Polycrystalline Thin Films: Structure, Texture, Properties and Applications

MATERIALS RESEARCH SOCIETY SYMPOSIUM PROCEEDINGS VOLUME 343

Polycrystalline Thin Films: Structure, Texture, Properties and Applications

Symposium held April 4-8, 1994, San Francisco, California, U.S.A.

EDITORS:

Katayun (Katy) Barmak

Lehigh University
Bethlehem, Pennsylvania, U.S.A.

Michael Andrew Parker

IBM Storage Systems Division
San Jose, California, U.S.A.

Jerrold A. Floro

Sandia National Laboratories
Albuquerque, New Mexico, U.S.A.

Robert Sinclair

Stanford University
Stanford, California, U.S.A.

David A. Smith

Stevens Institute of Technology
Hoboken, New Jersey, U.S.A.

MATERIALS RESEARCH SOCIETY
Pittsburgh, Pennsylvania

Single article reprints from this publication are available through
University Microfilms Inc., 300 North Zeeb Road, Ann Arbor, Michigan 48106

CODEN: MRSPDH

Published by:

Materials Research Society
9800 McKnight Road
Pittsburgh, Pennsylvania 15237
Telephone (412) 367-3003
Fax (412) 367-4373

Library of Congress Cataloging in Publication Data

Polycrystalline thin films : structure, texture, properties, and applications : Symposium
    H, April 4–8, 1994, San Francisco, California, U.S.A. / editors, Katayun (Katy)
    Barmak, Michael Andrew Parker, Jerrold A. Floro, Robert Sinclair, David A. Smith
        p. cm.—(Materials Research Society symposium proceedings ; v. 343)
        Includes bibliographical references and index.
        ISBN 1-55899-243-X.
        1. Thin films—Congresses. I. Barmak, Katayun II. Parker, Michael Andrew
    III. Floro, Jerrold A. IV. Sinclair, Robert V. Smith, David A. VI. Materials
    Research Society VII. Series: Materials Research Society symposium proceedings ;
    v. 343.

QC176.82P65 1994   94-29191
530.4'175—dc20   CIP

Manufactured in the United States of America

Contents

PART I: PHENOMENOLOGY OF
MICROSTRUCTURAL EVOLUTION IN THIN FILMS

*Invited Paper

PART II: REACTIONS, TRANSFORMATIONS,
AND DIFFUSION AT INTERPHASE INTERFACES
AND GRAIN BOUNDARIES

*Invited Paper

PART III: POLYCRYSTALLINE MAGNETIC THIN FILMS

III. A. GRAIN STRUCTURE, TEXTURE, AND EPITAXY IN THIN FILM MAGNETIC MEDIA

III. B. GRAIN STRUCTURE, TEXTURE, AND EPITAXY IN MAGNETIC MULTILAYERS

*Invited Paper

*Invited Paper

PART V: MECHANICS AND MECHANICAL PROPERTIES OF POLYCRYSTALLINE THIN FILMS

PART VI: POLYCRYSTALLINE METALLIZATION

PART VII: POLYCRYSTALLINE SEMICONDUCTOR FILMS

*Invited Paper

Preface

This volume is a compilation of more than 100 papers presented during a symposium on polycrystalline thin films held as part of the 1994 MRS Spring Meeting in San Francisco. Polycrystalline thin films are found in a variety of products that, taken together, comprise a major commercial market. This highly interdisciplinary symposium brought researchers working with different classes of materials (metals, ceramics, semiconductors, etc.) together into a single forum. The symposium emphasized the relationships between film microstructure and the observable properties, within the context of technologically relevant applications. The success of the symposium, as evidenced by the attendance at various sessions and the comments of the attendees, marks the need for future symposia dealing with polycrystalline thin films.

Katy Barmak
Michael Parker
Jerrold Floro
Robert Sinclair
David Smith

June 1994

Acknowledgments

This symposium was made possible through the financial assistance of the following sponsors:

- IBM Storage Systems Division
  - IBM Fellow Program
  - Analytical Services, San Jose
  - Disk Development
- JEOL USA
- Philips Electronics
- Gatan Inc.
- Komag Inc.
- Balzers
- Thermionics
- Rigaku USA

Katy Barmak acknowledges the support of NSF grant DMR-9308651 and the Department of Materials Science and Engineering, Lehigh University. Katy Barmak further extends a personal thanks to the conference chair, Dr. James Harper, and gratefully acknowledges the administrative assistance of Ms. Maxine Mattie. Michael Parker acknowledges the support of IBM Fellow, J. Kent Howard. Jerrold Floro acknowledges the support of Sandia Labs through Department of Energy Contract DE-AC04-94AL85000.

And finally, the symposium organizers wish to thank the MRS publication staff for their capable assistance in the preparation of this volume.

MATERIALS RESEARCH SOCIETY SYMPOSIUM PROCEEDINGS

MATERIALS RESEARCH SOCIETY SYMPOSIUM PROCEEDINGS

*Prior Materials Research Society Symposium Proceedings
available by contacting Materials Research Society*

PART I

Phenomenology of Microstructural
Evolution in Thin Films

GRAIN GROWTH IN POLYCRYSTALLINE THIN FILMS

CARL V. THOMPSON
Dept. of Materials Science and Engineering, Massachusetts Institute of Technology, Cambridge, MA 02139

ABSTRACT

The performance and reliability of polycrystalline films are strongly affected by the average grain size and the distribution of grain sizes and orientations. These are often controlled through grain growth phenomena which occur during film formation and during subsequent processing. Abnormal rather than normal grain growth is most common in thin films, and leads to an evolution in the distribution of grain orientations as well as grain sizes, often leading to uniform or restricted crystallographic orientations or textures. Surface and interface energy minimization and strain energy minimization can lead to development of different textures, depending on which is dominant. The final texture resulting from grain growth depends on the film thickness, the deposition temperature, the grain growth temperature, the thermal expansion coefficients of the film and substrate, and the mechanical properties of the film, as well as other factors.

INTRODUCTION

While epitaxial films have traditionally been the topic of more intense and extended discussions than polycrystalline films at MRS meetings, polycrystalline films are used in a larger and more varied number of applications. They are used as decorative and protective coatings; they are used for catalysis; and they are used in electronic, magnetic, and optical devices and systems. In all of these applications, the performance and reliability of polycrystalline films depend on their average grain size, their average crystallographic texture or orientation, the distribution of grain sizes, and the distribution of grain orientations. These, in turn, are all strongly affected by the conditions under which the film is formed, and by post-formation processing.

This review will begin with a brief general discussion of the processes which control the grain size and orientation distributions in polycrystalline films. This introductory discussion will be followed by a more focused overview of grain growth phenomena in films. Finally, a still more detailed review of recent developments in the understanding of texture evolution during grain growth in films will be given. It will be seen that grain growth phenomena can have a dominant role in controlling the final structure and properties of polycrystalline films.

FILM FORMATION

Whether polycrystalline films form through vapor, liquid, or solid phase processes, the fundamental kinetic phenomena involved are crystal nucleation, growth, and coalescence. While the latter can also involve grain boundary motion, if it does not, the grain size and grain size distribution in an as-formed polycrystalline film will be a function only of the nucleation and growth processes. These phenomena can be relatively easily simulated in two-dimensions [1, 2], as illustrated in Figure 1. In this

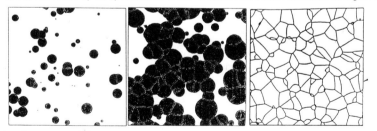

Figure 1:   Simulation of 2D nucleation and growth of crystals. The nucleation and growth rates are constant so that a Johnson-Mehl structure results (Reference 1).

Mat. Res. Soc. Symp. Proc. Vol. 343. ©1994 Materials Research Society

case nucleation occurs at a constant rate, I, and at random locations. Crystals are also assumed to grow isotropically at a constant rate, G. This results in the well-known Johnson-Mehl structure [3], for which it has been shown [4] that the final average grain "diameter" $\bar{d}$ is 1.448 $(G/I)^{2/3}$. Here and subsequently, d can be thought of as the diameter of a circular grain with the same 2D area as an actual grain. While the simple relation among $\bar{d}$, I, and G given above applies only for the Johnson-Mehl structure, it is generally the case that $\bar{d}$ increases with increasing G for a given I, even if G is not constant or isotropic.

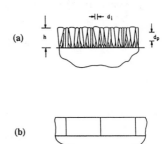

Figure 2:   Schematic cross-sectional views of two types of as-deposited microstructures in polycrystalline films.

Once a continuous film has formed, thickening can lead to several fundamentally different microstructures. These have been widely discussed in the context of so-called zone models [5-9]. Two possible structures are shown in Figure 2. In the first, Figure 2a, film growth continues through incorporation of new material into grains preexisting in the initial continuous layer. However, in the case illustrated in Figure 2a, some grains are occluded by others during film thickening, so that the in-plane grain size varies through the thickness of the film, being smallest at the bottom. No significant grain boundary motion occurs during thickening of films with this final structure. These non-equiaxed columnar structures are common for high melting temperature materials [5, 6] such as refractory metals and semiconductors [7] deposited at relatively low temperatures.

The equiaxed structures shown in cross-section in Figure 2b and in perspective in Figure 3, have an essentially constant grain size through the film thickness, and almost all of the grain boundaries are continuous through the thickness of the film. These conditions can be met if no grain occlusion occurs during film and grain thickening. However, the films in Figures 2b and 3 have grain sizes comparable to the film thickness, suggesting that grain boundaries have moved, and grain growth has occurred as the film has thickened [11]. This would have to be the case if the initial nuclei spacings are small compared to the final film thickness, as is usually expected. Equiaxed columnar structures such as those of Figures 2b and 3 are commonly observed in lower melting point materials such as fcc metals.

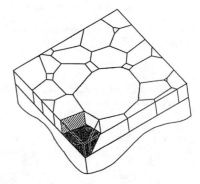

While the phenomena discussed above are observed and normally visualized in the context of film formation from the vapor phase, they also occur in solid phase formation of polycrystalline films. For example, the simple description embodied in the discussion of Figure 1, appears to apply to the crystallization of amorphous Co_2Si [12, 13], and the more general description given above for nucleation growth and coalescence should apply to crystallization processes in general. The general description appears also to apply to polycrystalline films formed through interfacial reactions [14, 15].

Figure 3:   Sketch of a polycrystalline film with an equiaxed columnar microstructure.

Films with as-deposited or as-formed microstructures of the type shown in Figure 2a will usually undergo grain growth when heated, evolving toward structures of the type shown in Figures 2b and 3. This grain growth phenomenon, which can occur during film coalescence, during film thickening, and during post-formation annealing, will now be described in more detail.

GRAIN GROWTH

Grain growth is a process through which the average grain size of *a fully crystalline* material increases through the motion of grain boundaries. As illustrated in Figure 4, this occurs through the shrinkage and eventual elimination of some grains, and the growth of others. The driving force for this process is the reduction of the total grain boundary area, and the corresponding reduction of the excess energy associated with the grain boundaries. Grain growth is a *competitive* process in that some grains grow at the expense of others. It is not to be confused with *crystallization,* in which new grains nucleate and grow in an amorphous matrix, of which Figure 1 could be an illustration, or *recrystallization,* in which *new* grains nucleate and grow in a crystalline matrix. Grain growth does not involve nucleation of new grains.

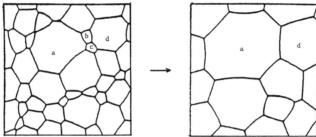

Figure 4: Top view illustration of a film undergoing grain growth. Grain boundaries move, causing grains a and d to grow and grains b and c to shrink and disappear. The average grain size increases.

Grain growth has been studied extensively in bulk materials [16], in metallic sheets, and more recently, in thin films [17]. There are two main categories of grain growth phenomena, normal grain growth and abnormal grain growth. *Normal grain growth* refers to a process in which the average grain size, $\bar{d}$, increases such that the time-rate-of-change of $\bar{d}$, is proportional to $1/d$, so that

$$\bar{d} - \bar{d}_o \; \alpha \; t^{1/2}, \tag{1}$$

where $\bar{d}_o$ is the average initial grain size and t is time. Normal grain growth is also considered to be characterized by a steady state behavior in which the shape of the grain size distribution function is time-invariant, or self-similar when scaled by $\bar{d}$.

Abnormal grain growth is any grain growth process which is not normal. It generally results when growth of a subpopulation of grains is favored over the growth of the main population. It therefore generally occurs when additional driving forces favor growth of a specific set of grains, and/or when additional drag forces impede the growth or shrinkage of a specific set of grains. A specific case of abnormal grain growth is *secondary grain growth,* in which only a few grains grow at the expense of a static matrix of grains with unmoving boundaries [18]. For reasons to be discussed shortly, abnormal and secondary grain growth are very common in thin films, but normal grain growth is not. As will also be discussed, abnormal grain growth in thin films leads to an evolution in the average crystallographic orientations of the grains, as well as the average grain size and the distribution of grain sizes.

MODELING OF GRAIN GROWTH

A number of analytic grain growth models have been developed which lead to the behavior described by equation (1). These can be viewed as models for normal grain growth [16, 20-23]. These models can generally be applied to grain growth in 2D systems, as well as 3D systems. In both cases, expressions of the form of equation (1) are obtained, where the proportionality constants are different for 3D and 2D systems. These models only consider the effects of grain boundary energy reduction in driving grain growth, and they only apply to the evolution of populations of equiaxed grains.

Films with structures such as the one shown in Figure 2a will undergo grain growth involving boundary motion in three dimensions. However, the structure is far from equiaxed, so that analytic models do not apply in this growth regime, and equation (1) is not expected to apply. Once the structure of Figures 2b, 3, and 4 develops, the grain structure has taken on a quasi-2D character, and further grain growth should occur predominantly through grain boundary motion in directions lying in the plane of the film. In this regime, analytic models for 2D normal grain growth might be expected to apply. A well-known model of this type is due to Hillert [19, 22], who argued, based on the analysis of Mullins [23] and Von-Neuman [24], that a given grain of "radius" r (r = d / 2) will grow or shrink according to

$$\mathrm{d}r \,/\, \mathrm{d}t \;=\; \mu \left(\frac{1}{\bar{r}} - \frac{1}{r} \right), \tag{2}$$

where $\bar{r}$ is the average grain radius for the system, μ is the product of the grain boundary mobility and γ_{gb} the grain boundary energy. The dependence of dr / dt on $\bar{r}$ described in this equation is in agreement with the observation that grains which are larger than average *tend* to grow, and grains which are smaller than average tend to shrink. By requiring that

$$\frac{\partial f}{\partial t} \;=\; \frac{- \partial \left[f \left(\dfrac{\mathrm{d}r}{\mathrm{d}t} \right) \right]}{\partial r} \tag{3}$$

Hillert also considered how the grain size distribution function f(r, t) would evolve with time. He determined self-similar distribution functions for both 2D and 3D growth. For these functions, equation (3) also leads to the behavior described in equation (1). In using equations (2) and (3) and finding self-similar grain size distribution functions, Hillert drew on an analogy between grain growth and the mean field models for precipitate coarsening developed by Lifshitz and Slyozov [25] and Wagner [26]. In this analogy, treatment of 2D grain growth is mathematically equivalent to the treatment of interface-limited coarsening of a population of isolated discs, as illustrated for discs of fixed thickness in Figure 5. It should be emphasized though, that Hillert did not imply that the *mechanisms* of these two coarsening processes were the same. Equation (2) results from a consideration of the geometry of growth of an individual grain or cell in a cellular structure, and equation (3) is simply a continuity equation.

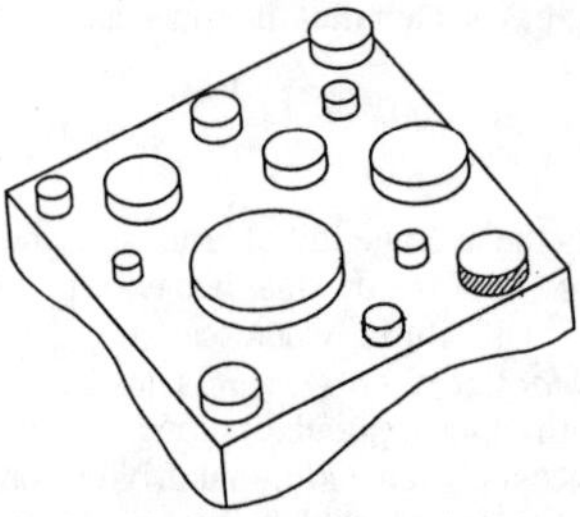

Figure 5: Array of coplanar discs of constant height. Disc coarsening is analogous to grain coarsening.

A number of computer simulations of grain growth, especially of 2D grain growth, have also been developed in recent years [16]. Of these, the two types of models which have been used to specifically treat grain growth in thin films are the Potts model [7, 27-29] and a front-tracking model due to Frost and co-workers [2, 30-34]. The latter involves generation of cellular structures, e.g., through simulation of nucleation and growth [1, 2], followed by simulation of boundary motion through incremental motion of points on the boundaries (see Figure 6), followed by motion of the points where the grain boundaries meet, known as triple junctions. For simulation of 2D grain growth, in each time step boundary points are moved a distance proportional to the local grain boundary curvature, κ, in the direction of the local grain boundary normal, where the distance is defined by a velocity, υ, given by

$$\upsilon \;=\; \mu \, \kappa . \tag{4}$$

If all the boundaries are taken to have the same energy, triple junctions are moved so as to retain a force balance, requiring that the boundaries meet at 120°. This simulation leads to the behavior shown in Figure 4, and the behavior characteristic of normal grain growth, as described earlier.

6

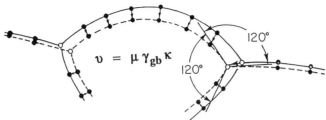

$$v = \mu \, \gamma_{gb} \, \kappa$$

Figure 6: Illustration of front-tracking approach for computer simulation of 2D grain growth. Points on boundaries are moved with velocities proportional to the local curvatures: triple points are moved to balance forces (Reference 31).

MODELING OF GRAIN GROWTH IN THIN FILMS

Surface and Interface Energy Effects

Thin films are *not* two dimensional. They have finite thickness, and they have top and bottom surfaces. The energies of the top and bottom surfaces of a grain in a free-standing film, γ_s, depends on the crystallographic orientation of the grain, relative to the orientation of the surface. If the film is on a substrate, the energy of the film-substrate interface, γ_i, also depends on the crystallographic orientation of a grain. In a polycrystalline film there will usually be a subpopulation of grains which have orientations which lead to lower than average values of γ_s and γ_i. This subpopulation of grains will be favored during growth, leading to abnormal grain growth. Unless all the grains in a film have the same surface and interface energies, normal grain growth cannot occur. Interface and surface energy minimization during abnormal grain growth in free-standing films or films on amorphous substrates will, in principle, lead to the development of a uniform *texture* (i.e., with all grains having the same crystallographic directions normal to the plane of the film) [18]. Interface energy minimization during abnormal grain growth in polycrystalline films on single crystal substrates will, in principle, lead to dominance of *epitaxial* orientations.

Clearly surface and interface energy differences are an important component of the driving force for grain growth in thin films. These have been accounted for in modifications of the Hillert model [36-38], and in some of the computer simulations of thin film grain growth [29, 32]. The means of accounting for these energies can be visualized by considering the special case illustrated in Figure 7, which shows a single "interior" grain, surrounded by another "exterior" grain. The boundary between the interior/exterior grains has two radii of curvature, the in-plane radius of curvature r_i, corresponding to the curvature κ considered in 2D models, and a curvature perpendicular to the plane of the film, r_p. If the interior and exterior grains have surface energies γ_s and $\bar{\gamma}_s$ and interface energies γ_i and $\bar{\gamma}_i$, respectively, force balances at the top and bottom surfaces require that

$$\frac{1}{r_p} = \frac{\left(\Delta\gamma_i + \Delta\gamma_s\right)}{h \, \gamma_{gb}} \equiv \Gamma , \qquad (5)$$

where h is the film thickness, $\Delta\gamma_s = \bar{\gamma}_s - \gamma_s$, and $\Delta\gamma_i = \bar{\gamma}_i - \gamma_i$. The rate at which the interior grain will grow or shrink will therefore be

(a)

(b)

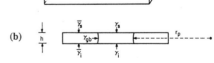

Figure 7: A single interior grain surrounded by another grain: (a) perpendicular view; and (b) cross-section.

$$\upsilon = \dot{r} = \mu \left(\frac{1}{r_i} + \frac{1}{r_p} \right) = \mu \left(\kappa + \Gamma \right),$$ (6)

where Γ is defined in equation (5). If Γ is zero, the grain will shrink. However, if Γ is negative, and $|\Gamma|$ is larger than κ. the grain will grow. This will happen only if $\gamma_s + \gamma_i$ is sufficiently smaller than $\bar{\gamma}_s + \bar{\gamma}_i$.

The Hillert model can be modified by considering the effect of Γ on coarsening of discs, and drawing on the analogy with grain growth [36-38]. A revised form of the growth law results when this is done, so that $\dot{r}$ is a function of surface and interface energy of a grain relative to the average surface and interface energies of the population of grains, as well as the size of the grain relative to the average grain size. The evolution of the distribution of grain sizes *and orientations*, $f(r, \theta, t)$, can then be predicted by satisfying equation (3), using the modified growth law, as described elsewhere in this proceedings [38].

In modifying the front-tracking computer simulation described earlier, it is necessary only to replace equation (4) with equation (6), and to assign initial values of $\gamma_s + \gamma_i$ to individual grains. The evolution in the average surface and interface energy of the film can be followed by keeping track of the values of $\gamma_s + \gamma_i$ for the surviving grains. Results of simulations of these types are described in Reference 32.

Grain Growth Stagnation

Grain growth in sheets and films often stagnates when the average grain size, $\bar{d}$, is 2 to 3 times the film thickness h [39-41]. Mullins [40] has argued that this occurs due to the formation of surface grooves where grain boundaries meet the surface of a film. This effect has been accounted for in the front-tracking simulation for thin film grain growth by assuming that boundaries are trapped and their velocities go to zero when the driving force for motion falls below a critical level,

$$\upsilon = \mu \, \Delta F \qquad \Delta F > \Delta F_{cr}$$
$$\upsilon = 0 \qquad \Delta F \le \Delta F_{cr},$$ (7)

where $\Delta F = \kappa$ when grain boundary energy alone is considered, and where $\Delta F = \kappa + \Gamma$ when surface and interface energy effects are also included. When only grain boundary energies are considered, it can be argued [31, 40] that $\Delta F_{cr} = \kappa_{cr} = \gamma_{gb} / (h\gamma_s)$ for uniform γ_s. Given this, it has been shown that 2D grain growth, driven by grain boundary energy elimination, stagnates when $\bar{d} \sim 3$ h, as observed experimentally. It can also be shown that the grain size distribution of the stagnant structure is lognormal, as is also often observed experimentally [41, 42].

Mullins [40] pointed out that boundaries between grains with sufficiently different surface energies can escape stagnation. This would lead to *secondary* grain growth in which a few grains with orientations which have low surface and interface energies (low Γ's) grow into a matrix of stagnated grains, as illustrated in Figure 8. In this case, secondary grain growth would lead initially to a bimodal

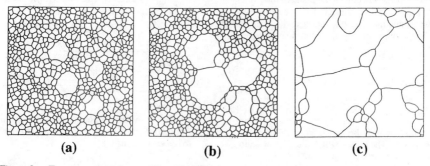

      (a)   **(b)**   **(c)**

Figure 8: Top view of a polycrystalline film undergoing secondary grain growth (simulator, Reference 33).

grain size distribution, as in Figure 8b, in which the large grains have restricted crystallographic orientations (e.g., texture). Eventually, once the secondary grains grew to impingement, a monomodal grain size distribution would result, and all the grains would have only the restricted orientations. This behavior has also been observed in a number of experimental systems [11, 17, 43].

Simulation of surface-energy-driven secondary grain growth, as illustrated in Figure 8, can be accomplished using the front-tracking model with equation (7) and with $\Delta F = \kappa + \Gamma$ (Reference 32).

Strain Energy Effects

In addition to surface and interface energy effects, another important aspect of real thin films that can lead to abnormal or secondary grain growth is associated with the fact that they are often attached to substrates which have different thermal properties than the films themselves. If a film with thermal expansion coefficient α_f is deposited on a substrate with thermal expansion coefficient α_s, at a temperature T_{dep}, when it is heated to a temperature at which grain growth occurs, T_{gg}, it will be subjected to a biaxial strain, ε, which can be approximated by

$$\varepsilon \cong \left(T_{gg} - T_{dep} \right)\left(\alpha_f - \alpha_s \right) = \left(\Delta T \right)\left(\Delta \alpha \right), \tag{8}$$

when α_f and α_s are only weakly temperature dependent. If the strain is elastically accommodated, the resulting strain energy density W_ε is given by

$$W_\varepsilon = M_{hkl} \, \varepsilon^2, \tag{9}$$

where M_{hkl} is an effective biaxial modulus whose value, for elastically anisotropic materials, depends on the crystallographic orientation of the grain relative to the plane of the strain [44]. Therefore, if we again consider the grains of Figure 7, and assume that they are attached to a substrate, for finite ΔT, they will have a difference in strain energy density, ΔW_ε, given by

$$W_\varepsilon = (M_i - M_e) \, \varepsilon^2, \tag{10}$$

where M_i and M_e are the effective biaxial moduli of the interior and exterior grains, respectively. This elastic strain energy can also contribute to the driving force for grain boundary motion, so that in equation (7), ΔF should be given by [34, 45-48]

$$\Delta F = \kappa + \Gamma + F_\varepsilon, \tag{11}$$

where $F_\varepsilon = \Delta W_\varepsilon / \gamma_{gb}$.

In the above discussion, it has been assumed that the strain due to differential thermal expansion has been elastically accommodated. The yield stress of a grain in a thin film is also a function of its crystallographic orientation [44, 49-51], as well as its size and the film thickness [49-52]. When a film yields, F_ε will be a function of the yield strains, as discussed in another paper in this Proceedings [34].

Minimization of the strain energy density of a biaxially strained film favors the growth of a subpopulation of grains with restricted crystallographic orientations. It therefore leads to abnormal grain growth, just as surface and interface energy minimization do. Strain energy minimization can also lead to secondary grain growth [53], when grooving-induced stagnation occurs. As in surface and interface energy driven grain growth, strain energy minimization leads to an evolution in the average crystallographic orientations of the grains in a film (e.g., to texture development). However, it is important to recognize that *surface, interface, and strain energy minimization do not necessarily favor the same orientations.* In fact, for fcc metals, surface energy minimization generally favors grains with (111) textures, while strain-energy-minimization generally favors other orientations. In fcc films in which strain is elastically accommodated, (100) texture is favored. In yielded films, the effective modulus is lowest for grains with (100) texture, but the yield strains are lowest for other orientations [49-51]. In this case the favored orientation is a function of the degree of the elastic anisotropy.

9

It can be seen, then, that abnormal and secondary grain growth can lead to different textures, depending on whether surface and interface energy minimization, or strain energy minimization dominates. These effects can be simulated using equations (7), (9), and (11) along with a model which accounts for the grain size, film thickness, and grain-orientation-dependence of the yield strain [51]. This is done for the case of no stagnation in another paper in this volume [34]. The case of strain, surface, and interface energy driven growth with stagnation has also been treated, but is not yet published [53]. Surface and interface energy minimizing textures, e.g., (111) textures in fcc metals, are favored in thin films, with small h, and at small strains, or ΔT's. Larger strains or ΔT's favor strain-energy-minimizing orientations, e.g., (100) textures in fcc metals. The observed texture of abnormal or secondary grains can therefore be a function of the difference between the deposition temperature and the grain growth temperature, as well as of the film thickness. This is shown for experimental results in Figure 9 for polycrystalline Ag films deposited on (100) Ni single crystal films. After grain growth, the Ag films

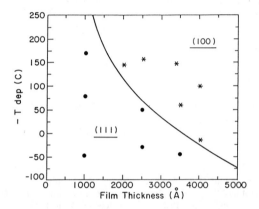

Figure 9:   Results from experiments showing the texture resulting from grain growth in initially randomly oriented polycrystalline Ag films deposited on single crystal (100) Ni films on MgO. Ag films were deposited at T_{dep} and had thickness h (Reference 48).

have either (111) or (100) epitaxial orientations, depending on the deposition temperature and the film thickness. Also shown in this figure is a line corresponding to the condition

$$\Gamma = F_\varepsilon.$$

calculated for the driving forces for growth of (111) and (100) grains, using calculated values of γ_s and γ_i, and known values of the elastic constants for Ag, as well as the thermal expansion coefficients for Ag and the MgO substrate on which the single crystal Ni and polycrystalline Ag films were grown [47, 48]. Above this line, (100) texture is expected, as observed, and below it, (111) is expected, also as observed. Similar curves can be calculated for other films on other substrates, given estimates of the surface and interface energies, and the catalogued elastic and thermal properties of the film and substrate.

<u>Summary of the Effects of Grain Growth on Texture Evolution</u>

Surface energy and interface energy minimization dominate in thin films undergoing grain growth at low strain, e.g., at temperatures near the deposition temperature. Surface-energy minimization is generally expected to favor orientations with close-packed surfaces, e.g., (111) textures for fcc metals and diamond cubic elemental semiconductors.

Minimization of elastically accommodated strains favors orientations with minimum effective biaxial moduli, e.g., (100) textures in fcc metals. Strain energy minimization can dominate when strains due to differential thermal expansion are high, e.g., due to large differences in the deposition and grain growth temperatures.

The yield strains of grains are a function of their size, as well as their crystallographic orientations, and the film thickness. If a material is elastically isotropic, as Al very nearly is, the strain energy driving force for grain growth, F_{ε}, is nonzero only after yielding occurs. In this case, grains with low yield strains are favored [45, 50]. For fcc metals the yield strain of (110) grains is lower than that of (111) and (100) grains, and it has been proposed [50] that this leads to the observation of (110) textures in some cases [54].

As grain growth proceeds, the yield strain of all grains decreases, so that F_{ε} decreases, and eventually becomes small compared to Γ. Therefore, even in those cases in which strain-energy minimization dominates initially, *un*impeded grain growth should eventually lead to a condition in which any remaining grains with surface-energy minimizing orientations should again be favored [34]. However, stagnation phenomena, such as those due to surface grooving or solute drag [55], can, and probably often do, lead to the "freezing-in" of strain energy minimizing orientations.

SUMMARY

Polycrystalline films form through the coalescence of independently nucleated crystals. Grain boundary motion can occur during this coalescence process, even if the film forms at relatively low temperatures. (Evidence for grain boundary motion during deposition of Au films has been observed, even when they were deposited at 20 percent of their absolute melting temperature!) Grain growth can also occur during post-deposition processing. Normal grain growth is not expected in thin films. Instead, abnormal grain growth occurs, leading to an evolution in both the grain *size* distribution and grain *orientation* distribution. The crystallographic textures resulting from abnormal grain growth depend on whether strain, surface, or interface energy minimization dominates. This, in turn, depends on the film thickness, the deposition temperature and the grain growth temperature, as well as on whether surface grooving, solute-drag, or other phenomena are impeding growth of some of the grains. The complex interplay of these conditions and phenomena can be investigated through use of computer simulations of grain growth in films.

ACKNOWLEGMENTS

The author would like to thank Roland Carel, Brett Knowlton, and Harold Frost for discussions and comments. This work was supported by NSF Contract #DMR-9001698 and SRC Contract #93-SP-309.

REFERENCES

1. H.J. Frost and C.V. Thompson, *Acta Metall.* **35**, 529 (1987).
2. H.J. Frost and C.V. Thompson, *J. Elec. Mat.* **17**, 447 (1988).
3. W.A. Johnson and R.F. Mehl, *Trans. Am. Inst. Min. Engrs.* **135**, 416 (1939).
4. E.N. Gilbert, *Ann. Math. Stat.* **33**, 958 (1962).
5. B.A. Mouchan and A.V. Demichisin, *Fiz. Met.* **28**, 83 (1969).
6. J.A. Thornton, *Annu. Rev. Mater. Sci.* **7**, 239 (1977).
7. D.J. Srolovitz, *J. Vac. Sci. Technol.* **A4**, 2925 (1986).
8. C.R.M. Grovenor, H.T.G. Hentzell, and D.A. Smith, *Acta Metall.* **32**, 773 (1984).
9. D.A. Smith and A. Ibrahim, *MRS Symp. Proc.* **317**, 401 (1994).
10. E. Kinsbron, M. Sternheim, and R. Knoell, *Appl. Phys. Lett.* **42**, 836 (1983).
11. C.C. Wong, H.I. Smith, and C.V. Thompson, *Appl. Phys. Letts.* **48**, 335 (1986).
12. K.N. Tu, D.A. Smith, and B.Z. Weiss, *Phys. Rev.* **836**, 8948 (1987).
13. D.A. Smith, K.N. Tu, and B.Z. Weiss, *Ultramicroscopy* **30**, 90 (1989).
14. E. Ma, C.V. Thompson, and L.A. Clevenger, *J. Appl. Phys.* **69**, 2211 (1991).
15. K.R. Coffey, L.A. Clevenger, K. Barmak, D.A. Rudman, and C.V. Thompson, *Appl. Phys. Lett.* **55**, 852 (1989).
16. H.V. Atkinson, *Acta Metall.* **36**, 469 (1969).
17. C.V. Thompson, *Ann. Rev. Mat. Sci.* **20**, 245 (1990).
18. C.V. Thompson, *J. Appl. Phys.* **58**, 763 (1985).
19. M. Hillert, *Acta Metall.* **13**, 227 (1965).
20. N.P. Louat, *Acta Metall.* **22**, 721 (1974).
21. O. Hunderi and N. Ryum, *J. Mater. Sci.* **15**, 1104 (1980).

22. P.K. Wu, *MRS Symp. Proc.* **317**, 559 (1994).
23. W.W. Mullins, *J. Appl. Phys.* **27**, 900 (1956).
24. J. von Neumann, "Metal Interfaces, ASM, Cleveland, Ohio, p. 108 (1952).
25. I.M. Lifshitz and V.V. Slyozov, *J. Phys. Chem. Solids* **19**, 35 (1961).
26. C. Wagner, *Z. Elektrochem.* **65**, 581 (1961).
27. M.P. Anderson, D.J. Srolovitz, G.S. Grest, and P.S. Sahni, *Acta Metall.* **32**, 783 (1984).
28. D.J. Srolovitz, M.P. Anderson, P.S. Sahni, and G.S. Grest, *Acta Metall.* **32**, 793 (1984).
29. D.J. Srolovitz, G.S. Grest, and M.P. Anderson, *Acta Metall.* **33**, 2233 (1986).
30. H.J. Frost, C.V. Thompson, C.L. Howe, and J. Whang, *Scripta Metall.* **22**, 65 (1988).
31. H.J. Frost, C.V. Thompson, and D.L. Walton, *Acta Met. et Mat.* **38**, 1455 (1990).
32. H.J. Frost, C.V. Thompson, and D.L. Walton, *Acta Met. et Mat.* **40**, 779 (1992).
33. D.T. Walton, H.J. Frost, and C.V. Thompson, *Appl. Phys. Letts.* **61**, 40 (1992).
34. R. Carel, C.V. Thompson, and H.J. Frost, "Computer Simulation of Strain Energy and Surface- and Interface-Energy on Grain Growth in Thin Films," *MRS Symp. Proc.,* Spring 1994, forthcoming.
35. C.V. Thompson, J. Floro, and H.I. Smith, *J. Appl. Phys.* **67**, 4099 (1990).
36. C.V. Thompson, *Acta Metall.* **36**, 2929 (1988).
37. J.A. Floro and C.V. Thompson, *Acta Met. et Mat.* **41**, 1137 (1993).
38. J.A. Floro and C.V. Thompson, "Mean Field Analysis of Abnormal Grain Growth Driven by Interface-Energy Anisotropy," *MRS Symp. Proc.,* Spring 1994, forthcoming.
39. P.A. Beck, M.L. Holtzworth, and P.R. Sperry, *Trans. Am. Inst. Min (Metall.) Eng.* **180**, 163 (1949).
40. W.W. Mullins, *Acta Metall.* **6**, 414 (1958).
41. J.E. Palmer, C.V. Thompson, and H.I. Smith, *J. Appl. Phys.* **62**, 2492 (1987).
42. B.M. Tracy, P.W. Davies, D. Fanger, and P. Gartman, "Microstructural Science for Thin Film Metallizations in Electronics Applications," ed. by J. Sanchez, D.A. Smith, and N. DeLanerolle (TMS, 1988), p. 157
43. H.-J. Kim and C.V. Thompson, *J. Appl. Phys.* **67**, 757 (1990).
44. W.D. Nix, *Met. Trans.* **20A**, 2217 (1989).
45. C.V. Thompson, *Scripta Met. et Mat.* **28**, 167 (1993).
46. C.V. Thompson, J.A. Floro, and R. Carel, *Modelling of Coarsening and Grain Growth* (The Metallurgical Society: Warrendale, PA), 205, 1994.
47. J.A. Floro, R. Carel, and C.V. Thompson, *MRS Symp. Proc.* **317**, 419 (1994).
48. J.A. Floro, C.V. Thompson, R. Carel, and P.D. Bristowe, "The Competition Between Strain and Interface Energy During Epitaxial Grain Growth in Ag Films on Ni (100)," *J. Mat. Res.,* in press.
49. P. Chaudhari, *IBM J. Res. Develop.* **197** (1969).
50. J.E. Sanchez, Jr. and E. Arzt, *Scripta Metall. Mater.* **27**, 285 (1992).
51. C.V. Thompson, *J. Mat. Res.* **8**, 237 (1993).
52. R. Venkatraman and J.C. Bravman, *J. Mater. Res.* **7**, 2040 (1992).
53. R. Carel, C.V. Thompson, and H.J. Frost, unpublished research.
54. H. Longworth and C.V. Thompson, *J. Appl. Phys.* **69**, 3929 (1991).
55. Y. Hayashi, H.J. Frost, C.V. Thompson, and D.T. Walton, *MRS Sym. Proc.* **317**, 431 (1994).

TOWARD REALIZING THE STRUCTURE–PROPERTY LINK IN POLYCRYSTALLINE THIN FILMS

C.S. NICHOLS
Department of Materials Science & Engineering, Cornell University, Ithaca, NY

1 Abstract

Many materials for engineering applications are used in polycrystalline form and contain grain boundaries with a range of structures and properties. However, most research on grain boundaries to date has focussed exclusively on symmetric coincidence site lattice interfaces. To go beyond descriptions for these simple interfaces and thence to an aggregate of grains and grain boundaries in a polycrystal will require a new approach. Here we discuss two models for properties of polycrystalline materials, including their advantages and drawbacks, and indicate the microstructural variables available to optimize properties.

2 Introduction

The goals of any theory of the structure–property link in polycrystalline systems are to predict properties for a given microstructure and to define microstructures for optimal properties. A polycrystalline microstructure consists of grains and grain boundaries as well as solutes and second phases. Polycrystalline materials have been described traditionally in terms of constitutive laws formulated in the context of mean-field theories. Familiar examples are the strain rate in diffusional Coble creep, the Hall-Petch relation for plastic yield stress, and the Fuchs' expression for resistivity. Within a mean-field approach, all grain boundaries are assumed to have identical properties. As a consequence, constitutive laws predict that properties of a polycrystal depend on a single, isotropic measure of the microstructure, most often the average grain size, d. This approach has enjoyed remarkable success, despite the evidence of bicrystal experiments that properties of grain boundaries depend on boundary structure.[1]

We address the problem of understanding properties of polycrystalline materials that are governed by grain boundaries by explicitly including boundary-to-boundary variation in properties.[2, 3, 4, 5, 6] At first, only a binary distinction between boundary types is considered so that boundaries are, for example, either "high" or "low" energy. A polycrystalline microstructure is then described by the "clusters" of grains that are connected by like-property boundaries. Constitutive laws are reformulated with cluster size as the relevant measure.

We begin with a brief discussion pointing out differences in macroscopic properties of polycrystalline systems. We next review the features of mean–field models for polycrystalline systems before going on to examine percolation models. Examples are included in the context of both models. We conclude with some comments on open questions.

3 Properties of polycrystalline systems

A chemically and compositionally pure polycrystalline material consists of an aggregate of grains and grain boundaries. A useful way of classifying properties of these systems is to

Mat. Res. Soc. Symp. Proc. Vol. 343. ©1994 Materials Research Society

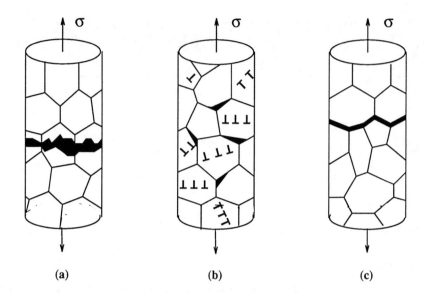

Figure 1: Schematic diagrams of polycrystalline materials with cylindrical geometry loaded in uniaxial tension along the cylinder axis. (a) Brittle transgranular fracture with no plastic deformation as an example of a property dependent only on the properties of the grains; (b) plastic deformation effected by dislocation motion within the grains followed by pile up at the grain boundaries with eventual fracture along the boundaries as an example of a property dependent on the properties of both the grains and grain boundaries; and (c) brittle intergranular fracture with no plastic deformation as an example of a property dependent only on the properties of the grain boundaries.

focus on the possible ways of coupling the two microstructural elements, of which there are three: (1) grain–grain; (2) grain–grain boundary; and (3) grain boundary–grain boundary.

Various fracture modes can be used to illustrate these classes. In Fig. 1a is shown a polycrystalline cylinder under uniaxial tension. Brittle fracture of the sample, with little or no plastic deformation, may propagate solely by a transgranular path, in which case the grain boundaries play no role (class (1)). Fig. 1b shows the same configuration of sample and load, but now plastic deformation occurs prior to fracture. Dislocations move within the grains and pile up at the grain boundaries, where stress is concentrated. Fracture occurs intergranularly (class (2)). In the final case, Fig. 1c, brittle intergranular fracture, with no plastic deformation, depends only on the grain boundary properties (class (3)). The classes of properties are summarized in Table 1, where other examples are included.

In addition to classifying properties by coupling microstructural elements, it is also instructive to subdivide those categories into structure–insensitive and structure–sensitive properties. In this paper we will focus on properties in the grain boundary–grain boundary coupling class. We will also ignore the effects of second phases and solute, although they can be incorporated into this scheme.

Structure–insensitive-grain–boundary–dependent properties depend on the *average* of

Table 1: Properties of polycrystalline materials classified according to how the two microstructural elements, grains and grain boundaries, couple together.

Coupling	Sample properties
Grain–grain	Brittle transgranular fracture
Grain–grain boundary	Dislocation pile-up followed by intergranular fracture; Nabarro-Herring diffusional creep
Grain boundary–grain boundary	Brittle intergranular fracture; electromigration; superconducting-to-normal transition

all grain boundaries and are not particularly sensitive to local fluctuations or defects. Examples of structure–insensitive properties include electrical conductivity and elastic moduli. Specifically, imagine altering the position of one or two grain boundaries in a polycrystal. The elastic moduli, for example, before and after the change are identical. This is not to say that two polycrystals with entirely different microstructures do not have different elastic properties, however. Traditional mean–field models are only strictly valid for structure–insensitive properties.

On the other hand, *structure–sensitive* properties are related to breakdown phenomena. Such properties are governed by a few grain boundaries and, hence, there are large sample–to–sample fluctuations. Examples of structure–sensitive properties are brittle fracture, dielectric breakdown, electromigration failure, and the superconducting–to–normal transition.

This distinction between classes of properties has been recognized by the physics community; structure–insensitive properties are termed "average properties", while structure–sensitive properties are termed "extreme properties."

4   Models for properties of polycrystalline systems

4.1   Mean–field models

Within mean–field models, polycrystalline materials consist of a collection of isotropic grains surrounded by identical boundaries. As a result of all boundaries being identical, there is a single length scale in constitutive relations, usually the average grain diameter, d.

Any property P of a polycrystalline ensemble can be expressed in the form[6]

$$P = P_o + \bar{P}d^m \tag{1}$$

where P_o describes the contribution of the grain interior, $\bar{P}$ is an average value of the property concerned, d is the average grain diameter, and usually $-3 < m < -\frac{1}{2}$. As a specific example, if P is yield stress, P_o is the Peierls stress, $\bar{P}$ is a parameter that accounts for boundary dislocation transmission resistance, and m is $-\frac{1}{2}$, then the Hall-Petch equation is recovered. According to Equation 1, the only microstructural parameter available to optimize properties is the average grain size.

Besides being in conflict with numerous bicrystal experiments, mean-field models can fail in a number of other ways. The averaging implicit in a mean–field treatment means that such models are inadequate for a microstructure in which any dimension of a grain is comparable to the sample size, the orientation distribution of crystallographic axes is not random, or the grain morphology is not equiaxial. Finally, constitutive laws of the form

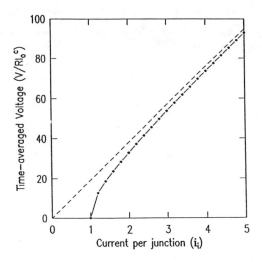

Figure 2: The calculated $I - V$ characteristics of a 20×20 array of Josephson junctions with a single critical current. The dashed line shows the expected behavior for an Ohmic system.

of Equation 1 fail for extreme properties because they rely on averages of grain boundary properties.

As an example of an application of mean–field theories, consider the current-voltage characteristics of a polycrystalline superconducting $YBa_2Cu_3O_{7-\delta}$ thin film. Bicrystal experiments of Dimos *et al.*[7] found that single grain boundaries have current-voltage characteristics similar to Josephson junctions. A mean–field polycrystalline thin film thus consists of an array of Josephson junctions in which all boundaries have identical critical current. Equations for the array critical current as a function of the input voltage can be set up and solved. See Ref.[2] for more details.

The time-averaged voltage as a function of the current per junction is shown in Fig. 2. The functional form for the $I - V$ curve is identical to that for a single junction. The total voltage drop is equal to the number of junctions in the direction of the input current, n, times the individual junction voltage drop, V_{ij},

$$V_{total} = nV_{ij} = nR(i_i^2 - 1)^{\frac{1}{2}} \tag{2}$$

for $i_i > 1$. There is a lossless current flowing through the array for $i_i \leq 1$. R is the Josephson shunt resistance and i_i is the current per junction. The junction critical current is normalized to 1.

Comparing Equation 2 to Equation 1, we can see that the grain interior does not contribute to the $I - V$ characteristics ($P_o = 0$) and since all boundaries are identical, $\bar{P}$ is the same resistance and current per junction. Furthermore, if the grain size is changed for a *constant* sample size, the only variable that changes in Equation 2 is the factor n Finally, the $I - V$ characteristics are identical for both hexagonal and square junction arrays. Grain shape and grain size therefore have little bearing on the superconducting $I - V$ characteristics within a mean-field model.

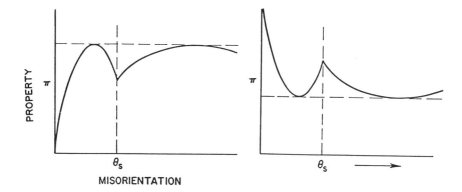

Figure 3: Schematic figures for the behavior of properties such as (a) energy, electrical resistance, and segregation, which are minimized at both low-angle and special grain boundaries, and (b) critical current and dielectric breakdown field, which are greatest for special boundaries. The behavior of a general boundary is indicated by the horizontal dashed line.

4.2 Percolation models

Because of the shortcomings of mean–field models as discussed above, we propose percolation models as an alternative. Similar to mean–field models, polycrystalline materials within percolation models are characterized by isotropic grains, but now we admit variation in grain boundary properties. Two types of dependences of properties on an interface degree of freedom are found. These are shown in Figure 3, where θ_m, the misorientation angle, characterizes the structure of the interface. Extrema in properties appear in these figures for low–angle and special grain boundaries on the one hand and for general boundaries on the other hand. We simplify the experimental results and assume there are only two types of grain boundaries in the system.

Constitutive laws are more difficult to codify for this model, but they will in general depend on p, the fraction of low–angle and special boundaries, on ξ, the percolation correlation length, and on the limiting value of the property of interest, P_{min}, so that

$$P = f(|p - p_c|, \xi, P_{min}) \tag{3}$$

where $|p - p_c|$ is the distance from the percolation threshold and f is some unspecified function. From Equation 3, we have more flexibility in optimizing properties of polycrystals. For example, by manipulating the types of grain boundaries in the system through processing choices, we can adjust p and by manipulating their spatial arrangement, we can adjust ξ.[8] Eq. 3 thus defines limits on the properties of polycrystalline materials. There is some experimental work exploiting these principles in fcc materials.[9]

Let us return to the example of a polycrystalline superconducting $YBa_2Cu_3O_{7-\delta}$ thin film. Besides the fact that superconducting grain boundaries behave like Josephson junctions, Dimos et al.[7] also found that such boundaries show a marked and distinct variation in critical current density on boundary misorientation angle, θ_m. Symmetric [001] tilt boundaries for which $\theta_m \leq 10°$ had a critical current density about 25 times larger than boundaries for which $10° < \theta_m \leq 45°$ (45° is the maximum symmetry-allowed for a tetragonal unit cell). This observation makes this system an ideal candidate for percolation models.

We can model the $I - V$ characteristics similarly as for the mean-field model, except that now we replace randomly a fraction $(1 - p)$ of the boundaries with junctions of low critical current, $I_c(low)$, to represent high misorientation angle boundaries. A schematic diagram of a realization of a small system is shown in Fig. 4. We have performed careful numerical calculations of the relation between array critical current, defined here as that value of the current for which the time-averaged voltage is greater than a chosen (small) amount, and p.[2] From these calculations it was found that

$$[I_c(array) - I_c(low)] \propto (p - p_c)^n \tag{4}$$

where a best-fit slope of the results gives $n = 1.23 \pm 0.1$. But, from standard percolation theory,

$$\frac{1}{\xi} \propto (p - p_c)^\nu \tag{5}$$

where $\nu = 1.33$. The two exponents, n and ν, lie within estimated error bounds and thus,

$$[I_c(array) - I_c(low)] \propto \frac{1}{\xi} \tag{6}$$

so that the array critical current is inversely proportional to the percolation correlation length. Or, the array critical current is proportional to the number of current–carrying paths through the sample. This relationship has been observed experimentally in aluminum wire networks[10], but has not been tested for disordered Josephson junction arrays to the best of our knowledge.

The scaling relationship in Equation 6 verifies our intuition regarding the dependence of critical current on system connectivity: the more connected paths of high critical current boundaries across the sample, the higher the sample critical current. Including boundary–to–boundary variation in a property yields a constitutive law that depends on a length scale that is generally *larger* than an average grain diameter. What's more, the correlation length depends nonlinearly on the percolation fraction of high current boundaries (Equation 5). Both of these dependences are quite different from predictions of mean–field models. In particular, note that the sample critical current tends to the limiting value, $I_c(low)$, as $p \to p_c$. This is an indication that the critical current of the sample depends sensitively on just a few grain boundaries and hence is an extreme property. Indeed, one can imagine that fluctuations in the position of a few boundaries can dramatically alter the sample critical current.

Constitutive laws within a percolation framework indicate limits on properties. The minimum critical current for a polycrystalline thin film is found for a sample containing *all* high angle boundaries (low critical current). In order to increase the sample critical current, it is necessary to increase not only the fraction of high critical current boundaries, but to optimize their spatial arrangement as well. We have studied here purely *random* systems, but textured systems, in which like–property boundaries are aligned preferentially, can provide enhanced properties. One group has reported that high critical currents may be obtained in samples with high–aspect–ratio grains.[11] The theory of non–random percolation systems, however, is not as well understood as random systems.

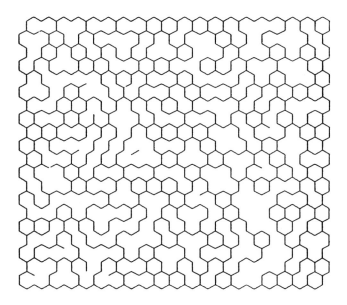

Figure 4: A realization of a two-dimensional polycrystalline microstructure. Low misorientation angle boundaries (high critical current) are present with a probability p while high misorientation angle boundaries (low critical current) are present with a probability $(1-p)$. The former boundaries are indicated on the figure by the presence of a line while the latter boundaries are indicated by the absence of this line. A hexagonal template has been used to generate this microstructure; boundaries therefore consist of segments with this symmetry. Dead-end boundaries are high–angle boundaries connected to two low–angle boundaries.

5   Closing remarks

We have given a brief overview of macroscopic properties of polycrystalline systems and discussed two models for describing and predicting these properties. Mean–field models have been successful largely because of their simplicity and the fact that, until recently, the only experimentally accessible microstructural variable has been the average grain diameter, d. Even so, there is broad scatter in the data supporting mean–field constitutive laws.[12, 13, 14]

As an alternative, we propose percolation models for describing and predicting properties of polycrystalline materials. Such models have the advantage that they make a distinction between properties of different boundaries. We have discussed only a binary distinction in properties, but this can be broadened into a multi–state classification and polychromatic percolation theory applied to obtain constitutive laws.

In any case, when boundary–to–boundary variation in properties is taken into account, the microstructure is then characterized in terms of *clusters* of grains. A new length scale emerges in constitutive laws that is the cluster linear dimension. We have chosen the percolation correlation length, ξ, as this measure. Cluster size depends nonlinearly on the fraction of like–property boundaries in the system and hence should be reflected in constitutive laws.

Our discussion has focussed throughout implicitly on columnar thin–film microstructures. It is in this context that theory and experiment converge most readily. Extension of percolation models to three–dimensional systems is in principle straightforward. However, it is not without computational and visualization problems. Furthermore, two–dimensional columnar microstructures offer more rigorous grounds for identification and development of microstructures with optimal structure–sensitive properties since the probability of finding a connected path is smaller in two dimensions than it is in three dimensions.

A number of outstanding questions remain, however. For example, we have chosen the misorientation angle as an appropriate measure for differentiating boundary properties. We might ask whether there are other or better measures. Further, we need to consider the experimental accessibility of the cluster size. Using electron backscattering techniques, however, it is now possible to determine relative grain orientations and hence attempt to identify grain clusters. Finally, we need to examine how to include extrinsic effects such as impurities or second phases into this methodology. To do so will require knowledge of how impurities or second phases alter the properties of individual boundaries. Such knowledge is currently scarce.

References

[1] T. Watanabe, Res. Mechanica **11**, 47 (1984).

[2] C.S. Nichols and D.R. Clarke, Acta metall. mater. **39**, 995 (1991).

[3] C.S. Nichols, R.F. Cook, D.R. Clarke, and D.A. Smith, Acta metall. mater. **39**, 1657 (1991).

[4] C.S. Nichols, R.F. Cook, D.R. Clarke, and D.A. Smith, Acta metall. mater. **39**, 1667 (1991).

[5] D.A. Smith and C.S. Nichols, in *Structure & Property Relationships for Interfaces*, eds. J.L. Walter, A.H. King, and K. Tangri (ASM, 1991), pgs. 345-360.

[6] D.A. Smith and C.S. Nichols, Materials Science and Engineering, in press.

[7] D. Dimos, P. Chaudhari, and J. Mannhart, Phys. Rev. B **41**, 4038 (1990).

[8] The two quantities, p, and ξ, are not unrelated. See Eq. (4).

[9] G. Palumbo, K.T. Aust, U. Erb, P.J. King, A.M. Brennenstuhl, and P.C. Lichtenberger, Phys. Stat. Sol. (a) **131**, 425 (1992).

[10] J.M. Gordon, A.M. Goldman, and B. Whitehead, Phys. Rev. B **38**, 12019 (1988).

[11] J. Mannhart and C.C. Tsuei, Z. Physik B **77**, 53 (1989) and D.R. Clarke, T.M. Shaw, and D. Dimos, J. Am. Ceram. Soc. **72**, 1103 (1989).

[12] R.W. Rice, Proc. Br. Ceram. Soc. **20**, 205 (1972).

[13] S.J. Bennison and B.R. Lawn, Acta metall. **37**, 2659 (1989).

[14] R.F. Cook, Mater. Res. Soc. Symp. Proc. **78**, 199 (1987).

ORIENTATION IMAGING OF THE MICROSTRUCTURE
OF POLYCRYSTALLINE MATERIALS

BRENT L. ADAMS
Brigham Young University, Department of Manufacturing Engineering,
Provo, Utah 84602

ABSTRACT

Recent developments, coupling the scanning electron microscope with image processing and crystallographic analysis, now make it possible to automatically index many thousands of lattice orientations exposed on section planes in polycrystalline materials. Backscattered Kikuchi diffraction patterns obtained from high-gain SIT and CCD cameras are analyzed using the Hough transformation (Cartesian coordinates - polar coordinates) in order to identify diffraction band widths and interplanar angles. From this basic information local lattice orientation can be determined. Information from raw data sets (in excess of 100,000 single orientations in some examples) can be used to construct orientation imaging micrographs which emphasize certain aspects of the exposed field of lattice orientations. Thus, features of the spatial placement of lattice orientation (including grain boundary misorientation, microtexture, and connectivity of the microstructure) are readily studied. In this paper these techniques are reviewed, and recent explorations of the connectivity of grain boundary misorientation structure are presented for an interesting nickel-chromium-iron alloy.

INTRODUCTION

Polycrystals present the investigator with microstructures extremely rich in detail, and highly stochastic in character. This paper describes new experimental technique and methodology for exploring these microstructures. The focus is upon an intermediate scale of inquiry, ranging from approximately 10^{-7} meters upwards to a (macroscopic) scale encompassing many thousands of crystallites or grains.

The new microscopy is called Orientation Imaging Microscopy (OIM), owing to the fact that image contrast is obtained directly from measurements of the gradients of local lattice orientation. More specifically, Backscattered Kikuchi Diffraction (BKD) patterns are analyzed to determine local lattice orientation from small localized regions

23

of the sample surface. As described later, OIM employs an ordinary scanning electron microscope, operated in stationary beam mode in conjunction with highly sensitive camera technology, to interrogate the flux of backscattered electrons emanating from a crystalline region which can be smaller than 2 X 10^{-7} meters. Recently developed algorithms achieve a rapid indexing of the BKD patterns, such that when coupled to stage (or beam) motion automation it becomes possible to scan sample surfaces with a probe which is extremely sensitive to the precise lattice orientation at each scan point. Contrast is then formed, from the raw data of such scans, by examining the spatial gradients of lattice orientation present in the surface microstructure.

OIM presents an attractive coupling of several aspects of microstructural inquiry of interest for polycrystals. Revealed in the measured gradients of lattice orientation is the size, shape and connectedness of crystallites (grains) that is the usual aim of ordinary optical and electron-optical microscopy. (The difference is, in our experience, that a small fraction of interfaces are missed by ordinary techniques, presumably because some grain boundary (crystallographic) structures simply do not give rise to contrast conditions in ordinary techniques.) However, since the lattice orientation of each crystallite is known by OIM, it is natural to correlate morphological features with crystallographic data. It is this coupling which makes OIM an important new tool for the study of polycrystalline microstructures.

First, the method of BKD analysis is described in further detail; then OIM is defined as a sequence of mappings which reveal contrast in gray scale (or color) based upon knowledge of lattice orientation gradients in the polycrystal. Finally, an example of the application of OIM to the study of intergranular stress corrosion cracking in a nickel-chromium-iron alloy is described.

THE METHOD OF BACKSCATTERED KIKUCHI DIFFRACTION

When a focused beam of electrons enters a crystalline phase a shower of back-scattered electrons emanates in all directions within a small interaction volume (typically of diameter ~ 4 X 10^{-7} meters at 20 KeV in metals). The energy of a small fraction of these electrons is reduced only slightly from their initial values, even though their momentum is altered dramatically. Some of these backscattered electrons find their way back through the surface where the flux pattern they form clearly marks the effects of diffraction by the occurrence of bands in the pattern. These banded patterns are commonly known as 'high-angle Kikuchi patterns.' Reference to 'high-angle' comes from the fact that optimal efficiency for the backscattering process occurs when the outward sample normal is at large angles to the incident beam direction (typically 110 degrees). Although the term Backscattered Kikuchi Diffraction patterns, or BKD patterns is used throughout this paper, the terminology Electron Back Scattered diffraction Patterns (EBSP's) is also currently in widespread use. For a review of the basic physics of BKD image formation the reader is referred to the early paper of Blackham, Alam and Pashley [1].

BKD patterns are known to contain a wealth of information about the local structure of the crystallite, but mainly the lattice orientation is of interest here. A measure of the overall strength of contrast in the BKD image is also easily obtained, which facilitates certain aspects of interpretation. Lattice orientation refers to the rotation/inversion required to bring a global reference lattice into coincidence with the local lattice of the crystallite, ignoring any small (elastic) distortions which may be present. It is common to use the symbol g(r) to refer to the lattice orientation associated with position r . The reader will note that it is not possible to assign lattice orientation to every position r in the polycrystal. Some points fall in regions where the lattice has been disturbed, such as near grain boundaries or triple junctions. At such a point it may be appropriate to associate the two, three or more lattice orientations which are present in the local neighborhood.

Using an electron-sensitive photographic plate positioned inside the vacuum chamber of a scanning electron microscope, Venables and Harland [2] reported obtaining high quality BKD patterns. Approximately one decade later, Dingley and Baba-Kishi described the interrogation of these patterns by a high-gain television camera focused upon a phosphor screen located in the vacuum chamber [3]. This novel advance led to interactive computer systems allowing the operator to rapidly index BKD patterns for lattice orientation. In this mode the operator is required to identify two or more known crystallographic zone axes or bands in the pattern. A skilled operator can index ~100 patterns an hour using this approach; when more data is required, however, this process becomes very fatiguing, and thus it is rather prone to indexing errors.

When indexing is required from a very large number of locations on the sample surface, manual indexing is rather impractical. Since 1990 several methods have been introduced to replace manual indexing with an automated methodology. Juul-Jensen and Schmidt [4] detected bands with an algorithm which scanned the BKD image line-by-line, finding the local intensity maxima. Wright and Adams [5] developed a scheme in which the 'Burns' algorithm is used to extract edges from the local intensity gradients present in the patterns. More recently, the identification of peaks in the Hough-transformation (Cartesian coordinates -> polar coordinates) with lines or bands in the BKD pattern has been exploited by Lassen, Conradsen and Juul-Jensen [6] and Kunze, et. al. [7] for rapid indexing.

Kunze, et. al. [7] found that indexing via the Burns and Hough algorithms is equally robust and reliable, although the Hough algorithm is now more widely used because it is less sensitive to properties of the crystal phase. Using current generation work station computers (~80 MIPS), implementation of the Hough algorithm requires approximately one second per pattern. The angular precision typically achieved is ~ one degree relative misorientation. The image quality parameter enables the operator to discard data which is believed to be susceptible to error.

It is known that all physically distinctive lattice orientations belong to the homogeneous space $O(3)/G$, which denotes the quotient space of right cosets of elements of the 3-dimensional orthogonal group, $O(3)$, with elements of the point symmetry subgroup of the lattice, G. (The reader who is not familiar with group theory may wish to consult any elementary text, such as the work of Hammermesh [8].) Points which fall on an interface between abutting grains belong to a homogeneous product space $O(3)/G \times S^2 \times O(3)/H$ where S^2 denotes the unit sphere of possible directions for a unit vector, G denotes the symmetry subgroup of the lattice occurring on the S^{2-} side of the interface tangent plane, and H is the symmetry subgroup of the lattice occurring on the S^{2+} side of the interface tangent plane (G and H need not reflect the same lattice phase). Thus, whereas points falling in crystallite interiors require specification by three independent parameters (e.g., euler angles, axis-angle, etc.), the crystallographic description of a grain boundary requires eight (two sets of three for each of two lattices, and two parameters defining the tangent plane normal). This expansion continues when triple junctions and quadruple points are considered.

Because of the enormous size and complexity of the space representing all grain boundary structure, it is common to consider various subspaces, wherein several of the free parameters are held constant. The crystallographic description of grain boundaries, and representations of grain boundary microstructure on subspaces were discussed in a recent paper by Adams and Field [9]. It is shown that fundamental stereological relations provide a direct association between measurements on section planes (i. e., measurements on $O(3)/G \times S^1 \times O(3)/H$), and the grain boundary texture (frequency of occurrence) on $O(3)/G \times S^2 \times O(3)/H$. Since OIM cannot yet distinguish lattice phase in most cases, OIM is typically limited to measurements on $O(3)/G \times S^1 \times O(3)/G$ to explore grain boundary texture on $O(3)/G \times S^2 \times O(3)/G$.

ORIENTATION IMAGING MICROSCOPY

The idea behind orientation imaging is not new ... it was suggested, in part, at least two decades ago in the work of Gotthardt, et. al. [10]. The potential of orientation scanning as a new form of microscopy, however, could not be realized until rapid automated indexing became a reality. The configuration for OIM, in is now described.

High quality BKD patterns are obtained using an silicon integrated target (SIT) camera coupled to a fiber-optic bundle and phosphor. Use of a fiber-optic bundle removes most of the distortion present in conventional camera lenses. (High quality images have also been obtained with a charge coupled device (CCD) camera.) The control unit for the camera is capable of averaging a variable number of video images (up to 128 frames); the choice is usually dictated by the quality of the BKD images, and is a function of atomic number, strain level in the sample, and the quality of mechanical and electropolishing procedures. This averaged pattern is compared by image subtraction with a stored background image obtained by scanning over a large number of grains on the surface. This image subtraction effectively removes intensity gradients in the image which are not fully removed by the camera control unit.

A sample holder is fixed upon a mechanical x-y stage, which is mounted such that the outward normal of the sample makes an angle of 110 degrees from the direction of the electron beam. The flat sample surface is stationed parallel to the plane of motion of the stage such that x-y motions remain in eucentric focus. The piezoelectric 'inchworm' stage is capable of on-demand motions of as small as 10^{-7} meters, with a total range of travel of .027 x .026 meters. Reproducibility over the entire range of travel is approximately 2.5 x 10^{-7} meters. The camera and stage control units are interfaced to a workstation operating at ~85 MIPS performance. This computer provides central control for both the stage and video control units via an RS232 interface. A frame capture board is installed in the computer which captures BKD images from the camera control unit. The indexing algorithms currently require ~ 1.2 second for indexing; thus approximately 3,000 patterns can be indexed each hour.

Adams, et. al. provided a formal definition of orientation imaging in terms of a sequence of three mappings [11]. Such a formal definition is not required for most purposes; here an informal description is used which carries the essence of the methodology. Local orientation measurements are obtained on a regular array of points positioned on flat polished sections. A representation of this array of points, appropriately scaled, is introduced into the computer graphics system. Overlapping this array of points, the investigator is allowed to introduce various graphical units, which are pixels or combinations of pixels differentiated by color or gray scale. Color and gray scale are assigned on the basis of the local details of lattice orientation, thus, these graphical units are used to form a visualization of some aspect of the microstructure. An example of this process is described in the next section.

APPLICATION OF OIM TO STUDY ISCC DAMAGE IN
NICKEL-CHROMIUM-IRON ALLOYS

It is known that nickel-chromium-iron alloys are vulnerable to intergranular stress corrosion cracking (ISCC) when subjected to stress in water at temperatures of ~300 °K. A set of OIM's for Alloy 600 Inconel are shown in Figure 1. These images illustrate several important aspects of OIM.

In the leftmost OIM is shown the image formed directly from the image quality index associated with the BKD patterns. Each pixel has been assigned a gray-level based upon the overall brightness of the diffraction pattern. Specifically, the image quality index is taken to be the sum of peak intensities associated with the seven highest peaks in the Hough space, which represents (roughly) the level of brightness for the seven brightest bands in the BKD image. Small variations of image quality with lattice orientation are very evident; these contrast the grains and their twins. Larger variations of image quality occur at grain boundaries where the natural lattice disturbance gives rise to poorer image quality and darker gray scales.

Also noted are dark spots (dark patches of poor image quality) decorating some grain boundaries. These are associated with the passage of an intergranular stress corrosion crack through the microstructure, from top to bottom of the OIM. The origin of these spots is not fully understood; the darker image quality of these spots suggests an interpretation as local regions of plasticity related to local difficulties of crack passage.

The OIM based solely upon image quality is comparable in information content to ordinary methods of optical and electron-optical microscopy. While the size and morphology of the microstructure is revealed, the underlying basis for the contrast is not coupled to lattice orientation. The centered and rightmost OIM's illustrate how these ordinary views of microstructure are now coupled with lattice orientation.

In the centered OIM, superimposed on the image quality pattern, darkened lines have been added to emphasize the location of grain boundaries which are characterized by abutting lattices of disorientation in excess of 10 degrees. Specifically, all pairs of adjacent measurement points in the BKD scan are interrogated for lattice misorientation. The absolute minimum disorientation is computed from among the multiplicity of possible rotations bringing the two lattices into coincidence. If this disorientation is larger than 10 degrees, then in the graphical representation of the microstructure a dark line segment (the graphical unit) is placed between the two scan points. This line segment is of length equal to the spacing between the scan points, and it is placed as a perpendicular bisector relative to the two scan points. These perpendicular bisectors connect together to form a visualization of the grain boundary network.

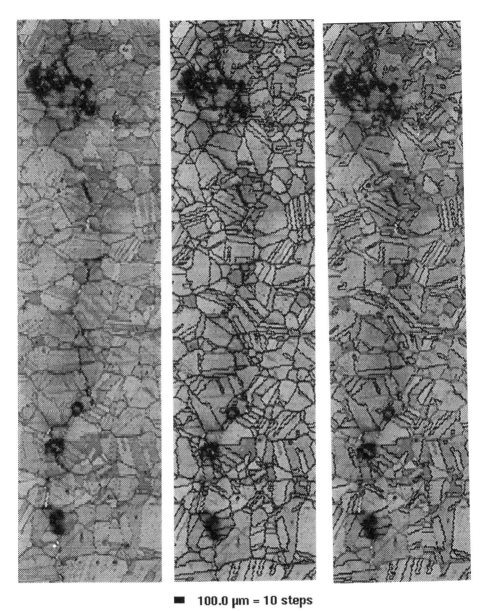

■ 100.0 μm = 10 steps

Figure 1. Orientation Imaging Micrographs of Alloy 600 Inconel showing the microstructure near ISCC damage. The left image is formed from the image quality. The center image highlights grain boundaries of lattice misorientation > 10°. The right image highlights grain boundaries whose misorientations lie within 8.66° of Σ3.

Such an image is called an orientation imaging micrograph (OIM) because precise lattice orientation information is employed in its construction. In the centered OIM the investigator knows that the dark boundaries are formed by lattice disorientations of more than 10 degrees. This particular criterion was chosen by the investigator. It is evident from a comparison with the image quality OIM on the left that the ISCC damage proceeds along the highlighted boundaries.

The rightmost OIM shows a more restrictive choice for imaging the grain boundaries. In this case all boundaries whose misorientation lie within 8.66 degrees of the $\Sigma 3$ coincidence misorientation (i. e., according to Brandon's criterion [12]), are highlighted with dark lines in the same manner as previously described. This forms a sub-network of the high-angle network highlighted in the centered OIM. Careful examination of these OIM's suggests that ISCC does not occur on boundaries highlighted in the rightmost OIM. (It is known that ISCC does not generally occur on $\Sigma 3$ boundaries in this material.)

Ordering Relations on $O(3)/G \times S^2 \times O(3)/G$

The ex-situ record of crack advance illustrated in Figure 1 provides an opportunity to explore ordering relationships for the local resistance to ISCC as a function of the crystallographic structure of the boundary. An outline of the approach is described here for single-phase polycrystalline materials.

Consider any arbitrary triple junction occurring along the intergranular crack. Suppose that the crack enters the triple junction from grain boundary A, and progresses along grain boundary B, but not along grain boundary C. Let S denote crystallographic structure in a grain boundary. Thus, S_A, S_B and S_C will denote the crystallographic structures for each of the three boundaries forming the triple junction; each belongs to $O(3)/G \times S^2 \times O(3)/G$.

We note that structure S_B has less resistance to ISCC than structure S_C (for the particular imposed temperature, stress-state, corrosive environment, etc.) Let R(S) denote the 'resistance' to ISCC in boundaries of structure S. We shall not be more precise about R, except to indicate that it is a real-valued function on the subspace $O(3)/G \times S^2 \times O(3)/G$, and R(S) ≥ 0 for all S.

The surface area (per-unit-volume), dS_V, of grain boundary structures lying in a neighborhood of measure dS about S is given by the real-valued grain boundary texture function $f(S) \geq 0$ such that $dS_V = f(S) \, dS$ [9]. To a first-order approximation we should expect the product of the resistance function and the texture function, $R(S) \cdot f(S)$, when integrated over $O(3)/G \times S^2 \times O(3)/G$, to be a reasonable estimate of the overall resistance of the microstructure to ISCC damage. However, it is not possible to directly measure R(S) from the ex-situ record. What is measurable is a sequence of ordering relations of the form $R(S_B) < R(S_C)$ for each triple junction in the ex-situ damage record. If a sufficiently dense record of such ordering relationships is available, then it is in principle possible to order any arbitrary partition of the entire manifold $O(3)/G \times S^2 \times O(3)/G$. From such an ordering the investigator is free to invent an auxiliary function, $R^*(S)$, which satisfies a system of ordering relationships $R^*(S_B) < R^*(S_C)$ in congruence with the true resistance function, $R(S)$. Thus, the integration of the product $R^*(S) \cdot f(S)$ over $O(3)/G \times S^2 \times O(3)/G$ defines an objective function for evaluating the overall resistance of microstructures to ISCC damage, in the first-order approximation. (It is, of course, to be emphasized that such a first-order approach may be entirely inadequate if the spatial placement of grain boundary texture is not approximately random. When placement is correlated, then the issue of connectedness is central, and the first-order approximation would be expected to fail.)

The problem with the approach just described is a difficult one, and well explored. Because the sample is opaque it is not yet possible to determine the full boundary plane orientation from the ex-situ record. Seven of the eight crystallographic parameters defining interface structure are directly measured using BKD analysis, but the eighth parameter (the angle of inclination the boundary makes relative to the surface normal) is obscured by the opacity. Thus, convenient measurements are possible only on $O(3)/G \times S^1 \times O(3)/G$. What is required is a connection between such measurements and the full problem on $O(3)/G \times S^2 \times O(3)/G$. This is the problem of opacity.

Two methods of stereology are known to deal with the problem of opacity. The first is a direct approach. The missing 'eighth' parameter can be measured directly by removing a thin layer of material of constant thickness (serial sectioning) and recording the translational shift of the grain boundary. The experimental challenge of serial sectioning is not a trivial one, but in principle, combining OIM measurements with serial sectioning enables a direct exploration of the ordering relationships of interest. The second method relies upon rigorous principles of mathematical stereology connecting information on several (oblique) section planes to estimate the full eight-dimensional function, f(S). This second approach has been reviewed and evaluated in the paper of Adams and Field [9].

SUMMARY

The technique and methodology of orientation imaging microscopy (OIM) as a new tool of microscopy has been reviewed. OIM is shown to link the morphological features of microstructure with lattice orientational attributes. This union of information reveals many important new insights into polycrystalline microstructures and their relationship to properties.

ACKNOWLEDGMENT

The author wishes to acknowledge financial support for this work through a grant from the Office of Basic Energy Sciences of the U. S. Department of Energy.

REFERENCES

1. M. N. Blackham, M. Alam and D. W. Pashley, Proc. Roy. Soc. Lond. **A221**, 224 (1953).
2. J. A. Venables and C. J. Harland, Phil. Mag. **27**, 1193 (1973).
3. D. J. Dingley and K. Baba-Kishi, Scanning Electron Microscopy **2**, 383 (1986).
4. D. Juul-Jensen and N.-H. Schmidt in Recrystallization '90, edited by T. Chandra (The Metallurgical Society, Warrendale, PA, 1991) p. 219.
5. S. I. Wright and B. L. Adams, Metall. Trans. **24A**, 759 (1992).
6. N. C. Lassen, K. Conradsen, and D. Juul-Jensen, Scanning Microscopy **6**, 115 (1992).
7. K. Kunze, S. I. Wright, B. L. Adams and D. J. Dingley in Symposium on Microscale Textures of Materials, edited by B. L. Adams and H. Weiland (Textures and Microstructures **20**, Gordon and Breach Publishers, Brussels, 1993) p. 41.
8. M. Hammermesh, Group Theory and its Application to Physical Problems, Addison Wesley Publishing Co., Reading, MA, 1962) p. 29.
9. B. L. Adams and D. P. Field, Metall. Trans. **23A**, 2501 (1992).
10. R. Gotthardt, G. Hoschek, O. Reimold and F. Haessner, Texture **1**, 99 (1972).
11. B. L. Adams, S. I. Wright and K. Kunze, Metall. Trans. **24A**, 819 (1993).
12. D. G. Brandon, Acta metall. **14**, 1479 (1966).

GRAIN GROWTH SUPPRESSION AND ENHANCEMENT
BY INTERDIFFUSION IN THIN FILMS

ALEXANDER H. KING and KAREN E. HARRIS

Department of Materials Science & Engineering, State University of New York at Stony Brook, Stony Brook, NY 11794-2275, U. S. A.

ABSTRACT

Grain structure and grain growth in thin metallic films are important because of their effects on properties such as yield strength, electrical resistance and electromigration resistance. Since almost all thin films are used in contact with a substrate and many also have contacts with overlayers, it is important to consider how interactions with other materials affect the grain growth process. In this paper we consider the effects of diffusive interactions. We will show that interdiffusion often accompanies grain growth and that it can result in a number of novel grain boundary reactions, driven by a variety of effects. Using TEM techniques, we demonstrate cases of grain growth suppression and grain growth enhancement resulting from interdiffusion of solute atoms in gold thin films. The reasons for the observed effects will be considered with a view to providing a fundamental understanding of the types of systems that might be expected to exhibit the various phenomena.

INTRODUCTION

The effects of solute additions upon grain boundary migration have been recognized for many years. Wherever systematic studies have been performed, it has been found that solute additions reduce grain boundary mobility, and often very severely [1]. This is so well accepted that large differences in grain boundary mobilities or velocities, measured in nominally identical materials, frequently are either ignored or ascribed to small variations in impurity content.

Although the "solute drag effect" is very well known, there have been no attempts to use it as a means of grain growth control. The reasons for this are not difficult to find: the greatest level of control is provided by the use of dopant-level impurity concentrations in materials of ultra-high purity levels that are hard to maintain in commercial processes. Adding solute often has other complicating effects on such properties as resistivity, corrosion resistance, strength and phase stability. And finally, for thin film applications, it is more frequently desirable to increase the rate of grain growth than to decrease it.

Nevertheless, there are reasons to pursue the idea that grain growth may be affected by solute additions. Thin film applications always involve structures with large surface-to-volume ratios so there is high potential for deliberate or accidental contamination of the material. This may be detrimental to the material performance or may be turned to good use. In this paper we are concerned with the effects of solute infusion on the development of microstructure.

A typical device comprising thin film structures is made through a complex process involving several heating steps. The materials are exposed to temperatures of $T_m/2$ or greater during deposition or in post-deposition treatments. At temperatures from $T_m/2$ down to about $T_m/4$, grain boundary diffusion dominates any mass transfer, and can easily provide for diffusion distances equal to the film thickness, within the process time. For example, the time taken to penetrate a 1 μm film of aluminum, via the grain boundaries, at 150°C, varies from about a second for gallium to about an hour for copper. Thus any thin film polycrystalline material may undergo diffusive mixing with adjacent materials via the grain boundaries, during normal processing.

The interactions between thin films and diffusing solutes are rather complicated. In this paper we first discuss the nature of some of the interactions and then describe some preliminary experiments aimed at elucidating which effects are the most important, for a given set of circumstances.

Mat. Res. Soc. Symp. Proc. Vol. 343. ©1994 Materials Research Society

EFFECTS OF GRAIN BOUNDARY DIFFUSION

Grain boundary diffusion can have several effects upon the microstructure of a thin film, and we need to understand the underlying reasons for them. Many processes occur simultaneously, interact and compete. In order to keep our discussion reasonably simple, we will presume that we are dealing with diffusion in the Harrison "type C" regime [2], in which diffusion occurs only in the grain boundary and lattice diffusion is almost completely frozen out. We will, however, consider cases in which there is a small degree of "leakage" from the grain boundary into the lattice. Shortly after the onset of diffusion, the situation is as indicated in Figure 1. At least four different phenomena can result from the introduction of solute into the grain boundaries of an initially "clean" material.

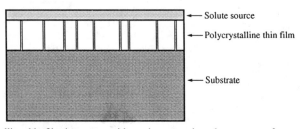

Figure 1: A polycrystalline thin film in contact with a substrate and a solute source, after commencement of diffusion in the Harrison type C regime. The solute source may also be a vapor, or the substrate itself. The film has solute-rich grain boundaries and clean grain interiors, even though the solute does not segregate to the boundaries under equilibrium conditions.

Interfacial Energy Effects

As solute is added to a grain boundary, the boundary energy is expected to change and it may either increase or decrease. Decreases in energy are usually associated with segregating systems, but the starting point here is somewhat different than is usual in the case of a segregation study. The driving force for interfacial segregation is the difference in system energy between the segregated and unsegregated material, and includes a contribution from the changing energy of the solid solution in addition to that from the change in interfacial free energy. For the present case we are only concerned with the change in energy of the grain boundary, since the solute is provided from an external source.

Nevertheless, solute segregation is found to depend upon the crystal misorientation and the boundary plane orientation [3,4]. Since the solid solution energy is unchanged by the grain boundary parameters, it is concluded that the variation of solute segregation derives from differences in the effect of the solute on the interfacial energy. Thus when solute is added to grain boundaries from an external source, the energies of the interfaces change by different amounts.

Consider a triple line initially at equilibrium, as shown in Figure 2: the equilibrium position of the line may be changed by adding solute to the three interfaces. In some cases the triple line may become unstable, and Jahn and King [5] have observed the solute-diffusion-induced wetting of a grain boundary by the third grain at a triple junction. If the "wetting grain" is larger than the mean (or has more than six sides) the wetting phenomenon will locally enhance grain growth. On the other hand, if the wetting grain is the smallest of the three, it will retard grain growth locally by reducing the rate of grain shrinkage. If the energy changes are randomly distributed among the grain boundaries in a polycrystal, small grains and large grains have equal probabilities of wetting adjacent grain boundaries so the effect of solute-induced-wetting on the overall grain growth is determined only by the distribution of triple lines. Since large grains have many triple junctions, there should be little net effect on their growth rates - the effects will average out. Small grains, with relatively few triple junctions, may be induced either to shrink or expand. Since either of these effects will result in grain annihilation, the net effect will be enhanced grain growth. This process may re-

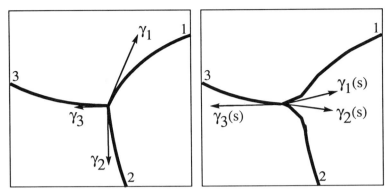

Figure 2: Equilibrium configurations of a triple line in pure material and after the addition of solute. It is possible for one of the boundaries to be wet completely by the third grain.

duce a tendency toward abnormal grain growth because it enhances growth more effectively in regions of more uniform grain size.

In many cases, large-angle grain boundaries are described as the superposition of low-energy interfaces and small-angle boundaries represented by dislocation arrays. The low-energy interfaces typically correspond to coincidence site lattices (CSLs) and the superposed small-angle boundaries may be made up of lattice dislocations or line defects with other Burgers vectors that are defined by the geometry of the interface itself [6]. Placing a solute layer in the grain boundary can increase the energy of the dislocation array by increasing the local elastic moduli, since the strain energy of the dislocation array is proportional to the shear modulus. The strain associated with the dislocation array is localized within a region of width approximately equal to the dislocation spacing [7]. Thus if the dislocation array (or small angle boundary) can move away from the solute-hardened special (or coincidence) interface by a distance larger than the dislocation spacing, the strain energy is reduced to its original level. The situation is as shown schematically in Fig. 3, and this behavior has been observed in the copper-zinc system by Jahn and King [5]. The diffusion of solute into the grain boundaries effectively drives the dissociation of large-angle near-coincidence boundaries into large-angle exact coincidence boundaries and associated small-angle boundaries. This reduces the grain size and may have additional effects upon the mobilities of the grain boundaries. Its effect would be to reduce the measured grain growth rate, and might even give rise to net grain shrinkage.

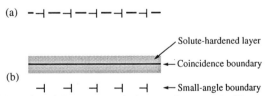

Figure 3: Schematic illustration of the structure of a large-angle, near-coincidence grain boundary (a) in pure material and (b) after addition of solute to the boundary. The strain energy is reduced by placing the dislocation array outside the elastically stiffened material.

Diffusion Induced Grain Boundary Migration (DIGM)

DIGM is an extensively documented phenomenon in which solute fluxes within the grain boundaries (as opposed to static solute concentrations) are accompanied by the sideways motion of the boundaries themselves [8]. In many cases, sharply curved interfaces result as the boundaries bulge into the neighboring grains, leaving behind solute enriched or depleted material. Pan and Balluffi [9] have demonstrated that the formation of sigmoidally curved boundaries can lead to the creation of new small grains when they "pinch off" as indicated in Figure 4. If sufficient driving force is available, new grains can also be created without the requirement for a pre-existing grain boundary, in a process called diffusion-induced recrystallization (DIR) [10].

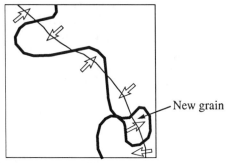

Figure 4: The creation of new grains by DIGM, as demonstrated by Pan & Balluffi [9]. The original and final grain boundary positions are indicated by thin and thick lines, respectively, and the directions of grain boundary migration are indicated by open arrows.

The free energy of mixing is not sufficient to permit the formation of bulges and/or new grains with very small radii of curvature, but the formation of sub-micron grains has been observed during DIGM in thin-film specimens [9]. It is now acknowledged that this is a result of the directed migration of the grain boundaries deriving from coherency strain effects [11]. Coherency strains result from the alloying or de-alloying of the material adjacent to the grain boundary. Two adjacent grains can have different strain energy deriving from the same concentration and atomic misfit, because of the elastic anisotropy of the material. The difference in elastic strain energy then drives an atomic flux toward the lower-energy grain, moving the boundary in the direction of the higher-energy grain. It has been shown that boundary migration velocities do vary as the square of the misfit as predicted [12], and that the direction of migration is always toward the grain presenting the elastically harder orientation [13]. Undersized and oversized solute atoms are equally able to drive DIGM.

At small driving forces, grain boundaries tend to undergo DIGM in a relatively planar form and this does not tend to lead to the creation of new small grains. Under these conditions, then, DIGM is likely to accelerate grain growth by providing an additional, if randomly directed, driving force for grain boundary migration. When the driving force is high, however, the creation of new grains becomes possible and DIGM can contribute to a net reduction in the grain growth rate by providing a means of replenishing the population of small grains.

Stress Effects

Adding further complication, thin films are never formed in a stress-free state. In general, thermally evaporated films on substrates are formed in a state of planar tensile stress while sputtered films tend to embody planar compressive stresses. Unrelieved stress in a thin film can affect the progress of grain growth, because the grain boundaries are narrow regions of less dense material than the crystal matrix [14]. As the grain size increases, the amount of this "grain boundary material" decreases and the mean density of the specimen increases. This can lead to the generation of tensile stresses in constrained specimens, which are superimposed upon the residual stresses from the growth process. This increases the strain energy in evaporated films and reduces it in sputtered films, with corresponding effects on the driving force for grain growth.

36

Grain boundary diffusion affects the generation and/or relief of thin film stresses that normally accompanies grain growth, in several ways. First, the tensile stress generated during grain growth is usually calculated on the basis of the model indicated in Figure 5. If the film thickness remains constant and the film is constrained from shrinking, the film develops a tensile strain corresponding to an increase in mean linear intercept grain size, m:

$$d\varepsilon_{11} = d\varepsilon_{22} = xd\left(\frac{1}{m}\right) \tag{1}$$

where x is the linear expansion per grain boundary, and is assumed here to be uniform. The energy associated with these strains can, however, be removed by a process essentially similar to Coble creep, in which atoms diffuse from the film surface to the grain boundaries allowing the film to relax in its plane while shrinking in thickness.

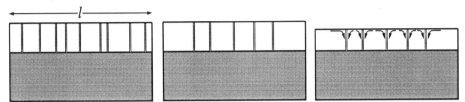

Figure 5: Schematic illustration of the creation of tensile stress by grain growth and its removal by diffusion. The fully relaxed length of the thin film is initially $l = n_0(m + x)$ where n_0 is the number of grains in length l, m is the mean linear intercept grain size and x is the average linear expansion at each grain boundary. If the film thickness remains constant, the product $n_i m$ is essentially constant and the film would shrink by $\Delta l = (n_0 - n_i)x$, if it were unconstrained. The resulting stress can be rapidly relaxed by diffusive fluxes as indicated in the final figure.

Chaudhari considered the effect of increasing strain energy and decreasing interface energy during grain growth [14]. Below a critical initial grain size, the combination of the two changes in energy produces a minimum in the film energy vs. grain size plot, and the stress generated during growth will result in grain growth stagnation. For an initial grain size larger than this critical size, there is no minimum in the energy curve and the generated tensile stress will not suppress growth. Capillarity limits on grain growth [15] occur at larger grain sizes than Chaudhari's critical size.

When a source of a second element is provided at the surface of the film, the free energy of mixing can drive grain boundary diffusion and generate stress in the film in a process distinct from diffusion resulting from film stress. The introduction of misfitting atoms certainly introduces coherency strains that drive DIGM as described above, but more significant stresses may arise from the Kirkendall effect which can force additional material into the film or deplete atoms from it by injecting vacancies. These will cause the generation of large compressive or tensile stresses, respectively, and may compensate or enhance any pre-existing stress in the film.

If grain growth occurs during film deposition it can provide a means of stress relief. The extent of grain growth is greatest for the material nearest the substrate, whether it results from a heated substrate or from the heat deposited in the film by the condensing material, since the first-deposited material is exposed to the high temperature for the longest time. This part of the film then carries the smallest residual stress and the free surface carries the greatest [16] resulting in a tendency for film curling which is revealed if delamination occurs.

Texture Effects

Texture development is commonly observed during grain growth in thin films and, in particular, FCC metals exhibit a tendency to develop a $\langle 111 \rangle$ "wire texture" which appears to result from the relatively low surface energy associated with close packed planes. $\langle 111 \rangle$-oriented grains have low energy and therefore enjoy a growth advantage which exhibits itself in the development of texture, and especially in the form of abnormal grain growth. If all of the grains in a specimen are ⟨111⟩-oriented there is no growth advantage and abnormal grain growth can be suppressed [17]. When the surface of the film is in contact with another material, rather than being exposed to a vacuum, it is the interfacial energy rather than the surface free energy that determines the formation of this type of texture.

Surface contamination and/or bulk alloying can affect the shape of the surface energy Wulff-plot, and thus alter the propensity of a thin film to adopt a texture. And since a surface source is pre-requisite for grain boundary diffusion, any experiment used to investigate diffusion is inevitably complicated by this effect.

If surface-energy-driven texture formation is suppressed, so too is secondary grain growth in thin films. Under these conditions, the process of grain growth tends to stagnate when the grain size is approximately equal to the film thickness, because of a balance of capillarity forces [14].

EXPERIMENTAL PROCEDURES

Thin films of gold were prepared by thermal evaporation onto unheated rocksalt substrates. A nominal film thickness of 25nm was used for all of the experiments. The films were removed from the substrates by immersing them in distilled water, and they were caught on TEM grids made of aluminum, silver or copper. The grids therefore served both to support the films and as the source of the diffusing solute. One set of films was collected on gold grids to serve as a control sample.

Films, supported on grids, were annealed at 150°C. The microstructures of the films before and after this anneal were compared by inspecting them in a TEM.

RESULTS

As deposited, the films had small equiaxed grains and a slight ⟨111⟩ fiber texture. During the 150°C anneal, grain growth took place in all specimens, but the microstructures of the diffusion specimens could easily be distinguished. The mean grain sizes (mean equivalent cylindrical diameter) were smaller for the copper and silver diffusion specimens and larger for the aluminum diffusion specimen than for the control specimen. Figure 6 shows the microstructures of the original film, the control specimen, and the three diffusion specimens after 24 hours of annealing. The control had a mean grain size of 86nm, while the aluminum, silver, and copper had mean grain sizes of 92nm, 52nm, and 54nm, respectively.

TEM conical dark field imaging revealed that the control specimen had microscopic texture variations. The shapes of strong and weak texture regions that developed during annealing reflect the shape of topographical variations on the substrate surface. A detailed discussion of the development of these texture variations and their dependence on substrate topography has been presented elsewhere [17]. Strongly and weakly textured regions of the control specimen have different microstructures. There is a broader distribution of grain sizes in the weakly textured regions than in the strongly textured regions as a result of abnormal growth of ⟨111⟩ oriented grains. After the 24 hour anneal, the weakly and strongly textured regions had mean grain size of 99nm and 72nm respectively.

Texture variations were not observed in any of the diffusion specimens. The strength of the ⟨111⟩ texture present in these specimens was estimated by comparing the brightness of the {220} ring with the other rings of the diffraction pattern. The aluminum diffusion specimen had a preferred ⟨111⟩ orientation; the {220} ring was considerably brighter than the other rings. Abnormal growth of ⟨111⟩ oriented grains took place during annealing, but these grains were evenly distributed, not localized in patches as in the control specimen. The strong texture observed in regions of the control specimen was not observed in the aluminum diffusion specimen. The copper and silver diffusion specimens had very narrow distributions of grains sizes, and no abnormal grains were

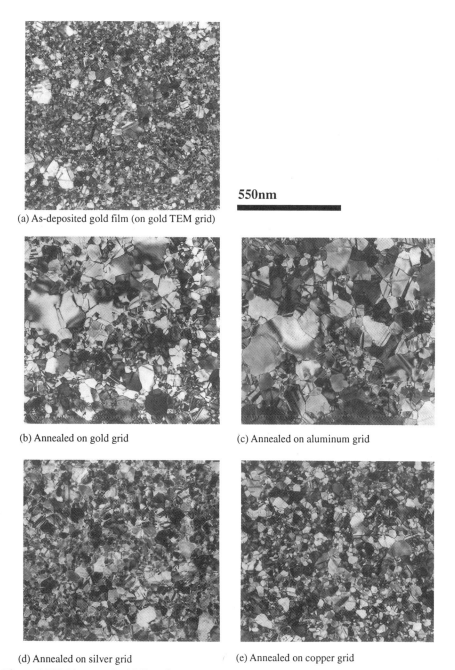

550nm

(a) As-deposited gold film (on gold TEM grid)

(b) Annealed on gold grid

(c) Annealed on aluminum grid

(d) Annealed on silver grid

(e) Annealed on copper grid

Figure 6: TEM images of gold film microstructure. (a) is the "as deposited" film and the other images are films annealed at 150°C for 24 hours on grids of various materials, as indicated.

observed. The diffraction pattern of the copper diffusion specimen showed a slightly preferred $\langle 111 \rangle$ orientation, and the silver diffusion specimen had nearly random orientation.

Figure 7 shows low magnification images of silver and copper films that were annealed for 20 hours. The specimens were tilted approximately 35° to enhance the contrast. The images show that both films are rumpled. The control and aluminum diffusion specimens both appeared relatively flat. The rumpling of the films is particularly pronounced around the grid bars and derives from the increasing specimen volume as solute diffuses into the film faster than gold diffuses out.

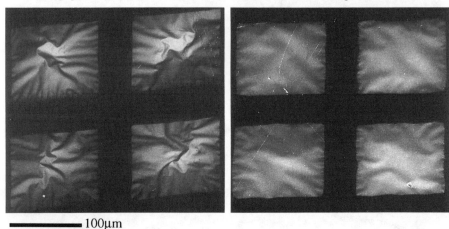

100μm

Figure 7: Low magnification TEM images of films annealed on silver (left) and copper (right) grids. Extensive rumpling of the films is caused by their growth by absorption of atoms from the grids.

DISCUSSION

Our experiments indicate that different diffusing species have differing effects upon grain growth in thin films. In the case of aluminum, grain growth appears to be accelerated, while copper and silver substantially retard it. It is of interest to try to understand why these effects occur, so they can be predicted when untested solute/solvent systems are encountered. The most obvious difference between aluminum and the other solutes, of course, is that aluminum is much less soluble in gold. Although electron diffraction produced no evidence for the formation of second phases, it is a relatively insensitive technique. Imaging, likewise, produced no evidence of precipitation. We base our discussion upon the phenomena introduced in the section on the effects of grain boundary diffusion.

Interfacial Energy

Grain boundary energy effects generally tend to slow grain growth, as described above. These may therefore contribute to the reduction of the grain growth rate in the presence of copper or silver, but do not offer any indication of why aluminum accelerates it. The same effects that we have considered here may also be involved in the nucleation of "conventional" DIGM.

Diffusion Induced Grain Boundary Migration

DIGM is thought to arise because of misfit strain effects and, as described above, it is likely to result in reduced rates of grain growth resulting from the nucleation of new grains. The atomic misfit of aluminum, relative to the gold lattice, is -0.76%, and lies between that of silver (+0.14%)

40

and copper (-11.37%), both absolutely and in magnitude. It is therefore unlikely that the observed effects on grain growth are related in a simple way to lattice mismatch. The total strain energy deriving from mismatch, however, is the product of the atomic misfit and the solute concentration; and the solute concentration will be affected by the grain boundary diffusivity. Silver diffuses about three orders of magnitude faster than copper in the grain boundaries of gold [18] so the higher silver concentrations can easily make up for the smaller misfit and yield a similar level of coherency strain energy. There are no reliable data for the diffusivity of aluminum in gold grain boundaries, although it appears to be a relatively slow diffuser. The similarity of the effects of copper and silver can thus be rationalized in terms of a misfit (or localized stress) argument, but the effect of aluminum in accelerating the grain growth is still puzzling.

Under extreme circumstances, DIGM may also result in grain refinement and there is some evidence of that at very short times in some of our experiments. DIGM is a well established phenomenon during the diffusion of both copper and silver in gold [8].

Stress

The effects of tensile stress in the film can easily be eliminated from consideration. Both silver and copper cause the films to grow, as demonstrated by the "rumpling" of the films shown in Fig.7. This is caused by a net flux of atoms into the film, deriving from the Kirkendall effect. The growth of the film relieves tensile stress and therefore any tendency of the tensile stress to retard grain growth. The grain sizes at the beginning of our experiment also exceed the estimated value of Chaudhari's limit so growth would not, in any case, be halted by this effect although it might have been retarded.

Texture

The texture development in the films appears to be markedly affected by the availability of a solute source. It is not clear whether this is a result of the altered grain growth process, or if it contributes to the changes in the grain growth rate. Texture development may be affected by the differential effects of solute on different boundaries in the specimen [19], but the existence of texture also provides driving force for grain growth under some circumstances. Surface-energy driven growth of $\langle 111 \rangle$-oriented grains certainly contributes to abnormal grain growth in pure gold films, and we observe many such grains in the aluminum-infused specimens in this study. Abnormal grain growth is suppressed in the presence of copper or silver sources.

Our findings for pure gold films have been reported elsewhere [17]. It was shown that the films can develop two distinctly different, co-existing textures. One is a very precise $\langle 111 \rangle$ texture with a mean deviation of only 1.5°; the other a less precise texture with a mean deviation of about 5°. Abnormal grain growth occurs only in the less precisely textured regions of the film. In the present study, the film on the aluminum grid developed a uniform texture qualitatively similar to that of the "less precise" pure gold texture developed by abnormal grain growth. The grain growth in the aluminum-exposed film was, however, more significant than in the "less-textured" pure gold film and the quantitative details of the texture and grain growth interdependence remain to be established, as does the mechanism by which aluminum suppresses the very sharp texture found in the pure gold films.

The film exposed to copper developed only a very slight texture and the film exposed to silver had no discernible change from the orientation distribution present in the unannealed gold. Neither of these films exhibited abnormal grain growth. It is not yet clear whether the lack of texture derives from the altered grain growth behavior, or from effects of the solute upon the surface free energy that drives abnormal grain growth.

A Final Comment

In many cases the goal of grain size control in thin films is the elimination of grain boundaries to reduce electromigration failure rates. Our experiments indicate that grain growth is enhanced if a "less precise" texture is developed in a thin film. The grain growth rate is smaller both for a very precise texture and in the absence of texture. One might then conclude that the less precise texture is desirable. Unfortunately, we have also found that this is the structure most prone to the formation of hillocks, under certain conditions [17]. For films attached to substrates, other considerations (most notably stress) may be important if solute sources are used to control grain growth.

SUMMARY

Solute diffusion exerts powerful effects upon grain growth. In some cases the effects appear through the nucleation of new grains and result in retardation of the average grain growth rate. In other cases the effects appear to be related to the control of texture and its consequent effect on the available driving force for abnormal grain growth. Depending upon the solute, grain growth in the film can be enhanced or suppressed.

ACKNOWLEDGMENT

This work was supported by NSF grant number DMR-9204589.

REFERENCES

1. D.A. Smith, C.M.F. Rae and C.R.M. Grovenor in Grain Boundary Structure and Kinetics, Ed. R.W. Balluffi, ASM, Metals Park OH (1980) p.337

2. L.G. Harrison, Trans. Faraday Soc., **57**, 1191 (1957).

3. J. Hu and D.N. Seidman, Scripta Metall. Mater., **27**, 693 (1991).

4. W. Swiatnicki, S. Lartigue-Korinek, A. Dubon and J.Y. Laval, Mater. Sci. For. **126-128**, 193 (1993).

5. R.J. Jahn and A.H. King, Acta Metall. Mater., **40**, 551 (1992).

6. W. Bollmann, G. Michaut and G. Sainfort., Phys. Stat. Sol. (a), **13**, 637 (1972).

7. W.T. Read and W. Shockley, Phys. Rev. **78**, 275 (1950).

8. A.H. King, Int. Mater. Revs. **32**, 173 (1987).

9. J.D. Pan and R.W. Balluffi, Acta Metall. **30**, 861 (1982).

10. Li Chongmo and M. Hillert, Acta Metall. **29**, 1949 (1981).

11. M. Hillert, Scripta Metall. **17**, 237 (1983).

12. W.H. Rhee, Y.D. Song and D.N. Yoon, Acta Metall. **34**, 2039 (1986).

13. F.S. Chen, G. Dixit, A.J. Aldykiewicz, Jr. and A.H. King, Met. Trans. A **21**, 2363 (1990).

14. P. Chaudhari, J. Vac. Sci. Technol., **9**, 520 (1971).

15. W.W. Mullins, Acta Metall., **6**, 414 (1958).

16. D. Goyal and A.H. King, J. Mater. Res., **7**, 359 (1992).

17. K.E. Harris and A.H. King, J, Elec. Mater., to be published.

18. I. Kaur, W. Gust and L. Kozma, *Handbook of Grain and Interphase Boundary Diffusion Data*, University of Stuttgart, 1989.

19. J.W. Rutter and K.T. Aust, Acta Metall., **6**, 375 (1958).

THE GEOMETRY OF GRAIN DISAPPEARANCE
IN THIN POLYCRYSTALLINE FILMS

S. J. TOWNSEND* AND C. S. NICHOLS**
Departments of *Physics, and **Materials Science and Engineering
Cornell University, Ithaca, New York 14853

ABSTRACT

During grain growth, shrinking columnar grains in thin-film polycrystalline microstructures eventually reach sizes comparable to the film thickness. Due to surface drag, the sides of such grains may bow inward rather than remaining flat through the bulk of the film. The grain boundaries delimiting such small shrinking grains may become unstable long before the surface of the shrinking grain reaches zero area. We report simulation results demonstrating such an instability in the limit of infinite surface drag. This may lead to extremely rapid disappearance of 4- or 5-sided grains, such as have been recently observed in *in situ* hot-stage TEM experiments on aluminum thin film polycrystals.

INTRODUCTION

Interest in thin polycrystalline films stems partly from their technological importance in applications such as integrated circuit interconnects and protective coatings. Since the properties of thin polycrystalline films are profoundly effected by their microstructure, we ultimately seek a complete understanding of microstructural time evolution. Progress in this area requires both experimental and modelling work. This paper emphasizes modelling work prompted by previous experiments[1].

Thin film microstructures with columnar grains are usually considered to be effectively two-dimensional. However, this picture can break down when the diameter of a grain (measured in the plane of the film) is comparable to the film thickness. This geometry occurs in two important cases: columnar grain growth and state-of-the-art integrated circuit interconnects. Grain disappearance events can be viewed as the hallmark of grain growth, since it is the redistribution of the volume of grains that disappear that allows the *average* grain size to increase. In columnar thin film grain growth, any shrinking grain must reach a size comparable to the film thickness before disappearance. In state-of-the-art interconnect lines the width of the line is about 10 times the thickness or less. Due this confined geometry, the size of columnar grains in a microstructure spanning the width of the line must typically be comparable to the film thickness.

FRAMEWORK FOR 3D COLUMNAR GRAIN MODELLING

Our general framework for columnar three dimensional grain modelling considers an aggregate of grains, each spanning the film thickness (Figure 1). The state of the system is taken to be the configuration of all grain boundaries. For each state there is a computed excess free energy, and the system evolves through minimizing this excess free energy. The free energy contributions cataloged by Stüwe[2] can be accommodated in this framework, including the grain boundary energy, energies from volume free energy densities in the grains, and surface energies of the exposed faces. The triple junction line energy, as

43

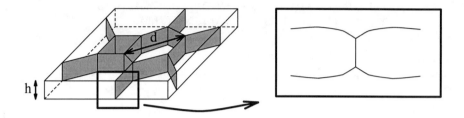

Figure 1: Schematic section of a thin film of thickness h containing a 6-sided grain of diameter d. Grain boundaries are shown in grey and *flattened* for clarity.

highlighted by Morgan and Taylor[3], can also be included. A strength of this framework is that it allows the direct modelling of the grain boundaries and their evolution using a finite element method. Computer simulations can be used to probe the sorts of system behavior characteristic of various combinations of energy contributions, and to compare to experimental data. For our analysis we use a simple grain boundary model, where grain boundary free energy is the only important term, this energy is isotropic, and all boundaries have the same mobility.

Grain boundary behavior is also effected by the presence of the free film surfaces. The evolution of the free surfaces is coupled to the boundary evolution at the grain-grain-air trijunction lines, where grain boundaries meet the film surfaces. In our framework the free surfaces are not explicitly modelled, so approximations to the surface effects must be added to the behavior of the boundaries where they meet the film surfaces. As also noted by Génin, Mullins, and Wynblatt[4], the free surface evolution and its coupling to the grain boundaries would ideally be explicitly simulated.

One way to incorporate surface effects is to consider an effective drag on the boundary motion at the surface. There are two limits: no surface drag and infinite surface drag. The no-drag limit is characterized by effective two-dimensionality, since there is nothing to break the symmetry through the thickness of the film. The von Neumann rule is expected to apply [5], predicting that the rate of area change of a grain is proportional to its number of sides minus six. In the infinite surface drag limit, the lines where the grain boundaries meet the free film surface are pinned making the overall grain boundary configuration static and area-minimizing. In this limit, the predicted shape for a "cylindrical" thin-film grain is the shape taken on by a soap-film stretched between to coaxial rings, known as the catenoid[6].

CASE STUDY: COLUMNAR GRAIN DISAPPEARANCE

In-situ TEM observations of grain growth in Aluminum thin films have found some unusual columnar grain disappearance events[1]. These unusual events involve the abrupt disappearance of four and five–sided grains with no resolved intermediate states. The grain size-to-thickness ratio in these case is on the order of one, suggesting that the bulk of the film might be important.

Our corresponding model simulates individual shrinking grains. Since the grain boundary configuration near the abruptly disappearing grains appeared nearly static before the abrupt disappearances, we use the infinite surface drag limit, in which the grain boundaries are pinned where they meet the film surfaces. This limit should also apply in quasi-static cases where the pinning lines of the boundaries at the film surfaces adjust slowly with respect to boundary migration rates.

Computational Method

Single grains were modelled via their delimiting grain boundaries in 3, 4, and 5–sided cases. The grain boundaries between adjacent neighbor grains are included and terminated at "posts" spanning the film thickness. This allows the grain boundary triple junction equilibria to effect the shape of the grain. The symmetrical positioning of the posts causes the neighbor-neighbor boundaries to stay radial and corresponds to an additional approximation that removes the influence of the microstructural environment on the grain-of-interest. The boundaries delimiting the grain of interest are held fixed where they intersect the film surfaces. Two cases of pinning line geometry were investigated: straight, and circular arcs meeting at 120°.

Grain diameter d is the diameter of a circle in a film surface passing through the points where the grain boundary trijunctions meet the film surfaces, and h is the film thickness. The energy of a configuration of grain boundaries is taken to be proportional to the total grain boundary area. In this approximation, minimal shapes for configurations with the same aspect ratio $\alpha \equiv d/h$ differ only in overall scale. This scale was set so that the volume of a grain of constant cross-section through the film is exactly one.

For each number of sides and pinning geometry, we varied the aspect ratio and carried out surface area minimizations using Surface Evolver[7]. In each geometry, there is found to be a critical aspect ratio, below which a stable solution minimizing total grain boundary area ceases to exist. In order to achieve good convergence, iterations must be carried out with a relatively coarse mesh representing the surface before refining the mesh. Slightly above the critical aspect ratio, however, the representational error from a coarse mesh can lead to the boundaries pinching off in the center of the film, falsely indicating an aspect ratio below the critical value. To overcome this problem, an additional constraint of fixed grain volume was introduced. The constraint volume becomes an additional input, and the minimization returns the pressure as an additional output. This pressure is the negative of the energy change per volume swept out by the boundaries if the constraint were released. Positive pressure thus indicates a surface that would shrink without the constraint, negative pressure indicates a surface that would expand, and zero pressure indicates a stable surface.

Typical coarse meshes had 48 triangular facets per grain side, with 8 edges along each triple line. Standard Evolver iterations moving all vertices simultaneously via a step-optimizing gradient descent method were carried out until the fractional energy change per iteration fell below 10^{-7}. The mesh was then refined by subdividing each facet into four by connecting the side midpoints. Iterations were then resumed until the same level of convergence was re-attained. Next, Evolver's conjugate gradient mode was employed until the fractional change fell below 10^{-12}. Additional steps were then carried out if pressure had not also converged. This was necessary when the boundary conditions led to nearly stable surfaces (i.e. with very small pressures).

For each number of sides and pinning geometry, there is a stability curve $\alpha_s(V)$ as shown in Figure 2 for the 4-sided, arc pinning case. For a given constraint volume V,

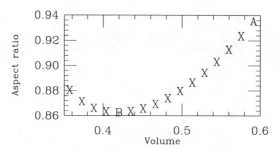

Figure 2: Computed stability curve $\alpha_s(V)$ in the 4-sided, arc-pinning case.

$\alpha_s(V)$ is the aspect ratio of the stable boundary configuration with that volume. Points on this curve correspond to stable grain boundary configurations (see Figures 3A and 3B). The minimum point on the stability curve (labeled B in Figure 2) has the critical aspect ratio α_c. Computationally, this critical aspect ratio is found a using a two dimensional search of pressure $P(V, \alpha)$ with respect to constraint volume V and aspect ratio α. Each evaluation requires a minimization run as described above. The critical aspect ratio α_c is found by applying Brent's minimization algorithm[8] to $\alpha_s(V)$. The value of $\alpha_s(V)$ at each volume V requested by the minimization routine is found as the root of $P(\alpha, V)$ using Brent's method[8].

Results and Discussion

The computed critical aspect ratios for 3, 4, and 5–sided grains in each of the two pinning geometries (Table 1), follow the expected trend of decreasing with increasing numbers of sides. In addition, the similarity of the 5–sided grains in the two pinning geometries, stemming from the nearness of pentagonal angles to 120°, leads to similar critical aspect ratios.

It is valuable to compare an analytical approximation suggested by H. J. Frost[9] to the computed results. In the approximation, the sides of a grain with circular arc pinning are taken to be sections of the corresponding catenoid surface. The coaxial ring constraints of this catenoid are separated by the film thickness and have the same radius of curvature as the pinning lines. A grain is then assumed to become unstable when this corresponding catenoid would become unstable. The results of this approximation are shown in Table 1. The approximate values fall systematically below the corresponding circular arc pinning computed points because the approximation neglects triple line equilibrium. A grain

Grain type	No. of sides		
	3	4	5
Straight line pinning	1.55	0.943	0.581
Circular arc pinning	1.16	0.863	0.566
Frost approximation	0.871	0.552	0.268

Table 1: Critical aspect ratios, α_c

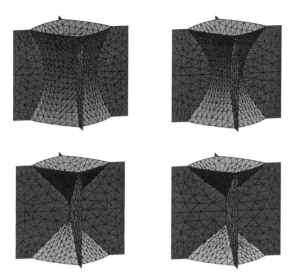

Figure 3: A (top-left) and B (top-right) are structures for points A and B in Figure 2. C (bottom-left) and D (bottom) are speculative states beyond the instability.

with catenoid section sides has interior dihedral angles less than 120° at its waist. With equilibration, the waist of such a grain would be forced to shrink to make these angles 120°, moving the grain boundary configuration closer to the instability (or pushing it past). Thus the simulated grains, which respect triple junction equilibrium, reach the instability at larger aspect ratios than their Frost-approximation counterparts.

Simulation Relevance

Through the simulations, we developed a description of anomalous disappearances as depicted in Figure 3. We imagine a grain as in Figure 3A shrinking quasi-statically as the pinning lines pull inward slowly. When it reaches the critical point (Figure 3B) there are no statically stable configurations of smaller aspect ratio. We envisage the grain boundaries then rapidly passing through configurations depicted *schematically* in Figures 3C and 3D as the central grain disappears. This view is the grain boundary analog of the "popping" of a catenoid-shaped soap film when its rings are pulled too far apart.

Suggestively, computed critical aspect ratios are similar to the observed aspect ratios (~ 1) of grains disappearing abruptly. However, it should be noted that the simulations do not include the possibility depinning of the boundaries at film surfaces. Just before the onset of the instability, the static simulations predict that the grain boundaries of the shrinking grain would meet the surface at sizable angles measured relative to the film normal (See Figure 3B). The well-known mechanism of pinning via thermal surface grooves (inset to Figure 1) probably cannot anchor a boundary that meets the pinning line at such angles[6]. Though it is not hard to imagine additional pinning owing to oxides layers, impurities at the surface, and other factors, more detailed investigations are needed. Dynamic simulations with high but finite surface drag should also shed light on the extent to which the underlying geometric instability in the static case effects the dynamic case.

CONCLUSIONS

Due to recent experiments, a new view of thin-film grain growth is emerging including abrupt disappearance events. We have proposed a novel mechanism for such events that depends crucially on the thickness of the film. Our analysis predicts that such events would be frequent in "young microstructures" where the average grain aspect ratio is on the order of one. Such events are also expected to be more common in confined geometries, such as state-of-the-art interconnect lines, that require any columnar grains to have small aspect ratio.

More generally, we have proposed a framework that is part of a shift toward investigating individual events in grain growth. This is important since some properties, such as electromigration failure, depend sensitively on the types of grain boundaries and their specific arrangement. In the context of such properties, individual events that happen to effect particularly important boundaries can have a large effect on the properties of the microstructure. Similarly, individual events are expected to be more important in confined geometry where the effected boundaries from a single event could be a significant fraction of the boundaries traversing the sample.

ACKNOWLEDGEMENTS

We would like to thank E. A. Holm and H. J. Frost for helpful discussions. We are indebted to K. A. Brakke whose public domain program Evolver was invaluable. This work was supported by a DOE Computational Science Graduate Fellowship(SJT) and an IBM Faculty Development Award(CSN) and was carried out in part at Sandia National Laboratories. This work made use of MRL Central Facilities supported by the National Science Foundation under Award No. DMR-9121654.

---

[1] C. S. Nichols, C. M. Mansuri, S. J. Townsend and D. A. Smith, *Acta metall. mater.* **41**, 1861 (1993).

[2] H. P. Stüwe, in *Recrystallization of Metallic Materials*, edited by F. Haessner (Riederer-Verlag, Stuttgart, 1978).

[3] F. Morgan and J. E. Taylor, *Scripta Metallurgica et Materialia* **25**, 1907 (1991).

[4] F. Y. Génin, W. W. Mullins and P. Wynblatt, *Acta metall. mater.* **40**, 3239 (1992).

[5] J. A. Glazier and D. Weaire, *J. Phys.: Condens. Matter* **4**, 1867 (1991).

[6] W. W. Mullins, *Acta met.* **6**, 414 (1958).

[7] K. E. Brakke, *Experimental Mathematics* **1**, 141 (1992).

[8] W. H. Press, S. A. Teukolsky, W. T. Vetterling, and B. P. Flannery *Numerical Recipes in C*, 2nd ed. (Cambridge University Press, New York, 1992).

[9] H. J. Frost (private communication).

COMPUTER SIMULATION OF STRAIN ENERGY AND SURFACE- AND INTERFACE-ENERGY ON GRAIN GROWTH IN THIN FILMS

R. Carel, C. V. Thompson, and H. J. Frost[1]; Department of Materials Science and Engineering, MIT, Cambridge, MA; [1]Thayer School of Engineering, Dartmouth College, Hanover, NH.

ABSTRACT

We have simulated strain energy effects and surface- and interface-energy effects on grain growth in thin films, using properties of polycrystalline Ag (p-Ag) on single crystal (001) Ni on (001) MgO for comparison with experiments. Surface- and interface-energy and strain energy reduction drive the growth of grains of specific crystallographic orientations. The texture that will result when grain growth has occurred minimizes the sum of these driving forces. In the elastic regime, strain energy density differences result from the orientation dependence of the elastic constants of the biaxially strained films. In the plastic regime, strain energy also depends on grain diameter and film thickness. In p-Ag/(001) Ni, surface- and interface-energy minimization favors Ag grains with (111) texture. In the absence of a grain growth stagnation, the texture at later times is always (111). However, for high enough strains and large enough thicknesses, the strain energy driving force can favor a (001) texture at early times, which reverts to a (111) texture at later times, once the grains have yielded.

INTRODUCTION

Grain growth in thin films rapidly results in a grain structure with an average grain size larger than the film thickness. The grain boundaries fully traverse the thickness of the film and the microstructure is columnar. Further grain growth occurs primarily through boundary motion in the plane of the film, so that grain growth in thin films can be modeled as a quasi two-dimensional process [1-5].

This paper presents a two-dimensional simulation of grain growth which has been modified to include surface effects including the effects of grain boundary grooving as well as two orientation dependent driving forces: surface- and interface-energy and strain energy. Surface- and interface-energy and strain energy need not favor the same crystallographic orientations in the film. We have investigated the competitive effects of these driving forces on the evolving average crystallographic orientation of the film.

DRIVING FORCES FOR GRAIN GROWTH

In a thin film on a substrate, grains with different crystallographic orientations have different surface- and interface-energy. The energy of the top surface of a grain is determined solely by its texture. In the case of a single crystal substrate, the film substrate interface energy also depends on the texture of the grains, but depends on the in-plane orientation of the grain lattice with respect to the substrate lattice, and the orientation of the interface plane as well. This leads to free energy differences between grains with different textures and/or in-plane orientations, which can drive secondary grain growth [6, 7]. The surface- and interface-energy driving force is expressed by

$$\Gamma = \frac{\Delta\gamma}{h\gamma_{gb}},\qquad(1)$$

where h is the thickness of the film, γ_{gb} is the grain boundary energy per unit boundary area and $\Delta\gamma$ is the difference in surface- and interface-energy between the two grains meeting at the grain boundary.

Thin films on substrates are often deposited at a temperature below the temperature at which grain growth occurs. Because the film and the substrate in general have different thermal expansion coefficients, the film, when annealed from the deposition temperature to the grain growth temperature, will be submitted to a strain in the plane of the film given by,

$$\varepsilon = \int_{T_{dep}}^{T_{gg}} \left(\alpha_f(T) - \alpha_s(T)\right)dT \equiv \left(\alpha_f - \alpha_s\right)\left(T_{gg} - T_{dep}\right),\qquad(2)$$

where α_f and α_s are the thermal expansion coefficients of the film and the substrate respectively, and T_{dep} and T_{gg} are the deposition and grain growth temperatures, respectively. The latter expression in equation (2) results when α_f and α_s are weakly temperature dependent.

Due to elastic anisotropy, grains with different textures will have different in-plane elastic constants. In the elastic regime, this leads to a texture dependent state of stress, and therefore texture dependent difference of the strain energy density. A grain with a texture (hkl) submitted to a biaxial strain ε has a strain energy density

$$W_\varepsilon = M_{hkl}\varepsilon^2, \tag{3}$$

where M_{hkl} is an orientation dependent effective biaxial modulus which can be derived from the stiffness tensor of the material [8, 9].

If the stress in the grain is high enough, yielding will occur. The yield stress of a polycrystalline thin film is expected to depend on both the texture [8, 10] and the geometry (grain thickness and grain diameter) [8, 11-13] of the individual grains. A simple model for the yield stress of individual grains in a polycrystalline thin film can be developed [14] by considering the propagation of a dislocation from the surface of a grain throughout the thickness of the grain. This leads to a yield stress approximately given by

$$\sigma_y \cong \frac{\sin\phi}{b\cos\phi\cos\lambda}\left(\frac{2K_s}{d\sin\phi} + \frac{K_b}{h}\right), \tag{4}$$

where d and h are the grain diameter and the film thickness respectively, and where λ is the angle between the Burgers vector and the film plane normal and ϕ is the angle between the normal to the slip plane of the grain and the film plane normal. K_s and K_b are the energy per unit length of the segments of the dislocation at the side and bottom of the grain, respectively [8, 12].

The strain energy density of a yielded grain of texture (hkl) is then

$$W_\varepsilon = \frac{\sigma_y^2}{M_{hkl}}, \tag{5}$$

where σ_y is given by equation (4) and depends on both the texture and geometry of the grain. When yielded, the state of stress and strain energy density of a grain depend on both the orientation and the geometry of the grain, and therefore vary from grain to grain in the film. This leads to free energy differences for grains with different orientations and/or diameters. This energy can contribute to the driving force for grain growth [10, 11].

The general expression for the strain energy density difference between two grains is given by the appropriate combination of equation (3) and (5) as none, one, or both grains can have yielded. The driving force for grain growth due to the strain energy density difference is:

$$F_\varepsilon = \frac{\Delta W_\varepsilon}{\gamma_{gb}}, \tag{6}$$

where ΔW_ε is the strain energy density difference between the two grains meeting at the grain boundary.

The total local grain boundary velocity is proportional [2-4] to the sum of the driving forces for grain growth,

$$v = \mu(\kappa + \Gamma + F_\varepsilon) \tag{7}$$

where μ is a mobility constant containing the grain boundary energy. Equation (7) represents the most general form of the local velocity law applied to grain boundary points in our simulations.

SIMULATIONS TECHNIQUES

Ag/Ni/MgO system

For the purpose of comparison with experiments [9, 15], we have used parameters characteristic of the Ag/(001)Ni/(001)MgO system in our simulations. The interface energy of Ag on (001) Ni as a function of in-plane orientation has been calculated for this system by Dregia et al using embedded atom potentials and molecular static relaxation of (001)Ag‖(001)Ni and (111)Ag‖(001)Ni twist boundaries [16, 17]. For these boundaries, the deepest minimum for the sum of the surface and the interface energy corresponds to a (111)Ag‖(001)Ni twist boundary with [110]Ag‖[001]Ni in-plane alignment. This orientation will be referred to as (111);0°. The minimum of the sum of the surface energy and the interface energy for the (001)Ag‖(001)Ni twist boundaries occurs for a 26° twist away from the cube on cube orientation, i.e. 26° from [001]Ag‖[001]Ni, referred to here as (001);26°. In our simulations, we use a simplified version of the energy functions calculated by Dregia et al (Figure 1), with fewer in-plane orientations.

The strain energy density of (001) Ag grains and (111) Ag grains in the elastic regime is [8, 9]:

$$W_\varepsilon (001) = M_{100} \, \varepsilon^2, \qquad W_\varepsilon (111) = M_{111} \, \varepsilon^2 \qquad (8)$$

where ε is the magnitude of the biaxial stress, $M_{100} = 76$ GPa and $M_{111} = 174$ GPa, when W_ε is given in GJ/m^3. The magnitude of the biaxial strain ε was chosen to be representative of the strain arising from differential thermal expansion between MgO and Ag, as described by equation (2), for a deposition temperature between 77K and 300K and a grain growth temperature of 693K [9, 15].

The yield stress of (001) and (111) Ag grains is given by equation (4). (111) grains always have a higher yield stress than (001) grains, but for a film under biaxial strain, they will also have a higher state of stress since the elastic constants of (111) grains are higher than those of (001) grains. For this system, surface- and interface-energy reduction favors the growth of grains in the (111);0° orientation and strain energy density reduction favors the growth of (001) grains. When both driving forces are accounted for in the simulations, they compete to determine the orientation and the microstructure of the film.

Initial Condition

We started with a Johnson-Mehl [18] initial structure comprised of 9,943 grains. We allowed normal grain growth, without driving forces other than due to κ, up to normalized time τ=0.1, when 7,732 grains remained. The resulting structure is characteristic of the steady-state normal grain growth regime where a parabolic growth law is obtained as well as a self similar grain structure [2]. At time τ=0.1, half of the grains were randomly chosen and assigned a (100) texture. The other grains were assigned a (111) texture. Grains with a given texture were then randomly assigned an in-plane orientation, and therefore an interface energy, thereby ensuring that all possible in-plane orientations for a given texture are equally represented in the initial grain structure. Once the thickness of the film was chosen, the average initial grain diameter d was chosen such that it scaled with the film thickness h, typically d/h=1. Simulation of further grain growth included the additional driving forces due to strain, surface and interface energies. More details about the simulation techniques can be found elsewhere [2-5, 20].

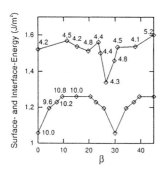

Figure 1: Sum of the surface- and interface-energy for two interface boundaries of Ag on (001)Ni as a function of misorientation angle (based on references 16 and 17). The top curve is for (001) Ag ‖ (001) Ni, the bottom curve corresponds to (111) Ag ‖ (001) Ni. The surface fraction of grains in a given orientation, as initialized at τ = 0.1, is indicated above each data point.

SIMULATION RESULTS

Figures 2a-2c shows the evolution of a 1000 Å thick structure submitted to a strain of 0.004. The (001) grains are rapidly consumed by the (111) grains and for τ greater than 20, there are only (111) grains left in the structure. The rapid decrease of surface fraction of (001) grains and the corresponding increase of the surface fraction of (111) grains is due to the lower relative magnitude of the strain energy driving force and the surface- and interface-energy driving force. For these values of strain and thickness, the surface- and interface-energy driving force supersede the strain energy driving force from the beginning of the simulation, resulting in an advantage for (111) grains, which can grow and consume the (001) grains. The relative surface fraction of grains in the (001);26° and (111);0° orientation is shown in Figure 2-d. Because those orientations correspond to the minima for each texture, grains in other orientations are rapidly consumed. Even though all orientations are equally represented at time τ=0.1, the structure can be considered to be composed of grains in only these two interface energy minimizing orientations as early as τ=8. For τ greater than 20, the structure is only constituted of grains in the (111);0° orientation and the surface- and interface-energy driving force is exactly zero. Further grain growth is driven only by grain boundary curvature and the relatively small strain energy driving force existing between yielded grains.

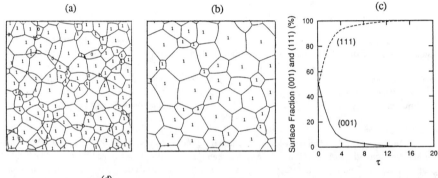

(a) (b) (c)

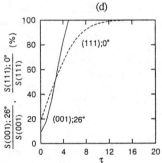

(d)

Figure 2: Simulated structure evolution for a 1000 Å thick Ag films subjected to a strain of 0.004. (a) structure at τ = 5, (b) structure at τ = 20, (c) surface fraction of grains with (001) and (111) texture as a function of τ, (d) surface fraction of grains in the (001);26° and (111);0° orientations as a percentage of the surface fraction of grains with (001) and (111) textures respectively.

Figure 3a-3c shows the evolution of a 1000 Å thick structure subject to a strain of 0.0066. For those values of strain and thickness, the structure initially develops a (001) texture which eventually reverses to a (111) texture. At early times, the average grain diameter is small and all grains are in the elastic regime. A strain of 0.0066 is sufficient for the strain energy driving force to overcome the surface- and interface-energy driving force if the grains are in the elastic regime, resulting in the rapid growth of (001) grains. However, when the grains reach the size at which they yield, as described by equation (4), their strain energy density becomes a rapidly decreasing function of grain size, as described by equation (5). Furthermore, the strain energy density difference between yielded (001) and (111) grains is also a rapidly decreasing function of grain size. The surface- and interface-energy driving force is grain-diameter-independent so that there is a grain diameter regime for which the surface and interface-energy driving force overcomes the strain energy driving force. When that regime is reached, the (001) grains stop growing and the texture reverses from (001) to (111). As in the case of a strain of 0.004, the in-plane orientation selection of the grains is determined by the minima of each energy curve so that for times greater than $\tau=12$, almost all the grains are either in the (001);26° or the (111);0° orientations.

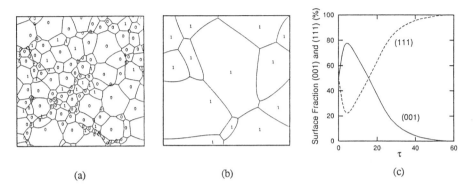

(a)   (b)   (c)

Figure 3: Simulated structure evolution for a 1000 Å thick Ag film subjected to a strain of 0.0066. (a) structure at $\tau = 5$, (b) structure at $\tau=55$, (c) surface fraction of grains with (001) and (111) texture as a function of τ.

DISCUSSION AND CONCLUSIONS

For the system considered here, surface- and interface-energy anisotropy and strain energy density anisotropy do not favor the same orientations. Depending on the applied strain and the film thickness, either (111) or (001) orientations dominate at early times. The texture developing at early times can be predicted by comparing the magnitude of the surface- and interface-energy driving force and the strain energy driving force. This is done by equating equations (1) and (6), which defines a line in the (ε,h) plane:

$$\varepsilon = \sqrt{\frac{\Delta\gamma}{(M_{111} - M_{100})}} \frac{1}{\sqrt{h}}, \tag{9}$$

where $\Delta\gamma$ can be taken as $\gamma(100);26^O]-\gamma(111);0^O]$ since these orientations rapidly consume all others. If no yielding of the grains is considered, the texture of the structure is determined by the quantity $\varepsilon\sqrt{h}$. If $\varepsilon\sqrt{h}$ is greater than $\sqrt{\Delta\gamma / (M_{111} - M_{100})}$, the texture of the structure is (001) at all times. If this is not the case the texture of the structure is (111). However, if yielding is allowed, the structure will always be (111) at later times, due to the grain-diameter-independent surface- and interface-energy driving force which always overcomes the strain energy driving force once the grains are yielded and have large enough diameters. There is therefore, under some circumstances, a grain size/film thickness "window" in which strain energy minimization can dominate in texture selection. Stagnation phenomena, such as due to grain boundary grooving [19, 20], can lead to stagnation of grain growth at a stage in which strain energy minimization textures are dominant [21]. Otherwise, surface and interface energy minimizing orientations will eventually become dominant as grain growth (and yielding) proceeds. The texture of the structure therefore depends, in general, on three parameters: biaxial strain, film thickness and time. The biaxial strain, in turn, is influenced by such things as the relative thermal expansion coefficients of the film and substrate, and the difference between the deposition temperature and the grain growth temperature. The range of strains and film thicknesses that lead to a (001) texture is reduced as the texture is observed at increasing time, as illustrated on Figure 4. The predicted dependences of the dominant texture on the strain and film thickness are consistant with experiments [9, 15].

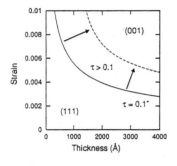

Figure 4: Dominant texture of the films when observed at different times, as a function of thickness and applied strain.

ACKNOWLEDGMENTS

Our thanks to Derek Walton and Nicholas Tung Chan for their help with the computer simulator. This work was supported by the NSF and the SRC, the latter through contract # 93-SP-309.

REFERENCES

1.   H. V. Atkinson, Acta. Metall. **36**, 469 (1988)
2.   H. J. Frost, C. V. Thompson, C. L. Howe, and J. Whang, Scripta Met. **22**, 65 (1988)
3.   H. J. Frost, C. V. Thompson, and D. Walton, Acta Metall. Mater. **38**, 1455 (1990)
4.   H. J. Frost, C. V. Thompson, and D. Walton, Acta Metall. Mater. **40**, 779 (1990)
5.   H. J. Frost, Y. Hayashi and C. V. Thompson, MRS Symp. Proc. **317** (1994)
6.   C. V. Thompson, J. A. Floro, and Henry I. Smith, J. Appl. Phys. **67**, 4099 (1990)
7.   H. J. Kim and C. V. Thompson, J. Appl. Phys. **67**, 757 (1990)
8.   W. D. Nix, Met. Trans. **20A**, 2217 (1989)
9.   J. A. Floro, C. V. Thompson, R. Carel, and P. D. Bristowe, submitted to J. of Mat. Res.
10.   J. E. Sanchez, Jr. and E. Arzt, Scripta Metall. Mater. **27**, 285 (1992)
11.   P. Chaudhari, IBM J. Res. Develop., 197 (1969)
12.   L. B. Freund, J. Appl. Mech. **54**, 553 (1987)
13.   R. Venkatraman and J. C. Bravman, J. Mat. Res. **7**, 2040 (1992)
14.   C. V. Thompson, Scripta Met. et Mat. **28**, 167 (1993)
15.   J. A. Floro, R. Carel, and C. V. Thompson, MRS Symp. Proc. **317** (1994)
16.   Y. Gao, S. A. Dregia, and P. G. Shewmon, Acta Metall. **37**, 1627 (1989)
17.   Y. Gao, S. A. Dregia, and P. G. Shewmon, Acta Metall. **37**, 3165 (1989)
18.   W. A. Johnson and R. F. Mehl, Trans. Am. Inst. Min. Engrs **135**, 416 (1939)
19.   W. W. Mullins, Acta Metall. **6**, 414 (1958)
20.   H. J. Frost, C. V. Thompson, and D. Walton, Acta Metall. Mater. **38**, 1455 (1990)
21.   C. V. Thompson, H. J. Frost, and R. Carel, unpublished.

ASYMPTOTIC BEHAVIOR IN GRAIN GROWTH

DAVID T. WU
Center for Materials Science, Los Alamos National Laboratory, Los Alamos, NM 87545

ABSTRACT

Hillert's model of grain growth consists of a drift term in size space that leads asymptotically to a distribution function and a growth exponent not often observed. Later theories introduce a diffusion term that is either assumed to dominate the drift term or a correction to it. This paper shows that the lower order drift term alone determines asymptotic grain growth behavior. A possible conclusion is that experimental results may need to be reinterpreted.

1. INTRODUCTION

Grain growth is the process by which the average size of grains in a polycrystalline material increases. Unlike computer models, which simulate the detailed behavior of individual grains, analytic theories for grain growth generally take a statistical approach based on the continuity equation

$$\frac{\partial f}{\partial t} = -\frac{\partial j}{\partial R},$$ (1)

where f is the distribution function for grains of size R at time t and j is the current in size space. Grains are assumed to be equiaxed, so that R is the radius of a sphere having equal volume, and parameters such as the number of sides of a grain are ignored. A "model" for grain growth consists of writing down an expression for j.

Hillert [1] started with the assumption that the boundary between two grains will migrate with a velocity proportional to its curvature. Additional assumptions about the behavior of grains in a mean field led to the expression

$$j_{Hillert} = vf$$
$$v = M_0 \left(\frac{1}{R_*} - \frac{1}{R} \right),$$ (2)

with v the drift velocity in size space and R_* a time-dependent critical radius. Since the current has no other terms, this is known as a drift theory. Hillert was able to show asymptotically that the distribution function becomes self-similar when plotted in terms of the reduced variable R/R_* and that R_* grew as $t^{1/2}$. Unfortunately, experimental results do not generally support his predictions. Measured distributions tend to be broader with a gentler cutoff at the large grain sizes, and the fitted growth exponents show a spread between $1/4$ and $1/2$ [2].

Louat [3] has developed a theory in which curvature plays no role. He contends that grain growth occurs through the random motion of grain boundaries and introduces a diffusive current

$$j_{Louat} = -D_0 \frac{\partial f}{\partial R}.$$ (3)

Even though his model has been shown to violate volume conservation [4] , Louat paved the way for modification of Hillert's theory. Hunderi and Ryum [5], for example, have suggested that grain growth contains elements of both deterministic drift and random walk. This paper considers the asymptotic behavior of Hillert's model with a diffusional correction:

$$j = j_{Hillert} + j_{diffusion}. \tag{4}$$

Since Louat's diffusion current is inadmissible, the question remains of what form the diffusive term should take.

2. DERIVATION OF THE DIFFUSION TERM

2.1 Grain growth as a discrete process

Analytic theories of grain growth usually classify grains in terms of the size R, but we are free to change this variable to any function of R. In the case of three-dimensional grain growth, for example, using the number of atoms k in a grain to replace R is just as valid if not more natural. There is, however, one essential difference between the two representations: R is a continuous variable, whereas k is discrete. We shall see shortly that treating grain growth as random walk in a discrete space actually determines the form of the diffusion current when continuum approximation of the rate equation is taken.

Given an assembly of grains, let us construct the distribution function H_k, the number of grains with k atoms. Since grains always shrink or grow by the exchange of atoms, we can model growth with transitions between bins (Fig. 1). In the general case, that of biased random walk, the forward and backward transition coefficients are not the same, the difference of which leads to a non-zero drift velocity [6]. Here we are concerned with unbiased random walk, for which the forward and backward coefficients are equal. Calling the transition rate $\mu_k H_k$, where μ_k is the transition coefficient, and taking into account the flux from neighboring bins, we obtain the rate equation for the population at bin k

$$\frac{dH_k}{dt} = \mu_{k-1}H_{k-1} - 2\mu_k H_k + \mu_{k+1}H_{k+1} \tag{5}$$

2.2 Continuum approximation

In order to obtain a partial differential equation, let us replace k by a continuous variable V,

$$V = \varepsilon k$$
$$\varepsilon = \text{volume of one atom,} \tag{6}$$

representing the volume of a grain, and introduce continuous distribution function h and diffusion coefficient D:

$$h(V,t) = \frac{H_k(t)}{\varepsilon}$$
$$D(V) = \varepsilon^2 \mu_k. \tag{7}$$

Substituting these functions into the rate equation and performing a Taylor expansion for all terms, we obtain an infinite order partial differential equation [7], which, to leading order in ε, is just the diffusion equation

$$\frac{\partial h}{\partial t} = \frac{\partial^2}{\partial V^2}(Dh). \qquad (8)$$

Since Hillert's flux is always written in terms of R, we should transform this equation to the same coordinate system. It can be shown [7] that the general result for grain growth in n dimensions is the diffusion equation

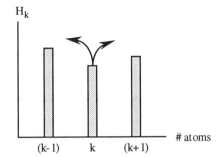

Fig. 1 The distribution function H_k is the number of grains having k atoms. The arrows represent the shrinking or growing of k-atom grains.

$$\frac{\partial f}{\partial t} = \frac{\partial}{\partial R}\left(\frac{1}{R^{n-1}}\frac{\partial}{\partial R}\left[\frac{D}{R^{n-1}}f\right]\right), \qquad (9)$$

which implies that the correction to Hillert's current in equation (4) should take the form

$$j_{diffusion} = \frac{1}{R^{n-1}}\frac{\partial}{\partial R}\left(\frac{D}{R^{n-1}}f\right). \qquad (10)$$

Mulheran and Harding [8] have given a physically motivated theory for the diffusion coefficient D. They argue that D must scale with the number of atoms on the surface of a grain

$$D = D_0 R^{n-1}. \qquad (11)$$

We will use this diffusion coefficient in the following section.

3. ASYMPTOTIC ANALYSIS

We are now ready to consider the behavior of the full rate equation

$$\frac{\partial f}{\partial t} = \frac{\partial}{\partial R}\left(M_0\left[\frac{1}{R} - \frac{1}{R_*}\right]f\right) + \frac{\partial}{\partial R}\left(\frac{1}{R^{n-1}}\frac{\partial}{\partial R}\left[\frac{D}{R^{n-1}}f\right]\right). \qquad (12)$$

While the following dimensional analysis is not rigorous, a more detailed argument leads to the same result [7]: Taking Mulheran and Harding's choice for D, we see that the terms in equation (12) have dimensions

$$\frac{1}{t} \sim \frac{1}{R^2} + \frac{1}{R^{n+1}}. \qquad (13)$$

Since the average size increases with time, the diffusion term will eventually become negligible relative to the drift term for $n = 2$ or 3. This shows that asymptotically the solution of the enhanced equation is identical to Hillert's.

4. DISCUSSION

Given that Hillert's predictions do not agree with experimental results and that his asymptotic solution is unchanged by a diffusional correction, we can perhaps conclude that Hillert's theory requires major revision. Alternately, we could argue that the *interpretation* of experimental results needs further consideration. This viewpoint can be made defensible by looking at the variation of the average grain size with time.

Using simple dimensional analysis once again, we can assert that at some short time when the average size is sufficiently small, the diffusion term actually dominates the drift term. In this limit the average size must be growing like $t^{1/(n+1)}$, in contrast with the asymptotic limit when it varies as $t^{1/2}$. There must exist a transition zone during which the exponent increases smoothly between the values $1/(n+1)$ and $1/2$ (Fig. 2). Because the decay toward steady state is governed by a power law rather than by an exponential, the transient regime may be considerable. If an experiment is not carried out for a sufficiently long time, it is possible to mistake the transient for the steady state. This is of course only one explanation for the scatter of growth exponents observed experimentally.

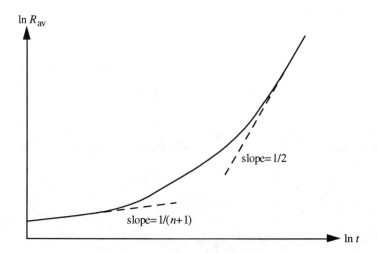

Fig. 2. Schematic variation of the average grain size with time. The slope of the plot gives the growth exponent, which varies smoothly through the transition zone between the values $1/(n+1)$ and $1/2$.

5. CONCLUSION

We have shown that adding a diffusion term to Hillert's theory of grain growth has no effect on the asymptotic solution. However, the transient solution now contains a transition zone governed by slow power law decay. Unless an experiment is followed for a sufficiently long time, it is possible to mistake the transient regime for the steady state.

ACKNOWLEDGMENT

We thank Dr. R. LeSar for discussions. This work was supported in part by an appointment to the U.S. Department of Energy Distinguished Postdoctoral Research Program sponsored by the U.S. Department of Energy, Office of Science Education and Technical Information, and administered by the Oak Ridge Institute for Science and Education.

REFERENCES

1. M. Hillert, *Acta metall.* **13**, 227 (1965).
2. H. V. Atkinson, *Acta metall.* **36**, 469 (1988).
3. N. P. Louat, *Acta metall.* **22**, 721 (1974).
4. A. Thorvaldsen, *Acta metall. mater.* **41**, 1347 (1993).
5. O. Hunderi and N. Ryum, *J. Mater. Sci.* **15**, 1104 (1980).
6. C. W. Gardiner, *Handbook of Stochastic Methods for Physics, Chemistry, and the Natural Sciences.* (Springer-Verlag, Berlin, 1985).
7. D. T. Wu (in preparation).
8. P. A. Mulheran and J. H. Harding, *Mater. Sci. Forum* **94-96**, 367 (1992).

MASS CONSERVATION IN GRAIN GROWTH

DAVID T. WU
Center for Materials Science, Los Alamos National Laboratory, Los Alamos, NM 87545

ABSTRACT

Several alternatives to Hillert's work on grain growth have been proposed. Unfortunately some of these theories have been shown to violate mass conservation. This paper introduces a way to enforce conservation for any model via a term representing self-interaction of the grain distribution.

1. INTRODUCTION

Analytic theories of grain growth are generally based on the continuity equation

$$\frac{\partial f}{\partial t} = -\frac{\partial j}{\partial R},$$ (1)

in which f is the distribution function for grains of size R at time t and j is the current in size space. Unlike computer models, which simulate the detailed behavior of individual grains, this description is statistical in nature. Grains are assumed to be equiaxed, so that R is the radius of a sphere having equal volume, and parameters such as the number of sides of a grain are simply ignored.

Hillert's model for j started with the assumption that the boundary between two grains will migrate with a velocity proportional to its curvature [1]. Additional assumptions about the behavior of grains in an average environment led to the expression for growth in n dimensions

$$j_{Hillert} = \alpha_0 \left(\frac{1}{R_*} - \frac{1}{R} \right) f$$

$$R_* = \frac{\int_0^\infty R^{n-1} f\, dR}{\int_0^\infty R^{n-2} f\, dR},$$ (2)

where R_* is a time-dependent critical radius. Ryum and Hunderi [2] have criticized this model for being axiomatic, as they could find little justification for it.

Louat [3], on the other hand, asserted that grain boundary motion is random. His expression for the current is independent of n:

$$j_{Louat} = -D_0 \frac{\partial f}{\partial R}.$$ (3)

In spite of reasonable agreement with experimental results, Louat's theory is fundamentally flawed. Hunderi and Ryum [4], among others, have pointed out that Louat's current violates mass conservation. Rather than asking whether a particular model conserves mass, the approach that Hunderi and Ryum took, this paper concentrates on *how* mass is conserved in growth.

61

Mat. Res. Soc. Symp. Proc. Vol. 343. ©1994 Materials Research Society

2. MASS CONSERVATION IN PHYSICAL SPACE

2.1 Global conservation

The total mass of an assembly of grains is

$$M \propto \int_0^\infty R^n f dR. \tag{4}$$

There is no question that any viable theory must conserve this quantity. By taking the time derivative of the total mass and using the continuity equation (1), we can show that the constraint on j is [5]

$$\int_0^\infty R^{n-1} j dR = 0. \tag{5}$$

This condition, however, is not sufficient to conserve mass because it allows for the existence of isolated sources and sinks. The principle of mass conservation states that mass is conserved locally, not just globally. Global conservation of course follows from local conservation.

2.2 Local conservation

By local mass conservation, we mean that given an arbitrary volume in physical space, mass in and out of that volume must be equal. Let us focus for now on the interaction between two grains in physical space (Fig. 1). Assume that the grain on the left is made of k atoms and is about to grow by the exchange of one atom (shaded) with its neighbor, which initially has k' atoms. Having made the exchange, the k grain has grown to a $(k+1)$ grain, but, at same time, the k' grain has shrunk to a $(k'-1)$ grain. Simply put, a grain never grows without making one of its neighbors shrink simultaneously. This is local mass conservation applied to grain growth. If we consider grain k in a real environment surrounded by several neighbors of varying sizes, the same conclusion holds: Grain k grows by making one of its neighbors shrink.

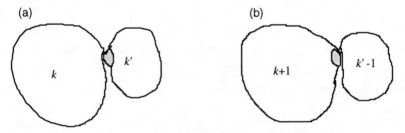

Fig. 1 Local mass conservation in physical space. A grain having initially k atoms grows by taking an atom from a neighboring grain. The neighbor loses an atom and shrinks from a k' grain to $(k'-1)$: (a). Before the exchange. (b). After the exchange.

3. MASS CONSERVATION IN SIZE SPACE

In this section we recast local mass conservation in the setting of size space . Only simple arguments are presented, as details are given elsewhere [5]. To make connection with the statistical description in size space, we must describe the process in Fig. 1 using probability distributions. Let us suppose that someone has constructed a model of grain growth defined by j. The current $j(R)$, its time-dependence assumed, tells us the rate at which grains of size R want to grow; but, as we have seen, a given R grain is surrounded by neighbors of arbitrary size R', which presumably want to grow at rate $j(R')$. The effective growth rate for grain R, $j_{eff}(R)$, must be less than $j(R)$ because its neighbors, in trying to grow, are taking atoms from grain R simultaneously. To calculate this correction, we must know the probability that a randomly chosen area of grain boundary is shared by grains R and R'. The general result is [5]

$$j_{eff}(R) = j(R) - \int_0^\infty dR' \, p(R \mid R') j(R') \left(\frac{R'}{R} \right)^{n-1} \qquad (6)$$

where $p(R \mid R')$ is the probability that, given a grain R', grain R is its neighbor. It is easy to show that j_{eff} conserves mass globally [5].

4. APPLICATION

In this section we apply local mass conservation to "derive" Hillert's model, thereby giving it the physical justification that Ryum and Hunderi found lacking. We use the same assumption as Hillert, that the boundary between two grains will migrate with a velocity proportional to its curvature, i.e.

$$j = -\frac{\alpha_0}{R} f \qquad (7)$$

(the minus sign is required because a boundary wants to move toward its center of curvature). This is a model in which all grains want to shrink to minimize surface area. If we assume mean field, i.e. that there are no size correlations so that the neighborhood of any grain looks the same, then $p(R \mid R')$ is proportional to the surface area of grain R:

$$p(R \mid R')\big|_{mean\,field} = \frac{R^{n-1} f}{\int R^{n-1} f dR} . \qquad (8)$$

Substituting equations (7) and (8) into the general expression (6) for the effective flux, we obtain

$$j_{eff} = \alpha_0 \left(\frac{\int_0^\infty R^{n-2} f dR}{\int_0^\infty R^{n-1} f dR} - \frac{1}{R} \right) f , \qquad (9)$$

which is precisely Hillert's current, equation (2).

5. CONCLUSION

We have shown that applying local mass conservation to grain growth leads to a correction term in the current representing the interaction between a grain and its surrounding. We also showed that Hillert's model of grain growth is equivalent to a mean field theory in which all grains want to shrink. The mass conserving backflux gives a physical justification for the critical radius in Hillert's theory.

ACKNOWLEDGMENT

We thank Dr. R. LeSar for discussions. This work was supported in part by an appointment to the U.S. Department of Energy Distinguished Postdoctoral Research Program sponsored by the U.S. Department of Energy, Office of Science Education and Technical Information, and administered by the Oak Ridge Institute for Science and Education.

REFERENCES

1. M. Hillert, *Acta metall.* **13**, 227 (1965).
2. N. Ryum and O. Hunderi, *Acta metall.* **37**, 1375 (1989).
3. N. P. Louat, *Acta metall.* **22**, 721 (1974).
4. O. Hunderi and N. Ryum, *Acta metall. mater.* **40**, 1069 (1992).
5. D. T. Wu (in preparation).

MEAN FIELD ANALYSIS OF ORIENTATION SELECTIVE GRAIN GROWTH DRIVEN BY INTERFACE-ENERGY ANISOTROPY

J. A. FLORO* AND C. V. THOMPSON, Department of Materials Science and Engineering, Massachusetts Institute of Technology, Cambridge, MA 02139.
*Currently at Sandia National Laboratories, Albuquerque, NM 87185-0350.

ABSTRACT

Abnormal grain growth is characterized by the lack of a steady state grain size distribution. In extreme cases the size distribution becomes transiently bimodal, with a few grains growing much larger than the average size. This is known as secondary grain growth. In polycrystalline thin films, the surface energy γ_s and film/substrate interfacial energy γ_i vary with grain orientation, providing an orientation-selective driving force that can lead to abnormal grain growth. We employ a mean field analysis that incorporates the effect of interface energy anisotropy to predict the evolution of the grain size/orientation distribution. While abnormal grain growth and texture evolution always result when interface energy anisotropy is present, whether secondary grain growth occurs will depend sensitively on the details of the orientation dependence of γ_i.

INTRODUCTION

The properties of polycrystalline thin films often depend not only on the average grain size, but also on the distribution of grain sizes and on the degree of preferred grain orientation. Orientation selective grain growth in thin films results in an increase of the average grain size with the concurrent development of preferred orientation. This is an *abnormal* grain growth process, i.e., no steady state grain size distribution can result during the transient period in which orientation evolution is occurring. *Secondary* grain growth, an extreme case of abnormal grain growth characterized by the development of a bimodal grain size distribution, is often observed experimentally [1-5] and generally results in a strong preferred orientation. Secondary grain growth can lead to much larger final grain sizes for a given thermal history than would be expected from normal grain growth [6].

Orientation selectivity arises from the presence of crystallographically anisotropic free energies which bias grain growth. The most ubiquitous of these is the anisotropy of the film/substrate interfacial energy and the energy of the film's free surface. We examine here how the functional dependence of the interfacial energy on the relative grain/substrate orientation promotes abnormal grain growth and the development of preferred orientation. A mean field model for grain growth which includes surface and interface energy anisotropy is numerically evaluated to determine the temporal evolution of the grain size/orientation distribution. We show that a bimodal grain size distribution results when the interface energy exhibits a discrete dependence on orientation, but generally does not result for other orientation dependencies. This result suggests that dramatic secondary grain growth, in which grains more than 2 orders of magnitude larger than the surrounding matrix are observed, requires suppression of normal grain growth, e.g., due to grain boundary grooving, in addition to interface energy anisotropy.

MEAN FIELD THEORY AND THE CONNECTION TO GRAIN GROWTH

Hillert [7] first applied the mean field coarsening formalism developed by Lifshitz and Slyozov [8] to grain growth. A variety of modifications of Hillert's approach have since been reported [9].

Mat. Res. Soc. Symp. Proc. Vol. 343. ©1994 Materials Research Society

Thompson extended the model to include the anisotropy of the surface and interface energy [10]. The physical situation of interest here is shown in Fig. 1. The grains are assumed to be columnar, with film thickness h, and each grain is characterized by its radius R, its surface energy γ_s, and its interface energy γ_i. Both γ_s and γ_i depend on grain orientation, described in general by three angles θ, α, and φ relative to a coordinate system defined for a specific substrate and interface orientation. This results in an orientation dependent mean field growth rate equation:

$$\frac{dR}{dt} = M\left[\frac{\overline{\gamma_s} - \gamma_s}{h} + \frac{\overline{\gamma_i} - \gamma_i}{h} + \gamma_{gb}\left(\frac{1}{\overline{R}} - \frac{1}{R}\right)\right], \qquad (1)$$

where $\overline{R}$ and $\overline{\gamma_{i,s}}$ are appropriately defined averages [10]. M is the grain boundary mobility and γ_{gb} is the grain boundary energy. Both are taken here to be constant. The effects of variable M and/or γ_{gb} have been treated using mean field approaches [11,12] and using simulations [13, 14]. This is an alternative source of orientation selectivity which requires that the initial matrix already be highly oriented in order to be observable. In equation (1), grains with lower than the average values of the matrix will tend to have larger growth rates, leading to development of preferred orientation. This effect increases with decreasing film thickness h.

To determine the evolution of the grain size/grain orientation distribution $f(R,\theta,\alpha,\varphi)$, we solve the size flux continuity equation,

$$\frac{\partial f}{\partial t} = -\frac{\partial(f\dot{R})}{\partial R}, \qquad (2)$$

where $\dot{R}$ is given by equation (1). Equation (2) is solved numerically in the transient regime, as describe in the next section

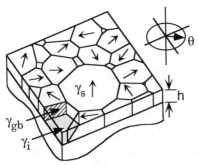

Figure 1. Continuous polycrystalline film of thickness h on a substrate.

DESCRIPTION OF THE ANALYSIS

We give only a brief description of the numerical techniques here. More details may be found in [14]. For simplicity we limit the analysis to one orientation degree of freedom such that $\gamma_i = \gamma_i(\theta)$ and $f = f(R,\theta,t)$. This corresponds to a film with all grains having the same crystallographic axis parallel to the substrate surface normal (uniform texture), but having random rotational orientations θ about that axis. The energy of the planar free surface, γ_s, will not vary with rotation about the surface normal and the first term on the right side of equation 1 will be zero. We thus subsequently refer to the interface energy $\gamma_i(\theta)$ simply as $\gamma(\theta)$. Numerical analysis was carried out by computer solution of equation 2 using the Lax-Wendroff finite differencing scheme [15].

We initially examine coarsening under two different functional forms of the interface energy vs. orientation function $\gamma(\theta)$. In the first interface energy function (IEF), $\gamma(\theta)$ varies logarithmically with θ, a common description of low angle boundaries. The second IEF is a step function in γ vs. θ, examined as an alternative approximation to a cusp in $\gamma(\theta)$. These two IEF's are shown in Fig. 2, where we plot the normalized quantity $(\gamma(\theta) - \overline{\gamma_0})/\gamma_{gb}$ on the vertical axis and where $\overline{\gamma_0}$ is the average interface energy of the system at t=0. The propensity of a system to undergo interface-energy-driven abnormal grain growth is given by a dimensionless scale factor $Z_0 = \overline{R}_0 \overline{\Delta\gamma}_0^{max}/h\gamma_{gb}$, where $\overline{R}_0$ is the initial mean particle radius and where $\overline{\Delta\gamma}_0^{max}$ is defined in Fig. 2. For convenience we introduce the normalized radius $\rho = R/\overline{R}_0$ and the normalized time $\tau = M\gamma_b t/\overline{A}_0$, where $\overline{A}_0$ is the initial mean particle area.

The initial particle size distribution $f(R,\theta,\tau=0)$ input to the program consisted of identical Hillert functions [7], assigned to each of the discrete orientation angles in the range $-\theta_m \le \theta \le -\theta_m$. The initial particle area per unit angle is constant in all cases discussed here. The particle size distribution, which is what is measured experimentally, is given by $f(R,t) = \sum_j f(R,\theta_j,\tau)$.

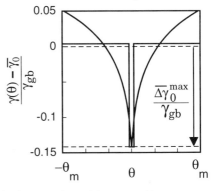

Figure 2. A comparison of the step and log interface energy functions (IEF's) used in the analysis.

RESULTS

Figure 3 compares the evolution of the grain size distribution for grain growth under the log and step IEF's with identical normalized driving force Z_0. It is immediately observed that coarsening under the step IEF produces a transient bimodal particle size distribution (secondary grain growth) while the log IEF results only in abnormal grain growth -- bimodality is not observed. Varying Z_0 by two orders of magnitude (e.g., by varying the film thickness) does not change this observation. Increasing Z_0 does increase the rate of abnormal/secondary grain growth, and, in the case of the step IEF, the degree of bimodality [14].

As abnormal/secondary grain growth occurs, an evolution in the overall crystallographic orientation of the system is also taking place. This is demonstrated in Fig. 4, which shows the grain area at each orientation θ for coarsening under the log and step IEF's. As grain growth progresses, the grain area at the cusp of the IEF's increases preferentially. Coarsening under the log IEF, shown in figure 4a, leads to development of a range of preferred orientations, whereas in Fig. 4b the preferred orientation due to coarsening under the step IEF clearly lies at $\theta<1°$. The shapes of the curves in Fig. 4 obviously follow directly from the shapes of the input IEF's.

In order to determine whether bimodal grain size distributions occur for IEF's intermediate in shape to the log and step functions, we use an empirically constructed interface energy function:

$$\gamma(\theta) = \frac{a \ln (1 + \frac{60}{\pi}\sin 3\theta) + b[1 - \exp(-\frac{\theta^n}{c})]}{a + b} . \qquad (3)$$

The first term in the numerator of equation 3 produces essentially logarithmic behavior except that the function obtains zero slope at θ_{max}. The second term in the numerator produces a simple exponential curve for $n=1$. For $n>2$, the curve becomes s-shaped, approximating a step function. As n is increased, the minimum of the IEF becomes flatter and the transition region between the bottom and top of the curve becomes narrower (more step-like). Decreasing parameter c decreases the width of the cusp. Parameters a and b weight the log-like and step-like contributions.

Since bimodality in the grain size distribution is a transient phenomenon, we need to construct a metric that quantitatively describes the degree of bimodality. Within the context of experimental observations of secondary grain growth, bimodality is large when a few grains grow very large

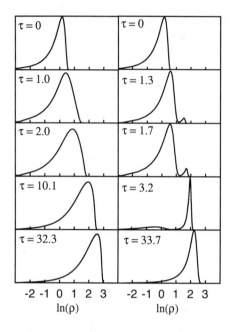

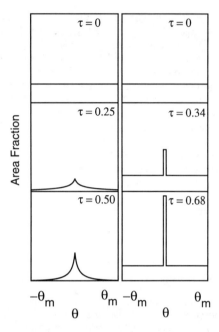

Figure 3. Time evolution of the grain size distributions (number of grains *vs.* normalized radius). Left column: coarsening under the log IEF. Right column: coarsening under the step IEF

Figure 4. Time evolution of the area fraction of the film *vs.* θ. Left column: coarsening under the log IEF. Right column: coarsening under the step IEF.

relative the majority of the grains. Bimodality is small, in this context, when all the grains have similar sizes, or when a few small grains exist within a field of much larger grains. Here we use a bimodality metric that is based on the product of the peak separation times the difference in peak heights. This metric, which satisfies the criteria above and is discussed in ref. [16], is monitored during the numerical analysis. The metric passes through a maximum during abnormal grain growth under any particular IEF and the grain size distribution corresponding to the maximum is output.

In Fig. 5, we vary the shape of the IEF and show the resulting distributions with maximum bimodality metric in each case. In Fig. 5a, we examine the effect of flattening the top of the IEF, i. e., progressively varying the IEF from a condition in which all orientations have different γ, to a condition in which all orientations away from the cusp have the same γ. The latter case is a reasonable description of real IEF's, i. e., orientations away from deep minima have roughly the same γ, with variations that are small compared with the depth of the minima themselves. We see that flattening the top of the IEF produces incipient bimodality, with the extension of a pronounced tail on the distribution to high radius. However, a fully separated secondary peak at large R is not obtained simply by flattening the top of the IEF. In Fig. 5b, we keep the top of the IEF flat, and consider n=1, which produces a sharp cusp, and n=6, which produces a cusp with a flat bottom

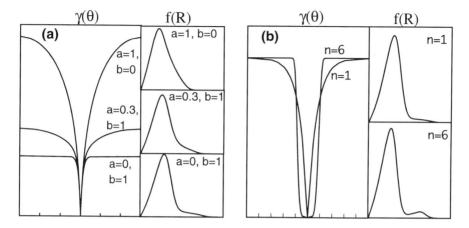

Figure 5. Interface energy functions $\gamma(\theta)$ (left panel of each figure), and the resulting distributions of maximum bimodality (right panel of each figure). (a) Effect of flattening the top of the IEF, with c=0.7, n=1,and (b) effect of flattening the bottom of the IEF, with a=0, b=1, c=0.7. In (b), only a narrow range of θ near the cusp is shown for clarity.

and a narrower transition region between the bottom and top of the IEF. A true bimodal distribution is observed in the latter case. A more extensive survey, varying parameters a, b, c, and n over a range of values and IEF shapes, results in the same conclusion -- only a step-like IEF produces secondary grain growth and true bimodality of the grain size distribution.

DISCUSSION

We find that the step IEF, a discrete, two-state approximation to the interfacial energy, produces secondary grain growth, characterized by a bimodal grain size distribution and well-defined preferred orientation. The continuously varying log IEF, a more realistic description of interfacial energies, results only in abnormal grain growth, characterized by a monomodal grain size distribution and a range of preferred orientations which evolves with time. Very similar behavior has been observed by Frost et al [18,19], who performed extensive simulations of grain growth using a boundary tracking model that incorporates the essential physical phenomena of grain growth in a direct (but computationally demanding) fashion. Frost *et al* used the step and log IEF's and followed the evolution of the total grain size distribution. They found that grain growth under the step interface energy function resulted in the occurrence of a transient bimodal grain size distribution [19]. However, grain growth under the log IEF with produced only monomodal grain size distributions. These results are in good agreement with our results using the simple mean field approach. From an experimental viewpoint these results are somewhat unexpected, in that *bimodality is frequently observed in real systems undergoing grain growth in the presence of surface or interface energy anisotropy* [1-5]. Inasmuch as all models of interphase boundaries predict smoothly varying energies as a function of orientation near the cusp, we do not believe that observation of secondary grain growth in real systems implies that a step-like IEF is present. Rather, the implication is that, within the context of this model, interface energy anisotropy alone does not produce dramatic secondary grain growth characterized by a highly bimodal size distribution. When secondary grain growth occurs in real systems, it is likely that an IEF such as that shown in Fig. 5b with n=1 is present. However, true bimodality apparently requires additional effects such as grain boundary grooving [17]. Grooving can cause stagnation of the misoriented "normal" grain matrix and promote selective breakaway growth of only a small number of grains with maximum driving force (those that lie at the minimum of the IEF).

CONCLUSIONS

Numerical analysis of a mean field model describing interface energy-driven grain growth in thin films has been presented. It was shown that interface energy anisotropy produces abnormal grain growth, with the concurrent development of preferred orientation as the film coarsens. However, only a step-like IEF produced secondary grain growth, i. e., a bimodal grain size distribution. Since interface energies in real systems are not expected to depend discretely on orientation, this result implies that secondary grain growth follows from a combination of interface energy anisotropy and additional effects such as stagnation of energetically unfavored grains due to boundary grooving.

ACKNOWLEDGEMENTS

The authors acknowledge Professor Harold Frost of Dartmouth University for his many useful suggestions and comments. This work was supported under NSF contract # DMR 9001698. Computation facilities were provided by Digital Equipment Corporation. J. A. Floro was supported in part by an AT&T Foundation Scholarship.

REFERENCES

1. J. E. Palmer, C. V. Thompson, and H. I. Smith, J. Appl. Phys. **62**, 2492 (1987).
2. C. V. Thompson, J. Floro, and H. I. Smith, J. Appl. Phys. **67**, 4099 (1990).
3. J.A. Floro and C.V. Thompson, in Thin Film Structures and Phase Stability, B.M. Clemens and W.L. Johnson, eds. (Mat.Res. Soc. Symp. Proc. Vol. **187**, Pittsburgh,PA, 1989), p. 273-8.
4. C. C. Wong, H. I. Smith, and C. V. Thompson, Appl. Phys. Lett. **48**, 335 (1986).
5. H. J. Kim and C. V. Thompson, J. Appl. Phys. **67**, 757 (1990).
6. C. V. Thompson, J. Appl. Phys. **58**, 763 (1985).
7. M. Hillert, Acta metall. **13**, 227 (1965).
8. I. M. Lifshitz and V. V. Slyozov, J. Phys. Chem. Solids **19**, 35 (1961).
9. H. V. Atkinson, Acta metall. **36**, 469 (1988).
10. C. V. Thompson, Acta metall. **36**, 2929 (1988).
11. G. Abbruzzese and K. Lucke, Acta Metall. **34**, 905 (1986).
12. H. Eichelkraut, G. Abbruzzese and K. Lucke, Acta metall. **36**, 55 (1988).
13. A. D. Rollett, D. J. Srolovitz, and M. P. Anderson, Acta metall. **37**, 1227 (1989).
14. H.J.Frost, Y.Hayashi, C.V.Thompson, and D.T.Walton, in Mechanisms of Thin Film Evolution, S. M. Yalisove, C. V. Thompson, and D. J. Eaglesham, eds. (Mat. Res. Soc. Symp. Proc. Vol. **317**, Pittsburgh,PA, 1994), p. 485.
15. J. A. Floro and C. V. Thompson, Acta metall. mater. **41**, 1137 (1993).
16. *Numerical Recipes*, William H. Press, Brian Flannery, Saul A. Teukolsky, and William T. Vetterling (Cambridge University Press, Cambridge, 1986), pp. 623-35.
17. J. A. Floro, Ph.D Thesis, MIT Dept. of Materials Science and Engineering, 1992.
18. W. W. Mullins, Acta metall. **6**, 414 (1958).
19. H. J. Frost, C. V. Thompson, and D. T. Walton, Acta metall. mater. **38**, 1455 (1990).
20. H. J. Frost, C. V. Thompson, and D. T. Walton, Acta metall. mater. **40**, 779 (1992).

GRAIN GROWTH KINETICS IN MICROCRYSTALLINE MATERIALS

V.G. SURSAEVA
Institute of Solid State Physics, Russian Academy of Sciences, Laboratory of Interfaces in Metals, Chernogolovka, Russia.

ABSTRACT

Certain experimental results are presented concerning grain growth in microcrystalline Ag films. Dark field **TEM** technique was used for the measurement of grain size, trijunction velocity and grain boundary mobility. We found that the activation energy for trijunction motion is 25.0 kJ/g.atom, and the activation energy for the grain boundary motion is 50.0 kJ/g.atom.

INTRODUCTION

Most of the films currently employed are polycrystalline aggregates, and their properties (mechanical, optical, magnetic, etc.) are essentially determined by the peculiarities of the grain structure formed during annealing. Despite numerous theoretical, simulation and experimental investigations, many significant aspects of grain growth are still poorly understood due to the enormous complexity of the collective grain boundary motion in a polycrystal.

It is common knowledge [1] that the mobility and surface tension of individual grain boundaries σ_b may differ substantially depending on the misorientation of neighbor grains. However, it has been well established that under certain conditions various materials irrespective of preliminary processing exhibit a particular universal behavior usually referred to as normal grain growth. It implies the power growth law of the mean grain size, self-similarity of grain size distribution and some other features.

The grain growth during recrystallization is realized through mutual motion of grain boundaries and trijunctions under grain boundary surface tension mainly σ_b. It is usually assumed that trijunctions do not drag the motion of boundaries and their role is reduced to maintaining thermodynamically equilibrium angles between them. The kinetics of grain growth is determined by the grain boundary mobility m_b and so-called parabolic law of grain growth takes place [2] :

$$R^2 - R_0^2 = 2A^*t \qquad (1)$$

R - grain size, R_0- initial grain size, t - annealing time, $A = \sigma_b^* m_b$ relation serves as an estimate for the proportionality coefficient A. In case the trijunction posses finite mobility m_j one should take account of the fact that the angles between the boundaries deviate from equilibrium ones, and due to this the trijunctions moves. But the trijunctions in this case drags the boundary migration.

As is shown in [3] the power of influence of trijunctions is determined by the ratio of the value of boundary mobility to that of trijunctions $\lambda = m_b / m_j$, having the length dimension, and by the grain size R. If $R >> \lambda$ the trijunction influence can be neglected, and when $R << \lambda$ the velocity of boundary migration is mainly determined by the trijunction mobility. And instead of Eq.(1) the following should be fulfilled:

$$R - R_o = B^*t \tag{2}$$

Here $B = \sigma_b^* m_j$, that is the grain size is linearly time dependent.

Apparently, no data on the trijunction mobility and , hence, no data about characteristic size of λ below which the trijunction recrystallization kinetics exists.

It is clear that such kinetics should be searched for mainly in _ systems with a small grain size. For instance, microcrystalline metallic films are such systems.

The purpose of my work is an attempt to describe the evolution microstructure in micricrystalline films during grain growth especially on earlier stages.

EXPERIMENTAL

Ag 99.999 was chosen as material for experiments. Microcrystalline 1000Å film, nontextured, were produced by vacuum evaporation on glass substrate with a sugar sublayer. In order to obtain a homogeneous microcrystalline structure the substrate temperature was maintained at 100C ^{o} . The mean grain size was 200-400 Å. The microdiffracton pattern of these films showed that the intensity of rings corresponds to randomly oriented grains. (Fig.1).

Thin films were annealed in a specially designed furnace in vacuum, the temperature being maintained within 1 C^o. Heating and cooling to the required temperature took a few seconds. The film were annealed and investigated in **TEM** on special supporting nets.

In our opinion for obtaining experimentally correct values the following requirements for the experiments should be fulfilled:

1.For representative data it is necessary to measure more than 500 grains;
2.Microcrystalline films should be nontextured and homogeneous;

The **TEM** based microdiffraction pattern give condensed information on the state of the films. Dark field **TEM** based technique was used for the grain size measurements (Fig.1). The dark field image is formed by deflecting electronic beam to different reflections and shows which grains contribute to a diffracted beam. The mean grain size, measured from dark field images, in different diffracted beam, was the same within experimental error.

RESULTS AND DISCUSSION

The mean grain size measurement was made twice on each film: before and after isothermal annealing. The annealing were done at 300 - 600 C ^{o},since it is known that at 700 C^o secondary recrystallization takes place[4]. Grain size growth increases at 300 C ^{o}already.The grain size is changing mainly at 400 - 500 C^o. The mean grain size increases up to 1000 Å. The grains have equilibrium shape. At T=500-600C°the mean grain size becomes more than film thickness.The time dependence of mean grain size for different annealing temperatures are presented in Fig.2. One would think experimental data may be described by parabolic law. However, the relative grain size distribution is steady-state within experimental error after annealing 1-5 hours only. (Fig.3.). The area distribution of grains (cumulative frequency) are linear for annealing time more than 1-5 hours only too (Fig.4.). The stright line in Fig.4 corresponds to the dependence

$$f(S) = 1/<S>*exp(-S/<S>) \qquad (3)$$

Parabolic law, conditions of self-similarity (Fig.3,4) are fullfiled for annealing more than 0.5-1hours only. We have normal grain growth in Ag film in late stage of grain growth only.

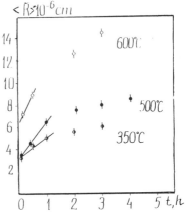

Fig.1.Dark field microcrystalline film image.
    Magnification 100.000.
Fig.2.The time dependence of the mean
    grain size in microcrystalline Ag film
    at different temperatures.

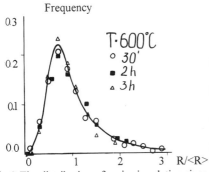

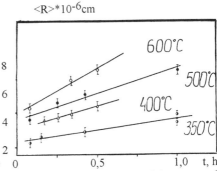

Fig.3.The distribution of grains in relative sizes.   Fig.5. The time dependence of the mean grain
                                                         size in early stage of grain growth.

73

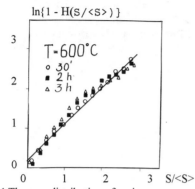

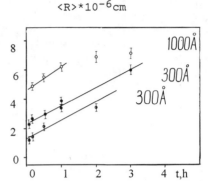

Fig.4. The area distribution of grains. (Cumulative frequency).

Fig.6. The time dependence of the mean grain size at T=350C° in films with different thicknesses.

The experimental data from initial time of recrystalization (0-1 hours) come out experimental error. We investigated more carefully the initial section of the mean grain size time dependence Fig.5 and found that it exhibits linear time dependence, not parabolic .We suppose linear time dependence is due to trijunction drag of the boundary motion. From the slope of this dependence we get the trijunction velocity - the value of coefficient B. Substituting the experimental values of the mean grain size obtained from the parabolic section (Fig.2) we find grain boundary mobility, the value of coefficient A. The estimation of the characteristic value of λ is given in Table I. These data agree with the experimental λ which were determined as the points of intersection of linear and parabolic dependence. Transition **a*** from the linear to parabolic section in the time dependence of the mean grain size may shift considerably to the side of long annealing times, depending on the initial mean grain size R_0.(Fig.6). As one may see from two dependence (for thickness film 300 A) even a transition to the parabolic law is not achieved. The value of the mean grain size corresponding to the linear-parabolic transition depends on the film thickness (Fig.6).

Hillert [5] regarded the grains as interacting spheres and wrote out the expression for the grain velocity as:

$$dR/dt = A/R \qquad (4)$$

Galina et al.[3] show that the trijunction velocity is R- independent and is determined by the product of the trijunction mobility and the surface tension .

$$dR/dt = B \qquad (5)$$

Since the boundary motion is drag by both trijunctions and boundaries, the expression for the grain boundary velocity can be given in the form

$$dR/dt = 1/(R/A + 1/B) = A/(R + A/B) \qquad (6)$$

$$(R + A/B)dR = Adt \qquad (7)$$

Integrating the both sides of the equation (7) we get

$$(R + R_0) + 2A/B = 2At/(R-R_0) \qquad (8)$$

An analysis of this expression shows that once in the experiments the boundaries are draged by both boundaries and trijunctions, the experimental results from Fig.2.,Fig.5,Fig.6. should be rectified in coordinates $(R + R_0)$ and $t/(R - R_0)$ on Fig.7.,Fig.8.,Fig.9. The parameters of the line give the values of boundary mobility A and trijunction velocity B . The values of A and B obtained by treating the data of Fig.2,5,6 agree well with the experimental data in Table I.The activation energy values for boundary motion were determined for straight and parabolic sections of time dependence of the mean grain size (Fig.10). For the linear section the activation energy was found to be 25.0 kJ/g.atom, for the parabolic one 50,0 kJ/g.atom.

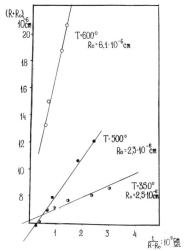

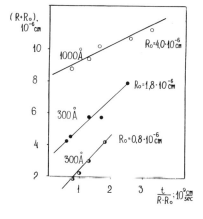

Fig.7.The experimental data on Fig.2 are presented in coordinates $(R+R_0)$ and $t/(R-R_0)*10^9$cm/sec.

Fig.8.The experimental data on Fig .5.are presented in coordinates $(R+R_0)$ and $t/(R-R_0)*10^9$cm/sec.

The grain growth activation energy, identified usually as the boundary migration energy, does not coincide with the activation energy of some diffusion process in solids[1] . As the experiment shows, however, random grain boundary activation energy is close to that of the boundary diffusion [5].

Particular attention should be paid to the section of the mean grain size time dependence with a linear function R(t). As has been mentioned the dependence displays such behavior when trijunction migration is the limiting link of the process [3]. Such dependence has been observed for the first time, and also the activation energy of this process first determined.

It should be noted that the activation energy is very low (25.0 kJ/g.atom), much below any diffusion process in Ag [5] known to us. This means that the migration of trijunction (in case the section under observation determined by it) proceeds as a non-diffusion process.

75

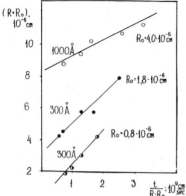

Fig.9.The experimental data on Fig.6 are presented in coordinates $(R+R_0)$ and $t/(R-R_0)*10^9$ cm/sec.

Table I
Results from grain growth kinetics in microcrystalline Ag films.

	350 °C	400 °C	500 °C	600 °C
$B_{exp}, \frac{cm}{sec}$	$3,5 \cdot 10^{-10}$	$5,9 \cdot 10^{-10}$	$9,2 \cdot 10^{-10}$	$1,7 \cdot 10^{-9}$
$A_{exp}, \frac{cm^2}{sec}$	$1,2 \cdot 10^{-15}$	$1,5 \cdot 10^{-15}$	$3,6 \cdot 10^{-15}$	$1,5 \cdot 10^{-14}$
$\lambda_{exp} = \frac{A}{B}, \mu m$	$0,34 \cdot 10^{-5}$	$0,48 \cdot 10^{-5}$	$0,58 \cdot 10^{-5}$	$0,8 \cdot 10^{-5}$
d^*_{exp}, cm	$0,40 \cdot 10^{-5}$	$0,53 \cdot 10^{-5}$	$0,65 \cdot 10^{-5}$	$0,88 \cdot 10^{-5}$

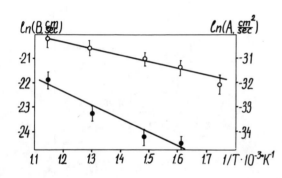

Fig.10.The temperature dependence of grain boundary mobility (•)and trijunction velocity (o) in Aq microcrystalline films.

REFERENCES

1.B.S.Bokstein, Ch.V.Kopezkii, L.S.Shvindlerman. Thermodynamics and Kinetics of Grain Boundaries in Metals, (Metallurgy, Moscow, 1986), p.224.
2.Recrystallization of metallic materials, edited by F.Haessner (Dr.Riederer Verlag,1978) p.351.
3.A.V.Galina, V.E.Fradkov, L.S.Shvindlerman, Fizika metallov i metallovedenie, V.63,1220 (1987).
4.E.I.Tochizkii.Crystallization and heat treatment of thin films, (Science and Tecnik,Minsk,1976) p.312.
5.I.Kaur, W.Gust, in Numerical Data and Functional Relationships in Science and Technology, edited by O.Madelung (Springer-Verlag Berlin Heidelberg,1990).

COMPUTATIONAL AND EXPERIMENTAL STUDIES OF GRAIN GROWTH

Tue T. Ngo*, Reza Ahmadi*, A. Alec Talin**, and R. Stanley Williams**.
*Department of Physics, University of California Los Angeles, Los Angeles CA 90024-1547.
**Department of Chemistry and Biochemistry, University of California Los Angeles, Los Angeles CA 90024-1569.

Abstract

This paper addresses the validity of the single particle growth law $r^n - r_0^n = Kt$ in describing the formation of 3D grains during the vapor deposition of thin metal films, especially with respect to the value of the exponent n. Computer simulations based on the Huygen's construction showed that the grain size distribution at full surface coverage did not depend significantly on the specific model chosen for initiating growth from nuclei. The particle size distributions obtained experimentally by STM measurements of thin Au films deposited on glass substrates agreed very well with the simulation results for n=2.

Introduction

Practical applications of thin films such as catalysts, high density photographic media, magnetic recording media, etc. rely on the unique properties of thin films, such as grain size distribution, grain orientations and film roughness. Improvement of the thin film performance for a specific application not only requires a detailed knowledge of the functionality required but also how to control film growth to produce the desired grain structure. Despite numerous computational and experimental studies, understanding and controlling film growth still remains a challenging problem.

Nucleation and growth of particles during physical deposition of films often occur preferentially at the defect sites on the surface of a substrate. The growth of a single particle is usually approximated by a power law of the form [1]:

$$r^n - r_0^n = Kt, \qquad (1)$$

where r is the linear size (radius) of the particle, r_0 is the critical size at which the growth law becomes valid, t is the growth time, K and n are constants. The value of n is dictated by the controlling transport mechanism. For example, if growth is controlled by diffusion of atoms on the sustrate surface and the driving force for atomic diffusivity is the Gibb-Thompson potential, then n is thought to be equal to 4 [2], but if the atoms perform random-walk diffusion, the atomic flux arriving at the perimeter of a growing particle is constant and n will take a value of 2 [3].

Models of growth by coalescence of droplets are valid for the deposition of low melting point materials, such as tin, gallium, etc., when the substrate is held at high temperature [4,5]. This growth condition allows liquid-like particles to coalesce upon contact with each other, and yields a droplet size distribution with striking scaling behavior [5]. On the other hand, if the deposited material forms solid-like particles (or grains) on a substrate held at relatively low temperature, the particles are immobile and grow until they contact each other and share a 'rigid' boundary. We here consider only the latter case and neglect growth by coalescence.

77

Simulations of 2D grain growth have been performed in the past by other authors [6]. Their studies were devoted primarily to the problem of recrystallization of thin amorphous films, which involves a post annealing process. Our kinetic problem of 3D particle growth shares an analogous formulation, i.e. single particle growth according to Eq. 1, and so can be treated by similar methods.

In the following section, we present selected results of our simulations to explore the behavior of a system of growing grains. We are particularly interested in the form of the particle size distribution functions, and will compare the results of the simulation to experimental data in a later section.

Computational simulations

The growth rate of a single particle is thought to obey a power law such as that of Eq. 1, where r_0 is the size of some minimum nucleus. After size r_0 has been reached, n is assumed to maintain a constant value. When r_0 is small (compared with r), it can be set to zero. In our simulation, particles were not allowed to move. We will followed the methods of simulation and analysis similar to those of Frost and Thompson [7] for 2D growth.

Two types of nucleation were considered in this work. In the first, all particles started to grow at the same time (this is the site saturation case described in Ref. 7). We began with a flat surface and defined a finite number of randomly distributed nucleation sites. All particles started growth simultaneously as hemispheres centered on these nucleation sites. After each time increment in the simulation, the new radius of each hemisphere is $r \propto t^{1/2}$. The hemispheres are truncated where they overlap with each other. This procedure is similar to the Huygen's Construction [8].

Figures 1a,b show two representative stages of the simulated growth for the site saturation case, with surface coverages of roughly 50%, and 98%, respectively. The specific parameters for the simulation are given in the figure captions. Initially, most particles can grow without interfering with each other, and Fig. 1a shows that most of them still retain their hemispherical shape. The average particle radius R = <r> of the collective system thus follows the same growth rate as that of the individual particle, i.e. $dR/dt \propto t^{1/2}$. At later stages, however, grains compete for the limited unoccupied area of the substrate surface (Fig. 1b). The mutual inhibition of growth results in a slower growth rate for R. Figure 2a, which is a plot of ln(R) vs. ln(t), indeed indicates a significantly slower growth rate for R as the system proceeds above 50% coverage. This general behavior should be independent of n.

Since all particles started to grow at the same time, the particles are uniform in size during the early stage of growth, and the initial distribution is narrow. As growth continues, the size distribution widens because of the mutual inhibition of particle growth. Figure (2b) displays the size distribution at 98% coverage, which also shows that when the surface is nearly covered, most particles are smaller than twice the average size and there is a distinct skewness to the distribution.

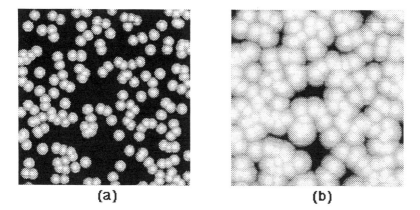

(a)   **(b)**

Figure 1. Simulation images at two different stages of site saturation growth. (a) 50% coverage, (b) 98% coverage. Total number of particles: 788. The expanded images presented here represent one-fourth of the total simulation area.

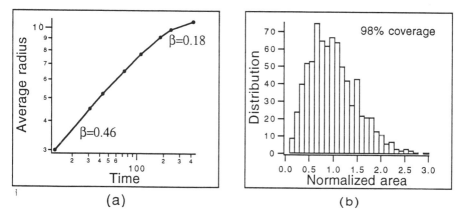

(a)   (b)

Figure 2. (a) The average size R of the system plotted as a function of time on a log-log scale. The dots correspond to 13%, 28%, 35%, 53%, 67%, 84%, 90%, 97% coverage. (b) The histogram of the size distribution of particles from figure 1b. The areas have been normalized with the average particle area set to unity.

In the second type of simulation, the total number of grains, the location of each grain, and the growth law were kept the same as in the previous case, but some particles were allowed to initiate growth earlier than others. Specifically, particles at different sites were born with the same probability at any given time. Thus the growth of an individual particle had the form $r \propto (t-t_0)^{1/2}$, with t_0 being the time when the particle was born. Particles were all born before the end of the simulation. This procedure is similar to the declining nucleation case described by Frost and Thompson [7]. Since nuclei were added at different times, one might expect that the distribution would be significantly broader than that in the site saturation case. However, both the images and the size distribution histograms for the declining nucleation growth model at 97% coverage were nearly identical to those in Figs. 1b and 2b, respectively [3] (and thus are not shown). This result is not too surprising if one considers that the radius is proportional to the square root of time and thus larger particles grow at a slower rate than smaller particles.

Experimental procedure and results

Several thin film samples of gold were grown for this study. The films were deposited in a high vacuum chamber (pressure~10^{-7}-10^{-8} Torr) onto glass substrates from a heated filament wrapped with gold wire. The flux (~10 Å per second) and film thickness were monitored by a quartz crystal monitor placed next to the substrate. Films were deposited with average thicknesses of 190 Å, 400 Å, and 1000 Å at room temperature. Each sample was removed from the chamber after deposition and immediately mounted on a Scanning Tunneling Microscope (STM) for imaging. Our STM [9], which is capable of atomic resolution, is a home-built unit that operates in air. The use of a STM allows us to clearly image grains with nanometer radii. The images are formed from a digital topograph, which is readily available for quantitative analysis.

Figures 3 a,b,c are the STM images of the 190 Å, 400 Å and 1000 Å thick Au films, respectively. As can be seen from these images, the average projected area of the grains increased with amount of Au deposited. The average particle size was taken to be the correlation length determined by calculating the autocovariance function (ACF), which is commonly employed to characterize film roughness [9,10,11]. The average grain size R plotted as a function of average thickness h (which is proportional to deposition time) for each sample is displayed on a log-log scale in Fig. 4a. The value of the slope β measured from this plot is 0.55. We thus obtain an empirical relationship

$$R \propto h^{0.55}, \tag{2}$$

which is close to the choice of n=2 in Eq. 1 that was used for the simulations above.

Discussion

In our simulations, the value of β continuously decreased as the surface coverage approached unity. In our experimental measurements, however, the value of β is about 0.5 (see figure 4a) even for depositions well beyond that which just cover the surface. Thus, in the range of Au film thicknesses studied (from 190 Å to 1000 Å thick), there may be significant grain overgrowth.

Figure 4b is a histogram of the size distribution (plotted as a function of area with the average area set to unity) of grains for the 1000 Å thick film. This plot shows that most grains are smaller than 3 times the average size, in agreement with the simulation results of Fig. 2b. The similarity of the two distributions shows that the Huygen's construction used above captures the essence of the grain growth process for Au films. The grain size in thin films has previously been observed to follow a log-normal distribution [6,12], but as yet there is no physical justification for such a distribution.

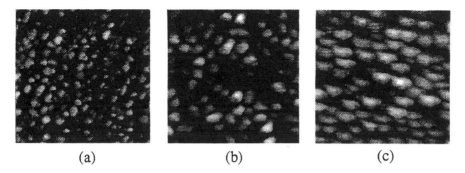

(a)   (b)   (c)

Figure 3. STM images of Au/glass samples with different average film thicknesses. The scan size is 300x300 nm^2. Tunneling current is 1 nA; bias voltage is -100mV. (a) 190 Å, (b) 400 Å, (c) 1000 Å, average Au film thickness.

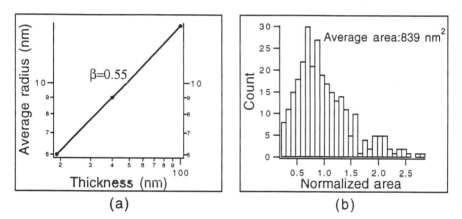

(a)   (b)

Figure 4. (a) The average grain size of each image in figure 4 plotted as a function of thickness on a log-log scale. The exponent value n as estimated from the slope of this plot is 2 (n~1/0.55). (b) Histogram of grain size distribution for the Fig. (3c). Total number of grains is 288. The horizontal axis is scaled with the normalized area a/A. (This histogram is tabulated from an image, not shown, which is four times the area of Fig. 3c).

        We have begun a program of computational and experimental studies aimed at elucidating the kinetic details of grain growth during thin film deposition. Our early simulation results indicate that within the Huygen's construction, the distribution of grain sizes at full coverage is not very sensitive to the details of the nucleation mechanism. Thus, careful experimental studies at low coverage may be required to determine the actual nucleation mechanism, but it may not be important for understanding the grain structure of thick films. The fact that the simulated and experimental grain size distributions are so similar gives us confidence that the simulation methods can yield important insight into the kinetics of film growth. Finally, our results indicate that the growth exponent n=2 represents the dominant grain growth mode for Au on glass.

Acknowledgements

This research was supported in part by the Office of Naval Research.

References

[1] P. Wynblatt and T-M. Ahn, Mater. Sci. Research, Vol. 10, Sintering and Catalysis, edited by G. C. Kuczynski (Ed.) (Plenum Press, New York ,1975) p. 83.

[2] I.M. Lifshitz and V. V. Slyozov, Sov. Phys. JETP **35**, 331 (1959); J. Phys. Chem. Solids **19**, 35 (1961).

[3] T. T. Ngo, R. Ahmadi, R. S. Williams, in preparation.

[4] A. Steyer, P. Guenoun, D. Beysens, and C. M. Knobler, Phys. Rev. A **44**, 8271 (1991).

[5] F. Family, P. Meakin, Phys. Rev. A **40**, 3836 (1989).

[6] H. V. Atkinson, Acta metall **36**, 469 (1988).

[7] H. J. Frost and C. V. Thompson, Acta metall. **35**, 529 (1987).

[8] See, for example, G. Carter and M. J. Nobes, Earth Surf. Process **5**, 131(1980).

[9] E. A. Eklund, Surface Science **285** (1993), p. 157.

[10] G. Rasigni, F. Varnier, M. Rasigni, and J. P. Palmari, and A. Llebaria, Phys. Rev. B **27**, 819 (1983).

[11] R. S. Williams and W. M. Tong, M.R.S. symposium proceedings **280**, 210 (1993).

[12] C. R. Berry, The theory of the Photographic Process, 4th ed., edited by T. H. James (Macmillan, New York, 1977) Chap. 3.

ON THE RELATIONSHIP BETWEEN CALCULATIONS AND MEASUREMENTS OF THE FREE VOLUME OF GRAIN BOUNDARIES

S. C. MEHTA AND D. A. SMITH
Department of Materials Science and Engineering, Stevens Institute of Technology, Hoboken, NJ-07030.

ABSTRACT

Grain boundary free volume, simply defined as the difference between the volume of a bicrystal and that of a single crystal containing an equal number of atoms, provides a good measure of average grain boundary coordination. Free volume is useful because (a) computer calculations suggest that the grain boundary free volume scales with the grain boundary energy and (b) experimental measurement of free volume may be relatively easier and more direct than that of grain boundary energy. The objective of this paper is to compare the predictions from computer models of grain boundary free volume with experimental measurements.

INTRODUCTION

The major characteristic which distinguishes a grain boundary from the bulk is a local reduction in coordination. This is the fundamental reason for the excess free energy of a grain boundary. In turn, the excess free energy is the thermodynamic basis for the differentiation that exists between the properties of the grain interior and those of the grain boundary. Grain boundary free volume (also known as excess volume or the grain boundary expansion) provides a global measure of the average decrement in the grain boundary atom coordination. Computer calculations suggest that the grain boundary free volume scales with the grain boundary energy [1]. Thermally activated phenomena such as grain boundary diffusion, boundary sliding, grain boundary migration, and other boundary dependent properties such as intergranular fracture strength, intergranular corrosion, electrical resistivity etc. have been shown to correlate well with the boundary free volume (or energy). The grain boundary free volume thus offers an accessible mechanistic link between structure and properties.

EVALUATION OF GRAIN BOUNDARY FREE VOLUME

Theoretical Approaches

Various theoretical techniques have been used to evaluate grain boundary free volume in metals; these include hard sphere modelling [3], computer simulation [4-10] and Sutton's analytical approach [11]. All these model the absolute zero grain boundary structures. Frost et al. [3,12] graphically constructed a variety of [001], [110] and [111] symmetrical tilt grain boundaries in f.c.c. metals. Rigid body translation was allowed parallel to the tilt axis. This simplified model predicts a trend of increasing free volume with increasing repeat distance at the boundary. In their early work, Pond et al. [4,5] used a central force potential to predict structures and free volumes of a variety of [001] and [110] symmetrical tilt grain boundaries in aluminum. A general trend of increasing boundary expansion with increasing repeat distance was observed. Later Chen et al. [6] also predicted free volumes for a series of [001] symmetrical

Mat. Res. Soc. Symp. Proc. Vol. 343. ©1994 Materials Research Society

tilt grain boundaries in Ni, Al and Ni₃Al using a local volume form of potential similar to the Embedded Atom Model (EAM). The data for [001] symmetrical tilt grain boundaries in Al showed a remarkable agreement with the free volume data of Pond et al. [5]. A linear relationship between the boundary energy and the free volume was predicted for Ni and Al. The volume expansions at particular boundaries in Ni₃Al were shown to be composition dependent. In a comprehensive study Wolf used zero temperature lattice statics simulation to obtain the structure-energy-free volume relationships for a variety of symmetrical, and asymmetrical tilt, twist and mixed boundaries in f.c.c. [7,8] and b.c.c. metals [9,13]. A Lennard Jones (LJ) potential and a many body EAM potential were employed and gave qualitatively similar results. The graphs of grain boundary energy and free volume as a function of angle showed sharp minima (cusps) at those values of rotation angles which left the boundary in purely symmetric or asymmetric tilt configurations. Further, the cusps in the free volume vs. angle curve were separated by a relatively large plateau region. The data available from the study suggest that the energy and the volume expansion for asymmetrical tilt boundaries are lower than those for symmetrical tilt grain boundaries with the same average interplanar spacing. These observations are consistent with the abundance of asymmetrical tilt grain boundaries observed in NiO bicrystals [16]. Simulation of grain boundaries in b.c.c. metals revealed similar trends. Wolf also obtained the structures and free volumes of incommensurate tilt, twist and mixed boundaries. Results for the incommensurate boundaries [17] were qualitatively similar except that the energies and free volumes of such boundaries were higher than their commensurate counterparts. The calculations also showed that, except for the most densely packed grain boundary planes, the relaxation of the interplanar spacings at the boundary exhibited an oscillatory pattern with the amplitudes of these oscillations increasing with decrease in interplanar spacing. In an alternate approach, Sutton [11] used Fourier analysis to elucidate the effect of boundary periodicity on the grain boundary energy and free volume for some commensurate and incommensurate boundaries. His results indicated that incommensurate symmetrical boundaries had a higher volume expansion than their commensurate counterparts in accord with Wolf's data for asymmetric tilt grain boundaries in f.c.c. metals. Further, the equilibrium expansion at a mixed boundary was predicted to be lower than that at a symmetrical twist boundary.

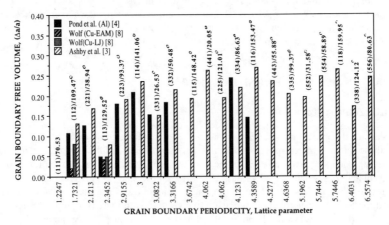

Fig. 1 A comparison of the results obtained by various theoretical approaches; the grain boundary free volume is plotted as a function of boundary periodicity for [1̄10] symmetric tilt grain boundaries in f.c.c. metals.

Fig. 1 shows a comparison of theoretically obtained grain boundary free volume data for a variety of $[1\bar{1}0]$ symmetric tilt grain boundaries in f.c.c. metals. It is apparent that (i) all the theoretical models predict zero expansion at the (111) coherent twin boundary; (ii) The free volume vs. misorientation curve has a deep cusp for the $\Sigma=11$, (113) symmetric tilt boundaries.

Experimental Approaches

There are four electron microscopical techniques predominantly used to measure free volume at a grain boundary. The first technique requires the measurements of the stacking-fault-like fringes [18] which arise at a near coincidence site grain boundary, imaged under common g conditions. The measured intensity profiles are compared with the theoretically simulated fringe profiles. This method allows the determination of a total translation vector, but is restricted to low Σ near coincidence site boundaries, since complications in image contrast interpretation arise when $|g|$ is large or in the presence of dislocations or moire effects. Pond [18] measured an expansion of 0.02nm ($\delta a/a = 0.049$) at the (121) twin boundary in Al.

The second technique involves measurements of the Fresnel fringe profiles at a boundary as a function of the objective lens defocus. This technique relies completely on the changes in local mean inner potential of the material at a boundary; a complication is that the local chemical composition and the atomic arrangement at grain boundaries both affect the mean inner potential. The interpretation of results depends on the comparison between the experimental fringe profiles and computer simulations. This method is generally applicable. However, the boundary has to be viewed edge on. Further, the boundary contrast depends very strongly on the verticality of the boundary, the beam convergence and the relative importance of refraction and pure Fresnel diffraction at the boundary. The uncertainty in the value of inner potential in the region near the boundary limits the accuracy of the method [19]. The Fresnel fringe profile has been shown to be sensitive to specimen artifacts such as grain boundary grooving [28]. Carpenter et al. [20] used the Fresnel fringe technique to study chemical segregation at the $\Sigma=13$, (510) <001> symmetrical tilt grain boundary in Si. They demonstrated the relative insensitivity of the Fresnel fringe technique to chemical segregation and how the technique can be used in conjunction with High Resolution Electron Microscopy (HREM) and nanospectroscopy to separate the structural and chemical effects at the interface.

The third method to measure the rigid body translation is a moire technique. The translation vector at the grain boundary can be determined by examining the displacement of the moire fringes formed by a doubly diffracted beam. This technique has been used to measure the free volume at the $\Sigma=3$, (112) twin boundary in gold [21] and the $\Sigma=3$, (111) twin boundary in stainless steel [19].

The fourth and most direct approach is lattice imaging using HREM [22-26]. This technique is restricted to the examination of boundaries which are parallel to the beam direction. It is important that the crystallographic orientation and thickness of the regions where the relative displacements are to be measured be identical. In spite of the point to point resolution 0.17 nm achievable with modern microscopes, the capability to image grain boundary structure is still restricted to low Σ grain boundary planes oriented along low index directions in most closely packed materials.

Fig. 2 shows the result of experimental free volume measurements superimposed on the theoretical data for a variety of symmetric and asymmetric tilt grain boundaries in f.c.c. metals.

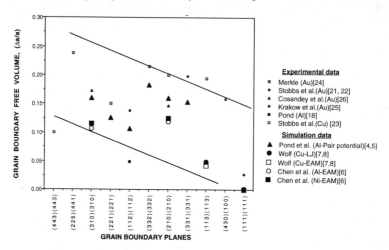

Fig. 2 Theoretical and experimental free volume data for a variety of symmetric and asymmetric tilt grain boundaries in f.c.c. metals.

All the theoretical free volume data can be enclosed in the region between two parallel straight lines with negative slope. This shows a trend of monotonic decrease in free volume with increase in interplanar spacing (or with decrease in grain boundary periodicity). A comparison of experimental free volume data for (112) boundaries in Au [21] and Al [18] show that this boundary has a significantly higher free volume in Au ($\delta a/a$=0.1380 in Au as against $\delta a/a$=0.049 in Al). Similar differences in free volume were observed for the (111) twin in Cu and Au ($\delta a/a$=0.0270 in Au as against $\delta a/a$=0.0028 in Cu) [21-23]. The experimental free volume data for (210) symmetric tilt grain boundaries in Au by Merkle ($\delta a/a$=0.2) [24] and Cosandey et al. ($\delta a/a$=0.1470) [26] do not match. For approximately the same average interplanar spacing, the asymmetrical (430)(100) boundary has a lower free volume than symmetrical (113)(113) tilt boundary in Au. This observation is consistent with the theoretical results [8]. A comparison of computer simulation results with the experimental data show that the experimentally measured free volumes are higher than the theoretically predicted values for almost all the boundaries. The Σ=11, (113) symmetrical tilt boundary, associated with a deep cusp in energy and theoretical free volume shows a large volume expansion in the HREM measurements [24]. The experimental measurements also indicate a finite free volume at the (111) twin boundary.

TEM OF A RANDOM BOUNDARY

Fig. 3(a) is a computer simulated Fresnel fringe profile of a model, hard sphere, random grain boundary in Cu. The crystal thickness is 20 nm in the direction of beam and the boundary is viewed edge-on. The simulation is performed using a very small objective aperture (5 nm^{-1}) so that diffracted beams do not contribute to the fringe contrast. All simulations have been performed using a multislice algorithm on EMS software. The microscope parameters used in the image simulation correspond to those for a 300 keV, Philips (CM30) Microscope. Fig. 3(b) shows the intensity profile of the same grain boundary.

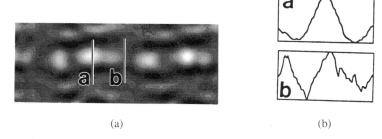

(a)   (b)

Fig. 3(a) Computer simulation of Fresnel fringes at a random grain boundary using a small objective aperture (5 nm^{-1}) and at -138 nm defocus. (b) Intensity profiles of the Fresnel finges across the grain boundary for the image in (a).

DISCUSSION

Discrepancies between theoretical and experimental free volume data could arise from a variety of factors. A major contribution to this discrepancy, however, comes from the neglect of entropic term in the theoretical computations. All the theoretical calculations for free volume apply to 0°K. The calculated entropy vs. misorientation curves indicate two pronounced minima for $\Sigma=5$, (013) and (012) boundaries (about 0.2 and 0.15 mJ/m^2/$^\circ$K respectively) [27]. Using these values, the entropic contributions to the free energy of the $\Sigma=5$, (012) and (013) grain boundaries; this at room temperature are 45 mJ/m^2/$^\circ$K and 60 mJ/m^2/$^\circ$K for the boundaries which constitute roughly 10 % and 7 % of the corresponding boundary free energies [24]. The effect of chemical segregation on the experimental free volume measurements must also be taken into account.

CONCLUSION

Capabilities for direct measurements using electron optical techniques have made measurement of grain boundary free volume more accessible. However, a large discrepancy between the thoretical and experimental values of grain boundary free volume is evident. Further, most of the theoretical and almost all experimental studies are restricted to chemically pure, low Σ, symmetrical and asymmetrical tilt boundaries in close packed fcc metals. There is clearly a need to develop an experimental technique which can be used to measure free volume at any general grain boundary and understand the effect of chemical segregation on the grain boundary structure. The Fresnel fringe technique appears quite promising for measuring the average grain boundary coordination and chemical segregation effects at a general boundary.

REFERENCES

[1] K. L. Merkle and D. Wolf, MRS Bull., 42, Sept. 1990.

[2] M. J. Weins, H. Gleiter and B. Chalmers, J. Appl. Phys. **42**, 2639, 1971.

[3] H. J. Frost, M. F. Ashby and F. Spaepen, Mater. Res. lab. Tech. Report, Div. of Appl. Sci., Harvard University, Cambridge, June, 1982.

[4] R. C. Pond, D. A. Smith and V. Vitek, Acta Metall., **27**, 235, 1979.

[5] R. C. Pond, D. A. Smith and V. Vitek, Acta Metall., **25**, 475, 1977.

[6] S. P. Chen D. J. Srolovitz and A. F. Voter, J. Mater. Res. **4**, 62, 1989.

[7] D. Wolf, Acta Metall. **37(7)**, 1983, 1989.

[8] D. Wolf, Acta Metall. **37(10)**, 2823, 1989.

[9] D. Wolf, Acta Metall. **38(5)**, 791, 1990.

[10] D. Wolf, Phil. Mag. A **62(4)**, 447, 1990.

[11] A. P. Sutton, Phil. Mag. A **62(4)**, 793, 1991.

[12] H. J. Frost, Scripta Metall. **14**, 1051, 1980.

[13] D. Wolf, J. Appl. Phys. **69(1)**, 185, 1991.

[16] K. L. Merkle and D. J. Smith, Ultramicroscopy **22**, 57, 1987.

[17] D. Wolf, J. Mater. Res. **5(8)**, 1708, 1990.

[18] R. C. Pond, J. Microscopy **116(1)**, 105, 1979.

[19] C. Ecob and W. M. Stobbs, J. Microscopy, **116**, 275, 1983.

[20] K. Das Chowdhury, R. W. Carpenter and M. J. Kim, J. Microscopy: The Key Research Tool, 61, March 1992.

[21] J. W. Matthews and W. M. Stobbs, Phil. Mag. A **36**, 373, 1977.

[22] W. M. Stobbs, G. J. Wood and D. J. Smith, Ultramicroscopy **14**, 145, 1984.

[23] G. J. Wood, W. M. Stobbs and D. J. Smith, Phil. Mag. A **50(3)**, 375, 1984.

[24] K. L. Merkle, Ultramicroscopy **40**, 281, 1992.

[25] W. Krakow, J. T. Wetzel and D. A. Smith, Phil. Mag. A **53**, 739, 1986.

[26] F. Cosandey, S. W. Chan and P. Stadelmann, Coll. de phys. **51**, C1-109, 1990.

[27] G. Hasson, J. Lecoze and P. Lesbats, Compt. rend. (Paris), **C271**, 1314, 1971.

[28] D. Rene' Rasmussen and C. Barry Carter, Ultramicroscopy **32**, 337, 1990.

EVOLUTION OF CRYSTALLINE MICROSTRUCTURE
IN GeTe THIN FILMS FOR OPTICAL STORAGE APPLICATIONS

M. LIBERA
Stevens Institute of Technology, Hoboken, NJ   07030

ABSTRACT

The bit-erase process in phase-change optical storage is based
on the amorphous to crystalline transformation. While there has
been significant progress developing compositions and
multilayered media for phase-change applications, quantitative
studies of the crystallization kinetics and microstructural
development are generally lacking. This paper describes work
quantifying crystallization in GeTe thin films. Microstructural
changes during isothermal annealing are measured using *in-situ*
hot-stage optical microscopy. This technique measures the
fraction crystallized, the number of crystallites, and
crystallite radii as a function of time. These data are
sufficient to deconvolute the individual contributions of
nucleation and growth. We find an Avrami exponent of ~4,
consistent with time-resolved reflection/transmission studies.
This exponent is due to 2-D growth at a constant rate plus
transient nucleation. The data are used in a kinetic model to
simulate non-isothermal crystallization during focused-laser
heating characteristic of the bit-erase process.

INTRODUCTION

Phase-change erasable optical data storage (1,2) uses
energy from a focused laser to reversibly switch a micron-sized
area in a multilayer thin film between the amorphous and
crystalline states. The storage layer is typically sandwiched
between two dielectric layers. Additional metal layers are also
often included. Together these layers control the coupling of
the incident laser to the film (3) and its thermal-conduction
properties (4,5). Storage-layer compositions are chosen from
systems where the optical properties are significantly different
between the crystalline and amorphous states.

The processes of reading, writing, and erasing a data bit
are illustrated schematically in figure 1. The storage layer is
typically amorphous in the vapor-deposited state. It is
crystallized in a process known as initialization by low-power
CW laser exposure. A data bit is then written by applying a ~50
nsec focused laser pulse to raise the local temperature of the
storage layer above its melting point. When the pulse ends, the
molten spot rapidly cools. Storage layer compositions are
chosen such that the operative cooling rate exceeds the critical
cooling rate for glass formation. The state of a bit can
subsequently be measured using low-intensity laser exposure to
measure the local reflectivity. A data bit is erased with laser
exposures that raise the local temperature of the storage layer
above its crystallization temperature. The amorphous region

89

crystallizes while the surrounding matrix remains relatively unaffected, and optical contrast is removed.

The materials problems associated with phase-change storage are many. There are issues related to phase transformations, thin-film adhesion, thin-film synthesis, thin-film stress, and conduction heat transfer, among others. These all bear on the performance of a particular multilayered recording medium as well as on its reliability.

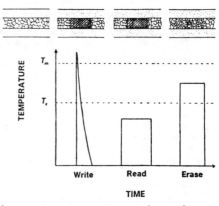

Figure 1 - Transformations in a dielectric/storage layer/dielectric phase-change recording medium.

Experimental study of such problems is difficult, however. The small length (~1µm) and time scales (~100 nsec) characteristic of this technology severely limit the nature and amount of collectable data. An alternate strategy is to study these problems under more accessible conditions, involving longer lengths and times, and then develop models to predict performance under conditions consistent with the application.

This paper describes research on one aspect of the phase-change problem: the bit-erase process. This amounts to understanding the crystallization kinetics and morphology. In the following sections we will briefly review the thin-film alloy systems used in phase-change media. We will then review our experiments studying crystallization of thin-film GeTe, a model phase-change system, under annealing conditions of long length and time scales by time-resolved reflection/transmission and hot-stage optical microscopy. This work identifies the Avrami exponent, the activation energies for crystallization and crystal growth, the Arrhenius growth law, and the nature of the nucleation process. These data are then used to model crystallization driven by focused pulsed-laser annealing characteristic of phase-change storage applications.

CHALCOGENIDE-BASED ALLOY THIN FILMS

The storage layer in phase-change media generally employs a multicomponent chalcogenide-based alloy film. Te is the most common chalcogenide ingredient. These materials are best able to satisfy the glass formation and crystallization requirements imposed by the technology while preserving sufficient changes in optical properties between the amorphous and crystalline states to give sufficient signal-to-noise (1). Chalcogenides and their alloys form glasses under rapid cooling rates (~10^8-10^{10} K/sec) because of their high melt viscosities. In contrast to liquid metals which generally behave as hard spheres, liquid

chalcogenides (S, Se, Te, Po) form covalently-bonded rings or
chains like simple polymers. The viscosity can be further
increased by alloying, and a number of chalcogenide-based alloys
have been amorphized by pulsed-laser melting. Crystallization
of such glasses presents a second challenge. Technological
constraints require that the bit-erase process be completed
within about 300nsec. Chen et al. (6) established that single-
phase alloys undergoing polymorphic crystallization tend to
satisfy this constraint. Two-phase alloys whose crystallization
is limited by diffusion largely do not. Most work now
concentrates on compositions at or close to those of a compound
or a pseudobinary compound (7-10). A perturbation of this
concept is being pursued by the Matsushita group (11) which
works with the GeTe-Sb$_2$Te$_3$ pseudobinary system where different
compounds within that system are fcc polytypes. The fact that
the various compounds bear structural similarities to each other
may lessen the effect of diffusion on the overall
crystallization kinetics of compositions in a two-phase field.

Our research has concentrated on alloys in the Ge-Te binary
system. The phase diagram is relatively simple (12). There is
a single compound - germanium telluride (GeTe) - which displays
limited solubility of Te. Both the Te-GeTe and GeTe-Ge
pseudobinaries are simple eutectics with very limited solubility
of Ge in Te and of Te in Ge. The following sections describe
work using 80nm Ge$_{48}$Te$_{52}$ films deposited on glass, carbon-coated
mica, or silicon substrates by coevaporation of pure Ge and pure
Te in an evaporator with a base pressure of 5x10^{-7} torr (13).

TIME-RESOLVED REFLECTION/TRANSMISSION

The kinetics of crystallization are most commonly modeled
using the Johnson-Mehl-Avrami (JMA) expression. The JMA
equation describes the fraction crystallized, $\chi(T,t)$, as a
function of temperature and time during isothermal annealing:

$$\chi = 1 - \exp[-Kt^n] \qquad \qquad \ldots [1]$$

the Avrami exponent, n, measures the time dependence of
nucleation and the dimensionality of growth. K is a
temperature-dependent rate function. This classical theory is
abundantly discussed in the literature (14). Subsequent
discussion here will assume familiarity with the JMA theory.

The most essential experimental quantity measured in an
Avrami analysis is χ. The transformation from a glass to a
crystal introduces changes in the enthalpy, electrical
resistivity, and volume, among other properties. Any of these
can be monitored to determine $\chi(T,t)$. In the case of
chalcogenide alloy films, time-resolved measurements of
reflection and transmission can be used to characterize
crystallization. This technique has been exploited by a number
of groups under conditions of focused pulsed-laser annealing as
well as furnace

annealing of bulk specimens (15-18). We have used the latter technique (13,19) where defocused red diode laser light with ~mW intensity is directed onto the surface of a specimen heated by a small furnace and the intensities of reflected and transmitted light is detected by photodiodes. Figure 2 shows typical data from a constant-heating-rate anneal at 20K/min. When the specimen crystallizes there is an abrupt decrease in transmission and a concurrent increase in reflection. Opposite changes occur during the crystallization of, for example, amorphous InSb (20).

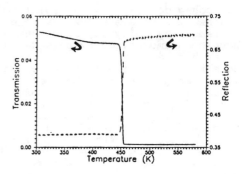

Figure 2 - Changes in reflection and transmission of an 80nm $Ge_{48}Te_{52}$ thin film during heating in air at 20K/min.

$\chi(T,t)$ can be derived from time-resolved data by fitting the linear portions of the reflected or transmitted signal before and after the transformation to straight lines, labelling each of these as $\chi=0$ and $\chi=1$, respectively, and then rescaling. A typical result is shown in figure 3 for isothermal annealing at 418K. These data have been plotted in the form $\ln[-\ln(1-\chi)]$ vs $\ln(t)$ to give a straight line whose slope corresponds to an Avrami exponent of $n=4.51(+/-0.05)$. Results from a series of constant-heating-rate experiments are shown in figure 4. A Kissinger analysis (21,22,23) of these data gives an activation energy characterizing crystallization of $Q_{\chi}=1.72(+/-0.03)eV$.

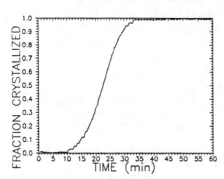

Figure 3 - $\chi(T=418K,t)$ for 80nm $Ge_{48}Te_{52}$ from time-resolved reflectivity data.

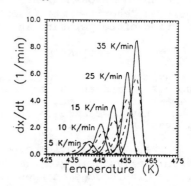

Figure 4 - crystallization rate data via time-resolved reflectivity (solid = exp.; dashed = calc.).

The literature is replete with similar analyses, perhaps employing different experiments, where sufficient data are collected to estimate an Avrami exponent and an activation energy. While certainly not effortless, such an analysis is relatively straightforward. All too often, however, interpretation is made by reference to a table such as that in Christian (14) which gives the exponent resulting from various growth dimensionalities, growth modes, and nucleation modes. The classical theory of transformation kinetics as represented by the JMA equation convolutes the effects of nucleation and growth. Trying to deconvolute these effects knowing only $\chi(T,t)$ is analogous to solving one equation in two unknowns. Additional experiments must be done to explore the individual contributions of nucleation and growth. This is particularly challenging under constrained time and length scales.

MEASUREMENTS OF MICROSTRUCTURAL DYNAMICS

A direct way to study the nucleation kinetics and the growth kinetics is to measure microstructural changes during crystallization by imaging. Smith et al. (21,22) have used TEM with *in-situ* heating to establish the kinetics characteristic of amorphous $CoSi_2$ thin-film crystallization. We have been unable to apply *in-situ* TEM to our GeTe films. The GeTe crystallite size is relatively large (~μms in diameter) for typical anneals, and operation of a TEM at magnifications low enough to give sufficient field of view (~x100's) is difficult. Moreover, our choice of substrates - 7059 glass, carbon-coated mica, and silicon - do not lend themselves well to *in-situ* analysis.

Figure 5 compares TEM micrographs of partially-crystallized films annealed: (a) in a furnace and subsequently removed from the substrate; and (b) in the TEM where the film is supported only by a Cu grid. The crystallite morphology for the *in-situ* experiment is highly dendritic, very unlike the disk-like appearance of the *ex-situ* annealed specimen. This difference is attributed to spatially inhomogeneous *in-situ* heating from the copper grid bars. *In-situ* TEM crystallization experiments designed to address crystallization kinetics are probably best done using amorphous films deposited onto amorphous electron-transparent supports such as SiO_2 or SiN windows in silicon.

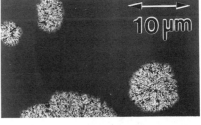
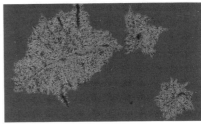

Figure 5 - bright-field TEM
a) furnace annealed;
b) *in-situ* TEM annealed

The morphology shown in figure 5a is typical of GeTe thin-film crystallization (26,27,28). In our films (80nm thick) selected-area electron diffraction shows that the crystallites grow quickly through the film (13). Crystal growth is predominantly two dimensional.

We have studied the nucleation and growth kinetics of these GeTe films using hot-stage optical microscopy at 200x (29). Isothermal experiments were done between 413K and 425K. Specimens were photographed every two minutes. Photographs were digitized by scanning into a Mac Quadra 950. Each image was studied using IMAGE V1.45 image-analysis software (30) to establish χ, the number of crystallites, N, and the crystallite diameters. The growth rate, G, was determined by measuring crystallite diameters for several times prior to impingement.

Figure 6 shows a typical sequences of images. χ(T=413K,t) data were collected by numerically evaluating the area-fraction crystallized in 33 such images. When plotted as $\ln(-\ln(1-\chi))$ against $\ln(t)$, these data give an Avrami exponent of n=4.0. A second identical experiment gives n=4.2. These results agree fairly well with our earlier time-resolved result of n=4.5. Figure 7 plots crystal size against time at 413K for 18 crystallites. The slope of each curve gives the crystal growth rate. The slope is constant. Growth appears to be controlled by an interface-limited mechanism, possibly atomic rearrangement at the interface, rather than by long-range diffusion. Growth rates were measured at several temperatures giving an Arrhenius plot ($\ln[G(T)]$ vs. $1/T$) with a straight line. G(T) can be modeled as $G(T)=G_o\exp[-Q_G/kT]$, where $G_o=5.86\times10^{14}$ m/sec and Q_G, the activation energy for growth, is 1.77(+/-0.14)eV agreeing with Q_χ=1.7eV via time-resolved reflection/transmission.

Figure 6 - time sequence of digitized (transmission) optical micrographs describing the crystallization of $Ge_{48}Te_{52}$ at 413K.

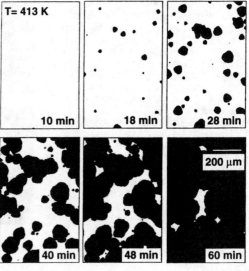

94

Figure 8 plots the number of crystallites, N=N(t), for isothermal annealing at 413K. A correction must be applied to these data before extracting a measurement of the nucleation rate. The specimen is finite, and the volume of amorphous material available for nucleation decreases as the transformation proceeds. One can show (29) that the nucleation rate, I, with units of # events/time, is:

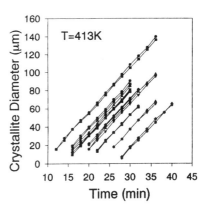

$$I(t) = \frac{dN(t)/dt}{[1-\chi(t)]} \quad ...[2]$$

Figure 9 shows I(t) calculated by this expression using our measurements

Figure 7 - Diameters of 18 crystallites during isothermal annealing at 413K.

of N(t) and χ(t). The noise in the curve is presumably due to the limited number of crystallites in this experiment. Figure 9 clearly shows that the nucleation rate increases with time. The data fall on a line given by a least-squares fit as: I(t)=0.4 + 0.1t (#/min). We attribute this result to transient nucleation. The parent amorphous phase is completely transformed before a steady-state nucleation rate, where I(t)=constant, is achieved.

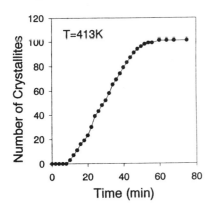

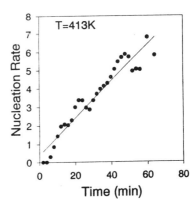

Figure 8 - Total number of crystallites, N(t), during isothermal annealing at 413K.

Figure 9 - I(T=413K,t) derived from N(t) by correcting for the diminishing amount of parent phase.

The transformation kinetics of these GeTe films can now be understood in terms of the individual contributions of nucleation and growth. Growth proceeds in 2-D at a constant rate. Transient nucleation occurs at a rate increasing linearly with time. The extended volume is: $V_{ex}(413K)=\eta I_o G^2 t^4=Kt^4$ where η is a shape factor (14). The characteristic JMA equation is thus: $\chi(T=413K)=1-\exp[-Kt^4]$. This expression can be generalized to other isotherms knowing the temperature dependence of the nucleation rate. We have thus far been unable to collect sufficient data for a confident analysis of $I(T,t)$. The strong temperature dependence of the transformation limits the range of temperatures over which meaningful $N(t)$ data can be collected.

JMA MODEL OF GeTe CRYSTALLIZATION

The Johnson-Mehl-Avrami equation applies to isothermal transformations with specific assumptions regarding the nature of the transformation. Henderson (31) has developed an analytical model relating χ and T for non-isothermal conditions where a specimen is heated at a constant rate. Greer (32) has developed a numerical JMA-type model where a non-isothermal anneal is approximated as a series of arbitrarily short isothermal anneals. To the best of our knowledge, no comparison has yet been made between these two models. We have used the Greer model to simulate the crystallization of GeTe films using our time-resolved data (n=4.5; Q=1.7eV). The dashed lines in figure 4 are calculated results. We have accounted for the temperature dependence of the crystallization reaction using the expression $K=K_o\exp[-nQ/kT]$. K_o is a free parameter fixed by matching the calculated peak temperature to the experimental peak temperature for the case of heating at 25K/min. The agreement between the model and experiments is fairly good over this range of heating rates (5-35 K/min).

We have used the Greer model to simulate the phase-change erase process. This amounts to modelling crystallization of a ~1μm diameter spot of amorphous GeTe surrounded by (non-participating) crystalline GeTe. We assume there is no temperature dependence of K_o, Q, or n. We employ a simple model for laser heating illustrated by figure 10. An objective lens focuses collimated laser light into a diffraction-limited spot with a Gaussian distribution of intensity at the film (33):

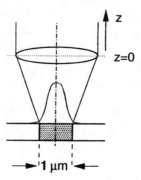

Figure 10 - The width of the gaussian laser intensity is controlled by lens defocus.

$$I(x,z) = [1/\alpha]\exp[-(x/w(z))^2] \qquad \ldots[3]$$

where x measures distance in the film plane from the center of the irradiated spot, z measures the defocus, and $w(z)$ is the beam waist (33) given as:

$$w(z)=w_o[1 + (\lambda z/\pi w_o)^2] \qquad \ldots[4]$$

w_o is chosen to give a diffraction-limited spot with FWHM of 0.5µm for a focused beam (z=0). The geometry is rotationally symmetric about the optic axis. We further assume the heating rate at position x is proportional to the laser intensity there and fix the proportionality constant by choosing $I(x=0,z=0)=5\times10^9$ K/sec based upon a more detailed model of laser coupling and heat flow in multi-layered films (4,34).

Figure 11 shows the results of one calculation for a fully-focused laser (z=0). Figure 11a shows the reduced temperature T/T_m ($T_m{}^{GeTe}\sim1000K$), as a function of distance from the center of the spot for several times. Figure 11b gives the corresponding fraction crystallized as a function of position. This calculation predicts that ~220nsec are required to fully crystallize a ~1µm diameter data bit. This time scale is consistent with the few-hundred nsec constraint imposed by the technology. However, the calculation also predicts that temperatures close to or exceeding the crystalline melting point are reached. This calculation suggests that the bit-erase process might be more accurately treated as a liquid-solidification process rather than as a glass-crystallization process.

A second point suggested by figure 11 is that focused laser heating during the erase process may affect the reliability of the recording medium. The center of the spot (x=0) is fully crystallized (erased) before the outer edge. In order to

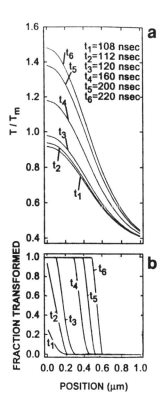

Figure 11 - (a) calculated temperature profiles and (b) fraction crystallized for focused (z=0) laser heating.

fully crystallize the edges,
the center is unnecessarily
heated which serves to only
degrade the multilayer
structure by thermal stress
or interfacial reaction. A
simple alternative is to
defocus the laser so the
incident energy is more
uniformly distributed over
the data bit. Figure 12
shows a calculation for a
defocus of 15μm. This
condition leads to more
uniform bit crystallization,
still within a reasonable
time, and the extremes of
temperature within the spot
are no longer reached.

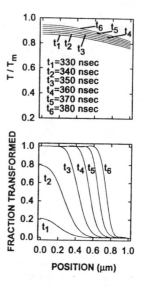

There is experimental
evidence supporting the
prediction that temperatures
exceeding ~0.5T_m are reached
during laser annealing.
Electron diffraction from
furnace-annealed and laser-
annealed amorphous Te-Ge-Sn
alloys have shown that solute
trapping can occur under
rapid heating (35). We used

Figure 12 - less
superheating and more uniform
temperatures are generated
with laser defocus (z=15μm).

a film of $Te_{36.3}Ge_{47.4}Sn_{16.3}$ which is amorphous as-deposited.
Furnace annealing for 20min at 623K produces a two-phase
microstructure with ~24vol% α-Ge and ~76vol% $(Ge_{55}Sn_{45})_{50}Te_{50}$.
Several different focused pulsed-laser annealing treatments all
produce a single-phase microstructure consisting only of the
telluride compound. The α-Ge is absent in the laser-annealed
specimens, and the excess Ge is accommodated by forcing Sn onto
the Te sublattice of the equilibrium $(Ge_xSn_{1-x})Te$ structure. A
typical grain size in the laser-crystallized structures is of
order 10-100nm. Interfacial velocities of order 0.1m/sec -
1.0m/sec are required to crystallize grains of this size in time
scales of order 100nsec. Such velocities are consistent with
predictions of solute trapping in highly supercooled liquids
(36).

DISCUSSION AND CONCLUSION

Experimental study of the bit-erase process is constrained
by the short time scales and length scales defined by the phase-
change technology. A solution is to do experiments with large
specimens for relatively long times where the transformation can
be better characterized. These data can then be used in a model
to predict behavior under less accessible conditions. Such an
approach would be valuable to other materials problems involving

short time and/or length scales such as rapid thermal annealing or various welding processes.

We have employed a very simple model describing non-isothermal crystallization kinetics. The model uses inputs - Avrami exponents, activation energies, etc. - generated from macroscopic experiments done over time scales of minutes and hours. One could improve its predictions by accounting for temperature-dependent changes in thermodynamic properties, for example, and for varying film/laser coupling as crystallization proceeds. An additional step which we have not yet made would use time-resolved reflection measurements during focused pulsed-laser annealing to address how well the model predicts the crystallization behavior at short times and small length scales.

A better model would predict microstructure. Just as our *in-situ* microscopy work provides a broader data set than the time-resolved reflection/transmission studies, a model that predicts microstructure is more useful than one that predicts only χ. In the case of phase-change storage, the grain size and grain orientation help determine the signal-to-noise ratio of a data bit. In the case of a weld, the strength and failure properties will be determined by such microstructural features as the grain size and precipitate distribution. In both cases, a JMA-based model would not address important issues concerning performance of a material in its application, since $\chi(T,t)$ does not specifically describe microstructure. More and better work needs to be done within the materials community to bring experimentally-based models with accurate predictive capabilities to real problems such as the phase-change storage technology discussed here.

ACKNOWLEDGMENT

Thanks are due to Martin Chen and Kurt Rubin of the IBM Almaden Research Center who provided important stimuli to and early collaboration on much of this work and to several graduate students - Q.M. Lu, K. Siangchaew, and B. Saltsman - from Stevens for their help with various aspects of this work.

REFERENCES

1. M. Libera and M. Chen, MRS Bulletin **XV**, 40-45 (April 1990).
2. D. Gravesteijn, C.J. van der Poel, P. Schulte, C. van Uijen, Philips Tech. Rev. **44**, 250-258 (May 1989).
3. A.E. Bell and F.W. Spong, IEEE J. Quant. Elec. **QE-14** (7), 487-495 (July 1978).
4. R. Kant, J. Applied Mechanics **55**, 93-97 (March 1988).
5. B.J. Bartholomeusz, SPIE Vol. **1078**, 179-188 (1989).
6. M. Chen, K. Rubin, and R. Barton, Appl. Phys. Lett. **49** (9), 502-504 (1986).

7. Proceedings of the Joint Symposium on Optical Memory and Optical Data Storage, July 5-9, 1993 (Maui) published as a special edition of the Japanese Journal of Applied Physics.
8. I. Morimoto, K. Furuya, M. Suzuki, and M. Nakao, SPIE Vol. **1663** Optical Data Storage (1992).
9. T. Yoshida, N. Akahira, S. Ohara, K. Nishiuchi, and T. Ishida, Jpn. J. Appl. Phys. **31**, 476-481 (1992).
10. K. Rubin, D.P. Birnie, and M. Chen, J. Appl. Phys. **71** (8), 3680 (1992).
11. N. Yamada, E. Ohno, K. Nishiuchi, N. Akahira, and M. Takoa, J. Appl. Phys. **69** (5), 2849-2856 (1 March 1991).
12. M. Hanson and K. Anderko, *Constitution of Binary Alloys*, 2nd ed. (McGraw Hill, New York, 1958).
13. M. Libera and M. Chen, J. Appl. Phys. **73** (5), (March 1993).
14. J. W. Christian, *The Theory of Transformations in Metals and Alloys*, 2nd ed. (Pergamon, Oxford, 1975).
15. C. van der Poel, J. Mat. Res. **3** (1), 126-132 (Jan/Feb 1988).
16. R. Barton, C.R. Davis, K. Rubin, and G. Lim, Appl. Phys. Lett. **48**, 1255-1257 (1986).
17. F.S. Jiang, J.C. Rhee, M. Okuda, and T. Matsushita, J. Non-Crys. Solids **95&96**, 533-538 (1987).
18. F. Ueno, Jap. J. Appl. Phys. **26**, Suppl. 26-4, 55-60 (1987).
19. M. Libera, T. Kim, L. Clevenger, and Q. Hong, Mater. Res. Soc. Symp. Proc. **V280**, ed. H. Atwater et al., 715 (1993).
20. J. Solis, K. Rubin, and C. Ortiz, J. Mat. Res. **5**, 190(1990).
21. P.G. Boswell, Scripta Metall. **11**, 701-707 (1977).
22. H.E. Kissinger, Analytical Chemistry **29**, 1702-1706 (1957).
23. E.A. Marseglia, J. Non-Cryst. Solids **41**, 31-36 (1980).
24. D.A. Smith, K. Tu, and B. Weiss, Ultramicr. **30**, 90-96 (1989)
25. D.A. Smith, P. Evans, and S. Koppikar, Mater. Res. Soc. Symp. Proc. **V321**, ed. M. Libera et al., 271-282 (1994).
26. K. Chopra and S. Bahl, J. Appl. Phys. **40**, 4171 (1969); J. Appl. Phys. **40**, 4940 (1969); J. Appl. Phys. **41**, 2196 (1970).
27. J. Aznarez and J. Mendez, Thin Solid Films **131**, 111 (1985).
28. T. Okabe and M. Nakagawa, J. Non-Crys. Sol. **88**, 182 (1986).
29. Q.M. Lu and M. Libera, submitted to J. Appl. Phys.
30. Image version 1.45, public-domain software (Mac) available from the National Institutes of Health
31. D.W. Henderson, J. Non-Cryst. Solids **30**, 301-315 (1979).
32. A. Greer, Acta Metallurgica **30**, 171-192 (1982).
33. O. Svelto, *Principles of Lasers* (Plenum, New York, 1982).
34. M. Libera, M. Chen, and K. Rubin, Mater. Res. Soc. Symp. Proc. **152**, ed. D. Poker and C. Ortiz (1989).
35. M. Libera, M. Chen, and K. Rubin, J. Mater. Res. **6** (12), 2666-2676 (1991).
36. M.J. Aziz, J. Appl. Phys. **53**, 1158-1168 (1982).

EFFECTS OF POST-DEPOSITION PROCESSING ON THE ULTIMATE GRAIN SIZE IN METASTABLE SEMICONDUCTOR THIN FILMS TO BE USED IN IR DETECTORS

SUSANNE M. LEE
Lawrence Livermore National Laboratory, 7000 East Ave., Livermore, CA   94550 and
Lawrence University, Department of Physics, P.O. Box 599, Appleton, WI 54912

ABSTRACT

   Through post-deposition annealing in a differential scanning calorimeter (DSC), we have manufactured both thin (200 nm) and bulk (8000 nm) single phase films of *crystalline* $Ge_{1-x}Sn_x$, using rf sputtering. The Sn concentrations produced ranged up to 31 at.%, well beyond the solid solubility limit of this system. There was a marked difference, in the as-deposited structure, between thick and thin films produced under the same deposition conditions. Quantitative models for both systems are given in this paper and were deduced from DSC measurements in conjunction with electron microscopy. The metastable crystalline state in the thin films formed by nucleation and growth from an amorphous phase; whereas in the thick films, the desired phase was already present in the as-deposited films and only growth of pre-existing grains was observed upon post-deposition annealing. When annealed to high temperature, the Sn phase separates from the alloys and we postulate here that it does so by nucleation and growth of $\beta-Sn$. With this hypothesis, the Sn separation in the 8000 nm thick films was accurately modeled by a two−mechanism process, however, in the 200 nm thick films, only one phase separation mechanism was necessary to accurately fit the data. Both models were corroborated by the subsequent melting behavior of the phase separated Sn which, though it varied depending on the sample being measured, always exhibited a melting endotherm starting 25-35°C lower than the bulk melting temperature of Sn. Speculation on the reasons for this are presented.

INTRODUCTION

   Semiconducting crystalline alloys of $Ge_{1-x}Sn_x$ are potentially useful in infrared to far infrared photodetectors for Sn concentrations greater than the equilibrium solid solubility limit $(x < 0.01)$ of the alloy. A wide variety of scientific fields ranging from medicine to astronomy need sensitive and inexpensive far infrared photodetectors for which crystalline $Ge_{1-x}Sn_x$ alloys have been predicted to be particularly well suited.[1] However these predictions are for $x > 0.20$ whereas the equilibrium solid solubility of Sn in Ge is less than one atomic percent. If these alloys are to be produced with the expected properties, they will have to be manufactured in a metastable form. Since the early 1950's this has been tried with only limited success: the thickness of the film is restricted to fewer than 200 nm or the Sn separates (in the form of $\beta-Sn$) from the alloy.
   With rf sputtering and post-deposition annealing in a Differential Scanning Calorimeter (DSC), I successfully manufactured single phase crystalline films ranging from 200 nm − 8000 nm. The mechanisms of formation are discussed in the following results section. For sensitive infrared detectors we need to optimize the crystal structure by forming as large a grain size as possible. Since this family of crystalline alloys is metastable, their maximum grain size, needed for sensitive photodetection, is determined by the onset of phase separation. For this reason we studied both the phase separation process(es) and the melting behavior of the phase separated Sn. We found that the total heat absorbed during the Sn melting, the onset temperature, and the shape of the Sn melting peak, provided crucial clues to understanding the phase separation transformation. This interdependence between *all* the phase transformations is critical to accurately analyzing each individual transformation and is a point I will emphasize in this paper.

Mat. Res. Soc. Symp. Proc. Vol. 343. ©**1994 Materials Research Society**

RESULTS

Two groups of materials will be discussed in this paper: the first set consists of 200 nm thick films of $Ge_{0.75}Sn_{0.25}$ and the second set of 8000 nm thick $Ge_{1-x}Sn_x$ alloys of varying composition. Both sets of films were deposited in an rf sputtering system and, immediately after deposition, their crystal structures were examined with electron diffraction (for the thin films) and x-ray diffraction (for the thick films). Both sets of films produced similar diffraction patterns: either diffuse rings in the electron diffraction or broad peaks in the x-ray diffraction, both of which results *indicate* that the as-deposited films were amorphous.

For the materials to have the desired infrared detection properties discussed above, the films must form a diamond cubic *crystal* structure. Thus we annealed both sets of alloys in a DSC to crystallize them and examine their "amorphous" to crystalline transformation. Despite the similarity of the diffraction results, the two sets of films produced markedly different DSC signals, leading us to deduce that their as-deposited structures were very different.

The 200 nm thick, as-deposited, 25 at.% Sn alloys, when annealed in the DSC, displayed two overlapping exotherms (see Fig. 1). A second heating scan of the same sample showed a Sn melting endotherm, indicating that one of the two exotherms during the first temperature scan corresponded to the phase separation of Sn from the alloy. Since we desired single phase material for the applications previously mentioned, it was necessary to somehow temperature separate the two exotherms observed during the first DSC scan. We therefore explored the phase space determined by the deposition variables and found that the structure of the as-deposited material was most sensitive to the externally applied dc bias. Varying the deposition conditions affected the amorphous structure in a manner that was unmeasurable by high resolution electron microscopy, however when annealed in the DSC a clear difference was observed: the two exotherms were still present, but were now well separated in temperature, unlike in the previous sample where the exotherms overlapped (see Fig. 2).

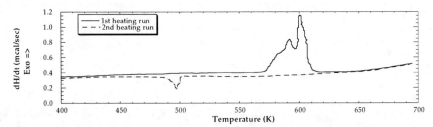

Figure 1: DSC anneal of an as-deposited 200 nm thick $Ge_{0.75}Sn_{0.25}$ sample showing overlapping exotherms during the first scan and Sn melting endotherm during the sample's second scan.

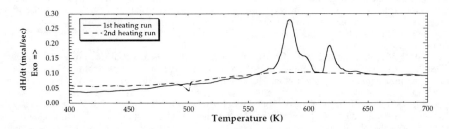

Figure 2: DSC anneal of a 200 nm thick $Ge_{0.75}Sn_{0.25}$ sample made under optimal deposition conditions. Compare the temperature separation of the two exotherms in this sample to the overlapping exotherms of fig. 1.

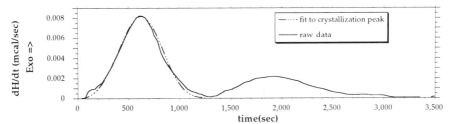

Figure 3: Isothermal anneal of the $Ge_{0.75}Sn_{0.25}$ sample whose scanned DSC spectrum is shown in fig 2. Note the time separated bell-shaped exotherms. The dashed curve is a JMA fit to the crystallization peak. The second bell-shaped exotherm corresponds to phase separation.

Electron diffraction was used to identify the first exothermic peak as corresponding to the amorphous to diamond cubic crystal transformation and the second peak as corresponding to the phase separation reaction. Thus by varying the processing techniques, the initial amorphous structure was altered in such a way as to affect the relative temperatures of crystallization and Sn phase separation, thereby permitting production of single phase polycrystalline $Ge_{0.75}Sn_{0.25}$.

To obtain information on the crystallization mechanisms, isothermal anneals were performed at approximately 10K below the onset of crystallization in the scanned measurement. Two bell shaped isothermal exotherms were observed, with the first again corresponding to crystallization and the second to Sn separation, see Fig. 3. We assumed a Johnson-Mehl-Avrami (JMA) type nucleation and growth for the crystallization transformation. JMA relates the rate at which the fraction of material is crystallized to the anneal time as follows

$$\frac{dH}{dt} = H_0 \, K \, n \, t^{n-1} \, e^{-K(T) \, t^n} \tag{1}$$

We fit the first exotherm to equation (1), where H_0 = the total heat released during the transformation, n = nucleation and growth parameter, and $K(T)$ = rate constant which is a function of temperature T, but not time t. We found $H_0 = 1.7 \pm 0.2$ kJ/mol, within measurement error of the measured area under the first exotherm of either the isothermal(1.6 ± 0.2 kJ/mol) or scanned runs (2.0 ± 0.3 kJ/mol); $K(540K) = 1.2 \times 10^{-10}$ sec$^{-3.5}$; and n = 3.5. An Avrami coefficient, n, between 3 and 4 corresponds to a decreasing nucleation rate if the growth of the grains is three dimensional.[2] To check this we examined the grain size of the diamond cubic crystals after the material had been annealed to just before the onset of phase separation. No phase separated Sn was observed and the grain size of the crystals varied from 50-500Å. That the final size of these crystals was less than the thickness of the film, together with the large variation in crystal sizes, indicated that we did indeed have three dimensional growth and could therefore interpret the nucleation parameter as corresponding to a decreasing nucleation rate with increasing time. To summarize, with this time and temperature separation of the two exothermic transformations, we produced, with no admixture of $\beta-Sn$, the diamond cubic crystalline $Ge_{1-x}Sn_x$ desired for infrared detection purposes.

Now that we had determined the deposition conditions which, with post-deposition annealing, would produce the desired crystalline state of the alloy, we applied these same conditions in depositing much thicker films, on the order of 4000-8000 nm. Such thicknesses make the determination of bulk electronic band structure properties much easier. As mentioned earlier, x-ray diffraction of these thicker films indicated an amorphous structure for the as-deposited material. We then annealed the films in the DSC to form the desired crystalline state, but the scanned DSC spectrum was very different than that of the thinner materials. Only one clear exotherm was observable with bumps on both the leading and trailing edges of the exotherm. The second scanned run, used to determine the baseline for the first scanned run, exhibited *two* Sn melting endotherms. A typical DSC spectrum from this group of films is shown in fig. 4 for the 8000 nm thick $Ge_{0.69}Sn_{0.31}$ samples.

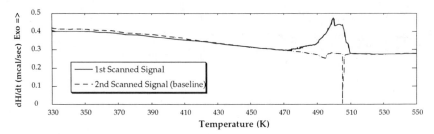

Figure 4: 10K/min scanned DSC spectrum of 8000 nm thick $Ge_{0.69}Sn_{0.31}$ showing a small bump on the leading edge of the exothermic signal of the first temperature scan. This small bump corresponds to the decaying exothermic signal shown in the fig. 5. Note the double Sn melting endotherms on the second scanned run used for baseline determination.

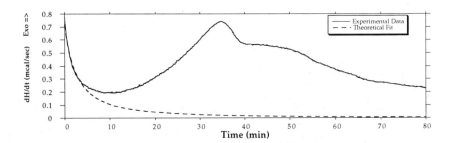

Figure 5: 450K isothermal DSC signal from 8000 nm thick $Ge_{0.69}Sn_{0.31}$ showing the initial decaying exothermic signal due to grain growth of pre-existing crystals followed by the exothermic peak(s) of Sn phase separation from the metastable, diamond cubic, crystalline alloy. The dashed curve is the theoretical fit to the experimental data.

To examine the various processes occurring during the first scan, isothermal anneals were performed on $Ge_{0.69}Sn_{0.31}$, an example of which is shown in Fig. 5. A decaying exotherm was observed during the first approximately 10 minutes of the isotherm in every thick sample, independent of Sn concentration. The decay was followed by a large, somewhat bell–shaped exotherm which corresponded to Sn phase separation, as determined by x-ray diffraction measurements. The bumpy shape of this latter exotherm will be discussed later. Concentrating on the decaying section of the exothermic signal, we annealed an as-deposited sample almost through the end of the decay, quenched it as quickly as possible (~360K/min), and remeasured the x-ray diffraction spectrum. Though the diffraction peaks were still broad, the corresponding grain size calculated from the width of the peak had increased by almost a factor of 2.5 from 18Å to 45Å, indicating that the decay corresponded to grain growth of existing crystals of a very small initial size (~18Å). Proceeding on this assumption, the isothermal decay was analyzed in a manner similar to Chen and Spaepen[3] using the standard grain growth equation for spherical grains of radius r and rate constant $K(T)$ assumed to be a function of temperature T only:

$$r^n(t) = (r_0)^n + t\,K(T) \tag{2}$$

Analyzing the decay shown in fig. 5, produced n = 1.7±0.1 which is close to the theoretical value of two obtained for ideal grain growth. Performing isothermal measurements at other temperatures produced an activation energy for the grain growth process of 0.5 ± 0.2 eV which is within the range of activation energies for grain growth in other materials determined by a variety of measurement techniques, including DSC. The interfacial grain boundary energy of the $Ge_{0.69}Sn_{0.31}$ crystals was deduced by combining equation (2) with the relation[4] between the evolution of the interfacial enthalpy (given by the area under the decaying exotherms as a function of time) and the enthalpic part of the interfacial tension γ_H

$$H(t) = g\gamma_H V/r(t) \qquad (3)$$

where g = 1.3 is a geometrical factor and V is the molar volume of the sample. From equations (2) and (3) and the DSC data, the interfacial grain boundary energy of the crystalline $Ge_{0.69}Sn_{0.31}$ alloys was calculated to be 113 mJ/m^2. Combining all these grain growth parameter values, the isothermal DSC output was simulated on the computer and the dotted curve in fig. 5 shows the result. The close agreement between the data and simulation justifies our hypothesis that the as-deposited material consisted of very fine–grained polycrystals and the DSC decay corresponded to growth of those small crystallites.

It is interesting to note that the enthalpic part of the interfacial grain boundary energy of the crystalline $Ge_{0.69}Sn_{0.31}$ is not much less than the grain boundary energy of β–Sn (164 mJ/m^2). The closeness of these two energies *may* indicate that the crystalline $Ge_{0.69}Sn_{0.31}$ boundaries are Sn-like, i.e. enriched in Sn, thereby enhancing the ease with which Sn separates from the alloy. If this is indeed the case, then understanding the process(es) by which Sn separates from the crystal is important in maximizing the crystalline grain size.

With this in mind, the materials were scanned through crystallization and beyond to examine the phase separation and subsequent Sn melting processes. Figure 6 shows a magnified view of the scanned DSC spectrum for the 8000 nm thick $Ge_{0.69}Sn_{0.31}$ sample. The large broad "bump" on the trailing edge of the sharp peak indicates that more than one process is occurring during phase separation. Upon heating the phase separated material a second time we should have observed a single melting endotherm resulting from the phase separated Sn. Instead, as is shown in fig. 6, *two* Sn melting endotherms were measured! This indicates that indeed there is more than one mechanism governing the Sn separation from the crystalline alloy, corroborating our earlier supposition deduced from the shape of the phase separation peak.

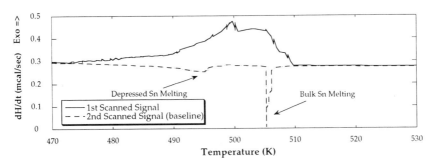

Figure 6: Magnified view of fig. 4, clearly showing, on the first temperature scan of the thick $Ge_{0.69}Sn_{0.31}$ sample, the bumpy bell-shaped exotherm, corresponding to grain growth and then phase separation of the Sn from the crystalline alloy. The second temperature scan of this sample distinctly shows two Sn melting endotherms.

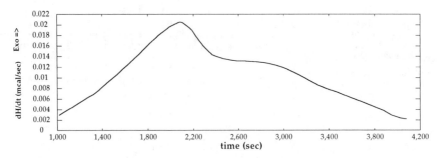

Figure 7: Isothermal DSC measurements of the thick $Ge_{0.69}Sn_{0.31}$ sample showing the complicated bell-shaped exotherm corresponding to phase separation. The presence of this "bumpy" exothermic peak indicates more than one phase separation process.

Isothermal DSC measurements were performed (see fig. 7) to possibly separate the multiple reactions occurring during the scanned anneals. As mentioned earlier, a decaying exothermic signal was initially observed, followed by a large, bumpy, but approximately bell-shaped, exothermic peak. The next set of measurements discussed here concentrate on this complicated exothermic *peak* which corresponded to Sn phase separation from the alloy, as determined from XRD measurements. The fact that the peak was approximately bell-shaped, indicated the presence of *at least* one *nucleation–and–growth* type process and, due to the bump on the trailing edge of this exothermic peak, more likely to *two* different processes.

A possibility for one of these phase separation mechanisms was spinodal decomposition. However this is a second order process and would not result in a *peak* in the rate of heat release. In addition, spinodal decomposition usually produces an interconnected pathway of phase separated material. To see if this occurred in our materials, we performed scanning electron microscopy (SEM) on the 8000 nm thick samples and scanning transmission electron microscopy (STEM) on the 200 nm thick samples, see fig. 8. From the electron microscope photographs, it is clear that the Sn separates into isolated spherical islands with the growth of these islands eventually becoming two-dimensional, since the STEM photographs show Sn disks whose diameters range from two to four times the thickness of the sample. The lack of inter-connected pathways together with the maximum in the rate of heat release observed during both the isothermal and scanned DSC measurements indicates that the Sn does *not* separate from the alloy by the second order process of spinodal decomposition, but rather by some sort of first order nucleation and growth processes.

To identify the *number* of phase separation mechanisms and their type, we combined the *isothermal* measurements with information obtained during the *scanned* DSC measurements. The shape of the isothermal phase separation curve indicates more than one Sn separation mechanism; the presence of *two* Sn melting endotherms in the *scanned* DSC spectrum further bolsters the hypothesis that *two* mechanisms are responsible for Sn separation from the crystalline alloy. Of the two melting endotherms, the lower temperature one starts 25K to 30K below the melting temperature of bulk Sn (505K). The measured heat of melting was significantly less than the heat needed to melt the original Sn concentration in the alloy.

Similar temperature– and enthalpy–depressed Sn melting endotherms have been observed by Jang and Koch[5] in ball milled $Ge_{1-x}Sn_x$ and the temperature depression was explained by Turnbull[6] to be the result of an atomic interfacial Sn layer coating the Ge crystals. In addition, Buffat and Borel[7] observed a significantly depressed melting temperature in materials consisting of clusters of atoms. Thus in our material, if the Sn separates into tiny clusters of Sn atoms, then we should see a depressed Sn melting temperature. If, on the other hand, the Sn separates into macroscopic islands, then we should observe an endotherm at the melting temperature of bulk Sn. Both types of endotherms were measured, see fig. 6, indicating that when the Sn separates it

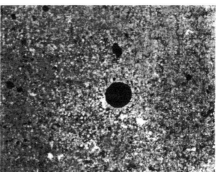

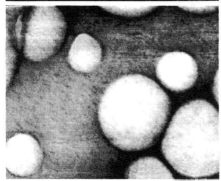

5UM 100KV BSD 0086/09

Figure 8: (clockwise from lower left) (a.) STEM photograph of 200 nm thick $Ge_{0.75}Sn_{0.25}$ showing disks of phase separated Sn ranging from 400 nm to 800 nm in diameter. (b.) SEM photograph of the 8000 nm thick $Ge_{0.69}Sn_{0.31}$ sample. The bright circles correspond to islands of pure Sn. (c.) Low magnification TEM micrograph of the sample in (a.), except annealed completely through the phase separation exotherm. Again the bright areas correspond to phase separated Sn. Note the *absence* of interconnected Sn pathways indicating spinodal decomposition. The large black circle in the middle of the photograph is due to a hole in the film.

forms both tiny atomistic clusters and larger crystals which have bulk melting properties. One possible conclusion is that the Sn first separates into the atomistic clusters and then by diffusion forms the larger crystals. The fact that the two melting curves are observed at the same time may just be a result of diffusion limited growth of the larger crystals.

To examine this hypothesis further, a sample of 4000 nm thick $Ge_{0.8}Sn_{0.2}$ was scanned to a high temperature to determine the onset and ending temperatures and shape of the phase separation curve. The shape was nearly identical to that of the sample whose scanned DSC spectrum is shown in fig. 6 and in which the interesting double Sn melting endotherms were subsequently observed. The only difference between the DSC scan of the 4000 nm thick $Ge_{0.8}Sn_{0.2}$ sample and fig. 6 was in the onset and ending temperatures of the phase separation exotherm, which was a result of the difference in Sn concentration between the two samples. After determining the onset temperature of phase separation to be 568±1 K, a fresh as-deposited sample of 4000 nm thick $Ge_{0.8}Sn_{0.2}$ was annealed to a series of temperatures within the phase separation regime, see fig. 9. The maximum temperature of the top two curves is lower than the onset of phase separation. The next curve, labeled 566 K, shows, at the highest temperature of the scan, the beginning of the phase separation exotherm. Up to this point no melting endotherm was observed indicating the absence of phase separated Sn in both the as-deposited sample and the first several temperature scans. Only after the first four anneals does a melting endotherm become clearly visible, as one would expect since the maximum temperature of the previous scan was approximately that of the onset temperature of phase separation.

Examination of this endotherm shows that its onset temperature is too low to correspond to bulk Sn melting, however the temperature and curve shape do correspond to that of the lower temperature endotherm (henceforth referred to as the first endotherm) in fig. 6. In the fifth and sixth curves of fig. 9, note the absence of the higher temperature endotherm (henceforth referred to as the second endotherm) observed in fig. 6, which occurred at the melting temperature of bulk Sn. As the maximum temperature of each anneal in fig. 9 was increased, the shape of the first endotherm changed and the second endotherm became visible. The end of the phase separation exotherm occurred at 630K as determined from the first temperature scan of the material. Thus the next to the last curve in fig. 9, in which the sample has been heated to 650K should show all the Sn phase separated from the GeSn alloy. Note the presence of both Sn melting endotherms. Heating this completely phase separated sample to a very high temperature (835K), cooling it, and then reheating through the Sn melting temperature produced a single endotherm at precisely the bulk Sn melting temperature. Correlating the "bumps" on the phase segregation exotherm with the appearance of the depressed and bulk Sn melting endotherms at different maximum anneal temperatures, indicates that there are *two* different paths or mechanisms by which the Sn separates from the crystalline alloy.

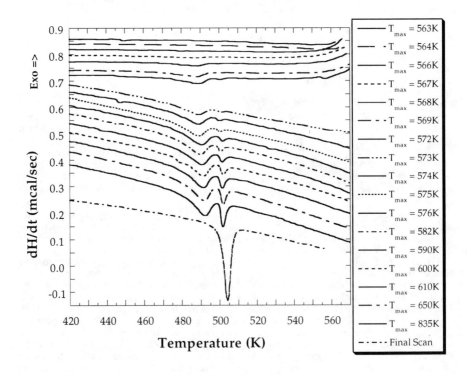

Figure 9: Successive 40K/min. scans of $Ge_{0.8}Sn_{0.2}$ showing the development of the two different types of phase separated Sn discussed in the text. Each scan was heated to a higher maximum temperature than the previous scan. For each curve, the maximum temperature of that scan is indicated in the legend of the plot. Each curve is shifted down from the previous one for clearer comparison.

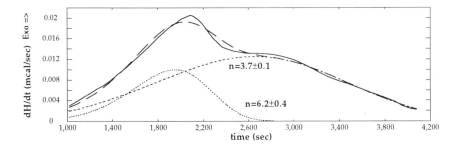

Figure 10: JMA fit to the 450K isothermal anneal of the thick $Ge_{0.69}Sn_{0.31}$ sample. The solid curve represents the experimental data and the long dashes correspond to the sum of the two fitted exothermic peaks shown by the short dashes and dots. The value of the parameters which produced the theoretical curves are: (first reaction) $n = 6.2 \pm 0.4$, $H_0 = 8.6 \pm 0.5$ mcal, $K(450K)=(3\pm7) \times 10^{-21}$; (second reaction) $n=3.7\pm0.1$, $H_0=25.5\pm0.7$ mcal, $K(450K)=(2\pm2)\times10^{-13}$.

With this information we deconvoluted the phase separation exotherm into two exotherms by assuming, for simplicity and from the STEM images, both processes to be some type of Johnson-Mehl-Avrami nucleation and growth. If two overlapping curves are each fitted to equation (1) using a general curve fitting routine such as that in Kaleidagraph©, the fit to the experimental data is exceptionally good (see fig. 10). The excellence of the fit indicates that the assumption of JMA type nucleation and growth for the tetragonal (β–Sn) phase is reasonable. However, the parameters that result from this fit are a little unusual: the Avrami exponent *n* corresponding to the first of the two fitted exotherms is 6.2±0.4. This *n* implies an *increasing* nucleation rate since the growth of this phase was three dimensional, as determined from STEM measurements. This first exotherm corresponds to the nucleation and growth of the phase which produced the depressed Sn melting endotherm. The meaning of this value of n is not understood at the moment. The Avrami exponent of the second curve, which corresponds in our model to the nucleation and growth of the phase causing the bulk-Sn melting endotherm, is n = 3.7±0.1 and has an associated rate constant which is almost eight orders of magnitude greater at this isothermal anneal temperature than the rate constant for the first exotherm. However, since the Sn ultimately forms two dimensional-like disks, the growth of this second phase must be more two dimensional and thus the value of 3.7 implies that the nucleation rate is decreasing.[8]

It is interesting to compare these values of *n* to the isothermal data of the 200 nm thick $Ge_{0.75}Sn_{0.25}$ samples, see fig. 11. In this material, the phase separation exotherm consisted of a smooth bell-shaped exotherm with no additional "bumps" on either the leading or trailing edges of the exotherm. If our explanations for the thicker material are correct, namely the presence of the double Sn melting implies two phase separation reactions with the earlier reaction corresponding to the lower temperature, broader melting endotherm, then we would expect a single β–Sn melting endotherm on the second heating scan of the thinner sample. This was indeed the case as can be seen in fig. 2. It is interesting to note that only the *lower temperature* (in comparison to the bulk melting temperature), *broad-shaped*, melting endotherm is observed. If we again assume JMA-type nucleation and growth for the phase separation reaction, we would expect to find an *n* close to 6 if our previous hypotheses from the thicker material are correct. Such an analysis on this thinner material produced an *n* of 6.4±0.4 and a rate constant of the same order of magnitude, $K(540K) = (4.9\pm0.9)\times10^{-22}$. Figure 11 shows the JMA fit to the data. The deviation of the fit from the experimental data is better than the accuracy of the DSC which is about 0.001 mcal/sec. The excellence of this fit corroborates our suppositions and preliminary conclusions from the thicker material.

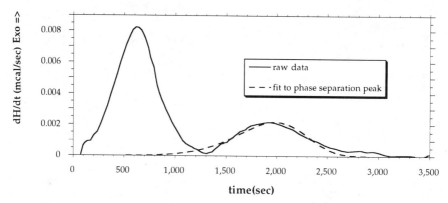

Figure 11: JMA fit to the 540K isothermal anneal of the 200 nm thick $Ge_{0.75}Sn_{0.25}$ sample. The solid curve represents the experimental data and the dashed curve corresponds to the fitted exothermic phase separation peak. The values of the parameters which produced the theoretical curve are: $n = 6.4\pm0.4$, $H_0 = 0.74$ kJ/mol, $K(540K) = (4.9\pm0.9) \times 10^{-22}$.

SUMMARY AND SPECULATIONS

We showed that by monitoring the relative crystallization and phase separation temperatures, we successfully determined the deposition conditions necessary to produce the appropriate amorphous structure for the 200nm $Ge_{0.75}Sn_{0.25}$ samples. Then when this amorphous structure was post-deposition annealed, it crystallized *without* simultaneous phase separation. Subsequent analysis of the shape of the crystallization exotherm as measured by a DSC together with TEM indicated three dimensional growth of the diamond cubic metastable crystalline form of the alloy with a decreasing nucleation rate.

Employing identical deposition conditions except for a longer deposition time and a slightly higher Sn concentration, 8000nm thick samples of $Ge_{0.69}Sn_{0.31}$ were produced. X-ray diffraction of these films exhibited the broad diffraction peaks associated with amorphous materials, however, isothermal DSC measurements provided startlingly different exothermic behavior than is associated with the crystallization of amorphous films. A large time decaying exotherm was measured which subsequent analysis showed corresponded *only* to grain growth of already pre-existing nanometer-sized crystals. Immediately following the exothermic decay, the transformation to the phase separated state was observed. The fact that the phase separation occurred so closely in time after the grain growth exotherm may be related to the fact that the interfacial energy of the metastable crystal grains, 113 mJ/m² as determined from analysis of the shape of the decaying exotherm, is just slightly smaller than the grain boundary energy of β–Sn (164 mJ/m²).

As mentioned earlier, the most desirable films for device applications are those with the largest grain size. To do this in these metastable $Ge_{1-x}Sn_x$ alloys requires controlling or delaying the phase separation process(es) and led us to concentrate on the mechanisms governing phase separation. By examining the β–Sn melting endotherms, of which there were two in the 8000 nm thick $Ge_{0.69}Sn_{0.31}$ samples, we deduced that there were two different paths by which the Sn separated from the metastable crystalline alloyed state. In this material, the Avrami nucleation exponent, n, was 6.2 ± 0.4 for the first of the two phase separation mechanisms corresponding to an increasing nucleation rate and 3.7 ± 0.1 for the second separation pathway. Since the growth of this second phase was two dimensional, the nucleation rate decreased as a function of time. Applying the same analysis to the phase separation exotherm in 200 nm thick $Ge_{0.75}Sn_{0.25}$ films produced evidence which corroborated our hypotheses deduced from the thicker material. The thinner samples exhibited only one depressed melting β–Sn endotherm leading us to conclude that there was only one pathway for the Sn to phase separate from the crystalline alloy. This inference was justified by the excellent fit of one exothermic nucleation and growth process to the phase separation peak. Assuming a JMA analysis, the Avrami nucleation constant was 6.4±0.4, exactly the same to within measurement error as the n for the first of the two exothermic phase separation reactions in the thicker material. This is consistent with our conclusion that this phase separation reaction is associated with the depressed melting endotherm. We associated this broad temperature-depressed endotherm with the melting of small clusters of Sn atoms and hence the n of 6.2-6.4 with the nucleation and growth of small clusters of Sn atoms. In addition the bulk melting endotherm observed in the thick material was associated with a second β–Sn separation process the n of which was 3.7±0.1. We therefore associated this second mechanism with the nucleation and growth of macroscopic β–Sn crystals.

There are two different pathways along which the Sn separates in the thick material and only one mechanism in thinner material. Working backwards from this information, we speculate that the crystals formed in the two types of materials are different in such a way as to lead to different pathways along which the Sn separates. Such a difference might be that the number of impurities or defects within the crystals is different in the two types of films. Indeed preliminary high resolution TEM measurements show that there are significantly fewer defects in the crystals which subsequently exhibit only the single, temperature-depressed, melting endotherm. Thus greater understanding of the mechanisms of how and where on an atomic scale the Sn separates might provide the information needed to control the as-deposited amorphous structure of the alloys such that the desired metastable crystal size is large enough to make the material useful for device applications.

ACKNOWLEDGMENTS

I thank Professor Katayun Barmak for many helpful discussions on both the grain growth analysis and the interpretation of the phase separation mechanisms.

REFERENCES

[1] S. Oguz, W. Paul, T.F. Deutsch, B.-Y. Tsaur, D.V. Murphy, Appl. Phys. Lett. **43**, 848 (1983).
[2] J.W. Christian, The Theory of Transformations in Metals and Alloys, 2nd ed., (Pergamon, Oxford, 1975).
[3] L.C. Chen and F. Spaepen, J. Appl. Phys. **69**, 679 (1991).
[4] L.C. Chen and F. Spaepen, ibid.
[5] J.S.C. Jang, and C.C. Koch, J. Mater. Res. **5**, 325(1990).
[6] D. Turnbull, J.S.C. Jang, and C.C. Koch, J. Mater. Res. **5**, 1731 (1990).
[7] Ph. Buffat and J-P. Borel, Phys. Rev. A **13**, 2287 (1976).
[8] K.R. Coffey, L.A. Clevenger, K. Barmak, D.A. Rudman, and C.V. Thompson, Appl. Phys. Lett. **55**, 852 (1989).

PLASMA EFFECTS ON THIN FILM MICROSTRUCTURE

MUNIR D. NAEEM[*,***], STEPHEN M. ROSSNAGEL[**] AND KRISHNA RAJAN[***]
[*] IBM Microelectronics Division, Hopewell Jn., NY 12533
[**] T.J. Watson Research Center, IBM Research Division, Yorktown Heights, NY 10598
[***] Materials Engineering Dept., Rensselaer Polytechnic Institute, Troy, NY 12180

ABSTRACT

We have studied the effects of low energy ion bombardment on thin copper films. Evaporated, sputtered and CVD copper films (~50 nm) were exposed to Magnetically Enhanced (ME) Ar plasmas. The microstructural changes (grain size) in the films were studied using Transmission Electron Microscopy (TEM).

Grain growth is observed in thin Cu films when the films are exposed to low energy (87 eV) Ar plasmas. The microstructural changes in sputtered and evaporated films are quite significant whereas the plasma bombardment has less effect on CVD films. These changes occur very rapidly and cannot be attributed solely to the thermal effects, especially at low RF power levels (500 W). The initial microstructure of the film has a significant effect on grain growth during plasma exposure.

INTRODUCTION

The microstructure of a material influences its electrical and mechanical properties. The effect of thin film microstructure on its electrical properties have been reported in the literature[1,2]. In semiconductor interconnect materials, the effect of grain size on electromigration is well documented[3]. The grain size of thin films is affected by processes such as thermal anneals during semiconductor processing. Ion implant, reactive ion etch (RIE) and routine sputter clean processes can also affect the film microstructure due to ion bombardment. The range of ion impact energies during these processes can vary from a few hundred eV (during RIE processes) to several KeV (during ion implant). The high energy (>10 KeV) ion bombardment effects on material surfaces have been reported[4,5,6]. Liu et al[7] and Atwater[8] have investigated grain growth in Cu and Au films during high energy ion bombardment. In RIE and sputter clean processes, the ion bombardment energies are usually low (<1 KeV). The effect of low energy ion bombardment on material surfaces is relatively unexplored. Our work focuses on the effect of low energy ion bombardment on thin Cu films during magnetically enhanced (ME) plasma processing.

EXPERIMENTAL PROCEDURES

Thin Cu films (~50 nm in thickness) were deposited on small pieces of special TEM membrane wafers attached to 125 mm wafers. The films were deposited using evaporation, sputtering and CVD techniques. The substrate on which the films were deposited was a thin Si_3N_4 film on Si wafers. The Si in the membrane areas of the wafer was etched out from the backside prior to the film depositions. No mechanical polishing or ion milling was used to

Mat. Res. Soc. Symp. Proc. Vol. 343. ©1994 Materials Research Society

ensure that effects observed in TEM microscopy are due to plasma processing only. The details of TEM membrane wafers are given elsewhere[9]. The evaporated films were deposited by electron beam heating of the Cu source. The deposition chamber was pumped down to the base pressure of ~ 2×10^{-7} Torr. The Cu film deposition rate was controlled at 0.5 nm/sec and no substrate heating was used. The substrate temperature due to E-beam heating was approximately 40°C. The sputtered films were deposited at room temperature and the substrate temperature for CVD films was approximately 180°C during deposition. The films were then processed in an ME Ar plasma reactor for 10, 20, 30 and 40 seconds at 500 W using magnetic field of 60 gauss. The Ar gas flow rate and pressure in the reactor were 100 sccm and 300 mTorr respectively. The cathode coolant temperature was 16°C. The ion impingement energies were simulated from the data collected by Langmuir probe. TEM sample temperature due to plasma heating was estimated by heat transfer calculations and from the experimental data collected by Tempe-dots[10]. A Tempe-dot covered with kapton tape was attached to the wafer next to the TEM sample. The wafers with a TEM sample and a Tempe-dot were processed in the plasma reactor. The plasma processed and the control samples were examined using TEM.

RESULTS AND DISCUSSION

The microstructures of the as-deposited evaporated film and the evaporated film exposed to ME plasma for 40 seconds at 500 W are shown in Fig 1. The microstructure of the as-deposited Cu film has a non uniform grain size distribution indicative of abnormal grain growth. The microstructure of the evaporated Cu film exposed to ME plasma for 40 seconds at 500 W is shown in Fig 1-b. Grain growth has occurred during plasma processing and the area occupied by large grains has increased. There is a large number of grains with twins in the plasma processed film. Fig 2 shows the microstructure of as-deposited and plasma processed (for 40 seconds at 500 W) sputtered films. As-deposited sputtered Cu film (Fig 2-a) also shows populations of small and large grains. The average diameter of small grains is relatively small as compared with the average diameter of large grains. As seen in Fig 2-b, the grain growth has taken place in sputtered films also as a result of exposure of these films to ME plasma. The area occupied by large grains has significantly increased due to ion bombardment during plasma process. The percentage increase in the area of large grains compared with control samples for sputtered and evaporated films is plotted in Fig 3. The plot is for the films exposed to ME plasmas for 10, 20, 30 and 40 seconds at 500 W. The microstructural changes in evaporated films happen in the first 10 seconds. The grain growth resulting from plasma exposure of sputtered films is slower in the beginning and the percentage area occupied by large grains is much higher for the samples processed for 40 seconds as compared with the evaporated films exposed to ME plasma for the same duration. The microstructures of as-deposited and plasma exposed (for 40 seconds at 500 W) CVD Cu films are shown in Fig 4. The grain size distribution of as-deposited CVD Cu film is more uniform (Fig 4-a) and there is no significant change in the microstructure of CVD film during plasma processing (Fig 4-b).

The ion impingement energies on the Cu surface are estimated[11] to be approximately 87 eV for the processing conditions described in the experimental procedures. The estimate[11] of TEM sample temperature is 319°C using 1-dimensional heat transfer model. The finite element analyses[11] for 2-dimensional heat transfer estimate the TEM sample temperature rise to be only 117°C which is closer to the temperature measured by the tempe-dot. The microstructural changes observed in evaporated and sputtered films during plasma processing take place very rapidly (in less than 40 seconds). Work by Gupta[12] et al shows that any noticeable grain growth

114

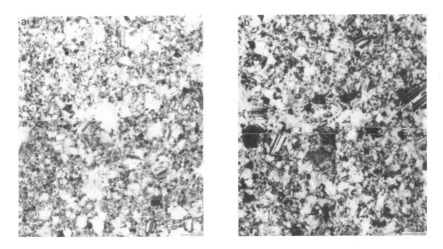

Fig 1 TEM micrographs showing microstructure of (a) as-deposited evaporated Cu film and (b) the evaporated Cu film exposed to ME plasma for 40 seconds at 500 W.

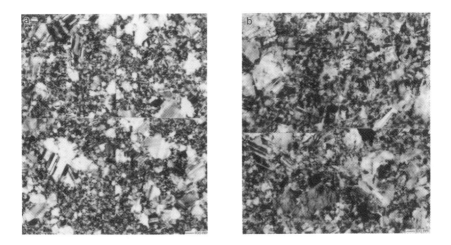

Fig 2 TEM micrographs showing microstructure of (a) as-deposited sputtered film and (b) the sputtered Cu film exposed to ME plasma for 40 seconds at 500 W.

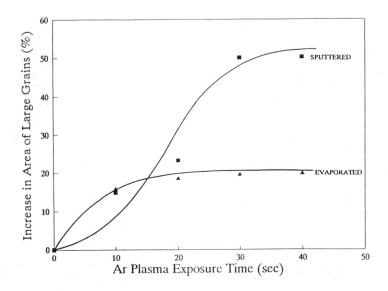

Fig 3 The comparison of increase in the area of large grains for evaporated and sputtered Cu films exposed to ME plasma at 500 W.

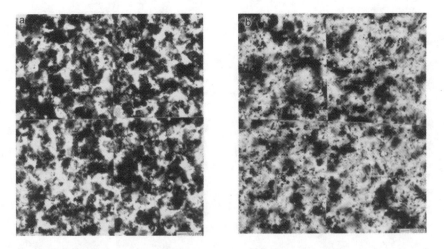

Fig 4 TEM micrographs showing microstructure of (a) as-deposited CVD film and (b) the CVD Cu film exposed to ME plasma for 40 seconds at 500 W.

in thin Cu films occurs only at temperatures higher than 200°C when the films are annealed for at least 900 seconds. In spite of the differences in initial microstructure of the films investigated in this work and those studied by Gupta[12] et al, the grain growth during plasma processing appears, in large part, not because of the thermal effects alone. The data presented here suggest that there are some non-thermal effects that lead to such a fast grain growth in Cu films during plasma processing. We are currently investigating the role of Cu surface modification due to plasma bombardment on grain growth observed in evaporated and sputtered Cu films. The effect of the initial microstructure of the film on subsequent grain growth during plasma processing is also very obvious. As seen in Fig 1, 2 and 4, the initial microstructure of Cu films depends on the deposition technique. The grain growth in evaporated and sputtered films is abnormal. Rollett[13] et al have shown in their simulation work that in a microstructure exhibiting abnormal grain growth, the grain boundary energy difference due to grain boundary orientations acts as a driving force for further grain growth. Thompson[14] has explored the role of surface energy during abnormal grain growth. Floro and Thompson[15] have shown that surface energy minimization leads to the preferential growth of grains of specific orientation in thin poly-Si and Au films. In general, the sputtered and the evaporated films have the higher tendency to undergo the grain growth due to the larger grain boundary density as compared with the CVD films. The CVD films are also believed to have relatively high amount of contaminants which can impede the grain boundary mobility. This is another reason why no noticeable change in grain size is observed in the plasma processed CVD films.

CONCLUSIONS

(1) Low energy ion bombardment during ME plasma processing leads to grain growth in thin Cu films.

(2) The extent of the plasma effects on these films depends on the initial microstructure of the film.

REFERENCES

[1]   Ohmi, T., Otsuki, M., Takewaki, T. and Shibata, T., **J. Electrochem. Soc.**, Vol: 139, No: 3, 922 (1992)

[2]   Lin, T. and Raman, V., **Mat. Res. Soc. Symp. Proc.**, Vol: 226, 203 (1991)

[3]   Walton, D. T., Frost, H. J. and Thompson, C. V., **Mat. Res. Soc. Symp. Proc.**, Vol:225, 219 (1991)

[4]   Auciello, O. in **Ion Bombardment Modification of Surfaces: Fundamental and Applications,** edited by: Auciello, O. and Kelly, R., Elsevier, New York (1984)

[5]   Wilson, I. H., Belson, J. and Auciello, O. in **Ion Bombardment Modification of Surfaces: Fundamental and Applications,** edited by: Auciello, O. and Kelly, R., Elsevier, New York (1984)

[6]   Rossnagel, S. M., Robinson, R. S. and Kaufman, H. R., **Surface Science**, Vol: 123, 89

(1982)

[7]   Liu, J. C., Li, J. and Mayer, J. W., **J. Appl. Phys.**, 67 (5), 2354 (1990)

[8]   Atwater, H. A., Thompson, C. V. and Smith H. I., **J. Appl. Phys.**, Vol: 64, 2337 (1988)

[9]   Naeem, M. D., Leary, H. J. and Rajan, K., **J. Electronic Mat.**, Vol: 21, No: 12, 1087 (1992)

[10]   Trade Mark product of **Omega Engineering**, Stamford Connecticut.

[11]   Naeem, M. D., **Ph.D. Thesis**, Rensselaer Polytechnic Institute, Troy, NY (1993)

[12]   Gupta, J., Harper, J. M. E., Mauer IV, J. L., Blauner, P. G. and Smith D. A., **Appl. Phys. Lett.**, Vol: 61, 663 (1992)

[13]   Rollett, A. D., Srolovitz, D. J. and Anderson, M. P., **Acta Met.**, Vol: 37, 1227 (1989)

[14]   Thompson, C. V., **J. Appl. Phys.**, 58 (2), 763 (1985)

[15]   Floro, J. A. and Thompson, C. V., **Acta metall. mater.**, Vol: 41, No: 4, 1137 (1993)

A STUDY OF THE SURFACE TEXTURE OF POLYCRYSTALLINE PHOSPHOR FILMS USING ATOMIC FORCE MICROSCOPY

R. Revay[†], J. Schneir[*], D. Brower[†], J. Villarrubia[*], J. Fu[*], J. Cline[**], T. J. Hsieh[†], and W. Wong-Ng[**]

[†] Optex Communications Corporation, Rockville, MD 20850

[*] Manufacturing Engineering Laboratory, National Institute of Standards and Technology, Gaithersburg, MD 20899

[**] Material Science and Engineering Laboratory, National Institute of Standards and Technology, Gaithersburg, MD 20899

ABSTRACT

Stimulable phosphor thin films are being investigated for use as optical data storage media. We have successfully applied atomic force microscopy (AFM) to the measurement of the surface texture of these films. Determination of the surface texture of the films is important for evaluating the effect of surface quality on optical scatter. In other thin film material systems it has been found that the surface "bumps" revealed by AFM correspond to grains in the film. This is not the case for the stimulable phosphor films used in our study. We have determined the grain size of our phosphor films by transmission electron microscopy (TEM) and x-ray diffraction (XRD). The grain size from TEM and XRD does not correlate with the size of the AFM surface "bumps." For example, in two of the five films studied, the XRD derived grain size varies by a factor of two but the size of the surface "bumps" remains the same. We conclude that the texture of the film surface is not directly determined by the grain size of the phosphor material.

INTRODUCTION

The Atomic Force Microscope (AFM) has developed into an important technique for determining the surface texture of insulating material. Using the AFM, spatial wavelengths on the order of 1 nm may be investigated. On these length scales the surface texture of polycrystalline thin films often depends on the film's grain size as well as the surface reactions that take place as the film is formed[1,2].

Stimulable phosphor thin films are being investigated for use as optical data storage media. Determination of the surface texture of the films is important for evaluating the effect of surface quality on optical scatter. We have successfully applied AFM to the measurement of the surface texture of these films. AFM images reveal that the surface texture consists of "bumps." In other thin film material systems it has been found that these "bumps" correspond to grains in the film. Hedge *et al*[1] have found that on LPCVD silicon films the features are related to the grain structure. Files-Sesler *et al*[2] have found good correlation between AFM

119

revealed surface topography and the grain size determined by TEM for silicon doped aluminum films. However, for copper doped aluminum, Files-Sesler and colleagues found no correlation between the AFM images and TEM determined grain size. Files-Sesler and co-workers warn that "care must be taken in assigning the term grain size to surface structure elucidated by AFM."

We have determined the grain size of our phosphor films by transmission electron microscopy (TEM) and x-ray diffraction (XRD). The grain size from TEM and XRD does not correlate with the size of the AFM surface "bumps." For example, in two of the five films studied, the XRD derived grain size varies by a factor of two but the size of the surface "bumps" remains the same. We conclude that the texture of the film surface is not directly determined by the grain size of the phosphor material.

SAMPLE MATERIALS

Thin phosphor (SrS host) films were deposited onto heated glass substrates using an electron beam evaporator. The substrate temperature was changed for each run to produce films of different grain size. The grain size was expected to increase significantly with elevated substrate temperature.

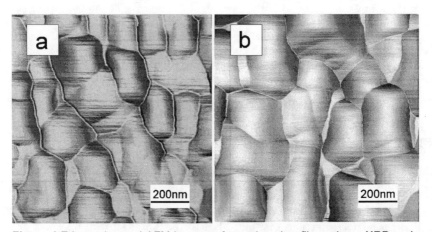

Figure 1 Edge-enhanced AFM images of two phosphor films whose XRD grain size estimates differed by a factor close to two as the temperature increased. The image with the smaller XRD estimate has slightly larger surface "bumps." **a**) Sample S631, 43 nm XRD estimate. **b**) Sample S638, 25 nm XRD estimate.

AFM IMAGING

All AFM images were obtained using a commercially available AFM[3,4]. Figure 1 shows AFM images of two films with XRD derived grain size that varies by a factor of two. The size of the surface "bumps" remains approximately the same.

The dependence of AFM feature size estimates on instrument parameters was studied in a full factorial experiment. Scan range, scan rate, tip force, and the control loop integral feedback parameter were varied. No statistically significant dependence was found on any of these parameters.

Effect of AFM Tip Size

The familiar images of atoms obtained by scanned probe microscopes might lead an incautious observer to conclude that lateral dimensions obtained from AFM images are trustworthy into the sub-nanometer size regime. Although this may be true for very flat specimens (especially when correction is made for inherent nonlinearities in piezoelectric scanners), interpretation of images of high aspect ratio specimens, like those containing grains with heights comparable to their lateral dimensions, is more problematical. The difficulty is illustrated in Figure 2b. Here, the surprising uniformity in size and shape of the "bumps" is explained by the fact that the image was taken with a damaged tip (Figure 2a). The surface features, each of which is smaller than the tip, produce multiple images of the tip. Such tip damage, in which the end of the otherwise pyramidal tip is removed leaving an approximately rectangular plateau, can occur if the tip collides with one or more sufficiently hard protrusions on the surface.

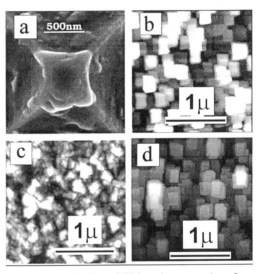

Figure 2 a) An SEM micrograph of a damaged AFM tip. b) An AFM image of phosphor film obtained with this tip. c) A similar area imaged with an undamaged tip. d) A computation of what the surface in (c) would have looked like had it been scanned with a pyramidal tip with a rectangular plateau of 100x165 nm on its end. Note the similarity to (b).

Figure 2b is an extreme example of a process that occurs even with an undamaged tip. The tip shape is "convolved" (strictly, dilated) with that of the surface.[5] The implication is that an image of a "bump" is enlarged compared to the actual "bump" by an amount comparable to the size of that part of the tip which comes into contact with it during imaging. The ends of our undamaged tips were approximately parabolic, as judged by SEM images, and 45 nm in radius as judged by the size of the smallest features in the image. For large features, 12 nm of distortion would be insignificant. For smaller ones, the effect of tip size must be taken into account. If the tip shape is known, those parts of the surface that were touched by the tip may be reconstructed and the surface feature sizes more accurately determined.

Figure 2c shows a similar area of the same sample, but imaged with an undamaged tip. Figure 2d is approximately the image which would have been obtained had the area shown in Figure 2c been imaged with a pyramidal probe with a 100 × 165 nm plateau on its end. This was calculated by dilation of the image with a model tip, as described in reference 5. Note the similarity to Figure 2b.

XRD

Powder diffraction data were collected on a diffractometer equipped with an incident beam monochromator, sample spinner and position sensitive detector. Data was collected from 20 to 150 degrees 2Θ using CuKα₁ radiation. Diffraction scans indicated the specimens were polycrystalline and displayed preferred orientation that varied in level and preferred direction with the substrate temperature. These data were further analyzed using commercially available software[6] to determine crystallite size and micro-strain of the specimens via a Williamson-Hall plot[7]. NIST SRM 660, LaB₆, was used to characterize the instrumental contribution to line broadening; i.e., the instrument profile function.

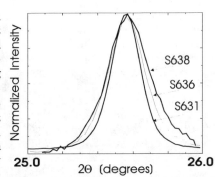

Figure 3 Line broadening of three of the films investigated. Film S631 was deposited at the highest substrate temperature, S638 at the lowest.

Both the split Pearson VII and pseudo-Voigt functions were used to refine the SRM 660 data; a superior fit was obtained with the pseudo-Voigt function. The analysis software models the dependence of the various profile terms on 2Θ so that it can generate an instrument profile at any 2Θ angle. The specimen contribution to the profile shape was then refined onto the instrument profile using a Lorentzian function. The integral breadths from the sample contribution were then used to generate a Williamson-Hall plot that allowed for the separation of the crystallite size and micro-strain contributions to the profile broadening.

We found that the model used to fit the instrument profile did not give a very good fit to our data. Furthermore, many of the diffraction lines from our specimens were of very low intensity due to the preferred crystallographic orientation observed. For these two reasons, we estimate the uncertainty of our grain size data to be no better than 20-30%. However, the reduction of line breadth with increasing temperature is clearly seen in Figure 3. This figure illustrates the line broadening of the (111) plane for three samples, S631, S636, and S638. Grain sizes estimated in this way varied from 25 to 43 nm($\pm$ 10 nm).

Figure 4 TEM cross section of film S637 showing the phosphor/substrate interface. The film was deposited at an intermediate temperature. The grains are elongated in the direction normal to the substrate.

TEM

Sections of the films imaged using a TEM showed grains that were not equiaxed. The grains are elongated in the direction normal to the substrate. It is clear from figures 4 and 5 that the grains are very strongly oriented. The films contain a high density of structural defects that could play a major role in limiting phosphor performance.

Figure 5 TEM cross section showing the top surface of sample S637. The AFM image of this film was similar to those shown in figure 1.

RESULTS AND CONCLUSIONS

The surface "bumps" observed in the AFM images are ~200nm in size. The XRD grain size estimate is ~30nm. The TEM estimate in the plane of the surface is ~50nm. It is clear that the absolute magnitude of the size of the AFM surface "bumps" is larger than the estimated grain size of the material.

We now address the question of the dependence of the size of the surface "bumps" on the grain size of the material. Figure 1 shows AFM images of two films with XRD derived grain sizes that vary by a factor of two. The size of the surface "bumps" remains approximately the same. We conclude that the texture of the film surface is not directly determined by the grain size of the phosphor material.

We can only speculate on the reason for the lack of correspondence between surface "bump" size and grain size. Some possibilities are: 1) differing sensitivity of the probes to sub-grain structure, 2) development of surface roughness based on kinetics of film growth independent of underlying grains, 3) surface specific chemistry which occurs during deposition.

ACKNOWLEDGMENT

We thank T. Estler, B. Polvani, and G. Storti for many useful discussions during the course of this work and for their help in interpreting the TEM micrographs. This work was funded in part by the Advanced Technology Program Office of the Department of Commerce.

[1]   R.I. Hegde, M.A. Chonko, and P.J. Tobin in *Evolution of Surface and Thin Film Microstructure*, edited by H.A. Atwater, E, Chason, M.H. Grabow, and M.G. Lagally (Mater. Res. Soc. Proc. **280**, Pittsburgh, PA, 1993) pp.103-108.

[2]   L.A. Files-Sesler, T. Hogan, and T. Taguchi, *J. Vac. Sci. Tecnol.* **A10(4)**, 2875-2879 (1992).

[3]   Digital Instruments Corporation Nanoscope II fitted with a stand-alone head.

[4]   Certain commercial equipment is identified in this report in order to describe the experimental procedure adequately. Such identification does not imply recommendation or endorsement by NIST, nor does it imply that the equipment identified is necessarily the best available for the purpose.

[5]   H. Gallarda and R. Jain, in *Integrated Circuit Metrology, Inspection, and Process Control V* (SPIE Proceedings **1464**, Bellingham, WA 1991) pp.459-473; G.S. Pingali and R. Jain, in *Proceedings of IEEE Workshop on Applications of Computer Vision*, (Los Alamitos, CA, 1992) pp. 282-289; D.J. Keller and F.S. Franke, *Surf. Sci.* **294**, 409 (1993).

[6]   "SHADOW", Materials Data, Incorporated, P.O.Box 791, Livermore,CA94551

[7]   G.K. Williamson and W.H. Hall, *Advances in X-ray Analysis*, **23** (Plenum Press, New York, 1953) pp.73-81

NANOCRYSTALLINE MICROSTRUCTURES
BY THIN-FILM SYNTHESIS METHODS

M. LIBERA AND D.A. SMITH
Stevens Institute of Technology, Hoboken, NJ 07030

ABSTRACT

Nanostructured materials are being extensively studied because
their ~1-100nm grain size can dramatically affect properties.
Most nanocrystalline synthesis methods produce particulate or
flake. The process of consolidation also allows coarsening,
contamination, and the introduction of porosity. The effect of
nanocrystallinity on mechanical properties must be deconvoluted
from these extrinsic artifacts. Most synthesis routes also
produce small quantities of material. Reproducibly making
enough specimens to explore more than a few properties is thus
difficult. This paper describes thin-film processes to produce
nanostructured materials. Thin-film deposition can easily
produce many specimens, free from extrinsic artifacts, with
identical composition and processing history. Many methods are
now well established to study a variety of thin-film mechanical
properties. We show examples of nanostructured films generated
by controlling deposition and/or post-deposition processing.

I. INTRODUCTION

Nanocrystalline materials have received tremendous
attention since their discovery in 1984 [1]. They have grain
sizes, d, several orders of magnitude finer than conventional
materials. Consequently, a nanocrystalline microstructure has
significant potential to affect many properties such as: yield
strength ($\sigma=\sigma_o+kd^{-1/2}$); volume diffusion-controlled creep
($\dot{\gamma}{\sim}D_vd^{-2}$); grain-boundary diffusion-controlled creep ($\dot{\gamma}{\sim}D_bd^{-3}$);
fracture toughness ($\sigma_F{\sim}k_Fd^{-1/2}$); and superplasticity ($\dot{\gamma}_{GS}{\sim}C_1d^{-2}$ or
$\dot{\gamma}_{GS}{\sim}C_1d^{-3}$).

The principal nanocrystal synthesis methods are gas
condensation [3], mechanical alloying [4], and chemical
precipitation [5]. These have been used in studies examining
microstructure and its stability, enhanced ductility, and
higher strength. While considerable progress has been made,
the mechanisms in many cases remain ambiguous. It is not
clear, for example, if the lateral extent of the grain-boundary
region in a nanocrystalline material is different from that of
a traditional large-grained material. There are conflicting
reports concerning the validity of the Hall-Petch model at
nanocrystalline length scales. A question has also surfaced
concerning the role of triple junctions in defining how
microstructure determines mechanical properties. Both the
extent to which properties are affected by a nanocrystalline
microstructure and the specific microstructural features
responsible for observed changes remain unclear.

Mat. Res. Soc. Symp. Proc. Vol. 343. ©1994 Materials Research Society

Two problems slow progress in this field. First, most
synthesis methods produce particulate or flakes. Consolidation
produces bulk specimens a few millimeters long and a few
hundred microns thick, but it introduces opportunities for
microstructural changes, contamination, and porosity. Most
measurements thus do not directly probe the effects of the
nanocrystallinity. The intrinsic properties of central
interest must somehow be deconvoluted from extrinsic artifacts.
Second, most synthesis routes produce relatively small amounts
of material. It is thus difficult to reproducibly make enough
specimens to explore more than a small subset of behaviors.
Such limitations have prompted controversy surrounding property
measurements as a function of grain size where a range of grain
sizes is produced by repeatedly exposing a single specimen to
coarsening anneals as opposed to producing a set of virgin
specimens having different grain sizes.

Thin-film science and technology offer an alternate means
by which to study many of the novel phenomena associated with
nanocrystalline microstructures. Since a consolidation step is
omitted, film deposition can produce fully dense material
without contamination or microstructural coarsening. For a
film $2\mu m$ thick with an average grain diameter of 20nm, the film
will have ~100 grains through its thickness. Like a bulk
material, the film properties will be determined by the average
behavior of this collection of grains. The mechanical
properties can be measured directly using a variety of
techniques developed over the past several decades by the thin-
film industry (6). The microstructure can be characterized
using the same diffraction, imaging, and microanalysis methods
applicable to bulk materials. A thin-film approach is also not
limited by insufficient quantities of nanocrystalline material,
since a single deposition run can produce a plethora of
specimens of identical composition and processing history.

Like other synthesis methods, ultrafine grain sizes can be
generated in thin films by controlling the nucleation of grains
and/or their subsequent growth. Nanocrystalline films can be
synthesized directly during deposition by exploiting the well-
established zone-model concepts of microstructural development
or by converting an amorphous film to a nanocrystalline
structure by some sort of post-deposition anneal.

II. DEPOSITION OF NANOCRYSTALLINE THIN FILMS

Zone models have been developed to describe the response
of thin-film microstructure to processing variables such as
substrate temperature, deposition rate, and base pressure,
among others. These variables control shadowing, surface and
bulk diffusion, and desorption, the extent of which affect the
film microstructure. Zone models have been generated for
evaporation (7,8) and sputtering (9). Figure 1 (8,10) shows a
zone model characterizing the evaporation of metal films. Zone
I corresponds to the case where the substrate is held at low
temperature $(T_s/T_m \sim 0.1)$. Within this regime, the adatom
mobility is restricted, and there is a large driving force for

crystal nucleation. Figure 1b summarizes grain-size data as a function of substrate temperature collected from a variety of metal systems. The grain size is about 10-25nm for $T_s/T_m < 0.15$.

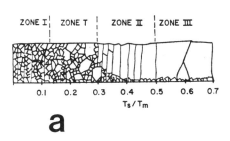

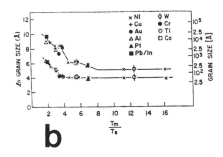

a   b

Figure 1 - a) zone model outlining microstructures as a function of substrate temperature; b) thin-film grain size for a variety of metals within Zone I conditions (after (8)).

An example of a Mo thin film deposited to a nanocrystalline microstructure is described by figure 2. This film was evaporated onto a single-crystal GaAs substrate at 0.1nm/sec in a system whose base pressure was ~10^{-7}Torr. The substrate temperature was T_s=643K giving a T_s/T_m ratio of ~0.22 (T_m^{Mo}=2890K). The diffraction pattern shows sharp Debye-Scherrer rings characteristic of a crystalline film with many randomly oriented grains. The dark-field micrograph was formed by imaging a portion of one of these diffracted rings and shows grains ~20nm in diameter.

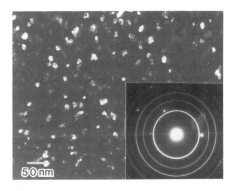

Figure 2 - Electron diffraction pattern (inset) and dark-field TEM image of an as-deposited nanocrystalline Mo film.

III. CONVERSION TO NANOCRYSTALLINE FILMS

In the limit of deposition under conditions where there is highly limited adatom mobility, amorphous thin films can be synthesized. This is most common in multicomponent systems and in compounds where the requirement of coordinated adatom motion to form crystals enhances the tendency to amorphize. Given sufficient time and temperature such structures crystallize by some form of nucleation and growth. A nanocrystalline microstructure can be generated during this conversion process if a high nucleation rate can be generated so the grain size is limited by impingement with other grains.

III.1 Conversion via furnace annealing

Furnace annealing heats an entire film at rates of ~1-20K/min. We have performed such experiments in a TEM with *in-situ* heating (11,12,13). This method enables direct study of the microstructural dynamics during annealing. Figure 3 shows one result from the crystallization of amorphous 40nm Cr_3Si films. This is a fully crystalline microstructure with equiaxed grains having d~25-50nm. Lu et al. [14] and Lui et al. [15] have produced nanocrystalline Fe-Mo-Si-B alloys by furnace annealing amorphous precursors.

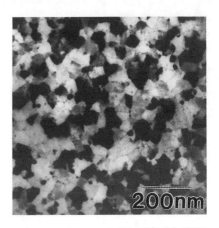

Figure 3 - Bright-field TEM image of Cr_3Si produced by furnace annealing an amorphous precursor.

There is a large chemical free energy driving the amorphous-to-crystal transformation which is absent when a fully crystalline microstructure coarsens by grain growth. The driving force for grain growth is related to the grain-boundary curvature. The curvature is quite high in a material with nanometer grain sizes. It quickly decreases with increasing d. There may be a threshold grain size above which coarsening is no longer significant at an annealing temperature sufficient for crystallization so the grain structure is effectively stabilized. This point deserves further exploration.

III.2 Conversion via annealing/ion irradiation

Irradiation by high-energy ions increases the point-defect population of a material. In the case of crystallization, TEM using the HVEM at Argonne National Laboratory [16,17] shows that ion irradiation can affect both the crystal nucleation kinetics and growth kinetics. Figure 4

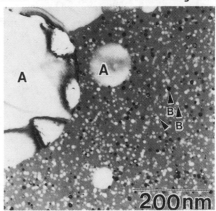

Figure 4 - Duplex crystallite size due to *in-situ* TEM annealing (grains A) and *in-situ* annealing+ion irradiation (grains B).

illustrates the impact that irradiation can have on film microstructure. This micrograph shows a 40nm $CoSi_2$ film subjected to a series of two *in-situ* TEM annealing treatments. The first was performed at T~430K with no ion irradiation and led to the formation of large widely-spaced crystals (grains "A"). The second was performed at T=380K with simultaneous irradiation. This conversion process generated the much finer grain structure (grains "B"), presumably because of an enhanced nucleation rate due to the extrinsic point-defect source.

III.3 Conversion via laser annealing

Laser annealing can be used to provide local heating with rapid heating rates ($\sim 10^6$-10^{10}K/min) depending upon the laser intensity and exposure time. Figure 5 describes the microstructural response of a 70nm film of $(GeTe)_{85}Sn_{15}$ to focused-laser heating. This film was deposited on a grooved disk substrate. The disk was spun at a linear velocity of ~6m/sec under focused krypton ion laser light. Figure 5 shows three tracks of the disk. Each was exposed to two passes of laser exposure at increasing power. 4.0mW incident power is insufficient to nucleate crystals while 4.5mW and 5.2mW exposure does induce crystallization. The grain size of the crystalline regions is of order 50-100nm. The fact that there appears to be little difference in the grain size between the 4.5mW and 5.2mW conditions supports the suggestion that the chemical free energy enables crystallization at a temperature low enough to prevent significant grain growth. We have also shown [18] that this crystalline microstructure is a supersaturated single-phase solid solution which will further decompose into a stable two-phase microstructure when subjected to additional annealing. We would expect such a duplex microstructure to resist coarsening even more effectively.

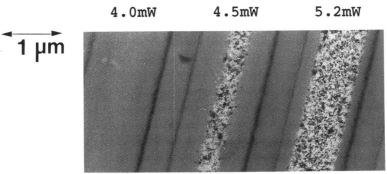

Figure 5 - Laser-annealed $(Ge_{50}Te_{50})_{85}Sn_{15}$ with nanocrystalline grain structure from an amorphous precursor (after (18))

IV. CONCLUSIONS

The discovery of nanocrystalline structures has raised a number of significant questions with important ramifications.

Are nanocrystalline materials fundamentally different from traditional large-grained materials? Is the average grain size a meaningful metric of mechanical behavior or should additional microstructural features such as triple-junction density and spacing or grain-boundary width be considered? Do traditional coarsening models apply to systems containing ultra-fine grain sizes where, once again, the high density of grain-boundary junctions may affect boundary mobility? Can one apply conventional concepts of Zener/Gladman boundary pinning and solute-drag effects to retard coarsening?

Thin-film science and technology is unlikely to replace the various existing and developing methods for synthesizing bulk quantities of nanocrystals. Nevertheless, there is a significant niche where thin-film methods can contribute to the understanding of many basic questions posed by nanocrystalline microstructures. This paper has shown several examples where nanocrystalline films can be generated with little difficulty by controlling the crystal nucleation and growth kinetics during or after deposition. Thin-film techniques can produce many samples with identical compositions free of extrinsic artifacts such as porosity and contamination. These can be combined with the array of well-understood mechanical-testing techniques now used throughout the thin-film community to experimentally address the many new questions regarding structure-property relations at nanometer length scales.

REFERENCES

1. R. Birringer, H. Gleiter, H. Klein, and P. Marquardt, Phys. Lett. 102A, (1984) 365.
2. H.J. Frost and M.F. Ashby, *Deformation-Mechanism Maps*, Pergamon Press, Oxford (1982).
3. R. Siegel, Ann. Rev. Mater. Sci. 21, (1991) 559.
4. C.C. Koch, Nanostructured Materials 2, (1993) 109-129.
5. L. McCandlish and B. Kear, Mat. Sci. & Tech. 6, 953 (1990).
6. *Thin Films - Stresses and Mechanical Properties IV*, ed. P. Townsend et al., Mat. Res. Soc. Symp. Proc. V308 (1993) and prior MRS Symposia on this subject.
7. B. Movchan and A. Demchishin, Phys. Met. Metallogr. 28, 83 (1969)
8. C.R.M. Grovenor, H.T.G. Hentzell, and D.A. Smith, Acta Metall. 32, 773 (1984).
9. J.A. Thornton, Ann. Rev. Mater. Sci. 7, 239 (1977).
10. H.T.G. Hentzell, C.R.M. Grovenor, and D.A. Smith, J. Vac. Sci. Tech. A2, 218 (1984).
11. D.A. Smith, K. Tu, and B. Weiss, Ultramicro. 23, 405(1987).
12. D.A. Smith et al, Mat. Res. Soc. Symp. Proc 321, 271 (1994)
13. M. Libera et al, Mat. Res. Soc. Symp. Proc. 280, 715 (1993)
14. X. Liu, J. Wang, B. Ding, Scr. Met. Mater. 28, 59 (1993).
15. R. Luck, K. Lu, W. Frantz, Scr. Met. Mater. 28, 1071 (1993)
16. C.W. Allen and D.A. Smith, Mat. Res. Soc. Symp. Proc. 201, 405 (1991)
17. J. Im and H. Atwater, Appl. Phys. Lett. 57, 1766 (1990).
18. M. Libera, M. Chen, and K. Rubin, J. Mater. Res. 7, 2666-2676 (1991)

TEXTURE DEVELOPMENT IN THIN NICKEL OXIDE FILMS

F. CZERWINSKI AND J.A. SZPUNAR
Department of Metallurgical Engineering, McGill University, Montreal, Que., H3A 2A7, Canada

ABSTRACT

X-ray diffraction was used to study the influence of ultra-fine dispersions of CeO_2 on the texture development and epitaxial relationships in oxide films, formed at high temperatures. The substrate orientation exerted the essential effect on microstructure and growth rate of oxide films on both pure as well as coated nickel. NiO grown on (100)Ni was polycrystalline with grains randomly oriented. Applying of CeO_2 resulted in a marked decrease of the oxide growth rate without significant changes in the texture. The NiO formed on pure (111)Ni exhibited strong (111) texture. During initial stages of oxidation the presence of CeO_2 caused the nucleation of randomly oriented oxide. However, the oxide developed during further exposure had the same character as that grown on pure (111)Ni face. X-ray measurements are compared with analysis conducted by TEM and electron microdiffraction. The role of chemically active element in inhibition of diffusion processes during the growth of oxides on both crystal faces is discussed.

INTRODUCTION

The growth of oxides on metallic substrates at high temperatures involves the diffusion of metal cations and oxygen anions through the oxide film. Of all oxides with practical importance in protecting against high temperature corrosion, the relative contributions of different transport processes have been most clearly established for NiO. At temperatures lower than about 1300 K, structural defects as grain boundaries and dislocations directly affect the NiO growth kinetics [1]. There is a good deal of evidence that the effectiveness of grain boundaries as easy diffusion paths can be markedly reduced by modification of NiO with elements which have a high affinity for oxygen [2,3]. Oxidation of polycrystalline nickel with a thin surface coating of CeO_2 dispersions showed the high anisotropy of oxide growth on differently oriented grains of the substrate, and on the different types of substrate grain boundaries [4,5]. Oxide growth on polycrystals is complex. Therefore, it is difficult to explain the role of substrate grain orientation in the growth of oxides modified by the reactive element. For preliminary experiments, Ni single crystal substrates have been chosen. The results obtained are the base for measurements on polycrystalline materials. By characterizing the effect of crystallographic orientation of substrate on growth rate and texture of oxides formed on selected crystal planes of Ni, one should find the consistent description of oxide growth on polycrystalline substrates modified with reactive element.

This study contains the analysis of oxide growth on (100) and (111) crystal faces of Ni superficially coated with CeO_2 dispersions.

Mat. Res. Soc. Symp. Proc. Vol. 343. ©1994 Materials Research Society

EXPERIMENTAL PROCEDURE

Thin oxide films were formed on (100) and (111) crystal faces of Ni with a purity of 99.999 wt. % supplied by Research Crystals, Inc. As a final step of substrate preparation the chemical polishing was conducted at temperature of 298 K in solution consisting of 65 ml acetic acid, 35 ml nitric acid and 0.5 ml hydrochloric acid. Crystallographic orientation of a specimen surface was determined by the Laue back reflection X-ray technique. The pure crystal surfaces were modified with 14 nm thick coatings composed of 5 nm CeO_2 particles derived from a sol-gel precursor [5]. Oxidation was conducted at temperature of 973 K and oxygen pressure of 5×10^{-3} Torr in an ultra high vacuum manometric system. Specimens were heated to the oxidation temperature at a rate of 200 deg/min in the presence of oxygen. The oxide growth rate was controlled by monitoring the oxygen uptake with a high sensitivity gauge. X-ray diffractions were obtained by a Rigaku rotating X-ray source. Texture was determined using a Siemens D-500 diffractometer equipped with a texture goniometer. Pole figures were measured using the reflection technique up to maximum tilt of the specimen of 80° in 5° in polar and angular intervals. The results were corrected for absorption and defocussing using a standard powdered specimen of NiO. The pole figures were normalized within the available interval of distribution of crystallographic planes and the intensity on the pole figures was shown using multiples of random intensities. X-ray measurements were supported by examinations of microstructure with a transmission electron microscopy (TEM) using selected area electron diffraction (SAD). Thin foils were prepared by the stripping of oxide films from substrates in saturated solution of iodine in methanol [4].

RESULTS AND DISCUSSION

Oxide growth rate

The oxidation kinetics for both pure and CeO_2 modified Ni crystal faces are shown in Fig. 1. The pure (100)Ni face oxidized at a higher rate and after 4 h exposure at 973 K the oxide was about 4 times thicker than that formed on (111)Ni. Observed differences support previous studies [6,7]. The reduction of the oxide growth rate caused by surface modification with CeO_2 depends on Ni face orientation. The comparison of the thicknesses of oxides grown on modified crystal faces after 4 h exposure indicate a higher reduction achieved for (100)Ni planes. The oxide formed on CeO_2 coated (100)Ni, as calculated from oxygen uptake, was about 2.5 times thinner than that formed on pure (100)Ni. At the same time, for a (111)Ni face, only a small difference was observed between thicknesses of pure and CeO_2 modified NiO (Fig. 1).

To analyse in detail the effect of CeO_2 dispersions on growth of NiO, the coefficient R was defined as:

$$R = k_{pi\ Ni}/k_{pi\ Ni/CeO2} \qquad (1)$$

where $k_{pi\ Ni}$ and $k_{pi\ Ni/CeO2}$ are the instantaneous parabolic rate constants for unmodified and CeO_2 modified samples respectively. The values of k_{pi} were calculated as:

$$k_{pi} = 2w\ (dw/dt) \qquad (2)$$

where w is oxygen uptake and t is time of oxidation. The calculations were performed numerically for time intervals dt = 60 s. The plot of R as a function of oxidation time is shown in Fig. 2. For (100)Ni the R values decrease with oxidation time after reaching the maximum at very early stages. This indicates that the role of the reactive element in oxide growth on this face is decreasing with time. Relatively low the growth rate of oxide on (111)Ni face allowed us to observe the changes in oxidation kinetics at the beginning of reaction, caused by the additions of the reactive element. The presence of CeO_2 particles on the (111)Ni promoted the nucleation and growth of NiO during initial stages of reaction. Therefore the measured growth rate of the oxide is even higher than on pure (111)Ni (R < 1). However, during further exposure, the R-value increases continuously suggesting that the reduction of the oxide growth rate is caused by CeO_2. After approximately 3 h of reaction at 973 K the effectiveness of CeO_2 coating, estimated by ratio R, is higher on (111)Ni than on (100)Ni. Since the diffusion through the oxide affects the growth rate, the presented kinetics suggest that there is a difference in the microstructure of oxide films. This difference is caused by variations in the number of easy diffusion paths for nickel and oxygen ions.

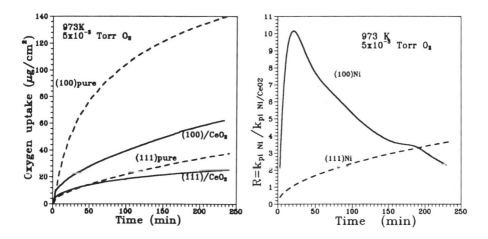

Fig. 1. Oxygen uptake versus time for pure and CeO_2 coated Ni crystal faces.

Fig. 2. Ratios R of instantaneous values of parabolic rate constants for pure and CeO_2 coated Ni crystal faces as a function of oxidation time.

Structure and texture

The X-ray diffraction patterns of oxide films formed after 4 h of oxidation at 973 K are shown in Fig. 3. Oxide with a thickness of 982 nm grown on pure (100)Ni face is polycrystalline. A similar diffraction pattern was determined for 437 nm thick oxide formed after coating with CeO_2. Trace quantities of CeO_2 introduced to the oxide from a 14 nm thick coating were just above the detection level. The diffraction patterns of oxides formed on pure and coated (111)Ni with thicknesses of 263 nm and 178 nm respectively, indicate a totally different microstructure.

Texture measurements revealed that the oxides grown on (100)Ni, both pure and CeO_2 coated, exhibited a random orientation of crystallites. The slight increase of the intensity of (111)NiO peak on the diffraction pattern after coating (Fig. 3), suggests the formation of weak (111) texture, but it is not supported by pole figure measurements. As is shown in Fig. 4a, the relative intensities in (111) pole figure of this oxide are characteristic for a random distribution of grain orientations. As was previously predicted from the diffraction pattern, a strong (111) texture is present in oxide formed on (111)Ni face. At this stage of growth, the X-ray study did not show the marked changes in NiO grain orientation caused by the addition of CeO_2. As is shown in Fig. 4b, strong (111) texture is also present in this oxide film.

The oxides examined by TEM using electron microdiffraction were generally thinner, especially for (100)Ni. The oxide films with thicknesses less than 200 nm formed on pure Ni crystals, examined with TEM, were epitaxially oriented with Ni substrate. The existing relationships were (100)NiO//(100)Ni and (111)NiO//(111)Ni. The presence of CeO_2 enhanced the nucleation of randomly oriented NiO on the outer surface of the coatings during their initial exposure to oxygen. SAD patterns of oxides stripped after the first few minutes of oxidation were composed of continuous rings for CeO_2 and NiO only. After longer exposure time, the faint spots which were produced on SAD pattern were essentially the same as that observed previously after the oxidation of pure crystal faces. This oxide, reproducing the Ni substrate structure, must be formed beneath the CeO_2 coating. Details regarding the TEM analysis are published elsewhere [8].

The comparison of the TEM study of the initial stages and the X-ray analysis of further growth indicates that oxide grain orientation is determined both by the orientation of the Ni surface and by the subsequent growth process. However, our studies revealed the crucial role of the crystallographic orientation of the substrate in texture formation. Thus, the initial epitaxially oriented oxide formed onto (100)Ni is transformed during growth to polycrystalline film with a random texture. After the same time of oxidation, the oxide on (111) face still mainly exhibited the <111> growth direction.

The measured difference in growth rate between oxides formed on pure crystal faces of Ni may be explained by the existing differences in oxide microstructure. The high density of grain boundaries in oxide grown on (100)Ni face is responsible for the higher oxide growth rate. X-ray examinations do not provide a direct explanation of the observed reduction of growth rate caused by CeO_2, partly because of the technique of texture determination used. The presence of the reactive element, introduces the complex depth-structure of oxide [5,8]. Therefore, the precise measurements are necessary for such a multilayer structure to observe the differences caused by the reactive element. The equipment for measuring the texture of thin layers of materials by using a glancing angle for X-ray beam is currently being tested in our laboratory [9].

Our experiments on thicker scales formed on both single crystal and polycrystalline substrates indicate the differences of oxide texture after modification with the reactive element [10]. Although no theoretical analysis of oxide grain growth and texture formation during oxidation is currently available, from analogy with metals one can predict that a small amount of ultra-fine reactive element oxide particles might influence the grain boundary geometry with consequences for grain boundary transport. The presence of grain boundaries in oxide, which have a low diffusivity, will affect the formation of the Ce-rich region in oxide by affecting the Ce ion diffusion. Subsequently, by affecting the transport of metal and oxygen ions, the grain boundaries will also affect the oxide growth.

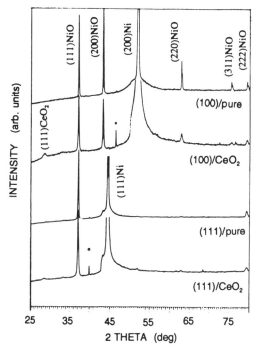

Fig. 3. X-ray diffraction patterns of oxides formed on pure and CeO_2 coated Ni crystal faces (* - Pt from thermocouple spot-welded to the specimen).

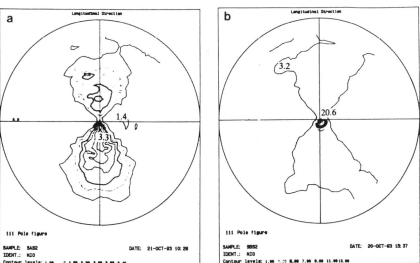

Fig. 4. (111) pole figures for oxides formed on CeO_2 coated Ni crystal faces: a) (100)Ni, b) (111)Ni.

CONCLUSIONS

The orientation of Ni substrate affects the microstructure, texture and growth rate of oxides both the pure and modified with the reactive element.

The (100)Ni oxidized at a higher rate than (111)Ni both before and after coating. For this crystal plane the higher reduction in oxide thickness caused by CeO_2 was achieved. However, the analysis of oxidation kinetics indicates the decreasing role of reactive element in oxide growth on this face, with time of oxidation.

Measurements of texture development and grain boundary geometry in oxides modified with reactive elements are still required. The strong texture may lead to a high proportion of grain boundaries having a low diffusivity. Such boundaries will affect the depth-distribution of CeO_2 in oxide and the level of Ce ions concentration in oxide grain boundaries. Thus, the texture will control the transport of metal and oxygen ions, i.e. the growth rate of oxide at high temperatures.

ACKNOWLEDGEMENT

The financial support of this work by the Natural Sciences and Engineering Research Council of Canada is gratefully acknowledged.

REFERENCES

1. A. Atkinson, R.I. Taylor and A.E. Hughes, Phil. Mag., A 45, 823 (1982).
2. D.P. Moon and M.J. Bennett, Mater. Sci. Forum, 43, 269 (1989).
3. D.P. Whittle and J. Stringer, Phil. Trans. R. Soc. Lond., A 295, 309 (1980).
4. F. Czerwinski and W.W. Smeltzer, in Oxide Films on Metals and Alloys, edited by B.R. Dougall, R.S. Alwitt and T.A. Ramanarayan (The Electrochem. Soc., Pennington, NJ, 1992) p. 81.
5. F. Czerwinski and W.W. Smeltzer, Oxid. Met., 40, 503 (1993).
6. N.N. Khoi, W.W. Smeltzer and J.D. Embury, J. Electrochem. Soc., 122, 1495 (1975).
7. R. Herchl, N.N. Khoi, T. Homma and W.W. Smeltzer, Oxid. Met., 4, 35 (1972)
8. F. Czerwinski and W.W. Smeltzer, in Microscopy of Oxidation, edited by S.B. Newcomb and M.J. Bennett (The Institute of Materials, London, 1993) p. 128.
9. J.A. Szpunar, S. Ahlroos and Ph. Tavernier, J. Mater. Science, 28, 2366 (1993).
10. F. Czerwinski and J. A. Szpunar, to be published.

MODELLING OF THE TEXTURE FORMATION IN ELECTRO-DEPOSITED METAL FILMS

D.Y. LI and J.A. SZPUNAR
Dept. of Metallurgical Engineering, McGill University, 3450 University Street, Montreal, PQ, Canada H3A 2A7

ABSTRACT

The texture formation during the electrodeposition process was simulated using a Monte Carlo technique. The simulation uses a two dimensional hexagonal lattice to map the microstructure of the deposit. The criteria for the texture formation was based on the minimization of the system's free energy. The anisotropy of surface-energy was taken into account. Since a metal's surface energy is influenced by hydrogen adsorption, the texture of metal deposits may vary with hydrogen co-deposition.

INTRODUCTION

The texture of electrodeposits has attracted long-term interest [1-5] because of its beneficial effect on anisotropic properties. However, the mechanism of the deposit's texture formation is not clear. Previous explanations of the texture formation fall into two major categories. One is called the two-dimensional nucleation theory [2]. The main idea of this theory is that the advance of a crystallographical plane is related to the formation of two-dimensional nuclei on the plane. Different planes require different overpotentials for the nucleation, which results in different rates of the two-dimensional nuclei formed on these planes. As a result, the deposit may be textured. However, the operating overpotential for metal deposition is around 10^2 mV which is much higher than the overpotentials required to generate the two-dimensional nuclei on different planes (e.g., a overpotential of 6 mV on the (110) and 8 mV on the (100) during Cu deposition [6]). Under such a relatively high operating overpotential, the two-dimensional nuclei could be generated on all planes at similar rates. Therefore, texture formation caused by the two-dimensional nucleation is not applicable. Disagreement between the theory and experiments have been found [3,4]. Another theory called the geometrical selection theory [1] suggests that texture formation is attributed to both the growth rates of crystallographical planes and deposit's surface morphology. However, this theory can not determine the surface morphology. Moreover, the observation of copper deposits shows [7] that the geometrical selection theory is not consistent because both top surfaces and side surfaces of small copper deposit platelets are close-packed faces, which does not agree with the theory.

This paper presents a Monte Carlo simulation approach to the study of texture formation during electrodeposition. The Monte Carlo simulation technique was suggested by Anderson et al [8], and was applied to simulate nucleation, grain growth and grain topology in recrystallization [9], VD film growth [10] and solidification processes [11]. Electro-deposition is a process in which mass transfer is accompanied by a charge transfer [12,13]. During the process, metal atoms are ionized at the anode and move to the cathode through the electrolyte. At the cathode, the ions are discharged and form a deposit. Formation of the deposit is also a nucleation and grain growth process [12-14]. It is therefore possible to apply the Monte Carlo technique to simulate texture formation and microstructural evolution in this process.

Mat. Res. Soc. Symp. Proc. Vol. 343. ©1994 Materials Research Society

THE MONTE CARLO SIMULATION MODEL FOR TEXTURE FORMATION

Process Controlling Parameter

Current density is often used as the parameter to control electrodeposition because it directly relates to the charge transfer and mass transfer, and thus relates to the deposition rate as described by Faraday's law [12]:

$$\frac{dh}{dt} = \frac{h}{t} = \frac{i_c \omega E_{el}}{60 \zeta} \qquad (1)$$

where h is the thickness of deposit, and E_{el} is the electrochemical equivalent of the metal. t, ζ and i_c are the deposition time, density of the metal and the current density, respectively. ω is the current efficiency. Current density is also of great importance to the microstructure of the deposit, since the current density strongly affects the nucleation frequency [15,16]:

$$F \propto \exp[-\sigma/(RTi_c^2)] \qquad (2)$$

where, σ is a constant. Thus, in the simulation, the current density is considered to be the process controlling parameter.

Factors affecting the Textural Development

When an ion arrives at the cathode surface, it looses its charge and diffuses until reaching a preferred site where this ad-atom can be incorporated into the metal lattice. The preferred site is where a minimum of the system's free energy is attained. *Therefore, those factors which are crystallographically anisotropic and related to the free energy of the system are considered to be the factors affecting the texture formation.* For instance, surface energy anisotropy might play an important role. When arriving at a conjunction of two grains, an ad-atom will tend to sit in the surface which has a lower surface energy and thus leads to a decrease in the overall surface energy. As a result, the boundary between the grains would move towards the grain which has a higher surface energy and its neighbour would therefore grow preferentially.

Implementation of The Monte Carlo Simulation

Microstructural and textural evolution of an electrodeposited coating can be simulated by adding atoms to random locations on the cathode surface and allowing them to diffuse until they reach low energy sites. However, direct modelling of this process is computationally intractable because it needs to take a huge number of atoms (e.g., 10^7 atoms) into account to simulate a surface as small as 1 μm^2. Thus, the present approach can not be an atomistic model but a quasi-continuum model that accounts for the statistical nature of the deposit growth.

In the Monte Carlo simulation, the continuum microstructure is mapped onto a discrete lattice. Each lattice site has the statistic behaviour of the continuum system. For instance, the interaction between two neighbouring sites with different orientations is the grain boundary energy rather than the interaction between two atoms.

A two dimensional hexagonal lattice containing 12000 lattice sites is used to map the microstructure at a cross section of the deposit. Each site is assigned a number, P, to represent the orientation of the grain to which the site belongs. In the simulated deposition, the lattice sites are occupied layer by layer. The occupation of a site is indicated by a change of its P value from zero to a positive number. The overall process is modeled in the following way. Initially, nuclei are generated at the surface of the cathode. The lattice is then scanned layer by layer from bottom to top. In the scanning of a layer, the site is randomly selected and its P value is tested. If $P=0$, the site will be occupied and its P value will change to a positive number. If $P>0$, meaning that the site has already been occupied, it is then skipped.

When the deposit grows through the occupation of lattice sites, the decision on which P number is selected is based on the change in the free energy of the testing site. The energy of the site is set by considering the interaction between the site and its neighbours:

$$H = -J \sum_{nn} (\delta_{P_i P_j} - 1) - J \sum_{mm}' \Delta(P_i \neq P_j) \qquad (3)$$

where J is one half of the bond energy between two neighbouring sites. P_i represents the P value of site i. $\delta_{P_i P_j}$ has the following values: $\delta_{P_i P_j} = 1$ if $P_i = P_j$, and $\delta_{P_i P_j} = 0$ if $P_i \neq P_j$. $\Delta(P_i \neq P_j)$ is a parameter representing the surface energy anisotropy. For each pair of sites (i,j), Δ is non-zero only when $P_i \neq P_j$ and one of the two sites is unoccupied so that the interface between these two sites is a surface (called a "surface bond"). $\Delta(P_i \neq P_j)$ depends on the P value of the occupied site, and the P is determined by the orientation of the grain to which the occupied site belongs.

The first term in equation (3) depends only on the number of bonds between the testing site and its neighbours. The surface energy anisotropy is considered in the second term. Sum $\sum$ is taken over all of the six nearest neighbour sites, while $\sum'$ is taken over only those sites which are connected to the testing site by "surface bonds". The bond energy between a pair of sites is 2J if the two sites have unlike P values, and zero if they have the same P. If the pair is connected by a "surface bond", the bond energy will be modified and becomes $2J(1-\Delta(P_i \neq P_j))$.

Initial-state energy, H_0, of the selected site is first calculated. Then, the P value of the site is changed to a P value of its nearest neighbours and a trial final-state energy, H_n, is calculated. The P values of all the nearest neighbours are calculated in the same way to obtain a series of trial H_n values, and the minimum H_n is chosen as the final-state energy of the site. If $\Delta H = H_n - H_0 < 0$, the new P value is retained. Otherwise this change in P value is not accepted. $\Delta H \geq 0$ usually occurs at sites on grain boundaries where the nucleation rate is high. Therefore, if $\Delta H \geq 0$, a fresh nucleus with a random orientation is generated in a probability described in eq.(2).

In the simulation, the scanning of one layer with 200 iterations is defined as one Monte Carlo step (MCS), which corresponds to a certain time interval. The deposition rate can thus be modelled using a different number (n) of MCSs in scanning each layer, i.e., $\Delta h/\Delta t \propto \Delta h/n$. The deposition rate is lower if the number of MCSs, n, is higher. Since the deposition rate is proportional to the current density (see eq.(1)), the current density can also be converted into n for the simulation. That is to say

$$i_c \propto 1/n \qquad (4)$$

The Effect of Hydrogen Co-deposition on Texture

Metal deposition is often accompanied by a simultaneous hydrogen co-deposition [12]. When the hydrogen co-deposition occurs, texture may change because the metal's surface energy can be lowered by hydrogen [17]. Since different planes have different adsorption abilities, this lowering of surface energy will lead to a change in the surface-energy anisotropy. Surface energies of lattice planes having lower reticular densities (i.e., loose-packed) decrease more rapidly with an increase in hydrogen content than those having higher reticular densities (close-packed) [1,18]. At present, accurate description of this relation is not available at present. However, a qualitative prediction could be made on the assumption that the change in surface energy of a plane is inversely proportional to its reticular density. Accordingly, $\Delta(P_i \neq P_j)$ in eq.(3) is replaced by $\Delta^*(P_i \neq P_j)$ which counts the effect of hydrogen adsorption:

$$\Delta^*(P_i \neq P_j) = \Delta(P_i \neq P_j) + \frac{\partial \gamma_s(C_H)}{\gamma_s} \cdot \frac{1}{D(P_i \neq P_j)} \qquad (5)$$

where $\Delta(P_i \neq P_j)$ is the anisotropy parameter without the hydrogen adsorption. $D(P_i \neq P_j)$ is the reticular density whose value depends on the orientation of the occupied site. $\partial \gamma_s(C_H)$ represents the decrease in the surface energy with the adsorbed hydrogen content.

RESULTS AND DISCUSSION

Texture and microstructural evolution in electrodeposition process

Six surface energies of iron have been chosen to model the development of six texture components. These orientations and corresponding surface energies, Δs, are given in Table I. The surface energies are calculated using the Lennard-Jones potential and the atom relaxation in the surface layer is taken into account[22]. Three deposition rates $(n = (\alpha/i_c) = 6, 13$ and $21)$ are used in the modelling. Simulated microstructures at the deposit's cross section have been illustrated in Fig.1. At low current density (i.e., larger n), the deposit is coarse-grained, while at high current density (i.e., smaller n), the deposit is fine-grained. In fact, at low current densities the deposition proceeds slowly, which provides time for the metal nuclei to grow

Table I. Δ Factors and Reticular Densities of Different Crystallographic Planes

Planes	Δ Factor	D
110	0.40	1.41
200 = 100	0.00	1.00
211	0.02	0.82
310	0.05	0.63
222 = 111	0.31	0.58
622 = 311	0.06	0.31

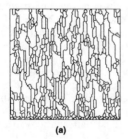 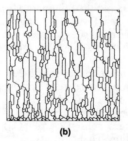 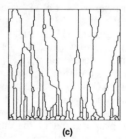

(a)   (b)   (c)

Fig. 1 Simulated microstructures at the cross section of deposit obtained at three deposition rates. Numbers of MCSs (for scanning one layer) corresponding to the three deposition rates are respectively (a) n=6, (b) n=13, (c) n=21.

and there is little scope for the creation of new nuclei. Therefore, the deposits are coarse-grained. Whereas at high current densities, the deposition rate is raised and new nuclei formation is enhanced. As a result, the deposits are fine-grained.

The volume fraction of the six differently oriented grains is presented in Fig.2. It is demonstrated that grains having lower surface energies have higher volume fractions. The {110} component, which corresponds to the lowest surface energy, is the strongest. It can also be seen that the {110} component becomes weaker when the deposition rate increases. This is because at high deposition rates, nucleation rate rises, and therefore the selective grain growth is deteriorated.

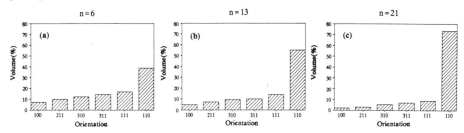

Fig. 2 Volume fractions of differently oriented grains simulated at deposition rates (a) n=6, (b) n=13, (c) n=21.

The effect of Hydrogen co-deposition on texture development

The effect of hydrogen co-deposition on the texture development in an iron deposit was simulated. In the simulation, $\partial \gamma_s(C_H)$ is taken from Petch's results [17]. Reticular densities [19] of the six planes are listed in Table I. The simulation demonstrates that iron deposit changes its texture from {110}, through {111}, to {311} type with an increase in the content of the adsorbed hydrogen (Fig.3).

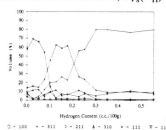

Fig. 3 The texture of iron deposit varies with the content of adsorbed hydrogen.

The Influence of Current Density, pH and Temperature

The content of adsorbed hydrogen at the cathode surface is proportional to the hydrogen coverage θ [20] which increases with increasing hydrogen overpotential (i.e., more negative) [20,21]:

$$\theta \propto \exp(-\lambda \eta_H/RT) \qquad (\eta_H < 0) \qquad (6)$$

where, λ is a constant. The hydrogen overpotential is the difference between the cathode potential (E) and the hydrogen equilibrium potential (E_H^{eq}) [13]:

$$\eta_H = E - E_H^{eq} = E + \frac{RT}{F} pH \qquad (7)$$

where, F and R are the Faraday's constant and gas constant.

(i). The Effects of pH

It is seen, from eq.(6) and (7), that hydrogen overpotential decreases (less negative) with an increase in pH, and this leads to a decrease of θ, which results in less hydrogen adsorption by the cathode. As a result, the texture of the iron deposit will change in the following way: {311} → {111} → {110}.

(ii). The Effect of Bath Temperature

A rise of temperature also reduces the hydrogen adsorption. This effect is caused by (1) a decrease in the absolute value of the hydrogen overpotential and (2) an increase in hydrogen desorption. Consequently, the texture of iron deposit will change in this way: $\{311\} \rightarrow \{111\} \rightarrow \{110\}$.

(iii). The Effect of Current Density

An increase in the current density corresponds to an increase in the cathode potential (i.e., more negative), which leads to a rise of hydrogen overpotential (see eq.(7)). For instance, the hydrogen overpotential on Fe is 0.40 V at 0.1 A/dm^2, 0.53 V at 1 A/dm^2 and 0.64 V at 10 A/dm^2 [12]. Therefore, the increase in current density will raise the content of adsorbed hydrogen. As a result, the texture will transform in the direction: $\{110\} \rightarrow \{111\}$ $\rightarrow \{311\}$.

The texture of iron deposits and its variation with the deposition condition have been measured. Agreement between the simulation and the experiment was found [22].

SUMMARY

A Monte Carlo model was used to simulate the textural formation during electrodeposition process. It was demonstrated that the texture formation was attributed to the minimization of the system's free energy. Although the texture type is affected by the surface energy anisotropy, the current density can affect the deposit's microstructure and the intensity of the deposit's texture. The change in texture type due to the hydrogen co-deposition was also considered. Factors influencing the hydrogen co-deposition and the texture type were analyzed and discussed.

REFERENCES

1. A.K.N. Reddy, J. Electroanal. Chem., 6, 141 (1963).
2. N.A. Pangarov, J. Electroanal. Chem., 9, 70 (1965).
3. J.R. Park and D.N. Lee, J. Korean Inst. Metals, 12, 243 (1976).
4. I. Epelboin, M. Froment and G. Maurin, Plating, 56, 1356 (1969).
5. D.Y. Li and J.A. Szpunar, J. Electr. Mater., 22, 653 (1993).
6. D. Postl, G. Eichkorn and H. Fischer, Z. Phys. Chem., 77, 138 (1972).
7. H.J. Pick, G.G. Storey and T.B. Vaughan, Electrochimica Acta, 2, 165 (1960).
8. M.P. Anderson, D.J. Srolovitz., G.S. Grest and P.S. Sahni, Acta Metall., 32, 783, 793, 1429 (1984), 33, 509 (1985).
9. D.J. Sorolovitz, G.S. Grest and M.P. Anderson, Acta Metall., 34, 1833, 2115 (1986).
10. D.J. Srolovitz, J. Vac. Sci. Technol., A4(6), 2925 (1986).
11. J.A. Spittle and S.G.R. Brown, J. Mater. Sci., 23, 1777 (1989).
12. E. Raub and K. Müller, Fundamentals of Metal Deposition, Elsevier Publishing Co., Amsterdam, 1967.
13. N.V. Parthasaradhy, Practical Electroplating Handbook, Prentice-Hall, Inc., Englewood Cliffs, New Jersey, 1989.
14. Frank C Walsh and Maura E Herron, J. Phys. D: Appl. Phys., 24, 217 (1991).
15. J.O'M. Bockris and G.A. Razumney, Fundamental Aspects of Electrocrystallization, Plenum Press, New York, 1967.
16. René Winand, Fundamentals and Practice of Aqueous Electrometallurgy (short course), The Metallurgical Society of CIM, Montreal, PQ, Canada (Oct. 20-21, 1990), P.7.
17. N.J. Petch, Phil. Mag., 1, 331 (1956).
18. G. Okamoto, J. Horiuti and K. Hirota, Sci. Papers Inst. Phys. Chem. Res., (Tokyo), 29, 223 (1936).
19. Gösta Wranglén, Acta Chemica Scandinavica, 9, 661 (1955).
20. R.N. Iyer, H.W. Pickering and M. Zamanzadeh, J. Electrochem. Soc., 136, 2463 (1989).
21. C.D. Kim and B.E. Wilde, J. Electrochem. Soc., 118, 202, (1971).
22. D.Y. Li and J.A. Szpunar, to be published.

A MICROSTRUCTURAL COMPARISON OF Cu(In,Ga)Se$_2$ THIN FILMS GROWN FROM Cu$_x$Se AND (In,Ga)$_2$Se$_3$ PRECURSORS

ANDREW M. GABOR*, J. R. TUTTLE*, D. S. ALBIN*, R. MATSON*, A. FRANZ*, D. W. NILES*, M. A. CONTRERAS*, A. M. HERMANN**, R. NOUFI*
*National Renewable Energy Laboratory, 1617 Cole Blvd., Golden, CO 80301
**University of Colorado, Boulder, CO 80309-0390

ABSTRACT

We fabricated CuInSe$_2$ and Cu(In,Ga)Se$_2$ thin films by two different pathways using physical vapor deposition. In the first we formed a Cu-Se precursor and then reacted it with a flux of (In,Ga) + Se. These films had large grains but were too rough for optimal device performance. In the other pathway, we first formed a smooth precursor of (In,Ga)$_2$Se$_3$ and then exposed it to a flux of Cu+Se. We overshot the optimal film composition to allow recrystallization of the film by a secondary Cu$_x$Se phase. We then consumed the excess Cu$_x$Se in a third stage deposition of (In,Ga) + Se. The recrystallization step increased the grain sizes, and the resulting films remained smooth. Photovoltaic solar cells made from these films have produced the highest total-area efficiencies of any non-single-crystal, thin-film solar cell.

INTRODUCTION

Thin films of CuInSe$_2$ (CIS) and related semiconductor alloys are among the most promising materials for use as light absorbing layers in thin-film photovoltaic (PV) solar cells. Significant advances have been made in cell efficiency recently with the alloying of CIS with the higher band-gap CuGaSe$_2$ (CGS) to form Cu(In,Ga)Se$_2$ (CIGS) [1,2]. This alloying raises the band gap of the absorber to a better match with the solar spectrum, and by varying the Ga content as a function of film depth, one can engineer graded-band-gap structures to aid cell performance. A common method of deposition has been coevaporation from the elemental sources. This has been successful, but a simpler approach where fewer elements are deposited at a time would be attractive from a manufacturing perspective. Selenization of the metallic precursors by H$_2$Se gas or Se vapor is potentially scalable, but problems associated with film adhesion and phase separation of Cu-In-Ga intermetallic compounds complicate this chemical pathway. Partly motivated by the difficulties of selenization, we have chosen a different simplifying approach where we prevent Cu-(In,Ga) compound formation by separating the deposition into stages where we coevaporate Cu+Se or In+Ga+Se. In doing this, we can benefit from the properties of the precursor and intermediate or secondary phases. Although we use physical vapor deposition (PVD) as flexible tool to explore different pathways, the results should apply to various manufacturing scenarios.

In an open system with sufficient Se activity, compositions of Cu, In, and Se tend to fall on the Cu$_2$Se - In$_2$Se$_3$ pseudobinary tie line. The Cu$_2$Se - In$_2$Se$_3$ pseudobinary phase diagram (see Fig. 1) shows that for Cu-rich compositions ([Cu] > [In]), the excess Cu will exist in the Cu$_2$Se phase. However, with high enough temperatures and Se activity, this Cu$_2$Se could be converted to a liquid Cu$_x$Se phase (x<2). Different groups have taken advantage of the liquid-phase-assisted growth potential in this two-phase region by initiating the deposition with a Cu-rich flux of Cu+In+Ga+Se to form a large-grain precursor [2,4,5]. These groups then introduce a Cu-poor flux to consume the excess Cu$_x$Se and grow additional CIGS quasi-epitaxially onto the precursor. The final composition for a good device must be slightly Cu-poor, because any remaining Cu$_x$Se will short the device. Well developed models for such growth commencing from the Cu-rich regime have already been presented [5,6].

For this work, we pursued two main paths toward the formation of CIGS. For pathway #1 we took the idea behind the Cu-rich precursor methods to its logical extreme and started with a precursor of only Cu+Se. We then reacted it with a flux of In+Ga+Se to form CIGS. For pathway #2 we started with an (In,Ga)$_2$Se$_3$ precursor and exposed it to a flux of Cu+Se. In a

143

variation of the latter method, we added enough Cu during the second stage to bring the composition into the Cu-rich regime in order to recrystallize the small CIGS grains with the aid of the liquid Cu_xSe. We then retreated into the Cu-poor phase region through a third-stage deposition of In+Ga+Se. This variation has yielded record-efficiency CIGS cells [1]. Currently, the best cells fabricated by this technique have achieved efficiencies > 16% [7], which represent the highest total-area conversion efficiencies of any non-single-crystal solar cells. The majority of the paper will focus on this approach. The device structure and fabrication are presented elsewhere [5]. The depositions are performed on soda-lime glass substrates coated with a 1-μm film of sputtered Mo.

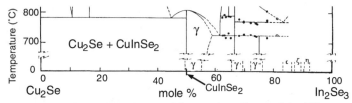

Fig. 1. The Cu_2Se-In_2Se_3 pseudobinary phase diagram (after [3])

Cu_xSe PRECURSOR METHOD

We formed CIS by coevaporating Cu+Se at a substrate temperature of 570°C and by then reacting the precursor at the same temperature with a flux of In+Se. The resulting film, as shown in Figure 2, exhibits enhanced grain growth with some grains > 10 μm wide. However the film is rough, and the voids between grains shorted the finished devices. We conclude that too much liquid phase can result in wetting problems during growth. Films with better coverage can be formed by variations on this method where most of the In is reacted with the Cu-Se at lower temperatures or where a seed layer of In-Se is first deposited. However, the resulting films are still too rough for optimal device performance, and the grains are not always densely packed.

The same enhanced mass transfer that allows this liquid-phase-assisted growth to form large grains, also allows the formation of an uneven grain structure. These results are consistent with those found from the codeposition of Cu+In+Se, where the more Cu-rich the precursor, the rougher the film. From a device standpoint, rough films have a larger effective p-n junction area over which detrimental dark forward current can flow, thus lowering the efficiencies. As is described below, films with a smooth surface and grain sizes > 2 μm can be achieved by limiting the film exposure to this Cu-rich regime to a short recrystallization step.

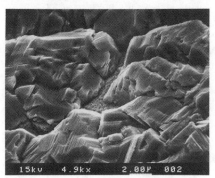

Fig. 2. SEM micrograph of a CIS film grown using the Cu_xSe precursor method

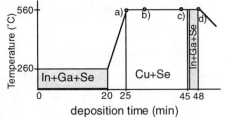

Fig. 3. Time-temperature profile for the three-stage $(In,Ga)_2Se_3$ precursor method

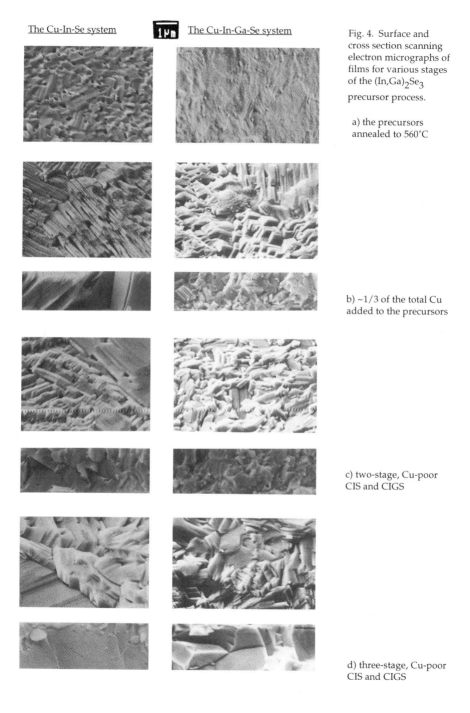

The Cu-In-Se system   **1 μm**   The Cu-In-Ga-Se system

Fig. 4. Surface and cross section scanning electron micrographs of films for various stages of the $(In,Ga)_2Se_3$ precursor process.

a) the precursors annealed to 560°C

b) ~1/3 of the total Cu added to the precursors

c) two-stage, Cu-poor CIS and CIGS

d) three-stage, Cu-poor CIS and CIGS

(In,Ga)₂Se₃ PRECURSOR METHOD

A typical time-temperature plot for the three-stage variation of pathway #2 is shown in Figure 3. We probed the recipe by stopping the deposition at different points for both CIS and CIGS formation. Some of these points are a) right before deposition of Cu, b) after adding ~1/3 of the total Cu, c) within the second stage at a good Cu-poor composition for device fabrication, and d) the finished three-stage film at approximately the same composition as in c). The Se rates were held at three times the atomic flux of any Cu, In, or Ga being deposited. When no Cu, In, or Ga were being deposited, the impingement rate of Se was held at ~ 15 Å/s, as calculated for a sticking coefficient of 1.

SEM micrographs of these films are shown in Figure 4. The appearance of the Ga-free and Ga-containing films are notably different. In the Ga-free case, domains 5-20 μm in width form upon reacting Cu with the precursor. Each domain has a layered morphology, where the orientation of the layering is different within different domains. Upon reaching the desired γ phase of CIS (see Fig. 1), the film appears to have lost this layered structure and is instead composed of tightly packed grains <1 μm in diameter. A distinct improvement in grain size is apparent in the finished film after the recrystallization is complete, with grains sometimes wider than the film thickness of ~2.5 μm. The Ga-containing films have somewhat different morphologies. The surface of the (In,Ga)₂Se₃ film has less surface structure than is the case for In₂Se₃, and the layered morphology is less apparent with the addition of Cu. A similar large-grain structure is found in both CIS and CIGS finished films. The surface morphologies of the finished films are sensitive to small variations in processing conditions. In some cases the full size of the grains is apparent, while in other cases the surface structure obscures the outline of the underlying grains. These films are significantly smoother than films grown by other techniques [1].

The finished films possess compositional nonuniformities as a function of depth, which affect the electrical and optical properties of the finished devices. Both CIS and CIGS films have a surface layer containing less Cu than the bulk. A surface phase of CuIn₃Se₅ was previously reported [8] for CIS films made by other techniques. Figure 5 displays the ratio of [Cu]/([In]+[Ga]) as a function of sputter depth into both CIS and CIGS films (which yielded 13.2% and 15.9% total-area efficiencies) as determined by x-ray photoelectron spectroscopy (XPS). The data clearly show compositions more Cu-poor than [Cu]/([In]+[Ga]) = 1/3 for the first hundred Å of the CIS film. Therefore a phase with lower Cu content than CuIn₃Se₅ exists at the surface (see Fig. 1). The measured composition continuously approaches the bulk value, implying the relative percentage of different phases varies throughout the depth of this surface region. However, spatial non-uniformities across the film surface could possibly smear out any actual phase-separated behavior perpendicular to the substrate. These surface phases can have higher band gaps and a different conductivity type than the bulk [8].

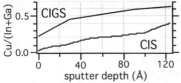

Fig. 5. The ratio Cu/(In+Ga) as determined from XPS compositional data as a function of sputter depth into CIS and CIGS films

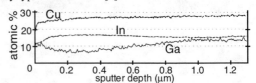

Fig. 6. Atomic compositions as determined by Auger electron spectroscopy as a function of sputter depth into a 2.7-μm-thick CIGS film that yielded a 16.1% efficient device

The Ga-containing films possess an additional compositional nonuniformity. During the deposition of the Cu, a gradient in the Ga content spontaneously forms with Ga decreasing toward the free surface. This grading in Ga content corresponds to a grading in the band gap that can aid electron transport in the device. We achieved the best device performance by adding a sufficiently high percentage of Ga in the third stage to form a dip near the surface in the resulting Ga profile (see Fig. 6).

146

The films pass through different Cu-poor phases during the second stage of the process. For example, Figure 7 shows the x-ray diffraction spectra for the Cu-In-Se films (b) and (c) of Figure 5. The extra peaks and the slight peak shifts for the spectrum of film (b) indicate the existence of one or more Cu-poor phases which no longer exist when enough Cu is added to bring the film to the composition of (c). Figure 8 shows an Auger depth profile for this same film (b). The film consists of two regions, with the back containing a higher percentage of Cu than the front. The compositions in the back and the front correspond closely to the compositions within the γ' and γ'' phase regions (see Fig. 1). As Cu is added, the film may progress through phases of increasing Cu content, the conversion occurring from back to front surfaces. We have not yet seen such a separation in the Ga-containing case.

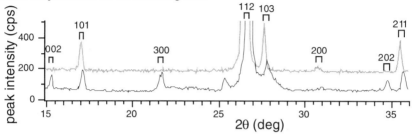

Fig. 7. X-ray diffraction spectra (Cu-Kα radiation) of CIS films (b) and (c) from Figure 5. Film c) is vertically offset by 100 counts per second for clarity.

In one possible pathway for the conversion, the Cu and Se delivered to the surface first react to form a liquid Cu_xSe phase. The Cu_xSe then reacts with In_2Se_3 and later with various Cu-poor phases to eventually convert the film to CIS. Other parallel pathways can also exist. Conversion of In_6Se_7 to CIS by exposure to a flux containing only Cu has been demonstrated [9]. However, when we omit Se from the second stage of the deposition, significant In loss occurs. Se-poor In_xSe phases, such as In_2Se, are known to be volatile at the substrate temperatures used. The Cu may react with the In_2Se_3 and the subsequent Cu-poor phases to release $In_xSe(g)$, possibly together with Se. If the $In_xSe(g)$ does not then react, it can move toward the surface and escape, resulting in In loss from the film.

Depending on the kinetics of the Cu_xSe formation and on the diffusivity of Cu into the film, this pathway can be followed even when Cu is coevaporated with Se. The presence of the Se flux could then prevent loss of In by reaction with the $In_xSe(g)$ as it moves toward the surface, forcing it back to a solid phase, which can then feed back into the reaction with Cu or Cu_xSe. Whichever pathway or pathways are actually followed, the growth proceeds through a significant restructuring of the lattice. An increase in film thickness of > 40% between films (a) and film (c) indicates that we cannot view the growth as mere insertion of Cu and Se into an already established In_2Se_3 lattice.

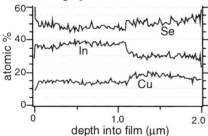

Fig. 8. Atomic compositions as determined by Auger electron spectroscopy as a function of sputter depth into Cu-In-Se film (b).

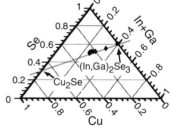

Fig. 9. A ternary phase diagram for the CIGS system. 20-kV electron probe microanalysis data from Cu-In-Ga-Se films from this study are plotted.

We assume here that Ga can partially substitute for In in the above reactions. The compositional data for CIGS supports this assumption through good fit to a pseudobinary tie line connecting Cu_2Se and $(In,Ga)_2Se_3$ [10]. Figure 9 shows the fit to such a tie line of several CIGS films grown for this study with $[Ga]/([In]+[Ga]) < 0.3$. The Ga profile resulting from the second stage deposition could be explained by the decomposition of $(In,Ga)_2Se_3$ releasing a more In-rich vapor phase that moves toward the surface before it resolidifies, leaving behind more Ga-rich phases at the back.

CONCLUSIONS

Although the liquid Cu_xSe is an excellent tool for recrystallization and liquid-phase-assisted growth when it is present in small amounts, too much of this phase can result in unacceptable roughness in the finished films. Films grown from $(In,Ga)_2Se_3$ precursors retain the smoothness of the precursor, and can attain large grain sizes through a recrystallization step. The films pass through different intermediate phases during growth, and the conversion may be proceeding from back to front. A Ga profile conducive to device operation results from the conversion, and the third stage of the process is successful in manipulating the Ga content as well as the phase compositions in the critical surface region of the films.

ACKNOWLEDGEMENTS

We thank J. Dolan, D. Du, A. Duda, A. Johnson, and A. Mason, for technical assistance. A. Gabor thanks A. Zunger for helpful discussions, and Associated Western Universities Incorporated for fellowship support. This work was performed at NREL under Contract No. DE-AC02-83CH10093 to the U.S. Department of Energy.

REFERENCES

[1] A. M. Gabor, A. M. Hermann, J. R. Tuttle, M. A. Contreras, D. S. Albin, A. Tennant, and R. Noufi, Proceedings of the 12th NREL Photovoltaic Program Review Meeeting, edited by R. Noufi (Amer. Inst. Phys., New York, 1993) (in press).
[2] J. Hedström, H. Ohlsén, M. Bodegård, A. Kylner, L. Stolt, D. Hariskos, M. Ruckh, and H. Schock, Proceedings of the 23rd IEEE Photovoltaic Specialists Conference, Louisville, KY, (IEEE, New York, 1993), p. 364.
[3] M. L. Fearheiley, Solar Cells 16, 91(1986).
[4] W. S. Chen, J. M. Stewart, W. E. Devaney, R. A. Mickelsen, and B. J. Stanbery, Proceedings of the 23rd IEEE Photovoltaic Specialists Conference, Louisville, KY, (IEEE, New York, 1993), p. 422.
[5] J. R. Tuttle, M. Contreras, A. Tennant, D. Albin, and R. Noufi, Proceedings of the 23rd IEEE Photovoltaic Specialists Conference, Louisville, KY, (IEEE, New York, 1993), p. 415.
[6] R. Klenk, T. Walter, H. W. Schock, and D. Cahen, Advanced Materials 5, 114 (1993).
[7] M. A. Contreras, et. al., (to be published).
[8] D. Schmid, M. Ruckh, F. Grunwald, and H. W. Schock, J. Appl. Phys. 73, 2902 (1993).
[9] J. Kessler, et. al., Proceedings of the 12th European Photovoltaics Solar Energy Conference, April 11-15, 1994, Amsterdam (to be published).
[10] J. L. Hernández-Rojas, M. L. Lucía, I. Mártil, G. Gonzalez-Díaz, J. Santamaria, and F. Sánchez-Quesada, Appl. Phys. Lett. 64, 1239 (1994).

Stability of Ruthenium-Silica Bilayer Structures

Zara Weng-Sieh*, Tai. D. Nguyen**, & Ronald Gronsky*

* Department of Materials Science & Mineral Engineering, University of California Berkeley and Materials Science Division, Lawrence Berkeley Laboratory, Berkeley, CA 94720.
** Center for X-ray Optics, Lawrence Berkeley Laboratory, Berkeley, CA 94720.

ABSTRACT

The microstructural evolution of ruthenium-silicon dioxide bilayer structures upon annealing is studied using transmission electron microscopy. $SiO_2/Ru/SiO_2$ structures, with thicknesses of 2/1/2 nm, 4/2/4 nm, 8/4/8 nm, and 20/10/20 nm, are formed by magnetron sputtering and annealed at 300 or 600°C. As-deposited films have grain sizes on the order of the Ru film thickness. After annealing at 600°C, significant grain growth is observed for all thicknesses, such that the final grain sizes are approximately 3 to 20x greater than the original film thickness. The largest increase in the average Ru grain size is observed for the 2 nm thick ruthenium film possibly due to the coalescence of Ru grains. The coalescence of the Ru particles in the 1 and 2 nm thick films results in the formation of lamellar Ru grains, which disrupts the contiguity of the Ru film. In all other cases, the increase in grain size is attributed to normal grain growth, but the formation of anomalous spherical grains is also observed.

Introduction

Thin metal-oxide structures are found in a variety of novel applications such as x-ray mirrors, integrated circuits, and catalysts.[1-3] In these applications, reactions between the metal-oxide structure may result in a dramatic change in the microstructure and possible degradation of the resulting properties. In such applications, characterization of the structures can be difficult. Therefore, understanding the basic microstructural evolution of simple model structures may provide insight into the control of the more complicated structures. In many of the aforementioned systems, unambiguous determination of the driving forces for the changes observed in the microstructure is difficult. For example, in x-ray mirrors,[2] studying reactions in a model bilayer structure is simpler and can provide information on the intrinsic behavior of the periodic multilayer structure. In catalyst structures, the interfaces between metal particles supported on silica substrates are difficult to characterize due to the random orientation of the metal particles on the support and the small size of the metal particles.[3] Using simple model sandwich structures, the stability and reactivity of these interfaces are more easily studied.

Experimental Procedures

Thin films of ruthenium sandwiched between silicon dioxide layers were prepared by magnetron sputtering onto standard 3 mm copper TEM grids.[2] Silica was sputtered from an amorphous silica target. The thicknesses chosen for silicon dioxide-ruthenium-silicon dioxide ($SiO_2/Ru/SiO_2$) sandwiches were 2/1/2 nm, 4/2/4 nm, 8/4/8 nm, and 20/10/20 nm. Sandwich films were annealed at 300 or 600°C for 30 minutes. High resolution electron microscopy was performed using a Topcon 002B microscope and electron energy loss spectroscopy was performed using a JEOL 200CX microscope fitted with a Gatan 666 parallel detector. Grain size distributions for about 100 particles per sample were measured by using the Microscopy Image program.[4] This image analysis program enables the determination of the maximum and minimum diameters of each particle. The particle diameter is then given by the average of the minimum and maximum diameter for each particle.

Results

The as-deposited $SiO_2/Ru/SiO_2$ sandwiches were continuous in nature, and the ruthenium films exhibited average grain sizes close to the initial ruthenium film thicknesses. Upon annealing at 300 or 600°C, a change in the microstructure of the sandwich structures is observed, indicating instability of the as-deposited films. In all cases, Ru grain growth is observed, but the mode of growth differs, depending upon the initial film thickness and the temperature of the anneal.

Figure 1 shows the morphology of the Ru grains resulting from annealing at 600°C for 30 minutes for the a) 1 nm, b) 2nm, c) 4nm, and d) 10nm Ru layers. The Ru particles in thinner films (1 and 2 nm thickness) exhibit spherical and lamellar particle morphologies, of which the lamellar morphology is more prevalent. Clearly, the contiguity of the Ru film is destroyed after annealing. The average grain size of the 1 nm thick Ru film increases to 11 nm while the average grain size of the 2 nm Ru film increases to 37 nm.

Also observed in these films are anomalous spherical Ru grains clustered in an oriented manner. Diffraction patterns of these clusters clearly indicate a textured microstructure with an [0001] preferred orientation perpendicular to the film, evident by the missing 0002, 101, and 102, etc. reflections. In some cases, an addition diffraction ring (and lattice fringe spacing) corresponding to 1.8Å lattice spacing is also observed. This additional 1.8 lattice spacing cannot not be uniquely identified as arising from ruthenium-silicide, ruthenium-oxide, or ruthenium-glass structures although electron energy loss spectra taken from these grains only show inner shell ionization edges of Ru (M4,5 edge at 279 eV),

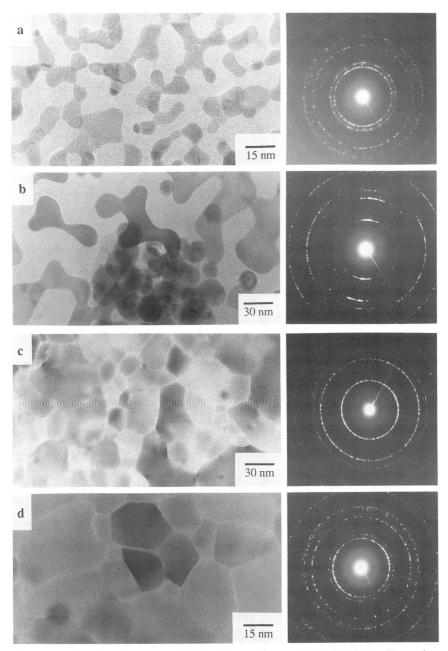

Figure 1: Plan-view images and diffraction patterns of $SiO_2/Ru/SiO_2$ films after annealing at 600°C for initial Ru film thicknesses of a) 1 , b) 2 , c) 4 , and d) 10 nm.

Si (L2,3 at 99 eV), and O (K at 532 eV).   Strong Moire patterns arising from the spherical particles are also observed, suggesting the presence of overlapping grains.

For the Ru films with initial thicknesses of 4 and 10 nm, annealing at 600°C for 30 minutes results in grain growth while maintaining the films' contiguity (figures 1c and 1d). In both cases, the increases in average grain size are not as large as the increase for the thinner Ru films. The average grain size of the 4 nm thick Ru film increases to 20 nm while the average grain size of the 10 nm Ru film increases to 30 nm. Images of these films revealed smaller spherical particles located within the normal grains. Since the sizes of these grains were close to that of the smaller normal grains, a single mode particle distribution is still observed.

Annealing of the Ru films at the lower 300°C temperature still provides enough thermal activation to encourage the growth of the Ru grains, but unlike the case of the 1 and 2 nm Ru films annealed at 600°C, the Ru films remain contiguous after annealing at 300°C. The average grain size increases from the initial film thickness to 4 nm for the 1 nm Ru film (or by a factor of 4), to 8.4 nm for the 2 nm film (or by a factor of 4.2), to 5.4 nm for the 4 nm film (or by a factor of 1.4), and to 11.2 nm for the 10 nm film (or by a factor of 1.2).

Anomalous spherical grains are also observed in these images. However, unlike the spherical grains found in the Ru films after annealing at 600°C, the average spherical grain sizes for the 300°C annealed Ru films are larger than those of the continuous film grain sizes. Again, these films exhibit Moire patterns in addition to an anomalous 1.8Å fringe spacing.

In all cases, the particle size distribution appeared to follow either a lognormal distribution [5] or the distribution described by Louat.[6] Since Louat's distribution is weighted towards smaller particles, only the films with a high density of smaller spherical grains (4 and 10 nm annealed at 600°C) appear to obey the Louat distribution better than the lognormal distribution.

Discussion

The grain size distribution of the Ru grains in the films for as-deposited samples, and those annealed at 300 and 600°C, is summarized in Figure 2. Grain growth is expected in metal films upon annealing to reduce the interfacial or grain boundary areas. After the average particle diameter surpasses 2-3 times the initial film thickness, further grain growth is limited by the specimen thickness effect.[7] Abnormal grain growth has also been observed when the grain size exceeded 2 to 3 times the initial film thickness.[8]

In general, for low temperature annealing, the Ru films seem to exhibit normal grain growth, following the grain size distribution of the as-deposited films. The particles in the 1 and 2 nm Ru films are approximately 4 times greater than the initial film thickness, which suggests that these films may be in the final stages of normal grain growth. Similarly, the particles in the 4 and 10 nm films annealed at 600°C are 5 and 3 times greater than the initial film thickness, respectively, which indicates that these films are also in the final stages of normal grain growth. On the contrary, particles in the 4 and 10 nm films annealed at 300°C are only 1.3 and 1.1 times greater than the initial film thickness, respectively, which implies early stages of normal grain growth.

Figure 2: Particle Size as a function of Initial Film Thickness and Annealing Temperature

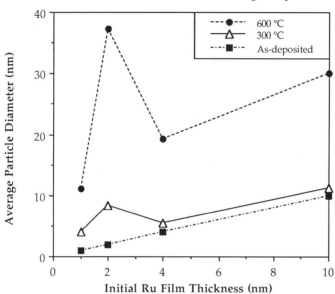

The largest increase in particle size is found in 2 nm Ru film annealed at 600°C. This unusual increase in the average particle size is attributed to the formation of the lamellar morphology of the Ru particles resulting from the strong coalescence of the Ru particles. The driving force for this coalescence of the Ru particles is the reduction of the total interfacial area and effectively the free energy of the system.[5] Such a dramatic change in the morphology and size of the Ru grain structure may require a large amount of thermal energy in addition to an unstable structure. A relatively smaller increase in the 1 nm Ru film particle size is observed, probably due to the smaller initial volume of the Ru film which limits the number of Ru atoms available for grain growth.

The presence of anomalous spherical grains for all films may be attributed to the uptake of Ru by the silicon dioxide and subsequent precipitation of Ru above or below the plane of the original Ru film. The apparent preferred orientation exhibited by the spherical Ru grains may be attributed to the formation of the low free energy interfaces with the silicon dioxide.[9] The strong texture exhibited by the spherical grains may be caused by the formation of a near-continuous network of the spherical grains, possibly seeded by a near-continuous network of lamellar grains.

Conclusions

The microstructural evolution of $SiO_2/Ru/SiO_2$ sandwich structures after annealing at 300 and 600°C has been examined. The formation of lamellar grain morphologies is observed for the thinner initial film thicknesses (1 and 2 nm) after annealing at 600°C. For all other Ru thicknesses, normal grain growth is observed. The formation of anomalous spherical grains is also found and is attributed to reactions between the ruthenium and the surrounding silicon dioxide. This model sandwich structure provides a unique system to investigate the structural stability and chemical reactivity of ruthenium-silicon dioxide interfaces.

ACKNOWLEDGEMENTS

Access to the facilities at the National Center for Electron Microscopy is gratefully acknowledged. The first author, Z. Weng-Sieh, acknowledges support through a Noyce Foundation Fellowship. This work is supported in part by the Division of Materials Science, Office of Basic Energy Sciences of the U.S. Department of Energy under contract No. DE-AC03-76SF00098.

REFERENCES

1. S. P. Murarka and M. C. Peckerar, Electronic Materials- Science and Technology, Academic Press, Inc., San Diego, 1989.
2. T. D. Nguyen, R. Gronsky, and J. B. Kortright, MRS Proc. Vol. 230, 109 (1991).
3. A. T. Bell, in Catalyst Design- Progress and Perspective, ed. L. L. Hegedus, Marcel Dekker, NY, 1988.
4. Microscopy/ FFT Image Program, version 1.35, 1991.
5. R. W. Cahn, Physical Metallurgy, North Holland, Amsterdam, 1970.
6. N. P. Louat, Acta. Metall., 22, 712 (1974).
7. J. E. Burke, Trans. AIME, 180, 73 (1949).
8. C. V. Thompson, J. Appl. Phys., 58, 763 (1985).
9. E. Grantscharova, Thin Solid Films, 224, 28 (1993).

Ag FILMS DEPOSITED BY IONIZED CLUSTER BEAM DEPOSITION

Zhong-Min Ren, Yuan-Cheng Du, Zhi-Feng Ying, Xia-Xing Xiong, Mao-Qi He, and Fu-Ming Li
State key joint laboratory for material modification by laser, ion and electron beams, Fudan University, Department of Physics, Shanghai, 200433, China

ABSTRACT

Ionized cluster beam deposition (ICBD) technique has been used to deposit Ag films on both Si(111) and Si(100) substrates. Sizes of clusters in ionized cluster beam are found to distribute in a range of 100-600 atoms/cluster. X-ray diffraction (XRD), and α-step profile methods are used to analyze the properties of Ag films. As a comparison, Ag films deposited by conventional evaporation are also investigated. Highly textured Ag films with strong (111) orientation on Si(111) have been obtained at high accelerating voltage Va=4kV. The crystallinity and surface flatness of Ag films can be improved by ICBD at high accelerating voltages.

1.Introduction

For the ionized cluster beam deposition (ICBD) technique, as suggested by Takagi et al.[], clusters of solids could be formed by condensation of supersaturated vapor atoms produced by an adiabatic expansion process into a relatively high vacuum region through a nozzle. The clusters are first partially ionized by an electron impact, then energy could be added to the ionized cluster beam by applying an acceleration voltage before reaching the substrate. The properties of the films could be controlled by adjusting the accelerating voltage, the electron current for ionization, the temperature of the crucible and the size of nozzle.

The ICBD technique has two remarkable advantages. The first is the low effective charge-to-mass ratio which eliminates space charge problems. The second is the reduced binding of atoms, by which upon collision of the cluster with the substrate, the atoms are more easily dislodged from the cluster for the purpose of surface diffusion, and it leads to the possibility of film depositions at substrate temperatures significantly below those of conventional evaporation technique. ICBD can offer films with remarkable high thermal stability, high density, strong adhesion, a low impurity level, and a smooth surface due to the effects of cluster ion bombardment[].

In this paper, we presents the results of our experiments of Ag films deposition using the ICBD technique. By changing the acceleration voltage, the qualities of Ag films can be improved as shown by the results of the analyses of X-ray diffraction (XRD), α-step profile for surface roughness measurements, and ellipsometry for studies of optical properties.

2.Experimental

The ICBD experimental build-up is schematically shown in Fig.1. The

155

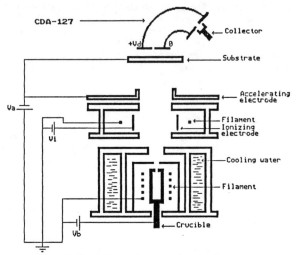

Fig. 1. ICBD experimental set-up.

crucible, with a cylindrical nozzle of 1mm length and 1mm diameter on the top, is heated by filament around it to keep a 1-10 Torr Ag vapor pressure in it during deposition. The electron current Ie for ionization is kept at 50mA while the acceleration voltage Va is varied from 0 to 4kV. The deposition rate is kept at about 12 nm/min. For comparison to those deposited by ICBD, a graphite crucible with a 10 mm orifice is used to produce an effusive beam of Ag atoms for conventional evaporation without ionizing and accelerating process to deposite the films.

The size distribution of Ag clusters is analyzed by using a cylindrical deflector analyzer CDA-127 as shown in Fig.1. The X-ray diffraction (XRD) is performed on a Siemens D500 diffractometer utilizing Cu K_α radiation. The measurements of thickness and surface roughness R_a of Ag films are performed by Tencor Instruments α-step 200 profilometer with tip radius of 12.5 micrometer with vertical resolution of 5 Å. For R_a measurements, the software in the profilometer instrument calculates the center-line through a trace. It then computes the average deviation from the center-line. The average deviation from the center-line defines the arithmetic average roughness R_a.

3.RESULTS AND DISCUSSION

Energies of ionized Ag clusters are analyzed using a CDA-127 analyzer. Based on the assumption of adiabatic expansion in the process of Ag clusters formation, there is a linear relationship between V_d and the size N of a cluster:

$$N=K*V_d$$

where K is a constant decided by the geometric structure of the analyzer and the crucible temperature. Figure 2 shows the distribution of clusters sizes. Although the signal is weak due to little amount of ionized clusters in the

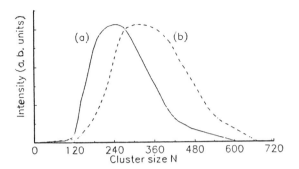

Fig. 2. Sizes distribution in Ag cluster beam. Crucible temperature
is 1500 K for (a), and 1800 K for (b).

ionized beam, Ag clusters could be found to distribute in a range of 100-600
atom/cluster. Our results support some papers published[]. Also shown in Fig.2
is that as the crucible temperature T_c increases, there is an increase in
cluster sizes. Briefly, the distributions of cluster sizes are also analyzed
using a retarding field analyzer (RFA), which shows good agreement with the
results from the CDA-127. About 1-4% atoms in the beam including both single
atoms and clusters are ionized, as indicated by the RFA method.Since
remarkable differences exist in many theoretical and experimental results[]
including our work presented here, obviously further studies of the formation
and measurements of clusters are most needed.

Fig.3 illustrates the X-ray θ-2θ diffraction (XRD) patterns from Ag films
deposited on Si(111) by ICBD. As a comparison, the XRD pattern of a Ag film
made by conventional evaporation is also shown in Fig.3.(a). For Fig.3.(a),
several orientations such as (111), (200), (220), and (311) appear, which
indicate that Ag film is polycrystalline. In Fig.3.(b), the Ag film deposited
with a neutral ICBD beam exhibits mainly the existence of (111) together with
(200) orientations, much the same as Fig.3.(a). The little difference between
Fig.3(a) and (b) is probably caused by the presence of little amount of
clusters in the Ag beam[18] for (b), while no cluster exists in the beam for

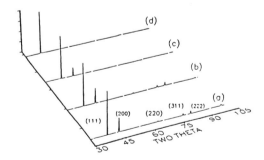

Fig. 3. XRD patterns of Ag films deposited on Si(111) substrates by
(a) conventional evaporation, (b) ICBD at Va=0V, (c) ICBD at Va=2000V,
and (d) ICBD at Va=4000V.

Tab.1. Average roughness R_a of Ag films

Thickness of films (Å)	R_a (Å) for Conventional evaporation	R_a (Å) for ICBD Va=0V	R_a (Å) for ICBD Va=kV	R_a (Å) for ICBD Va=4kV
1200	40	45	30	25
1800	45	55	40	33

(a). When ionizing and accelerating the Ag beam to a certain degree, highly textured Ag films are obtained, as shown in Fig.3.(d) where the acceleration voltage is 4kV. The acceleration voltage is 2kV for Fig.3.(c) in which the intensity of (200) orientation has been reduced by applying an acceleration field. From Fig.3, we could see that good crystallinity of Ag layers have been grown at large acceleration voltages. The strong (111) texture of the Ag films found at 4 kV as shown in Fig.3.(d) is suggestive of single crystallinity. The Ag(222) reflection in both (d) and (c) is very weak and could hardly be distinguished from the diffraction pattern. In Fig.3.(d), the full width at half-maximum (FWHM) for Ag (111) peak is 0.30 while that for substrate Si(111) peak is 0.48 larger than that for Ag(111) probably due to crystal damages of silicon substrates caused by Ag clusters/atoms' bombardment. We could conclude from Fig.3 that ionizing and accelerating components of a cluster beam streaming from a crucible with an initial vapor pressure of a few Torr through a small diameter nozzle alter the properties of films that are deposited. Ionic charge and kinetic energies added to the evaporated cluster beam are beneficial to the structure perfection of the grown films. ICBD could be recognized as an advantageous epitaxy technique when approperate ionizing and accelerating conditions are adopted.

In the case of Ag films on Si(100) substrates, the crystal properties could also be improved by applying high accelerating voltage. But it not easy to obtain highly textured Ag films as obtained on Si(111) substrates, probably due to lattice mismatch in this orientation.

The average surface roughness R_a of Ag films prepared by ICBD is investigated using an α-step profilometer, as shown in Table.1. The average roughness R_a of substrates before deposition is less than 5Å. For Tab.1, in the case of ICBD, as acceleration voltage increases from 0 to 4kV, R_a of Ag films is reduced to about 44% for 1200Å thick films or to about 40% for 1800Å thick films. These surface roughness improvement could be explained by the elevation of atomic migration energy added to single atoms and atoms broken from clusters at collison with the substrate, supplied by the applying of a high accelertaing voltage. The high migration energy of atoms is a striking feature of the ICBD technique.

The flatness of Ag films deposited by conventional evaporation technique is found to be better than that deposited by ICBD at Va=0V (i.e. neutral Ag cluster beam). This could be explained by the fact that the Ag films by ICBD at Va=0V contain many unbroken Ag clusters whose sizes are relatively large, and it is these clusters which makes the flatness of Ag film worse than that made by conventional evaporation.

In the SEM images which are not presented here, the sizes of Ag crystal grains are enlarged as accelerating voltage increase from 0 to 4000 V. Meanwhile, the effects of substrates on crystal properties of Ag films could be eliminated by applying high accelerating voltage, indicated by ellipsometry measurements. The following AES depth profile measurements show that the resistance to the diffusion of impurities could also be improved under ICBD conditions, especially at high V_a.

4.CONCLUSION

An ionized cluster beam deposition (ICBD) technique is used to deposit Ag films on Si(111) and Si(100) substrates. The cluster size distribution in the deposition beam is found to be in a range of 100-600 atoms/cluster. Highly textured Ag films on Si(111) are obtained by ICBD at Va=4kV. As accelerating voltage increases from 0 to 4kV, the average surface roughnesses R_a is reduced to 44% for 1200Å thick Ag films or to 40% for 1800Å thick films. ICBD has many advantages over conventional evaporation. The crystallinity and the surface roughness of the film could be improved by increasing the acceleration voltage.

REFERENCES

1. T.Takagi, I.Yamada, and A.Sasaki: J.Vac.Sci.Technol. **12**(6) 1126(1975)
2. I.Yamada and T.Takagi: IEEE Trans.Electron.Devices **ED-34**(5) 1018(1987)
3. T.Takagi: J.Vac.Sci.Technol. **A2**(2) 382(1984)
4. I.Yamada: Vacuum **41**(4-6), 889(1990)
5. W.Knauer: J.Appl.Phys. **62**(3), 841(1987)
6. D.Turner and H.Shanks: J.Appl.Phys. **70**(10), 5385(1991)
7. F.K.Urban and A.Bernstein: J.Vac.Sci.Technol. **A9**(3) 537(1991)

MONOCRYSTALLINE AND POLYCRYSTALLINE THIN FILMS FORMED BY COBALT ION IMPLANTATION IN THE ORGANIC SUBSTRATE (POLYESTER)

A.L.STEPANOV, R.I.KHAIBULLIN, S.N.ABDULLIN, YU.N.OSIN AND
I.B.KHAIBULLIN
Kazan Physical-Technical Institute, Dept. of Radiation Physics, Kazan, Russia

ABSTRACT

The structure and phase composition of thin films formed by 40 KeV cobalt ion implantation into organic substrate (polyester) were studied by transmission electron microscopy in conjunction with electron diffraction. Varying current density and dose implantation over the range 0.3×10^{16} - 2.4×10^{17} cm^{-2} we obtained island-like cobalt films of different type as well as labyrinth-like structure at the highest dose value. The granulometric and morphologic parameters were derived from the micrographs of the investigated films. Both amorphous state and α-Co crystalline lattice of cobalt granules were established from electron diffraction patterns of synthesized films. Along with discontinuous films, we formed monocrystalline plates of α-phase cobalt under the determined implantation regimes and conditions. Cross-section images of synthesized films showed that films are of about 300 Å thick and buried at the depth of 150 Å from the principal surface of the polyester.

INTRODUCTION

A method of ion-beam implantation is widely used to form the buried magnetic thin film for date storage into polymeric matrix [1, 2]. In this paper we present the result of investigation of structure and phase composition of thin films formed by 40 KeV cobalt ion implantation in new type of organic substrate - polyester networks.

EXPERIMENT

Polymer plates were irradiated at room temperature with cobalt ions of 40 KeV energy in doses ranging from 3×10^{16} cm^{-2} to 2.4×10^{17} cm^{-2} on ion accelerator ILU-3. The ion beam was scanned across the sample surface with an overage current density from 2 to 8 $\mu A/cm^2$.

The investigation of image of implanted layers of polymers were performed by transmission electron microscopy (TEM) on TESLA-BS500. The specimens for cross-section TEM were prepared by using microtome LKB.

RESULTS

It was established that in the near-surface layers of polymer a dispersed cobalt films are formed at the lowest and moderate dose values. Figure 1 exhibits micrographs of different films consist of an ensemble of spherical (Fig. 1a and 1b), needle-like (Fig. 1c) grains as well as of particles with regular polyhedral form close to the cubic form (Fig. 1e). Both samples with discontinuous island-like film and formation of labyrinth-like structures (Fig. 1d) are observed at the highest dose value. The uniform regular-shaped background in the micrograph (Fig. 1f) enables one to suppose that a continuous cobalt plates are formed in this sample.

The granulometric characteristics of the samples island-like films are given in Table 1. The measure estimating the discontinuous films were the linear parameters of granules - namely, the Fide diameter, which is determined by the distance between the tangents to the contour of the image of the particle drawn parallel to the chosen direction. The representation of each island-like film as approximately consisting of particles with regular form and the same size is described by a mean arithmetical diameter - d_{md} (Table I). In this Table the values of root-mean square deviation (dispersion) of particles - (σ) which determines the deviation of the size of particles from the mean size in the given film, are presented as well.

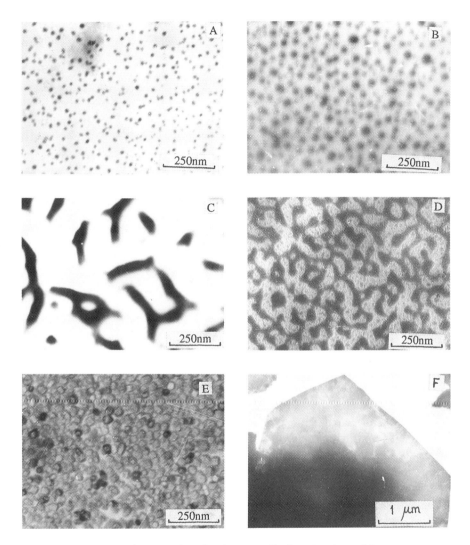

Fig.1 The electron micrographs of polyester Co-ion implanted layers:
    a.- sample N1, D=0.3x10^{17} cm^{-2}, j=4 mA/cm^2;
    b.- sample N2, D=0.9x10^{17} cm^{-2}, j=4 mA/cm^2;
    c.- sample N3, D=1.5x10^{17} cm^{-2}, j=4 mA/cm^2;
    d.- sample N4, D=2.4x10^{17} cm^{-2}, j=4 mA/cm^2;
    e.- sample N5, D=1.2x10^{17} cm^{-2}, j=2 mA/cm^2;
    f.- sample N6, D=1.8x10^{17} cm^{-2}, j=8 mA/cm^2;

Table I. The granulometric and morphologic parameters of the island-like films.

N sample	Type of particles	d_{md}(Å)	σ(Å)	η(%)	d_{max}(Å)
1	Spherical	149	45.1	30	228
2	Spherical	165	62.0	37	360
3	Needle-like	1208	1048.9	87	3952
5	Cubic	232	70.3	30	456

The calculation of the value of the mean linear size and mathematical dispersion enables one to estimate the extent of polydispersion of the sample by its variation coefficient (η) -

$$\eta = (\sigma / d_{md})100\% .$$

In Table I, the characteristic quantity for the island-like film d_{max} corresponding to the conventional maximal diameter of particles for their minimal quantity in one dimensional class not less then 1% is given as well.

Also the data of electron spectroscopy of cross-section of films showed that the synthesized cobalt films are of about 300 Å thick and occur at the depth of 150 Å from the principal surface of the polymer substrate.

For all films studied the electron diffraction measurements were carried out. The results of investigation of the implanted layers of organic compounds for the samples 1-3 and 5 (Fig.2) showed a system of narrow rings in the diffraction picture corresponding to the polycrystalline state of the material. The wide diffusion rings in the diffractograms for sample 4 correspond to the amorphous state of the material. The diffraction is obviously due to the particles formed under implantation in organic material and shown in Fig.1.

The interpretation of experimental diffraction lines and the corresponding interplane distances in the polycrystalline structures was carried out by their comparison with the reference values for pure cobalt and its compounds with chemical elements hydrogen, carbon and oxygen comprising the polymers. The analysis of these values enables one to conclude that under implantation of organic compounds with cobalt ions either a pure α-Co phase (samples 1 - 3) or a mixed α-Co and β-Co lattice (sample 5) is formed.

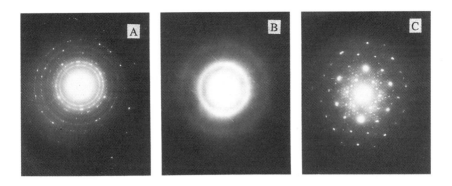

Fig.2 The electron microdiffraction patterns of the films presented in the Fig.1: a.- sample N2; b.- sample N4; c.- sample N6.

For sample 6 (Fig.1f) with no separate granules observed the diffraction picture consists of point reflections evidencing the formation of a monocrystalline film plates in the implanted polymer. The identification of the diffraction spectrum enabled us to determine the α-Co lattice corresponding to this spectrum.

DISCUSSION

As is clear from the presented results that the size and shape of cobalt grains depend from implantation parameters.

The generation of cobalt germs which can be localized on the structure and radiation defects of organic matrix is taking place at the early stage of implantation. The next implanted Co-ions diffuse to cobalt germs and produce crystal growth of germs. This process is responsible for formation of island-like films consist of spherical particles at low dose values. From Fig.1a and 1b it follow that size of spherical particles and particle density of film increase as soon as dose ranges from 0.3 to $1x10^{17}$ cm^{-2}. At dose value is well over 10^{17} cm^{-2} the cobalt clusters and particles may unite with each other and coalesce into larger lengthened particles. Both the film consist of need-like grain at the dose value of $1.5x10^{17}$ cm^{-2} and labyrinth-like at the higest dose value are formed structure as result of coalesced processes.

The current density plays an important role in ion-synthesis of films into polyester matrix. It is suggested that network structure of polymer influence on the process of formation of cubic particles at low current density (Fig.1e). At the highest current density a monocystalline plates with sizes not less then 2 μm are formed. The physical nature of this phenomenon isn't established. However it is known that the highest current density of Co-ion and high-dose implantation give rise to the large longitudinal stresses in the implanted layers. We belive that the orientational crystallization of implants can be induced by this internal stresses.

Data of Table I (σ and η) show a very high extent of uniformity concerning the size of regular spherical and cubic particles (samples 1, 2 and 5). It is of primary importance for example for the stability of magnetic characteristics in granular magnetic films[3]. It is interesting to note that sample 3 is characterized by a distinct geometrical anisotropy of the particle form. The similar parameters of the films may occur decisive when manufacturing a high-coercive magnetic material[3].

CONCLUSION

As follows from the above result, high-dose Co-ion implantation in polyester networks as matrix enable to form the thin granular films of cobalt consist of ensemble of fine particles with predetermined form and size as well as monocrystalline thin-films cobalt plates.

Reference

1.Kazufund Ogawa, U.S. Patent No. 4 751 100 (14 Jun 1988).
2.V.Yu.Petukchov, V.A.Zhikharev, N.R.Khabibullina and I.B.Khaibullin, Visokochistye veshestva 3, 45 (1993).
3.G.Bate, Proc. IEEE,74, 1513 (1986).

PART II

Reactions, Transformations, and Diffusion at Interphase Interfaces and Grain Boundaries

KINETICS OF INTERFACE REACTIONS
IN POLYCRYSTALLINE THIN FILMS

RÜDIGER BORMANN
Institute for Materials Research, GKSS Research Center, 21502 Geesthacht, Germany

ABSTRACT

A quantitative description of the nucleation and growth kinetics determining the interface reactions between two phases is given. Particular focus is given to the understanding of the mechanisms causing the distinct phase selection occurring during the early stages of interface reaction. Thereby the microstructure of the parent phases and the new phases formed are also taken into account and discussed with respect to their influence on the phase selection. The conclusions are applied to the results of two model systems, Al-Ni and Zr-Ni.

INTRODUCTION

Whereas the early stages of decomposition of homogeneous alloys have been studied intensively in recent years [1], there is little research on the quantitative understanding of the early stages of phase formation during the reaction of two elements (or more generally: two alloys). Although this area is technologically important for the processing of intermetallic compounds and for application of materials with heterostructures in the submicron range as well as for basic research, a quantitative description of phase formation is still not available.

In contrast to the later stages of phase reactions which are solely determined by the thermodynamics of the alloy system, the phase reaction in the early stages is mainly governed by the kinetics of the reaction process. By choosing appropriate reaction conditions this often results in a distinct phase selection in the early stages of phase reaction. Thereby only one of the possible intermetallic compounds or even a metastable phase which does not exist in the equilibrium phase diagram is formed. The latter occurs particularly in cases where large driving forces for the phase reaction exists. An impressive result of the kinetic influence is the formation of an amorphous phase during the solid-state reaction between an early and a late transition element [2].

This article summarizes the current quantitative understanding of the kinetics of interface reactions by taking into account also microstructural effects such as grain boundaries in the parent and in the formed phases. Particular focus is given to the understanding of the phase selection often occurring in the early stages of interface ractions. The conclusions will be applied to the results of two model systems, Al-Ni and Zr-Ni, which have been investigated in great detail.

PHASE FORMATION IN THE INTERFACE

For interface reactions to occur, a thermodynamic driving force for phase reaction must exist which originates from the negative free energy of mixing of the elemental components or from the negative heat of compound formation. In such cases the artificially prepared interface boundary represents a highly non-equilibrium state. It will decrease its energy by the formation of solid solutions due to interdiffusion, by the nucleation of intermetallic compounds or by the formation of a metastable amorphous phase. Which process will occur first depends on the kinetics and the 'time scales' [2] of each process. In this respect, the nucleation as well as the growth of each phase have to be taken into account. In principle, both processes can result in a distinct phase selection in the early stages of phase reactions and therefore will be described quantitatively in the following sections.

Mat. Res. Soc. Symp. Proc. Vol. 343. ©1994 Materials Research Society

NUCLEATION IN THE INTERFACE

As the formation of solid solutions by interdiffusion does not require nucleation, it is generally favored with respect to those processes which require nucleation, such as the formation of intermetallic compounds. However, in particular in systems with large driving forces for compound formation, the nucleation rates can be extremely high and may not limit the kinetics of compound formation. In order to quantify nucleation rates of compound formation, a nucleation theory for compound formation at interfaces is required.

As 'planar nucleation' of a new phase γ along the interface between the two parent phases α and β is unreasonable due to the large interface area to be created, it is assumed that the new phase nucleates locally in the interface (Fig. 1). Nucleation in the interface is not only favored by energetic effects related to the gain in energy of the α/β interface but also by a large probability to obtain concentration fluctuations which may serve as incipient nuclei of the new phase γ. As a first approximation, the classical theory of nucleation can be applied which calculates the free energy barrier of nucleation, ΔG^*, and the critical size of the nucleus, above which it can grow by gain of free energy. In equilibrium, the angles between the interfaces ω_i (cf. Fig. 1) and the radii of the nucleus γ are determined by the interface energies between γ and the parent phases:

$$\frac{\sigma_{\alpha\beta}}{\sin\omega_\gamma} = \frac{\sigma_{\alpha\gamma}}{\sin\omega_\beta} = \frac{\sigma_{\beta\gamma}}{\sin\omega_\alpha} \tag{1}$$

where σ_{ij} are the i/j interface free energies (i,j = α, β, γ).

By applying equation (1) to the classical nucleation theory, the equilibrium shape of the nucleus, the nucleation barrier and the critical size of the nucleus can be calculated if all the interface free energies are known. However, in most cases interface free energies have not been determined for the interfaces of interest or are difficult to calculate. Therefore, one has to to consider approximations which apply to the system which are of current interest.

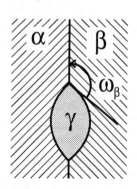

Fig. 1:
Nucleation of a new phase γ in the interface between the parent phases α and β.

Due to the negative heat of mixing of the components, the free energy of the α/β interface is lowered by the chemical contribution from the atomic interaction energy (see also below). Therefore it is reasonable to assume that $\sigma_{\alpha\beta} \ll \sigma_{\beta\gamma}$ and $\sigma_{\alpha\beta} \ll \sigma_{\alpha\gamma}$, in particular when the α/β interface is coherent or semicoherent. In case where $\sigma_{\alpha\gamma} \approx \sigma_{\beta\gamma}$, the quantitative description developed for grain boundary nucleation [3] can be applied. The shape of the nucleus then is a symmetrical doubly-spherical lens and

$$\cos\frac{\omega_\gamma}{2} = \frac{\sigma_{\alpha\beta}}{2\sigma_{\alpha\gamma}} \tag{2}$$

The free energy barrier for nucleation amounts to

$$\Delta G^* = \left(1 - \cos\frac{\omega_\gamma}{2}\right)^2 \left(2 + \cos\frac{\omega_\gamma}{2}\right) \frac{16\pi}{6} \frac{\sigma_{\alpha\gamma}^3}{\left(\Delta G_\gamma + \Delta G_\varepsilon\right)^2} \tag{3}$$

where ΔG_γ is the (chemical) driving force and ΔG_ε the elastic strain free energy.

For large negative heats of mixing, the chemical contribution to the free energy of the α/β interface can even compensate for the structural contribution. This seems to be a reasonable assumption for systems which exhibit small differences in atomic radii and therefore form coherent interfaces. Typical examples are Ni/Al, Nb/Al and Ti/Al interfaces. But also for other systems, such as Ni/Zr and Co/Zr interfaces and some of the silicide forming systems, the energy of the α/β interface is most probably negligible or can, in the initial state, even be negative due to a strongly attractive interaction of unlike neighbouring atoms. If $\sigma_{\alpha\beta} = 0$ and $\sigma_{\alpha\gamma} = \sigma_{\beta\gamma} = \sigma$, then the critical nucleus is a sphere of radius r* given by

$$r^* = \frac{-2\sigma}{\Delta G_\gamma + \Delta G_\varepsilon} \tag{4}$$

Further, the free energy barrier for nucleation is

$$\Delta G^* = \frac{16\pi}{3} \frac{\sigma^3}{\left(\Delta G_\gamma + \Delta G_\varepsilon\right)^2} \tag{5}$$

which corresponds to the nucleation barrier for homogeneous nucleation. This can be understood by considering that in the case of $\sigma_{\alpha\beta} = 0$ the nucleation in the α/β interface is no longer energetically favored by a gain in interface free energy as usual in heterogeneous nucleation. Therefore, the quantitative description of the (classical) theory of homogeneous nucleation can be applied which should allow a reasonable estimation of critical radius and the nucleation barrier in systems with large negative heat of mixing or compound formation.

For the quantitative description of the phase selection during nucleation, the rate of steady state nucleation I_γ, has to be considered, which is determined (exponentially!) by the nucleation barrier ΔG^*, to build up the nucleus interface and the growth barrier ΔG_a, for the formation of the nucleus [3]:

$$I_\gamma = Z \, N \, \exp(-\Delta G^*/kT) \, A^* \, v \, \exp(-\Delta G_a/kT) \tag{6}$$

where Z is the Zeldovich factor [1], N the number of atomic sites per unit volume on which the nucleation could have started, A* is assumed to be equal to the total number of atomic sites on the interface of the nucleus and v is the atomic vibration frequency of an atom at the interface.

Therefore the most important parameters determining the phase selection during nucleation are the activation free energy of growth ΔG_a, the interface energies σ, and the chemical driving force ΔG_γ. In order to evaluate a possible phase selection in the early stages of interface reaction by nucleation, reasonable estimates of these parameters have to be obtained.

The activation energy of growth ΔG_a

As the atomic processes related to the formation of a critical nucleus are mainly determined by diffusional jumps, the activation energy of growth can be approximated by the activation energy of diffusion or, if not available, by the activation energy of the growth of the formed γ phase in the planar growth regime, which in most cases is more frequently available from investigations of the later stages of interface reactions. However, in contrast to homogeneous nucleation in bulk systems, the atomic processes required to form the nucleus in the interface are not unambiguous as atoms not only from the parent phases but also from the α/β interface can participate in the growth of the nucleus.

From investigations on the later stage of phase growth it has been concluded that the precipitates formed in the α/β interface first grow to coalescence within the α/β interface [4]. These results indicate that growth of the nucleus is preferred in the direction of the interface emphasizing the importance of atomic mobility in the α/β interface. Therefore it is reasonable to assume that the activation energy of volume diffusion is only an upper estimate for the activation energy of the nucleus formation. (Similar considerations have been pointed out for heterogeneous nucleation in grain boundaries [5].)

The interface free energy σ

The interface free energy consists of two contributions: the chemical contribution related to the (chemical) atomic interaction energy, and the structural contribution which originates from the free energy of structural defects associated with semicoherent and incoherent interfaces.

The chemical contribution to the interface free energy was calculated for systems with positive heats of mixing by Becker [6] using a simple pair interaction model. In these cases the interface free energy scales with the square of the concentration difference across the interface Δc, and the pair exchange energy ε, which is positive in systems with positive heats of mixing:

$$\sigma \sim \varepsilon \, \Delta c^2 \qquad (7)$$

More recently the pair interaction model has also been applied to systems with negative heats of mixing [7], by considering different pair exchange energies for each phase. For interfaces in equilibrium, the interface free energy in a first approximation scales linearly with the concentration difference across the interface:

$$\sigma \sim \Delta c \qquad (8)$$

The pair exchange energies required to calculate the interface free energies can be determined from the thermodynamics of the system by applying the CALPHAD method [7-9]. The interface free energies thereby derived are in good agreement with those calculated from precipitation studies of homogeneous nucleation [10]. In particular, the small interface free energies of ≤ 0.1 J/m^2 observed in systems where intermetallic compounds precipitate from a supersaturated solid solution can be understood by the small concentration differences across the compound/matrix interface and the small differences between the pair exchange energies in solid solution phase and the intermetallic compound.

Further, the pair interaction model allows the calculation of the (chemical) free energy contribution of artificially prepared non-equilibrium interfaces. Due to the energetic preference of unlike atoms in systems with negative heats of mixing, interfaces between the elemental components in such systems exhibit negative chemical free energies of the interface which decrease when the heat of mixing become more negative. Therefore, in these non-equilibrium

172

interfaces the total interface free energy can be negligible or even negative, when the structural contributions to the interface free energy are small.

In the case of incoherent interfaces a structural contribution has to be considered which is related to the defects caused by the atomic mismatch across the interface. These defects can substantially raise the interface free energy, typically up to the free energies of large-angle grain boundaries (≈ 1 J/m^2). Therefore, in case of incoherent interfaces, the interface free energy is difficult to determine as quantitative free energies even for grain boundaries in pure metals for different crystal orientations are hardly available.

As an amorphous phase represents thermodynamically a highly undercooled liquid (at least close to the glass-transition temperature [11]), the structural contribution to the interface free energy of a crystal/amorphous interface may be approximated by models developed for crystal/liquid interfaces. By means of such models, a correlation between the entropy of fusion and the crystal/liquid interfacial free energy has been proposed [12]. However, as the difference in configurational entropy between the crystal and the undercooled alloy liquid substantially decreases with decreasing temperature [11], the question arises, whether this still holds for highly undercooled alloys liquids. Nevertheless, due to the lack of long-range structural constraints in the undercooled liquid, it is reasonable to assume that the structural contributions to the free energy of the crystal/amorphous interfaces are small compared to those of incoherent crystal/crystal interfaces. For high-energy incoherent interfaces it becomes possible that $\sigma_{\alpha\beta} >$ $\sigma_{\alpha\gamma} + \sigma_{\beta\gamma}$ and no nucleation barrier exists for the precipitation of the amorphous phase. This would result in a 'wetting' of the α/β interface by the amorphous phase. A similar situation can occur when high-energy grain boundaries of the parent phases form triple junctions with a (low-energy) α/β interface. This also would allow the amorphous phase to nucleate heterogeneously at those defects without an energy barrier and to grow along the grain boundaries of the parent phases.

The elastic strain free energy ΔG_ε

Elastic strain free energy can originate from atomic mismatch across the interface and scales e.g. with the difference in atomic radii $\Delta r/r$ in case of coherent interfaces. The strain free energy caused by atomic mismatch across the α/β interface would decrease the nucleation barrier for the precipitation of any new phase, whereas atomic mismatch between a precipitate and the parent phases increases the nucleation barrier. In addition, strain can also be developed when the new phase requires a different volume than the parent phases from which it has been formed [3]. However, in most cases, the strain free energy developing during nucleation is small compared to the chemical driving force, and most probably does not cause a distinct phase selection in the early stages of interface reaction.

The chemical driving force ΔG_γ

In order to calculate the chemical driving force for the nucleation of a new phase in the α/β interface the free energy curves of the solid solution phases and of the intermetallic compounds must be known. These can be determined by applying the CALPHAD method [8,9,11] to all available thermodynamic data (including the phase diagram) of the alloy system. In systems with large negative heats of compound formation the driving force for the nucleation of a new phase from the pure elements is large and equals to the free energy of compound formation at the reaction temperature. However, in most systems interdiffusion occurs first or simultaneously to compound formation[13], thus substantially decreasing the initially available driving force for compound formation. Further, it should be noted that the free energy of compound formation as calculated by the CALPHAD method refers to bulk materials. In a small nucleus the chemical long-range order may not be fully developed due to interfacial constraints thus increasing its free energy with respect to the bulk value. Therefore, the free energy of a chemically ordered intermetallic compound (as derived from the CALPHAD method) should be viewed as a lower boundary for the free energy of the compound nucleus.

APPLICATION OF THE NUCLEATION THEORY TO THE EARLY STAGES OF COMPOUND FORMATION

In the following sections the theory described above will be applied to two well-studied systems, Al-Ni and Zr-Ni. The former is a typical system where a distinct phase selection occurs in the early stage of interface reaction and only one out of several equilibrium intermetallic compounds is formed. The latter system is a typical system where a metastable phase, in this case an amorphous phase, is formed prior to the nucleation of the stable intermetallic compounds. The application of the nucleation theory to both systems will be undertaken in order to evaluate whether the nucleation kinetics can cause the experimentally observed phase selection during the early stages of interface reaction.

The Al-Ni system

The first compound which is formed during the reaction at the Al/Ni interface is the Al-rich phase Al_3Ni [14-16] if the annealing is performed at low temperatures of about 400-600 K. After Al_3Ni has grown to a certain thickness or the annealing temperature is increased, the second Al-rich phase Al_3Ni_2 nucleates and grows [17]. However, before the Al_3Ni phase is formed, interdiffusion occurs at the Al/Ni interface forming a supersaturated fcc solid solution [16].

Fig. 2 shows the free energy curves of the phases in the Al-Ni system. It demonstrates the strongly negative free energy of mixing of the fcc solid solution. Thus due to the preceding interdiffusion, the driving forces for the formation of intermetallic compounds are substantially reduced. For example, in case of Al_3Ni the driving force ΔG_γ amounts to about 5 kJ/g-atom if Al_3Ni is precipitated from the supersaturated fcc solid solution. Neglecting the strain free energy

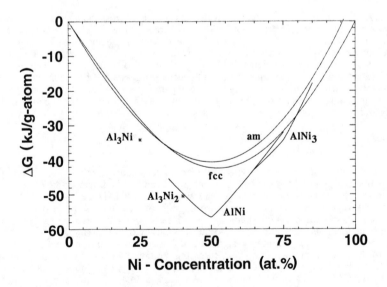

Fig. 2: The free energy curves of the Al-Ni system calculated by the CALPHAD method [18].

ΔG_ε, and taking an interface free energy of $\sigma = 0.1$ J/m^2, the free energy barrier for nucleating Al$_3$Ni amounts to about 0.3 eV corresponding to about 8 kT (k: Boltzmann constant) for a reaction temperature of 400 K. Thus, a nucleation barrier exists for Al$_3$Ni at temperatures where phase formation is frequently observed if the preceding interdiffusion is considered.

Therefore, heterogeneous nucleation at microstructural defects of the interface such as triple junctions with the grain boundaries of the parent phases become important. This conclusion is in agreement with calorimetric data and electron microscopy investigations performed on Al/Ni multilayers [16]. These results demonstrate that a nucleation barrier exists for the Al$_3$Ni formation and that the density of the heterogeneous nucleation sites depends on the grain size of the parent phases.

However, the results cannot explain why in the early stages of interface reaction the nucleation of Al$_3$Ni is preferred with respect to the formation of Al$_3$Ni$_2$ or other more Ni-rich intermetallic compounds. Therefore the nucleation rates of the competing intermetallic compounds have to be evaluated, which are determined by the free energy barrier for nucleation and the free energy barrier of growth. With respect to the nucleation barrier, Fig. 2 shows that the driving force e.g. for precipitating Al$_3$Ni$_2$ from the fcc solid solution is higher than for Al$_3$Ni, suggesting a lower free energy barrier for nucleation in case of Al$_3$Ni$_2$. On the other hand, the activation energy of growth determined in the later stages of interface reaction amounts to 1.2 ± 0.2 eV in case of Al$_3$Ni [15] and 2.0 ± 0.2 eV for Al$_3$Ni$_2$ [17]. Thus the preferred nucleation of Al$_3$Ni probably originates mainly from the pronounced lower value of ΔG_a with respect to the other intermetallic compounds.

In conclusion, the results show that due to the formation of an fcc solid solution an energy barrier exists at low temperatures for the nucleation of intermetallic compounds. Therefore, in agreement with the experimental observations, the microstructure of the parent phases at the interface becomes important and a phase selection during nucleation can occur, favoring Al$_3$Ni most likely by a lower growth barrier with respect to the more Ni-rich compounds.

Zr-Ni system

The Zr-Ni system is one of the most studied transition-metal systems which form an amorphous phase prior to intermetallic compounds during the early stages of interface reaction [2]. The thermodynamic origin of the easy formation of the amorphous phase is obvious from Fig. 3, which displays the free energy curves in the Zr-Ni system. It shows that the amorphous phase exhibits a very high stability, which for compositions in the middle concentration range is even higher than the stability of the various solid solutions and that the free energy of formation is comparable for the amorphous phase and the intermetallic compounds. As in the Al-Ni system, interdiffusion occurs prior or simultaneously to the formation of the first intermetallic phase, in this case the amorphous phase [13], thus decreasing the chemical driving force for nucleation. At low temperatures, Ni is more mobile in Zr than vice versa leading mainly to a supersaturation of the Zr crystal [2]. However up to now, it has remained unclear whether this interdiffusion results in a supersaturated hcp solid solution (as observed in the Zr-Co system [20]) or whether interdiffusion is fast enough to form a supersaturated fcc solid solution which is the more stable structure for higher supersaturations of the Zr.

As can be seen from Fig. 3, in both cases the initial driving force of about 40-50 kJ/g-atom for the formation of an amorphous phase or an intermetallic compound is substantially reduced. Taking a supersaturation of about 10 at.% in the hcp Zr (as observed in the Zr-Co system [20]) the driving force for the nucleation of the amorphous phase amounts to less than 5 kJ/g-atom. For a free energy of the amorphous/crystalline interface typically of 0.1 J/m^2 and by neglecting the strain contribution, a nucleation barrier ΔG^* for the amorphous phase of about 0.4 eV results, corresponding to over 10 kT for typical reaction temperatures of 400 K. Thus, a nucleation barrier exists most probably also for the formation of the amorphous phase and heterogeneous nucleation at defects in the Zr/Ni-interface becomes important.

This conclusion is supported by the observation that on defect-free single crystals of Zr the amorphous phase is not formed prior to the nucleation of the first intermetallic compound, usually ZrNi [20,21]. Only if high-energy defects exist in the Zr/Ni-interface, such as triple

175

junctions with high-angle grain boundaries, the amorphous phase is formed. In these cases, the nucleation of the amorphous phase is favored and its formation is often observed already during the preparation of the Zr/Ni-interface. Correspondingly, calorimetric investigations on multilayered polycrystalline Zr/Ni films do not indicate a nucleation barrier for amorphous phases [22].

With respect to the nucleation of the competing intermetallic compounds, the formation of the amorphous phase can be favored by its low interface free energy to the parent phases. However, in transition-metal systems, the intermetallic compounds may form low energy interfaces at least to one of the parent phases, thus decreasing their total interface free energies to values comparable to that of the amorphous nucleus. Therefore, it seems more reasonable that the amorphous phase formation is mainly favored by its growth barrier which is substantially lower than for the intermetallic compounds: In case of the amorphous phase, the later stages of phase reaction can be described by an activation energy of about 1.1 eV which is related to the fast diffusion of the late transition metal through the amorphous phase [2,22]. In contrast, growth of the intermetallic compounds requires a relatively high activation energy typically of about 2.0-2.4 eV for Zr systems [2,22]. This difference in activation energies of growth seems to dominate the competing nucleation rates of the amorphous phase and the intermetallic compounds, in particular at low reaction temperatures and in case of heterogeneous nucleation.

The formation of the amorphous phase further reduces the driving force for the nucleation of the intermetallic compounds (Fig. 3). Therefore, nucleation of the first intermetallic compound is usually associated with the existence of a nucleation barrier. This leads to a (slight) backgrowth of the amorphous phase, as been observed in the Zr-Ni system [21].

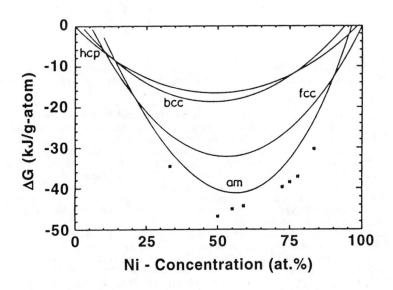

Fig. 3: The free energy curves of the Zr-Ni system calculated by the CALPHAD method. * indicate the free energy of the various intermetallic compounds (taken from Ref. 19)

PHASE SELECTION DURING GROWTH

As described in the previous chapter, nucleation of the new phase occurs in the interface of the parent phases. In case of a nucleation barrier, heterogeneous nucleation at preferred sites in the interfaces has to be taken into account. Subsequently, the precipitates grow to coalescence and form a continuous layer.

The growth of these precipitates will initially be determined by their interface velocities and later by the atomic transport to the interface or through the precipitated phase. Even the growth of the continuous layer can in principle be limited still by the interface velocities or by the atomic diffusion through the layer. Assuming that several different phases have nucleated at the parent interfaces, a distinct phase selection can also occur during further growth of the competing phases leading to a disappearance of initially nucleated phases as a consequence of different growth kinetics.

The kinetics of such a phase selection during growth in planar diffusion couples have been described by Gösele and Tu [24]. They consider the growth of two competing intermetallic compounds γ and δ between the saturated solutions of the parent phases α and β (Fig. 4). The growth of γ and δ then is determined by the particle fluxes $j(\gamma)$ and $j(\delta)$ through each phase, which initially is limited by interface reactions (i.e. the rate at which the atoms can be accommodated at the interface), and later by the diffusion through the phase. Only if the ratio of the particle fluxes lies in a certain range can both phases grow simultaneously. This range is determined by the composition ranges of the competing phases only. If, however, the particle flux e.g. through γ is much higher than through δ, then γ will grow at the expense of δ which can even result in the disappearance of δ. Under the latter conditions, it is reasonable to assume that δ has not even been formed in an isothermally treated diffusion couple and that only γ will be observed in the early stages of interface reaction. According to Gösele and Tu, such a phase selection is caused by the kinetics of the interface reactions which limit the increase of the particle flux through δ for very small thicknesses.

In the later stages of phase reaction, the thickness of the initially formed γ phase has increased, thus the particle flux through γ decreases and, above a critical thickness, δ can grow simultaneously. In addition, the particle flux through γ can be altered due to the formation of pores at the interfaces or if one of the parent phases is consumed. Both cases would favor the formation of δ or some other new phase.

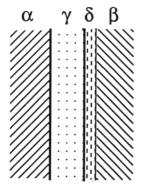

Fig. 4:
Planar growth model of the diffusion couple

The existence of a critical thickness for γ can, in principle, explain the experimental observation that the formation of intermetallic compounds in diffusion couples can occur in a well defined sequence. However, in metallic systems the model has not yet been quantitatively confirmed. This originates mainly from the difficulty of determining quantitative values for the interface reaction constants in the initial linear growth regime in which growth kinetics are interface-controlled. Experimentally, this regime turns out to be limited to layer thicknesses of a few monolayers for metallic systems [2,20] and therefore may occur only at stages of phase formation where coalescence of the precipitates to a continuous layer is not completed. This impedes the experimental determination of interface reaction constants and limits the predictive value of the model of Gösele and Tu.

On the other hand in metallic systems, the question arises whether the interfacial reaction barrier which limits the growth rate of small layers needs to be considered at all in in order to explain a phase selection during growth. Assuming the validity of the linear diffusion equations also for very small layer thicknesses, d'Heurle has pointed out that also large differences in the diffusion rates between competing phases can result in a suppression of compound formation in the early stages of phase growth [25]. Mathematically, diffusion rates become infinite for decreasing layer thicknesses. However, as a layer thickness below a monolayer is physically unreasonable, this limits the maximum growth rate attainable in the initial state of phase growth. As an example, if the diffusion constants of two phases differ by a factor of 1000 and the one layer would have grown to 100 nm, then the other would have to remain below 0.1 nm in order to allow simultaneous growth of the other phase. So from a physical point of view it would not exist yet. (As a first approximation the thicknesses of the different layers scale as the respective diffusion constants [25].)

In addition, the interface free energy of the very thin layers has to be taken into account, which results in a critical thickness of the layer below which it is energetically unstable. Only if the layer thickness given by the relative diffusion constants exceeds this critical thickness will the layer grow. This adds an additional constraint for the maximum growth rate of the competing phases.

Therefore, for compounds with large ratios in the diffusion constants, phase selection in the early stages of growth can occur also in the absence of an interface reaction barrier. In these cases, the phase selection can be predicted in a quantitative way if diffusion constants are available.

In this respect it should be noted that for crystalline layers, the flux through each phase can in principle be determined by volume diffusion or by grain boundary diffusion. Thus, a small crystallite size would also favor the growth of the compound emphasizing the importance of the microstructure for the growth kinetics and the phase selection [26-28].

In the model described above, it is assumed that the parent phases are saturated and equilibrated with the compounds formed in the interface. However, in the early stages of phase growth this is often not achieved due to the limited transport kinetics in the parent phases. In these cases, the atomic flux into the parent phase and in particular into the grain boundaries of the parent phases can result in a kinetic destabilize the compound formed at the α/β interface. Briefly speaking, with respect to the kinetic competition of phases during growth, the parent phases and their grain boundaries have to be considered as additional competing phases, which alter the flux balance of the compounds formed at the interface. (For a detailed description see Ref. 27 and 28 and references therein) These effects can be of great importance for the phase selection in polycrystalline diffusion couples with small grain sizes, as the parent phases and their grain boundaries can be supersaturated well beyond the equilibrium value in the initial state of phase reactions [20].

APPLICATION OF THE GROWTH MODELS

In the following the growth models described above will be applied to the Al-Ni and the Zr-Ni system in order to evaluate whether in addition to nucleation kinetics the observed phase selection can also originate from the growth kinetics of the competing phases.

The Al-Ni system

The growth of the two competing phases, Al_3Ni and Al_3Ni_2, have been investigated for different temperatures [15-17]. Within the experimental resolution no linear growth regime was observed. Instead the growth of the phases follows a parabolic time dependence.

At low temperatures the growth of Al_3Ni is substantially faster than Al_3Ni_2 or other intermetallic compounds of the Al-Ni system. This is reflected in the lower activation energy of growth (see above), which is related to the fast Al diffusion. However, the prefactor for the growth constant is much larger in case of Al_3Ni_2 favoring the growth of Al_3Ni_2 at higher temperatures with respect to Al_3Ni [17]. Therefore, after the formation of Al_3Ni_2 at higher reaction temperature, Al_3Ni_2 grows partly at the expense of Al_3Ni. Nevertheless, at low temperatures, Al_3Ni is clearly favored kinetically with respect to other competing phases allowing a distinct phase selection not only by its high nucleation rate but also by its enhanced growth kinetics.

The Zr-Ni system

As in the Al-Ni system, no linear growth behaviour has been observed for the amorphous phase, which can be partly due to its formation during the preparation of the diffusion couple [2,23]. The growth rate of the amorphous phase is determined by the low activation energy of Ni diffusion, favoring the formation of the amorphous phase in particular at lower reaction temperatures (see also above). Therefore, as in the Al-Ni system, the phase selection can be due to the favorable nucleation rate of the amorphous phase as well as due to the enhanced growth kinetics with respect to the crystalline intermetallic compounds.

SUMMARY AND CONCLUSIONS

The experimentally observed phase selection during the initial stage of interface reactions can be caused by differences in the nucleation rates as well as by the growth kinetics of the competing phases.

Due to the low energy of the parent interface, (classical) homogeneous nucleation theory is a reasonable approximation for a quantitative description of the steady-state nucleation rates of competing phases. By using interfacial energies derived from a pair interaction model and free energy driving forces calculated from the thermodynamics of the system a nucleation barrier can then be estimated in a relatively simple approach. The results show, that due to interdiffusion preceding the nucleation of a new phase, a nucleation barrier can exist even in systems with large free energy of formation for intermetallic compounds. Therefore, heterogeneous nucleation at defects in the interface has to be taken into account, and the microstructure of the parent phase can influence the phase selection. Besides the nucleation barrier which can be significally decreased by prefered heterogeneous nucleation sites, the nucleation rate is substantially determined by the activation energy of growth. As energy differences in the growth barrier seem to be even larger than differences in the nucleation barrier, the phase selection during nucleation may be completely determined by the growth barrier of the competing phases, in particular in polycrystalline diffusion couples.

The growth kinetics of competing phases are determined by the fluxes of diffusing species. Thereby phases with high diffusivities are favored and can suppress other phases in the early stages of interface reaction. As, at low reaction temperature, the activation energy of growth is again the dominant factor, this turns out to be one of the key parameters determining the phase selection in the early stages of interface reaction. However, in order to improve the predictive value of this conclusion, a microscopic understanding of the growth barrier and the differences in the diffusion constants of the competing phases would be desirable.

In this respect, it is important to note that the microstructure of the parent phases and of the formed phases can substantially influence the activiation energies of growth of the new phases. Further research should therefore not only consider the bulk properties of the phases but should also take into account their microstructure and the related interface kinetics in order to get a comprehensive description of the mechanisms causing the phase selection in the early stages of interface reactions.

Acknowledgements

The author would like to thank C. Michaelsen, S. Wöhlert, W. Sinkler, R. Wagner, F. Gärtner and R. Busch for valuable discussions and comments.

References

[1] R. Wagner and R. Kampmann, in Materials Science and Technology, edited by R.W. Cahn, P. Haasen and E.J. Kramer, Vol. 5 (VCH Publishers, New York, 1991), p. 211
[2] W.J. Johnson, Prog. Mater. Sci. 30, 81 (1986)
[3] R. Doherty, in Physical Metallurgy, edited by R.W. Cahn and P. Haasen, Vol. 2 (Elsevier Sicence Publishers B.V. Amsterdam 1983), p. 933
[4] K.R. Coffey, L.A. Clevenger, K. Barmak, D.A. Rudman, and C.V. Thompson, Appl. Phys. Lett. 55, 852 (1989)
[5] J.W. Christian, The Theory of Transformations in Metals and Alloys (Pergamon Press, Oxford 1975), p. 450
[6] R. Becker, Ann. d. Physik 32, 128 (1938)
[7] R. Bormann and C. Borchers, presented at the Frühjahrstagung 1991, Münster, Germany (unpublished)
[8] L. Kaufman and H. Bernstein, Computer Calculations of Phase Diagrams (Academic Press, New York 1970)
[9] H.L. Lukas, J. Weiss, and E.T. Henig, CALPHAD 6, 229 (1982)
[10] C. Borchers and R. Bormann, presented at the Frühjahrstagung 1991, Münster, Germany (unpublished)
[11] R. Bormann and K. Zöltzer, phys. stat. sol.(a) 131, 691 (1992)
[12] F. Spaepen, Acta Met. 23, 729 (1975)
[13] C.V. Thompson, J. Mater. Res. 7, 367 (1992)
[14] E.G. Colgan, M. Nastasi, and J.W. Mayer, J. Appl. Phys. 58, 4125 (1985)
[15] J.C. Liu, J.W. Mayer, and J.C. Barbour, J. Appl. Phys. 64, 651 (1988)
[16] E. Ma, C.V. Thompson, and L.A. Clevenger, J. Appl. Phys. 69, 2211 (1991)
[17] J.C. Liu, J.W. Mayer, and J.C. Barbour, J. Appl. Phys. 64, 656 (1988)
[18] R. Bormann (unpublished)
[19] F. Gärtner and R. Bormann, J. Physique, Colloque 4, 95 (1990)
[20] R. Busch, S. Schneider, F. Gärtner, R. Bormann, and P. Haasen, these proceedings
[21] A.M. Vredenberg, J.F.M. Westendorp, F.W. Saris, N.M. van der Pers, and Th.H. de Keijser, J. Mater. Res. 1, 774 (1986)
[22] W.J. Meng, C.W. Nieh, E. Ma, B. Fultz, and W.L. Johnson, Mater. Science and Engg. 97, 87 (1988)
[23] G.C. Wong, W.L. Johnson, and E.J. Cotts, J. Mater. Res. 5, 488 (1990)
[24] U. Gösele and K.N. Tu, J. Appl. Phys. 53, 3252 (1982)
[25] F.M. d'Heurle, in Diffusion in Ordered Alloys and Intermetallic Compounds, edited by B. Fultz, R.W. Cahn, and D. Gupta (TMS Warrendale, Pennsylvania 1993)
[26] C. Michaelsen, S. Wöhlert, and R. Bormann, these proceedings
[27] K.R. Coffey and K. Barmak, Acta Metall. et Mater., accepted for publication
[28] K.R. Coffey and K. Barmak, these proceedings

INVESTIGATING THE ROLE OF GRAIN BOUNDARIES IN INTERFACE REACTIONS

FRANÇOIS M. d'HEURLE,* PATRICK GAS,** and JEAN PHILIBERT.***
* IBM Research Center, PO 218, Yorktown Heights, NY 10598, USA.
** CNRS (URA 443), Univ. St Jérôme, Case 511, 13397 Marseille, France.
*** Laboratoire de Métallurgie, Bt. 413, Univ. Paris-Sud, 91405 Orsay, France.

ABSTRACT

The paper is built around a nonexhaustive review of the literature on the role of grain boundaries in reactive phase formation. Examples are chosen to illustrate these effects in silicide and oxide growths, and later on in metal-metal interactions. A short section deals with the effect of grain boundaries and grain boundary adsorption of impurities on the kinetics of growth and on the morphology of the growing layer. Some attempts at understanding the mechanisms of phase growth from the tracking of isotopes are briefly analyzed.

INTRODUCTION

The process of the formation of a new phase resulting from the interaction of two adjacent phases has attracted much attention (e.g. Refs. 1-3). In the formation of amorphous phases, grain boundaries have been reported[4,5] to play a definite role. It can be assumed that growth always follows some type of linear-parabolic kinetics, in which the first stage (possibly linear) is usually not very well understood. What is the importance therein of grain boundaries remains unanswered. Attempts to incorporate the kinetic and thermodynamic effects of grain boundaries are being made[6] and will be presented elsewhere in this symposium[7]; these will not be considered any further here. The main focus of this paper will be centered on that stage of interaction when the new phase has been formed and it grows through a diffusion controlled mechanism. While it appears simple to distinguish the relative participations of lattice and grain boundary diffusions, this is not really the case, especially if one wants to be precise. There are several reasons for such difficulties: 1) even if one considers the most simple case of a single phase growth, the activation energy for growth is not the same as that for diffusion, the reason being that if one uses Fick's law in the analysis, one can obtain the correct diffusion coefficient (ignoring that the value obtained is only an average of values that change as a function of composition) only if one knows the variations of the limits of composition as a function of temperature; that is rarely available. 2) for many intermetallic compounds the relative values of lattice and grain boundary diffusions and of their activation energies are not known. Thus even if the previous difficulty (1) is eliminated one cannot easily determine whether the activation energy that has been obtained is characteristic of grain boundary or lattice phenomena. This difficulty is compounded by the fact that in many intermetallic compounds the activation for lattice diffusion assumes remarkably low values because of the presence of constitutional defects - e.g. AlNi[8 – 10]. A review of the available experimental evidence on the role of grain boundaries in the growth of compounds such as silicides, as well as oxides shall be presented. Because of the difficulties mentioned above, some attempts were made to put in evidence the role of grain boundaries through the use of tracers; these will be analyzed. The dominance of grain boundary diffusion in the growth of a layer will not usually change the time ($t^{1/2}$) dependence of the thickness, except if the grain size

181

changes as a function of time. If the grain size increases linearly with the thickness of the layer the time dependence should become $t^{1/3}$. Values of the time exponent from 1/2.4 to 1/3.3 were found in the investigation of several intermetallic phases[11] that form in the system Al-Ti. With grains growing[12] as $t^{1/2}$ compound formation would be proportional to $t^{1/4}$; which apparently introduces extra complications in the intractable kinetics of growth for $TiSi_2$.

SILICIDES

Although silicides might not be the most important intermetallic compounds to be grown by solid state reactions, they have become interesting for the electronic industry, and as such have received a great deal of attention in recent times. As a result they are rather better known than other compounds and deserve to be treated first. Apparently, the first documentation[13,14] of the role of grain boundaries in the formation of a silicide concerns Ni_2Si. If one reacts a bilayer of Ni and Pt (on top) with a Si substrate, one observes the formation of Ni_2Si first. But once all the Ni has been consumed one observes the presence of Pt, presumably in the form of a silicide, at the Si/Ni_2Si interface. In view of the low solubility of Pt in Ni_2Si (some 0.2 at %) as detected by backscattering, one concludes that the Pt diffused through Ni_2Si via grain boundaries. Since that occurs at the same temperature and within the same time scale as the formation of Ni_2Si, it is safe to conclude that Ni_2Si actually grows via grain boundary diffusion of Ni. More recently, it has been observed[15] that when Cu is deposited over a layer of NiSi formed on a Si substrate, subsequent heat treatment results in the formation of Cu_3Si at the interface between the substrate and the Ni silicide, which leads to the same conclusion as in the precedent example. It is conceivable that a high flux of Pt in one case, or Cu in the second, could be obtained in spite of the low solubilities, if the Pt and Cu atoms were to migrate as interstitials as Ni does[16] in Si, but such interstitial diffusion can be excluded before hand for compounds with high coordination numbers.

Direct confirmation[17] that the growth of Ni_2Si (known from marker experiments[18] to proceed via the diffusion of Ni atoms) occurs by grain boundary diffusion came only in the last few years. This important result required an independent measurement of the diffusion rates of radioactive ^{63}Ni tracers in the lattice and along the grain boundaries of melt-cast Ni_2Si samples. The respective activation energies were found to be 2.48 eV and 1.75 eV. It is clear that this latter value is quite compatible with the activation energies reported[19,20] for the growth of Ni_2Si in thin film form: 1.5 eV - 1.6 eV. These results are illustrated in Fig. 1. Quite a similar procedure has now been followed[21] for

Figure 1. Diffusion coefficients of Ni_2Si. a) the diffusion coefficient of Ni in the Ni_2Si lattice and b) in the grain boundaries. The diffusion coefficients from the growth of Ni_2Si c) in "lateral" formation[18] and d) in "normal" formation[19]. From Ref. 17

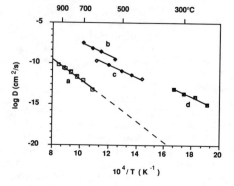

Figure 2. The effective diffusion coeffi- cients, including Co and Si, lattice and grain boundaries, as a function of the grain size. It is seen that the data points for thin film growth (400°C to 600°C) and bulk (800°C to 1000°C) agree quite well with the coefficients for 0.01 μm and 10 μm respectively. From Ref. 21.

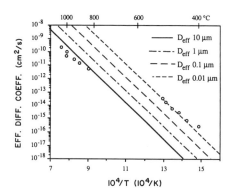

CoSi$_2$. The diffusion coefficients of Co were determined by means of radioactive ^{60}Co, the commensurate coefficients for Si were assumed to be the same as those of radioactive ^{68}Ge. On the basis of the values thus obtained and knowing the free energies of for- mation of CoSi$_2$ from the reaction of CoSi and Si, it is possible to estimate the rates of growth of CoSi$_2$ as a function of its grain size. The data are presented in Fig. 2, where one sees that the experimental rate of growth matches quite well calculations for re- spective grain sizes of 10 μm (bulk) and 0.01 μm (thin films), in agreement with that which is experimentally observed.

A tabulation[22] of the activation energies Q for the growth of different silicides in thin film form shows that the values scale as: $Q(eV) = T_m/1000$ (where T_m is the melting temperature, see Fig. 3). That is quite in keeping with the relation that obtains for grain boundary diffusion in pure metals. For CoSi$_2$ the activation energies for growth[23,24] have been reported: 2.3 eV for thin films (450°C to 600°C) and 2.4 eV for bulk (800°C to 1050°C). In view of the melting point of CoSi$_2$, 1550 K, it is surprising that the two values should be so close to one another. Since these values are much higher than what one would obtain from Fig. 3, one might be inclined to believe that they correspond to lattice diffusion. But that would violate the results presented in Fig. 2, which shows unmistakably the important role of grain boundaries. The point here is to emphasize the

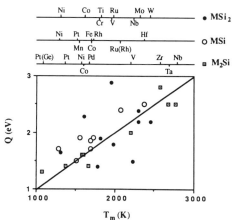

Figure 3. The activation energies (in eV) for the formation of various silicides as a function of their melting temperature (in K). From Ref. 22.

difficulties to conclude about the transport path from just growth experiments alone. A problem with $CoSi_2$ is that the activation energies[21] for grain boundary and lattice diffusion are anomalously close to one another, and both appear rather high with respect to the melting point. Similar difficulties are encountered with the formations of Cu_3Si and Cu_3Ge (respective melting points 1132 K and 1022 K) that have been studied[25,26] both in thin films at about 250°C and in bulk at about 500°C. The thin film experiments yield similar activation energies 0.95 eV, which seem to agree with the estimation made above ($T_m/1000$). However, in bulk experiments the activation energy for the growth of Cu_3Ge is 1.08 eV (well within experimental errors from the previous value), but microscopic determination of the large grain size leads the authors[25] to conclude that this represents lattice diffusion. The same authors find two activation energies for the growth of Cu_3Si 1.82 eV at high temperatures (lattice) and 1.04 eV at low temperatures (grain boundaries). The picture becomes even more confusing if one considers that in Cu_3Sn the lattice diffusion coefficients[27] for Cu and Sn are only 0.85 eV and 1.2 eV, respectively. Can these data be reconciled?

OXIDES, IMPURITY EFFECTS

Greater attention has been paid to grain boundary effects in oxide growth than in the formation of silicides and other compounds. General information on the subject is provided in Ref. 28. Discussions[29] about the role of short circuit diffusion during the growth of oxides appeared already in the early 1960's. We shall provide some detail here about NiO which has been extensively[30 – 32] studied. It was noticed[30] that the rate of growth of NiO increases as the grain size of the oxide decreases. While growth is known to occur via Ni motion, the activation energy for growth is smaller and the rate larger than what can be accounted for by the lattice diffusion of a ^{63}Ni tracer, while the growth data appear to fit the grain boundary diffusion[32] of Ni. Thus, the idea that Ni oxidation proceeds via grain boundary diffusion of Ni has received[28] wide acceptance. Yet, assent to this is not universal: the presence of pores[33] in the oxide layer renders the interpretation of growth kinetics quite problematical. What is quite certain is that the observed rates of growth require the presence of short circuit diffusion paths.

In discussing oxide growth and the role of grain boundary diffusion, Cr_2O_3 deserves special attention for at least three reasons. It is technologically important as a protective layer against oxidation of a great variety of industrial parts. Also very extensive work has been done on the role of impurities in controlling grain boundary diffusion. Finally there exists[34,35] good information on the diffusion of both Cr and oxygen in Cr_2O_3. The oxidation of pure Cr remains something of a clouded subject (see Ref. 1, pp. 389 - 401); it seems that oxidation proceeds via grain boundaries with a counter current of oxygen and Cr, the latter probably playing the most important role, in agreement with tracer studies[35] of grain boundary diffusion. In chromia (as well as in other oxide layers) the addition of oxygen active elements, Y and other rare-earth elements, Zr and Hf, has several beneficial effects: it improves adhesion and decreases the rate of oxidation. This is of such importance that it was the object[36] of a whole symposium. Implantation of Y (2×10^{16} at/cm^2) reduces[37] the oxidation rate of a Co-Cr (55-45) alloy at 1000°C by a factor of 100, in pure oxygen. Since it was previously shown[38] that Y segregates at the grain boundaries of Cr_2O_3, one draws the conclusion that adsorption of Y at the grain boundaries reduces the rate of grain boundary diffusion of Cr. Moreover, experiments[32] with Au markers indicate that in the absence of Y the chromia scale grows at the upper surface (via Cr diffusion) but in the presence of Y growth occurs at the metal-oxide interface, via oxygen diffusion. Although it is not immediately clear why Y should affect the diffusion of Cr but not that of oxygen, since in oxides the diffusion of anions and cations will generally depend on different types of defects, such an asymmetric behavior is conceivable. As with the theory of the grain boundary growth of NiO, a few words of caution are desirable: other mechanisms have been proposed (see Ref. 39). The last one[38] of these, in chronological order, proposes that diffusion requires the creation of defects at the oxide-metal interface, and that the reduction of oxidation rates in the presence of Y is due to a blocking of the process of defect creation. Space restrictions do not allow the examination of Al_2O_3. The behavior[40] of this latter oxide is not too different from that of Cr_2O_3.

TRACERS, SOME THEORETICAL POINTS

The use of tracers, radioactive or otherwise, during growth might provide precious information about growth mechanisms and the role of grain boundaries. When the tracer is not mobile, the interpretation is simple. If Ni is oxidized[24] first in ordinary ^{16}O and then in ^{18}O, most of the ^{18}O is found afterwards in the surface oxide, showing that oxygen is not mobile, and consequently that oxidation occurs via Ni motion. But that provides no information about the role of grain boundaries. An idealized representation of the conditions that obtain with a moving marker is shown in Fig. 4. If a layer A*-A (Fig. 4a) is made to react with B to form a compound and following reaction the positions of the two A isotopes is inverted (Fig. 4b), then the compound has grown via grain boundary diffusion of A. This will happen at low temperature, for idealized pure C (according to the classification of Harrison) type diffusion, with A alone being mobile. In practice with some lateral diffusion of A from the grain boundaries into the lattice of the already formed compound, the profiles are likely to be much less clear than in Fig. 4b, the more so as growth proceeds for increasing periods of time. Note that if both A and B are mobile, following reaction A* (in A*B) shall be found on both sides of A (Fig. 4c), with the relative amounts giving a measure of the respective mobilities of A and B (again assuming pure type C diffusion for both A and B). A very thorough mathematical analysis of this problem, for all three types (A, B and C) of diffusion is found in Ref. 41. One obtains a simpler version if one considers[42] only lattice diffusion. In either case, what matters is the ratio of the driving force ΔG for growth to kT, which measures the driving force for isotope mixing, and the time duration of the

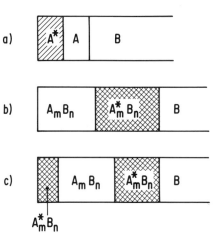

Figure 4. Sketch illustrating the position of an isotopic marker A* as a result of phase (A_mB_n) growth via grain boundary diffusion. a) Initial condition. b) After phase growth with A and A* alone being mobile (C type). c) With both A (and A*) and B being mobile (again type C diffusion); the amounts of $A^*_mB_n$ on the right and the left measure the respective mobilities of A and B atoms. According to Ref. 41.

185

experiment. For small $\Delta G/kT$ ratios mixing occurs more rapidly than growth, one obtains what the authors of Ref. 41 call equilibrium, the distribution of the tracer isotope is uniform throughout the growing layer. On the contrary for large ratios one obtains a drift condition in which growth so predominates that the tracer remains immobile as an inert marker. Whether one considers Refs. 41 or 42, the in-situ (during growth) use of isotope shall deliver information only in a restricted set of circumstances (between lower and upper values of $\Delta G/kT$), and only for experiments limited to short periods of times.

When grain boundary diffusion[43] predominates, one can encounter conditions where light and heavy isotopes are inverted. The idea is that the ratios of the diffusion coefficients for two isotopes are not the same for diffusion along the grain boundaries and in the lattice. At the limit where the diffusion coefficients in the grain boundaries would be the same, it is clear that an inversion should occur since the light element will be lost by lateral diffusion in the lattice faster than the heavy one. Mathematically one obtains a relation where the ratio of the two isotopes varies as $\exp\left[(D^*/D^{*\prime})^{1/4}(1 - (D^{**}/D^* \times D^{*\prime}/D^{**\prime})^{1/4})\right]$, where D^* and D^{**} are the respective lattice diffusion coefficients and the prime quantities refer to the grain boundary values. The maximum effect, of the order of $1/8 \times (M^{**} - M^*)/M$ (where the two M's refer to the isotopic masses), is bound to be very small. Applying[44,45] such an idea to the analysis of phase growth is certainly ingenious. However, the effects are so small that in the absence of extraordinary precautions they remain hidden behind analytical artifacts (see Ref. 46). In the initial derivation[43] the longitudinal force (along the grain boundary) was the same as the lateral force (in the lattice), e.g. the diffusion of two isotopes of Au in Ag (Fig. 5). In compound growth not only is the longitudinal force generally quite big, but the lateral force is exceptionally small, since it results only from the disproportionation in the concentrations of the two isotopes, respectively in the grain boundaries and in the lattice. It is not sure that this effect is sufficiently taken account in the derivations found in Ref. 44.

Figure 5. a) In solid solutions (type Ag - Au) the forces (arrows) on the diffusing atoms are equal, either along the grain boundaries or in a direction normal to this into the lattice. b) This is not true in compound formation.

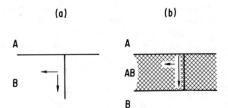

MORPHOLOGY, GRAIN SIZE, METAL - METAL REACTIONS

The morphology of the grown layers is a function of the mode of growth. In the already mentioned case of the growth of NiO by thermal oxidation it has been observed[30] that the grain size of the oxide depends on the state of the Ni substrate, with the grain size being smaller on highly cold worked Ni than on well annealed Ni. An extreme example[47] is the growth (via the dual motion of Ni and Al ions) of the spinel $NiAl_2O_4$ by the reaction of polycrystalline NiO with single crystal Al_2O_3; the new phase is a single crystal next to the alumina reactant and polycrystalline next to NiO. A similar effect is illustrated[48] in Fig. 6, where Nb_3Sn grows from the reaction of Nb with Nb_6Sn_5. The compound layer (a) grown next to the Nb has a small grain size, that next to Nb_6Sn_5 (b) has a large grain size. The ratio of the thicknesses a/b 1.4 implies that

Figure 6. Sketch illustrating the growth of Nb_3Sn from the reaction of Nb_6Sn_5 with Nb, via the diffusion of 3 atoms of Sn. While two moles of the product are formed on the right from the decomposition of Nb_5Sn_6, three are formed on the left from the reaction of Sn with Nb. This results in two product layers distinguishable by their grain size. According to Ref. 48.

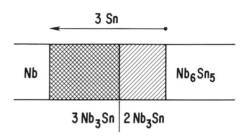

Nb_3Sn grows via Sn motion (a rare exception[10] to the ordered Cu_3Au rule), and the deviation, 1.4 instead of the ideal 1.5, if not simply an experimental error is a measure of the amount of phase growth due to Nb transport ($D_{eff.Nb}/D_{eff.Sn} = 1/12$). All of these observations imply some degree of epitaxy between the reactants and the product phase.

The fact that grain boundaries act as fast diffusion paths in so many reactions affect the geometrical character of the interfaces, which from first principles in the case of binary reactions at least should remain planar, since two phase regions are excluded. An example of such effects of grain boundaries on the interface is provided in Fig. 7, where it is seen that a growing layer[30] of NiO is thicker in the vicinity of the grain boundaries. This effect is purely kinetic: surface energy equilibrium would require that the interface discontinuities, observed within the Ni at the oxide grain boundaries, be inside the oxide. For a more general evaluation of such effects see Fig. 2.16 in Ref. 1.

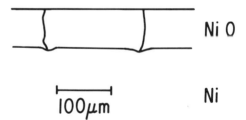

Figure 7. Tracing from a cross section photograph in Ref. 30, showing an interface instability due to enhanced Ni diffusion at the grain boundaries of Ni oxide.

Grain boundary (in the growing compound layer) diffusion is supposed to dominate many metal-metal interactions where the reacting materials are polycrystalline to begin with. The growth rate of Ni_3Al was reported[47] to be faster at 1000°C than at 1250°C because of the grain size difference. The growth of Ag_2Al below 200°C ($t^{1/2}$), with an activation[50] energy of 0.85 eV, is probably controlled by grain boundary diffusion. As with NiO (see above), the rate decreases as the grain size of the reacting Ag is increased. It is quite likely that activation energies[51] of 1.03 eV and 1.2 eV, respectively for the formations of Au_2Al and $AuAl_2$ are due also to grain boundary diffusion. For other such examples see Ref. 52. One important point, however, is that the presence of grain boundaries in the reactants may affect the morphology and even the mode of growth of the resulting compound. The growth ceases to be planar, the reaction occurs initially preferentially along the grain boundaries of one of the reactant. The reactions in Al-Nb, Al-Ag, Al-Hf and Al-Ti (respectively, Refs. 6, 50, 53, 54) display such a behaviour. An

Figure 8. Backscattering spectra illustrating the growth of WSi_2 in bilayers Si/Mo-W. Ignore the Mo part of the spectra between channels 250 and 300. The shape of the W part of the spectra at channel 375 shows that the formation of WSi_2 proceeds at a fast rate at the surface of the W layer, requiring Si diffusion, presumably via grain boundaries, through that layer. Temp. 750°C, time 0, 0.5, 0.75, 1 and 2 hr. From Ref. 56, Fig. 6.

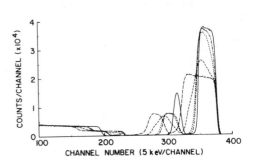

extreme example is illustrated in the reaction[55] of polycrystalline W with single crystal Si. In Fig. 8 it can be seen that the reaction to form WSi_2 proceeds at least as fast at the free surface of the W film as at the inner interface. This requires the grain boundary transport of Si through the W film. Similar observations[56,57] were made in the formations of $TaSi_2$ and $Ti(Si-Ge)_2$. The addition of impurities, that reduce grain boundary diffusion in one or the other of the reactants, eliminates or limits this anomalous behaviour. In certain electronic applications one wants a very thin and continuous layer of an intermetallic, $TiAl_3$ or $HfAl_3$, in the middle region of an Al thin film conductor. With pure Al it is extremely difficult to obtain such a continuous layer because of the preferential[53] grain boundary formation; this, however, becomes much easier[58,59] with the addition of Cu, that is otherwise known[60] to reduce grain boundary diffusion in Al. In the same way, the anomalous surface reaction of Ti with Si-Ge disappeared when the Ti film became saturated[57] with nitrogen. Similar observations have been reported[61] for Al-Ti; the surface formation of $TiAl_3$ disappears in the presence of residual oxygen in the annealing atmosphere. The effect of impurities in causing the formation of smoother interfaces is somewhat paradoxical, since the addition of a third element to a two component system opens the door to the formation of two phase regions, and subsequently very irregular interfaces. It is clear that the situation here is one where kinetic considerations completely dominate the conditions of thermodynamic equilibrium. An analysis of the effect of impurities in reducing grain boundary diffusion is found in Ref. 62. In general the addition of a third element may be expected to reduce[30] the grain size of the product layer either because of a solute drag effect or because of the formation of precipitates, both of which usually hinder grain growth. So-called differences between bulk and thin film reactions, often encountered in the literature, can be shown to be quite trivial. However, thin film reactions are usually carried at low temperatures where grain boundary diffusion should dominate. Yet one should always be careful: in the bulk formation of $TiAl_3$ (melting point about 1630 K) the activation energy was found[11] to be 1 eV, but it is about twice as big[58] in thin films.

Before terminating this section on the effects of grain boundaries in the reactants on phase formation one must mention again the formation[5] of amorphous phases such as result from the reaction of polycrystalline Zr with Ni. If the Zr reactant is a single crystal no amorphization reaction[63] is observed. But if a Ni single crystal[64] is coupled to polycrystalline Zr the reaction proceeds. The effect is probably related to what is discussed in the previous paragraph. It has been proposed[65] that nucleation of a new phase becomes difficult in sharp chemical potential gradients (see also Ref. 3). Grain boundary diffusion of Ni in Zr could decrease the gradient in a direction normal to the interface, which would become impossible with single crystal Zr.

CONCLUDING REMARKS

One finds throughout this paper, as a leitmotiv, the theme that it is extremely difficult to judge from kinetics alone, and without the help of fairly extensive corollary investigations, what is the role of grain boundaries as short circuit diffusion paths during phase growth. There are perhaps two exceptions. If the growth is expressed as a power function of time, a time exponent smaller than 1/2 is likely to reveal grain boundary diffusion. The other exception occurs when growth is started with a bilayer of one of the components, ideally two isotopes but extensions are conceivable; if growth results in an inversion of the isotopes it must have proceeded via grain boundary diffusion. Further insight can be obtained with the use of isotopes (in situ, during reaction), but the conditions where such use will be informative are restricted. The best results are obtained when isotope diffusion is studied independently from growth and the diffusion coefficients thus derived are compared to the growth data. Grain boundaries usually play a somewhat passive role, simply providing a path for diffusion to take place, but they may affect the morphology of interfaces. In some cases of epitaxy a single crystal reactant may cause the product phase to be monocrystalline. There exist records of a positive correlation between the grain size(s) of the reactants and that of the compound film resulting thereof, occasionally causing that film to be split into two distinguishable layers. Principally because of the segregation of impurities on grain boundaries, and of the resulting decrease in grain boundary diffusion, trace amounts of alloying elements can significantly affect the process of compound formation.

ACKNOWLEDGEMENTS

One of us (F. d'H.) is happy to acknowledge fruitful discussions of this grain boundary topic with Terje Finstad. We are all indebted to the many colleagues and friends who generously shared with us ideas and results.

REFERENCES

1 F. J. J. van Loo, Prog. Solid St. Chem. **20**, 47 (1990).

2 V. I. Dybkov, J. Phys. Chem. Solids **53**, 703 (1992).

3 J. Philibert, Defects and Diffusion Forum **95-98**, 493 (1993).

4 A. Vredenberg, J. Westendorp, F. Saris, N. van der Pers, and T. de Keijser, J. Mater. Res. **1**, 774 (1986).

5 K. Samwer, Physics Reports (Review Section of Physics Letters) **1**, 161 (1988).

6 K. Barmak, K. R. Coffey, and D. A. Rudman, J. Appl. Phys. **67**, 7313 (1990).

7 K. R. Coffey and K. Barmak, this symposium.

8 A. J. Bradley and A. Taylor, Proc. Roy. Soc. **A159**, 56 (1937).

9 S. Shankar and L. L. Seigle, Metall. Trans. **9A**, 1467 (1978).

10 F. M. d'Heurle and R. Ghez, Thin Solid Films **215**, 19 (1992).

11 F. J. J. van Loo and G. D. Rieck, Acta Metall. **21**, 61 and 73 (1973).

12 Y. L. Corcoran, A. H. King, N. de Lanerolle, and Bonggi Kim, J. Electron. Mater. **19**, 1177 (1990).

13 T. G. Finstad, J. W. Mayer, and M.-A. Nicolet, Thin Solid Films **51**, 391 (1978).

14 T. G. Finstad, Thin Solid Films **51**, 411 (1978).

15 J. Li, Y. Sacham-Diamand, and J. W. Mayer, Materials Science Reports **9**, 1 (1992).

16 R. D. Thompson, D. Gupta and K. N. Tu, Phys. Rev. **B33**, 2636 (1985).

17 J. C. Ciccariello, S. Poize, and P. Gas, J. Appl. Phys. **67**, 3315 (1990).

18 K. N. Tu, W. K. Chu, and J. W. Mayer, Thin Solid Films **25**, 403 (1975).

19 C. R. Zheng, E. Zingo, J. W. Mayer, G. Majni, and G. Ottaviani, Appl. Phys. Lett. **41**, 646 (1982).

20 C.-D. Lien, M.-A. Nicolet, and S. S. Lau, Phys. Status Solidi **a81**, 123 (1984).

21 T. Barge, Ph. D. Thesis, Université Saint Jérôme, Marseille, 1993.

22 P. Gas and F. M. d'Heurle, Appl. Surf. Sci. **73**, 153 (1993).

23 C.-D. Lien, M.-A. Nicolet, and S. S. Lau, Appl. Phys. A **34**, 249 (1984).

24 C.-H. Jan, C.-P. Chen, and Y. A. Chen, J. Appl. Phys. **73**, 1168 (1993).

25 S. Q. Hong, C. M. Comrie, S. W. Russell, and J. W. Mayer, J. Appl. Phys. **70**, 3655 (1983).

26 J. G. M. Becht, F. J. J. van Loo, and R. Metselaar, React. Solids **6**, 45 and 61 (1988).

27 H. Bakker, in *Diffusion in Crystalline Solids*, edited by G. E. Murch and A. S. Novick (Academic Press, New York, 1984) p. 189

28 P. Kofstad, *High Temperature Corrosion* (Elsevier Applied Science, London, 1988).

29 W. W. Smeltzer, R. R. Haering, and J. S. Kirkaldy, Acta Metall. **9**, 880 (1961).

30 D. Caplan, M. J. Graham, and J. Cohen, J. Electrochem. Soc. **119**, 1205 (1972).

31 A. Atkinson, R. I. Taylor, and P. D. Goode, Oxid. Met. **13**, 519 (1979).

32 A. Atkinson and R. I. Taylor, Philos. Mag. **A43**, 979 (1981).

33 B. Lesage and M. Déchamps, Defect and Diffusion Forum **95-98**, 1077 (1993).

34 A. C. S. Sabioni, B. Lesage, A. M. Huntz, J. C. Pivin, and C. Monty, Philos. Mag. **A66**, 333 (1992).

35 A. C. S. Sabioni, A. M. Huntz, F. Millot, and C. Monty, Philos. Mag. **A66**, 351 and 361 (1992).

36 *The Role of Active Elements in the Oxidation Behaviour of High Temperature Metals and Alloys*, edited by E. Lang (Elsevier Applied Science, London, 1989).

37 K. Przybylski, A. J. Garratt-Reed, and G. J. Yurek, J. Electrochem. Soc. **135**, 509 (1988).

38 G. J. Yurek, K. Przybylski, and A. J. Garratt-Reed, J. Electrochem. Soc. **134**, 2643 (1987).

39 B. Pieraggi and R. A. Rapp, J. Electrochem. Soc. **140**, 2844 (1993).

40 J. Philibert and A. M. Huntz, in *Microscopy of Oxidation*, edited by S. B. Newcomb and M. J. Benett (The Institute of Materials, London, 1993) p. 253.

41 Yu. M. Mishin and G. Borchardt, J. Phys. III (France) **3**, 863 and 945 (1993).

42 S.-L. Zhang, F. M. d'Heurle, and P. Gas, Appl. Surf. Sci. **53**, 103 (1991).

43 N. L. Peterson, Intl. Metals Rev. **28**, 65 (1983).

44 J. P. Stark, Acta Metall. **31**, 2083 (1983).

45 E.-S. M. Aly and J. P. Stark, Acta Metall. **32**, 907 (1984).

46 P. Gas, K. Zaring, B. G. Svensson, M. Östling, C. S. Petersson, and F. M. d'Heurle, J. Appl. Phys. **67**, 2390 (1990).

47 H. Schmalzried, *Solid State Reactions* (Verlag Chemie, Weinheim, 1981) p. 14.

48 W. L. Neijmeijer and B. H. Kolster, Z. Metallkde **81**, 314 (1990).

49 M. M. Janssen, Met. Trans. **4**, 1623 (1973).

50 J. E. E. Baglin, F. M. d'Heurle, and W. N. Hammer, J. Appl. Phys. **50**, 266 (1979).

51 S. U. Campisano, G. Foti, E. Rimini, S. S. Lau, and J. W. Mayer, Philos. Mag. **31**, 903 (1975).

52 J. E. Baglin and J. M. Poate, in *Thin Films: Interdiffusion and Reactions*, edited by J. M. Poate, K. N. Tu, and J. W. Mayer (John Wiley, New York, 1978) p. 305.

53 R. F. Lever, J. K. Howard, W. K. Chu, and P. J. Smith, J. Vac. Sci. Technol. **1**, 158 (1977).

54 P. Maugis, G. Blaise, and J. Philibert, Mater. Res. Soc. Symp. Proc. **237**, 679 (1992).

55 C. S. Petersson, J. E. Baglin, F. M. d'Heurle, J. Dempsey, J. Harper, C. Serrano, and M. Y. Tsai, in *Thin Film Interfaces and Interactions*, edited by J. E. Baglin and J. Poate (The Electrochemical Society, Pennington N. J., 1980) p. 290.

56 J. E. Baglin, J. Dempsey, W. Hammer, F. M. d'Heurle, C. S. Petersson, and C. Serrano, J. Electron. Mater. **8**, 641 (1979).

57 O. Thomas, S. Delage, and F. M. d'Heurle, J. Mater. Res. **5**, 1453 (1990).

58 M. Wittmer, F. LeGoues, and H.-C. W. Huang, J. Electrochem. Soc. **132**, 1450 (1985).

59 M. Wittmer, H.-C. Huang, and F. LeGoues, Philos. Mag. A **53**, 687 (1986).

60 F. M. d'Heurle and M. Gangulee, in *Nature and Behavior of Grain Boundaries*, edited by Hsun Hu (Plenum Press, New York, 1973) p. 339.

61 X.-A. Zhao, F. C. T. So, and M.-A. Nicolet, J. Appl. Phys. **63**, 2800 (1988).

62 D. Gupta, in *Diffusion Phenomena in Thin Films and Microelectronic Materials*, edited by D. Gupta and P. S. Ho (Noyes Publications, Park Ridge, NJ, 1988) p. 1.

63 A. Vredenberg, J. Westendorp, F. Saris, N. van der Pers, and T. de Keijser, J. Mater. Res. **1**, 774 (1986).

64 K. Pampus, K. Samwer, and J. Böttiger, Europhys. Lett. **3**, 581 (1987).

65 A. M. Gusak and K. P. Gurov, Solid State Phenomena **23 & 24**, 117 (1992).

A UNIFIED APPROACH TO GRAIN BOUNDARY DIFFUSION AND NUCLEATION IN THIN FILM REACTIONS

K.R. COFFEY[*] and K. BARMAK[**]
[*]IBM, Storage Systems Division, 5600 Cottle Road, San Jose, CA 95193
[**]Department of Materials Science and Engineering, Lehigh University, Bethlehem, PA 18015

ABSTRACT

An alternative model is proposed to extend the conventional view of diffusion under a concentration gradient in a grain boundary phase of width δ. The conventional model is well developed and readily applied to the thickening kinetics of polycrystalline product phases in binary diffusion couples, however it is not readily extended to other phenomena of interest in thin films, i.e., the nucleation and growth of the product phase crystallites prior to formation of a product phase layer. In the alternative model presented here, non-equilibrium thermodynamics is used to define the chemical potentials, μ_i^I, for each atomic specie in the grain and interphase boundaries of a polycrystalline diffusion couple. The chemical potential difference for each specie between the bulk phases of the diffusion couple is partitioned between the driving force for grain boundary diffusion and that for interfacial reaction. This partition leads to a characteristic decay length that describes the spatial variation of μ_i^I. Numerical calculations of μ_i^I are used to show that boundary diffusion favors heterogeneous nucleation. Product nucleation in thin film reactions is seen to be similar to precipitation from a bulk solid solution.

INTRODUCTION

While this paper will be primarily a theoretical development, the model for thin film reaction kinetics that will be developed was motivated by much experimental work that cannot be reconciled within other frameworks. Existing models of solid-solid reactions have found difficulty reconciling observations of nucleation limited first phase formation with the large driving force that would be expected from the assumption of local equilibrium at interfaces[1-4]. Moreover, no previous models have made attempts to understand such basic phenomena as product phase grain size or the details of product phase composition variations during the reaction[4,5]. In this paper we will present a model that explains the heterogeneous interphase boundary nucleation of the first phase and provides a new approach to selection of phases by reevaluating the "effective" driving force, $\_g^{eff}$ for nucleation of a product phase. Since in

193

this model _g^{eff} is determined by grain and interphase boundary diffusion, the grain structure of the reactant and product phases is seen to be fundamental to the reaction process.

Grain (and interphase) boundary diffusion is expected to be the dominant long range atomic transport mechanism in polycrystalline thin film reactions[6]. Currently, circumstances where grain boundary diffusion is dominant are modeled as "Type C" kinetics where diffusion in the bulk-like interior of the grains is frozen out by the relatively low annealing temperature. In this paper we will consider product phase formation under such conditions. "Type C" grain boundary diffusion is driven by a concentration gradient in a grain boundary "phase" of finite width[7,8]. In this paper, atomic transport in the lattice and in the boundaries will be described by using the chemical potentials of the atomic species in place of their concentration. The advantage of casting the grain boundary interdiffusion in terms of chemical potentials is that it allows comparison of the driving force for diffusion with the driving force for other reaction processes of interest, specifically, those for interfacial reaction and product phase nucleation. Furthermore, a division of bulk and interface properties similar to that of classical nucleation theory can be used, i.e., interfaces will be strictly two dimensional with no thickness and hence no volume. Both grain and interphase boundaries will be regarded as kinetic pathways for atom transport rather than equilibrium phases. This view of grain boundaries as kinetic pathways without volume is similar to that used in an experimental determination of grain boundary diffusivity[9]. The issue of penetration of grain boundary transported material into the adjacent grains, described as "Type B" kinetics, is in the present paper treated as interfacial reaction and bulk diffusion controlled transport. For simplicity the effect of stress and interfacial curvature on the chemical potential gradients along diffusion paths will be neglected here, but the model can be extended to include both of these effects.

TRANSPORT MODEL

In irreversible thermodynamics, all atom transport is modelled as a result of gradients in chemical potential:

$$J_A = - M_A (d\mu_A / dz) \qquad (1)$$

where J_A is the flux of atomic specie A, $d\mu_A/dz$ is the gradient of its chemical potential in the direction z, M_A is its mobility and cross terms have been neglected[10]. The correspondence between equation (1) and the conventional diffusion equation ($J_A = - D_A dC_A/dz$, where D is the diffusion coefficient and C the concentration) is well known.

Atom flux is across an interface (for example, an interphase boundary) without transport in the plane of the interface is shown schematically in Figure 1. The transport across a planar

interface is described by discrete versions of equation (1) in
which the discontinuity in chemical potential is considered to be
the driving force for transport across that interface. For the
case of an interface between two phases α and β, in a two
component system of species A and B, the A atom flux across the
interface is defined as:

$$J_A^{\alpha-\beta} = T_A^{\alpha-\beta} (\mu_A^\alpha - \mu_A^\beta) \qquad (2)$$

where $J_A^{\alpha-\beta}$ is the areal atom flux density (in atoms/cm²-sec) for
component A across the interface between binary phases α and β;
μ_A^α and μ_A^β are the chemical potentials (in eV/atom) for A in the
α and β phases respectively, immediately adjacent to the
interface; and $T_A^{\alpha-\beta}$ is a kinetic coefficient (in atoms²/cm²-sec-
eV) for this atom transfer (atom release or capture) process. A
similar equation can be defined for the B atom flux. When the
phases separated by the interface have substantially different
compositions, the velocity of interface motion, v, is
proportional $J_A^{\alpha-\beta}$, which in turn is limited by the lattice
(bulk) diffusional flux, J_A^β required to prevent build up of A
near the interface (J_B^β is assumed negligible). In this simple
picture, for the interfacial transport (reaction) process to be
the rate limiting step, it is necessary for the lattice diffusion
to be much faster than the interfacial transport. The more
common assumption is the opposite, that lattice diffusion is
sufficiently slow so as to allow the boundary to be considered
near equilibrium[11]. As we will see, the introduction of
lateral transport obviates the need to compare interface velocity
to bulk or lattice diffusion transport.

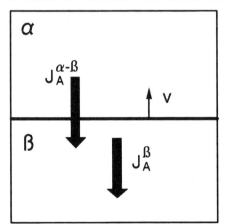

Fig. 1 - Transport across an
interphase boundary and bulk
diffusion away from the
interface.

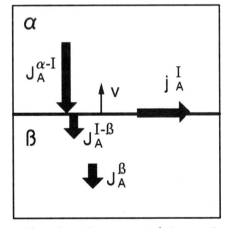

Fig. 2 - Transport into, out
of, and along an interphase
boundary, and bulk diffusion
away from the interface.

We can model grain or interphase boundary diffusion, i.e., lateral diffusion, with irreversible thermodynamics using equations of a form similar to equation (1):

$$j_A^I = - m_A^I (d\mu_A^I/dx)$$ (3)

where j_A^I is the linear flux density in the interface having units of (atoms/cm-sec) and m_A^I, the lateral mobility of A in the interface, has units of (atoms^2/sec-eV). Equation (3) is the analog of the conventional grain boundary diffusion equation cast now in terms of chemical potential gradient rather than concentration gradient. The conventional grain boundary diffusivity, D, is defined with a thickness given to the grain boundary, δ. The product, $D\delta$, is then the physically measured quantity and the flux, $D\delta$ dc/dx, is a linear flux density. The chemical potentials in the interface, μ_A^I and μ_B^I, are distinct from the chemical potentials of the adjacent bulk phase grains (μ_A^α, μ_A^β, μ_B^α, and μ_B^β). Their interaction with the bulk chemical potentials will be described below.

Figure 2 depicts the combined transport along and across grain and interphase boundaries. When lateral diffusion along (in the plane of) the grain and interphase boundaries is considered simultaneously with transverse transport across the boundaries, the flux across the interface is not conserved. Differences between the fluxes in the bulk phases to and from the interface, shown as $J_A^{\alpha-I}$ and $J_A^{I-\beta}$ in Figure 2, can be accommodated by grain boundary transport of atoms laterally in the plane of the interface, j_A^I. Significantly, interphase boundary motion can now occur without the occurrence of bulk diffusion in either adjacent phase, i.e., v can be independent of J_A^β and the reaction proceeds solely by interfacial transport across and diffusion along the interface. Physically, this describes a reaction that proceeds entirely by the mechanism of "type C" grain boundary diffusion. In this picture, for the interfacial transport (reaction) process to be the rate limiting step, it is only necessary for the grain and/or interphase boundary diffusion to be much faster than the interfacial reaction.

To model this situation the single transverse flux density of equation (2), $J_A^{\alpha-\beta}$, is now replaced by two (possibly unequal) interface areal flux densities, into and out of the interface:

$$J_A^{\alpha-I} = T_A^{\alpha-I} [\mu_A^\alpha (x,y) - \mu_A^I(x,y)]$$ (4a)

$$J_A^{I-\beta} = T_A^{I-\beta} [\mu_A^I (x,y) - \mu_A^\beta(x,y)]$$ (4b)

where the superscripts now refer to transport (release or capture) of atoms between the bulk phases and the interface plane, and x and y are the in-plane coordinates. For simplicity it will be assumed that the x and y dependences are similar, and therefore the y coordinate will be dropped and the x coordinate exclusively will be used to describe the in-plane position.

We can consider the thermodynamic driving force for reaction in a thin film system to be partitioned between the mechanisms of equations (3) and (4a&b). A further mathematical development is of interest because it defines a fundamental length scale to this partition. We can link lateral and transverse atomic transport by requiring that atomic flux is conserved for an area element of the interface. This conservation of flux is consistent with the lack of thickness and volume of the boundary. Flux conservation for an area element of a grain or interphase boundary is shown in Figure 3 and gives the gradient in linear flux densities along these boundaries as:

$$dj_A^I / dx = J_A^{\alpha-I} - J_A^{I-\beta} \qquad (5)$$

Equations (5), (4a) and (4b) can be combined with equation (3) to provide a second order differential equation which defines the interface chemical potential $\mu_A^I(x)$:

$$- d^2\mu_A^I / dx^2 + \mu_A^I \{ (T_A^{\alpha-I} + T_A^{I-\beta}) / m_A^I \}$$
$$= (\mu_A^\alpha(x)T_A^{\alpha-I} + \mu_A^\beta(x)T_A^{I-\beta}) / m_A^I \qquad (6)$$

A similar equation applies to the B atom fluxes.

Equation (6) is an inhomogeneous differential equation requiring a unique solution for each pair of functions, $\mu_A^\alpha(x)$ and $\mu_A^\beta(x)$, used to describe the bulk phase grains adjacent to the interface. It should be noted that equation (6) is not explicitly time dependent, as it describes the steady state transport properties of the interface (assumed to have no volume). The time dependence is implicit in the functions $\mu_A^\alpha(x)$ and $\mu_A^\beta(x)$ which change as the composition of the bulk grains adjacent to the interface changes. The homogeneous differential equation:

$$- d^2\mu_A^I / dx^2 + \mu_A^I \{ (T_A^{\alpha-I} + T_A^{I-\beta}) / m_A^I \}$$
$$= (\mu_A^\alpha{}_o T_A^{\alpha-I} + \mu_A^\beta{}_o T_A^{I-\beta}) / m_A^I, \qquad (7)$$

obtained by taking $\mu_A^\alpha(x)$ and $\mu_A^\beta(x)$ as the constants $\mu_A^\alpha{}_o$ and $\mu_A^\beta{}_o$, is not useful in describing the time evolution of the system as diffusion proceeds. However, the homogeneous equation can be used to describe the initial conditions for a thin film diffusion couple and provide insight into the interaction between lateral and transverse interfacial transport.

The solution to equation (7) for the chemical potential in the interface in a region where the chemical potentials in the bulk at the interface are constant with position is given by:

$$\mu_A^I(x) = C_{A1} \exp(x/L_A) + C_{A2} \exp(-x/L_A) + \mu_A^I{}_o \qquad (8)$$

where C_{A1} and C_{A2} are constants determined by the continuity of chemical potential and lateral interfacial flux at the limits of the region, L_A is a decay length for μ_A^I, and $\mu_A{}^I_o$ is the constant of integration.

Both L_A and $\mu_A{}^I_o$ have physical interpretations of interest. $\mu_A{}^I_o$ is the chemical potential of the interface in the absence of lateral gradients in the interface and is given by:

$$\mu_A{}^I_o = (\mu_A{}^\alpha_o T_A{}^{\alpha-I} + \mu_A{}^\beta_o T_A{}^{I-\beta})/(T_A{}^{\alpha-I} + T_A{}^{I-\beta}) \qquad (9)$$

As can be seen from this equation, $\mu_A{}^I_o$ is the "weighted average" of the values of the bulk phases at the interface and is therefore always intermediate to the chemical potential of the adjacent phases. The parameter L_A represents a decay length for changes in the chemical potential of specie A in the interface, and is given by:

$$L_A = \sqrt{[m_A{}^I/(T_A{}^{\alpha-I} + T_A{}^{I-\beta})]}. \qquad (10)$$

L_A is the characteristic distance for the interaction between regions of different chemical potential along the interface, i.e., it is a length over which fast grain boundary diffusion allows non-equilibrium long range interaction between bulk phase regions. In thin films L_A can be large compared to grain sizes.

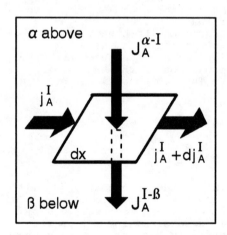

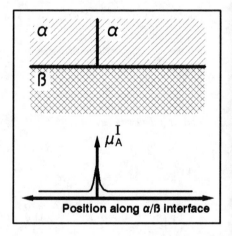

Fig. 3 - An area element of the α/β interface depicting conservation of atom flux across and along the interface.

Fig. 4 - The variation of $\mu_A{}^I$ in the α/β interface for the case of grain size much larger than the decay length.

It should be noted that L_A (and a similarly defined L_B) is a fundamental quantity dependent on the nature of the interface. To be exact, a unique value of L_A and L_B would need to be defined for each possible interface structure in a polycrystalline sample. However, in the remainder of this paper we will make the simplifying assumption of single values for L_A and L_B to present an application of the model. L_A is a steady state interaction length and is not time dependent in the usual sense of the characteristic distance for diffusion, $L = \sqrt{Dt}$. L_A is given by the square root of the ratio of grain boundary diffusivity to interfacial reaction rate, and thus describes the circumstances where these phenomena interact. In effect, the total chemical potential difference for species A, $\mu_A^{\beta} - \mu_A^{\alpha}$, is being partitioned between interfacial reaction barriers and grain boundary diffusion. When the interfacial reaction barriers are large ($T_A^{\alpha-I}$ and $T_A^{I-\beta}$ numerically small) or grain boundary diffusion is fast (m_A^I large) L_A is large. In thin film reactions where a product phase is formed, the interfacial reaction would be rate limiting when L_A is large compared to the product phase grain size or layer thickness and grain boundary diffusion would be reaction rate limiting when L_A is smaller than the grain size or product layer thickness.

THIN FILM REACTIONS - INTERDIFFUSION

The above transport equations are quite general and might be applicable to a variety of phenomena where the atomic transport is chemical potential driven and interfacial transport is the dominant mechanism. The transport model presented conserves atomic flux, but it does not conserve crystal lattice sites, which are assumed to be freely created and annihilated at interfaces. Although strain and its effect on free energy and chemical potentials is an important adjunct to this transport model, the focus will be on thin film reactions with large thermodynamic driving forces, and it will be assumed that the strain energies associated with the lack of conservation of lattice sites are negligible.

In the remainder of this paper, the above transport model will be applied to an idealized thin film diffusion couple (α phase primarily composed of atomic specie A and β phase of B) wherein the nucleation and growth of a product phase is expected to occur upon annealing.

The decay length parameter L_i allows the classification of different reaction microstructures as being bulk-like (grain size $\gg L_i$) or thin film (grain size $\ll L_i$), thus it is useful to consider some limiting cases. Figure 4 depicts the case where the α and β grain sizes are much larger than the decay length, L_i, thus there are no gradients in chemical potential over much of the interphase boundary and the interphase transport along the interface does not compete with transport across the interface The metastable equilibrium condition, $\mu_A^{\alpha} = \mu_A^I = \mu_A^{I_o} = \mu_A^{\beta}$, is rapidly achieved by interdiffusion across the interface. This situation is consistent with our understanding of bulk reaction in solids. Equation (8) would only need to be applied to regions close to where the α/α or β/β grain boundaries intersect the α/β interphase boundary.

Figure 5 depicts the opposite extreme, when the grain sizes are much smaller than the decay length. The interface chemical potential will be essentially constant around an individual grain, and will only vary over larger distances as one passes from the α phase to the β phase. Thus an α phase grain near the α/β interface will have essentially a single value for $\mu_A{}^I$ and for $\mu_B{}^I$ on all its interfaces with adjacent grains. When a reaction is occurring (not shown), i.e., product phase growth of a thin ($< L_i$) layer of grains between α and β , the reactant phase grains can dissolve into and the product phase grains precipitate out of a grain boundary "solution" whose composition changes only gradually across a phase boundaries. In this limit, the overall rate of product phase formation is interfacial reaction rate dependent, and this model predicts that the reactant phase grain sizes help determine the product phase growth rate as they define the available area of α and β phase interface within a distance $\approx L_i$ of the product phase grains.

An idealized example of a thin film reaction microstructure where the grain size is comparable to the characteristic distances L_A and L_B is shown in Figure 6, where it is assumed that $L_A = L_B = L$ is a constant for all interfaces. The α and β phase grains are shown as regular parallel slabs with the dimensions perpendicular to the interphase boundary and perpendicular to the plane of the page much greater than L (semi-infinite). The grain size in the third dimension, i.e., along the interphase boundary in the plane of the page, is equal to twice L. This geometry reduces the problem of finding the solution to equation (1.7) for the interfacial chemical potential

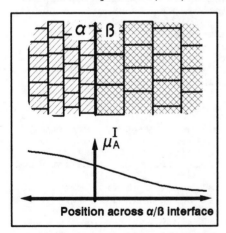

Position across α/β interface

Fig. 5 - The variation of $\mu_A{}^I$ with distance from the α/β interface as averaged over many grain boundaries in a plane parallel to the interface.

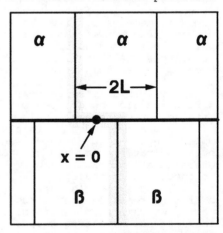

Fig. 6 - An example of thin film reaction microstructure with reactant phase grains as semi-infinite parallel slabs.

along simplified paths from the α to the β phase, with the parameter x being the distance along the path. An example of such a path is highlighted in Figure 6.

It is illustrative to present analytical and numerical solutions using this transport model for the geometry of Figure 6. For simplicity of calculation a regular solution model is used to describe the variation of chemical potential with composition of the α and β phases. The initial compositions of the two phases are taken to be the compositional extremes (99.999 at.% purity) and a regular solution parameter of 3kT was used.

To obtain the analytical solutions of equation (8) for μ_A^I and μ_B^I along the highlighted path three segments need to be considered. Along the α/α grain boundary portion the initial chemical potential in the bulk phase on both sides of the interface is given by $\mu_A{}^\alpha{}_o$, equal to 0.0 kT. Along the α/β interphase boundary portion the adjacent bulk phases are described by $\mu_A{}^\alpha{}_o$ and $\mu_A{}^\beta{}_o$ with the latter having a value of -8.5 kT for the initial purity and regular solution parameter chosen above. Along the β/β grain boundary portion, again, both adjacent bulk phases are identical and $\mu_A{}^\beta{}_o$ is -8.5 kT as given above. Equation (7) is then applied to each of these three path segments resulting in solutions of the form given by equation (8) for each region. A single solution for the entire path is then obtained by determining the constants C_{A1} and C_{A2} for each segment subject to the requirements that μ_A^I be bounded at large distances from the interfaces and be continuous at the segment junctions (triple points at +/- 1/2 L) and that the flux (-m $d\mu_A^I/dx$) be conserved at the segment junctions. Note that each α/α or β/β boundary is met by two α/β interphase boundaries at the triple points, so that for equal "m" values conservation of flux results in a factor of two discontinuity in $d\mu_A^I/dx$ across the junction. A similar analysis for the B atomic specie is omitted; because of the obvious symmetry, $\mu_B^I(x)$ is equal to $\mu_A^I(-x)$.

The solid curves in Figure 7 show the analytical solutions for μ_A^I and μ_B^I at a time prior to any interdiffusion across the interface, i.e., at t = 0. The vertical axis of the figure is the chemical potential in units of kT and the horizontal axis is the position along the highlighted path of figure 2 in multiples of L. Clearly the chemical potentials for both A and B along this path are much reduced from both their bulk values (μ = 0) and the α/β metastable equilibrium value of μ = -0.06 for the regular solution model chosen here. The dashed curve in Figure 7 gives μ_A^I (μ_B^I is here omitted from the figure for clarity) for a different set of boundary conditions corresponding to complete interdiffusion across the α/β interface, i.e., $\mu_A{}^\alpha = \mu_A{}^\beta = \mu_B{}^\alpha = \mu_B{}^\beta = -0.06$ along this interface at t = 0 (noting however that this boundary condition is non-physical). It is thus apparent that the kinetic reduction in chemical potentials of the interface occurs even when a local metastable equilibrium of the bulk phases on either side of the interface is hypothesized. In thermodynamic terms, the driving force for interdiffusion along grain boundaries that intersect the interphase boundary is greater than or equal to the driving force for interdiffusion across the interphase boundary. Fast diffusion in the interphase boundary requires that the competition between these processes be considered.

A more reasonable approach to interdiffusion across the interface is to calculate the system evolution from the initial condition of no interdiffusion in the bulk at t = 0. This is readily done by numerical finite element integration of equation (6) using the regular solution model to determine changes in $\mu_A^\alpha(x)$ and $\mu_A^\beta(x)$ that occur with diffusion[15]. The solid curve of Figure 8 shows the time evolution μ_A^I (and, by symmetry, μ_B^I) in the interphase boundary at x = 0 calculated using the transport model and assumptions given above. The dashed curve is for reference and is the result of a similar calculation constrained to disallow lateral transport in the grain and interphase boundaries and thus reflects the interfacial transport coefficients and bulk diffusivities chosen for this example. Time is given in units of L^2/vM_μ_{max}, where v is the atomic volume and $\_\mu_{max}$ is the maximum difference in chemical potential (equal to 8.5kT in the regular solution model used here for both α and β). The condition of t = 1 corresponds to the case where the characteristic bulk diffusion distance ($\sqrt{Dt}$, with D = vM_μ_{max}) is equal to L. As can be seen, the lateral transport significantly retards the evolution of the interface towards equilibrium.

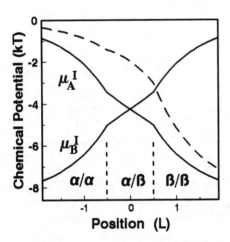

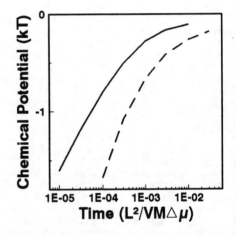

Fig. 7 - μ_A^I and μ_B^I along the highlighted path of Fig. 6. The solid curves are for the condition of no prior bulk interdiffusion across the α/β interface. The dashed curve shown for μ_A^I is for complete prior interdiffusion .

Fig. 8 - Result of a numerical caluculation of the time evolution of the chemical potential of μ_A^I at the central x = 0 position of figure 6. The units for time are such that at t = 1, L ≡ $\sqrt{Dt}$.

THIN FILM REACTIONS - NUCLEATION

Given this model for atomic transport, it is interesting to consider nucleation and phase selection. It is our premise that classical nucleation theory should be used with the driving force for nucleation defined by the interfacial chemical potentials[16]. When two polycrystalline layers are annealed, it is these interfacial chemical potentials that determine which phase will be able to nucleate and grow, in the same manner that the chemical potential of a homogeneous supersaturated solid solution determines nucleation and growth in bulk systems. In other words, this approach can be described as a grain boundary diffusion controlled precipitation model. The interface chemical potentials for A and B at each point in the interphase and grain boundary network define the free energy available for product phase formation. As can be seen from Figure 8, the interfacial chemical potentials along the interphase boundary will be initially low, increasing only as interdiffusion occurs in the adjacent grain boundary network. As the chemical potentials in the interphase boundary increase, they will be variable along the interphase boundary, μ_A^I larger near junctions of α/α grain boundaries with the interphase boundary and μ_B^I larger near the β/β junctions. These initially low and spatially variable chemical potentials define low and spatially variable driving forces for nucleation, which is expected then to be heterogeneous, and occur at adventitious sites.

Once a phase nucleates in the interface, it can act as a sink for species A and B, i.e., the new phase now drives the lateral diffusion. The formation of the nucleus causes a "capture zone" with a length scale of the order of L_A (or L_B) that prevents further nucleation in the zone similar to the case of thin film growth by vacuum deposition[17]. This characteristic length scale should then be apparent in the final microstructure as the size of the individual product phase crystallites when they coalesce into a contiguous product layer, provided that the density of the preferred nucleation sites is greater than L^{-2}. As specific examples, in the Nb/Al system this distance was found to be 20 - 30 nm for the formation of the first phase, $NbAl_3$. For the formation of Nb_3Al from a Nb/Nb_2Al interfacial structure a distance of 150 nm was observed[5,18].

CONCLUSIONS

In summary, a theory of thin film reaction kinetics has been developed which uses irreversible thermodynamics to model the atom transport occurring across interface reaction barriers and along grain and interphase boundaries. Grains of deposited reactant layers of α and β are assumed to initially contain the A and B species at the bulk chemical potentials of the α and β phases respectively. The resulting difference in chemical potential for each species (A and B) between the layers is considered to drive three processes. One is lattice diffusion, the other grain and interphase boundary diffusion, and the third, interfacial chemical reaction. Conservation of mass defines the boundary chemical potential and leds to a fundamental length scale, L, which describes the spatial variation of these potentials. The nucleation of a product phase is described as a process whose driving force is determined by these interfacial chemical potentials and which as a result competes for the atoms diffusing in the boundary. For reaction of layered thin films with small grain sizes, nucleation is predicted to occur primarily heterogeneously with a low driving force only after sufficient interdiffusion has occurred to raise the local A and/or B chemical potentials in the boundaries.

As a result of the incorporation of boundary transport, the model considers the effect of product and reactant microstructure to be fundamental to the reaction process and product phase formation. Moreover, it is argued that the characteristic length, L, should be determinable from the microstructure of the product phase.

REFERENCES

1. O. Thomas, C.S. Peterson, and F.M. d'Heurle, Appl. Surf. Sci. **53**, 138 (1991).
2. F.M. d'Heurle, J. Mater. Res. 3, 167 (1988).
3. K.R. Coffey, K. Barmak, D.A. Rudman, S. Foner, J. Appl. Phys. **72**, 1341 (1992).
4. K.R. Coffey, K. Barmak, D.A. Rudman, S. Foner, Mater. Res. Soc. Proc. **230**, 55 (1992).
5. K. Barmak, K.R. Coffey, D.A. Rudman, S. Foner, Mater. Res. Soc. Proc. **230**, 61 (1992).
6. R.W. Balluffi, and J.M. Blakely, Thin Solid Films **25**, 363 (1975).
7. L.G. Harrison, Trans. Faraday Soc. **57**, 1191 (1961).
8. D. Gupta, D.R. Campbell, and P.S. Ho, in <u>Thin Films - Interdiffusion and Reactions</u>, eds. J.M. Poate, K.N. Tu, and J.W. Mayer, Wiley Publishers, New York, 161 (1978).
9   J.C.M. Hwang and R.W. Balluffi, J. Appl. Phys. **50**, 1339 (1979).
10. H.B. Callen, <u>Thermodynamics and Thermostatistics</u>, John Wiley & Sons, 307 (1985).
11. J.S. Langer and R.F. Sekerka, Acta. Met. **23**, 1225 (1975).
12. U. Gösele and K.N. Tu, J. Appl. Phys. **53**, 3252 (1982).
13. R.W. Bené, J. Appl. Phys. **61**, 1826 (1987).
14. H. Muria, E. Ma, and C.V. Thompson, J. Appl. Phys, **70**, 4287 (1991).
15. K.R. Coffey and K. Barmak, Acta Met . et Mater., accepted for publication.
16. K. Barmak and K.R. Coffey, Mater. Res. Soc. Proc. **311**, 51 (1993).
17. R.A. Sigsbee, G.M. Pound, Advan. Col. Interf. Sci. **1**, 335 (1967).
18. K. Barmak, K.R. Coffey, D.A. Rudman, S. Foner, J. Appl. Phys. **67**, 7313 (1990).

PHASE FORMATION AND MICROSTRUCTURAL DEVELOPMENT DURING SOLID-STATE REACTIONS IN TI-AL MULTILAYER FILMS

CARSTEN MICHAELSEN, STEFAN WÖHLERT AND RÜDIGER BORMANN
Institute for Materials Research, GKSS Research Center, D-21502 Geesthacht, Germany

ABSTRACT

The phase selection which is generally observed in the early stages of solid-state reactions was studied using Ti-Al multilayer films as a model system.

Although all Ti-Al intermetallic phases have similar driving forces of about 30 kJ/g-atom, $TiAl_3$ is the only phase which is formed as long as the reactants are not consumed. The critical thickness beyond which a second phase is formed is larger than 100 μm. We found that the formation of $TiAl_3$ takes place by nucleation and growth, demonstrating that the driving force available for first-phase formation is considerably reduced by a preceding formation of solid solutions. Furthermore, we observed that nucleation continues at later stages, indicating that non-equilibrium conditions are maintained which are possibly influenced by grain-boundary diffusion. However, the phase selection is determined by the different growth velocities, being much higher for $TiAl_3$ than for all other phases.

INTRODUCTION

The early stages of phase formation during interface reactions of two elements with negative heats of mixing is of general importance in a wide variety of fields. A common feature is the large free energy available for phase formation, implying that all possible product phases should be able to form without any nucleation barrier. In contrast, a phase selection is often observed in the early stages but barely understood, and the first phase formed can even be a metastable phase. Furthermore, the occurence of nucleation barriers during first-phase formation has been found in a number of systems, such as Nb-Al[1] and Ni-Al[2], in spite of calculated nucleation barriers less than kT. Herein, the nucleation stage was observed to be kinetically separated from the following growth stage, giving rise to two distinct reactions although only a single product phase is formed. It has therefore been suggested[2,3] that interdiffusion precedes nucleation at an interface between two pure elements, leading to a reduction of the driving force available for product-phase formation. The phase selection can then be determined by differences in nucleation rates as well as growth velocities.

In this paper we report on our investigations of the phase formation in multilayered Ti-Al thin films. In the Ti-Al system various competing intermetallic phases exist which have comparable free energies of formation and small interfacial energies. Nevertheless, a pronounced phase selection in favor of $TiAl_3$ has been observed.[4]

Mat. Res. Soc. Symp. Proc. Vol. 343. ©1994 Materials Research Society

EXPERIMENTAL

The samples were deposited by triode-magnetron sputtering at an argon pressure of 3.7 x 10^{-3} mbar from elemental targets with rates of 0.5 - 1 nm/sec in an ultra-high vacuum chamber with a base pressure of about 3 x 10^{-9} mbar. Cu-K$_\alpha$ x-ray diffraction (XRD) measurements were performed using a Siemens D5000 diffractometer, equipped with a hot-stage chamber which allowed in-situ measurements at high vacuum. The differential scanning calorimetry (DSC) was carried out using a Perkin Elmer DSC-2. For the DSC experiments, the samples were deposited on 0.1 mm thin sapphire substrates which fit into the DSC pans. For the transmission electron microscopy (TEM) a Philips 400 T was used at a voltage of 120 kV.

RESULTS

The Ti-Al system contains the intermetallic phases Ti$_3$Al, TiAl, TiAl$_2$, Ti$_2$Al$_5$ and TiAl$_3$. All of them can structurally be viewed as superstructures of hcp Ti and fcc Al. In particular, the Ti$_3$Al phase represents a superstructure of the hcp Ti unit cell with a doubled lattice parameter in the basal plane, while TiAl, TiAl$_2$, Ti$_2$Al$_5$ and TiAl$_3$ are ordered superstructures based on the fcc unit cell of Al with a slight tetragonal distortion, the c-axis of these intermetallics being integral multiples of the 0.4 nm lattice constant of Al.

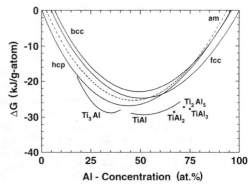

Fig.1 - Free energy curves of the Ti-Al system at 300°C, calculated by the CALPHAD method. Reference states are hcp Ti and fcc Al.

Fig. 1 shows the free energy functions of the system calculated by the CALPHAD method.[5,6] All intermetallic phases have similar free energies of formation which amount to about -30 kJ/g-atom. When not considering the ordered intermetallics, the phases with the lowest free energies are an hcp solid solution up to about 60 at.% Al, and an fcc solid solution for higher Al contents. At a composition of about Ti$_{40}$Al$_{60}$ the free energy of the amorphous phase is very close to those of the solid solutions.

Various Ti/Al multilayer films with different overall compositions and modulation wavelengths Λ were investigated by XRD, DSC and TEM. A typical x-ray pattern of an as-prepared multilayer film is shown in Fig. 2. Although the measurement is displayed with a square-root intensity scale which strongly enhances the visibility of diffraction peaks with small intensity, only two peaks were observed. These peaks correspond to the close-packed planes which have almost identical lattice spacings for Ti and Al. These results show that the sample exhibits a tremendous texture in growth direction, giving rise to a peak intensity and a rocking-curve width which is comparable to the single-crystal sapphire substrate. An evaluation of the peak width demonstrates that the grain size in growth direction is substantially larger than the layer thickness. Furthermore, multilayers with smaller Λ show satellites about the x-ray peaks typical for a (semi-) coherent superlattice, excluding the existence of an amorphous phase at the Ti/Al interfaces.[7] In contrast, plan-view TEM micrographs demonstrated that the in-plane grain size is about 50 nm, indicating a columnar microstructure.

Fig. 2 - X-ray pattern of a Ti/Al multilayer film with overall composition $Ti_{56}Al_{44}$ and a modulation period of 50 nm. Note the square-root intensity scale.

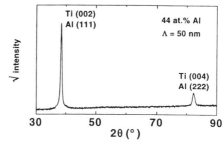

Fig. 3 - X-ray patterns of a Ti/Al multilayer with overall composition $Ti_{56}Al_{44}$ and a modulation period of 50 nm, taken in-situ during heating with a rate of 40 K/min in the range $37° \leq 2\theta \leq 40°$. Note the peak shift which is due to thermal lattice expansion. All measurements are displayed with the same intensity scale.

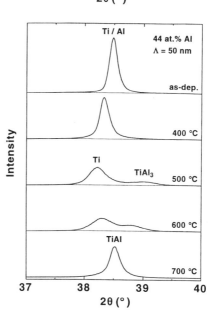

X-ray diffraction showed that the strong texture is maintained during solid-state reactions, Fig 3. A significant intensity was observed only at diffraction angles corresponding to the close-packed planes of the reactants as well as the product phases. This indicates that the orientation of the reactants is inherited to the product phase, which grows epitaxially, and maintains the coherency. As no other peaks were observed, a complete structure determination of the phases formed was impossible with our equipment. However, the diffraction intensity at angles corresponding to the close-packed planes which remained high during annealing enabled x-ray experiments to be performed with the same heating rates as used for the DSC scans, allowing a direct comparison of the results obtained by both methods. As the lattice spacings of the close-packed planes of the intermetallic phases in the Ti-Al system vary systematically with composition, an identification of the product phase composition was possible.

At a temperature of 400 °C, the intensity of the x-ray peak from the parent phases decreased and a small shoulder at higher angles developed. With further annealing this shoulder evolved in a peak which could be identified as $TiAl_3$. However, the TEM investigations showed that the $TiAl_3$ phase formed in the early stages did not exhibit the DO_{22} equilibrium structure but a metastable $L1_2$ modification. Finally, at 700 °C the x-ray peak position agrees with the γ-TiAl phase, indicating that the multilayer has been fully reacted to equilibrium.

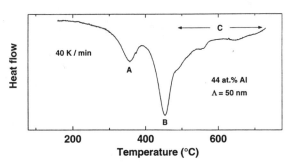

Fig. 4 - DSC trace of a Ti/Al multilayer film with overall composition $Ti_{56}Al_{44}$ and modulation period of 50 nm, taken at a heating rate of 40 K/min.

A typical DSC curve which was obtained upon annealing of a multilayered film is displayed in Fig. 4. This DSC curve exhibits two exothermal peaks A and B at about 350 °C and 450 °C, and a subsequent broad heat release C at higher temperatures. DSC traces taken from samples with the same periodicity Λ but different overall compositions showed the same two-stage reactions A and B up to about 450 °C for all specimens investigated, but the subsequent higher temperature heat flow C decreased with increasing Al content. The composition-independence of the reactions up to 450 °C indicates that the first two peaks A and B can be attributed to the phase formed before one of the reactants is consumed, while the subsequent heat release corresponds to the reaction into the equilibrium state. As the heat flow after the first two DSC peaks approaches zero for an overall composition of about 75 at.% Al, it can be concluded that the first phase formed must be $TiAl_3$, in agreement with the x-ray measurements.

Similar two-stage reactions which lead to the formation of a single product phase were also observed in a number of other systems, such as Nb-Al[1] and Ni-Al[2]. It has been shown that the first stage A can be interpreted as a nucleation from a fixed number of preferred sites and growth to coalescence until a contiguous layer is formed, while the second stage B can be attributed to the one-dimensional growth of this product phase layer. Following this model the area ratio of both DSC peaks should depend systematically on the modulation period. In fact, we observed that the second peak B decreases with decreasing Λ and is absent for $\Lambda = 30$ nm in the case of a 44 at.% Al sample. This result can be used to estimate the thickness of the $TiAl_3$ layer when coalescence occurs, leading to a value of about 10 nm. Furthermore we found that the $\Lambda = 30$ nm sample when heated beyond the first DSC peak (400 °C) produced the same x-ray pattern as the $\Lambda = 50$ nm sample after the second DSC peak (500 °C), confirming that both DSC peaks result from the formation of $TiAl_3$.

A TEM micrograph taken from a 1 µm Ti / 1 µm Al diffusion couple after annealing for 20 h at 400 °C is displayed in Fig. 5. It shows the columnar microstructure of the Ti layer, and larger Al grains with a diameter of about 1 µm. The product phase consists of many small grains, which size amounts typically to about 10 nm at the Ti interface and increases towards the Al interface, indicating a coarsening of the $TiAl_3$ grains after nucleation. The interfaces are not perfectly flat but have a roughness which corresponds to the product phase grain size. Apparently, the phase formation takes place via repeated nucleation at the Ti / $TiAl_3$ interface.

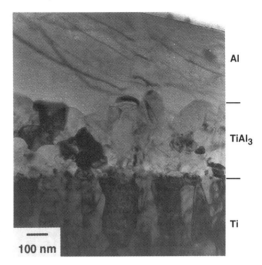

Al

—

TiAl₃

—

Ti

100 nm

Fig.5 - TEM micrograph of a 1 µm Ti / 1 µm Al diffusion couple after annealing for 20 h at 400 °C.

DISCUSSION

The observation of heterogeneous nucleation of even the first reaction product is quite surprising, as simple calculations using classical nucleation theory lead to nucleation barriers below kT using a free energy of formation of about 30 kJ/g-atom and an interfacial energy of 0.1 J/m^2, a value which appears reasonable for (semi-) coherent interfaces. Spontaneous nucleation of all possible product phases should therefore be easy, which should instantaneously result in a layered phase formation. However, as it has been pointed out in the Ni-Al system,[2] interdiffusion must precede phase formation, which leads to supersaturated solid solutions and to a considerable reduction of the driving force. In the case of Ti-Al, the formation of supersaturated hcp solid solutions in equilibrium with pure Al results in a driving force for the nucleation of TiAl$_3$ reduced to about 5 kJ/g-atom (Fig.1). For the reaction temperatures used, this corresponds to an energy barrier for nucleation of about 10 kT, indicating the existence of a nucleation barrier and favoring heterogeneous nucleation at microstructural defects at the Ti/Al interface.

The presence of nucleation barriers can lead to a distinct phase selection in the early stages of phase formation in case where the interfacial free energies σ differ for the competing phases. Although our measurements indicate that the interfaces involved are of low energy for all phases, we note that the nucleation barrier is proportional to σ^3 and furthermore enters exponentially into the nucleation rate. Therefore small differences in σ can result in considerable differences of the nucleation rate and can give rise to a dominant phase selection by nucleation. In this respect we note that the TiAl$_3$ phase which we observed in the early stages of the reaction exhibited an L1$_2$ structure with a lattice constant of about 0.4 nm. This L1$_2$ modification of TiAl$_3$ can be viewed as a (partially) ordered form of the fcc solid solution and should have a very low interfacial free energy to Al thus favoring its nucleation.

In addition, the following diffusion-controlled growth leads to a pronounced preference of TiAl$_3$ as well, as the growth constant of TiAl$_3$ is at least three orders of magnitude higher than that of all other intermetallic phases.[4,8] Annealing experiments performed on bulk diffusion couples showed that the critical thickness beyond which a second phase in addition to TiAl$_3$ occurs is larger than 100 μm, in agreement with calculations based on a pure competitive-growth model. Furthermore, dissociation experiments[9] demonstrated that phases other than $TiAl_3$ are consumed upon annealing when layered between Ti and Al.

The most interesting microstructural feature is that nucleation obviously continues during later stages of the reaction. This demonstrates that the initial driving force for nucleation is still present even after substantial growth of $TiAl_3$. In other words, a metastable equilibrium at the interface between Ti and $TiAl_3$ is not reached, but non-equilibrium conditions are maintained at the Ti interface, similar to the very early stages. This suggests that the Al supply is governed by a fast grain boundary diffusion through $TiAl_3$, and the microstructure of $TiAl_3$ becomes important for the reaction.

CONCLUSIONS

We found that during solid-state reaction between Ti and Al the phase formation takes place in two steps, the first being the nucleation from a fixed number of sites and growth to coalescence of $TiAl_3$, the second being the thickening of the same product phase. The observation of nucleation barriers in the early stages indicates that the driving force must be considerably smaller than expected from the free energy of formation of the intermetallic compounds. We conclude that highly supersaturated solid solutions must have been formed before the nucleation of a product phase takes place. Nucleation continues during later stages of the reaction, indicating that non-equilibrium conditions are maintained at the moving interface. However, the phase selection in favor of $TiAl_3$ is determined by the large differences of the growth velocities between the intermetallic phases.

ACKNOWLEDGEMENTS

We thank Katy Barmak, Michael Dahms and Wharton Sinkler for valuable discussions. The financial support of the Deutsche Forschungsgemeinschaft (Leibniz-Programm) is gratefully acknowledged.

REFERENCES

1. K. R. Coffey, K. Barmak, D. A. Rudman and S. Foner, Mat. Res. Soc. Symp. Proc. **230**, 55 (1992)
2. E. Ma, C. V. Thompson and L. A. Clevenger, J. Appl. Phys. **69**, 2211 (1991)
3. C. V. Thompson, J. Mater. Res. **7**, 367 (1992)
4. E. G. Colgan, Mater. Sci. Rep. **5**, 1 (1990)
5. H. L. Lukas and S. G. Fries, J. Phase Equilibria **13**, 532 (1992)
6. M. Oehring, T. Klassen and R. Bormann, J. Mater. Res. **8**, 2819 (1993)
7. B. M. Clemens and J. G. Gay, Phys. Rev. **B 35**, 9337 (1987)
8. K. Ouchi, Y. Iijima and K. Hirano, in Titanium '80 Science and Technology, ed. H. Kimura and O. Izumi, AIME New York 1980, p. 559
9. S. Wöhlert, Ph.D. thesis, University of Hamburg 1994

INTERFACE REACTION KINETICS
FOR PERMALLOY-TANTALUM THIN FILM COUPLES

A.J. KELLOCK*, J.E.E. BAGLIN*, K.R. COFFEY**, J.K. HOWARD**, M.A. PARKER** and D.L. NEIMAN*
* IBM Almaden Research Center, 650 Harry Rd., San Jose, CA 95120
**IBM Storage Systems Division, 5600 Cottle Rd, San Jose, CA 95193

ABSTRACT

Thermal interdiffusion mechanisms and kinetics have been studied for Ta-Permalloy ($Ni_{80}Fe_{20}$), Ta-Ni and Ta-Fe thin film couples, in the temperature range 300°C - 600°C. Interaction modes identified for the Ta-NiFe system include: fast diffusion of Ta into grain boundaries of NiFe, and nucleation and growth of Ni_3Ta, with consequent depletion of Ni in the remaining NiFe, and eventual segregation of Fe.

INTRODUCTION

Sensors for magnetic recording heads frequently employ thin films of tantalum in contact with permalloy ($Ni_{80}Fe_{20}$), where the Ta layer has been regarded as a diffusion barrier. However, when very thin metal layers are used, the performance of the magnetic layer may nevertheless be compromised by thermal interdiffusion during device processing.[1] A few Å thermal roughening of the interface has been reported[2] as a possible damage mechanism.

Comprehensive kinetic measurements have not been reported previously for this system. The binary phase diagrams[3] for Ni-Ta and Fe-Ta show several ordered alloy phases, but very low solid solubility for Ta in Ni or Fe.

This paper presents a preliminary report of new observations of the interfacial reactions between Ta and NiFe. So that the complex ternary case can be more readily understood, the binary systems of Ta-Ni and Ta-Fe have also been studied.

Three interaction processes are seen to dominate: (a) fast grain boundary diffusion of Ta through NiFe (or Ni or Fe); (b) amorphous or microcrystalline alloy phase formation; and (c) formation of ordered polycrystalline intermetallic phases at high temperature, with Fe segregation also observed in the Ta-NiFe case.

Evidently, these mechanisms are interdependent and concurrent. This makes a quantitative interpretation of the diffusion kinetics very complex, especially since different processes dominate at successive stages of heating, and grain growth[4] occurs in the host films simultaneously. Detailed analysis of the time- and temperature-dependences, and the implied activation energies, will thus be reserved for a future publication. Here, we present examples to illustrate the mechanisms at work, selecting annealing conditions where a simple process appears to dominate.

EXPERIMENT

Samples consisting of bilayers of (NiFe or Ni or Fe)(1000Å)/Ta(1000Å)/SiO$_2$/Si were sputter deposited on room temperature substrates, and annealed at relatively

Mat. Res. Soc. Symp. Proc. Vol. 343. ©1994 Materials Research Society

high temperatures and long times, with the objective of enabling subsequent prediction of low temperature kinetics. A second set of samples with the Ta layer on top and Ni or NiFe below was also run, to check for possible substrate- or surface- dependent effects. For brevity, this paper will report primarily on the series where the Ta layer was beneath the Ni or NiFe layer. The film thicknesses were chosen to ensure that material diffusing towards the surface or interface of each layer would be observable unambiguously using 2.3 MeV Rutherford Backscattering Spectrometry (RBS) for depth profiling. The amount of material diffused into either layer could then be determined with a sensitivity of $\sim 10^{13}$ atom/cm^2.

X-ray diffraction (XRD) was used to examine the microcrystalline structure of the films before and after annealing, to recognize ordered phase formation, and to estimate grain sizes in the films. Since RBS cannot sufficiently distinguish between Ni and Fe, Auger sputter profiling was used to monitor elemental diffusion in the Ta-NiFe system. Cross-sectional TEM of Ta-NiFe samples served to image the complex formation of layers and to identify their structure.

The samples were annealed in a tube furnace in purified flowing helium at temperatures from 300°C to 600°C, for periods ranging from 6 min to 128 hr. Particular attention was paid to ensuring an oxygen-free ambient by purifying the helium in Ti gettering furnaces en route to the sample tube.

RESULTS

Grain Boundary Transport of Ta in Ni and NiFe

For low temperatures or short times, the first observable effect was rapid diffusion of Ta through either Ni or NiFe to the free surface. Fig. 1(a) illustrates the typical RBS profile for samples in which the grain boundaries of Ni have, even after only the initial anneal, been uniformly populated with tantalum en route to the free surface, where Ta spreads laterally, as indicated by the growth of the RBS "surface" peak. Note that the Ta present between the surface and the Ta/Ni interface has a flat

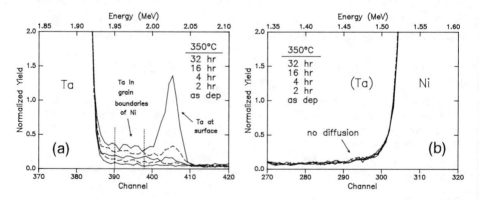

Fig.1. RBS spectra for the system Ni(1000Å)/Ta(1000Å)/substrate, showing interaction at 350°C. (a) Ta profile, characteristic of Ta diffusion through grain boundaries of Ni towards the free surface; (b) Ni profile, showing no equivalent movement of Ni into Ta.

profile in all the cases shown. This lack of appreciable concentration gradient within the Ni layer implies that the transport of Ta through the grain boundaries normal to the interface is much faster than any lateral diffusion of Ta within the Ni. RBS is very sensitive for high-Z elements, and the $\sim 1.5 \times 10^{14}$ atom/cm^2 Ta measured after 2 hr at 350°C represents approx. 1% of the available grain boundary volume (assuming a 5Å grain boundary width and a 200Å grain size (as deduced from XRD)). The movement of Ta through the grain boundaries is very rapid, and is difficult to observe with profiling techniques other than RBS in these early stages. An anneal of only 2 hr at 300°C already delivered Ta to the surface of the 1000Å Ni film. Surprisingly, Ni did not display a similar mobility within the Ta, as Fig. 1(b) illustrates. Shown on the same scale as the Ta profile of Fig. 1(a), the Ni profile showed no detectable change for the same thermal treatments. Similar behavior in every respect was observed for the Ta-NiFe system, although the rates of diffusion were slower. For example, to penetrate through a 1000Å NiFe film required 4 hr at 300°C.

Grain Boundary Limited Transport of Ta in Fe

The transport of Ta through Fe showed somewhat different characteristics from those observed for Ta in Ni or NiFe. There was no immediate diffusion to the surface, as indicated in Fig. 2(a). The diffusion of Ta seems to have progressed at comparable

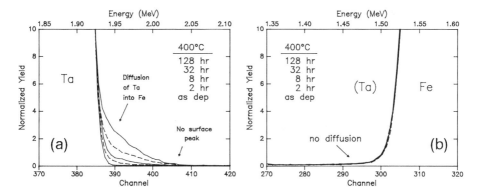

Fig.2. RBS spectra for the system Fe(1000Å)/Ta(1000Å)/substrate, showing interaction at 400°C. (a) Ta profile consistent with competition between lateral and normal diffusion of Ta into the Fe layer; (b) Fe profile, showing no equivalent movement of Fe into Ta.

rates, both laterally from the grain boundaries and normal to the film, towards the surface, indicating a type-B process with grain boundary limited kinetics. We observed no surface Ta peak at this temperature, unlike the cases of Ta-Ni and Ta-NiFe. Only when the Ta-Fe samples were heated above 450°C for 16 hr did Ta appear at the surface. The longest anneal of 128 hr at 400°C did not produce any evidence of compound phase formation between Ta and Fe (in contrast to the cases of Ni and NiFe, as we shall see). The only similarities between the Ta-Fe interaction and those of Ta-Ni and Ta-NiFe were the fast population of grain boundaries by Ta, and no observable movement of Fe or Ni into the Ta layer. The latter is evident from Fig. 2(b) which shows the RBS profile at the same scale as that of Fig. 2(a).

Formation of Compound Phases

When the Ta-Ni and Ta-NiFe systems were heated at high temperatures (> 400°C), new compound phases could be nucleated near the interface.

If the interaction between Ni and Ta was allowed to continue long enough, a compound phase formed, as indicated by the plateau in the spectra of Fig. 3. RBS spectrum simulation identified this plateau with the composition Ni_3Ta. XRD confirmed that this new phase was indeed polycrystalline Ni_3Ta. The same phase was observed to grow in the Ta-NiFe system. No such evidence for compound phase formation was found for the Ta-Fe system after heating at temperatures up to 500°C.

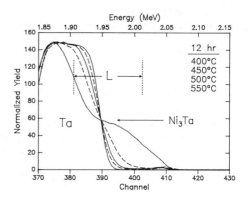

Fig.3. RBS profiles of Ta illustrating interaction with Ni after 12 hr at a variety of temperatures. Formation of a Ni_3Ta layer of width L ($\simeq$1000Å) is indicated for 12 hr, 550°C.

The initial stages of interaction observed in Fig. 3 (before the RBS plateaus form) may be indicative of amorphous or microcrystalline alloy formation.[5,6] TEM elongated probe micro-diffraction (EPMD) patterns[7] from our Ta-NiFe samples treated at high temperature showed background rings implying the presence of amorphous material.

In a similar fashion to the grain boundary transport of Ta, the rate of phase formation in Ta-Ni was faster than that in Ta-NiFe. If we define an interface layer width L as indicated in Fig. 3, then L = 400Å for Ta-Ni after 12 hr at 450°C, while L = 100Å for Ta-NiFe after the same anneal. We note also in Fig. 3 that the low temperature curves show the early stages of diffusion, i.e. grain boundary transport and surface peak formation, as well as interface phase formation. However, the RBS sequences displayed in this figure do not imply progressive phase growth from the Ni grain boundaries; instead, the nucleation and growth of a complete new compound phase layer at the interface is indicated by the developing plateaus. In addition, the slopes on the RBS profiles imply concomitant roughening of the interface region, consistent with nucleation and grain growth of the new alloy layer. Such roughening was also seen in the TEM images.

Fe Segregation in the Ta-NiFe System

RBS spectra from the Ta-NiFe system indicate that at $\geq$ 450°C, a new compound phase forms, and XRD has confirmed the presence of Ni_3Ta, (but no ordered Fe-Ta phase). In fact, the RBS spectra resemble those of Fig. 3 for the binary Ta-Ni case; however, since RBS does not produce separated profiles for Ni and Fe, we have used Auger depth profiling to address the question: What happens to the Fe in permalloy when the Ni_3Ta phase forms? Fig. 4. shows the analyses for three cases - (a) as deposited, (b) 36 hr at 500°C, and (c) 36 hr at 600°C. In case (b), Ni_3Ta was seen to form, excluding Fe in the process. The Fe concentration in the unreacted NiFe increases as the Ni migrates to form the new compound layer. We note also that some Fe has appeared at the Ta/Ni$_3$Ta interface. At the higher temperature, case (c), the

Fe has accumulated more strongly near this interface, and less remains near the original NiFe/substrate interface. The Ni$_3$Ta layer has become thicker, again excluding Fe from this region. The detailed mechanism for this process is obviously very complicated, and remains open to speculation.

In an attempt to elucidate this process, XTEM was performed on the sample that was profiled in Fig. 4(c), and the result is displayed in Fig. 5. The bright field image of Fig. 5(a) identifies four layers. We note the polycrystalline grain structure of the Ni$_3$Ta layer, and the amorphous or microcrystalline Fe-rich second layer. The EPMD patterns for the top two layers show the spot pattern from the polycrystalline Ta, which for the second layer is superimposed on the ring attributed to amorphous material. The grain structure is seen best in the dark field image of Fig. 5(c), and the EPMD pattern of this layer shows it to contain many defects, including twins. The micrograph of Fig. 5(b) is underfocussed to enhance a line of structures believed to be Kirkendall voids at the lower interface of the Ni$_3$Ta layer. These voids support the proposition that the Fe moves to the Ta/Ni$_3$Ta interface after most of the Ni has been consumed as Ni$_3$Ta. Finally, the last layer has a mixed composition, partially crystalline, including Ni, Fe and Ta.

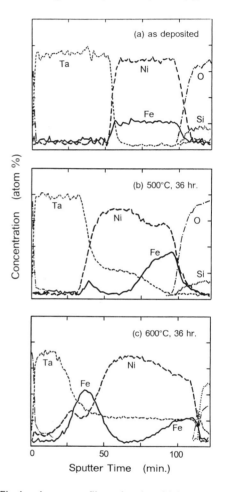

Fig.4. Auger profiles showing high temperature interaction for the system Ta(1000Å)/Ni$_{80}$Fe$_{20}$(1000Å)/substrate. (a) as deposited; (b) after 36 hr at 500°C; (c) after 36 hr at 600°C.

CONCLUSIONS

For the Ta-NiFe and Ta-Ni systems at 450 - 600°C, growth of a polycrystalline Ni$_3$Ta interfacial layer is observed. However, at ~ 300°C, this process would apparently yield only minimal layer growth, of the order of monolayers at the interface.

This is consistent with the interface roughening observed by x-ray reflectometry (7.4Å after 6.5 hr at 240°C) for Ta/NiFe by Huang et al.[2]

Perhaps of greater technological concern is the evidence of fast diffusion of Ta to populate the grain boundaries in permalloy films, which can be observed directly by RBS at temperatures as low as 300°C.

The Ta-NiFe interface clearly undergoes very dynamic change at moderate temperatures, involving many concurrent processes. This makes meaningful quantitative estimates of diffusivities or activation energies difficult to obtain. Such quantitative analysis will be the subject of a future publication.

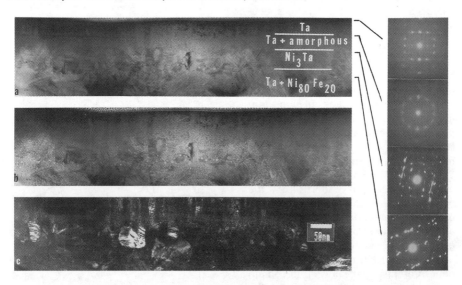

Fig.5. Cross-sectional TEM images for Ta/NiFe/substrate after anneal for 36 hr at 600°C. (see Fig. 4(c)). (a) bright field, showing 4 layers; (b) under-focussed to enhance the line of small defects (white dots) considered to be Kirkendall voids; (c) dark field, showing grain structure, especially in the Ni_3Ta layer. EPMD patterns for the individual layers are also shown.

References

1. I. Hashim, H.A. Atwater, K.T.Y. Kung and R.M. Valetta, J. Appl. Phys. **74**, 458 (1993).

2. T.C. Huang, J.-P. Nozieres, V.S. Speriosu, B.A. Gurney and H. Lefakis, Appl. Phys. Letters **62**, 1478 (1993).

3. Binary Alloy Phase Diagrams, 2nd ed. (ASM International, Materials Park, Ohio, 1991).

4. A. Gangulee, S. Krongelb and G. Das, Thin Solid Films **24**, 273 (1974).

5. M.A. Hollanders, C.G. Duterloo, B.J. Thijsse and E.J. Mittemeijer, J. Mater. Res. **6**, 1862 (1991).

6. W.L. Johnson, Prog. Mat. Sci. **30**, 81 (1986).

7. M.A. Parker, K. Johnson, C. Hwang and A. Bermea, Proc. Elec. Micros. Soc. Amer. **49**, 762 (1991).

GRAIN BOUNDARY DIFFUSION IN NiFe/Ag BILAYER THIN FILMS

M. GALL, J.G. PELLERIN, P.S. HO*, AND K.R. COFFEY AND J.K. HOWARD**
*Materials Laboratory for Interconnect and Packaging, University of Texas at Austin, BRC/MER Mail Code 78650, Austin, TX 78712-1100
**IBM Storage Systems Division, 5600 Cottle Rd, 808/282, San Jose, CA 95193

ABSTRACT

X-ray photoelectron spectroscopy (XPS) has been used to investigate grain boundary diffusion of Ag through 250 Å thick $Ni_{80}Fe_{20}$ (permalloy) films in the temperature range of 375 to 475 °C. Grain boundary diffusivities were determined by modeling the accumulation of Ag on $Ni_{80}Fe_{20}$ surfaces as a function of time at fixed annealing temperature. The grain boundary diffusivity of Ag through $Ni_{80}Fe_{20}$ is characterized by a diffusion coefficient prefactor, $D_{0,gb}$, of 0.9 cm^2/sec and an activation energy, $E_{a,gb}$, of 2.2 eV. The $Ni_{80}Fe_{20}$ film microstructure has been investigated before and after annealing by atomic force microscopy and x-ray diffraction. The microstructure of $Ni_{80}Fe_{20}$ deposited on Ag underlayers remained relatively unchanged upon annealing.

INTRODUCTION

Magnetoresistive head development for magnetic recording has concentrated on enhancing giant magnetoresistance (GMR). Recently, investigators have reported significant GMR by annealing sputter-deposited NiFe/Ag multilayer structures at temperatures between 300 and 400°C.[1] The range of layer thicknesses examined was 15 to 25 Å for NiFe and 10 to 40 Å for Ag. Whereas the as-deposited samples showed no evidence of GMR, a large, negative magnetoresistance was observed upon annealing. Samples exhibited 4 to 6% GMR at room temperature in fields of 5 to 10 Oe and therefore satisfy two important criteria in the design of magnetoresistive sensors for magnetic recording heads: a magnetoresistance greater than 2% and a sensitivity greater than 0.5% per Oe. The authors attributed the onset of GMR to discontinuities in the NiFe layers as a result of Ag grain boundary diffusion during annealing. However, direct evidence for grain boundary diffusion was not presented.

To investigate the nature of grain boundary diffusion in magnetic thin films, we have measured and modeled the surface accumulation of Ag through 250 Å thick NiFe films with x-ray photoelectron spectroscopy (XPS). By measuring the accumulation of diffusant on the NiFe surface as a function of time at a fixed temperature, the grain boundary diffusivity has been determined using a model developed by Hwang and Balluffi.[2] In addition, the effects of annealing on the NiFe thin film microstructure have been investigated.

METHODS

Experimental

The 250 Å NiFe / 100 Å Ag bilayer thin film structures were prepared by S-gun magnetron sputtering in a mixture of 4% H_2 and 96% Ar at a pressure of 3 mTorr with substrates at ambient temperature. Substrates were 3-inch Si (100) wafers with a 700 Å thick thermally grown oxide. Samples 16 x 16 mm^2 in size were prepared from these wafers. Further preparation of the as-received NiFe/Ag samples prior to permeation experiments consisted of Ar^+ sputtering to remove native oxides and carbides formed by exposure to air. Sputter cleaning was done in a sample introduction/preparation chamber at an Ar pressure of 5 x 10^{-5} Torr, with 30 mA ion gun emission current and 500 V Ar^+ beam energy. Prior to sputtering, the samples were degassed for 30 min at 150°C.

Permeation experiments were performed in a UHV system described elsewhere.[3] To increase surface sensitivity, the permeation experiments were performed with the sample oriented so that the angle between the surface normal direction and the analyzer was ~80°. Samples were annealed to temperatures of 475°C on a heating module that is mounted to the sample manipulator within the XPS analysis chamber.

Permeation experiments were performed by measuring the area of the $2p_{3/2}$ core level peak of Ni at -852.7 eV binding energy and the $3d_{5/2}$ core level peak of Ag at -386.0 eV binding energy. Also, the C 1s core level at -284.2 eV binding energy was monitored throughout the duration of each experiment. An inelastic background subtraction routine was used to calculate the core level peak areas during the data acquisition process.[4] Each of the core level peak areas was normalized to the acquisition time to quantitatively monitor the accumulation profile.

Analysis of Grain Boundary Diffusion

The accumulation of diffusant on a surface as a result of grain boundary diffusion has been modeled by Hwang and Balluffi.[2] Their model was developed within the conditions of type C kinetics where the barrier to bulk diffusion is high and mass transport within the thin film occurs only along grain boundaries.[5] As a coarse approximation, type C kinetics apply at temperatures less than ~35% of the film's melting point. In this study, annealing temperatures were in the range of 38% to 44% of the melting point. Due to relatively low diffusivities, higher temperatures had to be used to keep experiments within a reasonable timeframe.

Hwang and Balluffi applied Fick's 2nd law to the grain boundaries and the film's exit, or free, surface where diffusant accumulates and equated the material fluxes at the intersections of the grain boundary with both the diffusant source and exit surfaces. Two assumptions allow direct calculation of the grain boundary diffusivity from the measurement of surface accumulation as a function of annealing time and constant temperature. First, the diffusion of material on the exit surface is assumed to be much faster than diffusion in the grain boundary, so the amount of material accumulated on the exit surface is solely a function of time. Second, the capacity of the grain boundary is assumed to be much less than that of the exit surface. Consequently, a steady state concentration profile in the grain boundary is established and maintained. Combining these assumptions with that of an infinite diffusant source, the following expression for the surface concentration of permeated diffusant material as a function of time was derived:

$$C_s(t)/C_s(\infty) = 1-\exp(-h(t-t_0))$$ (1)

where C_s is the average surface concentration at annealing time t, t_0 is the time before permeated material reaches the surface, and h is a factor proportional to the grain boundary diffusivity. The factor h consists of the following terms:

$$h = (w_b/w_s)(2/d\ell) D_b$$ (2)

where w_b is the grain boundary width, w_s is the thickness of the accumulated layer on the exit surface, d is the grain diameter, ℓ is the film thickness, and D_b is the grain boundary diffusivity. While these equations have been successfully applied to a wide variety of materials systems, there is little information available for NiFe thin films.[6,7]

RESULTS AND DISCUSSION

Permeation experiments for the 250 Å NiFe/100 Å Ag thin film structures were carried out at temperatures of 375, 400, 425, 450 and 475°C. Background subtracted peak areas for Ni $2p_{3/2}$ and Ag $3d_{5/2}$ emission are plotted in Fig. 1 as a function of time at 450°C. The Ag $3d_{5/2}$ emission rises as Ag emerges from the grain boundaries onto the free surface and gradually reaches a plateau, while the Ni signal attenuates as the accumulation of Ag on the surface saturates. Superimposed on the Ag surface accumulation profile is a fitted curve in the form of Eq. (1) from which h can be found.

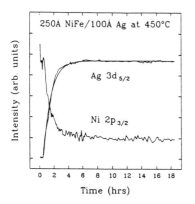
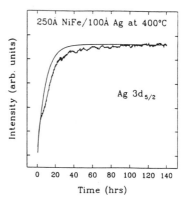

Figure 1. Background subtracted peak areas of Ni $2p_{3/2}$ and Ag $3d_{5/2}$ photoemission as a function of annealing time at 450°C.

Figure 2. Background subtracted peak area of Ag $3d_{5/2}$ photoemission as a function of annealing time at 400°C.

During each anneal, the amount of C on the NiFe surface was monitored. After correction for the different subshell photoionization cross-sections of each core level,[8] acquired data indicate relative amounts of about 50% NiFe and 50% C after sputtering. This corresponds to the presence of about one monolayer (ML) of C on the NiFe surface considering the escape depths of Ni $2p_{3/2}$, Fe $2p_{3/2}$ and C 1s core level electrons.[9] Even after extensive sputtering, the amount of carbon on NiFe surfaces could not be reduced below a typical value of about 1 ML. We attribute this fact to the presence of Fe, which most probably acts as a "getter" for C atoms adsorbed on the surface. This assumption can be supported by an observed shift of the metallic Fe $2p_{3/2}$ peak at -706.8 eV binding energy to -707.5 eV binding energy. Reported data for the Fe $2p_{3/2}$ peak in Fe_3C give a binding energy of about -708.1 eV.[10]

The background subtracted peak area for Ag $3d_{5/2}$ emission is plotted in Fig. 2 as a function of time at 400°C. We observed a deviation from the expected exponential increase in Ag $3d_{5/2}$ peak intensity. Also shown is a fit in the form of Eq. (1) where we fitted the initial exponential increase and the final saturation level to extract the parameter h. Deviations in this form were found at annealing temperatures of 400, 425 and 475°C. One of Hwang and Balluffi's assumptions is homogeneous surface accumulation of the diffusant. In this case a semilogarithmic plot of Ni $2p_{3/2}$ intensity vs. annealing time would show a straight line according to the attenuation formula:

$$I(d) / I(0) = \exp(-d / \lambda \sin\theta) \qquad (3)$$

where d is the overlayer thickness, λ is the inelastic mean free path of photoelectrons in the overlayer and θ is the photoelectron take off angle with respect to the sample surface. Figure 3 shows a semilogarithmic plot of the Ni $2p_{3/2}$ photoelectron signal as a function of annealing time at 450°C.

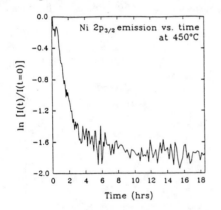

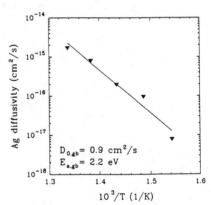

Figure 3. Semilogarithmic plot of Ni $2p_{3/2}$ photoemission as a function of annealing time at 450°C.

Figure 4. Calculated grain boundary diffusivities of Ag through 250 Å NiFe films.

Since the attenuation curve gradually flattens out with time, we can conclude that uniform surface accumulation throughout the duration of the experiment does not take place. From peak area calculations at the end of each anneal, we could conclude that the average amount of Ag on the sample surface at saturation is about one ML at higher annealing temperatures and less than one ML at lower temperatures. We believe that this premature saturation level is caused by surface energy effects. Although Ag, having a lower surface energy than Ni or Fe,[11] is expected to wet a free NiFe surface, the presence of 1 ML of C might influence the surface energy balance and therefore lead to premature saturation. This effect has been observed by Hwang[12] in the case of Ag accumulation on Au surfaces. Angle resolved XPS measurements and AFM surface roughness determinations support a model involving preferred Ag accumulation at the intersection of the exit surface with grain boundaries.

In order to determine grain boundary diffusivities using Eq. (2), we assumed that the initial exponential increase of the Ag $3d_{5/2}$ signal represents a uniform surface accumulation in its early stage. The ratio w_b/w_s was approximated to unity. The as-deposited film thickness of 250 Å was assumed to remain unchanged with annealing. Atomic force microscopy (AFM) and x-ray diffraction have been used to estimate the NiFe grain size on as-deposited and annealed NiFe/Ag bilayers. Prior to annealing, AFM revealed the as-deposited NiFe grain size to be approximately 250 Å in the lateral direction (parallel to the film surface). Analysis of diffraction peak widths gave a grain size of about 97 Å in the vertical direction (normal to the film surface). Upon annealing, the lateral grain size remained relatively unchanged while the vertical grain size increased to about 211 Å. For the calculation of grain boundary diffusivities, the lateral grain size has to be used and an approximate value of 250 Å was chosen. As expected, the NiFe layers showed a (111) fiber texture.

In Fig. 4, the calculated diffusivities are plotted versus 1/T for experiments conducted between 375 and 475°C. A straight line fit to the data yields a prefactor $D_{0,gb}$ of 0.9 cm^2/sec and an activation energy $E_{a,gb}$ of 2.2 eV. Compared to Ni grain boundary diffusion in Ni(75 wt%)Fe alloys, which occurs with an activation energy of about 1.2 eV,[7] the observed activation energy for Ag seems relatively high. Also, a low activation energy of about 0.92 eV for Ag grain boundary diffusion in pure Ni has been reported.[7] To our knowledge, data for Ag grain boundary in pure Fe is not available, but assuming an activation energy $E_{a,gb} \sim 1/2\ E_{a,bulk}$ and using reported values for $E_{a,bulk}$,[13] Ag grain boundary diffusion in pure Fe would occur with an activation energy of about 1.4 eV. Given the observed activation energy $E_{a,gb}$ of 2.2 eV for Ag grain boundary diffusion in $Ni_{80}Fe_{20}$ permalloy films, the effect of 20 wt% Fe in NiFe is significant.

According to Fisher's analysis of grain boundary diffusion,[14] the penetration depth x is given by:

$$x = A\ D_{gb}^{1/2}\ t^{1/4} \qquad (4)$$

where D_{gb} is the grain boundary diffusivity, t the penetration time and A a constant that can be found by measuring the time for the diffusant to reach the exit surface (incubation time). Eq. (4) was derived in the type B kinetics regime whereas our analysis assumes type C kinetics where lattice diffusion into the adjacent grains can be neglected. Therefore, we can give a lower limit for the penetration depth x using Eq. (4). The actual penetration depth will be higher since type C kinetics assumes no loss of diffusant into the adjacent grains. Hylton et al.[1] found the highest GMR sensitivity at annealing times of 10 min at 315°C for layered sandwich structures where the NiFe and Ag layer thicknesses were 20 Å and 40 Å, respectively. Using our results for A and D_{gb}, we obtain a penetration depth of about 18 Å, which is on the order of the NiFe layer thickness of 20 Å.

CONCLUSION

Permeation experiments made over the temperature range of 375 to 475°C indicate that grain boundary diffusion of Ag through NiFe does occur. The NiFe overlayer microstructure remains relatively unchanged with annealing. While the extraction of grain boundary diffusivities according to Hwang and Balluffi's model leads to reasonable results in the activation energy $E_{a,gb}$ and the prefactor $D_{0,gb}$, uncertainty remains regarding the applicability of this model. Given the data above, the penetration depth of Ag through NiFe under annealing conditions reported in Hylton et al.[1] can be estimated to be at least on the order of the NiFe layer thickness in the GMR multilayer structures. This supports the assumption that Ag grain boundary diffusion is the mechanism leading to the development of GMR in NiFe/Ag multilayer structures.

ACKNOWLEDGMENTS

This work was sponsored by the IBM SUR program under Agreement No. SJ92407.

REFERENCES

1.   T.L. Hylton, K.R. Coffey, M.A. Parker and J.K. Howard, Science **261**, 1021 (1993).
2.   J.C.M. Hwang and R.W. Balluffi, J. Appl. Phys. **50**, 1339 (1979).
3.   J.G. Holl-Pellerin, S.G.H. Anderson, P.S. Ho, K.R. Coffey, J.K. Howard, and K. Barmak, Mat. Res. Soc. Symp. Proc. Vol. **313**, 205 (1993).
4.   A. Proctor and P.M.A. Sherwood, Anal. Chem. **54**, 13 (1982).
5.   L.G. Harrison, Trans. Faraday Soc. **57**, 1191 (1961).
6.   D. Gupta, D.R. Campbell, and P.S. Ho in Thin Films - Interdiffusions and Reactions, edited by J.M. Poate, K.N. Tu, and J.W. Mayer (Wiley, New York, 1978), p. 161.
7.   I. Kaur and W. Gust, Fundamentals of Grain and Interphase Boundary Diffusion (Ziegler Press, Stuttgart, 1989).
8.   J.H. Scofield, J. Electron. Spectrosc. **8**, 129 (1976).
9.   M.P. Seah and W.A. Dench, Surf. Interface Anal. **1**, 2 (1979).
10. C.D. Wagner in Practical Surface Analysis (Second Edition), Vol 1: Auger and X-ray Photoelectron Spectroscopy, edited by D. Briggs and M.P. Seah (John Wiley & Sons, Ltd, 1990), p.607.
11. G.A. Samorjai, Chemistry in Two Dimensions: Surfaces (Cornell University Press, Ithaca, NY, 1981).
12. J. C.-M. Hwang, PhD Thesis, Cornell University, 1978
13. Landolt-Börnstein Numerical Data and Functional Relationships in Science and Technology, edited by O. Madelung and H. Mehrer (Springer-Verlag, Berlin, 1990), Group III, Vol. 26, p. 129.
14. J.C. Fisher, J. Appl. Phys. **22**, 74 (1951).

High resolution studies of the solid state amorphization reaction in the ZrCo system with the atom probe / field ion microscope

Susanne Schneider[1] *, Ralf Busch [1]** and Konrad Samwer ***

*   *I. Physikalisches Institut, Universität Göttingen, Bunsenstr. 9, D- 37073 Göttingen, Germany*
** *Institut für Metallphysik, Universität Göttingen, Hospitalstr. 3-7, D-37073 Göttingen, Germany*
*** *Institut für Physik, Universität Augsburg, Memminger Str. 6, D- 86135 Augsburg, Germany*

ABSTRACT

The atom probe/field ion microscope is introduced as a new powerful investigation device to study the early stages of the solid state amorphization reaction (SSAR). A bilayer of Zr and Co was condensed under UHV conditions on W wire tips and analyzed in a field ion microscope (FIM) combined with an atom probe (AP). The reaction of Co with Zr has been studied at room temperature. FIM pictures and AP analysis have shown that even at low temperatures an amorphous phase is formed at the Zr/Co interface and in the Zr grain boundaries. In these areas concentration profiles have been taken on a nanometer scale. Most likely, the extended solid solution of Co found in α- Zr grain boundaries causes the formation of the amorphous phase. Further, Rutherford backscattering spectrometry (RBS) suggests that even point defects and dislocations at the surface of an α- Zr single crystal are sufficient to initiate the SSAR between a polycrystalline Co layer vapour- deposited onto that single crystal.

INTRODUCTION

It has been shown that a number of metal/metal diffusion couples form an amorphous phase by SSAR [1]. The reaction is driven by a large negative heat of mixing and has mostly an asymmetry of the mobility of the reactants [2].

The diffusion controlled planar growth of the amorphous phase between the two elemental metals has been studied in the ZrCo system by electrical conductance measurements and cross-sectional transmission electron microscopy [3]. However, the very early stages of the SSAR deviate from a diffusion controlled type reaction and the amorphous phase formation mechanism must involve destabilization of the crystalline components. Due to the lack of appropriate investigation methods the very early stages of the SSAR are not well analyzed. It has been reported that the grain boundaries in the non- moving species are important for the formation of the amorphous phase [4].

This report attends to the formation mechanism of the amorphous phase in the ZrCo system. The investigations have been done by atom probe field ion microscopy (APFIM) on Zr/Co bilayers prepared on W tips at room temperature.

In addition, we report on diffusion studies done by RBS on a polycrystalline Co layer condensed on a α- Zr single crystal with different cleaning treatments of the crystal surface.

---

[1]present address: W.M. Keck Laboratory of Engineering, California Institute os Technology, Pasadena, CA 91125

Mat. Res. Soc. Symp. Proc. Vol. 343. ©1994 Materials Research Society

EXPERIMENTAL PROCEDURES

FIM tips are prepared by electropolishing thin W wires. The smoothness and shape of these substrate tips are proved by scanning electron microscope (SEM). The tips are cleaned with Ar ions under UHV conditions (base pressure $5 \cdot 10^{-11}$ mbar). A thin Cr layer of about 5nm is condensed onto the W wire to improve the adhesion of the Zr layer on the substrate. Then the Zr and Co layers are deposited onto the Cr layer at room temperature. To insure a uniform growth of the layers on the substrate the tips rotate during the evaporation. The growth rate is about 0.1nm/s. Two quartz balances are used to determinate the thickness of each layer.

After deposition the tips are transferred into the AP-FIM. The AP provides a chemical analysis with a high lateral resolution of about 2nm; it is used to measure concentration profiles at selected areas. Simultaneously, the surface structure of the layers is imaged by the FIM, providing a vertical atomic resolution. This combination provides a basis for distinction between crystalline and amorphous structure.

The (110) surface of a α- Zr single crystal was polished electrochemically to a surface roughness less than 0.3μm. Then the substrate was transferred to an UHV chamber and sputter cleaned by argon ions. After sputtering no carbon or oxygen contamination could be detected by in situ Auger electron spectroscopy (AES). A Co layer with a thickness of about 90nm was deposited onto the cleaned surface. The as- prepared sample was investigated by RBS with 0.9 MeV ^{4}He$^+$ ions. The channel and random techniques were performed to analyze the structure and the concentration profile near the Zr/Co interface. After annealing the crystal at 570K for 90min, RBS and X-ray analysis were applied again.

RESULTS

The FIM micrograph in fig.1 shows the transition from the amorphous ZrCo layer into the crystalline Zr layer. The picture is taken parallel to the growth direction of the metal layers. The sample was prepared at room temperature and has not been annealed. The size of the Zr grains can be estimated from this micrograph to be 8-12nm. This range is in good agreement with measurements on thin planar layers by X-ray and electron diffraction using the Scherrer formula [5]. The amorphous interlayer shows an irregular distribution of atoms well known from FIM micrographs of metallic glasses. The dark spot in the center is the probe hole of the AP.

The ladder diagram in fig.2 shows the AP analysis from this sample taken simultaneously. The concentration profile is shown from the top Co layer through the amorphous ZrCo layer into the Zr layer. The local Zr concentration is given by the slope of the curve. The depth scale is about 100 ions per nm [6]. The small amount of Zr found in the Co layer is within the range of the detection limit of the AP for this analysis. After 20nm depth, the Zr concentration c_{Zr} increases to 26(4)at.%. Another increase of c_{Zr} takes place 1.5nm deeper. Then c_{Zr} increases continuously to 75(7)at.% and finally steps to 100at.% Zr.

The FIM micrograph in fig.3 shows a vertical section of the layers after field evaporation. The Zr and the Co layer are shown beside the W substrate. The ladder diagram of an AP analysis in this Zr layer is plotted in fig.4. It shows an inhomogenous distribution of Co in the Zr layer. There are three different types of regions found in the Zr layer: type I are the pure Zr grains, type II are regions where, the Co content is about 5at.% due to an extended solid Co

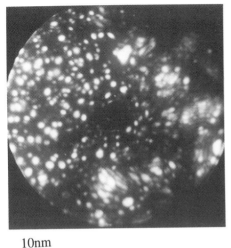

10nm

Fig.1: FIM micrograph from the transition of the amorphous interlayer into the poly-crystalline Zr layer

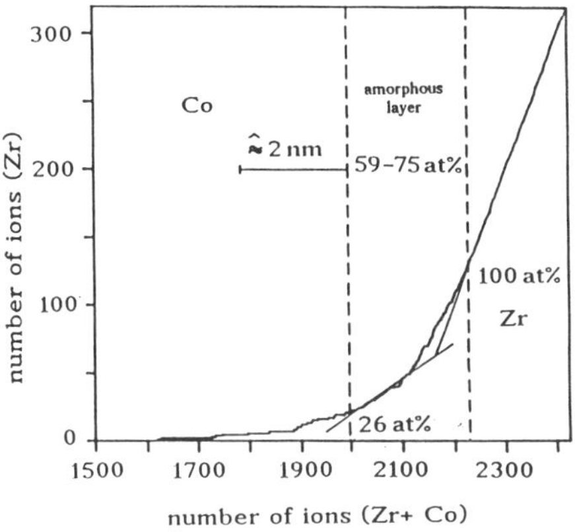

Fig. 2: Ladder diagram from the analysis of a Zr/Co bilayer. The depth analysis is made parallel to the growth direction.

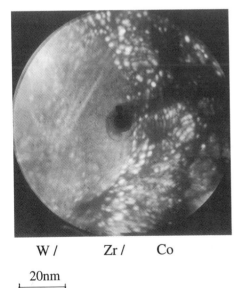

W /   Zr /   Co

20nm

Fig. 3: Vertical section of the layers after field evaporation

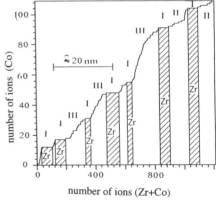

Fig.4: Ladder diagram from the analysis of a Zr layer perpendicular to the growth direction far away from the Co/Zr inter-face. TypeI: Zr grains; type II: Zr with about 5at.% Co; type III: Zr with about 40at.%

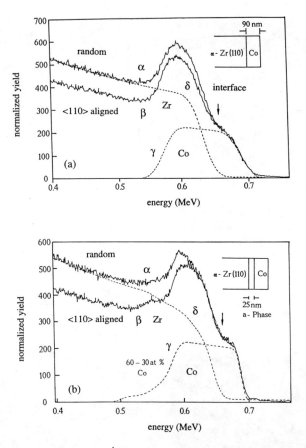

Fig. 5: Energy spectra for 0.9 MeV $^4He^+$ ions backscattered from the Co layer and a α- Zr single crystal as evaporated (a) and annealed at 570K for 90min (b). (α) Backscattering spectrum by random mode. (β) Backscattering spectrum by channeling mode. (γ,δ) Contribution of the Co and Zr layer to the random energy spectrum.

in Zr near the Zr grain boundaries, and finally type III are the amorphous regions with about 40at.% Co. The average Co concentration found in the Zr layer far away from the Zr/Co interface is about 8.8at.%.

Figure 5 shows the Rutherford backscattering spectra of a polycrystalline Co layer deposited onto an (110) α- Zr single crystal. The spectra are taken in random (α) and channeling geometry (β) from an as-prepared sample (fig.5a) and after annealing at 570K for 90min (fig.5b). The separated Zr and Co signals (δ, γ) are fitted to the random spectra (α). The arrows label the energy of the backscattered $^4He^+$ ions detected from the Zr/Co interface. In the channeling spectrum taken from the as- prepared sample the total yield from the aligned spectrum near the Zr/Co interface is reduced with respect to the random spectrum. However, the spectra for the annealed samples do not show any difference at the interface region. This means that the initial single crystalline structure of the α- Zr next to the interface has been disturbed. A long tail at the back edge of the calculated Co spectrum near an energy of 0.55MeV indicates a concentration range from 60-30 at% Co far behind the original interface. The change in the back of the Co signal in (5γ) indicates further an asymmetrical diffusion of into the Zr single crystal after annealing. The thickness of the disturbed region is about 25nm. Further, a small amount of less [2]than 5at.% Co diffuses into a depth of about 60nm. In additon thin film X-ray diffraction measurements do not show any Bragg peaks from crystalline phases. Those features give evidence for a new phase that has formed at the interface.

In agreement with measurements on polycrystalline Zr layers [3], we conclude that an amorphous ZrCo layer has been formed at the Zr/Co interface during the heat treatment.

DISCUSSION

The driving forces for the SSAR are the large negative heat of mixing and the decrease of interface energy by replacing the incoherent Co/Zr interface. It is evident from the present data that an amorphous region of about 2.5nm has formed at the Zr/Co interface even in samples which were not annealed. The discontinuities of c_{Zr} shown in fig.2 reveal the existence of two amorphous phases. CALPHAD calculations of the free energies in the ZrCo system have shown that the concentration of the amorphous phases according to the common tangent rule is different from the experimental data; therefore the earliest stage of the SSAR is far away from metastable equilibrium [7].

Our investigations confirmed that the Zr grain boundaries provide a path for the Co diffusion. It is known from many investigations that only Co is the moving species in this SSAR [3]. Vieregge et al. show that grain boundary diffusion of Co in α- Zr occurs even at room temperature [8]. However, in our experiment, the Co-rich regions are not confined to the grain boundaries. In some Zr grains (type II) we found an extended solid solution of Co; in other regions the amorphous phase has been formed already. Earlier investigations have concluded that the Zr grain boundaries are the nucleation sites of the amorphous phase and they are necessary for the formation of the amorphous phase [4].

It seems that the amorphous ZrCo phase forms first by supersaturation of the Zr grains near high angle grain boundaries. The high angle grain boundaries are favoured because of their large energies and open structure. Beyond a critical Co concentration the Zr lattice is destabilized by the supersaturation and the amorphous phase grows spontaneously. But the amorphization process can happen as well in low angle grain boundaries if the Co concentration exceeds the critical limit there. Our observations suggest that the defects in the grain boundaries of the non- moving Zr favor the supersaturation process and the formation of the amorphous phase. Even the relatively low concentration of point defects and dislocations at the surface of the Zr single crystal induced by the sputter cleaning process is sufficient to initiate the phase formation [9]. Only clean and undamaged single crystal Zr layer in contact with polycrystalline Ni layers show no phase formation at all, because no supersaturation of the moving species seems possible there [10].

CONCLUSION

In conclusion we have shown that AP- FIM is a powerful tool for investigations of the early stages of the SSAR. The results indicate the formation of two amorphous phases and show that the reaction is initiated at the Zr grain boundaries even at room temperature. We conclude that the amorphous phase formation is the result of a destabilization process of the Zr lattice by supersaturation with Co rather than a classical nucleation process. In comparision with previous work only a small defect concentration at the interface of the non- moving species is necessary for the destabilization process.

We would like to thank G. von Minnigerode and the late P. Haasen for helpful discussions and continuous support. We are grateful to W. Bolse for the RBS measurements. This work has been supported by Deutsche Forschungsgemeinschaft (DFG), Sonderforschungsbereich 345.

REFERENCES

[1]   W.L. Johnson, Mater. Sci. Eng. 97 (1988) 1.

[2]   K. Samwer, Phys. Rep.,161 (1988) 1.

[3]   H. Schröder and K. Samwer, J. Mater. Res. 3 (1988) 461.

[4]   J.C. Barbour, F.W. Saris, M. Nastasi and J.W. Meyer, Phys. Rev. B 32 (1985) 1363.

[5]   H. Schröder, K. Samwer and U. Köster, Phys. Rev. Lett., 54, 19 (1985)

[6]   R. Busch, PhD thesis, Universität Göttingen (1992)

[7]   R. Busch, F. Gärtner and S. Schneider, this proceedings

[8]   K. Vieregge and Ch. Herzig, Defect Diff. Forum 66-69 (1989) 811.

[9]   S. Schneider, R. Busch, W. Bolse and K. Samwer, J. Non-Cryst. Solids 156-158 (1993) 498

[10] P. Ehrhart, R.S. Averback, H. Hahn, S. Yadavilli and C.P. Flynn, J. Mater. Res.3 (1988) 1276.

THE EARLIEST STAGE OF THE SOLID STATE AMORPHIZATION
REACTION IN THE ZR-CO SYSTEM

Ralf Busch[1]*, Frank Gaertner*, Susanne Schneider[1]**, Rüdiger Bormann*** and the late Peter Haasen*

*Institut für Metallphysik, Universität Göttingen, Hospitalstr. 3-7, D-37073 Göttingen, Germany
**I. Physikalisches Institut, Universität Göttingen, Bunsenstr.9, D-37073 Göttingen, Germany
***Institute for Materials Research, GKSS Research Center, D-21502 Geesthacht, Germany

ABSTRACT

Based on atom probe field ion microscopy (AP/FIM) studies, electromotive force (EMF) measurements and CALPHAD calculations we discuss the earliest stage of the solid state amorphization reaction (SSAR) in Zr/Co-layers. The AP measurements show that two amorphous phases are formed at the Zr/Co interface from the early stages of the reaction. The metastable two phase field between these amorphous phases is shown by direct measurement of the chemical potential of Zr in amorphous co-sputtered ZrCo alloys by the EMF method. The comparison between the atom probe data and the CALPHAD calculation shows that the interfaces between the different layers are far away from metastable equilibrium in the beginning of the reaction. The amorphous phase formation at the Zr/Co interface and in the hcp-Zr grain boundary is preceded by a supersaturation of the hcp ZrCo solid solution that transforms polymorphically into the amorphous state.

INTRODUCTION

The amorphous phase formation in diffusion couples combining an early transition metal (ETM) like Zr, Ti or Hf with a late transition metal (LTM) like Ni, Co or Cu is attributed to the fact that the amorphous phase has a lower free energy than the original reactants and that the formation of the competing intermetallic phases is inhibited [1-3]. In this type of interdiffusion experiment the mobility of the LTM is orders of magnitude larger than the mobility of the ETM. For the ZrCo system it was shown that the amorphization is favored in the presents of grain boundaries and also line defects [4].

Investigations of the late stage of the amorphization reaction show a diffusion-controlled growth of the amorphous layers [5]. In this regime the concentration gradients across the amorphous layers and the concentration steps at the interfaces are expected to be determined by the metastable equilibria between the terminal solid solution layers and the amorphous layer. In order to interpret the profiles of composition and chemical potentials across the layers in the early stage of the reaction we combine high resolution atom probe studies with the determination of the chemical potentials of the amorphous phases using the EMF method and CALPHAD calculations [6-7].

EXPERIMENTAL

Zr/Co bilayers were deposited onto tungsten FIM tips under UHV conditions by electron beam evaporation. The samples were analyzed by AP/FIM. The AP provides a chemical analysis

---

[1]present address: W.M.Keck Laboratory of Engineering Materials, 138-78, California Institute of Technology, Pasadena, CA 91125

Mat. Res. Soc. Symp. Proc. Vol. 343. ©1994 Materials Research Society

across the interfaces with atomic resolution in depth and a lateral resolution of 2 nm. The sharpness of the interfaces and the exact composition changes can be detected.

The chemical potentials of Zr in amorphous Zr-Co meltspun ribbons and co-sputtered films were determined by measuring EMF as a voltage between the amorphous alloy and a Zr reference electrode in a galvanic cell. The experiments were undertaken in a temperature range between 340°C and 420°C. A molten salt mixture with a low melting point was used as electrolyte. The samples were checked before and after the measurement with respect to crystalline precipitates.

RESULTS

With the atom probe the concentration profile from the cobalt layer through the amorphous layer into the zirconium is determined. Fig.1.a shows two atom probe measurements for as deposited bilayers as ladder diagrams. Starting the analysis in the pure Co layer the composition increases to (25.5±3) at% Zr and (27±4) at% Zr, respectively. After approximately 1 nm a sharp step of composition up to (59±7) at% Zr is observed within the amorphous layer. The Zr concentration keeps on increasing up to (75±2) at% Zr until it reaches pure zirconium by another step [4]. Annealing of the bilayers for 1h at 250°C leads to a similar profile. However, the size of the concentration step within the amorphous layer decreases as shown in fig.1.b. For this later stage of the reaction the concentration steps from 22 at% Zr to 39 at% Zr (61-78 at% Co). The measurements reveal that there are two amorphous phases coexisting during this interdiffusion reaction. The phases are separated by a sharp interface.

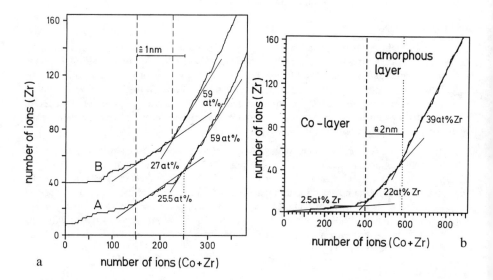

Fig.1.: AP analyses, visualized as ladder diagrams. The concentrations are calculated from the slope. a: two measurements going from the Co layer into the two amorphous layers in an as-deposited bilayer. b: after annealing for 1 h at 250 °C.

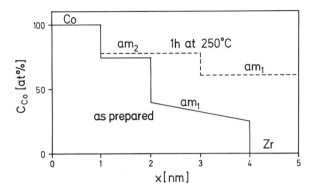

Fig.2: Concentration profiles across the Co/am$_2$/am$_1$/Zr interfaces

The composition changes across the layers are plotted in fig.2 as concentration profiles. The concentration steps at the interfaces decrease as the sample is annealed. The observed composition difference between the amorphous phases of the annealed diffusion couple shows a concentration step close to the expected metastable two phase field between these phases.

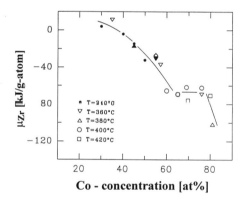

Fig.3: Chemical potential of Zr in amorphous ZrCo alloys with respect to hcp Zr.

The chemical potentials of amorphous Zr-Co alloys between 30 at% Co and 80 at% Co are determined by EMF measurements. After equilibrating the samples just below the crystallization temperature the chemical potentials are constant in a composition range between 63 at% Co and 77 at% Co. This metastable two phase field between the amorphous phases extends almost as far as the observed composition steps in the late stage of the interdiffusion experiment (dashed in fig.2). The observed step in the diffusion couple however is a little bit larger, because the interfaces in the interdiffusion experiment are still moving. This requires an additional chemical potential difference across the interfaces (see fig.5).

The thermodynamic functions of the Zr-Co system are determined by the calculation-of-phase-diagram (CALPHAD) method.

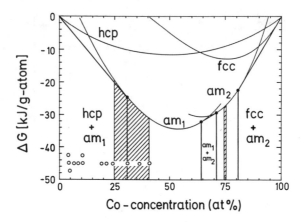

Fig.4: Free energy curves of the ZrCo system at 250 °C with respect to hcp-Zr and fcc-Co. Indicated are the homogeneity ranges of the two amorphous phases in as prepared samples (hatched) and the concentrations that were found in the Zr-grains next to different grain boundaries (circles).

The calculation is based on all available thermodynamic information concerning the Zr-Co system including phase diagram and calorimetric data. In particular the measured chemical potentials of the amorphous alloys are used to model the two amorphous phases. Fig.4 shows the calculated free energy curves for the terminal solid solutions and the two amorphous phases including the metastable two-phase fields. Indicated are the measured composition ranges of the two amorphous phases observed in the diffusion couple in the beginning of the reaction (hatched) and the compositions which were measured in different grain boundaries (circles). It is evident that the measured compositions in this early stage of the reaction cannot be described only by the metastable equilibria: 1, the measured Zr concentrations in the Zr-rich amorphous phase are higher, than expected using the equilibrium between hcp-Zr and the amorphous phase; 2, the observed homogeneity ranges of the amorphous phases are smaller than expected in a metastable equilibrium state according to the common tangent rule.

DISCUSSION

In the amorphous layer next to the zirconium and in certain Zr-grains next to grain boundaries of the hcp zirconium AP-measurements show a higher Zr concentration compared to the metastable equilibrium between the hcp solid solution and the amorphous phase (see fig.4). Other Zr-grains contain Co concentrations between 4 at% Co and 11 at% Co, which we attribute to solid solutions formed in the presence of the respective grain boundaries. These observations suggest that Co starts to diffuse into the Zr prior to amorphous phase formation, especially if we take the high diffusivity of Co in Zr into account. Supersaturation begins from the hcp Zr grain boundaries after segregation of Co due to the fast grain boundary diffusion of cobalt in zirconium [8], but also in the absence of grain boundaries because of the driving force according to the Gibbs free energies. Assuming classical nucleation with the maximum driving force, an amorphous nucleus should have an initial composition within the metastable homogeneity range of the amorphous phase. In the experiment we however observed zirconium concentrations that are much to high. This indicates that the initial stage of the SSAR is not a nucleation problem in the classical sense. More likely in the presents of defects the cobalt locally supersaturates the zirconium until it reaches its metastability limit and transforms polymorphically into the amorphous state. As observed in the earliest stage (fig.4) after the formation at the Zr/Co interface the amorphous phase tends to increase its Co concentration towards the metastable equilibrium. In the grain boundaries the phase formation strongly depends on the possible degree of segregation. Different grain boundary structures can explain the large range of Co concentrations observed in grain boundary regions. At those interfaces where the necessary supersaturation is not obtained, amorphous phase formation will not occur. This is the case for several interfaces we observed in our experiment, most likely small angle grain boundaries, but could also explain the absence of SSAR on undamaged Zr single crystals in contact with a late transition metal [9]. In contrast to amorphous phase formation the formation of an intermetallic crystalline compound requires nucleation.

Under the assumption that the Co-mobility is orders of magnitudes higher than the Zr mobility, the interdiffusion reaction can be discussed in terms of driving forces for the Co diffusion. In order to understand the restricted composition ranges of the amorphous phases that are found in the initial stage of the reaction, the measured concentrations across the interfaces are related to the calculated Co chemical potentials determined by the CALPHAD method. The course of the Co chemical potentials is visualized in fig.5.

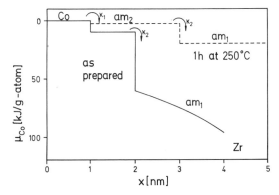

Fig.5: Chemical potential of cobalt across the different layers. Indicated are the different interface reaction coefficients κ.

In the initial stage of the reaction we find pronounced steps of the chemical potential at the different interfaces. This reveals that the interface reactions dominate the growth of the amorphous phases when the total thickness of the amorphous layer is small. Assuming a steady state of the fluxes across the interfaces and that the reaction is mainly controlled by the am_2/am_1 interface, the changeover thickness $x^*=M/\kappa$ between interface- and diffusion-controlled growth can be calculated, where M is the mobility of cobalt and κ_i are the interface reaction coefficients [10]. The flux across the am_2/am_1 interface $j=\kappa_2\cdot\Delta\mu_{inter}$ is equal to the flux through the Zr-rich amorphous phase am_1: $j= M_{am1}/x \cdot \Delta\mu_{layer}$. Using the data of fig.5. the changeover thickness is determined to be 2.6 nm. After annealing the sample for 1h at 250°C the step in chemical potential at the am_2/am_1 interface and the flux of Co across the interface decreases by a factor of three, caused by the increase of the layer thickness x_{am1}. From the reduced flux across the interface, the thickness of the am_1 layer after 1h annealing at 250°C can be recalculated. We obtain a value of 9 nm that is in good agreement with TEM and RBS investigations [11] [12]. The growth of the Co-rich (am_2) amorphous phase stays interface controlled even after annealing for 1h at 250°C. The changeover thickness x^* for diffusion limited growth of this phase is estimated to be larger than 5 nm, in spite of high Co mobility. This behavior can be explained by the small driving forces for the Co diffusion in the Co rich amorphous phase and a small interface reaction coefficient at the am_2/am_1 interface due to a large resistance for cobalt to cross the interface into the Zr-rich amorphous alloy.

SUMMARY

Concentration profiles across reacting Zr/Co bilayers are analyzed by AP/FIM in the early stage of the SSAR. A Co-rich and a Zr-rich amorphous phase separated by a sharp interface are formed. EMF measurements reveal that the phase separation also occurs in meltspun ribbons and co-sputtered thin films. The size of the measured metastable two-phase field between the amorphous phases corresponds to the observed concentration steps in the diffusion couple after annealing for 1h at 250 °C. In the Zr rich amorphous phase much higher Zr concentrations are found than the metastable equilibria require if the reaction is initiated by a nucleation process. This suggests that Zr and Co first locally form a solid solution in the interface region, which transforms polymorphically into the Zr rich amorphous phase. The thermodynamic functions and metastable equilibria between the amorphous phases and terminal solid solutions are determined with the CALPHAD method to study the initial stage of the observed SSAR in terms of driving forces for the interdiffusion. It is shown that the flux through the interface between the amorphous phases is relatively small. The growth of the Zr richer amorphous phase becomes diffusion-controlled from a thickness of 2.6 nm. The growth of the slower growing Co-rich phase stays interface controlled within the experimental time.

ACKNOWLEDGMENT

This research was supported by the Deutsche Forschungsgemeinschaft via SFB 345. The authors thank K.Samwer, L.A.Greer and the FIM group of Göttingen for valuable help and fruitful discussions.

REFERENCES:

1. W.L. Johnson, Prog. Mat. Sci. 30, **81** (1986).
2. B.M. Clemens, W.L. Johnson and R.B. Schwarz, J. Non-Cryst. Solids **61**, 817 (1984).
3. H. Schröder, K. Samwer and U. Köster, Phys. Rev. Lett. **54**, 19 (1985).
4. S. Schneider, R. Busch and K. Samwer, this proceedings.
5. H. Schröder and K. Samwer, J. Mat. Res. **3**, 461 (1988).
6. H.L. Lukas, J. Weiss and E.T. Henig, CALPHAD **6**, 229 (1982).
7. R. Bormann and K. Zöltzer, phys. stat. sol. (a) **131**, 691 (1992).
8. K. Vieregge and Chr. Herzig, Defect and Diffusion Forum **66-69**, 811 (1989).
9. W.Kiauka, C.van Cuyck and W.Keune, Mat.Sci. and Engg. **B12**, 273 (1992).
10. U. Gösele and K.N.Tu, J. Appl.Phys. **66**, 2619 (1989).
11. H.Schröder, PhD thesis, Göttingen, 1987.
12. K.Pampus J.Boettiger,B.Torp,H.Schröder and K.Samwer,Phys.Rev.B **35**, 7010 (1987).

ARTIFICIALLY SYNTHESIZED NON-COHERENT INTERFACES IN FE-TI
MULTILAYERS AND THEIR INFLUENCE ON SOLID STATE REACTIONS

Z.H.Yan*+, M.L.Trudeau+, A.Van Neste# and R.Schulz+*
D.H.Ryan*, P.Tessier*, R. Bormann~ and J.O.Ström-Olsen*

+ Materials Technology Dept., Hydro-Québec Research Institute,
Varennes, P.Q., J3X 1S1, Canada
* Centre for the Physics of Materials, McGill University,
Montreal, P.Q., H3A 2T8, Canada
~ Institute for Materials Research, GKSS-Research Center,
D-2054 Geesthacht, Germany
Metallurgy Department, Laval University, Ste-Foy, P.Q.,
G1K 7P4, Canada

ABSTRACT

The influence of the interfacial structure on the solid state
reaction products in Fe-Ti multilayers has been studied using
various preparation conditions and characterization techniques.
Sharp and diffused interfaces were produced by using either
sequential or co-evaporation in the interfacial region. The
reaction product, in the case of the sharp interface, is the bcc
supersaturated solid solution of Ti(Fe) while, in the case of the
diffused interface, an amorphous phase is formed. Therefore,
nucleating the amorphous phase at the interface by local co-
evaporation alters the reaction path observed in Fe-Ti
multilayers. The solid state reactions were studied using low and
high angle X-ray diffraction and Mossbauer measurements. The
results are discussed in light of recent thermodynamic
calculations on the Fe-Ti system.

INTRODUCTION

Evidences were given that lack of grain boundaries, triple
junctions or other interfacial defects could suppress the
formation of amorphous phases during solid state reactions of bilayers in spite of the existence of a thermodynamic driving force
for amorphization (1,2). Recent studies also revealed that the
nature of the end product during an interfacial reaction was very
sensitive to factors such as the annealing temperature and time,
the properties of the interfaces and of each component and the
impurities (3-5). The nucleation process also plays a crucial
role in the phase selection during the interfacial reactions,
especially when the competing phases have comparable
thermodynamic stabilities such as in the case of Ti-Fe. In this
system, the thermodynamic calculations indicate that the free
energy difference between the amorphous and the disordered bcc
solid solution is small (3). Special experiments have to be
designed to elucidate the kinetic aspects associated with the
nucleation of a given phase. Previous work showed that a bcc
phase grew upon annealing an Fe-Ti multilayer structure even
though the amorphous phase has a lower free energy over certain
concentration range (3). It is the purpose of the present work
to study even further, the influence of the interfacial
properties on the phase evolution.

The multilayers were deposited onto oxidized Si wafers or glass
substrates (Corning 7059) and subsequently annealed under a
vacuum better than 9*10-9 Torr. In order to investigate the
influence of the interface on the reaction, two kinds of
interfaces were synthesized. One was prepared by sequential

evaporation of each element (defined as sharp interface), and the other was artificially diffused by partially overlapping the evaporation of the two sources (defined as diffused interface). The temperature of the sample holder used during thermal treatments has been calibrated for improved accuracy.

EXPERIMENTAL RESULTS

The bcc phase has been formed at the interfaces in all cases (Ti55Fe45, Ti45Fe55 and Ti35Fe65) when sharp interfaces were fabricated. The structural evolution was characterized using high angle X-ray diffraction(Cu Ka), while the decay of the compositional modulation was monitored by the low angle diffraction. Fig. 1 shows the x-ray patterns of a Ti55Fe45 sample as a function of the annealing time at 400 °C. With the Ti as first layer, the as-deposited film is highly textured (Ti (002) and Fe (110)) with the compact planes parallel to the substrate. The as-deposited spectrum shows two groups of satellite peaks around hcp Ti (002) and bcc Fe (110), indicating partial coherency at the interface (6). This structural pattern begins to change upon annealing at 400 °C. The two groups of satellite peaks shift close to each other at first, then a new textured bcc crystalline phase grows as indicated by the presence of the strong bcc (110) peak and the absence of higher index reflections. The corresponding evolution of the superlattice peaks at low angles is shown in Fig. 2. The pair thickness, determined from the position of the peaks, is in agreement with the designed value of 60 Å. Below 400 °C, we

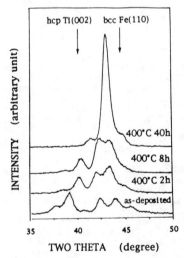

Fig. 1 High angle X-ray diffraction patterns of Ti55Fe45 multilayer sample before and after annealing at 400 °C.
(pair thickness: 60 Å)

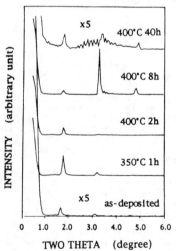

Fig. 2 Low angle X-ray diffraction patterns of Ti55Fe45 multilayer sample before and after annealing at 350 °C and 400 °C.
(pair thickness: 60 Å)

observed the enhancement of the superlattice peaks. Since there is no obvious structural change in the high angle portion of the spectrum, this enhancement probably reflects a structural relaxation and some modifications of the interface roughness. On the contrary, isothermal annealing at 400 °C induces a decay of the superlattice peaks indicating that the interfacial reaction is taking place. There is an increase in the second order peak for 8 hours of annealing, which is abnormal in view of the dependance of the diffraction intensity on the compositional modulation. This abnormal increase may originate from the interference between the newly formed bcc phases at each interface, on condition that the interlayers grow planarily and that the residual Ti and Fe layers shrink to nearly equal thicknesses. In this case, the bcc interlayers constitute a new set of quasi-periodic compositional modulation at half of its parent wavelength. The first order peak of the new superlattice superimposes over the second one of the parent structure. Since the Ti layer of the as-deposited Ti55Fe45 sample is thicker than that of Fe (Ti 39 Å / Fe 21 Å), the shrinkage of the Ti layers must be faster. After 8 hours, the superlattice peaks decrease again with further reaction. The X-ray diffraction also reveals that the formed bcc phase is also (110) textured suggesting that the bcc Fe layer provides the nucleation site for the growth of the interlayer. Therefore it is proposed that, in the interfacial reaction of Ti-Fe multilayers with sharp interfaces, the bcc phase epitaxially forms and grows planarily by asymmetric interdiffusion, Fe being the fast diffusing species.

The above experimental results could not be expected based on the calculated thermodynamic functions of the Ti-Fe system using the CALPHAD method (3). The calculated Gibbs free energy curves of the various phases are given in Fig.3 at T = 450 °C, the temperature near which the multilayers were subjected during annealing. It shows that the amorphous phase has lower free energy than the disordered bcc phase over a relatively wide range. This result originates from the chemical short range order which develops in the undercooled liquid (7). From this, one would expect the formation of the amorphous phase rather than the bcc phase during interfacial reactions assuming that the formation of the stable TiFe intermetallic compound is kinetically suppressed. We do observe, however, the formation of the ordered B2 TiFe compound but only after a long annealing time. Indeed, the (100) superlattice peak appears in the diffraction pattern after several hours. In the early stage of the reaction there is no evidence for the formation of the ordered phase. The absence of the amorphous phase could be due to a nucleation barrier while the partially coherent sharp interface provides the nucleation sites for the growth of the bcc phase.

To elucidate the kinetical aspects in the interfacial reaction in the Ti-Fe system, the sharp interfaces were destroyed by co-evaporation in the interfacial region. As vapor quenching produces amorphous Ti-Fe alloys from 25 to 80 at% Fe (8), this thin interlayer should be amorphous. Thus, we artificially synthesize non-coherent or diffused interface. Fig. 4 shows, as a function of the annealing time at 400 °C, the X-ray diffraction patterns of Ti45Fe55 multilayers with 12 Å diffused interlayer. No satellite peaks are visible around the main peaks at high angles, and the Fe bcc (200) peak is observed indicating weak texture. The disappearance of the satellite peaks is primarily caused by the loss of coherency at the interface because of the

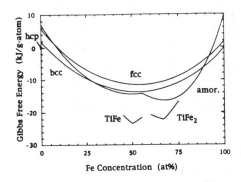

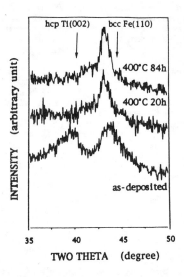

Fig. 3 Diagram of the calculated Gibbs free energy vs. concentration at T = 450 °C. The free energy of hcp Ti and bcc Fe chosen as reference states.

Fig. 4 High angle X-ray diffraction patterns of Ti45Fe55 multilayer sample with designed diffused interfaces.
(interlayer thickness: 12 Å)

amorphous interlayer. During annealing, the (110) diffraction peak of the bcc solid solution doesn't grow as in the previous case. Instead, a broad amorphous band is observed between 40 and 45 degrees. The end product, after 84 hours, is an amorphous phase with some bcc crystallites. With the diffused interfaces, the reaction path is altered. This shows that the structural evolution during thin film reactions can be strongly dependent on the nature of the interfaces.

When the thickness of the amorphous interlayer is reduced to 6 Å, the reaction path is also different. Fig.5 shows the Mossbauer spectrum of a multilayer sample with 6 Å diffused interface in the as-deposited state. In addition to the Mossbauer pattern of bcc Fe, the results reveal the existence of an amorphous phase whose quadrupole splitting and isomer shift agree with those reported in ref. (8). This confirms that the co-evaporated interlayer is amorphous but nothing is known on the uniformity of this layer. The structural evolution of this sample is shown in Fig. 6 as a function of annealing time at 400 °C. In contrast to the 12 Å diffused interface, the crystalline bcc phase forms upon annealing. The (110) diffraction peak grows even more rapidly than in the case of the sharp interfaces shown in Fig. 1. It is possible that the thickness of the amorphous interlayers is sub-critical and therefore would shrink upon annealing. More investigations are needed to study these phenomena.

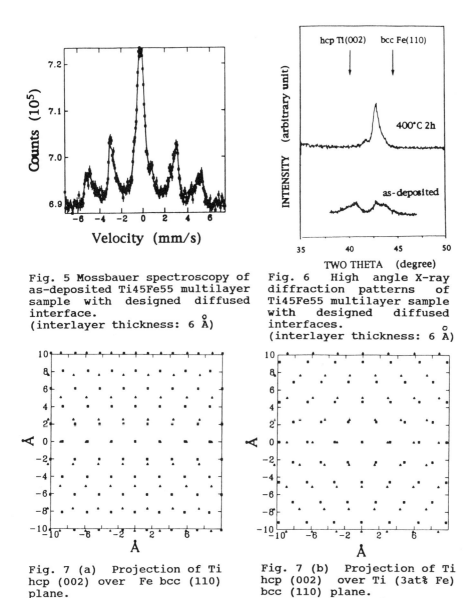

Fig. 5 Mossbauer spectroscopy of as-deposited Ti45Fe55 multilayer sample with designed diffused interface.
(interlayer thickness: 6 Å)

Fig. 6 High angle X-ray diffraction patterns of Ti45Fe55 multilayer sample with designed diffused interfaces.
(interlayer thickness: 6 Å)

Fig. 7 (a) Projection of Ti hcp (002) over Fe bcc (110) plane.
▲ hcp Ti,   ■ bcc Fe

Fig. 7 (b) Projection of Ti hcp (002) over Ti (3at% Fe) bcc (110) plane.
▲ hcp Ti, ■ bcc Ti (3at% Fe)

DISCUSSION

The above results indicate that the solid state reaction in Ti-Fe multilayers is strongly dependent on the interfacial conditions. The variations in the evolution path reflects the influence of

the nature of interface on the kinetics of the reaction. Fig. 7
(a) shows the projection of the Ti hcp (002) plane on Fe bcc
(110) using bulk lattice parameters. It emphasizes the
crystallographic relationship which exists in the as-deposited
state. We have found, using mechanical alloying, that 3 at% of
Fe in hcp Ti was sufficient to destabilize the hcp structure and
form bcc Ti(Fe)(9). In Fig. 7 (b), a Ti hcp (002) plane is shown
on top of a (110) plane of bcc Ti(3at%Fe). The lattice match is
much better indicating that the formation and growth of the bcc
interlayer must be accompanied by strain relaxation. The strain
energy can alter the relative stablity of the competing phases,
especially when they have comparable free energies such as in the
present case. Through the transformation of Ti from hcp to bcc
with Fe in solution, the lattice match is improved and the
strain is released. This provides an additional driving force
for the growth of the bcc solid solution. Moreover the bcc phase
can grow epitaxially-like from the parent bcc Fe layer, as
evidenced by the texture of the newly formed bcc interlayer.
Therefore the formation of Ti(Fe) bcc solid solution has
negligible nucleation barrier compared to the amorphous phase
which usually needs heterogeneous nucleation sites to form.

ACKNOWLEDGEMENTS:

The authors are grateful to Dr. L. Dignard-Bailey and Dr. C.
Michaelsen for discussions. The financial supports from NSERC
and FCAR and from Hydro-Québec are greatly appreciated.

REFERENCES:

1. A.M.Vredenberg, J.F.M.Westendorp, F.W.Saris, N.M. van der
   Pers and Th.H.de Keijser
   J. Mater. Res. 1, 774(1986)

2. B.M.Clemens
   Phys. Rev. B 33, 7615(1986)

3. Z.H.Yan, M.L.Trudeau, R.Schulz, R.Bormann, A.Van Neste
   and J.O.Ström-Olsen
   Mater. Res. Soc. Symp. Proc., vol 311, 79(1993)

4. W.L.Johnson
   Prog. Mater. Sci. 30, 81(1986)

5. J.J.Hauser
   Phys. Rev. B 32, 2887(1985)

6. B.M.Clemens and J.G.Gay
   Phys. Rev. B 35, 9337(1987)

7. R.Bormann, F.Gartner and K.Zoltzer
   J. Less-Commen Met. 145, 19(1988)

8. K.Sumiyama, H.Ezawa and Y.Nakamura
   Phys. Stat. Sol.(a) 93, 81(1986)

9. L.Zaluski, P.Tessier, D.H.Ryan, C.B.Doner, A.Zaluska,
   J.O.Ström-Olsen, M.L.Trudeau, R.Schulz
   J. Mater. Res. 8, 3059(1993)

SEGREGATION EFFECT AND ITS INFLUENCE ON GRAIN BOUNDARY DIFFUSION IN THIN METALLIC FILMS

A.N.ALESHIN*, B.S.BOKSTEIN**, V.K.EGOROV* AND P.V.KURKIN**
*Institute for Microelectronics Technology and High-Purity Materials, Russian Academy of Sciences, 142432 Chernogolovka, Moscow District, Russia
**Steel and Alloys Institute, Department of Physical Chemistry, Leninsky prosp., 4, 117936 Moscow, Russia

ABSTRACT

The diffusion in Au-Cu and Pt-Cu thin films has been studied by Rutherford backscattering spectrometry (RBS) under the kinetic regime "B" (within the temperature interval of 175-290°C) and "C" (room temperature). The 1,5-2,0 MeV He$^+$ RBS spectra were taken using 14-18 keV resolution. The RBS spectra were changed to depth-concentration profiles for both bulk and grain boundary (GB) diffusion. Under the kinetic regime "C" the absolute values of GB diffusion coefficients were obtained. Under the kinetic regime "B" the triple products δKD_b (δ is the GB width, D_b is the GB diffusion coefficient, K is the enrichment ratio) were obtained using Whipple's and Gilmer-Farrell's models. The activation energies for both GB diffusion and bulk diffusion were determined for Au-Cu system as well as for Pt-Cu system. The comparison between the data on the GB diffusion for kinetics "B" extrapolated to room temperature and the data on the GB diffusion for the kinetics "C" enables one to derive the product δK and to separate the contribution of segregation into the parameters of GB diffusion for systems under study.

INTRODUCTION

Theoretical model of GB self diffusion has been developed by Fisher [1] and Whipple [2] in which the GB was presented as a homogeneous isotropic slab with a width of δ. In this model a diffusion flux is directed along GB (parallel to the y axis) and furthermore, a loss of material from the GB into a grain bulk takes place. For example this lateral flux (parallel to the x axis) is important in the so-called kinetic regime "B" [3] which occurs when the condition $2(Dt)^{1/2} > 10\delta$ is validated (D is the bulk diffusion coefficient, t is the annealing time).
In according to Fisher's model at the GB region the following equalities must be fulfilled

$$c'(\delta/2 - 0) = c(\delta/2 + 0) \tag{1}$$

$$D_b(\partial c'/\partial x)\big|_{\delta/2-0} = D(\partial c/\partial x)\big|_{\delta/2+0} \tag{2}$$

where c' and c is the concentration inside and outside of the slab.

241

The treatment of experimental data in the framework of Fisher's solution enables one to derive the product δD_b. Gibbs [4] solved the Fisher's problem taking into account GB segregation in the form of Henry's isotherm

$$c' = Kc \qquad (3)$$
$$K = K_0 \exp(-H_a/kT) \qquad (4)$$

where K is the enrichment ratio, H_a is the heat of segregation and K_0 is the pre-exponential factor.

So, in the general case the activation energy for the GB heterodiffusion contains the energy of the interaction of the diffusant atoms with the GB. When studying the heterodiffusion the treatment of experimental data allows one to obtain the triple product $\delta K D_b$.

In accordance with Harrison's classification [3] there is the kinetic regime "C" when a loss of material from GBs into a grain bulk is absent and the rate of diffusion mass transport is characterized only by the GB diffusion coefficient D_b. The diffusion in the kinetics "C" is realized when the inequality $(Dt)^{1/2} < \delta/2$ is validated. The comparison between data on GB diffusion for the kinetics "B" extrapolated to temperature at which the kinetics "C" is realized and data on the GB diffusion for the kinetics "C" enables one to derive the product δK and to estimate the K value. An increased density of GBs in thin films as compared to massive samples facilitates the investigation of diffusion in the kinetic regime "C".

The diffusion in Au-Cu and Pt-Cu thin films has been studied in the present paper. The Au-Cu thin-film diffusion couple consists of the elements whose properties (surface tension, melting temperature) responsible for segregation ability are a like. In contrast, these properties of the elements forming the Pt-Cu thin-film diffusion couple are very different. This fact is the heart of the present investigation. Besides, the Au-Cu and Pt-Cu systems are suitable for RBS analysis.

EXPERIMENTAL

The Pt-Cu thin-film diffusion couples using in experiments when the kinetic regime "B" was under study have been obtained by magnetron sputtering through sequential deposition of copper and platinum on glass ceramic substrate at room temperature under a pressure 10^{-3} Pa. The thicknesses of the films were: Cu - 700 nm and Pt - 50 nm.

For the kinetic regime "C" the succession of deposition was changed: platinum was deposited firstly and copper was deposited after that. In this case the diffusion couple constituted the bimetallic film structure of the Pt and Cu layers respectively 210 and 400 nm thick. The circumstance was taken into account that such a succession of deposition provides the motion of the Pt atoms in a direction of free surface. This enables one to determine more precisely (as compared to the analysis at great depth) the Pt concentration profile, since in near-surface regions the higher depth resolution is reached.

The Au-Cu thin-film diffusion couples have been obtained by vacuum evaporation through sequential deposition of copper and gold on glass ceramic substrate with a deposition rate 3 nm/s under a pressure 10^{-3} Pa. The thicknesses of the layers were: Cu - 800 nm, Au - 60-120 nm (for the study of Cu diffusion in Au) and Cu - 200 nm, Au - 45-70 nm (for the study of Au diffusion in Cu).

The films were annealed in evacuated glass ampoules under a pressure close to 1 Pa. The Au-Cu thin films were annealed in 175-250°C temperature range (five annealing temperatures). The Pt-Cu thin film were annealed in 200-290°C temperature range (four annealing temperatures). Such temperature regimes provide the realization of the kinetic regime "B" for both the Au-Cu system and the Pt-Cu system. The kinetic regime "C" was realized for the diffusion of Cu into Au thin film and for the diffusion of Pt into Cu thin film at room temperature.

1,5-2,0 MeV ^{4}He$^+$ RBS spectra were obtained from the bimetallic films before and after annealing in a standard scattering configuration on a Sokol-3 ion beam analysis complex with a 160° scattering angle and 1,0 msr detector solid angle. The total resolution of the detection system varied from 14 to 18 keV (full width at half-maximum). The RBS spectra taken from the samples annealed in the regime of kinetics "B" were changed to concentration-depth profiles by a simplified algorithm [5] based on a single-element sample conception [6]. The use of this procedure for an element's profile in the bimetallic films is justified because the partial elemental RBS spectra before and after annealing did not overlap. The program RUMP [7] were used in our experiments for processing the spectra taken from samples in which the diffusion under the kinetics "C" was realized.

RESULTS OF DIFFUSION EXPERIMENT

Figure 1 shows the Pt concentration profiles plotted in Whipple's form for the diffusion of Pt in Cu flims under the kinetic regime "B". One can see two regions connected with both bulk and GB diffusion. To calculate D^{Pt} the method of middle gradient [8] has been applied. The parameters of the temperature dependence of D^{Pt} are expresed by the values: the activation energy for bulk diffusion $E^{Pt} = (1,50\pm0,05)$ eV; the pre-exponential factor $\delta D_0^{Pt} = (0,77\pm0,17)$ cm^2/s.

The calculation of the GB diffusion parameters was performed in accordance with Fisher-Whipple's model using Le Claire's relation [9]

$$\delta K D_b^{Pt} = 0,661(\partial \ln \bar{c} / \partial y^{6/5})^{-5/3} (4D^{Pt}t)^{1/2} \qquad (5)$$

where $\bar{c}$ is the mean concentration by x at a depth of y actually measured by RBS. The parameteres of the temperature dependence of the triple product $\delta K D_b^{Pt}$ can be expressed through the following values: the activation energy for the GB diffusion of Pt in Cu films $E_b^{Pt} = (1,25\pm0,05)$ eV; the pre-exponential factor $\delta K_0 D_{b0}^{Pt} = (1,73\pm0,32)\cdot10^{-8}$ cm^3/s.

The data on the bulk diffusion and the GB diffusion in Au-Cu thin films were described in great detail in [10,11]. The distinctive property of the diffusion process in the Au-Cu system is the higher-order value of D_b^{Cu} as compared with the D_b^{Au} value. In contrast to the Au concentration curves (Au concentration curves are described by Le Claire's relation) the Cu concentration curves have a large portion inside Au films where the concentration gradient is close to zero (Figure 2). The parameters for the GB diffusion of Cu into Au films were calculated on the basis of the Gilmer-Farrell's solution [12] by means of the so-called "universal curve". The values of the bulk diffusion coefficients D^{Cu} (and D^{Au} too) were calculated in the same manner as for D^{Pt}.

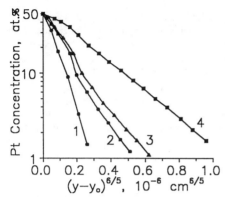

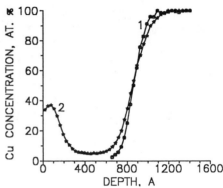

Fig1. The Pt concentration curves plotted in Whipple's form for the diffusion of Pt in Cu films under the kinetic regime "B": 1 - before annealing; 2 - 200°C for $1,5\cdot10^3$s; 3 - 260°C for $1,5\cdot10^3$s; 4 - 290°C for $6\cdot10^3$s.

Fig2. The Cu concentration curve for the diffusion of Au in Cu film under the kinetic regime "B": 1 - before annealing; 2 - 250°C for $3\cdot10^2$s;

The bulk interdiffusion coeficient D^{Au-Cu} as a function of temperature may be expressed by Arrhenius equation with the following parameters: $E^{Au-Cu} = (1,10\pm0,04)$ eV; $D_0^{Au-Cu} = (1,24\pm0,30)\cdot10^{-4}$ cm^2/s. The temperature dependence of the triple product $\delta K D_b^{Cu}$ in the Au-Cu system has the parameters: $E_b^{Cu} = (0,98\pm0,04)$ eV; $\delta K_0 D_{b0}^{Cu} = (5,8\pm0,9)\cdot10^{-10}$ cm^3/s.

As mentioned above the RBS spectra were processed by the program RUMP [7] when the diffusion regime "C" was under study. The program RUMP proceeding from the spatial distribution of the elements under study generates the model spectrum in such a way that to achieve matching of the experimental spectrum with the model one. The processing by the program RUMP reveals a very small peak in a RBS spectrum taken from the sample Cu/Au - 58 nm which was stored at room temperature during 5 months (Figure 3).In this experiment RBS spectra being taken every 2-3 weeks through this period. The peak was revealed at the channel region corresponding to scattering from Cu atoms at the free surface.

The appearance of this peak implies that Cu atoms penetrate through Au film along GBs and reach the sample surface. This low-temperature diffusion experiment can be described in terms of first appearance of diffusant on the sample surface t_0. In accordance with [13] in this case the value of D_b is expressed by formula:

$$D_b = Y_0^2/6t_0, \qquad (6)$$

where Y_0 is the film thickness.

The substitution of the values of Y_0 and t_0 into the equation (6) allows one to obtain the value of D_b^{Cu} equal to $(4,8\pm0,8)\times10^{-19}\ cm^2/s$.

In general case, when the kinetic regime "C" is realized the expression for the mean concentration $\bar{c}$ determined from RBS spectra has the form [14]:

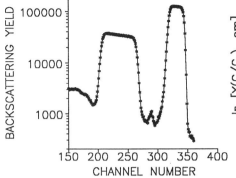

Fig3. The RBS spectrum taken from the Cu/Au - 58 nm thin-film diffusion couple stored during 5 months at room temperature (the kinetic regime "C").

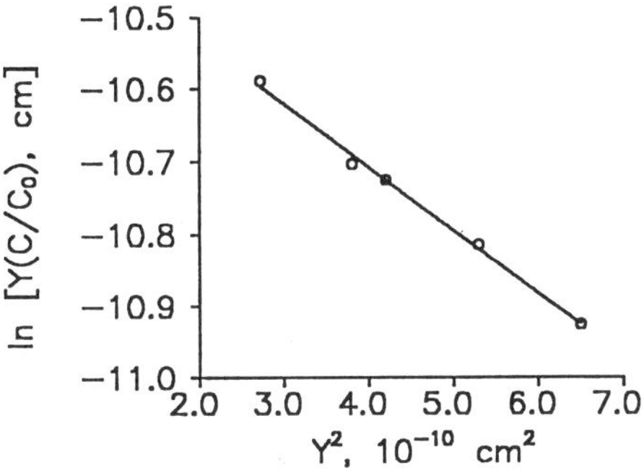

Fig4. The part of the Pt concentration curve plotted in accordance with expression (8) for the diffusion of Pt in Cu film under the kinetic regime "C" (room temperature). Each point in the plot corresponds to the processing step of the program RUMP.

$$\bar{c}(y,t) = \lambda\delta(c_0/2)erfc[y/2(D_bt)^{1/2}], \qquad (7)$$

where λ is the total GB length per unit area, c_0 is the initial concentration equal to 100 at.%.

As a rule the value of the product $\lambda\delta$ is unknown. The expression (7) may be transformed into:

$$\ln(y\bar{c}/c_0) = const - y^2/4D_bt \qquad (8)$$

when the inequality $y/2(D_bt) > 1$ is valid.

The use of the program RUMP allows one to separate the Pt concentration profile from the RBS spectrum taken from the sample which was stored at room temperature during 40 months. The processing of experimental data on the tail of the concentration profile by formula (8) enables one to determine the GB diffusion coefficient of Pt into Cu film at room temperature equal to $D_b^{Pt} = 2,8 \cdot 10^{-18} \ cm^2/s$

DISCUSSION

The extrapolation of the δKD_b^{Cu} parameter values found for the GB diffusion of Cu into Au films in the "B" regime to room temperature yields $\delta KD_b^{Cu} = 1,7 \cdot 10^{-26} \ cm^3/s$. Thus,

$\delta K=0{,}4$ nm. This value is close to Fisher's estimation of the GB width equal to 0,5 nm. Therefore, the K value approximates 1. Since Cu and Au are similar in many properties governing segregation ability such a value of K is realistic.

In contrast, the same procedure for diffusion data of Pt-Cu system yields $\delta K D_b^{Pt} = 1{,}4 \cdot 10^{-29}$ cm^3/s. Thus, in this case $K(25°C)=10^{-4}$ for $\delta=0{,}5$ nm. This value is very small. It is four orders of magnitude smaller than that has been obtained when the diffusion in Au-Cu thin films was under study. Such a value of K can be accounted for segregation phenomenon.

As mentioned above the experimentally determined activation energy for the GB heterodiffusion contains the energy of the interaction of the diffusant atoms with the GB H_a. Proceeding from the liquid model of GB structure [15] which takes into consideration the constitution of the phase diagrams one can conclude that for the Pt-Cu system $H_a^{Pt} > 0$ and consequently K < 1. The more precise calculation of the H_a^{Pt} value performed in according with the theory developed in [16] yields $H_a^{Pt}=0{,}24$ eV, and $K(25^0C) = 9 \cdot 10^{-5}$.

CONCLUSION

The estimation of the segregation contribution into the GB diffusion performed on the basis of measurements of the GB diffusion parameters for both the kinetic regime "B" and the kinetic regime "C" is fundamentally new.

REFERENCES

1. J.C.Fisher, *J.Appl.Phys.* **22**, 74, (1951).
2. R.T.P.Whipple, *Brit.J.Appl.Phys.* **54**, 1225, (1954).
3. L.G.Harrison, *Trans.Faraday Soc.* **57**, 1191, (1961).
4. G.B.Gibbs, *Phys.Stat.Sol.* **16**, K27, (1966).
5. A.N.Aleshin,B.S.Bokstein,P.V.Kurkin, *Poverkhnost* **1991** (5), 157.
6. W.K.Chu, J.W.Mayer and M.-A.Nicolet, *Backscattering Spectrometry*, (Academic Press, New York, 1978).
7. L.R.Doolitle, *Nucl.Instrum.Methods B*, **9**, 344, (1985).
8. P.M.Hall, J.M.Morabito, J.M.Poate, *Thin Solid Films* **33**, 107, (1976).
9. A.D.Le Claire, *Brit.J.Appl.Phys.* **14**, 351, (1963).
10. A.N.Alcshin, B.S.Bokstein, V.K.Egorov, P.V.Kurkin, *Thin Solid Films*, **223**, 51, (1993).
11. A.N.Aleshin, B.S.Bokstein, V.K.Egorov,P.V.Kurkin, *Defect and Diffusion Forum* **95-98**, 457, (1993).
12. G.H.Gilmer and H.H.Farrell, *J.Appl.Phys.* **47**, 3795, (1976).
13. P.H.Holloway and G.E.McGuire, *J.Electrochem.Soc.* **125**, 2070, (1978).
14. I.Kaur and W.Gust, *Fundamentals of Grain and Interphase Boundary Diffusion*, (Ziegler Press, Stuttgart, 1988), p.99.
15. K.Lucke and K.Dettert, *Acta Met.* **5** (11), 628, (1957).
16. N.H.Tsai, G.M.Pound, F.F.Abraham, *J.Catalys.* **50**, 200, (1977).

LIQUID METAL PENETRATION ALONG GRAIN BOUNDARIES

V. E. FRADKOV
Materials Engineering Department, Rensselaer Polytechnic Institute, Troy, NY 12180-3590, USA

ABSTRACT

Liquid metal grain boundary corrosion is discussed in terms of grain boundary etching profiles with equilibrium dihedral angles at the vertex of the grooves close to zero. It is shown that if the liquid solution is in equilibrium with the solid, then only grain boundary grooving occurs, producing small grooves growing in time as $t^{1/2}$. However, if the equilibrium cannot be reached, a long liquid filled canal develops along the grain boundary, rapidly propagating with constant velocity. To stop such rapid grain boundary corrosion certain measures should be taken to reach the equilibrium state. This explains, for example, why removal of oxygen from the Nb(s)-Li(l) system prevents rapid grain boundary corrosion of Nb.

INTRODUCTION

Contact of solid metal with a liquid one may bring on a phenomenon which is sometimes desirable, but most often harmful. Liquid filled canals appear and rapidly propagate along grain boundaries in solid metals. In many cases they are visible in an optical microscope, which implies a thickness of a few microns, and their propagation velocity can achieve several millimeters per hour. Stress is not necessary for such penetration, in contrast with the phenomenon of liquid metal embrittlement [1-3] . Liquid metal penetration has been observed in numerous solid-liquid metal pairs Nb-Li, Zn-Ga, Zn-Sn, Al-Sn, Al-In, Zn-Bi, Zn-Pb, Cu-Pb, Cu-Bi, and Al-Ga [4-18]. The phenomenon is rather subtle; sometimes a small change in the system is sufficient to stop it. For example, removal of oxygen eliminates penetration of liquid Li along the grain boundaries of Nb [19, 20]. The same effect was observed when a few percent of Zr was added to Nb.

Selective dissolution of grain boundaries is a well known phenomenon; it is widely used in metallography to achieve visualization of grain boundaries. Similar profiles appear when a polycrystal is in contact with liquid metal. However, liquid metal etching is peculiar, because the liquid-solid interface in this case has, as a rule, relatively low interfacial energy, γ_s, in comparison with the typical energy, γ_B, of a high-angle grain boundary. Hence, the equilibrium angle, φ_0, at the vertex of the groove (see Fig. 1) is not small, in contrast to the case of chemical or thermal etching, and may even reach its limit value of $\pi/2$, which corresponds to complete wetting. Therefore, the linearization used by Mullins [21, 22] to discuss the thermal etching groove is inapplicable. The shape and evolution of etching profiles at the values of φ_0 close to $\pi/2$ are studied in the present paper, focusing on the possibility of development and propagation of the liquid canals described above.

MATHEMATICAL FORMULATION OF THE PROBLEM

Liquid metal etching develops by dissolution of solid metal in the liquid one. That means that the solid metal atoms have to cross the liquid-solid interface driven by a chemical potential difference. Preferential etching of the grain boundary area is determined by the curvature of the surface, because the constant equilibrium value of the angle φ_0 is maintained at the vertex of the groove (Fig. 1). The chemical potential of the atoms on the curved surface is higher than that on the plane one by a value proportional to the mean curvature of the surface. We will consider the atom transfer through the boundary, with diffusional removal of the dissolved metal in the liquid phase being rather rapid. The equation describing the evolution of the surface shape is given by

247

Mat. Res. Soc. Symp. Proc. Vol. 343. ©1994 Materials Research Society

$$V_n = A \cdot K + V_0 \tag{1}$$

where V_n and K are the normal velocity and the local mean curvature of the interface, $A = \gamma \Omega \kappa$, where γ, Ω and κ are the interface energy, atomic volume, and kinetic coefficient of dissolution, respectively, and V_0 is the velocity of the plane surface receding due to dissolution. This velocity is positive if the liquid solution is undersaturated. In terms of variables x, y and φ (Fig. 1), eq.(1) can be written as

$$\left(\frac{\partial y}{\partial t}\right)_x \cos\varphi = A\left(\frac{\partial \varphi}{\partial x}\right)_t \cos\varphi - V_0, \tag{2}$$

where $\varphi = \arctan(\partial y / \partial x)_t$ is the angle between the tangent to the surface and the horizontal line. The first term in the r.h.s. of eq.(1) corresponds to the curvature of a cylindrical interface. For the sake of simplicity, crystallographic anisotropy is neglected.

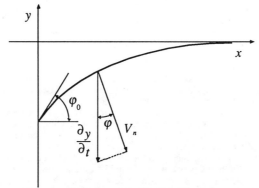

Fig. 1. Geometry of the liquid etching groove.

The boundary conditions for the eq. (2) are given by

$$\left.\begin{array}{l} \varphi\big|_{x=+0} = \varphi_0 \\ y\big|_{x=\infty} = 0 \end{array}\right\}, \tag{3}$$

The angle φ_0 may take any value in the interval $0 < \varphi_0 \leq \pi/2$.

CAPILLARITY MODE

Suppose that the liquid metal solution is in equilibrium with a planar surface of the crystal, i.e.,

$$V_0 = 0. \tag{4}$$

In this case, which is referred to as the capillary mode of liquid metal corrosion, grain boundary etching is caused by capillary forces alone. A similar situation was considered in [21, 22] for thermal etching grooves. For the capillary mode, eq.(2) with boundary conditions (3), has a scaling solution describing development of the groove on a plane surface, which can be expressed in the following form:

$$x = (At)^{1/2} \xi(\varphi); \quad y = (At)^{1/2} \eta(\varphi). \tag{5}$$

Equation (2) may now be written in terms of dimensionless functions $\xi(\varphi)$ and $\eta(\varphi)$ as

$$\eta \, d\xi - \xi \, d\eta = 2 d\varphi, \tag{6}$$

where
$$\tan\varphi = d\xi/d\eta. \tag{7}$$
This problem has a well known solution for the case of thermal grooving, where $\varphi_0 \ll 1$ [21], but here we are interested in the case characteristic for liquid metal etching, where $\varphi_0 \to \pi/2$. In this case eq. (7) can be written as
$$d\xi = (\pi/2 - \varphi)d\eta; \tag{8}$$
hence,
$$d\varphi = -\frac{d^2\xi}{d\eta^2}. \tag{9}$$
Using eq. (9) in eq. (6), one can obtain the solution of the problem which obeys the boundary condition $\lim_{\eta=-\infty}\xi = 0$ as
$$\xi \propto \int \exp(-\eta^2/4)d\eta. \tag{10}$$
Formally, eq. (10) implies that the depth of the groove is infinite. However, the width of the groove decreases exponentially fast with the depth, reaching the scale of interatomic distances. Therefore, the capillary mode cannot supply anything similar to the canals along grain boundaries described above. Instead, an ordinary groove appears and slowly grows as a square root of time, even for completely wetted grain boundaries with $\varphi_0 = \pi/2$.

ETCHING MODE

The situation with $V_0 > 0$ will be referred to as the etching mode of liquid metal corrosion. In this case eq. (2) has no scaling solution, and we cannot supply any closed analytical solution of it with the boundary conditions (3). However, a discussion of a related problem with other boundary conditions
$$\varphi|_{x=0} = \varphi_0/2; \quad \varphi|_{x=L} = -\varphi_0/2, \tag{11}$$
where the meaning of parameter L is clear from Fig. 2, will help us to understand the crystal surface shape evolution in the etching mode.
Equation (2) with boundary conditions (11) has a steady state asymptotic solution, which can be obtained in a closed analytical form [23] as
$$x(\varphi) = -\frac{A\varphi}{V_s} + \frac{V_0}{V_s}\frac{A}{\sqrt{V_s^2 - V_0^2}}\log\left(\frac{\sqrt{V_s^2 - V_0^2} - (V_s + V_L)\tan\dfrac{\varphi}{2}}{\sqrt{V_s^2 - V_0^2} + (V_s + V_L)\tan\dfrac{\varphi}{2}}\right),$$
$$y(\varphi,t) = -V_s t + \frac{A}{V_s}\log\left(\frac{V_s - V_0}{V_s\cos\varphi - V_0}\right) \tag{12}$$
where V_s is the velocity of steady-state motion, determined from the implicit formula
$$L = \int_{-\varphi_0/2}^{+\varphi_0/2} \frac{A\cos\varphi\, d\varphi}{V_s\cos\varphi - V_0}. \tag{13}$$
If $L \ll A/V_0$ then capillary forces substantially exceed the chemical driving force, and we return to the capillary mode (although in a different geometry) with $V \propto 1/L$. The case $L \ll A/V_0$ exhibits a rather different behavior from the capillary mode. The velocity of steady-state motion, V_s, in this case is independent of L:
$$V_s = \frac{V_0}{\cos(\varphi_0/2)}. \tag{14}$$

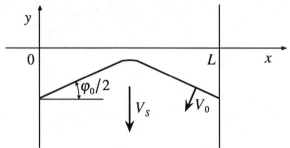

Fig. 2. Form of the steady-state solution.

The solution (8) corresponds to a shape of the surface containing two straight segments connected by a curve with characteristic size and curvature radius of about A/V_0.

Returning to the original problem of grain boundary etching, we note that for the grooves of size in excess of A/V_0, boundary conditions (3) may be obtained from (11) by a mere rotation by $\varphi_0/2$. The shape of the groove is depicted on Fig. 3, and the velocity of penetration is

$$V_s = \frac{V_0}{\cos(\varphi_0)}. \tag{15}$$

The vertex of the groove, like the point where the cutting blades of scissors meet, moves faster than the groove walls. Thus, the etching mode can supply a mechanism for development of a canal peculiar to rapid liquid metal corrosion along grain boundaries. For completely wetted boundaries, when φ_0 goes to $\pi/2$, the velocity of the groove propagation becomes infinite.

It is clear that the formally infinite velocity of the canal propagation V implies inapplicability of the chosen approximation. The kinetics of solution of the walls of the groove no longer determines the velocity of the process, which is determined by other factors — such as kinetics of the opening of the canal in the vertex, $i.e.$ the mobility of the junction of the three surfaces forming the groove, diffusional removal of the solute matter in the case of long canals, etc.

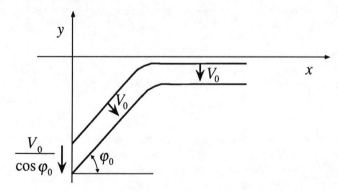

Fig. 3. Liquid metal etching groove at $L \ll A/V_0$.

CONCLUSIONS

The phenomenon of rapid liquid metal grain boundary corrosion may be explained by the analysis of grain boundary etching profiles for $\varphi_0 \to \pi/2$. It is essential that the length given by the ratio A/V_0 should be microscopically small, *i.e.*, the liquid solution should be far from saturation, because it explains the delicate nature of the phenomenon. For example, in the Nb-Li system rapid liquid metal corrosion can be stopped by removing oxygen. In the absence of oxygen, the solution of Nb in Li becomes saturated, and $V_0 = 0$, so the canal does not develop. In the presence of oxygen, insoluble niobium-lithium trioxides form [24], the solution becomes depleted and liquid lithium penetrates the niobium along its grain boundaries. The effect of Zr may be explained by the fact that Zr tends to bind to oxygen in the system [17, 24]. In some cases stress may have a similar effect, increasing the chemical potential of the solid and upsetting the equilibrium with the liquid solution [25].

It should also be pointed out, that the condition of complete wetting, that is necessary for the phenomenon under consideration, might depend on the temperature and other factors [26]. This condition also may not hold for all grain boundaries simultaneously [27], and then rapid corrosion may develop only at some grain boundaries and not at others.

ACKNOWLEDGMENT

I would like to thank E.A. Glickman for attracting my attention to the fascinating features of the phenomenon of liquid metal penetration, E.A. Brener for discussion of the mathematical formulation of the problem, and A.R.C. Westwood, N.F. Stoloff, and M.E. Glicksman for reading the manuscript and valuable discussions.

REFERENCES

1. N. S. Stoloff. *Metal-Induced fracture.* in *Environmrnt-Induced Cracing of Metals.* 1988. Kohler, Wisconsin: NACE-10.
2. A. R. C. Westwood: in Fracture of Solids, D. C. Drucker and J. J. Gilman, Ed. 1963, Interscience Publisyhers: New York, NY. p. 553.
3. W. Rostoker, J. M. McCaughey, and H. Markus, Embrittlement by Liquid Metals. 1960, Reinhold, NY.
4. V. W. Eldred, *Atomic Energy Research Establishment Rpt. X/R.* 1956, p. 1806.
5. N. Eustathopoulos, L. Condurier, J. C. Loud, and P. Desre. J. Cryst. Growth, 33(1), 105 (1976).
6. K. K. Ikeuye and C. S. Smith. Trans. AIME, 185 762 (1949).
7. E. M. Lyutyi, O. I. Eliseeva, and R. I. Bobyk. Soviet Materials Science, 26(November-December), 611 (1990).
8. A. Passerone, R. Sangiorgi, and N. Eustathopoulos. Scripta Met., 14(10), 1089 (1980).
9. A. Passerone, R. Sangiorgi, and N. Eustathopoulos. Scripta Met., 16(5), 547 (1982).
10. A. Passerone and R. Sangiorgi. Acta Met., 33 77 (1985).
11. J. H. Rogerson and J. C. Borland. Trans. AIME, 227 2 (1963).
12. E. Scheil and K. E. Schiessl. Zs. Naturfoorsch, 4a 524 (1949).
13. A. Passerone and N. Eustathopoulos. Acta Met., 30(7), 1349 (1982).
14. L. A. Onuchak, . 1975, Moscow State University, Moscow: Thesis
15. V. Y. Taraskin, V. Y. Goryunov, G. I. Denshchikova, and B. D. Summ. Phys. Chem. Mech. Mater., 68 643 (1965).
16. A. B. Pertsov, L. A. Pogosuan, B. D. Summ, and Y. V. Goryunov. Sov. Colloid J., 26 699 (1974).
17. J. R. Di Stefano, . 1964, University of Tennessee: Thesis
18. F. N. Rhines and B. R. Patterson. Metall. Trans. A, 13A 985 (1982).

19. E. E. Hoffman. USAEC Report ORNL - 2675 (1959).
20. E. E. Hoffman. USAEC Report ORNL - 2674 (1959).
21. W. W. Mullins. J. Appl. Phys., **28**(3), 333 (1957).
22. W. W. Mullins. Trans. AIME, **218**(April), 354 (1960).
23. A. V. Galina, V. E. Fradkov, and L. S. Shvindlerman. *Rapid Penetration Along Grain Boundaries*. Preprint of Institute of Solid State Physics, Chernogolovka, 1986.
24. A. V. Lapitsky and Y. P. Simanov. Sov. Vestnik MGU (Proc. of Moscow State Univ.), ser. Phys.-Math., **9(2)**, 67 (1954).
25. W. M. Robertson. Trans. AIME, **236** 1478 (1966).
26. E. I. Rabkin, V. N. Semenov, L. S. Shvindlerman, and B. B. Straumal. Acta Met., **39** 627 (1991).
27. E. I. Rabkin, L. S. Shvindlerman, and B. B. Straumal. Int'l J. of Modern Phys. B, **5**(No. 19), 2989 (1991).

CHEMICAL STABILITY OF REACTIVELY SPUTTERED AlN WITH PLASMA ENHANCED CHEMICAL VAPOR DEPOSITED SiO$_2$ AND SiN$_x$

JAESHIN CHO* AND NARESH C. SAHA**
*Compound Semiconductor-1, Motorola Inc., Tempe, Arizona 85284
**Semiconductor Analytical Laboratory, Motorola Inc., Mesa, Arizona 85202

ABSTRACT

We have studied the chemical stability of reactively sputtered aluminum nitride film with plasma-enhanced chemical vapor deposited SiO$_2$ and SiN$_x$ films. It was found that the PECVD SiO$_2$ film reacted with AlN to form aluminosilicate ($3Al_2O_3 \cdot 2SiO_2$) at the SiO$_2$/AlN interface after annealing above 550°C. The presence of Al-O bonds at the SiO$_2$/AlN interface was verified with x-ray photoelectron spectroscopy and x-ray induced Auger electron spectroscopy. The formation of aluminosilicate resulted in significant decrease in the wet etch rate of AlN layer. For SiN$_x$/AlN/Si layered structure, no interfacial reactions were detected at the SiN$_x$/AlN interface after annealing up to 850°C. These results confirm the thermodynamic predictions on the mutual stability of SiN$_x$/AlN and SiO$_2$/AlN.

INTRODUCTION

The AlN is a wide bandgap III-V compound which has unique properties: high thermal conductivity, chemical and thermal stability, high selectivity to fluorine-based etch plasmas, piezoelectric property, etc. AlN films have potential applications for high-frequency monolithic surface acoustic wave devices, because of their large electromechanical coupling constant and high acoustic velocity [1]. Also, the chemical and thermal stability as well as high electrical resistivity of AlN films make them suitable as insulating and passivating layers in semiconductor devices [2,3]. However, AlN is rarely used in semiconductor device fabrication because other aspects of AlN, such as the deposition technique for device quality film, its etch characteristics and chemical stability with other dielectric films are not thoroughly characterized.

In this study, the reactive sputtering technique was used to deposit device-quality AlN film. Chemical stability of AlN film were studied using layered structures of SiO$_2$/AlN/Si and SiN$_x$/AlN/Si. The x-ray photoelectron spectroscopy (XPS) and x-ray induced Auger electron spectroscopy (XAES) were used to characterize the interfacial reaction products, and these results were correlated with the wet etch characteristics of AlN film and of the layered structures.

EXPERIMENTAL PROCEDURES

The AlN films were reactively sputter deposited with Ar and N$_2$ gases using DC magnetron sputtering. The deposition conditions were adjusted to have amorphous AlN film. The optimum N$_2$ and Ar flow ratio was 80:20, and the power was 9 kW with deposition rate of 1000Å/min. The background pressure was maintained below 1.33×10^{-6} Pa to minimize the possible incorporation of oxygen into the film, and the deposition was performed without intentional heating of wafers. The refractive index of the as-deposited AlN film was around 2.0 with typical thickness uniformity of less than 1% across 4 inch wafers. The thickness of AlN used in this study was 300Å otherwise mentioned.

The etching of AlN was performed using either concentrated H$_3$PO$_4$ or dilute (1:10) NH$_4$OH:H$_2$O solutions. The dilution of the etchants and its temperature were optimized to have approximately 100-200Å/min of etch rate for the as-deposited AlN film. The bath temperature of H$_3$PO$_4$ acid was maintained at 50°C, and NH$_4$OH at room temperature. The thickness and refractive index of AlN were measured using ellipsometry before and after etching. In order to simulate the actual device fabrication, the wet etch rates of AlN film and of the layered structures, SiO$_2$/AlN/Si and SiN$_x$/AlN/Si, were measured after annealing and C$_2$F$_6$ plasma exposure. The

SiO$_2$ and SiN$_x$ films were deposited using plasma-enhanced CVD at 350°C. Annealing was done either in a conventional furnace in N$_2$ ambient or by rapid thermal process in Ar ambient.

The interfacial chemistry of SiO$_2$/AlN, SiN$_x$/AlN and AlN/Si was analyzed by x-ray photoelectron spectroscopy (XPS) and x-ray induced Auger electron spectroscopy (XAES) using Perkin Elmer 5400 spectrometer. The photoelectron spectra were excited by using Mg K$_\alpha$ x-rays (1253.6 eV) and the x-ray induced Al KLL Auger spectra by using Zr L$_\alpha$ x-rays (2040.4 eV). All samples were analyzed at 45° take-off angle with respect to the sample surface. The typical depth of analysis at this configuration was ~50Å.

RESULTS AND DISCUSSION

Etching characteristics of AlN

Fig. 1 shows wet etch characteristics of the 300Å thick as-deposited AlN film in concentrated H$_3$PO$_4$ acid as well as in (1:10) NH$_4$OH:H$_2$O solution. The etch rate of AlN film was ~100Å/min in H$_3$PO$_4$ acid and ~200Å/min in NH$_4$OH solution. It confirms that the AlN is soluble both in acid such as phosphoric acid, and in alkaline solutions such as ammonium hydroxide and positive photoresist developer [4].

The etch rate of AlN film can be a function of subsequent anneal temperature. Fig. 2 shows the etch rate of 300Å thick AlN film that was furnace annealed at 550°C for 30 min in N$_2$ ambient. While the etch rate of annealed AlN film in NH$_4$OH was the same as the as-deposited film (~200Å/min), the etch rate in H$_3$PO$_4$ acid was slowed down to 30Å/min. There was a slight decrease in the refractive index after anneal. The XPS analysis showed that the top surface of AlN was converted to 20-30Å thick aluminum oxide after annealing. It is believed that the observed discrepancy in the etch rate in NH$_4$OH vs. H$_3$PO$_4$ solutions after anneal resulted from the differences in the effectiveness of these solutions in removing the top aluminum oxide layer to initiate etching. It suggests that NH$_4$OH is more effective in removing surface Al$_2$O$_3$ layer than H$_3$PO$_4$ acid.

AlN film was known as an effective etch stop from fluorine-based etch plasmas due to the lack of volatile fluorine compounds with aluminum. A 30-50Å thick AlF$_3$ layer was formed on the AlN surface after exposed to C$_2$F$_6$ plasma as shown in Fig. 3 of the Al 2p XPS spectrum. It was found that the wet etch rate of C$_2$F$_6$ plasma treated AlN was significantly different from the as-deposited film as shown in Fig. 4. There was no etching action in H$_3$PO$_4$ solution due to insolubility of AlF$_3$ in H$_3$PO$_4$, while the etch rate in NH$_4$OH solution was the same as the as-

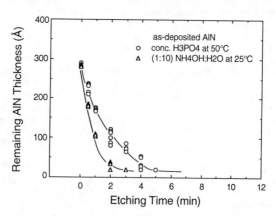

Fig. 1. Wet etch rate of 300Å-thick as-deposited AlN film in concentrated H$_3$PO$_4$ and (1:10) NH$_4$OH:H$_2$O solutions.

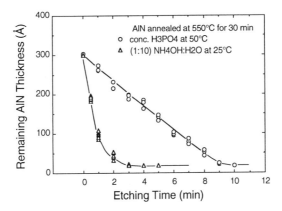

Fig. 2. Wet etch rate of annealed AlN film in concentrated H_3PO_4 and (1:10) $NH_4OH:H_2O$ solutions. Samples were furnace annealed at 550°C for 30 minutes in N_2 ambient.

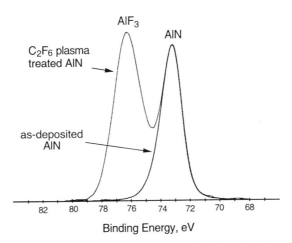

Fig. 3. Al 2p XPS spectra of C_2F_6 plasma-treated AlN surface showing the presence of 30-50Å thick AlF_3 layer.

deposited film. In order to break AlF_3 layer, higher etch temperature of H_3PO_4 acid was required, and once the top AlF_3 was removed the etching was proceeded with the same rate of as-deposited film.

Etching characteristics of layered structures

Since the AlN had consistent etch rates in dilute NH_4OH solution independent of both annealing history and the presence of AlF_3 layer as discussed above, the dilute NH_4OH solution was used to investigate the interfacial reactions of AlN with PECVD SiO_2 and SiN_x films. The layered structures of $SiO_2/AlN/Si$ and $SiN_x/AlN/Si$ were furnace annealed at 550°C for 30 minutes, and the SiO_2 and SiN_x layers were C_2F_6 plasma etched using underlying AlN layer as an etch stop. Then the wet etch rate of AlN layer in NH_4OH solution was evaluated. It was found that the

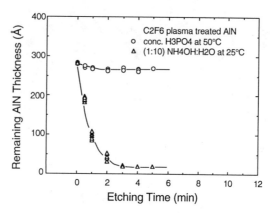

Fig. 4. Etch rate of C_2F_6 plasma treated AlN film in concentrated H_3PO_4 and (1:10) $NH_4OH:H_2O$ solutions.

AlN layer in SiN_x/AlN/Si sample had the same etch rate with the as-deposited film, i.e., completely cleared after 3 minutes of etching. However, for SiO_2/AlN/Si sample there was no etching action even after 30 minutes of etching. It is surprising because the previous studies showed that the etch rate of AlN was independent of both annealing and C_2F_6 plasma treatment. In order to have better understanding of AlN/SiO_2 interface, i.e., whether there was any interfacial reaction to slow down or block the etching, a sample with opposite layering, AlN/SiO_2/Si was prepared for XPS/XAES analysis. The reason for the opposite layering was to isolate the temperature history and possible oxidizing plasma exposure on AlN surface during SiO_2 deposition from thermally induced reactions during annealing.

2000Å thick PECVD SiO_2 film was deposited at 350°C on Si wafers and then 300Å thick AlN film was subsequently deposited on top of SiO_2. Then the samples were annealed at 550°C for 30 minutes. The XPS spectra showed that top AlN surface was covered with ~20Å of aluminum oxide layer. The formation of the surface oxide layer was presumably due to presence of residual oxygen in the annealing environment. This top Al_2O_3 can be easily wet etched with dilute NH_4OH solution as evidenced by XPS data. The AlN layer was then successively wet etched down to AlN/SiO_2 interface by controlling the wet etch rate deliberately using (1:100) $NH_4OH:H_2O$. Al 2p photoelectron peak and x-ray induced Auger transition (Al KLL) were acquired after each etch to monitor the AlN/SiO_2 chemistry. The XPS and XAES data were collected by using angle resolved technique to ensure that a continuous layer of AlN remained to protect the interface yet the interface chemistry was within the analysis depth.

Figure 5 shows four Al KLL XAES spectra from the AlN/SiO_2/Si sample as approaching towards the AlN/SiO_2 interface. The Al peak from the middle of AlN layer was mostly from Al-N bonds indicating that AlN is an effective oxidation barrier. The typical native oxide layer present on the AlN surface due to air exposure after NH_4OH wet etch was < 5Å (top curve of Fig. 5). As approaching towards the AlN/SiO_2 interface, Al KLL spectra showed presence of two peaks at the kinetic energies of 1386.7±0.2 eV and 1389.0±0.2 eV due to Al-O and Al-N bondings, respectively. The intensity of the 1386.7 eV peak continue to increase with increasing etch time. From the intensity ratio of the Al-O and Al-N peaks, the thickness of interfacial Al-O layer was estimated to be ~50Å. The depth profile using controlled wet etch showed distinct advantage over the sputter etch depth profile which when was used could not resolve the interface chemistry due to ion beam induced mixing and peak broadening.

The chemical stability of AlN with SiO_2 and Si_3N_4 has been predicted by Bhansali et al. [5] using thermodynamic calculations. The stable tie-lines are present from AlN to Si, Si_3N_4 and Al_2O_3 as shown in Fig. 6 of the ternary phase diagrams comprising the quaternary Al-Si-N-O system, and thus AlN is in equilibrium with Si substrate, Si_3N_4 and Al_2O_3. They also predicted

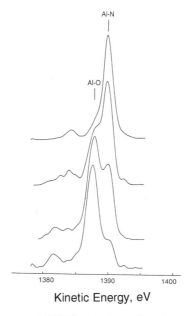

Fig. 5. Al KLL XAES spectra near AlN/SiO$_2$ interface taken from AlN/SiO$_2$/Si (300/2000Å) layered structure. Samples were annealed at 550°C for 30 minutes and the AlN layer was continuously wet etched using (1:100) NH$_4$OH:H$_2$O solution.

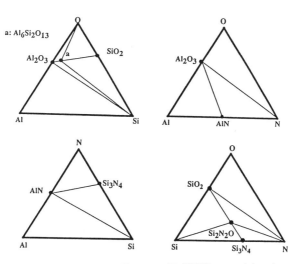

Fig. 6. The four calculated ternary phase diagrams (T=900K) comprising the quaternary Al-Si-N-O system from reference [5].

that SiO_2 and AlN are unstable with each other forming Al_2O_3 and/or aluminosilicate ($3Al_2O_3 \cdot 2SiO_2$) at sufficiently high temperatures. From our XPS and XAES analyses, the observed Al-O bonds at AlN/SiO_2 interface indicate the formation of Al_2O_3 or aluminosilicate. Since the phase present at the SiO_2/AlN interface was insoluble in NH_4OH solution, it suggests that the phase could be in the form of aluminosilicate which is insoluble in basic solutions [6]. The same XPS/XAES analysis was performed at the AlN/SiN_x interface from $AlN/SiN_x/Si$ layered structure using previously described method, and there was no interfacial reaction detected after rapid thermal annealing up to 850°C. It confirms the presence of stable tie line between Si_3N_4 and AlN as predicted from Al-Si-N ternary phase diagram in Fig. 6.

CONCLUSIONS

The chemical stability of reactively sputtered AlN film with PECVD SiO_2 and SiN_x films was characterized using XPS and XAES. For layered structure of $SiO_2/AlN/Si$, Al-O bonds were detected at SiO_2/AlN interface after annealing above 550°C, but for layered structure of $SiN_x/AlN/Si$ there was no interfacial reaction detected between SiN_x and AlN. The Al-O bonds at SiO_2/AlN interface are believed from the formation of aluminosilicate ($3Al_2O_3 \cdot 2SiO_2$) as predicted from the thermodynamic calculations, and this interfacial layer significantly decrease the wet etch rate of AlN in NH_4OH solution.

ACKNOWLEDGMENTS

The authors would like to thank Wayne Cronin, Kirby Koetz and Frank Barton for the deposition of AlN, and CS-1 for continuous support on this study.

REFERENCES

1. M.H. Francombe and S.V. Krishnaswamy, J. Vac. Sci. Technol. **8**, 1382 (1990).
2. R.D. Pashley and B.M. Welch, Solid State Electron. **18**, 977 (1975).
3. S.P. Kwok, J. Vac. Sci. Technol. **B4**, 1383 (1986).
4. T. Tada and T. Kanayama, J. Vac. Sci. Technol. **B11**, 2229 (1993).
5. A.S. Bhansali, D.H. Ko and R. Sinclair, J. Electronic Mater. **19**, 1171 (1990).
6. CRC handbook of Chemistry and Physics, 71st Ed. (CRC Press, Boston, 1990) p.4-42.

THE THERMAL STABILITY OF TI-AL-C PVD COATINGS

O. Knotek*, F. Löffler, L. Wolkers
Materials Science Institute, Aachen University of Technology, 52056 Aachen, Germany

ABSTRACT

While (Ti,Al)N coatings are well known and established in wear applications like drilling and milling, the corresponding TiC based metastable system Ti-Al-C is not well examined. This paper investigates the phase stability and the phase transition behaviour of post-deposition heat-treated Ti-Al-C coatings. The samples were annealed in a temperature range from 600-1400 °C in a high vacuum furnace. The influence of deposition parameters like coating compositions via the Ti/Al ratio and the metal/metalloid ratio on the properties were investigated. Another very important parameter in this investigation is the carbon carrier gas. Examined gases were methane, ethane, ethylene and ethine. The changes in microhardness, lattice parameters and particle sizes were analysed with respect to their temperature dependence.

The phase formation depends on the target composition and on the reactive gas. A formation of a metastable (Ti,Al)C phase can be observed. A higher C:H ratio of the reactive gas leads to better incorporation of carbon into the coating. An Al-loss independent of the coating composition begins 1000 °C. The microhardness increases with higher annealing temperature and decreases with higher Al-content.

INTRODUCTION

The phase formation and transformation of PVD-coatings is one of the most important influences on the properties of such materials. For example the metastable phase formation of the (Ti,Al)N coatings characterizes directly the field of applications [1,2]. The phase transformations of such thin film limit the application and especially the usable temperature range, because they very often cause a crack formation and other coating failures in thin films. This occurs for amorphous (Al-O-N, Si-N) and crystalline coatings [3]. In this paper, the phase tranformation and stability of crystalline Ti-Al-C coatings is described. The influence of the target composition, reactive gases and the coating composition is studied.

EXPERIMENTAL

The coatings were deposited with a Magnetron Sputter Ion Plating process from targets with the following compositions: Ti 100, Ti/Al 75/25, Ti/Al 50/50, Ti/Al 25/75 (at-%). The reactive gases used were methane, ethane, ethene and ethine. The samples were heat treated in a vacuum furnace in a temperature range from 600 to 1400 °C in 200 °C steps. After the annealing, they were examined with X-ray diffraction and microhardness measurements.

RESULTS

In Table 1 the phase formation as a function of target composition is given for substoichiometric Ti-Al-C coatings using methane as the reactive gas. A substoichiometric composition means that the metal/metalloid ratio is greater than 1. This phase transformation diagram must be read from lower to higher temperatures, because it describes irreversible transformations. Depending on the target composition, different phases are formed. The coatings deposited with the Ti-target show TiC formation, a phase which is stable up to 1400 °C. On the Al-side only an Al-phase is formed; carbon is dissolved in parts of the amorphous coating: annealings to temperatures under 1000 °C do not effect the phases; at 1000 °C the thin film is evaporated. The Ti/Al targets show a different behaviour. The Ti/Al 75/25 target forms the intermetallic phase Ti_3Al and a highly deformed TiC phase, which can be characterized as (Ti,Al)C. This phase is decribed more precisely by the lattice parameters as follows. This phase is stable up to 800 °C. Above this temperature the "deformation" of the TiC monocell diminishes. At 1200 °C only TiC is detectable. TiC is the only phase detected for the two other Ti/Al target compositions after the 1200 °C anneal. The Al-rich target forms $Al_{24}Ti_8$, the superlattice of Al_3Ti, and the 50/50 composition forms the H-phase, Ti_2AlC. At 1000 °C the formation of TiC starts in both coatings.

The phase formation changes drastically with an increased carbon content, as shown in Table 2. The coatings deposited with the Ti/Al 75/25 and 50/50 targets show a formation of

1400 °C	TiC	TiC	TiC	TiC	
1200 °C	TiC	TiC	TiC	TiC	
1000 °C	TiC	TiC Ti_3Al	TiC Ti_2AlC	TiC Al_3Ti	
800 °C	TiC	(Ti,Al)C Ti_3Al	Ti_2AlC	Al_3Ti	Al
600 °C	TiC	(Ti,Al)C Ti_3Al	Ti_2AlC	$Al_{24}Ti_8$	Al
200 °C	TiC	(Ti,Al)C Ti_3Al	Ti_2AlC	$Al_{24}Ti_8$	Al
Ti/Al-Ratio [at-%] →	100/0	75/25	50/50	25/75	0/100

(Left axis label: Annealing Temperature [°C])

Table. 1: Phase transition diagram of substoichiometric Ti-Al-C coatings

Annealing Temperature [°C]		$100/0$	$75/25$	$50/50$	$25/75$	$0/100$
	1400 °C	TiC	TiC	TiC	TiC	
	1200 °C	TiC	TiC	TiC	TiC	
	1000 °C	TiC	TiC (Ti,Al)C	TiC (Ti,Al)C	TiC Ti_2AlC	Al_4C_3
	800 °C	TiC	(Ti,Al)C	(Ti,Al)C	Ti_2AlC Al_3Ti	Al_4C_3
	600 °C	TiC	(Ti,Al)C	(Ti,Al)C	Ti_2AlC $Al_{24}Ti_8$	Al_4C_3 Al
	200 °C	TiC	(Ti,Al)C	(Ti,Al)C	$Al_{24}Ti_8$	Al
	Ti/Al-Ratio [at-%] →	$100/0$	$75/25$	$50/50$	$25/75$	$0/100$

Table. 2: Phase transition diagram for stoichiometric Ti-Al-C coatings

the metastable (Ti,Al)C phase, which is comparable to (Ti,Al)N, where some Ti in the monocell is replaced by Al. At 1000 °C Al-loss occured, so that at higher temperature anneal only TiC occurs, which is again the only phase on the Ti-side of the diagram. At the Al-rich side of the diagram, an $Al_{24}Ti_8$ phase is detected. After the 600 °C anneal Ti_2AlC is formed and is stable up to 1000 °C. At this temperature TiC is also formed, the only stable phase at high temperatures. The formation of Al_4C_3, the only known Al-C-phase, takes place at 600 °C and is stable up to 1000 °C. The influence of the reactive gases on the phase formation is given for Ti/Al 50/50 at-% targets. The differences between Table 3 and Table 1 and 2 are caused by the slightly lower reactive gas pressure used for this investigation. As a function of the reactive

Annealing Temperature [°C]		CH_4	C_2H_6	C_2H_4	C_2H_2
	1400 °C	TiC	TiC	TiC	TiC
	1200 °C	TiC	TiC	TiC	TiC
	1000 °C	TiC Ti_2AlC	Ti_2AlC	(Ti,Al)C	Ti_2AlC
	800 °C	Ti_2AlC	Ti_2AlC	(Ti,Al)C	Ti_2AlC
	600 °C	Ti_2AlC	Ti_2AlC	(Ti,Al)C	Ti_2AlC
	200 °C	Ti_2AlC	Ti_3Al	(Ti,Al)C	Amorph
	Target:Ti/Al 50/50	CH_4	C_2H_6	C_2H_4	C_2H_2

Carbon-Hydrogen-Gas

Table 3: Phase transition diagram of Ti-Al-C deposited with different reactive gases

gas, the following phases are formed: Ti_2AlC (methane), Ti_3Al (ethane), (Ti,Al)C (ethene) and amorphous (ethine). This enormous difference, which is caused by the different reactivities of the various carbon hydrogen gases, changes after the first annealing. Ti_2AlC is now formed in all coatings, only (Ti,Al)C is still stable. Annealings up to 1000 °C causes no detectable phase transformations. After the 1200 °C anneal, TiC is the only detectable phase. This means that there is an obvious Al-loss above 1000 °C. All Al-containing phases are decomposed.

Figure 1 shows the X-ray diffraction examined lattice parameters, of several TiC and (Ti,Al)C coatings and the change in lattice parameters indicate the deformation of the TiC monocell. The influence of the reactive gas on the lattice parameter of the TiC phase was examined. The carbon content of the plasma was the same for all depositions. This means that the different reactivity of the gases influences the film growth and the carbon incorporation. It is shown that for all gases, the lattice parameter of TiC is not reached, since there is not enough carbon incorporated, a known problem of the PVD processes. The lattice parameter decreases with increasing anneal temperature for all TiC coatings. The (Ti,Al)C shows an opposite behaviour. With higher annealing temperature the lattice parameter increases. At 1000 °C the lattice parameter of TiC is reached. Such a behaviour is known from the (Ti,Al)N coatings, the increase of the lattice parameter goes along with with an Al-loss in the metastable phase [4].

The particle sizes of the coating material have been investigated as well. In Fig. 2 the particle sizes of the TiC and the (Ti,Al)C phase are given as a function of the target composition. With increasing Al-content larger particle sizes were measured. This is valid up to

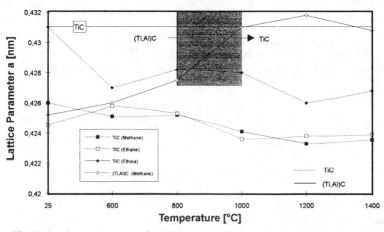

Fig. 1: Lattice parameter of TiC and (Ti,Al)C coatings in function of the annealing temperature

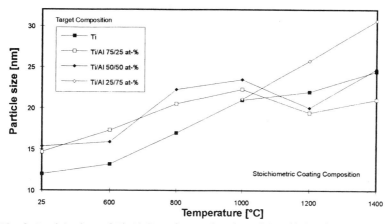

Fig. 2: Particle sizes of Ti-Al-C coatings as function of the Ti/Al ratio and the annealing temperature

1000 °C. Reasoning from the general Al-loss we concluded that te target composition did not influence the final product for the 1000 anneal. The particle sizes increase with increasing anneal temperature. An interrelationship between the carbon carrier gas and the particle size is not detectable.

The influence of the target composition and the annealing temperature on the microhardness is given in Fig. 3, showing, that with an increasing Al-content, the micro-hardness

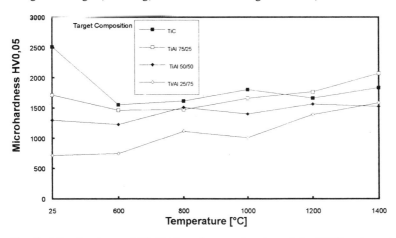

Fig. 3: Microhardness of Ti-Al-C coatings as function of the Ti/Al ratio and annealing temperature

decreases. The microhardness also increases with higher annealing temperatures. Such a behaviour is only known for coatings with new phase formations and transforma-tions [5]. We Speculate that the difference in microhardness which is determined by the disappearence of Al after the evaporation of the Al and the formation of the TiC. An influence of the carbon carrier gas on the microhardness can not be detected. The influence of Al on the microhardness is in contrast to the behaviour known for the comparable (Ti,Al)N coatings [6,7].

CONCLUSIONS

The phase formation in the system Ti-Al-C depends on the target composition and the reactive gases used. A metastable phase like (Ti,Al)C can be deposited. The formation of (Ti,Al)C depends on the reactive gas partial pressure and the reactive gas itself. To deposit this phase it is necessary to use a very high reactive gas pressure which is not far from target poisoning. In the cases where no (Ti,Al)C is formed through deposition, Ti_2AlC after a 600 °C annealing is formed. The reactive gases do not have a direct influence on coating properties like microhardness or particle size. However an influence on the lattice parameter can be detected. A higher C:H ratio leads to a better incorporation of carbon into the coating. Independent of the coating composition, the phase formation, and the carbon carrier gases used an Al-loss was observed at 1000 °C and ended after the 1200 °C anneal. In comparison to other metastable PVD coatings like (Ti,Al)N, the Ti-Al-C coatings have a lower thermal stability and a lower microhardness level. The coating structure is strongly influenced by the deposition parameters. Use of this material in typical PVD coating applications, like machining, seems to be inappropriate.

REFERENCES

[1] D.T. Quinto, G.J. Wolfe, P.C. Jindakl, Thin Solid Films, **153**, 19 (1987).

[2] O. Knotek, M. Böhmer, T. Leyendecker, J. Vac. Sci. Technol., **A4**, 2695 (1986).

[3] O. Knotek, F. Löffler, L. Wolkers; Proc. o. AMPT Conference **93**, Dublin, Aug. 1993, Vol.1, 427.

[4] O. Knotek, F. Löffler, L. Wolkers; Proc. o. AMPT Conference **93**, Dublin, Aug. 1993, Vol.2, 775.

[4] B. Wendler, K. Jakubowski; J. Vac, Sci. Technol., **A6**, 93 (1988).

[5] W.D. Munz J. Vac. Sci. Technol, **A4**, 2717, (1989).

[6] E. Vancouille, J.P. Celis, J. R. Roos, Thin Solid Films, **224**, 168 (1993).

MICROSTRUCTURAL DEVELOPMENT AT Ti-BASED ALLOY/COATED SiC INTERFACES

M. STRANGWOOD*, C.B. PONTON*, M.P. DELPLANCKE**, V. VASSILERIS** AND R. WINAND**
*The University of Birmingham, Faculty of Engineering, School of Metallurgy and Materials, Edgbaston, Birmingham B15 2TT, United Kingdom
**Université Libre de Bruxelles, Faculty for Applied Sciences, Metallurgy-Electrochemistry, 50 av. F.D. Roosevelt, 1050 Brussels, Belgium

ABSTRACT

The kinetics of formation of reaction layers at the interface between a Ti-based alloy (β-21s) and graphite blocks coated first, by CVD techniques, with a 100 μm layer of SiC and then either TiC or C were determined. The rate controlling step for reaction at 900°C and 10 MPa for up to 6 hours was found to be carbon diffusion through the reaction layer. The behaviour was found to be consistent with that of composite systems for prolonged heat treatment and the same growth behaviour was exhibited by both systems. Incubation times of 0.42 and 0.9 hours were determined for reaction layer growth in the TiC- and C-coated systems respectively. The reaction layer/Ti-alloy bond strength was good but all couples failed readily, especially if any C layers remained. The growth conditions of TiC coatings by reactive magnetron sputtering were determined for two different types of gas mixtures: CH_4/Ar and C_2H_2/Ar. The composition and structure of the films were extensively studied.

INTRODUCTION

The need for high specific strength and stiffness materials for aerospace applications, such as rotating components in the intermediate and high pressure compression stages of gas turbine engines, has stimulated the development of composite materials incorporating continuous SiC fibres in titanium alloy matrices. Many of the proposed applications involve fatigue-limited components under which situations the continuously reinforced material can give rise to crack bridging by the fibres and a decreasing crack growth rate as it extends, often leading to crack arrest [1]. This behaviour is dependent upon load transfer from the matrix to the fibre and so on shear strength of the fibre-matrix interface. This depends on the extent of any reaction between the fibre (or coating) and matrix, residual stresses developed on cooling and transformation stresses from solid-state transformations in the matrix, which will be affected by the reinforcement and any reaction between that matrix [2]. The extent of any chemical reaction will be dominated by high temperature processing of the material, which, for Ti-SiC composites, will be at temperatures around 900°C. The majority of Ti-SiC composites are available as processed panels, so that the initial development of any reaction layer cannot be determined, only their development with extended elevated temperature exposure. In addition, the fine nature of the fibres (100 or 140 μm diameter) means that there is considerable thermodynamic modification of the system through the Gibbs-Thompson effect. Therefore, in order to isolate the various effects and process, planar blocks of graphite coated with a 100 μm thick CVD layer of SiC, were further coated with C or Ti_xC and diffusion bonded to a titanium alloy. This arrangement allows the development of reaction layers under controlled conditions to be studied and compared with the published behaviour of composite systems.

Related to the preparation of the TiC coated model samples, a study of the synthesis of TiC by reactive magnetron sputtering was initiated. The structure and composition of the films were extensively studied as a function of the process parameters (power, gas mixture composition and pressure).

EXPERIMENTAL

Coating preparation

Ti$_x$C:H films of various thicknesses and compositions were prepared by d.c. reactive sputtering of 99,5% pure titanium with a commercial planar magnetron apparatus (Leybold LH560) using argon as the working gases and either methane or acethylene as the reactive gas. The purity of all the gas was at least 99.9%. The effects of the gas phase composition and of the current density were investigated. The concentration of the reactive gases ranged from 0% to 50%. While the experiments were carried out in the dynamic mode the total pressure was around 4×10^{-1} Pa corresponding to a total flux of 100 sccm (standard cubic centimeter per minute). The source-to-substrate distance was 57 mm and the target was 90 mm in diameter. In all experiments the substrates were kept at the ground potential and at room temperature.

The carbon films were deposited in the same device using pure argon as the sputtering gas and a graphite target. The current density and the pressure were respectively 8 mA cm^{-2} and 4.8×10^{-1} Pa.

The films were deposited on doped (100) silicon single crystals previously etched in concentrated hydrofluoric acid (HF 49%) for structural and composition analysis. Films were deposited onto the SiC covered graphite samples after cleaning in ultrasonic bath in ethanol and acetone.

The Ti$_x$C:H films, used in the diffusion couples described below, were prepared from a C$_2$H$_2$ 2 sccm / Ar 98 sccm gas mixture with a discharge current density of 8 mA cm^{-2}.

Composition of the films was investigated by Auger Electron Spectroscopy (AES) and X-ray Photoelectron Spectroscopy (XPS). The morphology and crystallinity of the films were determined by X-ray diffraction (XRD), high energy transmission electron diffraction (TED) and transmission electron microscopy (TEM). X-ray diffraction patterns were recorded with a Siemens diffractometer using the filtered Cu K$\alpha_1\alpha_2$ radiation.

Preparation of the Ti-alloy/C and Ti-alloy/Ti$_x$C couples

The coated blocks of SiC-coated graphite were degreased and interfaced with blocks of a metastable β-Ti alloy (β-21s, Ti-15 Mo-3 Al-2.7 Nb-0.2 Si-0.1 O$_2$ wt%), which had been solution-treated at 900°C, ground to a 1200 grit SiC-paper finish and degreased. The blocks were then placed between the recrystallised alumina anvils in the RF induction coil of a Theta Industries high speed deformation dilatometer. The specimens were then held at temperature, under a 10MPa load, for periods of up to 6 hours. After bonding, the specimens were argon quenched to room temperature under a decreasing load. The specimens were then polished and etched in Kroll's II reagent (2% HF, 10% HNO$_3$, 88% H$_2$O) prior to examination in a Jeol 6300 scanning electron microscope (SEM) operating at 20 kV. The extent of any reaction layer between the components of the diffusion couple was determined and compared with finite difference model for growth of Ti$_x$C layers [3].

RESULTS AND DISCUSSION

Coatings

The influence of two process parameters were investigated: the current density of the discharge and the composition of the gas phase, all other parameters (pressure, total flux, target-to-substrate distance, temperature) remaining constant. In all cases (CH$_4$ and C$_2$H$_2$) and at all current densities, the evolution of the composition of the deposited films as a function of the volume fraction id the gas phase was similar: at low carbon containing gas concentration, the carbon was incorporated in the films as carbide. As the carbon concentration in the gas phase increased, the titanium was completely carburized as can be deduced from the relative amplitude of the AES titanium peaks. An amorphous carbon (a-C) phase also appeared as the concentration of the C-containing gas was further increased. The value of the critical carbon concentration in the gas phase for the appearance of the a-C phase depended on the gas source and on the discharge current density. The carbon concentration in the films corresponding to this critical value was always close to the stoichiometric one. Simultaneously with the composition changes, the grain size decreased and the films became completely amorphous when the a-C phase appeared, as shown by TEM, TED and X-ray diffraction.

These results are consistent with those published in the literature [4,5,6] and can be interpreted by considering the species originating from the target, the mobility of the different species at the substrate surface and a d.c. plasma deposition process occuring simultaneously with the reactive sputtering one. This last factor and the implantation of carbonaceous species in the surface of the target are responsible for the formation of a surface compound on the target tha has a lower sputtering yield than the pure titanium. This explains the decreasing deposition rate with increasing carbon concentration in the gas phase. At high hydrocarbon concentration, the target surface is completely covered by a carbon layer and most of the sputtered material are C atoms. At the substrate surface, the deposition of the sputtered material is taking place simultaneously with the direct deposition of carbon from the plasma. This effect is especially important in the case of C_2H_2 gas source at low current density.

Increasing the current density increases the flux of atoms impinging and the energy of these atoms. It also results in a reduction in the fraction of molecular species sputtered from the target. Since molecules have lower mobilities than single atoms, the decrease in the number of molecules impinging the substrate results in an increase in the grain size. In the case of C_2H_2, an increase in the current density also increases the decomposition of the C_2H_2 molecules in the gas phase and at high current densities the C_2H_2/Ar plasmas give rise to similar films to CH_4/Ar plasmas.

A detailed description of the deposition experiments and results is given elsewhere[7].

Diffusion couples

The diffusion couples all exhibited poor strength, particularly when a carbon layer was encountered which agrees with previous push-out results [8]. The strength of any TiC layer with the titanium matrix was good and this material was always found on the metal side of the diffusion couple.

Specimen 1, for both coating types, exhibited a thin layer of reaction product at the metal/coated SiC interface, identified as Ti_xC, which corresponded with that (Ti_2C) detected in transmission electron microscopy (TEM) studies of Ti-alloy/SCS-6 composite systems aged at 900°C [9]. This confirmed that the model couple system was behaving in the same way as the composite. The reaction product was irregular and occurred as isolated allotriomorphs rather than a continuous layer. The measured thickness was then taken to be the maximum extent of the reaction product into the matrix, normal to the the metal/ceramic interface. The newly formed Ti_xC material for the TiC-coated SiC was detected as it formed with higher Mo and Nb contents from that deposited upon the SiC and so exhibited different contrast in the SEM. The use of short heat treatment times did not introduce any changes in the matrix adjacent to the interface, which remained single phase β.

Extension of the ageing time for specimens 2 and 3 resulted in the formation of a thicker, more continuous reaction layer, Figure 1 (a)-(d), although the nature of that layer did not alter. The micrographs in Figure 1 indicate increased formation of Ti_xC at the metal/ceramic interface, although this is still not a complete layer in the case of the C-coated SiC specimens. For the TiC-coated specimens, carbon diffusion into the matrix has occurred to a high enough level to stabilize the α phase, the extent increasing from specimen 2 (130 minutes), Figure 1(a), to specimen 3 (253 minutes), Figure 1(b). The same level of carbon penetration was not observed for the carbon-coated specimens.

Figure 2 illustrates the variation in reaction layer thickness for both sets of coated samples as well as that predicted based on growth of Ti_xC controlled by diffusion of C through that phase.

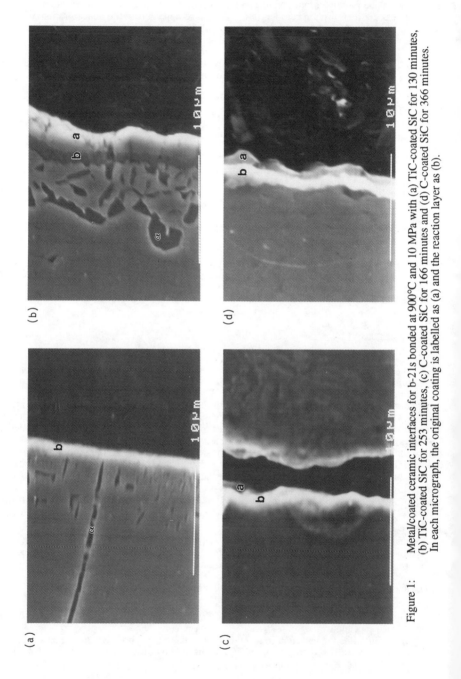

Figure 1: Metal/coated ceramic interfaces for b-21s bonded at 900°C and 10 MPa with (a) TiC-coated SiC for 130 minutes, (b) TiC-coated SiC for 253 minutes, (c) C-coated SiC for 166 minutes and (d) C-coated SiC for 366 minutes. In each micrograph, the original coating is labelled as (a) and the reaction layer as (b).

This is largely controlled by interfacial equilibrium compositions either side of the reaction layer, which can be predicted from thermodynamic models. The model prediction has been verified for ageing of composite materials with an initial reaction layer 0.8 μm thick and an overall diffusion distance (half the fibre separation) of 100 μm.

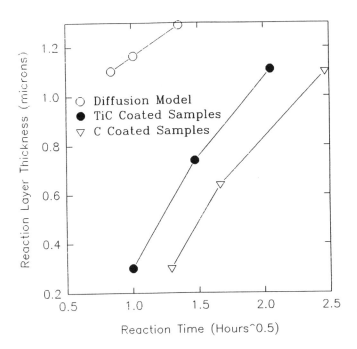

Figure 2:   Reaction layer growth kinetics as a function of time at 900°C under a stress of 10 MPa.

Comparison of the experimental data indicates very similar behaviour between the two coated systems. The increases in reaction layer thickness are parallel, indicating that the same rate controlling mechanism is acting in both cases. The rates are greater than that predicted by the diffusion model, but the diffusion couples are effectively semi-infinite (20 mm long), whereas the composite situation is finite. Once this is taken into account, it is apparent that the rate controlling step is, again, diffusion of carbon through the reaction layer. The TiC-coated material is underlain by SiC and so may not be thought to have a reservoir of C available for continued growth, but the deposited layer contains excess amorphous C, which would provide the necessary material, at least for the duration of these heat treatments. The kink in the growth curves may be considered to be due to experimental factors, but it is mirrored in both cases and corresponds to the region in which the allotriomorphs of reaction product are becoming more continuous. Hence an alternative view could be that the kink is a feature of the early stages of growth and arises as the contribution of matrix/reaction product interfacial diffusion is reduced through the formation of a more continuous layer, where transport is limited to bulk diffusion through Ti_xC. Extrapolation of the results to zero width indicates an incubation time exists for both types of coatings. Despite the preexistence of a layer of Ti_xC interfacing with the metal, 0.42 hours is required before significant growth into that matrix occurs. In the case of the carbon-coated specimen, then the incubation time is more than doubled to 0.9 hours as nucleation of the new interfaces is required raising the work of nucleation and hence the incubation time. Thus the initial stages of reaction zone formation can be studied and extended ageing gives growth behaviour consistent with the composite material behaviour. These results also indicate that for carbon-coated SiC fibres, neglecting the Gibbs-Thompson effect, processing times of around four hours at 900°C are necessary to produce the width observed in as-received composite plates.

CONCLUSIONS

During the reactive sputtering of titanium in CH_4 or C_2H_2, it was shown that the amount of carbon incorporated into the growing films depends on the formation of a compound at the surface of the target and on the relative importance of direct carbon deposition from the plasma. These two factors are influenced by the discharge current density. The morphology and crystallinity of the deposits are strongly influenced by the carbon content in the films; the presence of a-C phase results in a decrease of the TiC grain size.

The use of coated specimens in diffusion couples has been shown to indicate the initial stages of matrix/reinforcement reaction, whilst still duplicating composite behaviour at longer heat treatment times. The rate determining step has been confirmed as carbon diffusion through the reaction layer. Incubation times were determined for both TiC- and C-coated SiC samples interfaced with Ti-alloy matrix, indicating continued growth in the former case or nucleation and growth in the latter case.

ACKNOWLEDGEMENTS

The authors would like to thank Professors J.F. Knott and M.H. Loretto for provision of laboratory facilities, Professor J. Vereecken and Mrs N. Roose for the XPS analysis.

Financial support for this work was partially provided by The British Council and the French Community of Belgium.

REFERENCES

1. K.M. Fox, M. Strangwood and P. Bowen, 7th World Congress on Titanium, TMS-AIME, San Diego
2. P. Martineau, R. Pailler, M. Lahaye and R. Naslain, J. Mater. Sci. **19**, 2749 (1984)
3. K.M. Fox, M. Strangwood, 3rd European Conference on Advanced Materials and Processes, Euromat 93, Journal de Physique IV, Colloque C7, supplement au Journal de Physique III, volume 3, November 1993, pp 1699-1704
4. I.T. Ritchie, Thin Solid Films **72**, 65 (1980)
5. J.-E. Sundgren, B.-O. Johansson and S.-E. Karlsson, Thin Solid Films **105**, 353 (1983)
6. J.-E. Sundgren, B.-O. Johansson, S.-E. Karlsson and H.T.G. Hentzell, Thin Solid Films **105**, 367 (1983)
7. V. Vassileris, M.P. Delplancke and R. Winand, to be published
8. K.M. Fox, M. Strangwood and P. Bowen, Interfacial Phenomena in Composite Materials, Cambridge, to be published in Composites
9. M. Strangwood, K.M. Fox, P. Bowen and P.J. Cotterill, EMAG 93, Institute of Physics Conference Series No. 138: Section 8, Bristol, 1993, pp 391-394

SCANNING X-RAY DIFFRACTION: A TECHNIQUE WITH HIGH COMPOSITIONAL RESOLUTION FOR STUDYING PHASE FORMATION IN CO-DEPOSITED THIN FILMS.

T.I. SELINDER, D.J. MILLER, K.E. GRAY, M.A. BENO, AND G.S. KNAPP
Argonne National Laboratory, MSD-223, 9700 S Cass Ave., Argonne, IL. 604 39

ABSTRACT

Investigation of the formation of new metastable phases in alloy thin films requires ways of quickly determining the crystalline structure of samples with different compositions. We report a novel technique for acquiring structural information from films intentionally grown with a composition gradient. For example, binary metal alloy films were deposited using a phase-spread sputtering method. In this way essentially the entire composition range could be grown in a single deposition. By using a narrow incident x-ray beam and a translating sample stage combined with a position sensitive x-ray detector technique, detailed information of the metastable phase diagram can be obtained rapidly. Compositional resolution of the order of ±0.2% can be achieved, and is limited by the brightness of the x-ray source. Initial results from studies of phase formation in Zr-Ta alloys will be presented. Extensions of the analysis technique to ternary systems will be discussed.

INTRODUCTION

Even for binary systems, where virtually every compound has been studied and equilibrium phase diagrams established, there remain systems in which extended solubility, metastable-, and amorphous phase formation are unexplored. The development of material for use in, e.g., corrosive, and oxidizing environments requires that many alloy systems be investigated in detail. This task usually requires fabrication of a large number of samples with varying chemical composition, and subsequent application of analysis techniques. There is no conceivable way out of this dilemma if solely bulk specimens are considered.

Thin film co-evaporation, on the other hand, has proven versatile since films of different compositions can be simply fabricated by altering the relative deposition rates. Films themselves are used as protecting coatings and diffusion barriers but can, under adequate deposition conditions, be used as model systems for later bulk sample fabrication.[1] A feasible approach to rapidly obtain samples covering large portions of composition space is to use relatively large substrates, and the spread that occurs due to spatially separated deposition sources. The microstructure of the obtained films and the degree of metastability in the films can be varied by appropriately choosing deposition conditions, such as substrate temperature.

Once a quick sample manufacturing process exists, characterization limits fast surveying of the phase space. This becomes even more pronounced when the phase spread geometry is utilized. For example, the study of phase formation performed using XRD is a task that becomes rather tedious if high compositional resolution is required. In this paper a new technique, Scanning X-Ray Diffraction (SXRD) is described. It significantly reduces analysis times needed to acquire structural information connecting chemical composition with occurrence of various phases. The Ta-Zr binary system was chosen as a model for metastable phase diagram studies.

FILM DEPOSITION

The films were grown in a dual triode sputtering gun system, as illustrated in Fig. 1. The two guns were located next to one another. The target material was 12 mm diameter and 6 mm thick metal disks, bonded to a water cooled copper backing plate. The high vacuum chamber was cryogenically pumped with additional pumping provided by a titanium sublimation pump. Consequently, no pumping oils were in contact with this system, which was rough pumped using a dry rotary vane pump. The base pressure was better than 6 x 10^{-6} Pa.

Mat. Res. Soc. Symp. Proc. Vol. 343. ©1994 Materials Research Society

Sputtering was performed using 10 mTorr Ar introduced through the guns, close to the target surfaces. Target voltages of 200-400 V were used, resulting in sputtering powers of 30-60 W, per gun. Oscillating quartz sensors were used to monitor the deposition rates, which were in the order of 2-3 Ås^{-1}.

The substrates were r-cut sapphire wafers mounted on a stainless steel heater block. The block was heated using a resistive element and the temperature was controlled with a thermocouple located within the block. The substrate size was 12 mm wide by 95 mm long, and held at a distance of 48 mm from the target surfaces. The arrangement is shown to scale in Fig. 1.

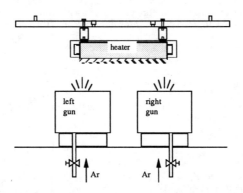

Fig. 1. Schematic drawing of the vacuum deposition chamber, displaying a side view of the phase spread geometry.

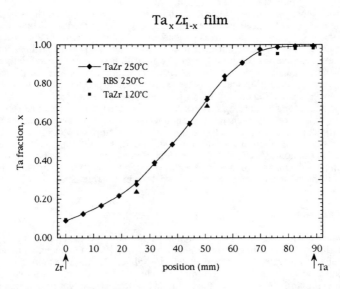

Fig. 2. Ta content x as a function of position on the substrate. Arrows indicate the locations of Zr and Ta target centers, respectively.

FILM ANALYSIS AND SCANNING X-RAY DIFFRACTION (SXRD)

X-ray fluorescence (XRF), and Rutherford Backscattering Spectroscopy (RBS) were used to obtain the composition as a function of position. Such a graph is shown for Zr-Ta alloy films in Fig. 2. The positions of the Zr and Ta target centers are indicated. Data points obtained by both RBS and XRF represent averages over app. 2×2 mm^2 regions. By interpolation a calibration curve of composition vs. substrate position can be drawn. For films grown at varying temperatures analysis repeatedly showed that the composition varied reproducibly and smoothly from one wafer to the adjacent ones.

For structural analysis a rotating Cu anode X-ray diffractometer was used. Variable slits were placed in the incident beam in order to restrict the analyzed specimen area to a narrow line perpendicular to the composition gradient, see Fig. 3. The diffracted beam was monochromatized and reflected out of the plane by a flat graphite analyzer crystal and detected in a position sensitive detector (PSD). The technique has been described in detail elsewhere,[2] and allows for simultaneous acquisition over approximately 5 degree range in 2θ. Ti-W-N and polycrystalline alumina samples were used to experimentally assess the acquisition efficiency and spatial resolution, respectively. Compared to a standard powder diffractometer, employing the Bragg-Brentano geometry (BB), it offers significantly faster data collection. This is evident in Fig. 4 where PSD and BB methods are compared. The scan speed was 2.5 times faster and the signal to noise ratio 10 times better in the PSD case. Taking these facts and x-ray intensity factors into account a 50-fold enhancement in efficiency was achieved.

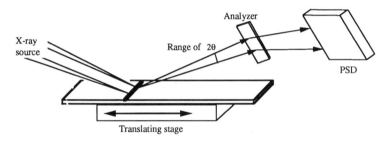

Fig. 3 A schematic illustration of the SXRD setup.

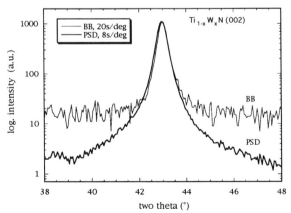

Fig. 4. Powder diffractograms showing the (002) peak of a Ti-W-N film, comparing the PSD diffractometer to the standard BB method.

In order to survey the phase spread film a means of moving the sample in the direction of the composition gradient was required. Scanning was accomplished by mounting the specimens on a translating sample stage, as indicated in Fig. 3. There is a trade-off between short acquisition times for each diffractogram and the spatial resolution, which both depend on the slit width. With this x-ray source acceptable count rates were achieved for widths down to 0.1 mm, which corresponds to a beam width W_0 less than 0.2 mm at the sample location. The actual spatial resolution W is, however, dependent on the diffraction angle. In Fig. 5 several diffractograms showing the alumina (116) peak at $2\theta = 57.5°$ are shown. The scans were acquired at different positions measured relative to the alumina wafer edge and the intensity actually drops to zero over a distance of less than 0.4 mm.

Application of the SXRD method is now demonstrated by effectively mapping the metastable phase diagram of Ta-Zr. Fig. 6 shows SXRD results obtained from a film quenched from vapor to a substrate held at 250 °C. Shown here is a transition between two crystalline states, one Zr rich and the other Ta rich. In order to show this as clearly as possible only a limited range of diffraction angles are shown here. From a full analysis, however, we have concluded that the two crystalline peaks correspond to metastable bcc solid solutions with unit cell sizes 3.55 and 3.36 Å, respectively. The incipient phase in-between shows no other diffraction than the broad feature shown in Fig. 6, and might be an amorphous phase. In the Ta rich end of the phase spread, shown graphically in Fig. 7, a peak assigned to (002) β-Ta[3] occurs at $2\theta = 33.7°$ in a very narrow range of 0.98<x<0.99. The film is two-phase in this region, the other phase being bcc Ta with a unit cell size of 3.32 Å, which is close to the bulk Ta value.

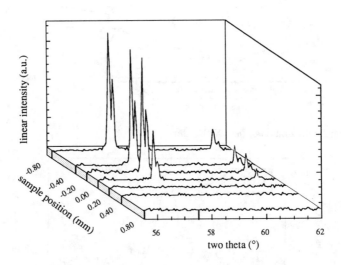

Fig. 5. A sequence of diffractograms showing the (116) peak of an alumina wafer. The sample position is given relative to the wafer edge.

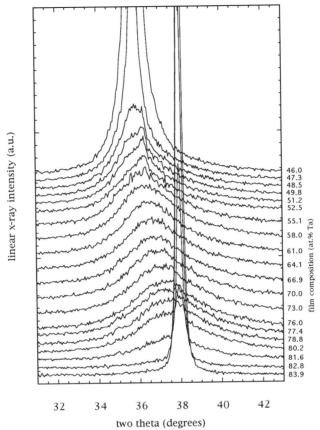

Fig. 6. SXRD obtained from a $Zr_{1-x}Ta_x$ film grown at 250 °C.

SUMMARY

The described phase spread growth technique is useful not only as a means of quickly obtaining a large number of samples. More importantly, since the relative rates vary smoothly, all compositions are covered continuously thus facilitating, for instance, determination of solubility limits and detection of intermediate phases. Accurate chemical analysis of the phase spread films is necessary to obtain the desired correlation between structure and composition.

To take full advantage of the rapid sample fabrication method, a means of fast structural characterization is essential. We have here reported of the SXRD-technique which is based on obtaining several consecutive diffractograms from different areas (compositions) of the sample. For reasonable analyzing times the spatial resolution is better than 0.4 mm at 57.5° diffraction angle. This may be converted to a compositional resolution which is dependent on the actual composition gradient, c.f. Fig. 2, but is in the present experiment on the average ±0.2 % absolute. The resolution could be improved either by reducing the composition gradient, i.e., use larger substrates, or by utilizing brighter x-ray sources, like synchrotrons. In such an experiment it would also be possible to analyze small spots on two-dimensional phase spreads in studies of phase formation in ternary systems.

Figures 6 and 7 illustrate examples of the versatility of SXRD in investigating phase formation in binary alloy systems. In the first case the possibly amorphous transition range between two metastable bcc structures was determined and in the second case the occurrence of non-equilibrium or impurity stabilized β-Ta was shown in a narrow composition range. At this point it should be pointed out, however, that once the interesting regions of the phase spread have been discovered, final phase identification and microstructural characterization should be undertaken, using, e.g., electron microscopy.

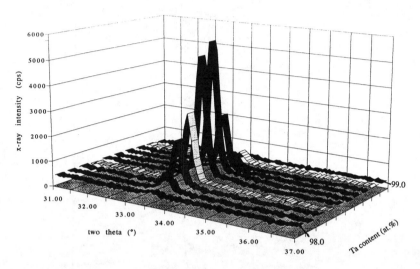

Fig. 7. Pseudo 3D graph showing SXRD from the same $Zr_{1-x}Ta_x$ film as in Fig. 6, in this case displaying the Ta-rich end of the phase spread.

In the present experiment the composition as a function of position on the substrate was determined in a separate system. A highly desirable further improvement of the technique should include the capability of simultaneous composition and diffraction analysis.

ACKNOWLEDGMENTS

This work was supported by the U.S. Department of Energy, Basic Energy Sciences-Materials Sciences, under contract #W-31-109-ENG-38. Moreover, the authors owe their thanks to J. Moser at University of Illinois at Champaign-Urbana for doing the RBS.

REFERENCES

1   N. Saunders and A P Miodownik, J. Mat. Sci. **22**, 629 (1987)
2   M.A. Beno, and G.S. Knapp, Rev. Sci. Instrum. **64**, 2201 (1993)
3   L. G. Feinstein, R. D. Hutteman, Thin Solid Films **16**, 129 (1973)

FINE STRUCTURE OF SPUTTERED Ni / Ti MULTILAYERED THIN FILMS STUDIED BY HREM

M. J. CASANOVE*, E. SNOECK*, C. ROUCAU*, J. L. HUTCHISON**, Z. JIANG*** and B. VIDAL***.
* Cemes - Loe, CNRS, B.P. 4347, 31055 Toulouse cedex, France.
**Department of Materials, University of Oxford, Parks Road, Oxford OX1 3PH, U.K.
***Faculté des Sciences de St. Jérôme, 13397 Marseille cedex, France.

ABSTRACT

Ni/Ti multilayered thin films can be efficient neutron guides and are therefore of great interest in neutron optics. Ni/Ti and NiC/Ti multilayers with various layer thicknesses were fabricated by magnetron sputtering and characterized by high resolution transmission electron microscopy (TEM). The TEM studies, performed on cross-sectional specimens, revealed that both kinds of layers were textured and showed coherence in the growth direction. The presence of a 2 nm thick amorphous zone at the Ni/Ti interface in the carbon free thin films was also confirmed. On the contrary, sharp interfaces were obtained in NiC/Ti multilayers. The fine structure of the different layers will also be reported.

INTRODUCTION

The study of metallic multilayer films is mainly concerned with their magnetic behaviour, of great interest in many technological applications. Some metallic systems are also good candidates for applications in neutron optics. Optical applications require multilayer structures in which the alternate layers present a high scattering contrast. However, a suitable neutron reflectivity cannot be achieved if the different interfaces possess significant roughness.

Different factors, such as the size of the grains or their orientation, can contribute to interface roughness. Moreover, interdiffusion of the different elements can lead to poorly defined interfaces. Finally, the thickness of the different layers also influences their structure and morphology. Careful examination of the microstructural features of the different layers and interfaces is then essential for a better understanding and an improved control of their properties.

The Ni/Ti system is one of the systems suitable for neutron optics applications. This magnetic-non magnetic transition metal system is also studied for its magnetic properties such as its magnetoresistance [1]. Because of the important lattice mismatch, coherent interfaces are not expected in this system. It has been reported that both nickel and titanium layers present an amorphous structure at small layer thicknesses (< 5 atomic planes) and a textured crystalline structure at larger thicknesses [2]. The presence of large grains in the crystalline layers induces some interface roughness, affecting the reflectivity of the films, [3-4]. The addition of light elements as impurities in the different layers was found to improve the reflectivity of the Ni / Ti films [3-4].

Sputtered Ni / Ti and NiC / Ti films of various thicknesses (where NiC stands for nickel with a small carbon content), were studied by high resolution electron microscopy (HREM). The microstructural features of the different layers were investigated and the influence of the incorporation of carbon in the nickel layers on the interface and layers morphology was analysed.

EXPERIMENTAL DETAILS

The Ni / Ti and NiC / Ti multilayer films were RF-magnetron sputtered at room temperature on (001) oxidized silicon wafers at a pressure of 2 mTorr. The sputtering power was 180 W for the Ni (NiC) layers and 300 W for the Ti layers. Prior to deposition the background pressure was

Mat. Res. Soc. Symp. Proc. Vol. 343. ©1994 Materials Research Society

Figure 1 : Low magnification [110]ₛ images of Ni/Ti and NiC/Ti multilayers showing the perfect regularity of the different layers as well as a suitable flatness of all the interfaces. The dark layers correspond to the Ni layers.

better than $1.5\ 10^{-7}$ Torr. More details of the preparation process as well as the results of x-ray diffraction performed on identical samples grown on float glass are given in [4].

Cross-sectional specimens for TEM investigations were prepared following the now standard method. Final thinning to electron transparency was achieved using argon ion-milling at liquid-nitrogen temperature. The high resolution TEM experiments were performed on Philips CM30/ST and JEM 4000 EX II electron microscopes with respective point resolution of 0.19 and 0.16 nm. The amorphous surface layer resulting from the oxidation of the silicon substrate prevents the influence of the silicon structure on the growth of the multilayer film.

RESULTS

The TEM specimens were tilted in order to be observed along the silicon [110] zone axis (labelled $[110]_S$). All the interfaces were then imaged edge-on. Figure 1 presents low magnification bright field images of Ni / Ti and NiC / Ti multilayers with respective layer thicknesses of 6 nm for the Ni and NiC layers and 7 nm for the Ti layers leading to a modulation period of 13 nm. The Ni layers appear dark in the figure while the bright layers correspond to the Ti layers. As can be seen, a perfect regularity in the stacking sequence and the layers thicknesses has been achieved in both cases. No significant difference in the interface roughness is revealed in these low magnification images.

The selected area electron diffraction pattern presented in figure 2 reveals a strong coherence in the growth direction. This diffraction pattern was taken in a thin region of the film far from the substrate. Ni layers grow with a set of {111} planes parallel to the substrate surface (d_{111}(Ni)=0.2 nm) as well as the Ti layers which grow in a fcc structure (d_{111}(Ti)=0.26 nm). As evident in the pattern, both Ni and Ti layers are textured, low intensity diffraction spots corresponding to the second set of {111} planes (when parallel to the electron beam) being present at about 71 ° of the {111} planes parallel to the substrate.

The Ti layers usually grow in the hcp structure, the (0002) planes lying parallel to the film surface ($d_{0002} = 0.24$ nm). However when a certain amount of hydrogen is included in the Ti layers, the x-ray diffraction scans show a Ti peak corresponding to an interplanar spacing of 0.26

nm (instead of the 0002 peak), as reported in [4]. This peak can be indexed either as $01\bar{1}0$ in the hcp structure or as 111 in the fcc one. In a study of Gd / W multilayer films [5], the presence of fcc crystallites within the Gd layers (usually hcp) was attributed to the presence of hybrid Gd,

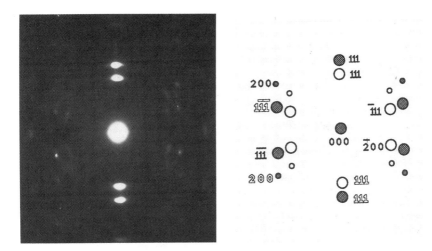

Figure 2 : Selected area [110]$_S$ diffraction pattern of a Ni (59 Å) / Ti (72 Å) multilayer taken far from the substrate. The strong reflections lie along the growth direction. For better understanding, the diffraction spots are schematized in the added drawing . Full circles are Ni spots, open circles correspond to Ti spots. Full and open digits refer to spots generated by twin-like domains.

hydrogen being incorporated during the sputtering process. Hydrogen being present as a residual gas in the deposition chamber and the Ti target targets being very reactive, the presence of hybrid Ti in our film cannot be excluded.

Another explanation for the occurrence of Ti fcc phase can be found in the literature, [6-7]. Indeed, in their study of Ni / Ti multilayers, the authors found that the Ti structure strongly depends on the Ti layer thickness. In the range of thicknesses 5 to 10 nm, the Ti structure is fcc, while both fcc and hcp phases are present in thicker layers.

A [110]$_S$ high resolution micrograph corresponding to a Ni (59 Å) / Ti (72 Å) multilayer film is presented in figure 3. The Ni layers are completely crystallized with extended crystallites. The fcc structure of the Ni layers with a set of {111} planes parallel to the substrate surface is well imaged in the figure.The titanium layers are also crystallized but with smaller, column-like crystallites. The small black arrows in the figure indicate three different Ti grains. Some amorphous zones are also found in these layers. Moreover Ti grains show higher misorientations than the Ni crystallites. The fcc (111) planes of Ni with an interplanar spacing of 0.20 nm and the fcc (111) planes or hcp (01$\bar{1}$0) planes of Ti with an interplanar spacing of 0.26 nm are the most visible in the figure. The highly contrasted Ti grain close to the centre of the figure clearly exhibits a fcc structure viewed along the [110] direction. An important feature in this image is the presence of an amorphous layer all along the interfaces (indicated by white arrows in the figure) . This layer is continuous and of constant thickness. Some interface roughness can also be seen. Clearly, a solid state amorphization reaction takes place in the Ni-Ti system as in other systems reported in the literature,[8]. Taking account of the better crystalline state of the Ni layers, it is assumed that Ni diffuses in the Ti layer and not the contrary. The order of the deposition does not seem to influence the solid state reaction, the amorphous interface layer being identical, and in particular of the same thickness, on both sides of the Ni layers.

Figure 4 presents a high resolution micrograph of a NiC / Ti multilayer film with a NiC and Ti layers thickness of 10 nm leading to a modulation wavelength of 20 nm. Here again we note the

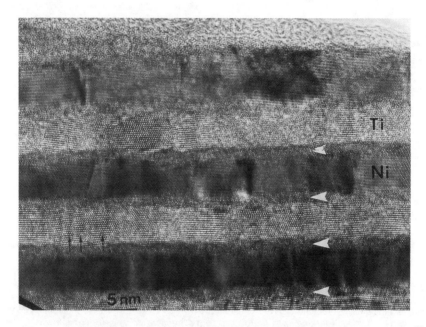

Figure 3 : [110]ₛ HREM micrograph of a Ni /Ti film revealing the presence of a continuous 2 nm thick amorphous interdiffusion zone at the interfaces.

strong texture of the Ni layers although there are not so well crystallized. The results of x-ray diffraction studies performed on NiC / Ti layers also show a degration of the crystallinity of the Ni layers,[3-4]. Moreover, the NiC crystallites present a higher misorientation than in the Ni / Ti films. The partial amorphization of the Ni layers is expected to smooth the Ni / Ti interfaces. Indeed the size of the Ni crystallites decreases and moreover, no continuous interdiffusion layer is present at the interfaces. On the contrary, sharp and well defined interface regions are obtained in the major part of the film. The important lattice mismatch between the two different layers does not lead to coherent interfaces. Numerous defects in both layers can be seen in the image, specially in the interface regions. No facetting of the layers surfaces was evident. The presence of carbon in the Ni layers seems to prevent the Ni diffusion in the Ti layers. As a result, the Ti layers appear more crystallized than in the Ni /Ti films.

A lattice image of one of the large Ti fcc grains is presented in figure 5. The multilayer was a Ni/Ti with a modulation wavelength of 13 nm. The Ni layers on both sides of the central Ti layers appear quite dark in this region while the fcc Ti grain displays a very high contrast. Accurate measurements of the angles between the different sets of planes were obtained on numerical diffractograms of the central region. The angle of 71° corresponds to the angle between two sets of {111} planes while the angle of 54 ° corresponds to the angle between a (111) and a (200) planes in the fcc structure. The interplanar spacings are respectively 0.26 nm for the (111) planes and 0.22 nm for the (200) planes. A slight difference between the interplanar spacing of the two sets of Ti {111} planes was found. The planes lying at 71° from the film surface plane are slightly compressed by the corresponding Ni (111) planes. The little arrow points at a twin boundary in the fcc Ti grain. Looking carefully at the figure, we see that the twin boundary presents several steps.

Numerous stacking faults were also found in the Ni layers.

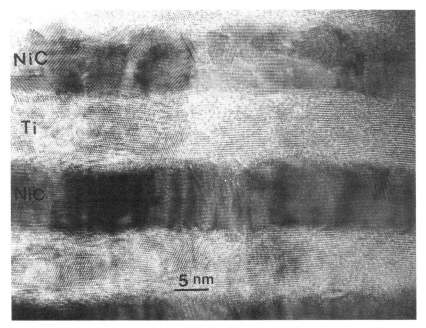

Figure 4 : [110] $_s$ HREM micrograph of a NiC / Ti multilayer film . Some interdiffusion is observed from place to place but it constitutes no more a continuous layer.

CONCLUSION

Cross-sectional TEM investigations of Ni/Ti and NiC /Ti multilayer films revealed some important differences and clearly showed the fine structural features of the different layers. In the investigated range of layer thicknesses, 5 to 10 nm, a crystallographic texture of the different layers was found. The Ni and NiC layers preferentially grow with a set of {111} planes parallel to the film surface and have a high degree of crystallinity. NiC layers however present some amorphous zones and a slight misorientation of the different crystallites. Ti layers are also highly textured although numerous amorphous regions were found in these layers in the Ni / Ti films. An unusual fcc structure of Ti crystallites was unambiguously confirmed by both diffraction and lattice imaging experiments. The presence of an amorphous interdiffusion zone at the Ni / Ti interfaces was revealed in HREM micrographs. This amorphous layer is continuous and of constant thickness all along the interfaces. The incorporation of carbon in the Ni layer seems to prevent the Ni from diffusing into the Ti layers and to flatten the interfaces.

ACKNOWLEDGMENTS

We are grateful to the Royal Society and the CNRS for a visiting fellowship for one of us (M.-J. C.) at the Department of Materials of the University of Oxford.

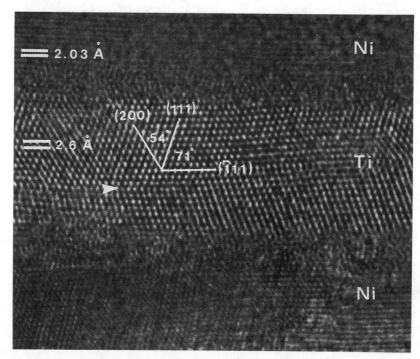

Figure 5 : Lattice image of a fcc grain in the titanium layer . [110]ₛ zone axis observation. The (111) Ti planes lie parallel to the substrate surface. The little arrow points at a stacking fault.

REFERENCES

1   A. Sdaq, J.M. Broto, H. Rakoto, J.C. Ousset, B. Raquet, B. Vidal, Z. Jiang, J.F. Bobo, M. Piecuch and B. Baylac, J. Magn. Magn. Mater. **121**, 409 (1993).
2   B. M. Clemens, Phys. Rev. B **33**, 7615 (1985).
3   O. Elsenhans, P. Böni, H.P. Friedli, H. Grimmer, P. Buffat, K. Leifer and I. Anderson SPIE **1738**, 130 (1992).
4   Z. Jiang, B. Vidal, M. Brunel, M. Maaza, F. Samuel, SPIE **1738**, 141 (1992).
5   A.K. Petford-long, P.E. Donovan and A. Heys Ultramiscroscopy **47**,323 (1992)
6   M. Porte, H. Lassri, R. Krishnan, M. Kaabouchi, M. Mâaza and C. Sella, Appl. Surf. Science **65-66**, 131 (1993)
7   S. Sella, M. Mâaza, M. Kaabouchi, S. El Monkade, M. Miloche and H. Lassri, J. Magn. Magn. Mat. **121**, 201 (1993).
8   G.A.Bertero, T.C. Hufnagel, B.M. Clemens and R. Sinclair, J. Mater. Res. **8**, 771 (1993).

Polycrystalline Magnetic Thin Films

Part A

Grain Structure, Texture, and Epitaxy in Thin Film Magnetic Media

THE EFFECT OF MICROSTRUCTURE ON THE MAGNETIC PROPERTIES OF THIN FILM MAGNETIC MEDIA

T. Yamashita, R. Ranjan, L.H. Chan, M. Lu, C.A. Ross, J. Chang and G. Tarnopolsky
Komag Inc. 275 S. Hillview Drive, Milpitas California, 95035

ABSTRACT

Magnetic recording media used today are based on sputtered cobalt alloy films with thicknesses in the order of 50 nm. As recording density is increased, the microstructure of the film must be controlled with increasing level of sophistication to achieve the magnetic properties necessary for good recording performance. Recording density has been increasing at the compound annual rate of 30%, and in recent years at a higher rate. Already, 1 Gb/in^2 has been achieved in the laboratory [1], and 10 Gb/in^2 is being contemplated [2]. In order to achieve such densities, microstructural characteristics such as grain morphology, size distribution, crystallographic orientation, and grain separation must be controlled with great precision, and their relationship to magnetic properties must be understood. The paper will describe the effect of sputter process conditions and the selection of magnetic alloys on the film microstructure and describe what might be required to achieve high recording densities. Particular attention will be paid to grain size distribution and grain separation. Grain separation is important for low noise performance of the media. Alloy selection and sputtering conditions can be manipulated to achieve different levels of separation between the grains.

INTRODUCTION

Sputtered thin film media has become the choice for use in small form factor hard disk drives. The recording density is being increased at a compound annual rate of 60% with the commercial introduction of magnetoresistive heads [3]. The 95 mm disk format recording density is approximately 500 Mb/in^2, with a 100 kbpi linear density and a 5000 tpi track density [4]. If the present trend continues, it is expected that drives with 1 Gb/in^2 recording density will become available in 1996. Further increases in recording density up to 10 Gb/in^2 are being contemplated [2,5]. In order to achieve such high densities, advances in media, head, mechanical and signal processing technologies must take place. Innovations in material systems and deposition technology are required. The film is expected to have high coercivity Hc (> 2500 Oe), low magnetization thickness product MrT (~ 0.5 memu/cm^2), and extremely low noise. The head will fly very low. A critical element in realizing the current and future media needs in achieving high recording density is to obtain an understanding of the relationship between film microstructure and the magnetic properties.

The following sections will review strategies used for the media in terms of alloy selection, film structure and microstructural manipulations. Examples that follow some of the outlined strategies are reported. The conclusion will discuss possible future direction for achieving much higher recording densities.

UNDERCOATS

An undercoat applied beneath the magnetic film is perhaps one of the most important contributors to the magnetic properties of the film. The undercoat can be used to mask irregularities such as contaminants that might exist on the substrate surface which adversely affect the nucleation and growth of the magnetic film. The substrate can provide potential sources for corrosion in the magnetic layer. Hence, the undercoat also serves as a barrier between the substrate and the magnetic film. However, the key feature of the undercoat used today is to manipulate the nucleation and growth of the magnetic layer and thereby achieve good uniformity of magnetic properties across the entire disk. The parameters that are affected are Hc, hysteresis loop squareness and noise properties, just to name a few. Chromium is the most predominant undercoat today. Originally reported by Lazarri et.al. in 1967 [6], the Cr undercoat increases the Hc of the cobalt-alloy film and provides high in-plane anisotropy in the film. A key feature of the Cr undercoat is that epitaxy is obtained between Cr and the HCP-cobalt lattice. The orientation relationship was determined by Daval & Randet in 1970 [7]. A large variety of cobalt alloys has been used with a Cr undercoat. In its current industrial form, Cr undercoat thickness is typically between 500 to 2000 Å, and it is deposited on a heated substrate. A high degree of epitaxy between the Cr and the magnetic layer is required in order to obtain high Hc and high hysteresis loop squareness. At deposition temperatures above 200°C, Cr grows with strong <100> vertical orientation, and an epitaxy between the Cr and Co films develops with the relationship of Cr<110> // Co<0001>, and Cr{100} // Co{11$\bar{2}$0}. The epitaxial relationship between pure Cr (a_0 = 2.885 Å) and pure Co (c = 4.069 Å, a_0 = 2.507 Å) is illustrated in Figure 1.

HCP-Co can have two directions of epitaxy with respect to the Cr lattice, since there are two <110> type directions on the Cr {100} plane [7]. If the CoCrTa/Cr film stack is deposited at temperatures above 200°C and in good vacuum, the Cr grains are typically much larger than CoCrTa grains [8], and on any given Cr grain with {100} surface, we have observed that there are several HCP-Co sub-grains with their c-axis oriented along the two orthogonal Cr<110> directions [8,9]. Graphical representation of the CoCrTa growth on Cr is shown in Figure 2, together with the actual high resolution TEM lattice image of the CoCrTa grains. Arrows in the TEM lattice image indicates the direction of c-axis of HCP-CoCrTa grains. The c-axis of the CoCrTa grains are generally always in the plane of the film. However, the c-axis orientation within the film plane is quite random since Cr grains themselves have random <110> directions in the plane of the film. Careful examination of the high resolution lattice images of CoCrTa film reveals that grain boundaries have a thin layer of amorphous material separating each grain [8,9] (see figure 2). These films have excellent low noise performance, therefore the grain boundary phase is thought to be effective in reducing exchange coupling between the grains. An interesting observation about the film is that grain boundary separation between each Co clusters (on each Cr grain) is larger than the separation between the individual Co-subgrains. The implication of this two-levels of grain separation is still not certain.

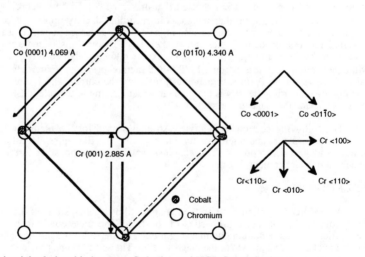

Figure 1: Epitaxial relationship between Cr lattice and HCP Cobalt lattice.

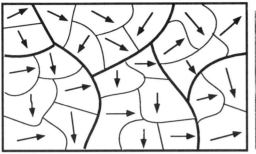 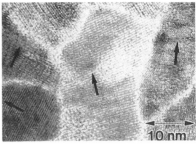

Figure 2: Schematic representation of Cr/CoCrTa film structure from the top surface (left), and actual high resolution TEM micrograph (right). Large lines represent Cr grain boundaries. Thin lines represent CoCrTa grain boundaries. Arrows represent the direction of Co c-axis. (Courtesy of T.Nolan, Stanford University)

Zhu et. al. recently showed by computer simulations that clustered particle structures with two levels of particle separation have higher hysteresis loop squareness than those with a single level of particle separation and yet still possess the same level of noise performance [10]. Although Zhu's theory is still yet to be proven experimentally, we have seen experimental verification of the observed scale of the grain isolation through correlation lengths obtained by uniform magnetization noise measurements [11].

A more unusual undercoat is the sputtered Ni_xP film reported by Yamashita et al. [12]. Typical phosphorus content is 10 - 20 wt. %. The sputtered film is amorphous as analyzed by x-ray and electron diffraction, and has a "grain-like" appearance as seen by high resolution SEM (HRSEM). TEM micrograph of the Ni_xP film by itself having nominally Ni_3P composition (P = 14.6 wt. %) is shown in figure 3a, and Ni_xP/CoNiCrPt film is shown in figure 3b. Ni_xP film is 30nm thick, and a CoNiCrPt film is ~ 400 Å thick (MrT = 3.0 memu/cm^2). The films were deposited at 25 mTorr by rf deposition on aluminum/NiP substrates. It can be seen in the micrograph that CoNiCrPt film has an isolated grain structure, which we believe is the main reason for the S/N improvement in the media.

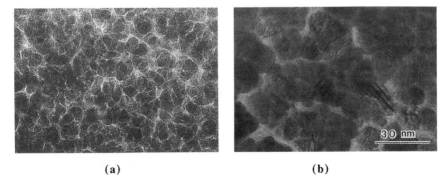

(a)   (b)

Figure 3: TEM micrographs of Ni_xP film by itself (a) and Ni_xP/CoNiCrPt film (b)

When CoPt-alloy films are deposited on Ni_xP film, Hc of the film increases with increasing Ni_xP film thickness up to about 700 Å. Then the effect saturates as shown in Figure 4. During this interval, the "grain" size of the Ni_xP film increases as shown by the HRSEM micrographs in Figure 5. A corresponding HRSEM micrographs of Ni_xP/CoNiCrPt films are shown in Figure 6. Noise properties of the media improves with Ni_xP thickness up to 500 Å and degrades with further increase in thickness. It is thought that there is an optimum Ni_xP grain size where one obtains the right combination of grain size and particle isolation. The shape and size of the Ni_xP appears to play a crucial role in the Hc and the noise properties one obtains for the Ni_xP/CoPt alloy films. Another factor is grain size uniformity and formation of grain clusters. With a Ni_xP thickness of approximately 300 Å, one obtains the best uniformity of grain size in the magnetic film and a minimum of grain clusters. This microstructure provides the best combination of noise performance and hysteresis loop squareness. The results indicate that the role of Ni_xP film on the magnetic layer is quite different in character than the Cr undercoat. Co grains on Ni_xP film has random orientations, and there is no eveidence of epitaxy between the

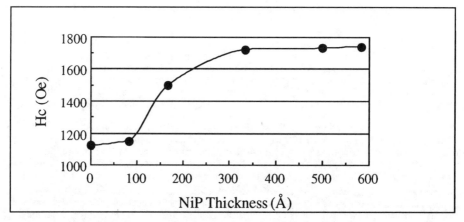

Figure 4: Hc vs. Ni_xP film thickness for CoNiCrPt alloy film

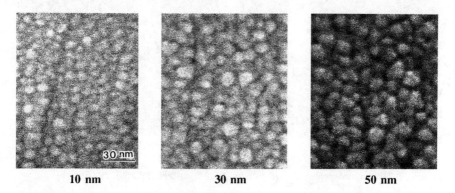

           10 nm   **30 nm**   **50 nm**

Figure 5: HRSEM micrographs of sputtered Ni_xP films of different thicknesses.

two layers. In fact, the Co film sputtered on Ni_xP film which has been exposed to air behaves rather similarly to a film deposited together in a good vacuum without air exposure. Only CoPt-alloys can be used for the Ni_xP undercoat, since the undercoat does not seem to enhance the crystalline anisotropy of the magnetic film. Random orientation of the Co grains does not seem to adversely affect the squareness of the film. S^* is typically between 0.8 to 0.9 for films with $MrT \sim 3.0$ memu/cm^2.

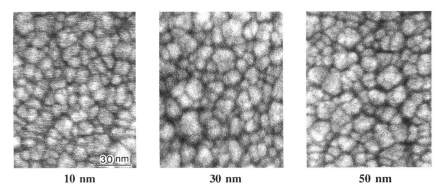

| 10 nm | 30 nm | 50 nm |

Figure 6: HRSEM micro graphs of CoNiCrPt films on Ni_xP undercoat with different thicknesses.

MAGNETIC ALLOY SELECTION

Choice of magnetic alloy for the sputtered film has traditionally centered on its Hc capability, magnetization (Ms), hysteresis loop squareness, noise and recording parameters and corrosion resistance. Other considerations include cost and ease of control over the sputtering process and target fabrication is also an important consideration. The most popular target fabrication method is vacuum casting which provides the highest purity. However, certain alloy compositions may be difficult to manufacture by vacuum casting. For example, DC magnetron sputtering requires that the magnetic field from the cathode penetrate through the target for the sputtering to take place. A cast target is usually rolled and forged in order to increase the permeability of magnetic flux through the target. Some alloy compositions may be too brittle for forging and rolling.

In the Cr undercoat process, CoNi and CoCrNi [13] were initially used. The most popular alloy at the present time, CoCrTa alloy, was first reported by Fisher et. al. in 1987 [14]. The need for higher Hc is generating more interest in Pt containing compositions such as CoCrTaPt [15] and CoCrPt [16]. High uniaxial anisotropy in the Pt alloy makes it possible to obtain high Hc with or without the Cr undercoat. Properties of CoPt film and its high Hc potential was first reported by Aboaf et. al. in 1983 [17]. CoPt has the distinction of having zero magnetostriction at the composition of 19 at. % Pt. The properties of CoPt alloys as a function of Pt content have been reported by Opfer et. al. [18].

One of the first considerations for an alloy is its magnetization. Various alloying components lower $4\pi Ms$ from the pure cobalt value of 18,000 gauss. Cr is usually added to the alloy to improve corrosion resistance and the noise performance [16]. Cr rapidly reduces magnetization so that at 25 at.% Cr, Ms is reduced to zero [19]. Lower magnetization leads to larger film thickness for a given magnetic remanance; it contributes to spacing loss and degrades recording performance. A minimum of about 10 - 12% Cr is usually required for corrosion protection. At this level of Cr, magnetization is reduced by one half. Alternative alloying elements that enhance corrosion protection without sacrificing the magnetization are desirable.

Low noise performance is one of the key properties of the film. Noise is reduced in thin film media by reducing inter-particle exchange interactions [20,21]. Spin wave Brillouin scattering analysis by Murayama et. al. showed 10 Å separation between the grains is sufficient to achieve exchange decoupling [22]. Noise reduction strategy for thin film media involves attempts to physically separate the grains through sputtering pressure [1], or from undercoat such as Ni_xP [12], or by phase segregation by Cr segregation [23] in CoCr and CoCrTa alloy films. It was shown by Nolan et. al. [8,9] that Cr/CoCrTa films clearly exhibit a segregation effect. However, CoNiCrPt and CoCrPt compositions with the same amount of Cr as CoCrTa do not show grain boundary segregation. There is still some mystery about why the CoCrTa alloy has good noise property. Ta is clearly a factor in the noise improvement, but with only 2 to 6 at. % Ta that is usually used for CoCrTa, there is not enough Ta in the alloy to explain the grain boundary segregation observed by high resolution TEM. Therefore, it is speculated that Ta may help Cr to diffuse to the grain boundary. Segregation of Cr in CoCr vertical films was speculated many years ago [24]. Since the Ms values observed in CoCr films deposited on heated substrates are often much higher than expected from bulk [25], it is reasonable to expect Cr segregation. With the availability of a new generation of TEM with high resolution EDS capabilities, it should be possible to determine the segregating species for the CoCr alloy films.

Another emerging alloy manipulation method involves addition of small amount of insoluble elements or species into the alloy. Insoluble species are such materials as oxides, borides, nitrides and carbides. If these species can be incorporated to the magnetic alloy, they should cause significant modifications in the sputtered film. Addition of boron [26], SiO_2 [27], and various oxides [28] into CoPt alloys have been reported. In some of these examples, a dramatic increase in Hc has been observed. Grain size reduction accompanies the Hc increase, together with grain isolation which helps the noise performance. As an example, TEM micrographs of CoNiPt film with and without SiO_2 addition is shown below in Figure 7. With the addition of only 2 molar % of SiO_2 into the alloy, the Hc can increase by several hundred oersteds, and it can be seen that grain size refinement and isolation occurs. Since future media will require smaller grain size and lower noise while maintaining high Hc, this direction in alloy manipulation will be sure to attract more attention in the future.

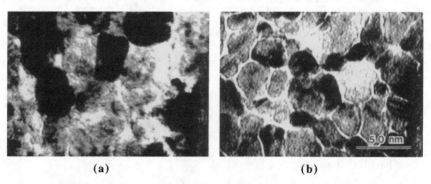

(a)   (b)

Figure 7:   TEM micrograph of CoNiPt film by itself (a) and with SiO_2 addition (b). Hc = 1600 Oe for (a) and 2000 Oe for (b). (Courtesy of A. Murayama, Asahi-Komag Co., Ltd.)

SPUTTER PROCESS

Some of the microstructural manipulations that are possible through sputter deposition are adjustment of sputter pressure, substrate bias, substrate temperature and deposition rate, to name a few. There is a considerable literature on the effect of the mentioned parameters on microstructure and magnetic properties. However, one important parameter which will become more significant as film thickness is reduced is the residual gas effect. Two examples will be described. First, the effect of base pressure on the growth of NixP/CoNiPt films is shown. Hc of the films deposited at different base pressures is shown in Figure 8. When base pressure is increased, Hc increases dramatically while squareness is reduced (not shown). Most of the base pressure effect is due to residual water vapor in the chamber. Corresponding HRSEM micrographs for two base pressure conditions are shown in Figure 9. These photographs along with many others have indicated to us that at high base pressure, the grain size uniformity of the film deteriorates, and a significant amount of clustering of the grains is observed (arrowed in figure 9a). The high base pressure media has lower noise, but the OW and PW50 are degraded due to loss in squareness. The effect shown in Figure 8 is reproducible by deliberate introduction of water vapor into the vacuum chamber during sputter deposition. When films become thinner for use in magnetoresistive head technology, the sensitivity of the film to changes in vacuum conditions will be increased. Therefore it is expected that vacuum conditions of the sputter system will have to be improved and more closely controlled in the future.

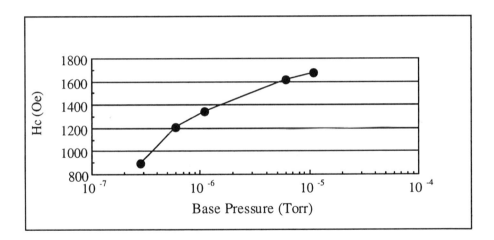

Figure 8: Hc vs. base pressure for NixP/CoNiPt film

Another interesting effect of residual gas is on the preferred growth orientation of the film. CoCr and CoPt alloys show a strong tendency to grow with c-axis out of the plane of the film. This tendency is a function of the undercoat and the magnetic alloy composition. An undercoat which forces the c-axis to lie in the plane of the film such as the Cr undercoat is probably an exception rather than a rule. Many undercoats, especially the amorphous films appear to show a strong tendency to promote vertical growth in Co films. Sputtered NixP films show similar tendencies. The alloy dependence of vertical growth is shown in Figure 10. Three different alloys were analyzed by x-ray diffraction. Films were deposited on aluminum/NiP substrates with sputtered NixP as undercoat. Sputtering pressure was 30 mtorr in all cases. Seehman-Bohlin glancing angle diffraction geometry [29] was used. Film thickness in all cases was approximately 600 Å. The general tendency is that films with higher Cr content exhibit a stronger (0002) reflection, which is an indication of the strength of vertical film growth. CoNiPt with no Cr in the film has the weakest (0002) reflection. Vertical film growth has strongly detrimental effect on the hysteresis loop squareness. Therefore steps must be taken to avoid it.

(a)   (b)

Figure 9:   HRSEM micrograph of $Ni_xP/CoNiPt$ film for two different base pressure deposition. (a) = $1x10^{-5}$ torr, (b) = 2×10^{-7} torr

An interesting method to break the tendency for vertical growth in CoPt alloy films is to dope the film during growth with a small amount of impurity gas such as nitrogen or oxygen. This is shown in Figure 11 for $Co_{78}Cr_{12}Pt_{10}$ film doped with three levels of nitrogen gas during sputtering. Even though the amount of nitrogen introduced during sputtering is very small, there is a significant decrease in the intensity of (0002) peak with increasing nitrogen doping. The ($10\bar{1}1$) peak is also reduced. The ($10\bar{1}1$) peak is being broadened, which is an indication of stacking fault formation. This was verified by TEM diffraction analysis of the films. Figure 12 is a set of electron diffraction patterns for a film of CoNiPt film with different amount of nitrogen added during deposition. The streaking or greater "fuzziness" of ($10\bar{1}1$) reflection is indicative of the formation of stacking faults in the grains (arrowed in figure 12). Hc of the film is lowered as more nitrogen is added as indicated in figure 13. Different alloys exhibit different response to the nitrogen gas. The control over the preferred growth orientation of the film is considered one of the important factors in the control of film microstructure.

FUTURE PROSPECTS

With the commercial introduction of the magnetoresistive (MR) heads, the MrT of the film is being reduced substantially. Typical MrT values that are being used today range in value from 0.8 to 1.5 $memu/cm^2$, compared to greater than 2.5 $memu/cm^2$ used in inductive applications. Using current alloy compositions, physical thickness of the media will be 200 - 300 Å for MrT of 1.0 emu/cm^2. Making the physical thickness of the media smaller is an important requirement for higher density since thinner media provides less spacing loss to the head. In the future, heads with a MR element using giant magnetoresistance effect will become available which will allow the MrT to be reduced at least another factor of two or more. Therefore, one of the key challenges will be to control the microstructure during the sputtering process for thinner magnetic media. This will be quite difficult. Figure 14 shows Hc and MrT plotted as function of CoCrTa sputter film thickness for a conventional heated Cr undercoat process. It can be seen that Hc starts to decrease rapidly below MrT of 1.0 $memu/cm^2$ for this

$Co_{86}Cr_{10}Ta_4$ alloy composition. It is apparent that grains are becoming superparamagnetic as thickness is reduced below 200 Å. One of the alternatives is to reduce the magnetization of the alloy so that greater physical thickness is available at low MrT range. This will allow better control for the sputter process. However, as heads are made to fly closer to the media, the physical thickness of the media will have to be reduced again.

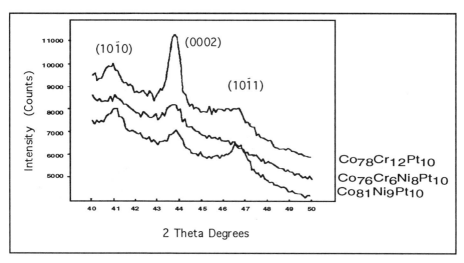

Figure 10: Glancing angle x-ray diffraction scan of $Co_{78}Cr_{12}Pt_{10}$, $Co_{81}Ni_9Pt_{10}$ and $Co_{76}Ni_8Cr_6Pt_{10}$ alloy films

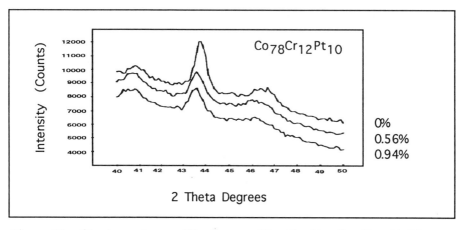

Figure 11: Glancing angle x-ray diffraction scan of $Co_{78}Cr_{12}Pt_{10}$ alloy film with different amount of nitrogen doped into the argon gas.

In Figure 15, the relative size of a bit as proposed for an areal density of 10 Gb/in^2 is indicated in the HRSEM micrograph of a Ni$_x$P/CoPt-alloy film surface. The size of the bit is 0.8 x 0.064 μm as proposed by Murdock et. al. for case #3 [2]. For this recording density there will be only a few hundred grains in each cell. This situation will be clearly unfavorable from the signal to noise point of view. The grain size will have to be reduced significantly so that more particles can be placed within each bit cell. The challenge for the future will be to learn how this grain size reduction can be accomplished without sacrificing any of the magnetic properties of the film such as Hc and loop squareness. Noise performance will be paramount, and it will have to be achieved by precise control over individual grain separation through alloy segregation or by physical separation. New alloy systems and deposition methods will have to be invented to achieve the increase in density that is being proposed.

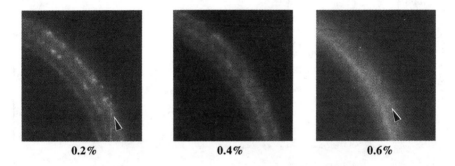

0.2%   0.4%   0.6%

Figure 12: Electron diffraction pattern of Co$_{77}$Ni$_9$Cr$_4$Pt$_{10}$ alloy film with different amount of nitrogen doped into the argon gas.

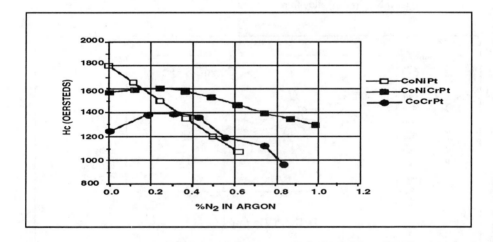

Figure 13: Hc vs. nitrogen content in the argon gas for Ni$_x$P/CoNiCrPt alloy.

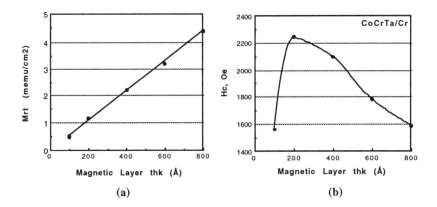

(a)   (b)

Figure 14:   Hc and MrT vs. film thickness for Cr/CoCrTa media. $T_{sub} = 200°C$.

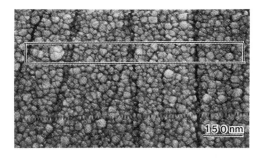

Figure 15:   Overlay of expected bit cell size for 10 Gb/in^2 over the HRSEM micrograph of of the Ni$_x$P/CoPt- alloy film.

ACKNOWLEDGEMENTS

Authors would like to thank Professor Robert Sinclair and Dr. T. Nolan of Stanford University for their help in TEM analysis.

REFERENCES

1. T. Yogi, C. Tsang, T.A. Nguyen, K. Ju, G.L.Gorman and G. Castillo,
    IEEE Trans. Magn., MAG-26, 2271 (1990)
2. E.S. Murdock, R.F. Simmons, and R. Davidson,
    IEEE Trans. Magn., MAG-28, 3078 (1991)
3. 1988 Head/Media Technology Review, R. Balanson ; Las Vegas Organized by Dan
    Mann Magnetics and Peripheral Research Corporation, 351
4. 1993 Head/Media Technology Review, L. Procker ; Las Vegas Organized by Dan Mann
    Magnetics and Peripheral Research Corporation, 351
5. NSIC/ National Storage Industry Consortium
6. J.P. Lazzari, I. Melnick and D. Randet, IEEE Trans. Magn. MAG-3, 205 (1967)
7. J. Daval and D. Randet, IEEE Trans. Magn. MAG-6, 768 (1970)
8. T. Nolan, R. Sinclair, R. Ranjan and T. Yamashita,
    J. Appl. Phys.73 (10), 15 May, 5117 (1993)
9. T. Nolan, R. Sinclair, R. Ranjan and T. Yamashita,
    J. Appl. Phys. 73 (10), 15 May, 5566 (1993)
10. X.G. Ye and J.G. Zhu, to be published, Paper # DB-04, 1993 3M Conf.
11. R. Ranjan, W.R. Bennett, G.J. Tarnopolsky, T. Yamashita, T. Nolan and R. Sinclair,
    to be published, paper #DB-09, 38th Annual Conference on Magnetism and Magnetic
    Materials, Nov. 1993
12. T. Yamashita, L.H. Chan, T. Fujiwara and T. Chen,
    IEEE Tran. Magn. MAG-27, 4727 (1991)
13. T. Yamada, N. Tani, M. Ishikawa, Y. Ota, K. Nakamura and A. Itoh,
    IEEE Trans. Magn. MAG-21, 1429 (1985)
14. R.D.Fisher, J.C. Allan and J.L. Pressesky,
    IEEE Trans. Magn. MAG-22, 352 (1991)
15. B.B. Lal, H.C. Tsai and A. Eltoukhy, IEEE Trans. Magn. MAG-27, 4739 (1991)
16. US Patent 4,789,598 K. Howard et al.
17. J. Aboaf, S. Herd and E. Klokholm, IEEE Trans.Magn. MAG-19,1514 (1983)
18. Opfer et. al. Hewlett Packard Journal, Nov. 1985
19. R.M. Bozorth, *Ferromagnetism*, D. Van Nostrand Company, March 1951
20. T. Chen and T. Yamashita, IEEE Trans.Magn. MAG-24,No.6, 2700-2705 (1988)
21. J.G. Zhu and H.N Bertram, J. Appl. Phys. 63, 3248 (1988)
22. A.Murayama, M. Miyamura, K. Miyata, and Y. Oka,
    IEEE Trans. Magn., MAG-27, 5064 (1991)
23. Y. Maeda and K. Takei, IEEE Trans.Magn. MAG-27, 4721 (1991)
24. T. Chen, G.B. Charlan and T. Yamashita,
    J.Appl. Phys., 54(9) 5103-5111, Sept. (1983)
25. J.E. Snyder, M.H. Kryder and P. Wynblatt, J. Appl. Phys., 67, 5172 (1990)
26. C.R. Paik, I. Suzuki, N. Tani, M. Ishikawa, Y. OTa and K. Nakamura,
    IEEE Trans.Magn., MAG-28, 3084 (1992)
27. M. Murayama, S. Kondoh, and M. Miyamura, to be published, 3M Conf. (1993)
28. T. Shimizu, Y. Ikeda, and S. Takayama, IEEE Trans.Magn., MAG-28, 3102 (1992)
29. P.A. Flinn, J. Vac. Sci. Technol. Nov/Dec. 1988 1749 - 1755

CORRELATION OF STRUCTURE AND PROPERTIES IN THIN-FILM MAGNETIC MEDIA

T. P. NOLAN*, R. SINCLAIR*, R. RANJAN**, T. YAMASHITA**,
G. TARNOPOLSKY**, AND W. BENNETT**
Department of Materials Science and Engineering, Stanford University, Stanford, CA 94305
Komag Inc. 591 Yosemite Drive, Milpitas, CA 95035

ABSTRACT

Correlation of detailed microstructural information obtained from a temperature series (25-275°C) of otherwise identical CoCrTa/Cr recording media with detailed recording performance measurements demonstrates two mechanisms for magnetic isolation.

Media were prepared by sputter deposition of 75 nm Cr underlayers and 60 nm $Co_{84}Cr_{12}Ta_4$ layers onto circumferentially textured NiP plated aluminum substrates using an Intevac MDP-250 machine.

High-resolution transmission electron microscopy (HRTEM) and high-resolution scanning electron microscopy (HRSEM) show columnar isolation by voiding, via the mechanism of shadowing during low temperature growth. Shadowing decreases with increasing temperature until voiding disappears at 150°C. At high temperature (275°C), large intergranular separation is observed, which has been attributed to elemental segregation. This separation decreases as temperature decreases, until it disappears at 150°C.

Detailed magnetic recording measurements show that the maximum normalized media noise power is observed at 150°C. It decreases as more isolated structures are formed by increasing or decreasing temperature. Uniform magnetization noise, a measurement related to crosstrack correlation length within a magnetization transition also shows this trend.

It is suggested that HRTEM shows two different microstructural features that control magnetic correlation lengths and resulting media noise.

INTRODUCTION

With the development of low noise recording channels and magnetic recording heads, the noise intrinsic to the magnetic recording medium becomes increasingly important. Much of this noise occurs at recorded transitions. Increasing recording densities exacerbate the problem because the number of transitions and the percentage of transition area per bit increase. Noise reduction requires control of the sharpness and shape of the magnetization transition, which is in turn controlled by the microstructure of the magnetic alloy. Unfortunately, the relationships between observed microstructural features, standard bulk magnetic measurements, micromagnetic structure and media noise are not fully understood.

The important microstructural features are difficult to analyze because of the multiple thin film layers and the extremely small size of grains and topographical features. Complete analysis requires a combination of techniques including HRTEM, HRSEM and electron diffraction.

Media can now be readily made that have nearly identical bulk hysteresis loop properties (eg. H_c), but significantly different magnetic recording (noise) performance. High noise is generally related to exchange coupled grains. Lower noise media are often presumed to have magnetostatically coupled grains or small clusters that may promote small magnetic vortex structures in transitions.[1,2] Recent efforts have lead to more direct measurements of media noise from high density written transitions. Among these is uniform magnetization noise (UMN).[3,4] It is obtained by first applying a saturating head field to an entire track on the medium. A reverse field equivalent to the remanence coercivity H_r is then applied to bring the remanent magnetization in the track to zero. After this, the micromagnetic structure throughout the track is believed to be similar to that at the center of a written transition. The noise power measured from this track, UMN, is correlated to the transition noise, and the length scale is thought to be related to the cross-track correlation length, the size of the independently magnetized particles or clusters in the magnetic film. UMN may be thought of as a measurement of micromagnetic structure. Here we correlate this information with microstructural features observed by TEM and HRSEM for a series of $Co_{84}Cr_{12}Ta_4$ films deposited on substrates heated between 25°C and 275°C.

EXPERIMENTAL PROCEDURE

Media were sputtered under 5 mtorr Ar in an Intevac MDP 250 sputter system. The Cr underlayer, $Co_{84}Cr_{12}Ta_4$ and C overcoat thicknesses are 75, 60 and 20 nm respectively. Ta concentration is larger than in previous work to enhance the grain separation effect.[5] The temperature series consists of samples deposited on circumferentially textured NiP/Al substrates heated to 25°C, 100°C, 150°C, 175°C, 200°C, 225°C, 250°C and 275°C. TEM images were obtained from back side ion milled disc samples using a Philips EM430ST operating at 300kV. SEM images were obtained from small coupons observed at 0° tilt using a Hitachi S900 FEG-SEM operating at 30kV. SEM samples were heated to 150°C to minimize carbon contamination during imaging. The TEM and SEM images appear different because they offer complementary information about the samples. Both contain information about the grain morphology, but TEM adds information about the crystallography, while HRSEM has stronger topographic contrast.

The detailed recording measurements described above were performed on the 25°C, 100°C, 150°C, 200°C and 250°C samples using a 0.35 μm thin-film nano-head flying at 3.6μ" (~90 nm). Integrated media noise and power spectral density (PSD) measurements were collected in the frequency interval 0.15<f<15 MHz, with background noise subtracted.[4,6,7]

RESULTS

Coercivity (H_c) and signal to media noise ratio at 60 kfci (SNR) are plotted versus temperature in Figure 1. H_c and SNR increase rapidly between 100°C and 200°C, before saturating at higher temperature. Mechanisms have been suggested for the apparent improvement in noise performance. Compositional inhomogeneity may increase as substrate temperature is raised, resulting in more separated magnetic particles.[8] Film stress may change with temperature, altering defect structures or bulk properties.[9] Evidence for these claims is limited.

HRTEM micrographs and corresponding diffraction patterns of temperature series samples are shown in Figure 2. The {1120} diffraction ring of the plan view diffraction patterns is used to study changes in crystallographic orientation with increasing temperature. At low temperature (Figure 2a) this diffraction ring is clearly apparent, corresponding to many grains with <1120> axes lying in the film plane and a c-axis growth direction. At 275°C (Figure 2f) this diffraction ring has completely disappeared, as is expected for a sample with a strong <1120> growth orientation. The diffraction results show that the transition to oriented growth occurs in the range from 100°C to 200°C.

Evidence of shadowing induced physical separation is observed in the thinnest regions of the 100°C sample. There are weak ≅10 nm features within grains, corresponding to the columnar structure. Evidence of topography is no longer observed at 150°C. Adatom mobility has apparently become sufficient to dominate shadowing, and grain separation is no longer obtained from this mechanism.[10] HRSEM results (Figure 3) confirm the result. The 100°C sample has ≅10-20 nm scale surface roughness that is not observed in the 175°C sample.

The 175°C SEM sample has weak contrast corresponding to a smooth surface, but there is evidence of slightly larger ~20-40 nm features. These are more apparent in the 250°C sample. The size scale is that of the crystallographic grains, and may be interpreted as marking grain boundaries. The TEM micrographs of high temperature samples (Figure 2) show apparently separated grains. Separation at 275°C can be quite large (~1 nm), but decreases with decreasing temperature, until at 175°C, separation is no longer observed. Largest separation is generally observed between CoCrTa grains nucleated from different Cr underlayer grains (black arrows), as determined by a misorientation <90°. Smaller, but still visible separation is sometimes observed between CoCrTa grains from the same underlayer grain (white arrows) that are misoriented by 0° or 90°. These observations extend the previously published growth model[5], as shown schematically in Figure 4.

The microscopy suggests that there is a magnetic particle separation mechanism for noise reduction at low and high temperature, so that a maximum media noise might be expected between 150°C and 175°C. This is in disagreement with the SNR data, that suggests a monotonic decrease in media noise with temperature.

SNR is, however, not an independent measure of noise. The signal effect can be removed by considering the integrated media noise power independently, as shown in Figure 5. There is a clear peak in the media noise at 150°C, just as predicted by the microscopy. It is apparent that the effect of media noise on the SNR measurement is simply overwhelmed by the monotonic increase in signal in the range 100°C to 200°C.

Previous results[11] suggest that the orientation change from random to c-axis in-plane orientation (observed here between 100 and 200°C) increases coercivity. This decreases the magnetic transition length and increases the output signal. It does not necessarily affect noise performance significantly.

The micromagnetic structure related measurement, UMN (Figure 6), shows the same basic trend as noise power, with a peak at 150°C.[8] The size scale of UMN is believed to be proportional to cross-track correlation length, and may be related to the size scale of flux vortices (ring structures) that produce transition noise.[3] Modeling and some observations of larger scale transitions have shown that magnetization at transitions settles into a turbulent array of ring structures.[1,2] The size of these structures is sensitive to the media microstructure and also to the noise performance of media. UMN estimated cross-track correlation length has the interesting feature that it is comparable to the grain size of the Cr underlayer (50 nm). This suggests that, although the 10-20 nm Co alloy grains are not magnetically independent, flux vortices could circulate within one Cr template grain. A possible mechanism for this is shown in Figure 7.

At high temperature, multiple CoCrTa grains grow on each Cr underlayer grain, with c-axis parallel to one or the other Cr <110> axis (Figure 4).[5,12,13] With the extremely strong growth orientation, it is common that neighboring grains lie with c-axes oriented as in Figure 7. Disregarding exchange energy, the magnetostatic energy from neighboring grains promotes the magnetization loop shown, just as is often observed on a much larger scale with 90° closure domains found in cubic materials.[14] It is likely that the formation of such a structure will depend on the magnetic exchange interactions between the separated clusters of 90° rotated grains and between the less separated individual grains within the clusters.

CONCLUSIONS

Microstructural separation occurs during deposition by two mechanisms:
1. Shadowing induced voiding at column boundaries.
2. High temperature grain boundary separation.

Lower noise media are promoted by microstructural separation produced by two mechanisms:
1. Shadowing induced noise reduction occurs for many low temperature Co alloy media.
2. CoCrTa has high temperature noise reduction too, giving a peak in media noise for that alloy for 150°C deposition temperature.≥

Grain separation appears to occur to different extents for grains nucleated from the same Cr underlayer grain and those nucleated from different Cr grains.

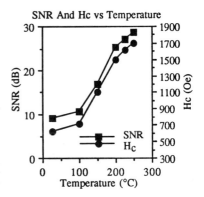

Figure 1: Coercivity and SNR (at 60 kfci) of temperature series samples.

REFERENCES
1. J. G. Zhu and H. N. Bertram, J. Appl. Phys., 63, 3248 (1988).
2. T. Chen, IEEE Trans. Magn., 17, 1181 (1981).
3. T. J. Silva and H. N. Bertram, IEEE Trans. Magn., 26, 3129 (1990).
4. G. J. Tarnopolsky, H. N. Bertram and L. T. Tran, J. Appl. Phys., 69, 4730 (1991).
5. T. P. Nolan, R. Sinclair, T. Yamashita, and R. Ranjan, J. Appl. Phys., 73, 5117 (1993).
6. R. Ranjan, W. R. Bennett, G. J. Tarnopolsky, T. Yamashita, T. P. Nolan and R. Sinclair, J. Appl. Phys., in press.
7. G. A. Bertero, H. N. Bertram and D. M. Barnett, IEEE Trans. Magn. 29, 67 (1993)
8. Y. Maeda and K. Takei, IEEE Trans. Magn., 27, 4721 (1991).
9. M. Lu, Q. Chen, J. H. Judy and J. M. Sivertsen, submitted for publication.
10. J. A. Thornton, Ann. Rev. Mater. Sci., 7, 239 (1977).
11. T. P. Nolan, R. Sinclair, T. Yamashita, and R. Ranjan, J. Appl. Phys., 73, 5566 (1993).
12. T. P. Nolan, R. Sinclair, T. Yamashita, and R. Ranjan, IEEE Trans. Magn., 29, 292 (1993).
13. J. Daval and D. Randet, IEEE Trans. Magn. 6, 768 (1970).
14. B. D. Cullity, Introduction to Magnetic Materials (Addison Wesley, Reading, 1972) Ch. 9.

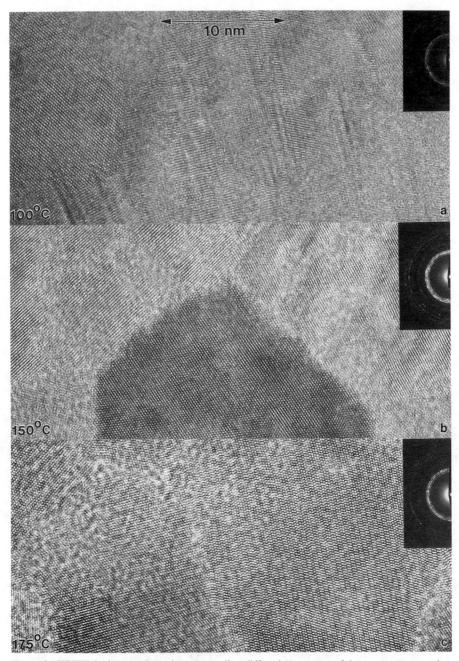

Figure 2: HRTEM micrographs and corresponding diffraction patterns of the temperatures series.

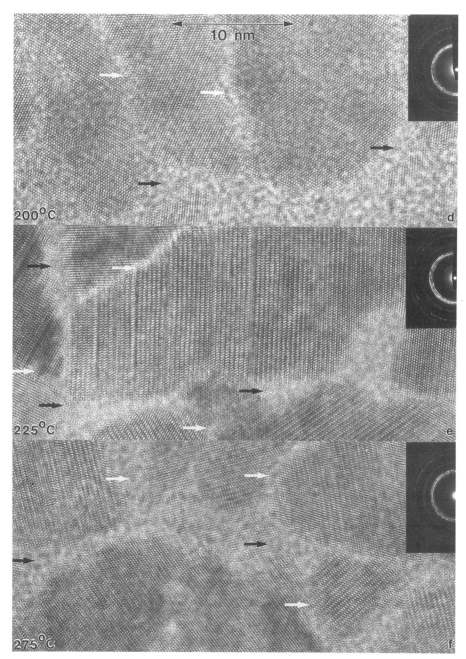

Figure 2: HRTEM micrographs and corresponding diffraction patterns of the temperatures series.

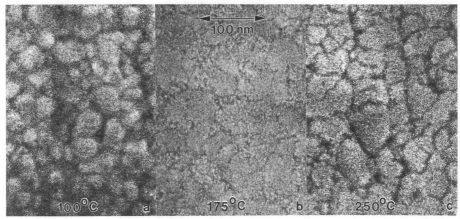

Figure 3: HRSEM images of temperature series samples.

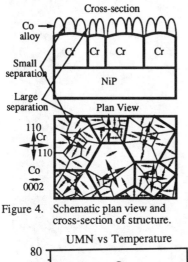

Figure 4. Schematic plan view and cross-section of structure.

Transition Noise vs Temperature (at 60 kfci)

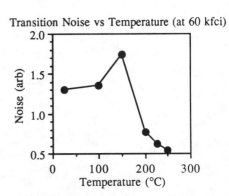

Figure 5: Media noise power of temperature series samples.

UMN vs Temperature

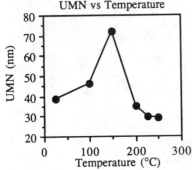

Figure 6: UMN of temperature series samples.

Schematic of Magnetic Rotation Within 1 Cr Underlayer Grain

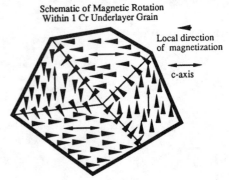

Figure 7: Diagram of microstructure promoting small flux vortices.

THE EFFECTS OF CROSS HATCH ANGLE (CHA) ON MAGNETIC PROPERTIES AND READ/WRITE CHARACTERISTICS OF CoCrPtTaB THIN FILM MEDIA

GA-LANE CHEN*, H. DO*, T.-A. YEH**, J. M. SIVERTSEN**, AND J.H. JUDY***
*HMT Technology, 1055 Page Ave., Fremont, CA 94538
**Department of Materials Science, University of Minnesota, Minneapolis, MN 55455
***Department of Electrical Engineering, University of Minnesota, Minneapolis, MN55455

ABSTRACT

Cross-hatch type texturing has been widely used to replace circumferential texturing in the thin film media industry for optimizing tribology performance. However, higher "cross hatch angle" (CHA) may degrade magnetic performance. The relationship between texturing cross hatch angle (CHA) and magnetic properties and their effects on the read/write characteristics of sputtered thin film media are discussed in this paper. As the CHA increases, both the orientation ratio (OR) and the squareness ratio "S(c)/S(r)" decrease. The film with the highest OR is located at CHA=9^0, which has the lowest media noise. The medium has highest signal/noise ratio when CHA is equal to 9^0. PW_{50} can decrease as much as 4 nanoseconds as CHA increases to 73^0. By calculating a ratio of the intensity of Cr (110) + CoCrPtTaB (0002) to the intensity of Cr (200), we are able to correlate the intensity ratio to magnetic properties and read/write characteristics.

INTRODUCTION

Due to the higher stiction and friction caused by too smooth a polished surface, the circumferential texturing process with Ra from 30Å to 70Å was introduced to reduce stiction. Tribologists have found that most circumferential texturing has higher take-off velocity (TOV) and causes higher stiction at head disk interface (HDI). Now, the cross-hatch type texturing has been widely used to replace circumferential texturing for optimizing tribology performance. However, too high a cross hatch angle and too rough a surface may degrade the magnetic performance. The impact of texture on the soft error rate (SER) on CoPtCr has been reported [1]. It is believed that surface roughness may partly determine SER performance, particularly at high linear recording densities where the probability of a scratch intersecting a written bit increases. The effect of disk cross hatch texture on the tribological performance has been studied [2], but its effect on the magnetics are not reported yet. Disks prepared with a cross hatch texture have a low take-off velocity and better tribological performance. For the magnetic alloy research, CoCrTa/Cr [3,4,5], CoCrPt/Cr [6], and CoNiPt [7] films have been widely studied. The CoCrTa media have very low noise but have difficulty for attaining coercivity exceeding 2000 Oe. With a appropriate Pt content for CoCrPt [8] and CoNiPt alloys, they can easily achieve coercivity in excess of 2000 Oe, but generally higher Pt content leads to higher manufacturing cost. Paik, et. al. have introduced CoCrPtB/Cr thin film media [9]. They demonstrated that Pt content can be reduced by the addition of boron in CoCrPt/Cr media. HMT has developed new CoCrPtTaB/Cr media which give a good range of high coercivity and low media noise. The effect of orientation ratio of circumferentially textured media on the signal-to-noise ratio has been studied [10]. However, no literature has been found to report the effect of cross hatch angles (CHA) on the magnetic and recording characteristics yet. Here, we have

303

developed 11 different kinds of texturing pattern to study the effects of cross hatch angles on the performance of sputtered CoCrPtTaB/Cr thin film media with CHA from zero to 73 degrees.

EXPERIMENTAL PROCEDURES

Al_2O_3 slurry is used to make 11 different mechanical textures on NiP/Al substrates. The cross hatch angles are 0-4, 9, 16, 25, 32, 43, 51, 61, 66, 68, and 73 degrees. All of 11 textured substrates have the same roughness Ra at 65 +/- 5 Å. After texturing, substrates were DC magnetron sputtered with the same Cr underlayer, CoCrPtTaB layer, and hydrogenated carbon overcoat. Then, PFPE lubricant was applied on the carbon surface to enhance CSS performance. Measurement of the hysteretic curve and magnetic properties was performed on a vibrating sample magnetometer (VSM). Both the circumferential and radial directions of the M-H curve for various textured disks were measured. Recording properties of CoCrPtTaB media were measured with a Guzik 501 read/write analyzer and a thin film head with a 70% slider and a gap length of 0.37 μm and pole widths of P1=11 μm and P2=9 μm. The coil had 34 turns. The flying height for measurement was 89-100 nm at a linear velocity of 6.4-7.6 m/sec. By using a Guzik 501 and a HP 5358A spectrum analyzer, the total media noise (Nm) was measured in a span of 0-20 MHz with a 30 KHz resolution bandwidth. Zone bit recording (ZBR) tests were performed on our CoCrPtTaB media. At the ID (r= 0.95") of disk, the low writing frequency (LF) and high writing frequency are 1.36 MHz (7.6 KFCI), and 5.44 MHz (30 KFCI) respectively. At the OD (r=1.33") of disk, the low writing frequency (LF) and high writing frequency are 2.13 MHz (8.5 KFCI), and 8.50 MHz (34 KFCI) respectively. The writing current was set at 25 mA for all radii of disk. X-ray diffraction of CuK_α x-ray by θ/2θ scan was used to study the crystal structure and texture orientation of CoCrTaPtB. By calculating the ratio of the intensity of Cr (110) + CoCrPtTaB (0002) to the intensity of the Cr (200) peak at 2θ=64.7⁰, we were able to correlate the intensity ratio to read/write characteristics.

RESULTS AND DISCUSSION

Fig. 1(a) is the coercivity in the circumferential direction Hc(c) versus cross hatch angles. Fig. 1(b) is the coercivity in the radial direction Hc(rad) versus CHA. Fig. 1(c) is the relationship between coercivity ratio Hc(c)/Hc(r) and CHA. The coercivity ratio decreases as CHA increases. The highest coercivity ratio is at CHA=9⁰ and the lowest of coercivity ratio is at CHA=73⁰. Fig. 2(a) is the product of magnetic remanance and thickness in the circumferential direction, $M_rt(c)$, versus CHA. Here, M_r is the magnetic remanance and t is the magnetic film thickness. The highest $M_rt(c)$ is at 0-4 degree and the lowest $M_rt(c)$ is at 66⁰. Fig. 2(b) is the M_rt in the radial direction, $M_rt(r)$, versus CHA. $M_rt(r)$ has the highest reading at CHA=73⁰ and $M_rt(r)$ has lowest value at CHA=16⁰. Fig. 2(c) is the ratio of $M_r(c)/M_r(r)$ versus CHA. The ratio of $M_r(c)/M_r(r)$ is almost a linear function of CHA. At CHA=9⁰, we have the highest ratio of $M_r(c)/M_r(r)$. At CHA=73⁰, the lowest value of $M_r(c)/M_r(r)$ was obtained. Fig. 3(a) is squareness ratio in circumferential direction S(c) vs CHA. In the range of CHA from 32 to 68 degree, S(c) is linearly decreasing as CHA increases. Fig. 3(b) is the squareness ratio in the radial direction S(r) versus CHA. S(r) has the lowest value at CHA=9⁰ and 16⁰. The highest S(r) is at CHA=73⁰. Fig. 3(c) is the ratio of S(c)/S(r) versus CHA. The highest ratio of S(c)/S(r) is at CHA=9⁰. The lowest ratio of S(c)/S(r) is at CHA=73⁰. The squareness ratio, S(c)/S(r), decreases as CHA increases. Fig. 4(a) is the coercive squareness ratio in the circumferential direction S*(c) versus CHA. S* is a common way to evaluate the strength of coupling in thin film

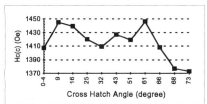

Fig. 1(a) Hc(c) vs Cross Hatch Angle

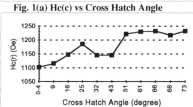

Fig. 1(b) Hc(r) vs Cross Hatch Angle

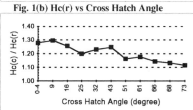

Fig. 1(c) Hc(c) / Hc(r) vs Cross Hatch Angle

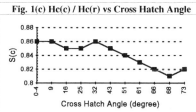

Fig. 3(a) S(c) vs Cross Hatch Angle

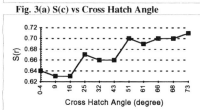

Fig. 3(b) S(r) vs Cross Hatch Angle

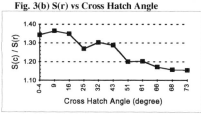

Fig. 3(c) S(c) / S(r) vs Cross Hatch Angle

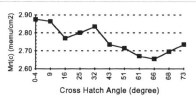

Fig. 2(a) Mrt(c) vs Cross Hatch Angle

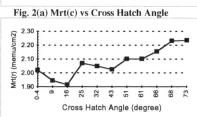

Fig. 2(b) Mrt(r) vs Cross Hatch Angle

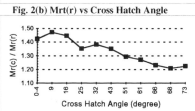

Fig. 2(c) Mr(c) / Mr(r) vs Cross Hatch Angle

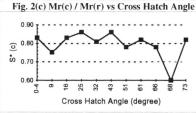

Fig. 4(a) S*(c) vs Cross Hatch Angle

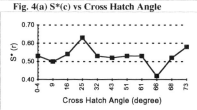

Fig. 4(b) S*(r) vs Cross Hatch Angle

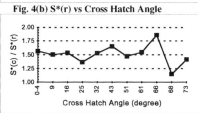

Fig. 4(c) S*(c) / S*(r) vs Cross Hatch Angle

305

media [11]. In general, cross hatch angles do not affect the exchange coupling. Fig. 4(b) is the coercive squareness ratio in the radial direction S*(r) versus CHA. Fig. 4(c) is the ratio of S*(c)/S*(r) versus CHA. The ratio of coercive squareness S*(c)/S*(r) is not a linear function of CHA. Fig. 5 gives the testing results for amplitude at low frequency (LFA) and high frequency (HFA) versus cross hatch angle. In general, LFA decreases as CHA increases. The highest LFA and HFA are located at CHA=9^0 and the lowest LFA and HFA are located at CHA=68^0. Fig. 6 gives the results of resolution vs CHA. At CHA=32^0, the highest resolution was obtained. At CHA=73^0, we obtained the lowest resolution. Fig. 7 is the measurement for signal-to-noise ratio (SNR) vs CHA. As CHA increases, SNR decreases. The highest SNR is at CHA=9^0 and the lowest SNR is at CHA=73^0. Since the disk with CHA=9^0 has the highest orientation ratio (OR), it may have the highest SNR [4]. High O.R. medium makes the transition sharp and high signal output as well as low noise. Therefore, the medium at CHA=9^0 showed the best signal-to-noise ratio (SNR). Fig. 8 is the relationship between half peak width (PW$_{50}$) and CHA. In the range of CHA from 32 to 68 degrees, PW$_{50}$ increases linearly as CHA increases. For the range of CHA at 0 to 25 degree, PW$_{50}$ is almost a constant. Fig. 9 is the measurement for bit shift versus CHA. For CHA=0 to 16 degree, bit shift is almost the same. At CHA=68^0, the worst bit shift was detected. Fig. 10 is the relationship between overwrite(O/W) and CHA. In general, overwrite (-dB) increases as CHA increases, which may be due to lower amplitude at higher CHA as indicated on Fig.5.

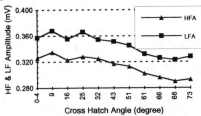

Fig. 5 HF & LF Amplitude vs CHA

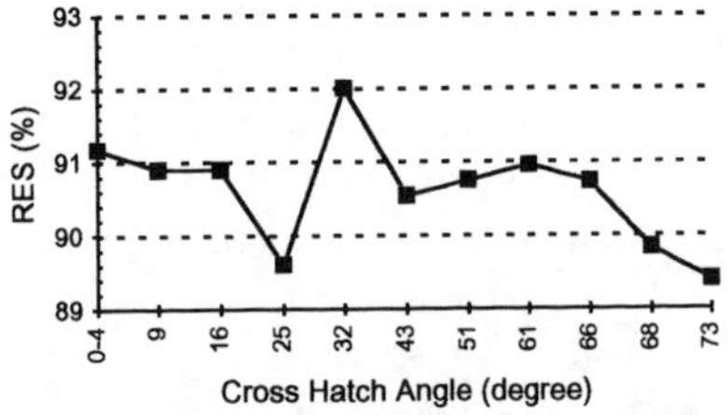

Fig. 6 Resolution vs Texture CHA

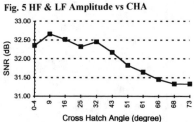

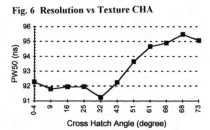

Fig. 7 SNR vs Texture CHA

Fig. 8 Half Peak Width vs Texture CHA

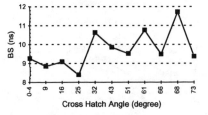

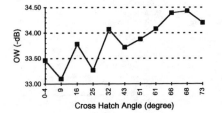

Fig. 9 Bit Shift vs Texture CHA

Fig. 10 Overwrite vs Texture CHA

The preferential direction of CoCrPtTaB for different CHA texturing is used to explain the mechanism for different magnetic properties and R/W characteristics. The primary difference for all of the x-ray spectra is in the peak intensity between the combined Cr (110) and CoCrPtTaB (0002) peaks and the Cr (200) peak. A comparison of the peak intensity differences is shown in Table 1. Fig. 11 is the X-ray diffraction spectra for disks with different cross hatch angles. The highest X-ray intensity ratio is found for the sample with cross hatch angles of 9^0, which has highest S/N ratio. The lowest intensity ratio is at CHA of 68^0, which has lowest S/N ratio and the worst PW_{50}. From a previous study [3], it was concluded that the higher the intensity ratio, the stronger the CoCrTa (0002) texture. In our cases, the CoCrPtTaB films have strong c-axis out-of-plane orientations as they were deposited on a substrate with a CHA of 9^0.

Table 1   The intensity ratio of Cr(110) + CoCrTa(0002) (peak at about 44.45^0) to Cr(200) (peak at about 64.7^0)

Texture I.D. (Cross Hatch Angle (degree))	1072 (0-4)	1074 (9)	1075 (16)	1076 (25)	1073 (32)	1077 (43)	1078 (51)	1079 (61)	1080 (66)	1082 (68)	1081 (73)
Intensity Ratio ($I_{Cr(110)+CoCrTa(0002)}/I_{Cr(200)}$)	1.24	1.84	1.29	1.14	1.40	1.40	1.31	0.97	1.16	1.04	1.35

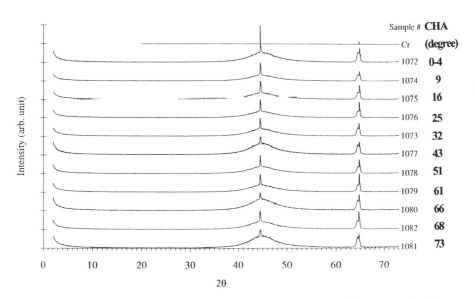

Fig. 11 X-ray diffraction spectra for all of eleven samples with different cross hatch angle.

SUMMARY AND CONCLUSIONS

The effects of cross hatch angle on the magnetic properties for CoCrPtTaB can be summarized as follows. (1) As CHA increases, both the coercivity ratio Hc(c)/Hc(r) and Mr ratio Mr(c)/Mr(r) decrease. This means that orientation ratio (OR) for media decreases as texturing cross hatch angle increases. (2) As CHA increses, S(c) decreases and S(r) increases. The squareness ratio S(c)/S(r) decreases too. (3) However, the coercive squareness S*(c), S*(r), and their ratio, S*(c)/S*(r) is not a linear function of CHA. The effects of CHA of texturing on the read/write characteristics can be summarized as follows. (1) Both of the amplitudes at low and high frequencies decreases as CHA increases. (2) The signal/noise ratio, SNR, decreases as CHA increses from 32 to 73 degrees. (3) For half peak width, PW_{50}, is almost a constant for CHA from 0 to 25 degrees; PW_{50} increases as CHA increases from 32 to 73 degrees. PW_{50} can degrade as much as 4 nano-seconds for a very high CHA texturing surface. At cross hatch angle CHA of 9 degrees, CoCrPtTaB magnetic films show the best orientation ratio OR, the highest readback signal at low and high frequencies, the highest signal/noise ratio SNR, and the lowest PW_{50}. X-ray diffraction results indicated the highest intensity ratio for Cr (110) + CoCrPtTaB (0002) to the intensity of Cr (200) peak at CHA equal to 9 degrees. This highest ratio of crystallographic texture may explain why the films have the best magnetic properties and the magnetic recording performance for a CHA at 9 degrees.

REFERENCES

[1] T.M. Reith, K.E. Johnson, A.C. Wall, and E.Y. Wu, JA-07, Digest of Intermag, 1992.
[2] J.K. Lee, A. Chao, J. Enguero, M. Smallen, H.J. Lee, and P. Dion, IEEE Transactions on Magnetics, Vol 28, NO.5, P. 2880, 1992.
[3] M.S. Miller, J.P. Walber, H.F. Erskine, B.M. Hagen, Y. Hsu, J.M. Sivertsen, J. H. Judy, and D.E. Speliotis, J. Appl. Phys, Vol. 70, P. 7518, 1991.
[4] H. Takano, T.T. Lam, J.-G. Zhu, and J. H. Judy, IEEE Transactions on Magnetics, VOL. 29, NO.6, P. 3709, 1993.
[5] Y. Deng, D. N. Lambeth, and D.E. Laughlin, IEEE Mag., VOL. 29, NO.6, P. 3676, 1993.
[6] M.F. Doerner, T.Y. Yogi, D.S. Parker, and T. Nguyen, IEEE Transactions on Magnetics, VOL. 29, NO.6, P. 3667, 1993.
[7] T. Chen, and T.Yamashita, IEEE Transactions on Magnetics,VOL. 24, NO.6, P. 2700, 1988.
[8] P. Glijer, J.M. Sivertsen, and J.H. Judy, J. Appl. Phys., Vol. 73, P. 5563, 1993.
[9] C.R. Paik, I. Suzuki, N. Tani, M. Ishikawa, Y. Ota, and K. Nakamura, IEEE Transactions on Magnetics, Vol. 28, NO. 5, P. 3084, Sep. 1992.
[10] M. Mirzamaani, M. Re, S.E. Lambert, A. Praino, T.S. Perterson, and K.E. Johnson, IEEE Transactions on Magnetics, Vol. 26, NO. 5, P. 2457, Sep. 1990.
[11] T.C. Arnoldussen, and L.L. Nunnelley, "Noise in Digital Magnetic Recording", World Scientific Publishing Co., 1992

THE MICROSTRUCTURE OF SPUTTERED Co-Cr MAGNETIC RECORDING MEDIA

ASTRID PUNDT*, RALF BUSCH** AND CARSTEN MICHAELSEN***
* Institut für Metallphysik, Hospitalstraße 3-7, D-37073 Göttingen, Germany;
** Keck Laboratory of Engineering Materials, California Institute of Technology, Pasadena, CA 91125;
*** Institut für Werkstofforschung, GKSS-Forschungszentrum, D - 21502 Geesthacht, Germany.

ABSTRACT

With the increase of magnetic storage capacity in view, sub grain decomposition in Co-Cr magnetic recording material is examined with the field ion microscope / atom probe (FIM/AP) on a nanometer scale. Concentration variations in sputtered 50 nm Co-20at% Cr and 3 μm Co-22at% Cr films are analysed. Layers with up to 50 nm thickness are sputtered directly onto tungsten FIM tips. FIM-tips of layers with up to 3 μm thickness are prepared by a lithography and etching technique. The measurements show decomposition on the scale of a few nanometers. The concentration amplitudes vary between 11 at% Cr and 42 at% Cr within 50 nm films. Within 3 μm films the concentration varies from about 10 at% Cr to 30 at% Cr. Furthermore, at grain boundaries 3-4 nm precipitates of about 40 at% Cr are observed. The observed concentrations depend on substrate temperature. The compositional inhomogeneity can be explained by a miscibility gap in the hcp phase solid solution which is induced by magnetism at low temperatures.

INTRODUCTION

In 1978 Iwasaki et al. [1] introduced sputtered Co-Cr thin films as perpendicular magnetic recording material. Magnetic measurements on Co-Cr films with Cr concentrations between 12 at% and 24 at% (Fisher et al. [2]) show a higher saturation magnetisation than expected for a homogeneous alloy. A reason for this behaviour could be a phase separation. Phase separation within grains might lead to an increase of magnetic storage density.

A lot of research concerning phase separation in Co-Cr magnetic material has been done. Investigations by means of transmission electron microscopy or x-ray diffraction are constrained because both the atomic size difference and the atomic scattering contrast is small [3]. In order to investigate the microstructure of Cr, Maeda et al.[4,5] etched the films to remove Co. Chrysanthemum like patterns were found in samples treated in this way. Compositional studies were performed by Parker et al.[6] using Mössbauer spectroscopy and Yoshida et al.[7] using nuclear magnetic resonance. To investigate microstructure and composition of the decomposed Co-Cr films at the same time, field ion microscopy including time of flight mass-spectrometry (FIM/AP) [8] is a powerful method. Chemical analysis can be done with nanometer resolution in the lateral dimension and on the atomic scale in depth. Additionally, the microstructure can be determined from FIM-image contrast.

EXPERIMENTAL

Sputtered Co-Cr magnetic layers of 50 nm and 3 μm thickness are investigated. The FIM/AP analysis requires samples in form of tips with radius of curvature between 50 nm and 200 nm. This specimen geometry is obtained by different preparation methods for the two types of layers. For the thinner films tungsten wires are first electropolished to a tip radius of about 20 nm [9]. Subsequently, a 50 nm Co-20at% Cr layer is sputtered onto these tips at an Ar pressure of $3.7 \cdot 10^{-3}$ mbar. The tip temperature is expected to increase to about 100 °C during deposition. In Fig.1 two scanning electron

309

microscopy (SEM)-micrographs of the same tungsten tip before and after sputtering procedure are superimposed. The thickness of the Co-Cr layer is determined to be 50 nm. The quality of the coating process can be checked by means of small rises on the tungsten substrate. They have to reappear on the surface of the layer (arrow). At the same time Co-Cr films are deposited on planar substrates to characterise the hcp structure by x-ray diffraction. From rocking curve measurements the full width of the half maximum of the (0002) line was determined to be 6.7°.

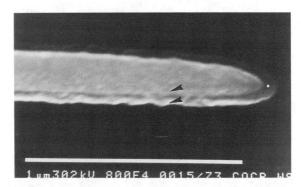

Fig.1: *Superimposed SEM-micrographs of the FIM-tip before and after deposition. The inner bright part is the substrate tungsten tip with radius of curvature of about 20 nm. Thickness of the Co-Cr layer is about 50 nm.*

In the second preparation method tips are prepared directly from 3 µm sputtered Co-22at% Cr films. The films are deposited on flat Cu coated Si-substrates using a Co-22at% Cr alloy target at Ar pressure of $1.1 \cdot 10^{-2}$ mbar. During deposition the substrate temperature was 200°C. Ribbons with square cross-section (0001 in plane) are produced using a lithography technique [10]. Finally, these ribbons are electropolished from both sides to prepare FIM-tips.

AP analyses are performed under UHV conditions with a tip temperature of about 54 K. FIM-micrographs are obtained using a gas mixture of $5 \cdot 10^{-6}$ mbar Ne and $1 \cdot 10^{-6}$ mbar He gas.

RESULTS

Atom probe analyses of 50 nm Co-20at% Cr films show regions with nominal concentration that coexist with regions indicating phase separation. In Fig.2 an analysis is plotted as a ladder diagram, i.e. the number of Cr atoms is plotted versus the total number of analysed atoms. In such a diagram the slope corresponds to the concentration of the analysed region. In distances of 1 nm the concentration changes from more than 42 at% Cr to less than 11 at% Cr. Since the analysis cylinder is of the same size as these 1 nm regions the real concentration can not be exactly determined [11]. In Fig.3 a corresponding FIM-micrograph is shown. Bright regions corresponding to low Cr concentration have the same size, as was determined by AP analyses.

Because combining the data into blocks destroys some information, we choose the ladder diagram as a sensitive method. To check the evidence of the concentration variations in the ladder diagram, the null hypotheses that the mean values of each two regions are equal is tested with a significance level of 1%, providing the distribution of the mean values in each region is a normal distribution [12,13]. Furthermore, we have determined the standard deviation $\pm 2\sigma = \pm 2\sqrt{c(c-1)/(n-1)}$ of each region (containing the concentration with a probability of 96%), with the mean concentration c and the number of ions n in the concentration region. Only regions, where the null hypothesis could be rejected and the ±2σ values (in brackets) do not overlap are indicated in the ladder diagram. The mean values of these regions are significant different and indicate a non-random Cr-distribution.

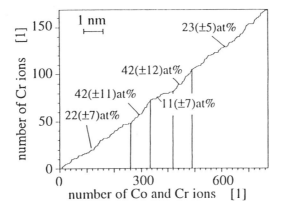

Fig.2: *Ladder diagram of the AP analysis of a 50 nm Co-20at% Cr layer. Concentration variations on a nanometer scale can be seen.*

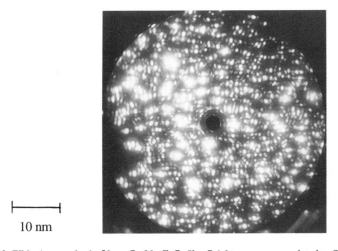

Fig.3: *FIM-micrograph of a 50 nm Co-20at% Cr film. Bright areas are correlated to Cr-poor regions.*

Atom probe analyses in 3 μm Co-Cr films indicate similar concentration variations on a 1 nm scale, as shown in Fig.4. Concentrations less than 8 at% Cr and more than 33 at% Cr are observed. A corresponding FIM-micrograph is shown in Fig.5. It also indicates 1-2 nm bright regions. Bright regions correspond to low Cr concentration, as measured by selected area AP-analyses. The radii of curvature of the FIM-tip and the resulting magnifications differs in two directions.

These results show a smaller concentration difference between the Cr-rich and Cr-poor regions within Co-Cr layers at higher temperature sputtered.

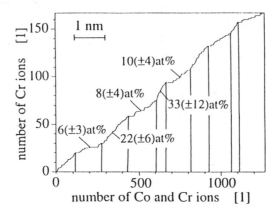

Fig.4: *Ladder diagram of the AP analysis of 3 μm Co-22at% Cr film. Concentration variations on a nanometer scale are evident.*

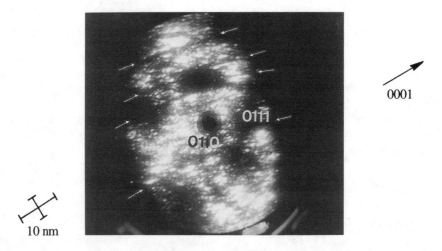

Fig.5: *FIM-micrograph of 3 μm Co-20at% Cr film. Bright regions correspond to Cr poor concentrations. Arrows indicate grain boundaries. Because the radii of curvature vary in two directions the magnification is different in these directions.*

In the 3 μm layers a second type of concentration variation can be found on a scale of about 8 nm, as shown in Fig.6. The concentration varies between 10 at% Cr and 30 at% Cr. The scale of this second type corresponds to the scale of the Chrysanthemum like structure found by Maeda et al..

Fig.7 shows the ladder diagram of a grain boundary analysis. It demonstrates the occurrence of a 3-4 nm precipitation with a composition of 40 at% Cr which is located at the grain boundary. In the vicinity of these precipitates Cr-poor regions of about 8 at% Cr are often observed.

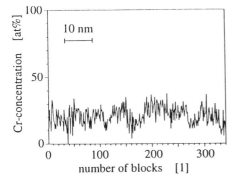

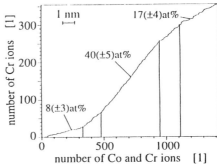

Fig.6: *Concentration variations in a 3 μm Co-20at% Cr film on a scale of about 8 nm. Each block contains 69 ions.*

Fig.7: *Ladder diagram of the grain boundary analysis of a 3 μm Co-20at% Cr film. Probing direction is close to <1122>.*

DISCUSSION

We discuss our results with respect to the calculated Co-Cr phase diagram of Hasebe et al.[14]. As shown in Fig.8, the calculated free energy curves of the hexagonal phase are extrapolated to low temperatures. At temperatures lower than 800 °C, magnetism induces a miscibility gap: the ferromagnetic Cr-poor solid solution is thermodynamically stabilised with respect to the paramagnetic phase. The AP measurements of the as sputtered layers show a phase separation into Cr-poor (ferromagnetic) and Cr-rich (paramagnetic) phases. Furthermore, the AP studies show an increase of the concentration difference with decreasing substrate temperature in good qualitative agreement with the free energy curves. As shown in Fig.8 this can be explained by the larger miscibility gap at lower temperatures. Thus we conclude that thermodynamics play an important role on phase formation during deposition.

As bulk diffusion is known to be very sluggish [15] and can be neglected for the temperatures present in our experiments, the decomposition must have taken place at the growing surface during deposition. The measurements suggest that small concentration variations on a nanometer scale are formed within the grains. The surface mobility is apparently most pronounced at grain boundaries (due to their open structure), leading to larger Cr enriched precipitates. We note that these precipitates were only observed in the 3 μm samples, because grain boundaries could not clearly be identified in the 50 nm samples (see Fig.3). Furthermore, composition variations on an 8 nm scale are observed which might correspond to the Chrysanthemum like pattern found by Maeda et al.[4].

The results indicate that a thermodynamic interpretation is a reasonable approximation to describe the experimental observation, although other parameters such as deposition rates or process pressure are expected to influence the kinetics of phase formation during deposition.

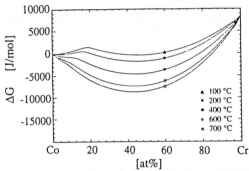

Fig.8:. *Free energy curves of the hcp phase calculated by Hasebe et al.[14]. We extrapolate these curves to lower temperatures. Magnetism induces a miscibility gap at temperatures lower than 800°C.*

SUMMARY

Concentration variations in as sputtered 50 nm and 3 μm Co-20at% Cr films are analysed with field ion microscope / atom probe (FIM/AP). The results show an already decomposed structure on various length scales: on a scale of 1-2 nanometers the concentration varies between 11 at% Cr and 42 at% Cr within 50 nm films. Within 3 μm films the concentration differs from 8 at% Cr to 34 at% Cr. Furthermore, variations on a scale of 8 nm are observed. At grain boundaries a 3-4 nm Cr enriched region of 40 at% Cr can be found. We conclude that the compositional inhomogeneity is caused by the miscibility gap in the hcp phase, which is induced by the magnetism occurring at low temperatures.

ACKNOWLEDGEMENT

The authors would like to acknowledge the support of their advisor, the late Prof. P.Haasen.

They also wish to thank Dr. K. Hono for kindly providing the lithography-samples and thank the Göttingen FIM-group for helpful discussions. This work was supported by the Deutsche Forschungsgemeinschaft via Sonderforschungsbereich 345.

REFERENCES

[1] S.Iwasaki, K.Ouchi, IEEE Trans.Magn., MAG-**14**(5), 849 (1978)

[2] R.D.Fisher, V.S.Au-Yeung, B.B.Sabo, IEEE Trans.Magn., MAG-**20**(5), 806 (1984)

[3] K.Kobayashi, G.Ishida, J.Appl.Phys.**52**(3), 2453 (1981)

[4] Y.Maeda, S.Hirono, M.Asahi, Jap.J.Appl.Phys. **24**(12), L951 (1985)

[5] Y.Maeda, M.Takahashi, IEEE Trans.Magn., MAG-**24**(6), 3012 (1988)

[6] F.T.Parker, H.Oesterreicher, E.Fullerton, J.Appl.Phys.**66**(12), 5988 (1989)

[7] K.Yoshida, H.Kakibayashi, H.Yasuoka, J.Appl.Phys.**68**(2), 705 (1990)

[8] G.P.Geber, T. Al-Kassab, D.Isheim, R.Busch, P.Haasen, Z.Metallkd.**83**, 449 (1992)

[9] R.Busch, S.Schneider (these proceedings)

[10] N.Hasegawa, K.Hono, R.Okano, H.Fujimori, T.Sakurai, Appl.Surf.Sci.**67**(1-4), 407 (1993)

[11] R.Wagner, Field-Ion Microscopy in Materials Science, in: Crystals, Vol.6, edited by H.C.Freyhardt, Springer Verlag, Berlin 1982

[12] E. Kreyszig, Statistische Methoden und ihre Anwendungen, Vandenhoeck & Ruprecht Verlag Göttingen 1965

[13] A.Pundt, D.Isheim (in preparation)

[14] M.Hasebe, K.Oikawa, T.Nishizawa, J.Jap.Inst.Met.,**46**(6), .577 (1982)

[15] A.Green, D.P.Whittle, J.Stringer, N.Swindells, Scr.Met.**7**, 1079 (1973)

STRESS, MICROSTRUCTURE AND MAGNETIC PROPERTIES OF THIN SPUTTERED Co ALLOY FILMS

C. A. ROSS, R. RANJAN AND J. CHANG
Komag Inc., 275 S. Hillview Drive, Milpitas, CA 95035

ABSTRACT
The internal stress and magnetic properties have been measured for cobalt alloy films deposited using a range of sputter conditions. The sputter conditions affect the internal stress and the microstructure, both of which contribute to the magnetic properties of the films. Stress influences the magnetic properties via magnetostriction, while the microstructural effect on coercivity comes primarily from the degree of separation and size of the grains in the film. The effect of stress on coercivity was investigated separately by applying external stresses to the films, and magnetostriction coefficients were derived for different alloys. The change in coercivity with sputter parameters indicates that in alloys such as $Co_{77}Ni_7Cr_4Pt_{12}$, the microstructure has the dominant effect on coercivity.

INTRODUCTION

Control of the magnetic properties of sputtered cobalt alloy films is important in the manufacture of rigid magnetic recording media. Hard disks are usually made from aluminum alloy substrates which are coated with a layer of electroless nickel phosphorus (NiP_x) which is polished and textured. An underlayer, magnetic layer and carbon overcoat are then sputtered onto the substrate. The magnetic layer is a cobalt alloy, typically 50nm thick with a coercivity of 1500 Oe or greater. The properties of the magnetic layer govern the recording behaviour of the disk, and the present trend is towards thinner films with high coercivity and squareness. Alloys used in rigid magnetic media are primarily made from cobalt, to give a high magnetisation, with additions of other materials including Cr, Pt and Ta to increase the coercivity. The underlayer is commonly Cr, in which case the Co alloy may grow with an epitaxial relationship to the underlayer (1), or it may be sputtered amorphous NiP_x (2) in which case there is no such relationship.

The magnetic properties are strongly influenced by the microstructure of the film and this is affected by the sputtering parameters, in particular by the sputtering pressure, power, substrate temperature and substrate bias. Any mechanical stress in the film will also influence the film's magnetic properties through magnetostriction. The film stress has an intrinsic component which is determined by the material and the sputter parameters, and an extrinsic component caused by thermal mismatch between the film and substrate. It is of interest to identify the effects of the stress and the microstructure on the magnetic properties of the film. In this paper, we will show how sputter parameters affect both the stress and magnetic properties of some cobalt-based alloy thin film media, and we will measure separately the effect of stress on the magnetic hysteresis loop by applying an external stress to the films.

RESULTS AND DISCUSSION

Influence of sputter parameters on stress and magnetic properties of $Co_{77}Ni_7Cr_4Pt_{12}$ films

The effect of film thickness, sputter pressure and sputter power on stress in rf-sputtered $Co_{77}Ni_7Cr_4Pt_{12}$ and NiP_x (x = 25at.%) is shown in Fig. 1. The samples were made in a load-locked r.f.-sputter system at base pressures of 10^{-6}torr. The sputter pressure varied between 6mtorr and 30mtorr and the power varied between 0.5kW and 1.5kW over a 20cm diameter target. Samples were made on thin glass slides which had been precoated with 30nm sputtered

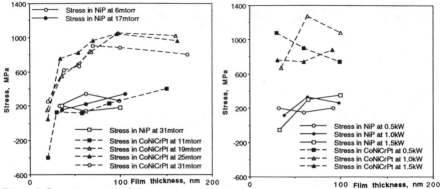

Figure 1. Stress in sputtered $Co_{77}Ni_7Cr_4Pt_{12}$ and NiP_x on glass as a function of sputter pressure and power.

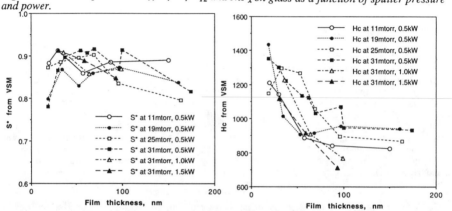

Figure 2. Coercivity and S* of sputtered $Co_{77}Ni_7Cr_4Pt_{12}$ on NiP_x on glass.

NiP_x to make them reflective and to provide a surface which more closely resembled a metallic disk substrate. The substrates were preannealed at 120°C for 10min, then their curvature was measured using a laser system. NiP_x or $Co_{77}Ni_7Cr_4Pt_{12}$ was then deposited onto the substrates and the curvature was remeasured in order to derive the film stress. The stress measurements are estimated to be accurate to ±10% for the thicker films and ±20% for the thinner films.

The NiP_x films showed tensile (positive) stresses of up to 350 MPa for the conditions used, without a systematic variation with sputter power or pressure but with generally a slight increase with thickness. The $Co_{77}Ni_7Cr_4Pt_{12}$, however, had a larger tensile stress of up to 1000 MPa, which also increased with thickness up to 100nm but had little variation with power. Films made with the lowest sputter pressure, 11mtorr, had a smaller tensile stress and the thinnest film had a compressive stress. A transition from tension to compression is expected as the sputter pressure is reduced, due to increased bombardment of the growing film (3).

VSM (vibrating sample magnetometry) was used to measure the magnetic hysteresis loops of the $Co_{77}Ni_7Cr_4Pt_{12}$ samples, from which the saturation magnetisation M_s, remanent magnetisation M_r, coercivity H_c, squareness S (=M_r/M_s) and S* (=1 - (M_r/H_c x dM/dH at H_c)) could be found. Some results are shown in Fig. 2. Many of the trends observed can be explained in terms of the effect of the grain separation (4), which governs the magnetostatic and exchange interactions between the grains. For example, the coercivity of the films decreases with increasing film thickness as the crystals grow larger and become less well separated, and is higher at high sputter pressure or low sputter power because the grains are better separated. Higher pressures promote shadowing and a more porous microstructure, while the coercivity increase at low power may be caused by increased sensitivity to residual gases at the lower deposition rates. The magnetic properties were isotropic in the film plane.

Measurements on $Co_{81}Ni_9Pt_{10}/NiP_x$ sputtered onto smooth electroless NiP_x on Al were also made. These films were deposited on 0.64 mm thick substrates which were preannealed at 250°C for 2h before sputtering. The films were made in a r.f. batch sputter system with a base pressure of 10^{-6} torr. NiP_x films alone made at 30mtorr and 1.0kW with thicknesses of 50nm had a tensile stress of 170 MPa, slightly smaller than that measured for the films on glass. Magnetic films of 50nm thickness had a tensile stress which varied between 260 MPa and 375 MPa as the sputter pressure increased from 15mtorr to 30mtorr. These values are lower than those measured on glass substrates presputtered with NiP_x, which varied between 200 and 800 MPa as the pressure increased from 11 to 31mtorr. The lower tension in the film on the aluminum substrate arises at least partially from the thermal mismatch difference between the substrates. Cooling by an estimated 100°C after sputter will give a compressive thermal component of 370 MPa for the film on aluminium but a tensile component of 150 MPa for the film on glass. Subtracting the thermal stress from the total stress shows that the intrinsic stress is similar for the two alloys.

We therefore expect $Co_{77}Ni_7Cr_4Pt_{12}$ or $Co_{81}Ni_9Pt_{10}$ alloy films on aluminum/electroless NiP_x/ sputtered NiP_x substrates to be in a tensile stress state of order 300 MPa if the films are made with relatively high sputter pressures, with some heating arising from the sputtering process. For glass disk substrates, higher tensile stresses are expected. This stress state will affect the magnetic properties of the films through the magnetostriction.

Influence of stress on the magnetic hysteresis loop of $Co_{75}Ni_7Cr_6Pt_{12}$, $Co_{76}Cr_{12}Pt_{12}$ and $Co_{84}Cr_{14}Ta_2$ films

It has been seen that sputter conditions affect microstructure and stress, and both of these factors influence magnetics. To isolate the effects of stress on the magnetic properties, an external stress can be applied to the film by heating or bending the substrate in situ in the VSM. Heating applies a biaxial thermal mismatch stress while bending the substrate in one direction adds a uniaxial stress.

$Co_{84}Cr_{14}Ta_2$, $Co_{75}Ni_7Cr_6Pt_{12}$ and $Co_{76}Cr_{12}Pt_{12}$ films were deposited onto smooth or textured aluminum/electroless NiP_x substrates with a sputtered Cr underlayer. The smooth substrates were polished using fine abrasive, while textured substrates were made by introducing circumferential grooves into a polished substrate. It is well known (5-8) that magnetic films sputtered onto Cr on a textured and heated substrate are anisotropic in the film plane, such that the circumferential coercivity (parallel to the grooves) is higher than the radial coercivity by a factor known as the coercivity orientation ratio, H_c OR. An orientation ratio for S* can similarly be defined. In contrast, films sputtered onto a smooth substrate are found to have H_c OR equal to 1.0. Some of the films in this study made on smooth substrates showed H_c OR significantly less than 1.0; this was attributed to the presence of fine radial scratches on the substrate resulting from the polishing process. The films made on textured substrates had H_c OR in the range of 1.4 - 1.8.

Samples of the three alloys were made on textured substrates using similar sputter conditions. The Cr underlayer was 150nm thick, sputtered at 10 or 12mtorr while the magnetic films were 54-65nm thick and sputtered at 5 or 6mtorr. Prior to deposition, the substrates were heated to 250°C. The effect of a radial external stress on the radial coercivity and S* is shown in Fig. 3. A uniaxial stress was applied to the films by cutting radial strips from the disks, applying an elastic strain by bending the strips and estimating the stress from the substrate curvature. Magnetic properties of the samples were measured by VSM in the stressed and unstressed states. As a compressive stress was applied, the coercivity increased implying a negative magnetostriction

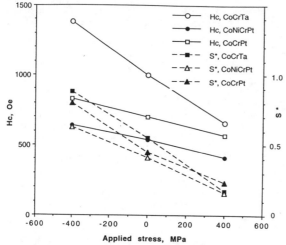

Figure 3. Effect of tensile stress on radial coercivity and S* of films on Cr underlayers

coefficient as expected (9,10). Both S and S* also increased with compressive stress and decreased with tensile stress.

If a uniaxial stress σ is applied in the film plane, the magnetostatic energy contribution (in an isotropic material) is given by $-(3/2)\lambda_s\sigma\cos^2\phi$, where λ_s is the magnetostriction coefficient and ϕ the angle between the stress and the direction of magnetisation, zero in the geometry of this experiment. The resulting change in coercivity ΔH_c can be estimated as being proportional to $\lambda_s\sigma/M_s$. The exact relation may depend on the degree of exchange coupling in the film (11) but a proportionality coefficient of 3 is often used. Values of M_s for the three alloys in this study were measured by VSM, and have the values 680, 990 and 800 emu cm^{-3} for the $Co_{84}Cr_{14}Ta_2$, $Co_{75}Ni_7Cr_6Pt_{12}$ and $Co_{76}Cr_{12}Pt_{12}$ respectively. From the rate of change of coercivity with stress and the M_s values, the magnetostriction coefficients are in the ratios 1:0.46:0.44 for the $Co_{84}Cr_{14}Ta_2$, $Co_{75}Ni_7Cr_6Pt_{12}$ and $Co_{76}Cr_{12}Pt_{12}$ alloys. A value of -30×10^{-6} has been measured for a $Co_{89}Cr_9Ta_3$ alloy (8) which can be compared to the -20×10^{-6} calculated for the $Co_{84}Cr_{14}Ta_2$ by applying the relationship $\Delta H_c = 3\lambda_s\sigma/M_s$.

Heating the samples adds a biaxial tensile component to the film stress of 3.7 MPa/K. Samples were heated in the VSM to temperatures of up to 150°C, and the effect of measurement temperature on the orientation ratios and circumferential coercivity and S* of three samples is shown in Fig. 4 and Fig. 5. The samples in Fig. 4 were the same ones used in Fig. 3, while those in Fig. 5 were made after adjusting the sputter conditions to yield a circumferential coercivity of 1200 Oe for all three alloys in the as-deposited state. In Fig. 5, the isotropic media, made on smooth substrates, have orientation ratios which remain stable or decrease (in the case of samples with H_c OR initially less than 1) with heating while for the oriented media, H_c OR increases with temperature. The change in stress in the films will affect both the radial and the tangential coercivity by the same amount, neglecting effects of topography on stress; this will alter the H_c OR for anisotropic films but not that of isotropic films. Changes in M_s are small over this temperature range, so the coercivity changes are believed to originate from the change in stress in the films. As expected from the negative magnetostriction and tensile thermal stress, the coercivity decreases with heating and the rate of change is highest for the $Co_{84}Cr_{14}Ta_2$ which has the highest magnetostriction. Over the temperature range studied, the ratios of the magnetostriction coefficients calculated from $M_s\Delta H_c$ are 1:0.50:0.55 for the $Co_{84}Cr_{14}Ta_2$, $Co_{75}Ni_7Cr_6Pt_{12}$ and $Co_{76}Cr_{12}Pt_{12}$ alloys respectively, in good agreement with the ratios calculated from the uniaxial bending. Assuming that the relation $\Delta H_c = 3\lambda_s\sigma/M_s$ holds, $\lambda_s = -28\times10^{-6}$ for the $Co_{84}Cr_{14}Ta_2$.

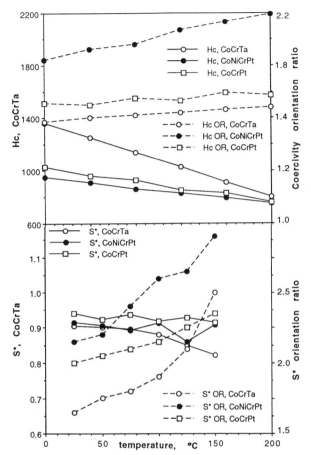

Figure 4. Effect of measurement temperature on circumferential coercivity, S* and orientation ratios of films on Cr underlayers

Returning to the $Co_{77}Ni_7Cr_4Pt_{12}/NiP_x$/glass experiment, the contributions of stress and microstructure to the coercivity can now be assessed. Comparing 50nm thick films sputtered at different argon pressures, a film sputtered at 31mtorr has a tensile stress which is 600 MPa higher than that of a film sputtered at 11mtorr. Taking $\lambda_s = -15\times10^{-6}$, the high pressure film might be expected to have a coercivity which is lower than that of the low pressure film by approximately 300 Oe. However, the coercivity of the high pressure film is actually 250 Oe higher than that of the low pressure film despite the stress difference, most likely due to its enhanced grain separation. This illustrates that in such alloys, the microstructure is a more important determinant of the magnetic properties than is the stress.

CONCLUSIONS

Sputter parameters affect the film microstructure and stress and both of these affect the magnetic properties of the film. The microstructural influences on magnetics are consistent with the degree

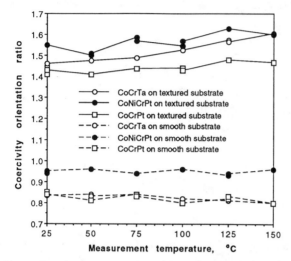

Figure 5. Coercivity orientation ratio of 1200 Oe films on smooth and textured substrates

of grain isolation, for instance higher coercivities are promoted by lower film thickness, higher sputter pressures and lower deposition rates. $Co_{81}Cr_9Pt_{10}$ and $Co_{77}Ni_7Cr_4Pt_{12}$ films made under conditions of high sputter pressure and low power have tensile stresses of order 300 MPa when sputtered onto aluminium/NiP_x but higher stresses of up to 1000 MPa when sputtered onto glass/NiP_x.

Stress affects the hysteresis loop through magnetostriction, so that application of an external stress can change the coercivity and squareness. Applying a tensile biaxial stress increases the orientation ratio of oriented media but not of isotropic media. Of three alloys investigated, the $Co_{84}Cr_{14}Ta_2$, which had the highest magnetostriction, showed the greatest sensitivity of magnetic parameters to heating or the application of a uniaxial stress. Its magnetostriction coefficient was approximately -28×10^{-6}, twice as large as that of $Co_{75}Ni_7Cr_6Pt_{12}$ and $Co_{76}Cr_{12}Pt_{12}$. According to these magnetostriction values, the microstructure has a dominant effect on the magnetic properties of the films compared to the effect of stress.

REFERENCES

(1) T.P. Nolan, R. Sinclair, R. Ranjan and T. Yamashita, *J. Appl. Phys.* **73(10)** 5117 (1993)
(2) T. Yamashita, L.H. Chan, T. Fujiwara and T. Chen, *I.E.E.E. Trans. Mag.* **27(6)** 4727 (1991)
(3) J.A. Thornton and D.W. Hoffman, *J. Vac. Sci. Technol.* **14**, 164 (1977)
(4) T. Chen and T. Yamashita, *I.E.E.E. Trans. Mag.* **24(6)** 2700 (1988)
(5) M.F. Doerner, P.-W. Wang, S.M. Mirzamaani and D.S. Parker, *Proc. Mater. Res. Soc.* **232** 27 (1991)
(6) M. Mirzamaani, K. Johnson, D. Edmonson, P. Ivett and M. Russak, *J. Appl. Phys.* **67(9)** 4695 (1990)
(7) E.M. Simpson, P.B. Narayan, G.T.K. Swami and J.L. Chao, *I.E.E.E. Trans. Mag.* **23(5)** 3405 (1987)
(8) A. Kawamoto and F. Hikami, *J. Appl. Phys.* **69(8)** 5151 (1991)
(9) D. Mauri, V.S. Speriosu, T. Yogi, G. Castillo and D.T. Peterson, *I.E.E.E. Trans. Mag.* **26(5)** 1584 (1990)
(10) J.A. Aboaf, S.R. Herd and E. Klokholm, *I.E.E.E. Trans. Mag.* **19(4)** 1514 (1983)
(11) J-G. Zhu, *I.E.E.E. Trans. Mag.* **29(1)** 195 (1993)

EFFECTS OF MICROSTRUCTURE AND CRYSTALLOGRAPHY ON THE MAGNETIC PROPERTIES OF CoCrPt/Cr THIN FILMS

G. Choe and S.J. Chung
Dept.of Metallurgical Eng., Chonnam National University, Kwangju, Korea 500-757

ABSTRACT

The effects of grain morphology and crystallography on the magnetic properties of CoCrPt/Cr thin films were investigated as a function of sputtering parameters of CoCrPt thin films. Cross sectional TEM exhibited strong separation of Co grains with increasing argon sputtering pressure, resulting in reduced exchange coupling and increased coercivity. Most Co grains sputtered at high pressure were separated by 10 to 20 Å thick boundaries, possibly due to the enhanced scattering of argon ions by sputtering particles and thus the observed higher content of Pt in the deposited films. The application of substrate bias greatly decreased the Co grain separation and the nonmagnetic grain boundary disappeared by -200 V bias sputtering, resulting in increased exchange coupling. The bias sputtered CoCrPt thin films decreased their grain size and increased epitaxial growth on Cr layer and also (10$\bar{1}$0) texture. The magnetic field required to saturate the film increased significantly with increasing substrate bias, possibly due to the induced residual stress in the bias sputtered films.

INTRODUCTION

There has been considerable progress in the improvement of areal density of longitudinal recording media. Areal density of 1- 2 Gb/in^2 has been reported[1,2] and fabrication of recording media having 10 Gb/in^2 has been recently attempted[3]. For high density recording applications, the recording media are required to exhibit high coercivity and remanent magnetization while maintaining low recording noise. CoCrPt thin films with Cr underlayers are strong candidates for ultra high density recording, since high coercivity can be easily obtained in CoCrPt/Cr bilayer films. In particular, the magnetic and recording characteristics of CoCrPt/Cr thin films were reported to be dependent on the deposition parameters that influence the film microstructure, crystallography and grain morphologies [4 - 7].

In this paper, we report the correlation between the grain morphologies and the magnetic properties of CoCrPt/Cr bilayer films by varying sputtering conditions of CoCrPt thin film. Various features including grain isolation, crystallography, composition, film stress, and defects, were strongly affected by increasing argon sputtering pressure and substrate bias. Plane and cross sectional transmission electron microscopy (TEM), and x-ray diffraction were employed to observe the microstructural and morphological features.

EXPERIMENTS

$Co_{68}Cr_{14}Pt_{18}$ and Cr thin films were deposited on cover glass substrates by rf magnetron sputtering. The deposition rates of CoCrPt and Cr layers were 13 Å/sec and 8 Å/sec, respectively. Sputtering conditions of 2, 000Å thick Cr layer were kept constant at 2 mTorr argon pressure, 150 °C substrate temperature and 0 V substrate bias, while CoCrPt films on the Cr underlayer were deposited under various argon pressure and substrate bias. Argon pressure varied from 2 mTorr to 20 mTorr and substrate bias varied from -50 V to -200 V during deposition of CoCrPt at 10 mTorr. All the Cr underlayers exhibited (110) bcc texture that was observed by x-ray diffraction. Magnetic properties were measured by a vibrating sample magnetometer (VSM) with a 10 kOe maximum field. Remanence-thickness product (Mrδ) of CoCrPt films was between 1.8 and 3.1 memu/cm^2. The crystallography was obtained using an x-ray diffractometer with Cu$K\alpha$ radiation. Specimens for plane view TEM observation were prepared by ion milling the films from the substrate side. The microstructural study of CoCrPt films was performed using TEM at 200 kV accelerating voltage. Cross-section TEM samples were prepared by conventional ion milling technique.

Mat. Res. Soc. Symp. Proc. Vol. 343. ©1994 Materials Research Society

RESULTS AND DISCUSSION

The effect of the Cr underlayer on the morphology of CoCrPt was investigated by sputtering CoCrPt film at 2 mTorr and 10 mTorr with and without Cr underlayers. Figure 1 shows the TEM micrographs and electron diffraction patterns of CoCrPt films. While the Cr underlayer is sputtered at identical sputtering pressures (2 mTorr), the grain size of the CoCrPt films sputtered at 2 mTorr is much bigger than that of the film deposited at 10 mTorr. The CoCrPt films grown at 2 mTorr exhibited very low Hc (100 Oe) without the Cr underlayer but increased their coercivity up to 440 Oe with a Cr underlayer due to the decreased Co grain size as shown in Figure 1-a. With increasing argon pressure, the Co crystal size decreased significantly. The CoCrPt films sputtered at 10 mTorr showed substantial grain isolation as showin in Figure 1-b. It is, however, noted that the Cr underlayer influence the grain size of CoCrPt films deposited at the same argon pressure. The grain growth of CoCrPt films sputtered at 2 mTorr is prohibited by Cr underlayer, whereas the Cr underlayer enhances the grain growth as well as the grain isolation for the CoCrPt films sputtered at 10 mTorr, as illustrated in Figure 1. This effect resulted in the increased coercivity for all the CoCrPt films sputtered at any sputtering pressure with Cr underlayer compared to those without Cr underlayer. Most Co grains were found to exhibit preferential hcp (0002) whose interplanar spacing is close to Cr bcc (110). It is observed that there are many more stacking faults in the CoCrPt film sputtered at 2 mTorr than in the films sputtered at 10 mTorr, possibly due to more induced residual stress of CoCrPt films sputtered at lower pressure. Crystal orientation of CoCrPt films was not affected by Cr underlayer as shown in electron diffraction patterns of Figure 1.

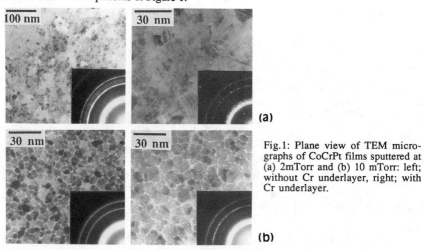

Fig.1: Plane view of TEM micrographs of CoCrPt films sputtered at (a) 2mTorr and (b) 10 mTorr: left; without Cr underlayer, right; with Cr underlayer.

Figure 2 shows the variation of coercivity, Hc, and squareness, S, as a function of CoCrPt sputtering pressure for CoCrPt/Cr bilayer films. Coercivity increases with increasing argon pressure and reaches a maximum value of 1880 Oe at 10 mTorr followed by a slight decrease with further increasing pressure. The squareness (S) and coercive squareness (S*) decreased slightly with increasing argon pressure. This result was expected since the reduced intergrain exchange coupling may increase Hc and decrease S and/or S*[8].

Cross-sectional TEM view (Figure 3) clearly revealed the grain morphological features of CoCrPt/Cr bilayer films. Columnar grain growth of Cr crystals sputtered at 2 mTorr is apparently observed. The CoCrPt films sputtered at 2 mTorr show columnar growth of continuous Co grains which are well connected each other (Figure 3-a). However, for the CoCrPt film sputtered at 10 mTorr, significant Co grain isolation is observed with well grown columnar structure on Cr underlayer. Most Co grains (80 Å - 150 Å) were separated by 10 to 20 Å thick boundaries. This result is attributed to the difference in collisional scattering of argon ions by sputtering particles owing to the change in sputtering pressure. Since an elevated argon pressure tends to enhance the scattering of argon ions with sputtered atoms, the deposition flux may become oblique, which may introduce the lower atomic mobility on the growing film and more columnar morphology. The observed compositional change of CoCrPt films strongly supports the above effect. As sputtering pressure increased from 2 mTorr to 20 mTorr, Pt content relatively increased by 6 at.% but Co content decreased by 7.5 at.% , since heavier Pt atoms are less influenced by atomic scattering at high pressure.

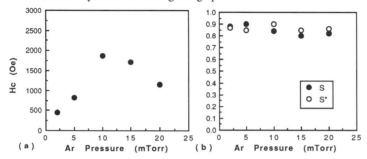

Fig.2 : Change in magnetic properties of CoCrPt films (with Cr underlayer) with Ar pressure.

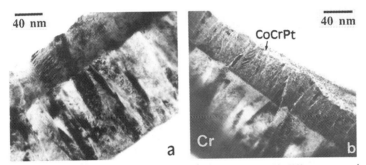

Fig.3 : Cross-sectional TEM micrographs of CoCrPt/Cr films with different sputtering pressure of CoCrPt : (a) 2 mTorr and (b) 10 mTorr.

The application of substrate bias during deposition of Cr underlayer [9] and Co alloy film layer [5,6,10] was reported to alter the grain growth of Co crystal and therefore the media noise. In this paper, we report the effect of substrate bias during deposition of CoCrPt films at 10 mTorr on the variation of grain morphology, crystallorgraphy and magnetic properties. Figure 4 shows change in plane view TEM micrographs of CoCrPt films without Cr underlayer as a function of substrate bias. As substrate bias increases, the observed Co grain isolation of CoCrPt films sputtered without substrate bias starts to decrease and almost disappears by -100 V substrate bias. On the other hand when the CoCrPt films were grown on Cr underlayer, as shown in Figure 5-a, the isolation of Co grains is still observed for the -100 V bias sputtered CoCrPt films while the grain size decreased with relatively thin boundaries compared to the unbias sputtered films. The grain isolation was completely removed after application of -200 V substrate bias, as shown in Figure 5-b. Combining this result with the above Figure 1 data, it is concluded that the Cr underlayer promotes the Co grain segregation and also inhibits the substantial grain growth of Co grains, therefore resulting in the increased Hc and reduced magnetic exchange coupling. It is also seen from electron diffraction pattern of -200 V bias sputtered film that the intensity of Co (10$\bar{1}$0) ring is dominant. The high substrate bias appears to increase the Co grain size by breaking the columnar boundaries, while the moderate

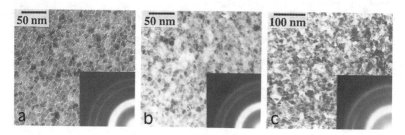

Fig.4: Plane view of TEM micrographs of bias sputtered CoCrPt films (without Cr underlayer): (a) -50V, (b) -100V and (c) -150V.

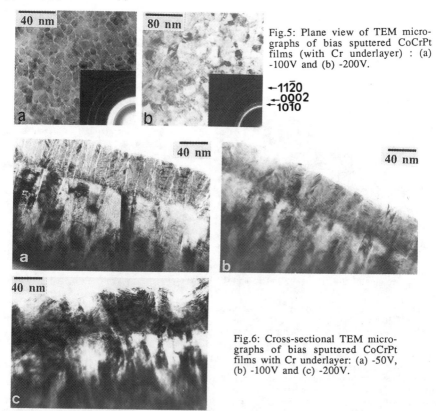

Fig.5: Plane view of TEM micrographs of bias sputtered CoCrPt films (with Cr underlayer) : (a) -100V and (b) -200V.

←11$\bar{2}$0
←0002
←10$\bar{1}$0

Fig.6: Cross-sectional TEM micrographs of bias sputtered CoCrPt films with Cr underlayer: (a) -50V, (b) -100V and -200V.

magnitude of substrate bias decreases the Co grain size.

The application of substrate bias played an important role in changing the Co grain morpholgy. Figure 6 shows cross-sectional TEM morphologies of CoCrPt/Cr films as a function of substrate bias. In contrast to the previous report [4,11], it is obvious that the separation of columnar Co grains decreases with increasing substrate bias and disappears by -200 V bias. In particular, the boundary layer between Cr and CoCrPt films becomes less distinct with the application of substrate bias and the epitaxial growth of Co grains on Cr underlayer is more enhanced. In addition, stacking fault and micro-twins were easily observed for the bias sputtered CoCrPt films. Electron diffraction pattern indicated the preferred growth of Co hcp (1010) of bias sputtered CoCrPt films, consistent with other report [4]. It appears that the applied substrate bias induces the bombardment of argon ions to the growing film and thereby increases substrate temperature, atomic mobility and film density, resulting in the dense film without non-magnetic grain boundaries.

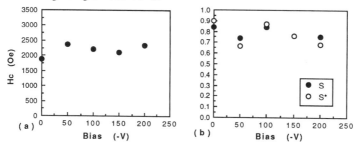

Fig.7: Change in magnetic properties of CoCrPt films (with Cr underlayer) with substrate bias.

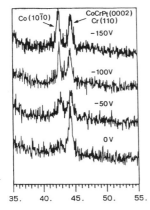

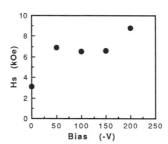

Fig.8: X-ray diffraction spectra of bias sputtered CoCrPt films with Cr underlayer.

Fig.9: Change in saturation field of CoCrPt films(with Cr underlayer) with substrate bias.

Figure 7 shows the change in Hc and S with substrate bias for CoCrPt/Cr films. Hc increases to 2400 Oe while S and S* slightly decrease, which is similar to the reported data [4]. The cause of this result was assumed to result from the reduced exchange coupling between magnetic grains [4,6,10,11]. However, as illustrated in the above cross-sectional TEM views of bias sputtered films, the magnetic grain isolation diminished by substrate bias, contrary to the previous reporter's expectation. The possible physical phenomenon that affects the magnetic properties of bias sputtered films is the increased compressive stress in the films and/or the preferred crystallographic orientation of Co (1010). The increased film stress may result in defect structures like the observed stacking faults and Co (1010) may align the c axis of crystal parallel to the film plane, resulting the increased coercivity. The composition of CoCrPt films was not affected by substrate bias. As a result, the reduced squareness of bias sputtered films may be attributed to the compressive stress of deposited film, indicated by the observed larger d-spacing of Co (1010) planes for CoCrPt films with higher substrate bias, as shown from x-ray diffraction spectra (Figure 8). It can be noted that the required magnetic field (Hs) to saturate

CoCrPt films significantly increases with increasing substrate bias. Figure 9 shows the change in Hs with substrate bias. The magnetization of CoCrPt films with random crystallographic orientation, as indicated earlier by x-ray diffraction and electron diffraction patterns, may be influenced by magnetostriction effect arising from the compressive stress of bias sputtered films. Since the intensity of compressive stress increases with increasing bias voltage, Hs increases with substrate bias. Magnetization reversal in the Co grains with preferred orientation may occur at lower applied field before the magnetization is totally reversed.

CONCLUSIONS

Effects of microstructure and crystallography on the magnetic properties of CoCrPt/Cr thin films were investigated by varying the sputtering conditions during deposition of CoCrPt layer. The deposition pressure strongly affected the Co grain size and the compositional variation by changing the collisional scattering of argon ions with sputtering particles. The Co grain isolation occurred with increasing pressure, resulting in increased coercivity and reduced magnetic exchange coupling. In dependent of sputtering condition of CoCrPt layer, Cr underlayer appeared to influence the Co grain size as well as the grain isolation. The application of substrate bias played an important role in changing the Co grain morpholgy and crystallography. The bias sputtered CoCrPt films increased epitaxial growth on Cr layer, stacking faults and also (1010) texture. The required magnetic field to saturate the CoCrPt films increased significantly with increasing substrate bias, possibly due to induced compressive stress in the deposited films.

ACKNOWLEDGEMENTS

We are grateful to Kim, Dongyoung of SKC for VSM measurements.

REFERENCES

[1] T.Yogi, C.Tsang, T.A.Nguyen, K. Ju, G.L. Gorman, and G. Castillo, IEEE Trans. Magn., **MAG-26**, 2271 (1990).
[2] M. Futamoto, F. Kugiya, M. Suzuki, H. Takano, Y. Matsuda, N.Inaba, Y. Miyamura, K. Akagi, T. Nakao, H. Sawaguchi, H. Fukuoka, T. Munamoto and T. Takagaki, IEEE Trans. Magn., **MAG-27**, 5280 (1991).
[3] E. S. Murdock, R.F. Simmons, and R. Davidson, IEEE Trans. Magn., **MAG-28**, 3079 (1992).
[4] P.Glijer, J.M. Sivertsen and J.H. Judy, J. Appl. Phys., **73**, 5563 (1993).
[5] T. Yogi, T.A. Ngyen, S.E. Lambert, G.L. Gorman and G. Castillo, IEEE Trans. Magn., **MAG-26**, 1578 (1990).
[6] M. Lu, J.H. Judy and J.M. Sivertsen, IEEE Trans. Magn., **MAG-26**, 1581 (1990).
[7] M.A. Parker, J.K. Howard, R. Ahlert and K.R. Coffey, J. Appl. Phys., **73**, 5560 (1993).
[8] J.G. Zhu and H.N. Bertram, IEEE Trans.Magn., **MAG-24**, 2706 (1988).
[9] J. Pressesky, S.Y. Lee, S. Dulan and D. Williams, J. Appl. Phys., **69**, 5163 (1991).
[10] Y. Deng, D.N. Lambeth and D.E. Laughlin, IEEE Trans. Magn., **MAG-28**, 3096 (1992).
[11] T. Yogi, T.A. Nguyen, S.E. Lambert, G.L. Gorman and G. Castillo, Mat. Res. Soc. Symp. Proc., **232**, 3 (1991).

EPITAXY AND CRYSTALLOGRAPHIC TEXTURE OF THIN FILMS FOR MAGNETIC RECORDING

DAVID E. LAUGHLIN,* Y.C. FENG, LI-LIEN LEE* AND B. WONG

Materials Science and Engineering Department, Carnegie Mellon University
*Data Storage Systems Center, Carnegie Mellon University, Pittsburgh, PA 15213

ABSTRACT

Thin films for magnetic recording are usually deposited on metallic Cr underlayers to control their crystallographic texture and other microstructural features such as grain size and crystal perfection. In this paper, the mechanisms involved in the production of thin films (both underlayers and magnetic layers) with specific crystallographic textures is reviewed and discussed. We first present an overview of a model for the development of the initial crystallographic texture of films grown on amorphous substrates. The textures that various films develop during growth are next discussed followed by a review of the role of epitaxy of Co-alloy films on Cr with underlayers and interlayers. The control of grain size of the magnetic film is also discussed in light of the underlayer grain size. Finally, the role of epitaxy on crystalline perfection and subgrain structure is discussed.

INTRODUCTION

It is well known that the crystallographic texture of thin films for magnetic recording has a profound influence on the recording properties of the films[1-6]. A schematic of a cross section of thin film recording medium is shown in Figure 1. In this paper we will discuss the development of the crystallographic texture of the underlayer film as well as the development of the crystallographic texture of the magnetic layer. Over the last decade or so a plethora of data has been collected and published concerning the crystallographic texture of both the underlayers and the magnetic layers.[7-17] In the main, the mechanisms involved in the development of crystallographic texture in the Cr underlayers depend on the interplay between the energetics and growth of the films, while the basic mechanism for the development of the texture of Co alloy magnetic thin films is one of epitaxy. In this paper we will discuss both of these general mechanisms of crystallographic texture formation. In the first section a model for the understanding of the development of either the (110) or (002) texture in Cr films grown on amorphous substrates will be presented and illustrated. In the second section we will summarize the various epitaxial relations observed to date between hcp Co alloy thin films and bcc Cr underlayers. In the third section we will briefly discuss the effect of underlayers and interlayers on the grain size and crystalline perfection of the magnetic layers.

I. DEVELOPMENT OF UNDERLAYER CRYSTALLOGRAPHIC TEXTURE

We start our discussion with the case of bcc Cr underlayers. It is well known that by changing the substrate preheating temperature two different crystallographic textures can be produced in Cr thin films. The normally obtained crystallographic texture is (110). However, under specific conditions, the (002) Cr texture can be obtained. These are the two crystallographic textures for Cr underlayers that are most frequently reported in the literature.

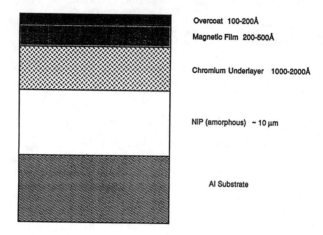

Overcoat 100-200Å

Magnetic Film 200-500Å

Chromium Underlayer 1000-2000Å

NIP (amorphous) ~ 10 μm

Al Substrate

Figure 1.   A schematic of the cross section of a typical hard disk media. Not to scale.

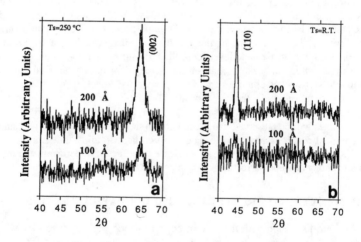

Figure 2.   X-ray diffraction patterns of Cr thin films deposited on glass at (a) 250°C,
(b) room temperature.

Each of these textures gives rise to a different texture for the subsequently deposited Co-alloy films: the (110) Cr texture gives rise to a $\{10\bar{1}1\}$ Co-alloy texture and the (002) Cr texture gives rise to a $\{11\bar{2}0\}$ Co-alloy crystallographic texture. The Co-alloy texture develops from the Cr underlayer by means of epitaxy. We will discuss this mechanism in section II, below. But first we discuss how the different Cr crystallographic textures develop on amorphous substrates.

There are two ways a non-epitaxial texture may develop during the formation of thin films: the texture may form primarily in the nucleation stage of the film or the texture may form primarily during the growth stage of the film.

Figures 2a and 2b display the X-ray diffraction patterns for 10 and 20 nm thick Cr films deposited at room temperature and 250°C respectively. The $\{002\}$ texture is observed for both the 10 nm and 20 nm film deposited at 250°C, while only the 20 nm film deposited at room temperature has the $\{110\}$ texture. From this, we see that the $\{002\}$ crystallographic texture which developed at 250°C is most probably formed during the nucleation stage of the film. However, the $\{110\}$ crystallographic texture of the Cr films seems to develop during the growth stage for the Cr films deposited under these process conditions. See also reference 18.

If the texture is determined during the nucleation stage, the relative values of three surface energies are important: the substrate/film, the substrate/vacuum and the film/vacuum energies.[19-20] In Figure 3, these values are denoted as S12, S2 and S1 respectively. Furthermore, since the film is crystalline, there are different values of S1, which depend on the crystallographic plane in question. For crystals with the bcc structure, the $\{110\}$ planes have the lowest surface energy since they are the closest packed ones. The $\{100\}$ planes have the next lowest surface energy for bcc structures. Looking at Figure 3 it can be seen that a simple force balance applied to these surface energies shows that if the value of S12 is small, the film will wet the surface, and form in the shape shown in Figure 3A. This would give rise to the $\{110\}$ texture for the growing film, since as indicated above the $\{110\}$ planes have the lowest surface energies in the bcc structure. On the other hand, if the value of S12 is large, the film will tend to be more equiaxed in shape. To minimize the surface energy of the film the (002) planes would be parallel to the substrate surface since this configuration maximizes the amount of $\{110\}$ surface area for the developing Cr island. This argument assumes that there is enough time (or energy) available for the film to obtain equilibrium or near equilibrium shapes. This appears to be the case for Cr films when the substrate is pre-heated. Thus the formation of the (002) crystallographic texture can be explained as a nucleation texture.

If there is not enough time for the equilibrium shape to form, the initial orientation of the Cr grains should be nearly random. Any texture that develops in this case would do so during the growth stage of the film. Figure 4 is a schematic showing the growth of two films with differing initial grain sizes. As the films grow, the balance of surface energies would predict that the grains with the $\{110\}$ surfaces would expand on growth. This can be seen again to be the result of a simple force balance, since any other grain orientation would have a surface with a higher energy parallel to the plane of the film. As shown in the Figure this would tend to expand the (110) oriented grain, hence lowering the overall surface energy. Note also that the film with the smaller grain size would obtain a strong $\{110\}$ texture at smaller thicknesses than the one with the larger grain size.

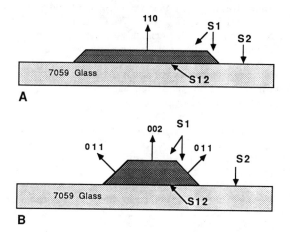

Figure 3.   Schematic view of the nucleation of islands of Cr on glass substrates for (a) small values of S12, (b) large values of S12. Note that in (b) the shape of the nucleus that minimizes the energy produces the (002) crystallographic texture.

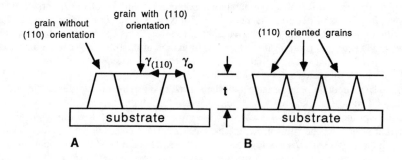

Figure 4.   Schematic view of the development of texture during growth of film for (a) large initial grain size, (b) small initial grain size.

As a general rule, the textures which develop during the growth of thin films are those with the closest packed planes parallel to the plane of the film. The growth textures of several structures are listed in Table I.

TABLE I

Growth Textures for Films

Structure	Space Group	Texture
bcc	Im$\bar{3}$m	{110}
fcc	Fm$\bar{3}$m	{111}
hcp	P$\frac{6_3}{m}$mc	(0001)
B2	Pm$\bar{3}$m	{110}
D03	Fm$\bar{3}$m	{110}

This approach to understanding the crystallographic textures of thin films grown on amorphous substrate has been discussed by Feng *et al.*[21]

II. DEVELOPMENT OF MAGNETIC LAYER CRYSTALLOGRAPHIC TEXTURE

The development of the crystallographic texture of the Co alloy films depends for the most part on epitaxial relationships between the Co alloy film and the underlayer. If no underlayer is used, the Co alloy films (which usually have the hcp structure) are grown on an amorphous substrate, such as glass or amorphous NiP films. The crystallographic texture that develops will be (0001). This is fine if the films are to be utilized in the perpendicular recording mode, but is of little use for longitudinal recording. The use of underlayers has been demonstrated to cause the c axes of the grains of the film to lie in or near the plane of the film.

As mentioned above, two crystallographic textures are well documented for Co alloy magnetic thin films. For Co alloys deposited on (110) Cr textured underlayers, the (10$\bar{1}$1) Co texture is obtained. Figure 5 shows that there is a good match between the atoms on these two planes. Figure 6 shows that the fit between atoms on the (10$\bar{1}$0) planes of Co and some of the atoms on the (110) planes of Cr is also good, but the density of atoms on the planes is very different. The extra atom in the Cr plane would raise the interfacial energy between these planes due to the lack of any atomic bonds across the interface for the extra atom. This orientation relationship in crystallographic texture is not usually reported.

The second well documented texture for the Co/Cr thin films is the (11$\bar{2}$0) Co parallel to the (002) Cr. As indicated above, the (002) Cr crystallographic texture is formed when the substrate is held at elevated temperatures. Figure 7 shows the atomic matching for the case of the (11$\bar{2}$0) of the Co alloy parallel to (002) of Cr. Once again, there is a good match in both the size of the projected unit cell and the number of atoms per unit area of both structures.

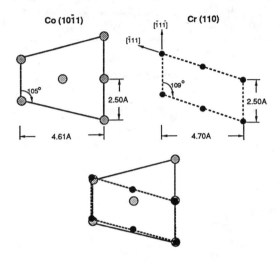

Figure 5   Schematic of the atomic positions of (10$\bar{1}$1) Co and (110) Cr planes.

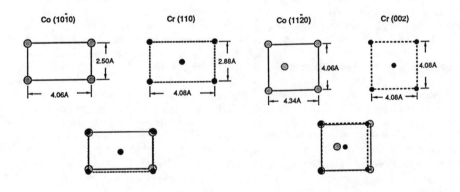

Figure 6
Schematic of the atomic positions of
(10$\bar{1}$0) Co and (110) Cr planes.

Figure 7
Schematic of the atomic position of
(11$\bar{2}$0) and (002)Cr planes.

In recent work, Lee et al.[22] have used NiAl as an underlayer for Co alloy films. The lattice parameter of this material is nearly identical to that of Cr, so it was expected that the same crystallographic texture would result. NiAl (which has the B2, bcc derivative structure) develops a weak (110) texture on deposition. See Table I. However, the Co alloy film develops the (10$\bar{1}$0) crystallographic texture, not the (10$\bar{1}$1) as with a Cr underlayer. See Figure 8. On closer examination of the X-ray pattern, it can be seen that a fairly strong (112) peak also exists for the B2 NiAl films. The atomic matching of (10$\bar{1}$0) Co with (112) of Cr and/or NiAl is shown in Figure 9. It can be seen to be an excellent fit. In fact, other workers[23-24] had reported the (10$\bar{1}$0) texture for Co alloys deposited on Cr, and an inspection of their diffraction pattern showed that their Cr film also had a fairly strong (112) peak. See also reference 25. We conclude that this texture: (10$\bar{1}$0) Co alloy parallel to (112) NiAl or Cr is a low energy one. Table II summarizes different orientation relationships that we have observed by means of electron microdiffraction. See references 17, 26 and 27.

TABLE II

Orientation Relationships of Thin Films

(10$\bar{1}$1) Co	//	(110) Cr
(11$\bar{2}$0) Co	//	(002) Cr
(10$\bar{1}$0) Co	//	(112) Cr
(10$\bar{1}$0) Co	//	(113) Cr

Thus the crystallographic textures that develop in Co alloy hcp thin films can be explained in terms of a best fitting of the atomic planes across the thin film interface. Hence the mechanism of the formation of the initial crystallographic texture of the magnetic layers is that of epitaxy. It should be noted, however, that as the hcp magnetic layers grow thicker, there is a tendency for them to obtain their low energy growth texture, namely the (0001). It should also be noted that different alloys of Co may have large enough differences in lattice parameter to change the plane of "best fit".

Other microstructural features are also affected by this epitaxy, and to these we now turn.

III. GRAIN SIZE AND CRYSTALLINE PERFECTION

Keeping these various mechanisms of the formation of crystallographic textures in mind, we turn to the application of them to several interesting features. Figure 10 is a schematic drawing which demonstrates the use of an interlayer between two magnetic layers. The identity of this interlayer can vary, though published works to date have used Cr as an interlayer as well as the underlayer. The interlayer serves to "break up" the magnetic layer, and may help to increase the signal to noise ratio of recording due to an effectively smaller "particle size" in the films. The interlayer may form epitaxially on the first magnetic film. If it does, it has a good chance of "passing on" the crystallographic texture of the first magnetic layer to that of the second magnetic layer. Cr interlayers have been shown to do this. See Figure 11.[28] Here the dark field

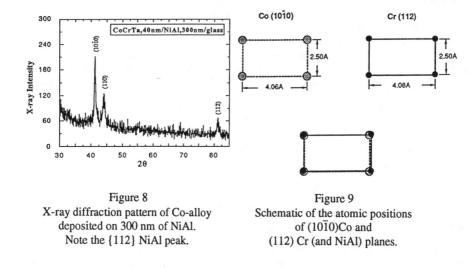

Figure 8
X-ray diffraction pattern of Co-alloy
deposited on 300 nm of NiAl.
Note the {112} NiAl peak.

Figure 9
Schematic of the atomic positions
of $(10\bar{1}0)$Co and
(112) Cr (and NiAl) planes.

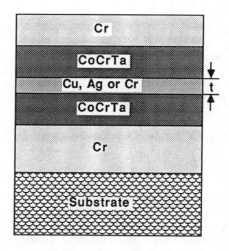

Figure 10.   Schematic of thin film configuration including an interlayer.

Figure 11. TEM cross section of multilayer film (a) Bright field, (b) Dark field.

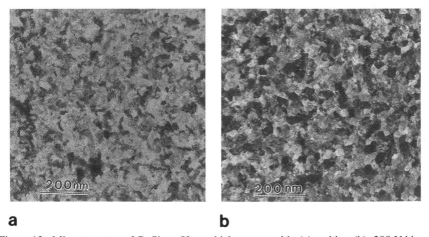

a   b

Figure 12. Microstructure of Cr films, 50 nm thick, grown with: (a) no bias, (b) -200 V bias. Notice that the grains in (b) are much more clearly defined.

micrograph clearly shows that several of the columns of the films have bcc and hcp regions that are all correlated to one another. Using other interlayers offers the opportunity of breaking up the crystallographic texture of the magnetic layers, as these metals are fcc in structure. Feng *et al.* have presented preliminary work on this and reported that the results are promising.[29]

Another feature of the use of interlayers is that it may help in the production of magnetic layers with smaller grain size. As reported earlier, the magnetic layer always has a grain size that is the same size or smaller than that of the Cr underlayer. When an interlayer is deposited on the first magnetic layer it will take on the grain size of the magnetic layer. Then, when the second magnetic layer is deposited on the interlayer, it will probably have a grain size that is smaller than that of the interlayer. This mechanism of producing magnetic layer thin films with small grain size is a topic of our current research.

One final interesting topic for discussion is the degree of crystallographic perfection of the underlayers and how that might affect the perfection of and hence the magnetic properties of the subsequently grown magnetic layer. Figures 12a and 12b show TEM microstructures of Cr grains that were deposited with and without applied bias, respectively. The Cr film formed with bias has a much more clearly defined grain structure than the one formed without bias. From high resolution TEM it can be seen that the Cr film formed with an applied bias during sputtering has fewer subgrain boundaries within its grains. This means that it has fewer defects present than the Cr film formed without bias. The magnetic layers which formed on top of these different underlayers will to some extent replicate these features, and thus have differing amounts of defects in them and hence have different magnetic properties. Thus, the control of certain extrinsic magnetic properties of magnetic thin films can be performed by careful control of the conditions by which their respective underlayers are produced.

IV. SUMMARY

A brief overview of various features of the development of crystallographic textures of thin films utilized for magnetic recording has been presented. The magnetic thin film usually obtains its initial crystallographic texture as well as grain size and perfection by means of epitaxy with its underlayer. Various important crystallographic textures between hcp Co alloy films and their metallic underlayers were also presented and explained in terms of the atomic arrangement of atoms across the epitaxial interface. The development of the crystallographic texture of the Cr underlayer was discussed. A model for the development of the initial textures during either the nucleation or growth stage of the film was presented. Finally the importance of epitaxy on other microstructural features of the magnetic thin film was discussed.

REFERENCES

1. J.P. Lazzari, I. Melnick and D. Randet, IEEE Trans. Magn. **3**, 205 (1967).
2. J.P. Lazzari, I. Melnick and D. Randet, IEEE Trans. Magn. **5**, 955 (1969).
3. J. Daval and D. Randet, IEEE Trans. Magn. **6**, 768 (1970).
4. G.-L. Chen, IEEE Trans. Magn. **22**, 334 (1986).
5. R.D. Fisher, J.C. Allen and J.L. Pressesky, IEEE Trans. Magn. **22**, 352 (1986).

6. S.L. Duan, J.O. Artman, J.-W. Lee, B. Wong and D.E. Laughlin, IEEE Trans. Magn. **25**, 3884 (1989).
7. B.C. Movchan and C.B. Demchishin, Phys. Met. Metallogr. **28**, 83 (1963).
8. J.A. Thorton, J. Vac. Sci. Technol. **11**, 666 (1974).
9. H.J. Lee, J. Appl. Phys. **57**, 4037 (1985).
10. T. Ohno, Y. Shiroishi, S. Hishiyama, H. Suzuki and Y. Matsuda, IEEE Trans. Magn. **23**, 2809 (1987).
11. D.P. Ravipati, W.G. Haines, and J.L. Dockendorf, J. Vac. Sci.Technol. **A5**, 1968 (1987).
12. D.M. Mattox, J. Vac. Sci. Tecnol., **A7** (3), 1105 (1989).
13. S.L.Duan, J.O. Artman, B. Wong, and D.E. Laughlin, J. Appl. Phys. **67**, 4913 (1990).
14. J. Pressesky, S.Y. Lee, S.L. Duan and D. Williams, J. Appl. Phys. **69**, 5163 (1991).
15. T. Yogi, T.A. Nguyen, S.E. Lambert, G.L. Gorman and G. Castillo, Mat. Sci. Soc. Symp. Proc. **232**, 3 (1991).
16. J. Lin, C. Wu and J.M. Silvertsen, IEEE Trans. Magn. **26**, 39 (1990).
17. K. Hono, B. Wong and D.E. Laughlin, J. Appl. Phys. **68**, 4734 (1990).
18. Li Tang and Gareth Thomas, J. Appl. Phys. **74**, 5025 (1993).
19. W. L. Winterbottom, Acta Metallurgica **15**, 303 (1967).
20. J. W. Cahn and J. Taylor, Phase Transformations '87, 545 (1988).
21. Y. C. Feng, D. E. Laughlin and D. N. Lambeth, (unpublished)
22. L.-L. Lee, D.E. Laughlin and D.N. Lambeth (unpublished).
23. Y. Hsu, Q. Chen, K.E. Johnson, J.M. Suvertson and J.H. Judy, Proceedings of the Second International Symposium on Magnetic Materials, Processes, and Devices of the Electrochemical Society, **92-10**, 95 (1991).
24. A. Nakamura and M. Futamoto, Jpn. J. Appl. Phys. **32**, Pt.2, L1410 (1993).
25. T. Hikosaka and R. Nishikawa, IEEE Trans. Magn. in Japan, **67**, 678 (1991).
26. T. Yogi, G. L. Gorman, C. Hwang, M. A. Kakalec and S. E. Lambert, IEEE Trans. Magn. **24**, 2727 (1988).
27. David E. Laughlin and Bunscn Y. Wong, IEEE Trans. Magn. **27**, 4713 (1991).
28. B.Y. Wong and D.E. Laughlin, Appl. Phys. Lett. **61**, 2533 (1992).
29. Y. C. Feng, D. E. Laughlin and D. N. Lambeth, submitted 6th MMM-INTERMAG.

ACKNOWLEDGEMENT

Work presented in this review has been sponsored by the Department of Energy (DE-FG02-90ER45423, Y.C.F.), the National Science Foundation (ECD 89-07068, L.L.L.) and Hitachi Metals Limited (B.W.). The government has certain rights to this material. We thank Prof. David N. Lambeth for many stimulating discussions.

ROLE OF THE EPITAXY RELATIONS IN THE CoCrPt THIN FILMS WITH BILAYER V/Cr UNDERLAYERS

PAWEL GLIJER*, JOHN M. SIVERTSEN* AND JACK H. JUDY**
The Center for Micromagnetics and Information Technologies (MINT)
*Department of Chemical Engineering and Materials Science
**Department of Electrical Engineering, University of Minnesota, Minneapolis, MN 55455.

ABSTRACT

The effects of grain-to-grain epitaxy conditions on the magnetic properties of high coercivity $CoCr_{13}Pt_{13}$ films with bilayer Cr/Cr, V/Cr and V/V underlayers were studied. V/Cr films were used in order to obtain underlayers that would differ from Cr only by a size of the lattice constant. Introduction of V instead of Cr as an underlayer changed the relative misfits between CoCrPt and the underlayer lattice and increased the amount of {0002} texture in the magnetic film. This effect led to a decrease in the in-plane coercivity. CoCrPt/V/Cr and CoCrPt/Cr/Cr films exhibited similar texture of the underlayer, the same microstructure and similar level of internal stress. Correspondingly the decrease in coercivity did not exceed 10% (from 2.7 kOe to 2.5 kOe). On the other hand CoCrPt/V/V film had more than 4 times smaller coercivity (0.6 kOe). In this case, change in the magnetic properties was caused not only by altered misfits but also by the change in microstructure, texture and state of stress in the film. Results obtained indicate that small changes in misfits between a magnetic layer and an underlayer can be effectively used to control coercivity only if other structural parameters remain unchanged.

INTRODUCTION

Grain-to-grain epitaxy due to lattice matching between a Co-based magnetic layer and a Cr-based film is a basis for using underlayers in longitudinal Co-based/Cr thin film magnetic recording media. Under certain conditions this matching may lead to in-plane orientation of c-easy axes in the Co-based layer causing improved in-plane coercivity and squareness. The in-plane or close to in-plane orientation of the c-axis can be achieved by matching Co{$10\bar{1}0$} and Co{$10\bar{1}1$} with {110} oriented Cr or by matching Co{$11\bar{2}0$} with {100} oriented Cr[1]. It was suggested that small changes in the Cr underlayer lattice constant may improve match between both lattices and lead to even better in-plane magnetic properties[2]. Such small changes in lattice constant may be achieved by proper alloying of the Cr underlayer. In the case of CoCrPt/Cr magnetic media alloying of the Cr underlayer with V increased coercivity for certain compositions of the underlayer[3,4]. These compositions correspond to minimized lattice misfits between magnetic layer and Cr-V underlayer. On the other hand, Hsu et al.[5] reported that in case of CoCrTa/Cr media, there was no difference between magnetic properties of the films deposited under different epitaxy relations, despite a significant difference in misfit values. Also, recent study of CoCrPt/Cr-V thin films[6] revealed that addition of V to Cr underlayer not only changes lattice constant of the underlayer but also affects the grain size, number of {100} oriented grains and mosaic spread of $\{11\bar{2}0\}$ texture in the CoCrPt layer. All of these factors may affect the coercivity of the medium to a high degree. It is evident that experiments are needed which will explain the relative importance of all the above factors.

The purpose of this paper is to determine the relative importance of small differences in lattice misfits (see Table I) between the underlayer and the magnetic layer in controlling magnetic properties of CoCrPt thin film media.

EXPERIMENTAL PROCEDURE

Magnetic films of ~270 Å thick $CoCr_{13}Pt_{13}$ with 650Å - 900 Å bilayer V/Cr underlayers were deposited on cover glass in a RF diode sputtering system with RF substrate bias. The V/Cr underlayers were used since, due to V/Cr epitaxy, they should differ from Cr underlayers only by the size of the lattice of the top layer. For comparison, pure Cr and V underlayers were also used. These underlayers were also sputtered as bilayers in order to factor out the effect of the interface in the comparative analysis. Deposition rates were 140 - 350 Å/min and were measured with ~15% relative error using an alpha-step profilometer. Film thickness was estimated through the deposition time and subsequently verified using cross-section TEM samples. The results of the latter measurement are given in Table II.

Magnetic properties of the samples were measured using a vibrating sample magnetometer (VSM) with 13 kOe maximum field. Magnetic hysteresis loops as well as remanent magnetization curves were obtained for each sample. The delta-M parameter[7] was calculated from the remanent magnetization curves and used to assess the level of magnetic intergranular interactions in the CoCrPt film.

The crystallographic texture of the films and stress assessment were determined using an x-ray diffractometer. The stress assessment was performed using a method based on the measurement of the difference in peak positions (θ_n) for a set of $\{hkl\}$ planes parallel to the film surface and peak position (θ_i) for the same set of $\{hkl\}$ planes inclined by an angle ϕ to the film surface[8]. In-plane stress is proportional to $[(\sin\theta_n/\sin\theta_i)-1]$, for constant ϕ; $\phi=15°$ was used in the measurements.

Surface morphology, surface roughness and grain size of the films were measured using an atomic force microscope (AFM). Two surface roughness parameters were measured: R_a- the mean roughness measured relative to the center plane[9]; R_z - 10 points mean - defined as the average difference in height between the five highest peaks and five lowest valleys. Areas of $\approx 4 \ \mu m^2$ were used for this measurement.

The microstructure of the films was studied using a transmission electron microscope (TEM) with an accelerating voltage of 300 kV. Plane view samples were prepared by chemically removing the glass substrate. In some cases, the underlayers were removed using ion-milling in a liquid nitrogen cooled stage.

Some complementary samples were prepared in order to supplement the structural and property analyses:

- CoCrPt films with V/V, Cr/Cr and V/Cr underlayers were deposited on thin silicon wafers for cross-section TEM analysis.
- V/V and Cr/Cr underlayers without magnetic coating were sputtered as reference samples for x-ray and microstructure analysis.
- CoCrPt (700 Å) films with Cr/Cr and V/Cr underlayers were sputtered on NiP/Al disk substrates in order to study the magnetic recording properties using a thin film head.

RESULTS AND DISCUSSION

Values for the lattice misfits between the CoCrPt magnetic layer V and Cr underlayers are given in Table I. The CoCrPt-hcp lattice parameters were calculated from x-ray diffraction spectra of a CoCrPt film deposited on glass; standard values for Cr and V lattice parameters were assumed[8].

Table I Misfits between CoCrPt layer and Cr,V underlayers, calculated as[10]

$$\% \text{misfit} = \frac{a_{CoCrPt} - a_{bcc}}{0.5(a_{CoCrPt} + a_{bcc})} \cdot 100\%, \text{ where a is a lattice dimension in epitaxial plane}$$

Epitaxy relation	misfit [%]	
	Cr	V
CoCrPt {10$\bar{1}$0}<010>// bcc{110}<100>	-11.7	-16.3
CoCrPt {10$\bar{1}$0}<001>// bcc{110}<110>	2.1	-2.6
CoCrPt {10$\bar{1}$1}<010>// bcc{110}<111>	2.7	2.0
CoCrPt {10$\bar{1}$1}<$\bar{1}$01>// bcc{110}<111>	-2.1	-6.7
CoCrPt {11$\bar{2}$0}<001>// bcc{100}<110>	2.1	-2.6
CoCrPt {11$\bar{2}$0}<1$\bar{1}$0>// bcc{100}<110>	8.6	3.9

It is evident that misfits increase when Cr is replaced with V for matching of CoCrPt planes with {110} plane of the underlayer. Since {110} was a dominant texture in the underlayers used CoCrPt/V tended to develop more of a perpendicular {0002} texture (Fig.1). The CoCrPt/Cr/Cr film (sample A, Table II) exhibited high in-plane coercivity of 2.7 kOe. Replacement of Cr layers with V led in sample D to a profound decrease in the in-plane coercivity and coercivity squareness S*. This decrease is certainly partially caused by a sharp increase in the {0002} texture of the CoCrPt film. However, the change in the texture does not explain the observed decrease in the perpendicular coercivity. CoCrPt films with V/Cr underlayers exhibit magnetic properties similar to those of sample A (Cr/Cr underlayer).

Table II. Magnetic properties and thickness of examined samples A,B,C,D.

Sample	Layer thickness	$H_c^{in-pl.}$ [Oe]	S*	$H_c^{perp.}$ [Oe]
A	270Å CoCrPt/ 420Å Cr/ 460Å Cr	2710	0.80	1200
B	260Å CoCrPt/ 370Å V/ 250Å Cr	2490	0.84	1210
C	265Å CoCrPt/ 250Å V/ 420Å Cr	2250	0.80	1170
D	280Å CoCrPt/ 350Å V/ 300Å V	640	0.52	470

In the case when texture of V/Cr was the same as in a Cr/Cr underlayer, both texture and properties of the magnetic layers were almost the same (samples A and B). The 10% difference in in-plane coercivity between samples A and B is most likely caused by the larger amount of {0002} texture in B and by the 30% difference in overall thickness of the underlayer.

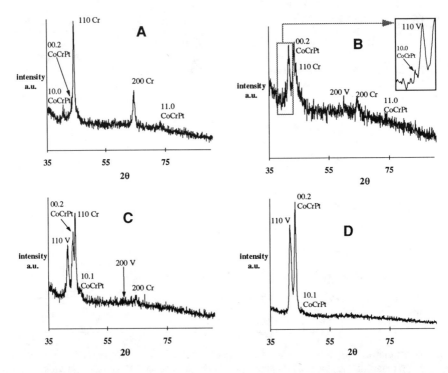

Fig. 1. X-ray diffraction spectra for CoCrPt thin films with bilayer underlayers; A-CoCrPt/Cr/Cr, B-CoCrPt/V/Cr (enlarged lower angle portion of smoothed spectrum is also shown), C-CoCrPt/V/Cr, D-CoCrPt/V/V.

It was estimated that the average stress in the CoCrPt film with Cr/Cr underlayer (A) was ~1.6 times larger than in the case of CoCrPt film with V/V underlayer (D) and both stresses were compressive. Stress levels in samples B and C seemed to be similar to the stress measured in sample A. However, in the case of these samples, the peak overlap makes results of the analysis uncertain. Delta-M curves may provide some insight into both stress level and magnetic interactions in the films (Fig. 2). The unusual shape of the delta-M curve for sample with V/V underlayer arises from the perpendicular orientation of the c-axes in the film[7]. Similar curves for samples A, B and C suggest similar type of interactions. Judging from both maximum slope and positive peak of the delta-M curves sample B should have the largest and sample C the smallest level of intergranular exchange interactions in the CoCrPt film. Noise measurements in thicker media with the same underlayers support this conclusion. Media with V/Cr underlayer (type B) had the highest media noise among the three and medium with V/Cr underlayer of type C had the lowest. It has been demonstrated that external compressive stress can increase the intergranular exchange interactions, leading to an increased positive peak of delta-M[11]. Therefore the high positive peak in case of sample B can be an indication of large compressive stresses.

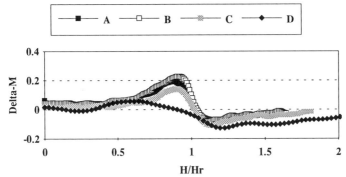

Fig. 2. Delta-M curves for CoCrPt films with bilayer underlayers.

Differences in the intergranular interactions may arise also from microstructural differences. AFM micrographs (Fig. 3) provide information about surface morphology and the microstructure of the films. Samples A and B have similar surface morphology. The grain size observed in the micrograph from sample A is ~150 Å, with some grains merged into 400 Å conglomerates. Deeper valleys around some groups of grains correspond probably to regions of separation between Cr underlayer grains and cause slightly larger roughness of the film. Roughly the same grain size can be observed on the micrograph from sample B. In sample C ~ 200 Å in diameter CoCrPt grains are merged into 1000Å - 1500 Å conglomerates. CoCrPt/V/V sample (D) has distinctly smaller grain size and roughness than the other three samples. Grains with diameters smaller than 100 Å can be distinguished in the surface plot from this sample.

A
CoCrPt/
Cr/Cr
R_a=10.0Å
R_z=87.4Å

B
CoCrPt/
V/Cr
R_a=7.8Å
R_z=73.6Å

C
CoCrPt/
V/Cr
R_a=8.5Å
R_z=72.2Å

D
CoCrPt/
V/V
R_a=2.7Å
R_z=27.1Å

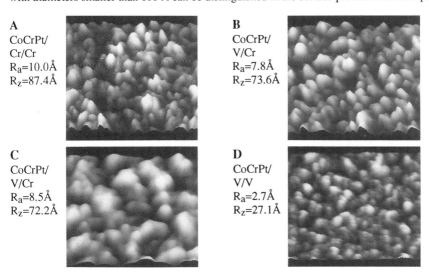

Fig. 3. AFM surface plots and roughness of CoCrPt films in examined samples. Scans cover area of 5000 Å x 5000 Å, the height scale is 150 Å.

Findings based on AFM images were confirmed and supplemented by TEM observations. In sample A, Cr underlayer grain diameters were 500 ± 120 Å with columnar CoCrPt grains with diameter of 150 ± 40 Å. In sample B, grains with diameters of 410 ± 150 Å were observed in both V and Cr underlayers and columnar CoCrPt grain diameters of 180 ± 65 Å could be distinguished in the micrographs. In sample C, grain diameters were: 430 ± 220 Å for Cr, 300 ± 100Å for V and 190 ± 60 Å for CoCrPt layers. CoCrPt/V/V film (sample D) exhibited much finer grain sizes than the other three samples. Grain diameters of 250 ± 80 Å were measured in V layers. Diverse grain size was observed in CoCrPt (140 ± 80 Å). CoCrPt grains did not exhibit columnar shape. Such small grain volumes can lead to superparamagnetic behaviour which may be partially responsible for the lower coercivity.

CONCLUSIONS

CoCrPt thin films with bilayer Cr/Cr, V/Cr and V/V underlayers were studied. It has been shown that replacement of the Cr underlayer with V led to increased amount of {0002} orientation in the CoCrPt film resulting from increased lattice misfits between CoCrPt and underlayer lattices. CoCrPt films with V/Cr and Cr underlayers exhibited similar structure when similar microstructure, state of stress and similar texture was achieved in the underlayers. In this case, the increase in the amount {0002} oriented grains in CoCrPt film with V/Cr underlayer accounted for less than 10% decrease in the coercivity. CoCrPt/V/V film had significantly different microstructure, stress level and texture of the underlayer, which led to different structure the magnetic layer and hence completely different magnetic properties.

These results suggest that change in lattice misfits by as much as 5% may be effectively used as a tool for controlling magnetic properties only if other structural properties of the film are not affected.

REFERENCES

1. D.E. Laughlin, B.Y. Wong, IEEE Trans. Mag., **27**, 4713 (1991).
2. J.K. Howard, R. Ahlert, G. Lim, J. Appl. Phys. **61**, 5579 (1987).
3. J.K. Howard, U.S. Patent No. 4,652,499 (24 March 1987).
4. T.A. Nguyen, J.K. Howard, J. Appl. Phys. **73**, 5579 (1993).
5. Y. Hsu, Q. Chen, K.E. Johnson, J.M. Sivertsen, J.H. Judy in Proceedings of the Second Symposium on Magnetic Materials, Processes, and Devices, edited by L.T. Romankiw and D. A. Herman, The Electrochemical Society Inc., 95, (1991).
6. M.A. Parker, J.K. Howard, R. Ahlert, K.R. Coffey, J. Appl. Phys. **73**, 5560 (1993).
7. I.A. Beardsley, J.-G. Zhu, IEEE Trans. Mag., **27**, 5037 (1991).
8. B.D. Cullity, Elements of X-ray Diffraction, 2nd ed. (Addison-Wesley Publishing Co., 1978) p. 506.
9. Reference manual for Nanoscope III, ver. 2.5, Digital Instruments Inc, 6780 Cartona Drv., Santa Barbara, CA 93117.
10. J.H. van der Merwe, J. Waltersdorf, W.A. Jesser, J. Mat. Sci. Eng. **81**, 1 (1986).
11. R. Ranjan, J. Chang, T. Yamashita, T. Chen, J. Appl. Phys. **73**, 5542 (1993).

EFFECTS OF SUBSTRATE TEMPERATURE ON MAGNETIC AND CRYSTALLOGRAPHIC PROPERTIES OF Co-Cr-Pt/Cr FILMS DEPOSITED BY LASER ABLATION

AKIRA ISHIKAWA*, YOSHIHIRO SHIROISHI* AND ROBERT SINCLAIR**
*Central Research Laboratory, Hitachi, Ltd., 1-280, Higashi-koigakubo, Kokubunji-shi, Tokyo 185, Japan
**Department of Materials Science & Engineering, Stanford University, Stanford, CA 94305-2205

ABSTRACT

The effects of the substrate temperature on the magnetic properties and crystal structure of $Co_{89}Cr_7Pt_4$/Cr films deposited by laser ablation were studied. A pulsed KrF excimer laser with a wavelength of 248 nm was focused on ablation targets in a vacuum chamber. The laser irradiation energy was estimated to 10 J/cm^2. Cr underlayer and $Co_{89}Cr_7Pt_4$ films were successively deposited on the Si substrate at a rate of 0.012 nm/pulse. At a substrate temperature lower than 350 °C, an in-plane coercivity of 30 nm thick $CoCr_7Pt_4$ film deposited on 50 nm thick Cr underlayer was 50~150 Oe showing no significant influence by the substrate heating. The coercivity at 450 °C, however, increased to 800 Oe showing a maximum at this temperature. X-ray diffraction analyses showed that the magnetic films formed at lower than 350 °C have fcc-Co phase, while the films deposited at 450 °C had hcp-Co phase with the c-axis oriented parallel to the substrate. At temperatures higher than 450 °C, the coercivity decreased concomitant with a disappearance of the oriented hcp-Co phase. The grain size of the films increased from 10 nm to 20~30 nm by increasing the substrate temperature. The increase of coercivity is probably due to annealing effects resulting in the grain growth of Cr underlayer and the phase transformation of the magnetic films to form the hcp-Co phase.

INTRODUCTION

Co-alloy films which are formed on Cr underlayers show high in-plane coercivity and so have been widely used for longitudinal magnetic recording media.[1-3] As the films are usually formed by Ar sputtering, they have a microscopically rough surface and granular structure with a grain size of 50~100 nm. This is due to a shadowing effect which is caused by the Ar gas introduced during the deposition.[4] The granular structure seems advantageous for the reduction of the media noise.[5] However, for the high density recording media, it is also important to form thin magnetic films with a homogeneous thickness and a small grain size compared to the recorded bit size.[6,7] Moreover, a microscopically smooth surface is preferable to allow a narrow head-media spacing.

Laser ablation deposition is one of the possible methods to form a homogeneous film with a smooth surface since the films can be formed in a vacuum.[8-10] It has also been reported that this method is suitable for close control of the film composition. Previous work showed that the Co-alloy film and Cr underlayer deposited by laser ablation at 250 °C had a smooth film surface with small crystalline grains of 10~20 nm.[11] However, the coercivity of these Co-alloy films was as low as 20~150 Oe since they had the fcc-Co phase.

In this paper, we report on the improvement of magnetic properties and a crystal structure change of the CoCrPt/Cr films deposited by excimer laser ablation by means of increasing the substrate temperature during the film deposition.

EXPERIMENTAL

The Si wafer substrate which had a native oxide layer was placed on the heating block which faced an ablation target at a distance of 40 mm.[11] The pulsed KrF laser with a

345

wavelength of 248 nm and a power of 500 mJ/pulse was focused by a quartz lens, and irradiated at an incident angle of 45 ° to a water-cooled ablation target. The pulse width of the excimer laser was about 30 ns and the repetition frequency was 10 Hz. The laser beam was scanned across the target surface to prevent local ablation. A Cr underlayer of 50 nm thickness and $Co_{89}Cr_7Pt_4$ film of 30 nm thickness were successively deposited onto the Si substrate at a deposition rate of 0.012 nm/pulse. The substrate was heated up to 600 °C during the deposition. The base pressure of the chamber was 3×10^{-4} Pa, while the pressure rose to 3×10^{-3} Pa during the substrate heating and laser ablation. Magnetic properties were measured by a vibrating sample magnetometer and a torque magnetometer with a maximum applied field of 13 kOe. The crystalline structure of the deposited film was analyzed by x-ray diffraction (primary x-rays: $CuK\alpha$) and transmission electron microscopy (TEM).

RESULTS AND DISCUSSION

The substrate temperature dependence on the in-plane coercivity of 30nm thick $Co_{89}Cr_7Pt_4$ films deposited on 50nm thick Cr underlayer is shown in Fig. 1(a). At a substrate temperature lower than 350 °C, the coercivity was 50~150 Oe showing no significant effect from substrate heating. On the other hand, the coercivity of the films deposited at 450 °C increased abruptly up to 800 Oe showing a maximum at this temperature. The maximum anisotropy field (H_{kmax})[12] determined by the rotational hysteresis loss for this film was about 10 kOe. At a substrate temperature higher than 450 °C, however, the coercivity decreased and no magnetization was observed for 600 °C. The H_{kmax} at 550°C was 2.5 kOe. The substrate temperature dependence on the in-plane saturation flux density (B_s) for the same samples in Fig. 1(a) is shown in Fig. 1(b). As the substrate temperature increased to 450 °C, B_s gradually decreased becoming zero at 600 °C. The decrease of B_s is probably due to the reaction between the magnetic layer and Cr underlayer as will be discussed in the TEM analyses. The substrate temperature dependence on the coercivity squareness (S*) for the same samples is shown in Fig. 1(c). As the substrate temperature increased to 550 °C, S* gradually increased to 0.8.

In order to analyze the mechanism of the coercivity improvements, the crystal structure of the $Co_{89}Cr_7Pt_4$/Cr films was analyzed. Fig. 2 shows the x-ray diffraction spectra for $Co_{89}Cr_7Pt_4$(30 nm)/Cr(50 nm) films deposited at the various substrate temperatures. As the temperature increased from room temperature to 250 °C, the intensity of the broad diffraction peak of fcc-Co(111) $(d_{111}= 0.2059$ nm) increased. There was no diffraction peak from Cr underlayer, possibly due to the small grain size as will be shown later. At 450 °C, the fcc-Co diffraction peak diminished at 450 °C and bcc-Cr(200) and hcp-Co(11$\bar{2}$0) diffraction peaks appeared. The crystal orientation possibly showed an epitaxial growth between Cr underlayer and $Co_{89}Cr_7Pt_4$ film, resulting in the c-axis of hcp Co parallel to the substrate. Therefore, the growth of bcc-Cr grains and the formation of hcp-Co phase with the c-axis lying in the film plane is possibly the reason for the increased coercivity at 450 °C as shown in Fig. 1(a). Above 520 °C, this crystal structure disappeared. Instead, four diffraction peaks of $CrSi_2$ and a peak which could not be indexed (d=0.1997nm) were observed. This is due to the reaction of the Cr underlayer and the Si substrate at the higher deposition temperature. The formation of the $CrSi_2$ seemed to suppress the epitaxial growth of bcc-Cr and hcp-Co crystalline with the decrease of the coercivity as shown in Fig. 1(a).

Fig. 3 shows the in-plane high-resolution TEM micrograph for the $Co_{89}Cr_7Pt_4$ film deposited at 250 °C. The cross-fringe image in the grain is characteristic of the fcc crystal structure. The high temperature Co phase seemed to be formed during the pulsed deposition owing to the high deposition energy (<100 eV) and the high instantaneous deposition rate (~1 µm/s) of the laser ablation.[13] This corresponds to insufficient grain growth of the $Co_{89}Cr_7Pt_4$/Cr films deposited at 250 °C. Fig. 4 shows the in-plane bright field TEM image of the $Co_{89}Cr_7Pt_4$ films deposited at 250 and 450 °C. The crystalline grain size of about 10 nm increased to 20~30 nm by increasing the substrate temperature. The grain boundaries become clearer for 450 °C and stacking faults (typical of hcp-Co thin film) could be seen within the grains. Diffraction rings showed the formation of hcp-Co phase at 450 °C. Fig. 5 shows the cross-sectional TEM image of the same sample as in Fig. 4. The Cr underlayer shows a

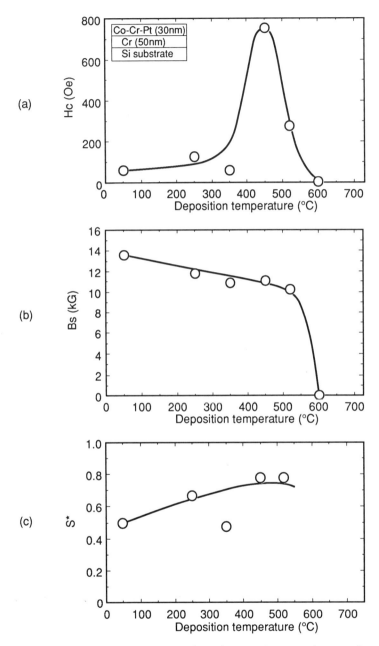

(a)

(b)

(c)

FIG. 1. The substrate temperature dependence on the magnetic properties of Co89Cr7Pt4/Cr films deposited by laser ablation; (a) In-plane coercivity; (b) Saturation flux density; (c) Coercivity squareness.

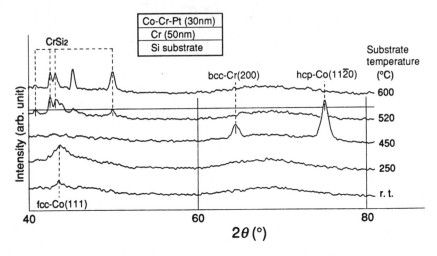

FIG. 2. X-ray diffraction spectra for CoCr7Pt4/Cr films deposited by laser ablation at various temperatures; primary x-rays: CuKα.

columnar structure with a grain diameter of 5~10 nm at 250 °C. The film surface was microscopically flat. By increasing the temperature to 450 °C, the grain size of Cr underlayer and magnetic film increased to 20~30 nm. The grain boundaries of the Cr underlayer were almost continuous with those of the magnetic film indicating an epitaxial growth between both films. The film surface became rough at 450 °C. The change of the film structure was possibly caused by the annealing effect during the pulsed deposition, resulting in atomic diffusion and grain growth. At the same time, the phase transformation from thermally unstable fcc-Co phase to the stable hcp-Co phase could occur. However, the high temperature annealing also brought about a reaction at the interface of the films as shown in the micrograph. The reaction layer between Si substrate and Cr underlayer appears to be $CrSi_x$ though no x-ray diffraction peak of $CrSi_x$ was observed in Fig. 2. The reaction layer between the magnetic film and the Cr underlayer might be non-magnetic, thus that decreasing the B_s of the sample as shown in Fig. 1(b).

It is known that evaporated Co films have a hcp structure.[14] This is probably due to the relatively low deposition energy (~ 0.1 eV) of the thermal evaporation. Consequently, it seems important to decrease the deposition energy of the laser ablation for the direct deposition of the hcp-CoCrPt film with high coercivity, a smooth film surface and good composition control.

CONCLUSIONS

The effects of the substrate temperature on the magnetic properties and crystal structure of $Co_{89}Cr_7Pt_4$/Cr films deposited by excimer laser ablation were studied. The coercivity could be increased to 800 Oe by increasing the substrate temperature up to 450 °C. At this temperature, the magnetic film had hcp-Co phase with its c-axis oriented parallel to the substrate. The grain size of the films increased to 20~30 nm by substrate heating with an increase of a surface roughness. The increase of the coercivity at this temperature is probably due to grain growth of Cr underlayer and the phase transformation of the magnetic films to form the hcp-Co phase.

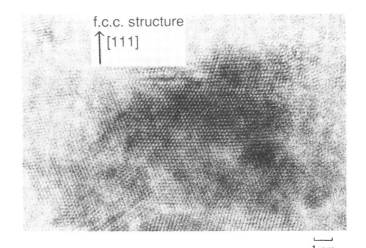

FIG. 3. In-plane high-resolution TEM micrograph of $Co_{89}Cr_7Pt_4/Cr$ film deposited by laser ablation at 250 °C.

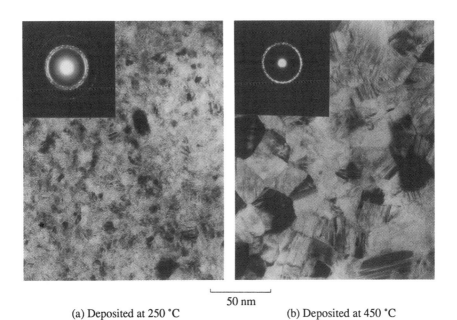

(a) Deposited at 250 °C   (b) Deposited at 450 °C

FIG. 4. In-plane bright field TEM micrograph of $Co_{89}Cr_7Pt_4/Cr$ films deposited by laser ablation at 250 and 450 °C.

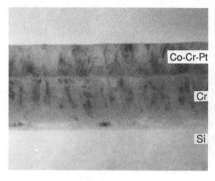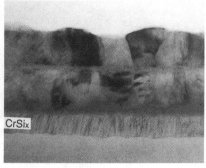

Co-Cr-Pt

Cr

Si CrSi$_x$

$\vdash\!\!\!\!\dashv$
20 nm

(a) Deposited at 250 °C   (b) Deposited at 450 °C

FIG. 5. Cross-sectional bright field TEM micrograph of Co$_{89}$Cr$_7$Pt$_4$/Cr films deposited by laser ablation at 250 and 450 °C.

ACKNOWLEDGMENTS

The authors would like to thank Dr. D. Laughlin of Carnegie Mellon University for helpful discussion on the film structure. We also thank Dr. M. Hiratani of Central Research Laboratory, Hitachi, Ltd. for discussions on the laser ablation.

REFERENCES

1. J.P. Lazzari, I. Melnick, and D. Randet, IEEE Trans. Magn. **MAG-3**, 205 (1967).
2. T. Ohno, Y. Shiroishi, S. Hishiyama, H. Suzuki, and Y. Matsuda, IEEE Trans. Magn. **MAG-23**, 2809 (1987).
3. I. Sanders, T. Yogi, J. Howard, S. Lambert, G. Gorman, and C. Hwang, IEEE Trans. Magn. **25**, 3869 (1989).
4. J. Thornton, J. Vac. Sci. Technol. **A4**, 3059 (1986).
5. T. Yogi, T. Nguyen, S. Lambert, G. Gorman, and G. Castillo in Magnetic Materials: Microstructure and Properties, edited by T. Suzuki, Y. Sugita, B.M. Clemens, K. Ouchi, and D.E. Laughlin (Mater. Res. Soc. Proc. **232**, San Francisco, CA, 1991) pp. 3-13.
6. E. Murdock, R. Simmons, and R. Davidson, IEEE Trans. Magn. **28**, 3078 (1992).
7. T. Yogi and T. Nguyen, IEEE Trans. Magn. **29**, 307 (1993).
8. J. Cheung and H. Sankur, CRC Critical Reviews in Solid State and Materials Science **15**, 63 (1988).
9. N. Cherief, D. Givord, A. Liénard, K. Mackay, O.F.K. McGrath, J.P. Rebouillat, F. Robaut, and Y. Souche, J. Magn, Magn, Mater. **121**, 94 (1993).
10. N. Cherief, D. Givord, O. McGrath, Y. Otani, and F. Robaut, J. Magn, Magn, Mater. **126**, 225 (1993).
11. A. Ishikawa, K. Tanahashi, Y. Yahisa, Y. Hosoe, and Y. Shiroishi, J. Appl. Phys. **75**, 1 (1994).
12. M.R. Khan, S.Y. Lee, J.L. Pressesky, D. Williams, S.L. Duan, R.D. Fisher, N. Heiman, and D.E. Speliotis, IEEE Trans. Magn. **26**, 2715 (1990).
13. G. Huber, MRS Bulletin **XVII** (2), 26 (1992).
14. J. Daval and D. Randet, IEEE Trans. Magn. **MAG-6**, 76 (1970).

HCP-FCC PHASE TRANSITION IN Co-Ni ALLOYS STUDIED WITH MAGNETO-OPTICAL KERR SPECTROSCOPY

A. CARL[*], D. WELLER[*], R. SAVOY[*] AND B. HILLEBRANDS[**]
[*]IBM Research Division, Almaden Research Center, 650 Harry Road, San Jose, CA 95120
[**]2. Physikalisches Institut, RWTH Aachen, Templergraben 55, 52056 Aachen, Germany

ABSTRACT

We have studied 100nm thick electron beam evaporated $Co_{1-x}Ni_x$ alloy films in the composition range $0 \leq x \leq 1$ using magneto-optical spectroscopy and magnetic anisotropy measurements. For films with $x \approx 0.2$ we have also investigated the dependence of these quantities on the growth temperature, which was varied in the range $27 \leq T_G \leq 408°C$. Both as function of the Ni content x and the growth temperature T_G we observe the anticipated hcp $\rightarrow$ fcc phase transition, e.g. by monitoring the magneto-crystalline anisotropy constant $K_{u,1}$, which changes continuously from values of $\approx 0.3 MJ/m^3$ for Co rich hcp alloys ($0 \leq x \leq 0.25$) to $\approx -0.05 MJ/m^3$ for Ni rich fcc alloy films, in good agreement with bulk literature data. The new and most striking result, however, is observed in the polar magneto-optical Kerr and ellipticity spectra, which were measured in the photon energy range $0.8 \leq h\nu \leq 5.5eV$. Changes by up to about 40% in Kerr rotation for films of constant composition and magnetization are observed when the structure changes from hcp $\rightarrow$ fcc. This demonstrates the sensitivity of magneto-optical effects to structural changes, making Kerr spectroscopy a useful electronic and physical structure probe.

INTRODUCTION

Investigations of the magneto-optical Kerr effect (MOKE) have found widespread interest in magnetic thin film research, not only in view of their importance for magneto-optical recording applications, but also since they provide fundamental information on magnetic as well as electronic properties of thin films [1]. Investigations include magnetic hysteresis loop measurements, usually at fixed wavelengths as well as spectroscopic investigations, typically within a photon energy range of $\approx 0.5 \rightarrow 5.5eV$. The sensitivity of MOKE spectra to details of the structure has recently been pointed out for Co-Pt alloys [2] and was most convincingly demonstrated for single crystalline 100nm thick films of fcc and hcp Co [3, 4], which can both be stabilized by seeded epitaxial growth in UHV [5]. Drastic changes of up to 50% in the Kerr rotation and Kerr ellipticity were found, depending on the crystal structure and/or crystallographic orientation [3].

Here we exploit the structural sensitivity of MOKE to investigate the hcp $\rightarrow$ fcc phase transition in 100nm thick polycrystalline $Co_{1-x}Ni_x$ alloy films. This is realized (i) by increasing the Ni content x for films grown at room temperature and (ii) by increasing the growth temperature of films with fixed Ni content $x \approx 0.2$.

Mat. Res. Soc. Symp. Proc. Vol. 343. ©1994 Materials Research Society

EXPERIMENT

The present $Co_{1-x}Ni_x$ alloy samples were prepared by simultaneous electron beam evaporation of Co and Ni in a high vacuum (HV) system with $\approx 10^{-8}$ mbar base pressure. During evaporation the background pressure reached into the 10^{-7} mbar region. Various samples with different compositions in the range $0 \leq x \leq 1$ and constant total film thickness $t = 100$nm were prepared. Typical evaporation rates were 0.02 to 0.1nm/s, monitored with quartz crystal balances. These films were grown at room temperature ($\leq 50°C$) onto 20nm thick Pt buffer layers, which were predeposited at 400°C onto fused silica substrates, resulting in strongly (111) textured films [6]. In addition, a series of $Co_{0.80}Ni_{0.20}$ films (constant composition) was grown at different growth temperatures between $27 \leq T_G \leq 408°C$. To protect the films against oxidation during subsequent ex-situ measurements, all films were capped with a 2nm Pt layer. After film preparation, the composition x and the actual film thickness t were obtained from x-ray fluorescence and microprobe measurements.

MO spectra (polar Kerr angle and ellipticity) were measured ex-situ in air and at room temperature in the photon energy range $0.8 \leq hv \leq 5.5eV$ with an automated spectrometer described elsewhere [7]. During these measurements the magnetization was saturated in the out of plane direction within magnetic fields of ±20kOe. Polar MOKE loops were recorded at room temperature at fixed HeNe laser wavelength (2eV photon energy) in fields of ±20kOe.

RESULTS AND DISCUSSION

Figure 1 shows the structural (and magnetic) phase diagram for the $Co_{1-x}Ni_x$ system in the whole composition range $0 \leq x \leq 1$, taken from Bendick et al. [8].

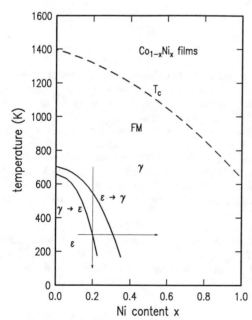

Figure 1 Structural and magnetic phase diagram for ferromagnetic $Co_{1-x}Ni_x$ alloys within $0 \leq x \leq 1$, showing the existence of a hcp (ε) $\rightarrow$ fcc (γ) phase transition at low Ni concentrations upon increasing and decreasing temperature (solid lines) after Bendick et al. [8]. The dashed line gives the Curie temperature T_c, which decreases from $T_c = 1390$K (Co) to $T_c = 632$K (Ni) with increasing Ni content x. The arrows indicate the two different routes which were chosen to cross the phase transition as function of (i) the Ni content x and (ii) the film growth temperature T_G.

Co$_{1-x}$Ni$_x$ forms ferromagnetic (FM) alloys at all compositions and the Curie temperature decreases from T$_c$ = 1390K (Co) to T$_c$ = 632K (Ni) as indicated by the dashed line in Fig. 1. At high temperatures (above ≈ 700K) Co$_{1-x}$Ni$_x$ alloys form solid solutions with fcc (γ) structure for all compositions. At lower temperatures and low Ni concentrations a martensitic phase transformation into the hcp (ε) phase is found with both decreasing (γ → ε) as well as increasing (ε → γ) temperature, as indicated by the corresponding solid lines. The horizontal and vertical arrows in Fig. 1 indicate the two routes we have chosen to cross the γ → ε phase transition: (i) by changing the Ni content x and (ii) by changing the growth temperature T$_G$.

Figure 2 shows the variation of the Kerr rotation θ$_K$ at 2eV photon energy as function of x. These data were obtained from polar MOKE loop measurements. The hysteresis loops indicate an in-plane easy magnetization direction for all films.

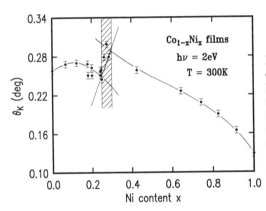

Figure 2 Saturation Kerr angle θ$_K$ obtained from polar MOKE hysteresis loops for a series of Co$_{1-x}$Ni$_x$ films as a function of the Ni content x. The strong increase of θ$_K$ within the narrow composition range 0.25 ≤ x ≤ 0.3 (hashed area) marks the position of the structural hcp → fcc transition in these alloys.

As one can see in Fig. 2, the saturation Kerr angle θ$_K$ shows a step like behavior with a sharp 15% increase in θ$_K$ right in the composition range where the hcp → fcc phase transition is expected to occur according to the bulk Co-Ni phase diagram (Fig. 1). This sharp rise is consistent with earlier reports of a much larger Kerr rotation of fcc (110) Co (−0.455° at 2eV) compared to hcp (0001) Co (−0.345° at 2eV) [3]. It can therefore be interpreted as directly reflecting the transition from hcp to fcc Co-Ni in the range marked by the hashed area.

It is interesting to note that the presently observed transition region (0.25 ≤ x ≤ 0.3) appears to be narrower than that expected from the bulk phase diagram (0.2 ≤ x ≤ 0.31). In order to investigate this point further, we have also determined the magnetic anisotropy fields H'$_{a,1}$ and H$_{a,2}$ from the hard axis MOKE loop measurements. These are related to the corresponding uniaxial anisotropy constants K'$_{u,1}$ and K$_{u,2}$ according to

$$H'_{a,1} = \frac{2K'_{u,1} + 4K_{u,2}}{M_s} \quad , \quad H_{a,2} = \frac{4K_{u,2}}{M_s} \tag{1}$$

where intrinsic quantities are obtained by adding the demagnetization field or energy, i.e. K$_{u,1}$ = K'$_{u,1}$ + 1/2μ$_0$M$_s^2$. Using the measured room temperature saturation magnetization M$_s$, we obtain the leading first order anisotropy constant K$_{u,1}$ as function of x as shown in Fig. 3.

$K_{u,1}$ decreases with increasing Ni content x from about $K_1 \approx 0.4MJ/m^3$ for a pure Co film to $K_1 \approx -0.05MJ/m^3$ for a pure Ni film. Both values are in good agreement with existing bulk literature data [9]. The sharp drop in $K_{u,1}$, marked with the hashed area in Fig. 3 is direct evidence for an hcp → fcc phase transition in the present Co-Ni alloys. It matches, within the experimental error, exactly the range inferred from the Kerr rotation alone, thus confirming the high sensitivity of MOKE to structural order in thin films.

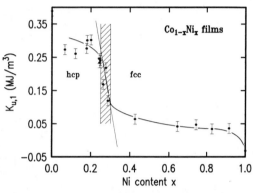

Figure 3 The anisotropy constant $K_{u,1}$ vs. Ni content x for the same series of $Co_{1-x}Ni_x$ films displayed in Figure 2. The hashed area ($0.25 \leq x \leq 0.3$) marks the hcp → fcc phase transition and matches the transition range found in Fig. 2.

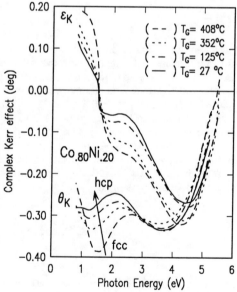

Figure 4 Growth temperature dependence of the complex magneto-optical Kerr effect (Kerr angle θ_K and Kerr ellipticity ε_K) for various $Co_{1-x}Ni_x$ films with $x \approx 0.2$ in the photon energy range $0.8 \leq h\nu \leq 5.5eV$. As indicated by the arrow, these results indicate a continuous transition from fcc Co-Ni to hcp Co-Ni as the film growth temperature is lowered.

We now turn to MOKE spectra of a series of $Co_{0.80}Ni_{0.20}$ films grown at different temperatures T_G. Figure 4 shows the dependence of the complex Kerr effect (Kerr angle θ_K and Kerr ellipticity ε_K) on T_G in the range $27 \leq T_G \leq 408°C$ for four different $Co_{0.80}Ni_{0.20}$ films. The MOKE spectra are quite different, especially in the infrared region near 1.5eV photon energy, resulting in changes of up to about 40%. The film grown at $T_G = 408°C$ clearly shows the characteristic features of a pure fcc Co film (see [3] for comparison), whereas lowering the growth temperature to room temperature (27°C) reveals the characteristic behavior found in hcp Co films [3]. Once again confirming the expected hcp $\rightarrow$ fcc phase transition. For a more quantitative discussion with respect to the position of the phase transition and details of the microstructure in these films, we refer to a future publication [10].

SUMMARY

We have demonstrated the sensitivity of magneto-optical effects to structural changes in the $Co_{1-x}Ni_x$ system. The hcp $\rightarrow$ fcc phase transition is accompanied by drastic changes in the magneto-optical Kerr effect (MOKE) and the magnetic anisotropy in these films. This suggests that MOKE can be used to delineate different phase regions in Co-Ni or other magnetic alloy systems.

ACKNOWLEDGEMENT

We thank A. Cebollada and M.F. Toney for various contributions. One of us (A.C.) gratefully acknowledges financial support from the Alexander von Humboldt-Foundation (AvH), Bonn, Germany.

REFERENCES

1. W. Reim and J. Schoenes, Ferromagnetic Materials, edited by K.H.J. Buschow and E.P. Wohlfarth (Elsevier Science Publishers B.V., VOL. 5, 1990), p. 133ff.

2. D. Weller, H. Brändle, R.F.C. Farrow, R.F. Marks, G. Harp, Magnetism and Structure in Systems of Reduced Dimensions, edited by R.F.C. Farrow et al. (NATO ASI series, Plenum Publishing Corp., New York, 309, 1993), p. 201: G.R. Harp, D. Weller, T.A. Rabedeau, R.F.C. Farrow, M.F. Toney, Phys. Rev. Lett. 71, 2493, (1993); G.R. Harp, D. Weller, T.A. Rabedeau, R.F.C. Farrow, R.F. Marks, Mat. Res. Soc. Symp. Proc. 313, 493, (1993).

3. D. Weller, G.R. Harp, R.F.C. Farrow, A. Cebollada, J. Sticht, Phys. Rev. Lett., 72, 2097, 1994.

4. T. Suzuki, D. Weller, C.-A. Chang, R. Savoy, T.C. Huang, B. Gurney and V. Speriosu, Appl. Phys. Lett. (to be published), 1994

5. G.R. Harp, R.F.C. Farrow, D. Weller, T.A. Rabedeau, R.F. Marks,
    Phys. Rev. **B48**, 17538, (1993).

6. R.F.C. Farrow, G.R. Harp, R.F. Marks, T.A. Rabedeau, M.F. Toney, D. Weller
    and S.S.P. Parkin, Journal of Crystal Growth, **133**, 47, (1993).

7. H. Brändle, D. Weller, S.S.P. Parkin, J.C. Scott, and P. Fumagalli, W. Reim,
    R.J. Gambino, R. Ruf, G. Güntherodt, Phys. Rev. **B46**, 13889, (1992).

8. W. Bendick and W. Pepperhoff, J. Phys. F: Metal Phys. **9** (11), 2185, (1979).

9. R.M. Bozorth, Ferromagnetism, (D. van Nostrand Company, INC., Princeton), p. 568.

10. A. Carl, D. Weller, M.F. Toney, R. Savoy and B. Hillebrands (unpublished)

PART III

Polycrystalline Magnetic Thin Films

Part B

Grain Structure, Texture, and Epitaxy in Magnetic Multilayers

ATOMICALLY LAYERED STRUCTURES FOR PERPENDICULAR MAGNETIC INFORMATION STORAGE

BRUCE M. LAIRSON*, M.R. VISOKAY**, R. SINCLAIR**, S.M. BRENNAN**, B.M. CLEMENS**, J. PEREZ*** and C. BALDWIN***.
*Materials Science Department, Rice University, Houston TX.
**Materials Science and Engineering Department, Stanford University, Stanford, CA.
***Censtor Corporation, San Jose, CA.

ABSTRACT

Layered structures can possess a high volumetric density of interfaces, which can result in novel magnetic properties. We report magnetic characteristics of two types of atomically layered structures, (001) oriented intermetallics with the CuAu(I) crystal structure and (111) oriented artificial multilayers. C-axis oriented superlattices of PtFe, PtCo, PdFe and PdCo with the layered CuAu(I) structure, possess large magnetic anisotropies and novel magneto-optic properties relative to the corresponding random alloys. These films are produced by annealing, with tetragonality and magnetic anisotropy developing in a fashion which depends on the initial bilayer period and annealing parameters. (111) oriented artificially grown Pd/CoCr multilayers are reported for perpendicular magnetic recording applications. Using a perpendicular contact probe transducer, the multilayers exhibit two times the readback signal of CoCr media. The multilayers can be made with low roughness suitable for contact recording, and can generate narrow readback pulses with large amplitude. Many aspects of Pd/Co performance, such as overwrite, can be predictably optimized for a given set of transducer properties. The reported results show that compositionally modulated structures can be made which have a wide variety of useful properties for perpendicular hard disk and magneto-optic recording.

INTRODUCTION

Dramatic changes in magnetic recording materials technology have occurred in the past three decades, which have allowed magnetic recording to double in storage density every two years over this period[1]. Much of the improvement in recording density which has occurred during this time has resulted from improvements in the materials used for recording. Hard disk media technology has progressed from thick, painted coatings to the present sputter deposited coatings in the 200Å-1000Å thickness regime. Magneto-optic technology benefited from the discovery of amorphous rare earth-transition metal alloys, which have nearly ideal properties for magneto-optic recording. For recording technology to continue to develop, new materials technologies will be required. Investigations of new thin film materials for magnetic recording applications are therefore of great interest. Here, we consider the application of layered thin film materials to two perpendicular recording technologies, magneto-optic recording and perpendicular hard disk recording. The media for these technologies must have magnetic anisotropy perpendicular to the film plane. In the case of magneto-optic recording, the media must also have low Curie temperatures and large Magneto-Optic Kerr Effect (MOKE) rotations. For perpendicular magnetic recording,

Mat. Res. Soc. Symp. Proc. Vol. 343. ©1994 Materials Research Society

media must have high saturation magnetizations, nanometer-scale surface roughness, and corrosion resistance.

CuAu(I) SUPERLATTICES FOR M-O RECORDING

We have recently reported a thin film processing technique which produces crystallographically oriented intermetallic compounds that are atomically layered parallel to the film surface[2]. These films possess high composition modulation in the direction out of the film plane. When a multilayer, stacked along the proper crystallographic axis, is annealed, the film "remembers" the initial stacking direction and forms an intermetallic compound with a unique axis along this direction. For example, a multilayer with bilayers consisting of [8 monolayers Pt(001)/ 8 monolayers Fe(001)], when annealed, restacks itself into the tetragonal intermetallic compound PtFe, with its c-axis oriented out of the film plane. This annealed structure has [1 monolayer Pt/1 monolayer Fe] comprising a bilayer.

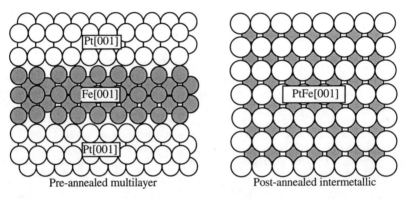

| Pre-annealed multilayer | Post-annealed intermetallic |

Fig. 1. Schematic of oriented ordered phase formation resulting from the annealing of epitaxial multilayers. The CuAu(I) structure on the right forms with its c axis preferentially aligned parallel to the surface normal.

Table 1
(001) Oriented Thin Film Intermetallics
Formed by Multilayer Annealing

Intermetallic Compound[2,3,4]	Magnetic Anisotropy K_u(ergs/cm^3)	Comments
PtFe(001)	8×10^6	
PtCo(001)	19×10^6	
PdFe(001)	6×10^6	
PdNi	----	Did Not Order (No Bulk Phase Known)

In effect, very small bilayer period superlattices, with the highest attainable composition modulation, can be formed by taking advantage of the thermodynamic driving force for ordering, with the final ordering direction determined by the high density of oriented interfaces in the original multilayer. The effect is shown schematically in Fig. 1. The ordering results from the free energy driving force for intermetallic phase formation, while the orientation results from the oriented initial interfaces. In a more complete drawing, one must also include the formation of other phases, other orientations and antiphase boundaries, none of which are shown in Fig. 1.

Table 1 lists the phases which have been attempted to date using this technique for multilayer films on MgO[001] single crystals substrates, along with their measured net perpendicular magnetic anisotropy. The measured anisotropy for PtCo(001) films is very large, about four times the magnetocrystalline anisotropy of pure Co. This measured anisotropy for the films is consistent with the measured perpendicular anisotropy which has been reported for single crystal PtCo. It would be of interest to attempt the production of known metastable and unknown phases, e.g., PdCo, PdNi, or MnAl. Some of the authors are currently exploring the production of metastable phases[3]. Ordered materials

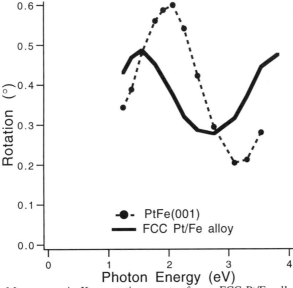

Fig. 2. Magneto-optic Kerr rotation spectra for an FCC Pt/Fe alloy and for ordered PtFe(001). A peak appears in the MOKE spectrum in the ordered alloy at an energy of 2.0eV.

this type are expected to have novel magnetic and electronic properties[5,6]. It is also noteworthy that large net perpendicular anisotropy is produced in films consisting only of Pt and Fe.

The magneto-optic rotation spectra for the ordered phases are quite different from those which have been measured for the equiatomic random alloys. For example, Fig. 2 shows a comparison between MOKE spectra for an epitaxial random FCC Pt-Fe alloy and the ordered PtFe phase. The Kerr rotation increases by 70% at 2.0eV[5]. It is not particularly surprising that the magneto-optic spectrum changes upon ordering of the alloy. While the alloy possesses cubic magnetic anisotropy, the intermetallic compound possesses large magnetic anisotropy along the c axis, as listed in Table 1. The electronic structure influences the magnetic anisotropy through the strong spin-orbit coupling of Pt. Changes in the magneto-optic spectra also occur for ordering in PtCo, PdCo, and PdFe[3]. However, it is notable that in some portions of the spectrum large peaks in the optical rotation occur, which are related to the development of a band structure upon ordering. The investigation of doping and the production of other ordered phases is of great interest in this context.

Magneto-optic recording uses thermomagnetic writing to produce reversed magnetic domains. In practice, writing is done with low power solid state lasers, requiring that the media layer have a Curie temperature less than about 300°C. Measurements were made of the PtFe magneto-optic Kerr rotation at various temperatures, shown in Figure 3. The extrapolated Curie temperature, 430°C, is near that of fully ordered bulk PtFe, 450°C. The Curie temperatures of these phases must be reduced by about 20% to employ them as magneto-optic media, either by alloying or changing their stoichiometry.

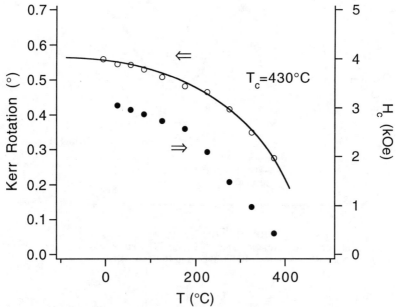

Fig. 3. MOKE rotation vs. temperature for PtFe(001). A Curie-Weiss best fit to the rotation yields a Curie temperature of 430°C.

Preliminary magneto-optic writing experiments were carried out in collaboration with F.J.A. den Broeder and H. van Kesteren. Epitaxial PtCo(001) films on MgO(001) with high anisotropy were prepared which had a total thickness of 320Å. A perpendicular MOKE hysteresis loop of one of these films is shown in Fig. 4a for a wavelength of 400nm. This film had a perpendicular remanence of 93% and a coercivity at room temperature of 2.9kOe. Figure 4b shows magneto-optically recorded reverse domains in this film, written in a reversing field of 625 Oe (50 kA/m). Well-formed marks are possible in the PtCo(001) films, but only at very high write power, about 200mW. This high write power is due to factors including the high bare film reflectivity, the high Curie temperature of PtCo, and the good thermal conductivity of the single crystal substrate. All of these must be addressed for continued development of these films as media.

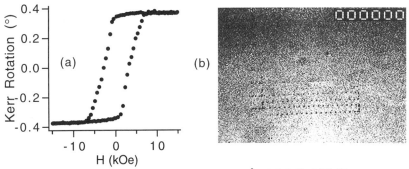

Fig. 4. a) Polar Kerr hysteresis loop at 400nm for a 320Å thick PtCo(001) film. b) Thermomagnetically recorded bits on this film. The bottom row of recorded domains is for a laser power of 210mW, the 3rd row from the bottom is for 196mW. The spacing between the recorded domains is approximately 5 microns.

Pd/Co MULTILAYERS FOR PERPENDICULAR MAGNETIC RECORDING

Perpendicular magnetic recording has undergone substantial development in the past two decades. Throughout this period, the preferred magnetic media materials have been hexagonal CoCr(0001) alloys, which have sufficient magnetocrystalline anisotropy to overcome the demagnetizing energy inherent to the thin film geometry. These alloys were particularly attractive because they could be prepared at large thicknesses with simple techniques, such as sputtering.

However, issues associated with the increasing areal storage density of magnetic recording limit the technological lifetime of CoCr(0001) based alloys as perpendicular recording media. The alloys exhibit a magnetically "dead" initial layer, they have relatively small anisotropy energies, and incomplete perpendicular remanence. As recording density increases and media thicknesses decrease, these issues will become increasingly important. Conversely, media thicknesses have decreased to a regime where the production of multilayer media is viable. For instance, to fabricate a 300Å multilayer media with a 10Å bilayer period requires only 30 bilayers, which can be deposited in several minutes.

Co/Pt and Co/Pd multilayers with no magnetic underlayer have been extensively studied for magneto-optic recording [7,8]. Some of the performance criteria for perpendicular magnetic recording are similar to those for magneto-optic recording, particularly for media noise, perpendicular anisotropy energy, and domain reversal dynamics. However, there are very substantial differences in media requirements for these technologies. Rather than measuring magnetization, magneto-optic storage technology measures Kerr rotation. Perpendicular magnetic recording requires large remanent out-of-plane magnetizations in the recording layer, a quantity which is normally minimized in magneto-optic recording. In addition, the Curie temperature, which is a key materials property for M-O recording, is relatively unimportant for perpendicular magnetic recording. Surface smoothness and corrosion resistance at the Angstrom level are vital to the functionality of perpendicular magnetic recording, yet are relatively unimportant for magneto-optic recording.

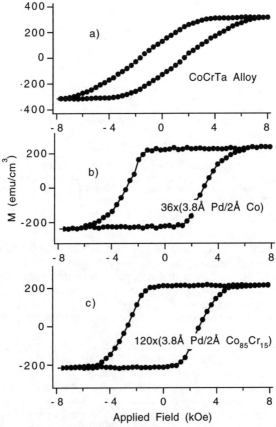

Fig. 5 Magnetic moment vs. perpendicularly applied field for a CoCrTa alloy media and for multilayer media.

We have studied the application of multilayer media to perpendicular magnetic recording using a contact single pole head with a NiFe flux return underlayer[9]. $Nx(Pd/Co_{1-x}Cr_x)$ multilayer thin films were prepared near room temperature by magnetron sputtering. The substrates were either NiFe plated Al disks for read/write performance studies or microscope slide witness samples for investigating structural and magnetic properties. The media disks were subsequently coated with approximately 100Å of carbon for wear protection and then lubricated. NiFe was employed as an underlayer material on the media disks to act as a flux return path for the recording head. Multilayers on glass and on NiFe underlayers were prepared simultaneously by successive deposition of Pd and $Co_{1-x}Cr_x$. To promote adhesion and remove some of the influence of the substrate, multilayers were sputtered onto a 10Å Pd seed layer. Generally it was found that coercivities measured from magneto-optic Kerr hysteresis loops for the multilayers on glass were similar to those for the multilayers on NiFe. Read/write evaluations were performed with perpendicular contact recording using a Censtor Flexhead(TM) operated in this study at a velocity of approximately 150 inches/sec. VSM measurements were made on a PAR Model 150 vibrating sample magnetometer. Magnetic anisotropy was determined from torque magnetometry performed on a Digital Measurement Systems magnetometer in an applied field of 15kOe. Surface roughness was measured using a Nanoscope III Atomic Force Microscope(AFM).

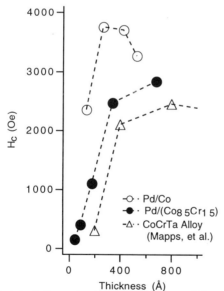

Fig. 6. Media coercivity vs. thickness comparing multilayers with conventional CoCrTa media, showing the development of high coercivity at small thicknesses for the multilayers.

Figure 5 compares hysteresis loops for Pd/Co$_{1-x}$Cr$_x$ multilayers with a conventional perpendicular CoCr alloy thin film. Figure 5a shows the out-of-plane hysteresis loop for a conventional CoCrTa thin film which has typical magnetization characteristics for perpendicular recording[5]. Figures 5b and 5c show out of plane hysteresis loops for multilayer films of (4Å Pd/ 2Å Co$_{1-x}$Cr$_x$)xN sputter deposited at room temperature. As can be seen in the figure, the squareness, remanent magnetization, and perpendicular coercivity are substantially higher for the multilayers than for conventional media. These features suggest that these multilayers may possess advantages for recording media applications relative to CoCr alloy films.

Fig.6 shows the coercivity of Pd/Co, Pd/CoCr, and CoCrTa media as a function of thickness. The CoCrTa data is taken from Mapps, et al[10]. The CoCrTa data displays the well-known "dead layer" effect, in which the coercivity extrapolates to zero at finite thickness, e.g. 200Å. The multilayers, on the other hand, show a much smaller dead layer, and the coercivity extrapolates to zero at very small media thicknesses, e.g. less than 50Å.

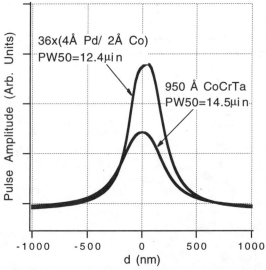

Fig. 7. Isolated pulse response of perpendicular recording head to a Pd/Co multilayer and a CoCrTa alloy thin film. The multilayer is approximately 275Å thick.

Fig. 7 shows the electrical readback signal vs. spatial frequency in thousands of flux changes per inch (kFCI) comparing a 950Å thick CoCrTa and a 275 Å thick (Pd/Co) multilayer. In this case the media were deposited onto NiFe-plated aluminum disks which provide magnetic flux return paths for the recording probe head. Reading and writing were performed with a Censtor Flexhead(TM) contact recording probe head[9]. Using the same recording transducer, the multilayer Pd/Co media showed two times the isolated pulse amplitude and a 20% smaller pulse width(12.4 μin vs. 14.5μin). At high linear densities, more than a factor of two improvement in the readback signal is obtained from the multilayers relative to the thicker conventional media. For instance, at 145kFCI, the track-averaged readback signal from the conventional medium is 140μV, compared to 380 μV for the multilayer. This improvement in resolution and readback signal is also apparent in the isolated pulse waveforms. It should be noted that relative signal amplitudes are dependent not only on the media selection, but on the appropriate matching of the media properties, such as the thickness and coercivity, to the recording head characteristics. A low write amplitude head, for instance, was unable to record domain walls into high coercivity multilayer media.

Media thicknesses have decreased to a regime where metal multilayers are viable magnetic media candidates. Pd/Co multilayers show great promise as perpendicular magnetic recording media. They possess a number of advantages over conventional CoCr alloy media, such as producing larger and narrower voltage pulses, having a much smaller dead layer, and being easily modified to match the characteristics of a given recording head. Multilayers can be produced which are sufficiently smooth for contact recording technology. In addition, they possess much higher anisotropy energies, are corrosion resistant, and many of their properties can be predictably tailored by adjusting deposition parameters, stoichiometry and layer thicknesses. Artificial multilayers do not require high temperature deposition to produce acceptable media properties.

CONCLUSIONS

Compositionally modulated materials offer the advantage of large magnetic anisotropies, which can be produced along the (111) or the (001) crystallographic axes. For some structures, this anisotropy is accompanied by other favorable properties, such as large magneto-optic rotations and saturation magnetizations. Their unique properties make them suitable candidates for a variety of applications, including perpendicular hard disk storage and magneto-optic data storage.

ACKNOWLEDGEMENTS

We wish to acknowledge useful discussions with H. Hamilton and F.J.A. den Broeder. We also thank H. van Kesteren for making and providing us with images of the written domain patterns in PtCo.

REFERENCES

[1]. K. Howard et al., these proceedings.
[2] B. M. Lairson, M. R. Visokay, R. Sinclair, and B. M. Clemens, Appl. Phys. Lett. 62, 639 (1993).
[3] M. R. Visokay, B. M. Lairson, B. M. Clemens and R. Sinclair, these proceedings.
[4] B. M. Lairson, M. R. Visokay, E. E. Marinero, R. Sinclair, and B. M. Clemens, J. Appl. Phys. 74, 1922 (1993).
[5] B. M. Lairson and B. M. Clemens, Appl. Phys. Lett. 63, 1438 (1993).
[6] Y. Suzuki and T. Katayama, MORIS'92 Proceedings 29 (1993).
[7] P. DeHaan, Q. Meng, T. Katayama, and J. C. Lodder, J. Magn. and Magn. Mat. vol. 113, 29, (1992).
[8] P. F. Carcia, A. D. Meinhaldt, and A. Suna, Appl. Phys. Lett. vol. 47, 178 (1985).
[9] B. M. Lairson, J. Perez, and C. Baldwin, to appear in Appl. Phys. Lett.
[10] D. J. Mapps, et al. J. Magn. and Magn. Mat. vol. 120, 305 (1993).

ON THE PERPENDICULAR MAGNETIC ANISOTROPY OF Pt/Co MULTILAYERS:
STRUCTURE-PROPERTY RELATIONSHIPS.

G.A. BERTERO and R. SINCLAIR
Department of Materials Science and Engineering, Stanford University, Stanford, CA 94305

ABSTRACT

A series of Pt/Co multilayers were sputter-deposited under various deposition conditions to promote structural changes resulting in marked differences in both the perpendicular anisotropy and magnetic coercivity. High resolution transmission electron microscopy was used in combination with other analytical techniques to study the structure of these films. It was found that the most important feature controlling the magnitude of the anisotropy is the interface sharpness. Conversely properties such as the quality of the (111) texture, grain shape or the defect structure, were found to be of secondary importance. The magnetic coercivity ranged from 0.6 to 7 kOe depending on the sputtering conditions and was found to depend strongly on the grain boundary structure rather than on the grain size.

INTRODUCTION

Co-based multilayers and in particular Co/Pt multilayers, have been found suitable as magneto-optic recording media at blue wavelengths [1,2]. In addition, they have recently been shown to offer advantages over Co-alloy thin film media for perpendicular magnetic recording [3]. An important property in these multilayer systems is their large perpendicular magnetic anisotropy, $K_\perp$. The origins of $K_\perp$ have been attributed by many investigators to a range of structural features, sometimes in contradiction to one another.

The work of Carcia et al. [4,5] showed that the perpendicular anisotropy of Xe-sputtered Co/Pt films was significantly larger than that of Ar-sputtered films and comparable in magnitude to films deposited by evaporation techniques. We chose a similar approach in an effort to identify the structural features leading to these large differences in $K_\perp$. We have sputter-deposited a series of multilayer films in a number of deposition ambients to promote structures yielding differences in both the perpendicular anisotropy and magnetic coercivity. High resolution transmission electron microscopy (HRTEM) and x-ray diffraction were used in combination with other analytical techniques to characterize these films. Particular emphasis was given to those structural features with potential influence on the magnetic anisotropy.

EXPERIMENTAL

The multilayer films were DC-magnetron sputter-deposited onto thermally oxidized silicon wafers at room temperature in Ar, Xe, and 50/50 Ar/Xe gas ambients. A 100Å thick Pt underlayer film is deposited at approximately 500°C to enhance the (111) crystallographic texture and control the grain size in the multilayer. Sputtering pressures were of 1.5 and 5.0 mTorr Ar, and 5 mTorr Xe. The sample-to-target distance during deposition was ~15 cm and the deposition chamber base pressure was ~ 4×10^{-9} Torr. Deposition rates were kept below 0.5 Å/s for accurate thickness control and a 6-second time delay was used between the end of the deposition of one layer and the beginning of the deposition of the next to ensure that no co-deposition takes place. Magnetic measurements were performed with a vibrating sample magnetometer (VSM) and a torque magnetometer. Kerr rotation measurements were performed in a magneto-optic spectrometer with a range of 350 to 1000 nm wavelength utilizing monochromatic light. Sample preparation for high resolution transmission electron microscopy (HRTEM) of cross-section specimens was performed utilizing a similar technique to that described elsewhere [6].

Mat. Res. Soc. Symp. Proc. Vol. 343. ©1994 Materials Research Society

RESULTS AND DISCUSSION

The Ar-sputtered films present, in general, lower coercivities and magnetic anisotropies compared with Xe-sputtered films. Figure 1 shows a plot of the magnetic coercivity and perpendicular anisotropy of Pt/Co multilayers with period (4.1 Å Co, 13.5 Å Pt)$_{12}$ as a function of sputtering gas in 5 mTorr ambients. This can also be seen in Figure 2 where magneto-optic hysteresis loops from the Ar and Xe-sputtered samples are shown. The respective saturation magnetizations are also indicated and it is seen that the Ar-sputtered film has a 25 % smaller Ms compared to the Xe-sputtered film, an effect that has frequently been observed in these films and will be discussed below.

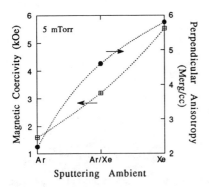

Figure 1: Magnetic coercivity and perpendicular anisotropy as a function of sputtering gas. Xe sputtering results in more coercive and anisotropic films

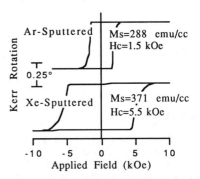

Figure 2: Magneto-optic hysteresis loops from Ar and Xe-sputtered films at 5 mTorr pressures. The saturation magnetization and perpendicular anisotropies are also shown.

The structure of these films was characterized in detail by HRTEM and x-ray diffraction. High resolution TEM images from Pt/Co multilayers deposited in 1.5 mTorr Ar, 5 mTorr Ar and 5 mTorr Xe ambients are shown in Figure 3. A large number of stacking defects can be seen in the 1.5 mTorr sputtered sample. These defects have been observed in films grown by other techniques and were expected because of the presence of Co (equilibrium form of Co is hcp and that of Pt is fcc at room temperature) and the relatively low stacking fault energy associated with closed packed structures. The high density and non-uniformity of these defects throughout the thickness makes the assignment of Co to fcc or hcp structure difficult. However, the 5 mTorr Ar and 5 mTorr Xe-sputtered films also shown in the figure present subgrain structures practically free of stacking defects. The crystal structures are clearly fcc and the (111) texture is strong in both cases. In this work, no increase in K⊥ was found in films with a high defect density, as was proposed by Chien et al. [7]. In fact, the 1.5 mTorr case which clearly had the highest defect density, had the lowest anisotropy of the three shown in Fig. 3.

The grain boundary structure of Ar and Xe-sputtered films are shown in Figure 4. In general, Xe sputtering resulted in rougher surfaces and more separated grains with considerable grooving and frequently disordered grain boundaries as discussed in ref.[8]. Ar sputtering, on the other hand, resulted in smother surfaces, reduced grooving, and sharper grain boundaries, particularly at lower sputtering pressures. The grain sizes, however, were comparable in all the films since these are fixed by the Pt underlayer grain size. The large differences in Hc can be better correlated in our films to the grain boundary structure and separation rather than to the grain size itself.

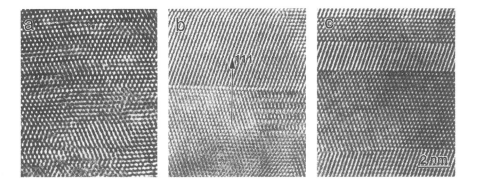

Figure 3: High resolution TEM images in cross-section from Pt/Co multilayers sputtered deposited in (a) 1.5 mTorr Ar, (b) 5 mTorr Ar, and (c) 5 mTorr Xe.

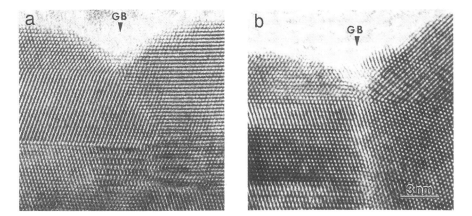

Figure 4: High resolution TEM images showing the grain boundary structure typical of Ar and Xe-sputtered Pt/Co multilayers at 5 mTorr pressures.

The effect of grain shape has also been found to be insufficient to explain the 3 to 4-fold difference in anisotropy between Ar and Xe-sputtered samples. In fact, sputtering at higher Ar pressures produces more separated grains similar to those seen in Xe-sputtered films. Such less magnetically coupled grains (as indicated by the larger coercivity) would presumably reduce the shape anisotropy factor thus reducing the demagnetizing contribution to the perpendicular anisotropy resulting in a larger $K\perp$. However, despite the coercivity increase, no appreciable gain in anisotropy is observed.

Carcia et al. attributed the differences in $K\perp$ to a better crystallographic texture resulting when sputtering with heavier ions [4]. In our films, x-ray diffraction indicates that the texture is very similar between the differently processed films. Figure 5 shows x-ray omega rocking curves at the (111) peak from the Ar, Ar/Xe and Xe-sputtered films. The full width at half maximum (FWHM) is relatively small for all the samples and the difference in FWHM between the films is negligible indicating a good and comparable degree of (111) texture. This is consistent with the TEM observations showing that the multilayer crystallographic texture is largely determined by that of the Pt underlayers. Low angle x-ray diffraction results for three films with the same nominal bilayer period are shown in Figure 6. The peak positions indicate equivalent periods, however, the multilayer peak intensities are considerably larger for the Xe-sputtered case than in the Ar-sputtered case despite the comparable background levels. This is consistent with the expectation of sharper interfaces in the Xe-sputtered samples and suggests a correlation between interface sharpness and magnetic anisotropy in these multilayers. Further evidence for the differences in interface structure can be found by measuring the saturation magnetization of Ar and Xe-sputtered films of equal periods. Figure 7 is a plot of Ms as a

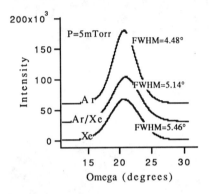

Figure 5: Omega x-ray rocking curves about the (111) multilayer peak for Ar, Ar/Xe, and Xe-sputtered films. The full width at half maximum (FWHM) for each peak is indicated.

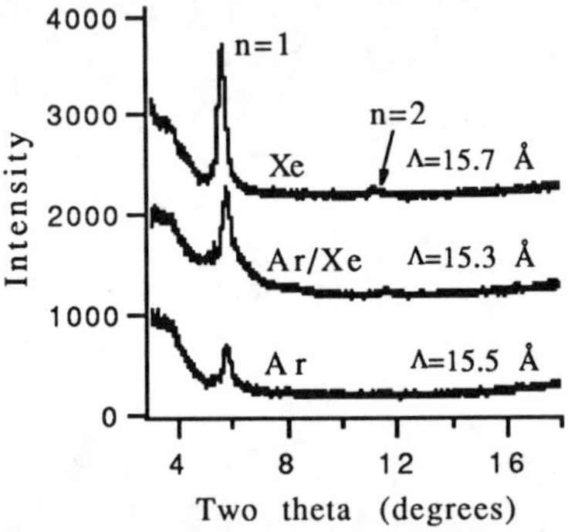

Figure 6: Low angle x-ray diffraction data from the Ar, Ar/Xe, and Xe-sputtered films. The bilayer periods, Λ, are indicated.

function Co layer thickness illustrating this effect. The difference is seen to decrease for thicker Co layers indicating that the origin is interfacial. We can expect as a consequence a lower Curie temperature in the Ar-sputtered film reflecting the differences in interface structure. This was confirmed by performing magnetic measurements at elevated temperatures where a difference of approximately 100 °C in Curie points, namely 308 °C and 410 °C, was found for the Ar and Xe-sputtered films respectively.

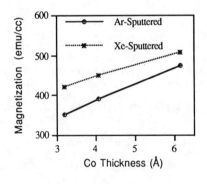

Figure 7: Saturation magnetization as a function of Co layer thickness in Co/Pt multilayers sputtered deposited in Ar and Xe at 5 mTorr pressures.

In order to study the effects of energetic bombardment in more detail during Ar sputtering we deposited a series of multilayers with periods Co/(Pt/Pd) and Co/(Pd/Pt), i.e., Pt is deposited always onto Co in the first case while Pd is deposited always onto Co in the other. In these samples, the Co layer thickness corresponds to two atomic monolayers, i.e., 4.1 Å, while the Pt and Pd layer thicknesses correspond to 3 atomic monolayers, i.e., 6.7 and 6.8 Å respectively. Backscattered energetic bombardment is expected to be largest during Pt sputtering and therefore the interfaces in each sample are expected to have different degrees of intermixing and roughness. Consequently, the properties of these two films should reflect the differences in the resulting structures. Figure 8 shows magneto-optic hysteresis loops taken at 400 nm photon wavelength

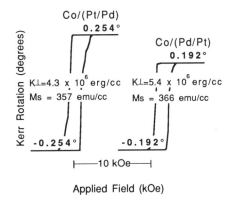

Figure 8: Magneto-optic hysteresis loops from Co/(Pt/Pd) and Co/(Pd/Pt) samples.

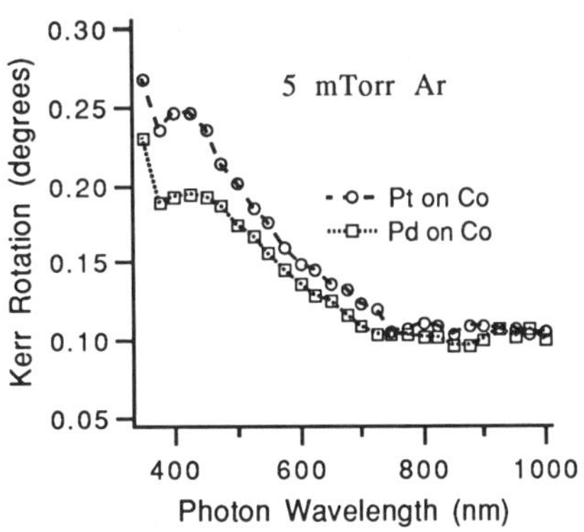

Figure 9: Kerr rotation as a function of photon wavelength for 5 mTorr Ar-sputtered Co/(Pt/Pd) and Co/(Pd/Pt) films.

from Co/(Pt/Pd) and Co/(Pd/Pt) multilayer films. The saturation magnetization and the resulting perpendicular magnetic anisotropy energies are also included. The Kerr rotation of the Co/(Pt/Pd) sample is larger by about 25 % than that of the Co/(Pd/Pt) sample despite their similar saturation magnetizations. This reflects the fact that in the Co/(Pt/Pd) case the Co is extensively intermixed with Pt giving rise to the larger Kerr rotation characteristic of Co/Pt multilayers near this wavelength. The Kerr rotation of Co/Pd multilayers at 400 nm is about 30 % less than in Co/Pt and therefore the Co/(Pd/Pt) results suggest a small degree of intermixing of Co with Pd in that sample as well. The magnetic anisotropy is largest for the Co/(Pd/Pt) film which is expected to have the sharpest interfaces.

The Kerr rotation as a function of photon wavelength for these two films is shown in Figure 9. The similar Kerr rotations at long wavelengths reflect the fact that the Co magnetization is similar in the two cases. The increasingly larger differences in Kerr rotation at shorter wavelengths reflect the characteristic contributions to the Kerr effect from the polarization of Pt and Pd. It is clear than in the Co/(Pt/Pd) multilayer it is mostly the Pt that is polarized indicating heavy intermixing of Pt with Co. The fact that this film presents in turn a smaller magnetic anisotropy is consistent with our findings in the Pt/Co multilayers and is additional evidence that intermixing is adversely affecting the magnitude of the anisotropy energy.

We have recently performed circular x-ray dichroism experiments in the Ar and Xe-sputtered Pt/Co films. Preliminary results indicate that the unquenched Co orbital moments in the Xe-sputtered multilayer are about three times larger than for the Ar-sputtered case in good agreement with the differences in K⊥ observed between the films. Unquenched orbital moments in the overall cubic structure of these films arise from the symmetry breaking effect of interfaces as proposed by Néel [9]. Intermixing homogenizes the structure resulting in a lower $K\perp$ [8]. In the limit, a homogeneous disordered fcc solid solution does not allow for uniaxial anisotropy. These magnetic dichroism results are then a strong confirmation of the correlation between the magnetic anisotropy and the sharpness at the atomic level of the interfaces in Pt/Co films.

SUMMARY AND CONCLUSIONS

We have sputter deposited a series of Pt/Co multilayers in a number of sputtering ambients to induce differences in the structure and thus properties of the films. Particular emphasis was given to the effects of structure on the magnetic anisotropy. We found that $K\perp$ is extremely sensitive to the interface sharpness. Intermixing of Co and Pt results in lower anisotropies. Conversely we found no correlations from features such as stacking faults and differences in the (111) texture as have been proposed elsewhere. The magnetic coercivity was found to depend strongly on the grain boundary structure and less on the grain size. Xe-sputtering, as opposed to Ar-sputtering, resulted in Pt/Co multilayers with more separated grains and less intermixed layers. As a consequence, the coercivity and perpendicular anisotropies are largest and comparable to that of films deposited utilizing less energetic evaporation techniques. Sputtering in a mixed Ar/Xe ambient results in a compromise between the two. The magneto-optic Kerr rotation, Θ_K, near 400 nm photon wavelengths is sensitive to the extent of Pt polarization in Pt/Co multilayers . Therefore, comparing multilayers with equal nominal periods, Θ_K is larger for the films displaying the largest degree of intermixing.

ACKNOWLEDGMENTS

The authors acknowledge the financial support of Kobe Steel USA. In addition we would like to express our thanks to Chul-Hong Park for performing the magnetic circular x-ray dichroism experiments.

REFERENCES

1. S. Hashimoto, A. Maesaka, K. Fujimoto and K. Bessho, J. Magn. Magn. Mater. **121** 471 (1993).
2. D. Weller, J. Hurst, H. Notarys, H. Brändle, R.F.C. Farrow, R. Marks and G. Harp, J. Mag. Soc. Jap. **17** 72 (1992).
3. B.M. Lairson, J. Perez and C. Balwin, Appl. Phys. Lett. (in press) (1994).
4. P.F. Carcia, Z.G. Li and W.B. Zeper, J. Magn. Magn. Mater. **121** 452 (1993).
5. P.F. Carcia, S.I. Shah and W.B. Zeper, Appl. Phys. Lett. **56** 2345 (1990).
6. J.C. Bravman and R. Sinclair, J. Electron Microsc. Tech. **1** 53 (1984).
7. C.J. Chien, R.F.C. Farrow, C.H. Lee, C.J. Lin and E.E. Marinero, J. Magn. Magn. Mater. **93** 47 (1991).
8. G.A. Bertero and R. Sinclair, J. Magn. Magn. Mater. (in press) (1994).
9. L. Néel, Journal of Physical Radium **15** 225 (1954).

MICROSTRUCTURE AND CHEMICAL ORDERING IN UHV-DEPOSITED, POLYCRYSTALLINE Co_xPt_{1-x} ALLOY FILMS FOR MAGNETO-OPTICAL RECORDING[*].

R.F.C. Farrow, D. Weller, M.F. Toney, T.A. Rabedeau[+], J.E. Hurst, G.R. Harp, R.F. Marks, R.H. Geiss, H. Notarys, IBM Research Division, Almaden Research Center, 650 Harry Road, San Jose, CA 95120-6099;
[+] Stanford Synchrotron Radiation Laboratory, Stanford CA 94305
[*] This work was supported in part by ONR Contract N00014- 92-C-0084.

ABSTRACT.

We report magneto-optical recording with 62 dB CNR (carrier-to-noise ratio) at 488 nm for quadrilayer structures comprising polycrystalline Co_xPt_{1-x} (x = 0.28) alloy films deposited on ungrooved, silicon nitride -coated glass discs, held at 300°C, by UHV evaporation. The key parameter controlling magnetic anisotropy of the films is substrate temperature during deposition. Polycrystalline films grown on amorphous silicon nitride have a large coercivity and full perpendicular remanence when grown at 300°C. Synchrotron X-ray diffraction data for polycrystalline films grown at 300 and 600°C confirm the presence of short-range compositional order, to the $CoPt_3$-$L1_2$ phase in both cases. Such ordering can introduce an anisotropy in the distribution of Co-Co pairs in the alloy and is a possible source of the magnetic anisotropy.

INTRODUCTION.

Magneto-optical recording using polycrystalline, Co_xPt_{1-x}(x~0.25) alloy films as media was first reported in 1992 [1,2]. The initial studies were for high-vacuum evaporated films. A limitation in recording performance resulted from the low coercivity (<1 kOe) of these films when grown on dielectric underlayers such as amorphous silicon nitride films or fused silica (quartz) substrates. This led to difficulties in writing stable marks for M-O recording and to a low (≲50 dB) carrier-to-noise (CNR) ratio. Farrow et al. demonstrated[3] that high-coercivity alloy films could be prepared on a variety of amorphous and crystalline substrates by depositing the alloy at temperatures near 300°C under UHV conditions. The coercivity and perpendicular remanence showed a peak near a substrate temperature of 300°C for polycrystalline and [111]-oriented epitaxial films. TEM examination of films grown onto amorphous carbon prelayers on mica substrates showed evidence for compositional ordering to the $CoPt_3$ ($L1_2$) phase for a substrate temperature of 400°C and for localized domains of this phase at 500°C. A correlation between compositional ordering and perpendicular magnetic anisotropy was suggested but the TEM data was unable to confirm any ordering at the temperature of maximum perpendicular remanence and coercivity (300°C). A CNR of 57.5dB (488nm) was reported for a quadrilayer structure comprising a 200Å polycrystalline alloy (x=0.32) film on an ungrooved glass disc. Here we report CNR measurements of 62 dB for an alloy of slightly lower Co content (x=0.28) and present cross-section TEM and synchrotron X-ray diffraction data for polycrystalline films.

EXPERIMENTAL TECHNIQUES.

The structures reported in this paper were prepared in a VG 80-M MBE system (VG Semicon Ltd.) using e-beam sources for both Co and Pt as described earlier[3]. Films

Mat. Res. Soc. Symp. Proc. Vol. 343. ©1994 Materials Research Society

for M-O recording were deposited onto a 400Å-thick film of sputtered silicon nitride (amorphous) on 3.5 inch diameter Corning 0313 glass discs. The Kerr loop was then recorded before the quadrilayer structure was completed by sputtering a 200Å-thick film of silicon nitride, followed by a 500Å -thick Al film. Alloy film thicknesses were measured by either electron microprobe or X-ray fluorescence techniques and are given as mean values. The accuracy of these thicknesses is about 5%.

MAGNETO-OPTICAL RECORDING.

Dynamic testing was performed on media comprised of an optimized quadrilayer structure deposited on flat glass using a 488nm dynamic tester described previously[4]. Test conditions were large marks (1 MHz tone), media velocity 10 m/s, 30 kHz resolution bandwidth, and write pulse width of 450 nS. Shown in Figure 1 are carrier (CRR), carrier-to-noise (CNR) and noise measurements made as a function of media read power for a structure comprising an alloy film ($x = 0.28$) 160Å-thick grown at 300°C. The Kerr loop for this film (see inset, Figure 1) recorded at 633 nm prior to quadrilayer completion, was square with a coercivity of 4.3 kOe. Excellent CNR values of $\simeq$62 dB are observed for read powers around 2 mW. High write powers ($\simeq$30 mW) and media bias fields ($\simeq$ 1 kOe) were required to produce these marks probably because of the high Curie point and large magnetization of these materials. The high Curie point is apparent in the read power dependence data as read powers > 6 mW can be used without significant degradation of the carrier. Also evident in the data is $\simeq$ 2 dB of write noise at 2 mW read power and the CNR values are media noise dominated. It should be possible to improve the CNR values by reducing write noise, media Curie points, and required bias fields eg. by further reducing Co content.

STRUCTURAL CHARACTERIZATION.

TEM studies were made on an alloy of $Co_{0.2}Pt_{0.8}$ deposited simultaneously on 400Å-thick silicon nitride on Si(111) and on 100Å amorphous carbon / Si (111) at 300°C. The alloy film, with silicon nitride and carbon films attached, was easily detached from the Si for plane-view TEM examination. Synchrotron X-ray diffraction studies were made of a film ($x = 0.24$) grown directly on fused silica (dynasil) at 300 and 600°C. Polar Kerr loops (recorded at 25°C and a wavelength of 633 nm) for these samples are shown in Figure 2. The films grown at 300°C have full remanence and coercivities $\gtrsim$4 kOe but coercivity and perpendicular anisotropy have collapsed for the film grown at 600°C.

Figure 3(a) shows a XTEM image of the film (Kerr data for which is shown in Figure 2 (a)) grown on silicon nitride / Si (111). The film is comprised of grains of width ~20-180Å often extending through the film (188 Å-thick). The film has no evident preferred orientation and the transmission electron diffraction pattern (Figure 3 (b)) shows a superposition of a ring pattern with spots from the silicon single crystal substrate. The ring pattern is that expected from a chemically-disordered, fcc alloy. A plan-view transmission electron diffraction pattern (Figure 3(c)) recorded from an alloy of similar composition, grown onto silicon nitride at 300°C also shows no evidence for superstructure lines. Such lines were observed[3] only from alloy films grown at higher temperatures ($\geq$400°C and above). In order to determine whether or not ordering was present at 300°C, grazing incidence X-ray scattering measurements[5] were made using synchrotron X-ray diffraction. Figures 4 (a) and (b) show data for polycrystalline films grown at 300 and 600°C. The film grown at 600°C shows strong, broad superstructure

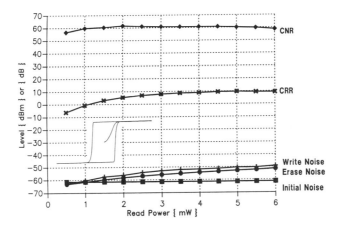

Figure 1.
Carrier (CRR), carrier-to-noise (CNR) and noise measurements made as a function of media read power for a structure comprising an alloy film ($x = 0.28$), 160Å-thick grown at 300°C. Inset shows Kerr loop for film (H_C 4.3 kOe, Rem. θ_K 0.30°) prior to quadrilayer completion.

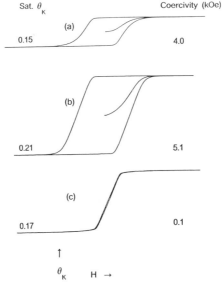

Figure 2.
Polar Kerr loops (recorded at 25°C and a wavelength of 633 nm) for samples characterized in this study.
(a) $x = 0.20$ alloy film, thickness 188 Å, deposited on 400Å silicon nitride / Si (111), 300°C.
(b) $x = 0.24$ alloy film, thickness 1108 Å, deposited on silica at 300°C.
(c) $x = 0.24$ alloy film, thickness 1172 Å, deposited on silica at 600°C.

377

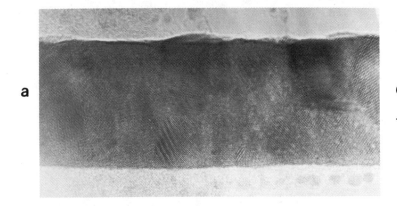

a

Co$_{0.2}$Pt$_{0.8}$

188±10Å

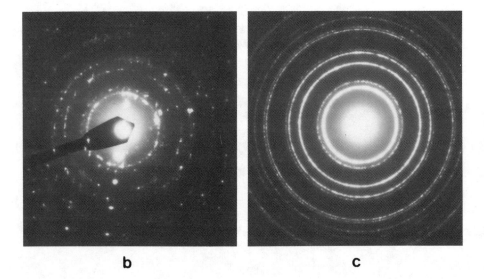

b   c

Figure 3.
(a) XTEM image of x = 0.20 alloy film, thickness 188Å, deposited on 400Å silicon nitride / Si(111), 300°C.
(b) cross-section transmission electron diffraction pattern of
x =0.20 film deposited on 400Å silicon nitride / Si(111), 300°C.
(c) plane-view transmission electron diffraction pattern of x =0.20 film deposited on 400Å silicon nitride / a-C / Si(111), 300°C. Note that the indices of the first 4 rings in the pattern are: (111), (002), (022) and (113). Superstructure rings would be present between (002) and (022) and between (022) and (113) but are not detectable in this pattern.

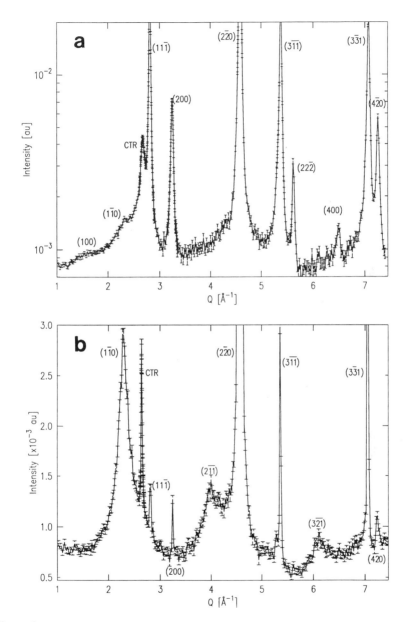

Figure 4
X-ray diffraction, in-plane scans:
(a) x = 0.24 alloy film, thickness 1108Å, deposited on silica at 300°C.

(b) x = 0.24 alloy film, thickness 1172Å, deposited on silica at 600°C

reflections: $(1\bar{1}0)$, $(2\bar{1}\bar{1})$ and $(3\bar{2}\bar{1})$ while the film grown at 300°C shows very weak and broad superstructure peaks: (100), $(1\bar{1}0)$. Note that the data for the 300°C film is plotted on a log scale to show the very weak superstructure peaks. The much weaker fundamental peaks: $(11\bar{1})$ and (200), compared with $(2\bar{2}0)$, in the 600°C film is due to strong (111) preferred orientation in this film. From the widths of the superstructure peaks we estimate the in-plane coherence lengths of ordering to be $\sim$14 Å (300°C) and $\sim$27Å (600°C). Thus ordering is short-range in both cases but a much larger fraction of the sample is ordered at 600°C.

DISCUSSION.

The possible role of anisotropic strain[6-8] and surface roughness[8] in controlling magnetic anisotropy in these polycrystalline alloy films was considered earlier. However, these parameters did not correlate with perpendicular anisotropy and are considered unlikely sources of magnetic anisotropy in our films. We have shown that short-range compositional ordering is present at the growth temperature (300°C) where perpendicular remanence and coercivity are peaked and also in the 600°C film, where the perpendicular anisotropy is absent. As pointed out [3] earlier, ordering to the CoPt$_3$ phase removes Co-Co nearest neighbor pairs. An anisotropic shape of the ordered regions may exist and control magnetic anisotropy by introducing a directional imbalance in Co-Co pairs. In the 600°C film, the shape of the ordered regions may be more isotropic, as a result of bulk diffusion, than in the 300°C film in which ordering will have a tendency to occur within the film plane due to surface diffusion. Consequently, a development of this model will require further structural characterization of the ordered domains and their interfaces with the disordered regions.

CONCLUSIONS.

Magneto-optical recording in Co_xPt_{1-x} (x = 0.28) alloy films has been demonstrated with high CNR. The magnetic anisotropy is controlled by growth temperature and we have confirmed that short-range compositional ordering is present in films with perpendicular anisotropy. Such ordering may be a source of magnetic anisotropy.

REFERENCES.

1. D. Weller, H. Brändle, G. Gorman, C.-J. Lin and H. Notarys, Appl. Phys. Lett., 61, 2726 (1992).
2. D. Weller, J. Hurst, H. Notarys, H. Brändle, R.F.C. Farrow, R. Marks, G. Harp, J. Magn. Soc. Japn., Suppl. S1, 72 (1993).
3. R.F.C. Farrow, R.H. Geiss, G.L. Gorman, G.R. Harp, R.F. Marks, E.E. Marinero, J. Magn. Soc. Japn., Suppl. S1, 140 (1993).
4. J.E.Hurst Jr., W.J. Kozlovsky, Jpn. J.Appl.Phys. 32, 5301 (1993).
5. P.H. Fuoss, S. Brennan, Annual Rev. Mater. Sci., 20, 365 (1990).
6. D. Weller, C. Chappert, H. Brändle, G. Gorman, R.F.C. Farrow, R. Marks, G. Harp, J. Magn. Soc. Japn., Suppl. S1, 76 (1993).
7. D. Weller, H. Brändle, C Chappert, J. Mag.Magn. Mat., 121, 461 (1993).
8. E.E. Marinero, R.F.C. Farrow, G.R. Harp, R.H. Geiss, J. A. Bain, B. Clemens, Mat. Res. Soc. Symp. Proc., 313, 677 (1993).

EFFECT OF INTERLAYERS UPON TEXTURE AND MAGNETIC PROPERTIES IN Co ALLOY MULTILAYER FILMS FOR LONGITUDINAL MAGNETIC RECORDING

M.R. VISOKAY[1,2], M. KUWABARA[1], H. SAFFARI[1], H. HAYASHI[1], R SINCLAIR[2] and Y. ONISHI[3]
[1]Kobe Steel USA, Applied Electronics Center, Palo Alto CA, 94304
[2]Department of Materials Science and Engineering, Stanford University, Stanford CA 94305
[3]Kobe Steel Limited, Technical Development Group, Kobe, Japan

ABSTRACT

CoCrNi bilayer films with Cr, Ti and Zr interlayers were deposited by DC magnetron sputtering onto ultra densified amorphous carbon substrates and characterized using cross-section transmission electron microscopy and vibrating sample magnetometry. The films are polycrystalline with a columnar microstructure. In the Cr interlayer case there is layer-to-layer epitaxy throughout a given column. Magnetic measurements showed a simple in-plane easy axis magnetic hysteresis loop. In the Zr and Ti interlayer cases epitaxy within a column was lost beginning with the first CoCrNi/(Zr/Ti) interface, resulting in a magnetic layer consisting of two crystallographically unrelated CoCrNi films. Magnetic measurements revealed a complex step-structure hysteresis loop for this case. Annealing the Zr interlayer film led to a significant growth of the amorphous layer due to a solid state amorphization reaction between the Zr and CoCrNi, which was accompanied by a decrease in the saturation magnetization.

INTRODUCTION

It has been shown that a significant reduction in media noise in longitudinal magnetic recording media can be attained by separating the magnetic medium into two or more magnetically decoupled layers, the result being a film with double (in the case of a bilayer) the number of independent grains[1-3]. To avoid degradation of the magnetic properties, the grains must be magnetically identical and as similar to the single layer case as possible. Since the film magnetic properties are intimately related to the crystallographic orientation of the magnetic portion of the film it is important to maintain similar orientation relationships between these layers. Therefore, in the case of a multilayer film it is important to use a non-magnetic spacer that maintains a consistent crystallography relative to the magnetic layers throughout the film. In most studies Cr has been used as the non-magnetic spacer layer, and the resulting magnetic properties and film crystallography have been well documented [1-4]. Since Cr is bcc, it is expected that using an hcp spacer material would modify the orientation relationships between the various layers. To this end hcp Ti and Zr, along with Cr, were used as spacers in CoCrNi bilayer films on a Cr underlayer in order to investigate the effect of interlayer composition upon the crystallography and magnetic properties of the resulting medium.

EXPERIMENTAL

The deposited film structure is shown schematically in Figure 1. The interlayers used were Cr, Ti, and Zr and the Co alloy was CoCrNi, with equal layer thicknesses in all cases. The films were deposited by DC magnetron sputtering in a 5 mTorr Ar ambient onto ultra-densified amorphous carbon (UDAC) substrates at a temperature of 200°C [5]. Vibrating sample magnetometry (VSM) measurements were performed using a maximum field of 12 kOe, applied in the film plane. All VSM measurements were done using samples of equal area. Cross-section transmission electron microscopy experiments were done on a Philips 430ST TEM operated at 300 kV, with a point resolution of around 0.19 nm. Cross-section TEM samples were made by standard grinding polishing and dimpling techniques, followed by ion milling at 77 K. Some samples were also annealed at 400°C in vacuum. Following the annealing, the samples were again characterized by VSM, and the Zr interlayer case was examined using cross-section TEM.

RESULTS AND DISCUSSION

The magnetic hysteresis loops of CoCrNi bilayer films with Cr and Zr interlayers are shown in Figure 2. As can be seen in Figure 2a, the film with a Cr interlayer has a simple hysteresis loop, with a coercivity of around 1 kOe. The film with the Zr interlayer (Fig. 2b), on the other hand,

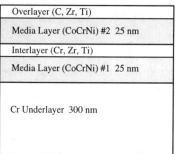

| Overlayer (C, Zr, Ti) |
| Media Layer (CoCrNi) #2 25 nm |
| Interlayer (Cr, Zr, Ti) |
| Media Layer (CoCrNi) #1 25 nm |
| Cr Underlayer 300 nm |

Figure 1. Schematic diagram of the film structure used in this study. The interlayers are Cr, Zr or Ti, the media is a CoCrNi alloy, and the substrate is ultra densified amorphous carbon (UDAC).

shows a distinct step structure. This step structure is consistent with the presence of two magnetic materials with different coercivities, one relatively large and the other significantly smaller [6]. For a small reverse field, the low coercivity layer switches polarity but the high coercivity one does not, leading to a net moment of nearly zero. As the reverse field is increased above the switching field of the high coercivity material, this layer reverses polarity and the net moment again increases. This process is shown schematically in Fig. 2b for several ranges of applied field. In the films examined, it is clear that the magnetic properties of the film with Cr interlayers are significantly different than those for which Ti or Zr were used.

The origin of this disparity in magnetic properties can be determined by careful microstructural characterization. Films had a columnar microstructure with a Cr underlayer grain size on the order

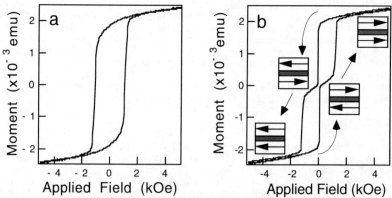

Figure 2. Magnetic hysteresis loops for CoCrNi bilayer films with a)Cr and b)Zr interlayers. The field was applied in the plane and the samples were of equal area. The film with a Cr spacer shows a simple hysteresis loop. The film with a Zr interlayer shows a step loop due to a difference in coercivities between the two layers. The insets show schematically how the directions of the moments in the two layers change with applied field.

of 50 nm. Several differences between films with Cr and Zr or Ti spacers can be seen immediately upon examination of cross-section bright field (BF) TEM images, shown in Figure 3. When a Cr interlayer is used (Fig. 3a) all CoCrNi/Cr interfaces are abrupt, while in the Zr interlayer case (Fig. 3b) distinct interfacial layers were observed at all CoCrNi/Zr boundaries (shown by arrows), but not between the CoCrNi and $Cr_{underlayer}$.

Another interesting difference between the two cases is the presence of similar defect structures in both CoCrNi layers for films with a Cr spacer, an example of which is shown by arrows in Fig. 3a, but not for those with Ti or Zr interlayers. If the CoCrNi grains were to grow in the same orientation on both the bottom and middle Cr layers it would not be unreasonable for a defect structure similar (though not necessarily identical) to that in the first layer to also be generated in the second. This suggests that a strong epitaxial relationship is present throughout a column for films with a Cr interlayer. Films with Zr or Ti spacers, on the other hand, do not show the same effect. In this case, defect structures in the grains in the two magnetic layers are independent, indicating that there is no shared orientation relationship with the interlayer. Additionally, the top media layer as well as the hcp Ti or Zr interlayers often show stacking defects parallel, rather than perpendicular, to the film plane, indicating (0002) (c-axis) growth out of the film plane. Microdiffraction patterns from the interlayer and top CoCrNi layers confirmed the presence of (0002) out-of-plane texture for these cases.

Figure 4 shows cross-section high resolution TEM (HREM) images of the magnetic layers in CoCrNi bilayer films with Cr and Ti spacers. For the Cr interlayer case (Fig. 4a), both CoCrNi/Cr boundaries are sharp with continuous lattice fringes crossing the interface. It can also be seen that the orientation of the top CoCrNi film is identical to that of the bottom one, confirming the presence of a strong epitaxial relationship within the column. Microdiffraction patterns taken from individual grains in the two layers also confirmed this relationship. The fact that the crystallographic orientation of the grains in the CoCrNi films is essentially identical means that the magnetic properties of the two layers should also be nearly identical, leading to the simple hysteresis loop shown in Fig 2a.

Examination of HREM images of films with Ti or Zr spacers, shown in Fig 4b for the Ti case, immediately demonstrates the nature of the observed interfacial layers. As can be seen from the image there is a thin amorphous layer on the order of 1 nm thick at all the CoCrNi/interlayer interfaces. No such layer was observed between CoCrNi and Cr underlayer in these films. The presence of this disordered region is not surprising since Co/Zr as well as many other similar materials systems such as Ni/Zr, Fe/Zr and Co/Gd are known to undergo a solid state amorphization reaction (SSAR) whereby a metastable amorphous phase is formed as an interfacial layer between two materials [7-10]. It has also been found in the Zr/Co system that this amorphization reaction can occur even during deposition, with no annealing [7]. The presence of this amorphous phase results in c-axis growth of the interlayer and CoCrNi film,

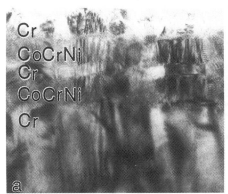
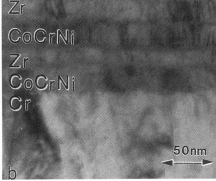

Figure 3. Bright field images of CoCrNi bilayer films with a) Cr and b) Zr interlayers. Note the presence of interfacial layers in the Zr interlayer case that are not present in the Cr interlayer case.

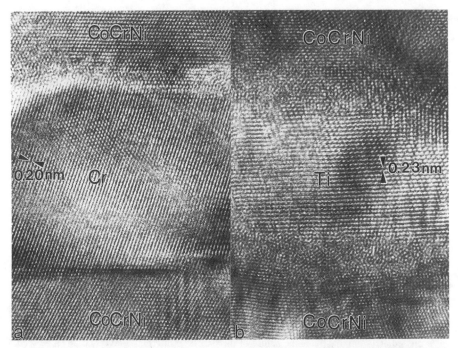

Figure 4. Cross-section HREM images of CoCrNi bilayer films with a)Cr b)Ti interlayers. No interfacial layer is seen in the Cr interlayer case and both Co alloy layers have the same orientation. In films with a Ti or Zr interlayer a thin amorphous interfacial layer is present, resulting in c-axis growth of the interlayer and media .

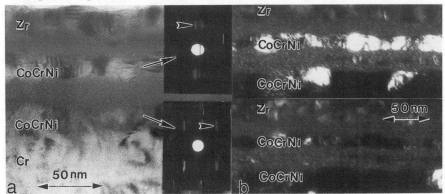

Figure 5. Cross-section a)bright field and b)dark field images of an annealed CoCrNi bilayer film with a Zr interlayer. Annealing led to a significant growth of the amorphous interfacial layer, which was only about 1 nm thick upon deposition. The insets to Fig. 5a are elongated probe microdiffraction patterns from the top and bottom media layers. The dark field images in Fig. 5b were formed using the diffraction spots indicated in the diffraction patterns.

Figure 6. Cross-section HREM image of an annealed CoCrNi bilayer with a Zr interlayer. A solid state amorphization reaction resulted in a thick amorphous layer between the CoCrNi layers. Note that the CoCrNi grains are in different orientations.

since these layers grow on what is essentially an amorphous substrate and hence adopt the most favored growth orientation for this situation, (0002).

The magnetic behavior shown in Fig 2 can now be explained in light of the observed difference in microstructure between films with Cr and Ti or Zr spacers. The coercivity of the bottom CoCrNi layer should be independent of the choice of interlayer, since the processing used to form that part of the film was identical in all cases. If a Cr spacer is used, the crystallographic orientation of the two CoCrNi layers is maintained at all the Cr interfaces, resulting in two independent but identical populations of magnetic grains. If a Ti or Zr spacer is used, the top CoCrNi film adopts an entirely different growth orientation than the bottom, resulting in two independent magnetic films with significantly different coercivities.

Considerable growth of the amorphous layer present in the as-deposited films with Ti and Zr spacers is expected upon annealing due to the continuation of the SSAR begun during deposition. Figure 5a shows a BF image of such a film annealed at 400°C which shows that growth of the amorphous layer occurred, but that neither of the original magnetic layers was entirely consumed. Note that the upper media layer was bordered by Zr at the top and bottom interfaces and therefore reacted more than the bottom layer. Consistent with the lack of amorphization reaction upon deposition, the $CoCrNi/Cr_{underlayer}$ interface in these films remained sharp, with no amorphous phase formation. Elongated probe microdiffraction (EPMD) patterns obtained by intentionally astigmating the condenser lenses in the TEM to result in a thin, long probe can be used to obtain diffraction information from a large number of grains in the various layers of a thin film structure, resulting in a "crystallographic depth profile" [11,12]. EPMD patterns from each of the media layers, shown as insets in Figure 5a, again confirm the orientation difference between the two. For the bottom CoCrNi layer, a significant ($11\bar{2}0$) out of plane component is present, while for the top a (0002) texture is observed. Dark field images taken using the diffracted beams indicated in the EPMD patterns are shown in Figure 5b. The image taken using the (0002) reflection from the top medium layer shows that most of the grains in that layer are strongly diffracting into that reflection, confirming the presence of strong (0002) texture. Some Zr grains in the overlayer are also seen to be strongly diffracting, indicating that they are also (0002) oriented. There are some small Zr grains remaining in the amorphous layer,

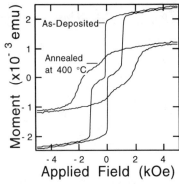

Figure 7. Hysteresis loops for Co alloy bilayer films with a Zr interlayer before and after annealing at 400 °C.

indicating that the amorphization reaction did not go to completion. These grains are roughly in the center of the former Zr layer, indicating that the reaction occurred at both interfaces. The image taken usingdiffracted intensity from the bottom media layer shows a nearly complete absence of intensity in the upper CoCrNi layer. A HREM image showing the detail of this film is shown in Figure 6. The interfacial region is clearly amorphous aside from some small residual Zr grains denoted by arrows. Also apparent in this image is the distinct difference in crystallographic orientation between the Co alloy layers which must necessarily have been present upon deposition.

The SSAR resulting from the annealing led to further changes in the magnetic properties of the Zr and Ti interlayer films. Figure 7 shows the magnetic hysteresis loop of the Zr interlayer film before and after annealing at 400 °C. The saturation magnetization shows a marked decrease due to the SSAR and the step structure present in the unannealed case persisted somewhat after annealing, though it was much reduced.

CONCLUSIONS

The crystallographic and magnetic properties of CoCrNi bilayer films with Cr, Zr, or Ti interlayers were examined using cross-section transmission electron microscopy and vibrating sample magnetometry. Films have columnar grains and in the Cr interlayer case there is layer-to-layer epitaxy throughout individual columns. This structure resulted in films with a simple in-plane hysteresis loop. This epitaxial relationship is absent in films with Zr or Ti interlayers, with the exception of the CoCrNi/Cr$_{underlayer}$ interface. Deposition of the Ti or Zr results in the formation of a thin interfacial amorphous layer due to a solid state amorphization reaction which causes (0002) growth of the top CoCrNi layer. The resulting film consists of two independent magnetic grain populations with significantly different coercivities and a complex step-structure hysteresis loop. Annealing of the Zr interlayer film resulted in an extensive solid state amorphization reaction and a decrease in the saturation magnetization.

REFERENCES

1. E.S. Murdock, B.R. Natarajan and R.G. Walmsley, IEEE Trans. Mag. **26** 2700 (1990).
2. S.E. Lambert, J.K. Howard and I.L. Sanders, IEEE Trans. Mag. **26** 2706 (1990).
3. M.M. Yang et al., IEEE Trans. Mag. **27** 5052 (1991).
4. M.R. Visokay, M. Kuwabara and H. Hayashi, J. Mag. Magn. Mater. **126** 131 (1993).
5. M. Kuwabara et al., J. Appl. Phys. **73** 6686 (1993).
6. H. Yamamoto et al., Mat. Res. Soc. Symp. Proc. **231** 217 (1991).
7. S. Schneider et al., J. Non-Cryst. Sol. **156-158** 498 (1993).
8. H. Schröder and K. Samwer, J. Mat. Res. **3** 461 (1988).
9. B.M. Clemens and R. Sinclair, MRS Bulletin **15** 19 (1990).
10. K. Yamamoto et al. J. Mag. Magn. Mater. **126** 128 (1993).
11. M.A. Parker et al., IEEE Trans. Mag. **27** 4730 (1991).
12. T.P. Nolan, R. Sinclair, R. Ranjan and T. Yamashita, J. Appl. Phys. **73** 5117 (1993).

MAGNETIC AND MAGNETO-OPTICAL PROPERTIES OF SPUTTER-DEPOSITED AND ANNEALED Co-Pt ALLOYS

SUNG-EON PARK*, PU-YOUNG JUNG*, KI-BUM KIM*, SEH-KWANG LEE**, AND SOON-GWANG KIM**

*Department of Metallurgical Engineering, Seoul National University, Seoul, KOREA
**Materials Design Laboratory, Korean Institute of Science and Technology, Seoul, KOREA

ABSTRACT

We have produced $Co_{1-x}Pt_x$ (X = 0.53 and 0.75) alloy films using DC magnetron sputtering and investigated their magnetic properties using vibrating sample magnetometry(VSM) and Kerr hysteresis loop tracer. The as-deposited Co-Pt alloy films show a strong in-plane magnetization. By annealing the alloy samples, we have identified that the magnetic properties are drastically changed. While the magnetic properties of the $Co_{0.25}Pt_{0.75}$ alloy films show no noticeable changes, the coercivity and the squareness of the $Co_{0.47}Pt_{0.53}$ alloy films are drastically increased after annealing. Transmission electron microscopy(TEM) and x-ray diffractometry(XRD) analysis showed that $CoPt(L1_0)$ and $CoPt_3(L1_2)$ ordered phases, respectively, are formed in each case with a strong (111) texture. We suggest that the perpendicular magnetic anisotropy in the Co-Pt system does not depend on the mere textureness of the layer but strongly depends on the arrangement of Co and Pt at an atomic scale.

INTRODUCTION

Co/Pt[1-7] and Co/Pd[2-4,8] multilayers have been intensively investigated as the next generation magneto-optical recording materials due to their advantages such as large Kerr rotation angle at short wavelengths and their anti-corrosive property. However, since these systems are typically composed of extremely thin Co layers of about 3 to 5 Å in between either Pt or Pd whose thickness is in the range of 8 to 20 Å, the thermal stability and the difficulty of manufacturing the samples remain as major disadvantages in these systems. The recent results of Lin et al.[9] and Weller et al.[10-12], demonstrating the perpendicular magnetic anisotroy and large coercivity in the as-deposited Co-Pt alloy films with a similar Kerr rotation angle to the corresponding multilayer systems, thus, brought a great attention in these systems in both practical and scientific reasons. Practically, these results demonstrate that we can prepare the thermally stable magneto-optical recording material with easy and reliable manufacturing method. Scientifically, it invoked an interest in identifying the origin of perpendicular magnetic anisotropy in these alloy systems.

The purpose of this paper is to identify the origin of perpendicular magnetic anisotropy in sputter deposited Co-Pt alloy films. We chose two different compositions of CoPt and CoPt$_3$ to identify the effect of composition on magnetic properties. Also, the samples were annealed in the furnace to observe the resulted changes in the magnetic properties.

EXPERIMENTAL PROCEDURE

The Co-Pt alloy films were sputter deposited using DC magnetron sputtering on both glass and oxidized Si(100) wafers. The background pressure was maintained below 2.0×10^{-6} Torr and Ar was used as a sputtering gas. Deposition was performed at 2,4, and 10 mTorr and substrate temperature was at room temperature, 150°C, and 300°C. $Co_{1-x}Pt_x$ alloy films with x = 0.50 and 0.75 were prepared to investigate the effect of composition on the magnetic properties. The composition of the alloy films were adjusted by varying the number of Pt chips mounted on a 6 inch Co target and the composition of the film was evaluated by using Rutherford backscattering spectrometry(RBS) and energy dispersive spectroscopy(EDS). Heat treatment of the samples was carried out at the annealing temperatures ranged from 400°C to 700°C in a N$_2$ atmosphere. To

387

prevent the oxidation of the films during annealing, 1000 Å of SiO_2 layer was sputter deposited on top of the films before annealing.

The magnetic properties were measured using VSM with magnetic fields up to 15 kOe and Kerr hysteresis loop tracer with AlGaAs solid-state laser at 780 nm wavelength. The structural analysis was carried out by XRD and TEM.

RESULTS AND DISCUSSION

Magnetic Properties of The $Co_{0.25}Pt_{0.75}$ Samples(As-deposited and Annealed)

Figure 1 shows the XRD patterns of the $Co_{0.25}Pt_{0.75}$ alloy films deposited at different Ar pressures at room temperature and 300°C. The data show that, irrespective of the deposition temperature and pressure, a single peak occurs at around $2\theta = 40°$. The d-spacing of this peak is about d = 2.227±0.003Å, which is commonly attributed to the (111) of disordered Co-Pt alloy. From XRD results, we can clearly note that the degree of the (111) texture of the as-deposited film is increased as the Ar pressure is decreased and also as the T_{sub} is increased.

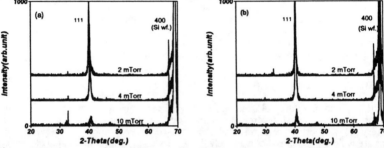

Fig. 1   XRD patterns of $Co_{0.25}Pt_{0.75}$ alloy films,   (a) T_{sub} = r.t., (b) T_{sub} = 300°C

In figure 2, we show the typical M-H hysteresis loops measured at in-plane and out-of-plane direction of the as-deposited $Co_{0.25}Pt_{0.75}$ alloy films deposited at 2 mTorr. It is shown that the coercivity H_c of the as-deposited film is always less than 200 Oe and easy magntization axis lies in-plane. Such a small H_c and in-plane magnetization were shown irrespective of the deposition pressure. Neither the deposition temperature or Kerr loop tracer data is shown in here, but no differences are observed in these samples.

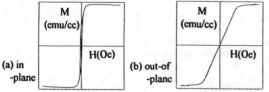

Fig. 2   The hysteresis loops
of the as-deposited
$Co_{0.25}Pt_{0.75}$ alloy films
(M_s = 304 emu/cc,
Saturation field is 10 kOe.)

Our results clearly show that in the sputtered alloy films of $Co_{0.25}Pt_{0.75}$, the deposition conditions such as the sputtering pressure and the substrate temperature, which affect the (111) texture of the film, do not induce any changes on the magnetic properties of these films. Our results thus appear contrary to the results of Lin et al.[9] and Weller et al.[10-12] who obtained a strong perpendicular magnetic anisotropy in the electron beam deposited $Co_{0.25}Pt_{0.75}$ alloy films. They indicated that the key to establishing a perpendicular anisotropy and a suitable coercivity is the slightly elevated deposition temperature to about 200~300 °C, and reported that perpendicular anisotropy seems to be related with a control of the crystallographic orientation of the as-deposited films to have a strong (111) texture. We note that the perpendicular anisotropy is also observed in sputtered films by Shoimi et al.[15] in the composition of around $Co_{0.6}Pt_{0.4}$. While they identified that magntic anisotropy in the Co-Pt alloy films is strongly dependent on the

Pt underlayer and DC bias, they also failed to identify that (111) texture is the sole contributor to the magnetic anisotropy. Indeed, it should be noted that they didn't observe magnetic anisotropy in these samples when the composition is close to $CoPt_3$.

To identify the effect of heat-treatment on the magnetic properties, we annealed the sample at 600 °C for up to 14 hours. There are no drastic changes in the magnetic properties after annealing. Figure 3 shows the XRD results of the as-deposited and annealed samples. No notable changes are observed in the XRD results except that all the diffraction peaks from the film moves to slightly higher angle indicating that the d-spacings of the film becomes smaller. Again we note that a strong (111) texture is maintained. By annealing, we note that (222) spacing of Co-Pt alloy changes from 1.120 Å to 1.103 Å which is similar to the (222) planar spacing of $L1_2$ ordered phase $CoPt_3$ (d_{222} = 1.106 Å). However, we can not conclude that ordered phase is formed by the annealing since this value is also close to the (222) planar spacing of the disordered $CoPt_3$ calculated from the Vegard's rule(d_{222} = 1.106 Å).

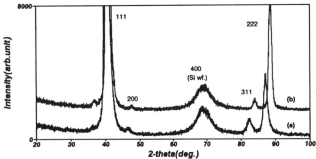

Fig. 3   XRD patterns of $Co_{0.25}Pt_{0.75}$ alloy films, (a) as-deposited, (b) 600°C 1hr. annealed

To verify the structural changes after annealing, we investigated the microstructural and phase changes of this sample by TEM(Fig. 4). Figure 4-a shows the bright field image and the selected area diffraction pattern of the as-deposited $Co_{0.25}Pt_{0.75}$ alloy film. The bright field micrograph clearly reveals an uniform distribution of grains with the size of about 200 to 300 Å. The selected area diffraction pattern shows ring-patterns of the FCC phase with an average lattice parameter of about 3.85 Å which is somewhat larger than the value calculated using Vegard's rule(a = 3.731 Å). Figure 4-b shows the bright field image and the selected area diffraction pattern of the $Co_{0.25}Pt_{0.75}$ alloy film annealed at 600 °C for one hour. In the bright field image, we note that the grain size distribution is not uniform. There exists large grains of about 2000 to 4000 Å in between the small grains of about 500 to 600 Å . The selected area diffraction pattern in Fig. 4-b shows the (110) ring patterns inside that of the (111). The occurrence of (110) ring pattern indicate that the ordered phase is formed by the annealing. The formation of ordered phase is also shown in the microdiffraction pattern(Fig. 4-c) where the weak (100) pattern spots which should not be shown in disordered FCC phase are shown. While EDS was attempted to identify any concentration diffrences in these two different sizes of grains, no differences were identified. We certainly believe that additional work is necessary to identify the origin of the formation of this non-bimodal distribution of the grains.

Our results thus clearly show that both the as-deposited and annealed samples of the $Co_{0.25}Pt_{0.75}$ alloy films do not show the perpendicular anisotropy irrespective of the deposition pressure and the deposition temperature in the ranges described above even though the film shows a strong (111) texture throughout the experiment. We also noted that the annealing of the samples to form an ordered phase of $CoPt_3$ does not induce a perpendicular anisotropy.

In view of our results, we are not clear how Weller et al.[10-12] and Lin et al.[9] obtained the perpendicular magnetic anisotropy in the composition of $CoPt_3$ by electron beam deposition. We have to stress that our films are deposited by DC sputtering system compared to their films deposited by electron beam deposition. However, we can certainly claim that the mere (111) textureness of the samples is not the main cause of the perpendicular magnetic anisotropy in the sputter deposited samples.

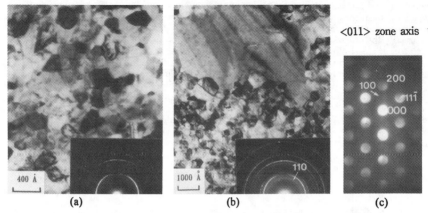

<011> zone axis

Fig. 4   TEM bright field images and diffraction patterns of $Co_{0.25}Pt_{0.75}$ alloy films,
(a) as-deposited, (b) 600°C 1hr. annealed and its microdiffraction patterns (c)

Magnetic Properties of The $Co_{0.47}Pt_{0.53}$ Samples(As-deposited and Annealed)

In order to study the effect of composition on the magnetic properties in the Co-Pt alloy film, we have sputter-deposited $Co_{0.47}Pt_{0.53}$ alloy films and annealed at 400 to 700 °C for up to 14 hours. Figure 5 shows the change of the H_c and the squareness defined as M_r/M_s in both parallel and perpendicular direction of the film. We again note that an in-plane anisotropy is present in the as-deposited sample and in the low temperature annealed samples. However, as the annealing temperature is increased, both the H_c and the squareness are incresed. The films show a parallel magnetization at low temperature, but not a simple parallel magnetization at high annealing temperatures. The rather drastic changes occurs at about 500°C. Torque measurements of the samples again clearly reveals these changes with the annealing temperature(Fig. 6). The easy axis of the sample annealed below 500 °C exists at 0, 180, and 360 ° but the location of easy axis isn't clear for the samples annealed above 500 °C.

Figure 7 shows the XRD patterns of the $Co_{0.47}Pt_{0.53}$ samples annealed at different temperatures. The lattice parameter measured from the XRD pattern is about a = 3.793 Å which is slightly bigger than the number calculated by the Vegard's rule(a = 3.749 Å). The as-deposited sample shows a strong (111) peaks with other minor peaks at (200), (311), (222) indicating the formation of a FCC phase with a strong (111) texture. Thus, we do know that the as-deposited sample forms a disordered FCC type Co-Pt phase. With annealing below 500 °C, all the peaks move slightly to the higher angles indicating the formation of a higher density of the film. However, annealing above 500 °C does not cause any shifts in the peak position. Instead, we note that the (200) and (311) peaks are separated into two peaks. This peak separation indicates the formation of tetragonal ordered phase of CoPt $L1_0$ and, can be indexed as (200) and (002) and (311) and (113), respectively.

The formation of ordered phase by the annealing above 500°C can be confirmed by the TEM investigation. Figure 8-a shows the bright field image and the selected area diffraction pattern of the as-deposited sample showing the uniform distribution of the grains of about 200 to 300 Å and a disordered FCC ring pattern. Figure 8-b shows the bright field image and both the selected area diffraction and microdiffraction patterns of the annealed sample(700°C, 1hr.). The occurrence of the forbidden (110) ring in the selected area diffraction pattern and the (110) spots in the microdiffraction pattern are evidence of the ordered phase formation(Fig. 8-c).

Out results thus clearly shows that the magnetic properties are noticeably changed for the $Co_{0.47}Pt_{0.53}$ samples with annealing by the formation of $L1_0$ ordered CoPt phase. This result is quite contrary to the case of $Co_{0.25}Pt_{0.75}$ where the formation of $L1_2$ ordered phase does not induce any changes in the magnetic properties. It has been well known that there is no magnetic easy axis in CoPt$_3$ ordered phase, and the shape anisotropy always dominates in determining the location of easy axis in thin films with cubic symmetry which is in-plane. However, for a CoPt ordered phase, the magnetocrystalline anisotropy effect is stronger than the shape anisotropy and the magnetic easy axis lies along the [001] direction of the grains. Indeed, the recent result of Lairson et al.[16-17] who obtained a strong perpendicular magnetic anisotropy from the (001) textured Co$_{0.5}$Pt$_{0.5}$ alloy film suggests that the magnetocrystalline anisotropy effect is dominant in CoPt ordered phase.

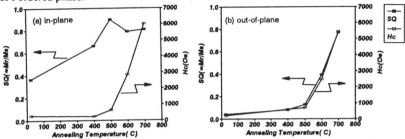

Fig. 5   The change of H$_c$ and squareness vs. annealing temperature in Co$_{0.47}$Pt$_{0.53}$ alloy films,   (a) in-plane   (b) out-of-plane

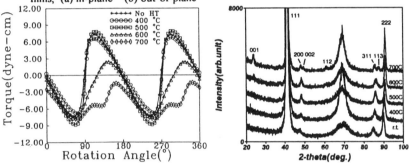

Fig. 6   The change of torque curves as annealing temperature

Fig. 7   XRD patterns of Co$_{0.47}$Pt$_{0.53}$ alloy films

Fig. 8   TEM bright field images and diffraction patterns of Co$_{0.47}$Pt$_{0.53}$ alloy films, (a) as-deposited, (b) 700°C 1hr. annealed and its microdiffraction patterns (c)

It is interesting to note the atomic arrangement in the CoPt and $CoPt_3$ ordered phases. In $CoPt_3$ phase, there does not exist any planes of Co in [110] projection. Co and Pt atoms are intermixed in any of the crystallographic planes. On the contrary, in the CoPt ordered phase, there exists separate Co planes and Pt planes of (001) in [110] projection. In other words, if we assume that there is an exchange induced magnetic moment between Co and Pt, then we can easily anticipate the formation of a strong magnetic anisotropy to the [001] direction. An interaction between a separate Co and Pt layer to induce a magnetization can be an explanation of why the CoPt ordered phase have a perpendicular anisotropy, like Co/Pt multilayer system as has been studied[1-7]. In this respect, we can also explain why we do not form a perpendicular magnetic anisotropy in the $Co_{0.47}Pt_{0.53}$ alloy samples. Our sample shows a strong (111) texture to the perpendicular direction of the sample surface, the easy axis [001] lies with an oblique angle about 54° to the film surface. Therefore, even though the perpendicular magnetic term is increased by the formation of $L1_0$ phase, we do not obtain a pure perpendicular magnetization.

CONCLUSION

In summary we have deposited the Co-Pt alloy films by DC magnetron sputtering at different sputtering pressures and at different substrate temperatures and investigated the relationships between the magnetic properties and the microstructures of the films. We have shown that the development of (111) texture is not the main cause of the formation of the perpendicular magnetic anisotropy in these samples. We have identified that the formation of $CoPt_3$ ordered phase does not induce a notable change in the magnetic properties of the film while the formation of CoPt ordered phase drastically induces a large magnetic term to the [001] direction in the film. The differences of the magnetic properties of $CoPt_3$ and CoPt ordered phase are caused by the atomic arrangement, namely, the existence of a distinct Co layer in between the Pt layer. We conclude that a strong magnetic anisotropy in the Co-Pt system is caused by the interaction between Co and Pt.

REFERENCES

[1] W. B. Zeper, F. J. A. M. Greidanus, P. F. Carcia
    IEEE Trans. Magn., 23 No. 3 3764 (1989)
[2] S. Hashimoto, Y. Ochiai and K. Aso, J. Appl. Phys. 66(10), 15 Nov. 4909 (1989)
[3] Noboru Sato, J. Appl. Phys., 64(11), 1 Dec. 6424 (1988)
[4] S. Hashimoto, Y. Ochiai and K. Aso, J. Appl. Phys., 67(9), 1 May 4429 (1990)
[5] Z. G. Li and P. F. Carcia, J.Appl. Phys., 71(2), 15 Jan. 842 (1992)
[6] Ping He, William A. Mcgahan, S. Nafis, John A. Wollam,
    Z. S. Shan, and S. H. Liou, J. Appl. Phys., 70(10), 15 Nov. 6044 (1991)
[7] C. J. Chien, R. F. C. Farrow, C. H. Lee, C.-J. Lin and E. E. Marinero
    J. Magn. Magn. Mat. 93 (1991) 47-52
[8] Douglas G. Stinson and S.-C. Shin, J. Appl. Phys. 67(9), 1 May 4459 (1990)
[9] C.-J. Lin and G. L. Gorman, Appl. Phys. Lett. 61(13), 28 Sep. 1600 (1992)
[10] D. Weller, H.Brandle, G. Gorman, C.-J. Lin and H. Notorys
    Appl. Phys. Lett. 61 (22), 30 Nov. 2726 (1992)
[11] D. Weller, H. Brandle and C. Chappert, J. Magn. Magn. Mat. 121 (1993)
[12] D. Weller, C. Chappert, H.Brandle, G. Gorman, R. F. C. Farrow,
    R. Marks, and G. Harp, MORIS '92 Proc. Dec.7-9
[13] D. Treves, J. T. Jacobs and Sawatzky, J. Appl. Phys. 46 No. 6, June 2760 (1975)
[14] S. Tsunashima, K. Nagase, K. Nakamura and S. Uchiyama
    IEEE Trans. Magn., 25 No.5 3761 (1989)
[15] S. Shiomi, H. Okazawa, T. Nakakita, Jpn. J. Appl. Phys. vol. 32 L315 (1993)
[16] B. M. Lairson, M. R. Visokay, E. E. Marinero, R. Sinclair and B. M. Clemens,
    J. Appl. Phys. 74(3), 1 August 1992 (1993)
[17] B. M. Lairson, M. R. Visokay, R. Sinclair, S. Hagstrom, and B. M. Clemens
    Appl.Phys.Lett., 61(12) 21 Sep. (1992)

MAGNETIC X-RAY CIRCULAR DICHROISM
IN FE CO PT MULTILAYERS

J.G. TOBIN* A.F. JANKOWSKI*, G.D. WADDILL*, P.A. STERNE**,*
*Lawrence Livermore National Laboratory, Livermore, CA 94550
**University of California, Davis, CA 95616

ABSTRACT

Magnetic x-ray circular dichroism in x-ray absorption has been used to investigate the ternary multilayer system, Fe Co Pt. Samples were prepared by planar magnetron sputter deposition and carefully characterized, using a variety of techniques such as grazing-incidence and high-angle x-ray scattering, Auger depth profiling and cross-section transmission electron microscopy. As previously reported, the Fe9.5Å Pt9.5Å exhibits a large dichroism in the Fe 2p absorption. Interestingly while the Co9.5Å Pt9.5Å has no measurable dichroism, the Fe4.7Å Co4.7Å Pt9.5Å sample has a dichroism at both the Fe 2p and Co 2p absorption edges. These and other results will be compared to slab calculation predictions. Possible explanations will be discussed.

INTRODUCTION

The last several years have witnessed a massive growth in the research and development of nanoscale magnetic materials. Perhaps the best review is provided by the Falicov Report[1] on "Surface, Interface, and Thin-Film Magnetism." Three general lessons can be derived from this report: (1) Magnetism is one of those special cases where fundamental research can directly lead to technological applications; (2) The key to understanding and manipulation of magnetic properties is the subtle yet overwhelming interplay of atomic geometric structure and local magnetic properties. For example, the giant magneto-resistance effect (GMR), which is already being explored for technological exploitation[2,3,4], appears to be intimately coupled to interfacial and thin film effects and probably will require elementally-specific probes for an explicit determination of the underlying causes[5,6,7]. This also appears to be the case for spin valves[8,9,10], another source of device miniaturization in read heads and magnetic sensors. [While it may eventually be found that these two effects are fundamentally connected, for now it appears that the GMR effect (up to 60%) is dependent upon an anti-ferromagnetic coupling through a non-ferromagnetic layer while the spin valve effect ($\leq$ 10%) is associated with an uncoupled ferromagnetic layer[9], which can be controlled externally.]; (3) The importance of probes with a direct spin-dependence. A very recent illustration of this is the development of the magnetic x-ray circular dichroism (MXCD) using x-ray absorption[11-15] and photoemission[16,17] as a probe of surface, monolayer, and multilayer magnetism. It is this advantage of elemental selectivity and spin specificity from MXCD x-ray absorption that we have to utilized, coupled to extensive structural characterization using techniques such as x-ray diffraction and transmission electron microscopy[18], plus spin-dependent slab calculations[19], to investigate Fe Co Pt magnetic multilayers.

SAMPLE PREPARATION AND CHARACTERIZATION

The Fe Co Pt multilayer samples are prepared using magnetron sputter deposition. The deposition chamber is cryogenically pumped to a base pressure of 1.3×10^{-5} Pa. A circular array of magnetron sources is situated 20 cm beneath an oxygen-free copper platen. The magnetron sources are operated in the dc mode at a 330–390 Volt discharge. An argon working gas pressure of 0.40 Pa is used at a flow rate of 15.5 cc min.$^{-1}$. The substrates are sequentially rotated over each source at 1.0 rev. min.$^{-1}$. The target materials are > 0.9994 pure. The polished Si substrates are cleaned with a procedure consisting of a detergent wash, deionized water rinse, alcohol rinse, and a N_2 gas drying prior to deposition. The substrates remain at a temperature between 293 and 306 K during the deposition. The sputter deposition rates, between 0.02 and 0.50 nm sec.$^{-1}$, are monitored using calibrated quartz crystals. The quartz crystals indicate the component layer thicknesses and the layer pair thicknesses, d_{FeCoPt}^{XTC}. The multilayer films are grown to a 0.2 μm thickness. Samples were futher characterized with x-ray diffraction, transmission electron microscopy and Auger depth profiling, as described elsewhere[18,20,21].

MXCD AND SLAB CALCULATION RESULTS

The MXCD measurements were performed at Stanford Synchrotron Radiation Laboratory (SSRL) on a spherical grating monochromator having the ability to generate soft x-rays with a high degree of circular polarization. This beamline (BL 8–2) is part of the UC/National Laboratories facilities at SSRL[22,23]. The absorption measurements were made in a total electron yield mode. Samples were magnetized *in situ* with a pulse coil, and the absorption was measured in remanence. For $3d$ transition metals, MXCD in x-ray absorption is observed as a polarization-dependent intensity variation in the L_2 and L_3 edges. A typical example of the Fe Co Pt multilayers is shown in Figure 1. The polarization dependence requires that the incident x-ray helicity (either parallel or anti-parallel to the direction of propagation) be aligned or anti-aligned with the sample magnetization[18,20,21]. When these vectors are perpendicular, the polarization dependence vanishes. The spectra in Figure 1 are for a grazing x-ray incidence angle of 80° from the sample normal. The solid curve is for a nearly anti-parallel arrangement of x-ray helicity and majority electron spin, and the dashed curve for a nearly parallel geometry. The intensity differences for for the L_2 and L_3 white lines are apparent. Our measurements demonstrate that the magnetization for the Fe Co Pt samples is in the plane of the multilayer films. It has also been shown that the relative strengths of the L_2 and L_3 absorption edges contain information about the spin-dependent density of states above the Fermi level and, therefore, about the spin and orbital magnetic moments of the material[18,20,21]. Interestingly in the case of Figure 1, there is an observable MXCD effect at both the Fe 2p and Co 2p edges. Preliminary analysis indicates for $Fe_x Co_y Pt_{9.5Å}$, little or nor MXCD effect if x=0 and $0 \leq y \leq 10$Å or if $0 < x < 4$Å and 2Å$\leq y \leq 4$Å. A significant MXCD effect is observable for y = 0 and 5Å$\leq x \leq 12$Å or 4Å$< x$ and 4Å$< y$. Further analysis, using branching ratio and sum rule approaches, is in progress[18,20,21].

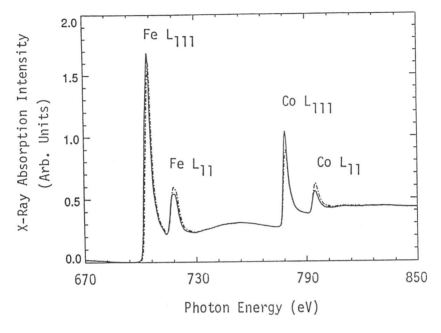

Figure 1. The x-ray absorption spectra of $Fe_{4.7Å}$ $Co_{4.7Å}$ $Pt_{9.5Å}$ using circularly-polarized x-rays. See text for details.

This raises an important question: Is the observation of an MXCD effect a true moment transfer from Fe to Co or only a proximity effect upon the Co from the nearby polarized Fe conduction states? To address this issue, we have begun a modeling of the system using a slab calculational method[19]. Preliminary results are shown in Figure 2. Here the structures are all fcc with a = 3.81Å, extracted from x-ray diffraction results. In this case large moments are seen for both Fe and Co in both the binary (x = 0 or y = 0) and ternary cases. Further calculations with adjustments of spacings will be necessary before any additional insight can be gained.

MXCD in Fe Co Pt Multilayers
Preliminary Slab Calculation Results
Magnetic Moments (in Bohr magnetons)

----	Pt	0.15		----	Pt	0.16
----	Pt	0.007		----	Pt	0.01
----	Pt	0.15		----	Pt	0.01
0000	Fe	2.82		----	Pt	0.16
0000	Fe	2.78		****	Co	1.92
0000	Fe	2.82		****	Co	1.88
				****	Co	1.88
				****	Co	1.92

----	Pt	0.15		----	Pt	0.16
----	Pt	-0.04		----	Pt	0.00
----	Pt	-0.04		----	Pt	0.02
----	Pt	0.15		----	Pt	0.15
0000	Fe	2.84		0000	Fe	2.83
0000	Fe	2.78		0000	Fe	2.77
0000	Fe	2.78		****	Co	1.85
0000	Fe	2.84		****	Co	1.94

Figure 2. Slab calculation results for $(Fe_{3ML}\ Pt_{3ML})n$, $(Fe_{4ML}\ Pt_{4ML})n$, $(Co_{4ML}\ Pt_{4ML})n$ and $(Fe_{2ML}\ Co_{2ML}\ Pt_{4ML})n$. ML stands for monolayer. 1 ML $\approx$ 2.2Å in these systems.

SUMMARY

We are using a combined approach based upon magnetic x-ray circular dichroism with x-ray absorption, structural characterization and theoretical simulations with slab calculations to probe the structure-property relationships in magnetic Fe Co Pt multilayers. Thickness dependences, template effects, and interfacial mixing are all crucial contributions that need to be isolated, controlled and understood. The issue of moment transfer versus proximity conduction band polarization is being investigated further.

ACKNOWLEDGEMENTS

Work was performed under the auspices of the US Department of Energy by the Lawrence Livermore National Laboratory under contract number W-7405-ENG-48. We wish to thank Karen Clark for her clerical support of this work. Conversations with Jim Brug were greatly appreciated.

REFERENCES

1. L.M. Falicov, D.T. Pierce, S.D. Bader, R. Gronsky, K.B. Hathaway, H.J. Hopster, D.N. Lambeth, S.S.P. Parkin, G. Prinz, M. Salamon, I.K. Schuller, and R.H. Victora, J. Mat. Res. 5, 1299 (1990).

2. G. Avalos, San Ramon Valley Times, Friday, August 20, 1993.

3. R. Pool, Science 261, 984 (20–AUG–1993).

4. T.L. Hylton, K.R. Coffey, M.A. Parker, and J.K. Howard, Science 261, 1021 (20–AUG–1993).

5. S.S.P. Parkin, Phys. Rev. Lett. 71, 1641 (1993).

6. A.C. Ehrlich, Phys. Rev. Lett. 71, 2300 (1993).

7. V. Grolier, D. Renard, B. Bartenlian, P. Beauvillian, C. Chappert, C. Dupas, J. Ferre, M. Galtier, E. Kolb, M. Mulloy, J.P. Renard, and P. Veillet, Phys. Rev. Lett. 71, 3023 (1993).

8. B.A. Gurney, V.S. Speriosu, J.P. Nozieres, H.F. Lefakis, D.R. Wilhoit, and D.U. Need, Phys. Rev. Lett. 71, 4023 (1993).

9. B. Dieny, V.S. Speriosu, S. Metin, S.S.P. Parkin, B.A. Gurney, P. Baumgart, and D.R. Wilhoit, J. Appl. Phys. 69, 4774 (1991).

10. B. Dieny, V.S. Speriosu, S.S.P. Parkin, B.A. Gurney, D.R. Wilhoit, and D. Mauri, Phys. Rev. B 43, 1297 (1991).

11. G. Schutz, W. Wagner, W. Wilhelm, P. Keinle, R. Zeller, R. Frahm, and G. Materlik, Phys. Rev. Lett. 58, 737 (1987); G. Schutz, M. Knulle, R. Wienke, W. Wilhelm, W. Wagner, P. Kienle, and R. Frahn, Z. Phys. B 73, 67 (1988); G. Schutz, R. Frahm, P. Mautner, R. Wienke, W. Wagner, W. Wilhelm, and P. Kienle, Phys. Rev. Lett. 62, 2620 (1989).

12. C.T. Chen, F. Sette, Y. Ma, and S. Modesti, Phys. Rev. B 42, 7262 (1990); C.T. Chen, Y.U. Idzerda, H.J. Lin, G. Meigs, A. Chaiken, G.A. Prinz, and G.H. Ho, Phys. Rev. B 48, 642 (1993).

13. J.G. Tobin, G.D. Waddill, and D.P. Pappas, Phys. Rev. Lett. 68, 3642 (1992).

14. Y. Wu, J. Stohr, B.D. Hermsmeier, M.G. Samant, and D. Weller, Phys. Rev. Lett. 69, 2307 (1992).

15. J. Stohr, Y. Wu, B.D. Hermsmeier, M.G. Samant, G.R. Harp, S. Koranda, D. Dunham, and B.P. Tonner, Science 259, 658 (29–JAN–1993).

16. L. Baumgarten, C.M. Schneider, M. Petersen, F. Schafers, and J. Kirschner, Phys. Rev. Lett. 65, 492 (1990).

17. G.D. Waddill, J.G. Tobin, and D.P. Pappas, Phys. Rev. B <u>46</u>, 552 (1992).

18. G.D. Waddill, J.G. Tobin, and A.F. Jankowski, J. App. Phys. <u>74</u>, 6999 (1993).

19. Hans L. Skriver, LMTD Method, Springer-Verlag, Berlin, 1984.

20. A.F. Jankowski, G.D. Waddill, and J.G. Tobin, Mat. Res. Soc. Symp. Proc. <u>313</u>, 227 (1993).

21. A.F. Jankowski, G.D. Waddill, and J.G. Tobin, J. Vac. Sci. Tech. A, (May 1994).

22. L.J. Terminello, G.D. Waddill, and J.G. Tobin, Nuc. Instrum. Meth. <u>A319</u>, 271 (1992).

23. K.G. Tirsell and V. Karpenko, Nuc. Instrum. Meth. <u>A291</u>, 551 (1990).

GMR, STRUCTURAL, AND MAGNETIC STUDIES IN NiFeCo/Ag MULTILAYER THIN FILMS

J.D. JARRATT AND J.A. BARNARD
Department of Metallurgical and Materials Engineering, The University of Alabama,
Tuscaloosa, Alabama 35487-0202.

ABSTRACT

Giant Magnetoresistance (GMR), crystal structure, and magnetic properties
have been investigated in a series of sputtered $Ni_{66}Fe_{16}Co_{18}$/Ag multilayer films
with induced uniaxial anisotropy. The film thickness ranges studied were 20 and 25
Å for NiFeCo, and 25 to 50 Å for Ag. GMR was only evident in the films after post-
deposition annealing. This onset of GMR is thought to be due to the breaking up of
the NiFeCo layers into ferromagnetic platelets or islands by the immiscible Ag
diffusing perpendicular to the film plane along the grain boundaries. The magnitude
and field sensitivity of the GMR was dependent on the annealing time and
temperature. High angle x-ray diffraction (HXRD) was used to reveal the overall film
structure and growth texture and low angle XRD (LXRD) was used to investigate the
quality of the multilayer structures. M-H hysteresis loops revealed in-plane uniaxial
anisotropy in as-deposited films which is eventually eliminated with annealing. The
easy axis squareness experiences a pronounced decrease with lower temperature
annealing, but then increases slightly with annealing temperature.

1. INTRODUCTION

It has recently been reported that highly field sensitive GMR is observed in
annealed NiFe/Ag multilayer films that show no as-deposited GMR [1]. This system
has produced the most sensitive MR values to date (1.2 %MR/Oe). The formation of
ferromagnetic platelets due to Ag diffusion on annealing is thought to be the cause of
the GMR. These multilayers can be termed 'discontinuous'. The platelet size and
their intralayer spacing allows for magnetization changes to occur in very small
fields. A very recent theoretical treatment of this phenomenon concludes that the
magnetostatic coupling between adjacent layer platelets and the spacing between
intralayer platelets is critical in achieving high MR magnitude and highly field
sensitive films [2]. A gap of 10 Å between intralayer platelets has been calculated to
give the highest MR magnitude and a gap of 6 Å has been calculated to result in the
maximum field sensitivity.

Earlier work in the NiFe/Ag multilayer system reports GMR in as-deposited
films of NiFe 12.5 Å/Ag 10.5 Å that exhibit a decrease in magnitude but an increase
in sensitivity upon annealing [3]. Their GMR mechanism was attributed to exchange
coupling between ferromagnetic layers [4]. This argument is supported by the
extreme sensitivity of the MR magnitude to Ag spacer layer thickness in the range
from 8 to 14 Å. Another report on NiFe/Ag multilayers revealed GMR in samples of
NiFe 350 Å/Ag 50 Å that were first annealed in oxygen to promote mixing and then
annealed in hydrogen to remove the oxygen species and leave behind Ag
precipitates in the NiFe layers [5].

We report here the first work on the $Ni_{66}Fe_{16}Co_{18}$/Ag multilayer system that
exhibits GMR only after annealing. We attribute this onset of GMR to the formation of
discontinuous NiFeCo layers by Ag diffusion, similar to the mechanism reported

Mat. Res. Soc. Symp. Proc. Vol. 343. ©1994 Materials Research Society

earlier [1]. The soft ternary alloy $Ni_{66}Fe_{16}Co_{18}$ has been shown to display very low magnetostriction and magnetocrystalline anisotropy and exhibits GMR when multilayered with Cu [6]. In this lab, highly field sensitive and low hysteresis GMR has been reported in the $Ni_{66}Fe_{16}Co_{18}$/Cu multilayer system [7].

2. EXPERIMENTAL METHODS

These films were DC magnetron sputter deposited at 100 W in 2 mTorr of ultra-high purity argon in a Vac-Tec Model 250 side sputtering system with a base pressure of 3×10^{-7} Torr onto Corning 7059 glass substrates. The geometry of the films is illustrated in Figure 1 and follows the arrangement suggested in [1]. The final Ag layer has been split in half to surround the NiFeCo layers allowing for even Ag diffusion through each NiFeCo layer. For the five bilayer samples n is four. A permanent magnet was positioned behind each substrate providing a 90 Oe parallel field resulting in in-plane uniaxial anisotropy in as-deposited films. The sputtering rates were determined from reference film step heights measured on a Dek-Tak IIA surface profilometer.

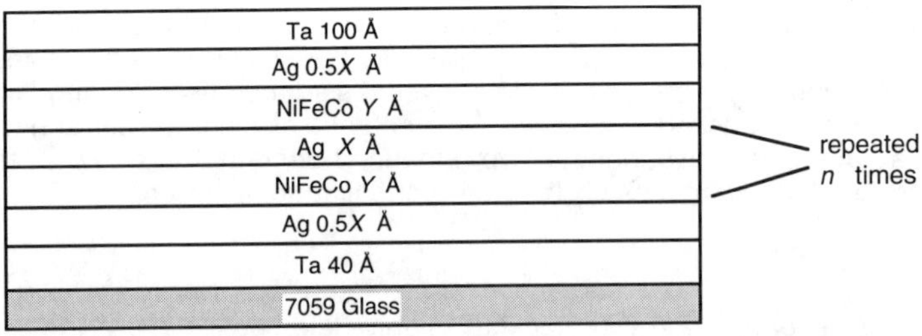

Fig. 1. Sputtered film configuration; $Ni_{66}Fe_{16}Co_{18}$/Ag multilayers.

Magnetic properties were measured on a Digital Measurement Systems VSM Model 880. X-ray diffraction was performed on a Rigaku D/Max-2BX XRD System with thin film attachment using Cu Kα (λ=1.54178 Å) radiation. The reported MR measurements were made by a 4-point probe assembly with the current and easy axis of the film both perpendicular to the applied magnetic field (i.e., hard-axis MR loops). Anisotropic MR (AMR) values were deduced by comparison with MR measurements made with the current and easy axis both parallel to the applied field (easy-axis MR loops). The annealing was done in a 1×10^{-6} Torr lamp annealing system. It took approximately 90 sec to reach the annealing temperatures and 1 hour to return to room temperature once the lamps were shut off. A magnetic field of 50 Oe was maintained in the films' easy axis direction during the anneal using a permanent magnet.

3. EXPERIMENTAL RESULTS

Figure 2 shows the hard-axis MR loops given by $d\rho/\rho=[\rho(H)-\rho(H_{sat})]/\rho(H_{sat})$ for an as-deposited sample and a series of annealing temperatures (10 min anneals, each conducted on a fresh sample). For clarity, only one half of each loop is shown. We observe no measurable MR in the as-deposited state. An initial MR peak is formed after annealing at 250 °C. A broader peak was found at 260 °C. Annealing

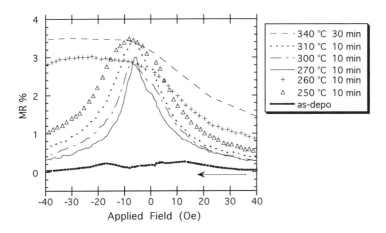

Fig 2. Magnetoresistance as a function of annealing conditions
for a (NiFeCo 25 Å/Ag 50 Å)x5 sample.

at 270 °C produced the sharpest MR profile. Higher temperature annealing gradually broadens the profile. The curious inversion in the 250 and 260 °C behavior was found for different Ag thicknesses. The general behavior was seen in both the 20 and 25 Å NiFeCo layer thickness samples with 5 NiFeCo layers.

Generally speaking, higher sensitivity but lower magnitude MR profiles were obtained in the 5 bilayer sample. Typical values after annealing at 270 °C are ~3% MR in a FWHM of 15 Oe giving a sensitivity of 0.2 %/Oe. The most sensitive hard-axis MR obtained was 0.26 %/Oe (2.94% in a FWHM of 11.16 Oe) for a (NiFeCo 25 Å/Ag 45 Å)x5 sample annealed at 270°C for 10 min. Over 4% GMR was observed after annealing ten and seven bilayer samples (NiFeCo 25 Å / Ag 50 Å) which saturated in 60~100 Oe. The extremely sensitive dependence on annealing temperature is in agreement with the results on NiFe/Ag [1]. Similar results were found in samples with 45 Å Ag spacers. Thinner Ag layer samples also revealed post-annealed GMR, but not with the field sensitivity and magnitude of the thicker Ag layer samples. The hysteretic nature of the MR loops is apparent. We find that the minimum hysteresis corresponds with maximum sensitivity. We also note that there was a general decrease in the films' overall resistivity, ρ, of ~30% with any annealing. Higher annealing times and temperatures were required to achieve the maximum GMR in the higher numbered bilayer systems, but lower MR sensitivity resulted. AMR values of ~0.1% were seen in these samples after easy-axis MR measurements were made. These profiles revealed higher field sensitivity than hard-axis profiles, but are not discussed here (this configuration is not conventionally used in MR head design).

The HXRD data (Fig. 3a) revealed a strong Ag (111) growth texture in the as-deposited state, but no peak splitting. A more complex XRD structure (characteristic of a layered structure) develops on annealing. The development of peak splitting in the high angle spectra only after annealing is surprising and may be due to the reduction of defects (associated with the sputtering process) present in the as-

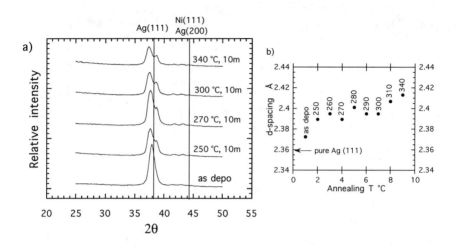

Fig. 3. a) High angle x-ray diffraction data and b) 'Ag(111)' d-spacing as a function of annealing temperature for (NiFeCo 25 Å /Ag 50 Å)x5 (10 min anneals).

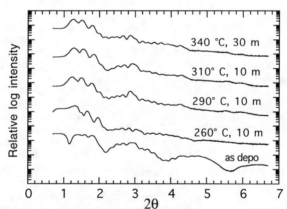

Fig. 4. Low angle x-ray diffraction data as a function of annealing conditions for (NiFeCo 25 Å/Ag 50 Å)x5.

deposited state. The as-deposited Ag(111) peak clearly splits in two with the dominant component shifting systematically to lower 2θ values on annealing. The d-spacing associated with this shifting peak is plotted as a function of annealing in Fig 3b. The angular position of the shoulder is stationary within experimental error.

As-deposited LXRD (Fig. 4) shows both the 'coarse' (low order Bragg) and 'fine' (Keissig fringe) structure features associated with a periodic composition modulation and finite total film thickness effects, respectively. Annealing at 260 °C reduces the strength and number of coarse structure LXRD diffraction peaks; the quality of the layering deteriorates with the initial low temperature annealing. This deterioration corresponds also to the onset of GMR and the presumed diffusion of the Ag through the NiFeCo layer to form a discontinuous but still layered structure.

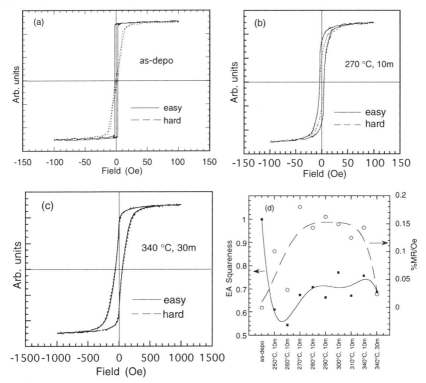

Fig. 5(a-c). The effect of annealing on M-H loops for (NiFeCo 25 Å/Ag 50 Å)x5 films; (d) easy axis squareness and MR sensitivity vs annealing treatment. (The curves are guides for the eye)

The M-H hysteresis loops exhibit well defined uniaxial anisotropy as-deposited that is reduced with annealing and finally eliminated after longer annealing times and higher temperatures (Fig. 5). This loss of in-plane anisotropy on annealing is consistent with a transition from continuous to discontinuous NiFeCo layers. Note that an aligning field was used during annealing. Also shown is a plot of the easy axis squareness and MR sensitivity both as a function of a series of anneals. The easy axis squareness decreases substantially after lower temperature anneals, but then tends to increase and reach a plateau with higher temperature annealing. This is in agreement with previously reported results [1]. There are large differences in the sensitivity between the first three low temperature anneals, but after the maximum sensitivity is reached at 270°C, there is a general decrease as the annealing temperature is increased. This maximum sensitivity sample is located in the middle of the increasing squareness trend with temperature.

4. CONCLUSIONS

We have investigated annealing induced GMR in $Ni_{66}Fe_{16}Co_{18}$/Ag multilayer films with spacer thicknesses well beyond that found in standard multilayer films which exhibit exchange coupling and as-deposited GMR. Annealing these films apparently breaks up the ferromagnetic layers by interdiffusing Ag through the grain boundaries thus forming discontinuous NiFeCo layers (platelets). The magnetization of these platelets can be rotated in modest fields resulting in a relatively high field sensitive MR of ~3% in ~11 Oe. A maximum of >4% MR can be obtained from further annealing but the field sensitivity is greatly reduced. This induced GMR is very sensitive to the annealing conditions suggesting an optimal platelet size and intralayer spacing. High angle x-ray diffraction of as-deposited samples reveals a strong Ag(111) growth texture. Peak splitting in HXRD is observed only after annealing. The quality of film layering, monitored by LXRD, degrades on initial low temperature annealing but then is essentially stable on further annealing up to 340 °C. Magnetic hysteresis loops reveal an as-deposited in-plane uniaxial anisotropy that decreases with annealing and easy axis squareness that rapidly decreases with low temperature anneals, but then slightly increases with further annealing.

ACKNOWLEDGMENT

This work is supported by NSF-DMR-9301648.
The use of the facilities of the Center for Materials for Information Technology at The University of Alabama, Tuscaloosa, Alabama, is gratefully acknowledged.

REFERENCES

[1] T.L. Hylton, K.R. Coffey, M.A. Parker, J.K. Howard, Science **261**, 1021 (1993).
[2] J.C. Slonczewski, J. Magn. Magn. Mater. **129**, L123 (1994).
[3] B. Rodmacq, G. Palumbo, Ph. Gerard, J. Magn. Magn. Mater. **118**, L11 (1993).
[4] M.N. Baibich, J.M. Broto, A. Fert, F. Nguyen Van Dau, F. Petroff, P. Eitenne, G. Creuzet, A. Friederich, J. Chazelas, Phys. Rev. Lett. **61**, 2472 (1988).
[5] M. Kitada, J. Magn. Magn. Mater. **123**, L18 (1993).
[6] M. Jimbo, T. Kanda, S. Goto, S. Tsunashima, S. Uchiyama, Jpn. J. Appl. Phys. **31**, L1348 (1992).
[7] S. Hossain, D. Seale, G. Qiu, J. Jarratt, J.A. Barnard, H. Fujiwara, M.R. Parker, J. Appl. Phys. **75** (10), in press (1994).

GIANT MAGNETORESISTANCE IN AS-SPUTTER-DEPOSITED Au/NiFe MULTILAYER THIN FILMS

W. Y. LEE, G. GORMAN, and R. SAVOY
IBM Almaden Research Center, San Jose, CA 95120

ABSTRACT

Giant magnetoresistance with low saturation fields (H_s's) is reported in Au and permalloy ($Ni_{0.82}Fe_{0.18}$) or Co-doped permalloy multilayer thin films as-deposited on Ta-overcoated Si and glass substrates. A $\Delta R/R$ as high as 4.0% with ≈ 25 Oe H_s was observed at 295 K for the film consisting of 10 layers of 24 Å Au/13 Å $Ni_{0.82}Fe_{0.18}$ deposited on a 3 Å Ta-overcoated glass at 50 °C. A H_s value as low as ≈ 20 Oe with a 15% smaller $\Delta R/R$ has been observed for the films with a thicker (e.g., 50 Å) Ta underlayer. Magnetic hysteresis loops of these films indicate the presence of antiferromagnetic exchange coupling between the $Ni_{0.82}Fe_{0.18}$ layers. This exchange coupling is much smaller for the multilayer films without the Ta underlayer, resulting in a 6x smaller $\Delta R/R$ and 10x larger H_s observed for these films. Results of x-ray diffraction analysis indicate stronger (111) texturing for the multilayer films with a Ta underlayer, consistent with the stronger antiferromagnetic coupling between the the $Ni_{0.82}Fe_{0.18}$ layers in the film. The addition of 2-10 % Co moderately increases the $\Delta R/R$ value, but also increases substantially the H_s (up to ≈ 200 Oe).

INTRODUCTION

The effect of giant magnetoresistance (GMR) has been observed in a large number of magnetic/non-magnetic metallic multilayers since its initial discovery in Fe/Cr superlattices (1, 2). The effect is generally attributed to the spin-dependence of electron scatterings at the interfaces. The magnetic layers in the superlattice are exchange-coupled through the non-magnetic spacer layers (3). The coupling oscillates between ferromagnetic and antiferromagnetic as the spacer layer thickness is varied (4). The antiparallel spin orientation in the successive magnetic layers results in a high electrical resistance of a multilayer in its antiferromagnetic state due to high electron scatterings at the interfaces. When the spins of the magnetic layers are aligned parallel to each other by a sufficiently high external field i.e., saturation field (H_s), the resistance of the multilayer attains the lowest value due to its greatly reduced electron scatterings. The discovery of GMR in magnetically coupled multilayers has attracted a great deal of interest in the possible use of such structures as magnetoresistive sensors. For low-field sensor applications, e.g., magnetoresistive read heads, a large resistance change at low magnetic fields is required. In this paper, we report GMR with low saturation fields in as-deposited Au and permalloy ($Ni_{0.82}Fe_{18}$) multilayer thin films. The effect of substrate temperature and Co addition to permalloy will be published elsewhere.

EXPERIMENTAL

Multilayer thin films of Au and permalloy or 2-10% Co-doped permalloy were sputter-deposited on Si and glass substrates at up to 300 °C, using a DC magnetron sputtering technique. A controller interfacing with two quartz crystal thickness monitors was used to sequentially open and close each target shutter to deposit the desired thickness in each layer. The Si and glass substrate typically were overcoated with a thin layer (1-50 Å) of Ta immediately prior to the deposition of the multilayer to enhance the (111) texture of the film as will be discussed later. The background pressure in the sputtering chamber is $\approx$ 5-8 x 10^{-8} Torr. The deposition rate is $\approx$10 and 20 Å/min. with a power input of 7 and 35 W to the

Mat. Res. Soc. Symp. Proc. Vol. 343. ©**1994 Materials Research Society**

Au and permalloy or Co-doped permalloy targets, respectively. During deposition, the substrate holder rotates at a rate of 30 rpm between the two targets to ensure the thickness uniformity of the films. The temperature of the substrate (T_s) was measured with a chromel-alumel thermocouple located at ≈ 1 mm away from the edge of the substrate holder. The substrate holder and thermocouple are enclosed in a quartz-lamp heater housing shielded with glazed Au-reflectors to minimize the heat loss and temperature gradient.

The resistivity of the film was obtained from an 1-inch diameter sample, using a commercial four point probe. The thickness of the film was determined with a commercial profilometer across a chemically etched step. The dependence of resistance of these films on DC magnetic fields (10 KOe maximum) was obtained using a low frequency (85 Hz) four point probe with the DC field applied parallel to the film, and with the current (0.5 mA) flowing in the direction parallel or transverse to the in-plane magnetic field. Magnetic properties were investigated at room temperature using a commercial vibrating sample magnetometer. The composition of the permalloy or Co-doped permalloy films were determined by a commercial electron microprobe. High angle X-ray diffractometry were used to study the crystal structure, while low angle X-ray reflectometry were used to study the integrity of the multilayer films.

RESULTS AND DISCUSSION

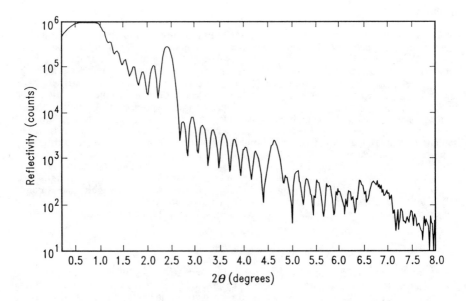

Figure 1. Low angle X-ray reflectivity patterns for a (24 Å Au/13 Å $Ni_{0.82}Fe_{0.18}$)$_{10}$ film deposited on a 50 Å Ta overcoated glass at 50 °C.

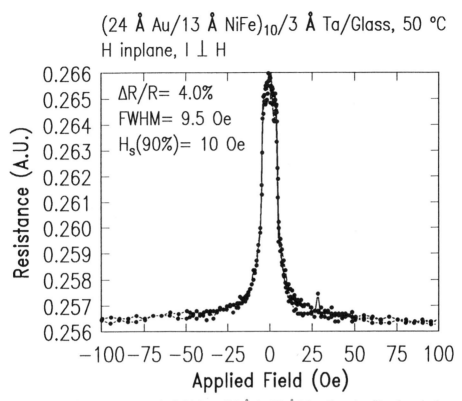

Figure 2. Resistance vs magnetic field for a (24 Å Au/13 Å $Ni_{0.82}Fe_{0.18})_{10}$ film deposited on a 3 Å Ta overcoated glass at 50 °C. The resistance were measured with the sensing current applied perpendicular to the in-plane magnetic field at 295 K.

The low angle X-ray reflectivity of a (24 Å Au/13 Å $Ni_{0.82}Fe_{0.18})_{10}$ film deposited on a 50 Å Ta overcoated (100) Si is given in Fig. 1. The interference fringes at 2.4 and 4.7 ° are first and second order "Bragg" peaks due to X-ray reflected from the 10 pairs of 24 Å Au/13 Å $Ni_{0.82}Fe_{0.18}$ layers in the film (5). The smaller fringes between the intensive "Bragg" peaks are related to the total thickness of the Au and $Ni_{0.82}Fe_{0.18}$ layers in the film. The individual layer thickness deduced from the data shown in Fig. 1 and X-ray fluorescence analysis is within 10 % of the nominal value. These results indicate no significant interface roughness and interdiffusion and thus well-defined interfaces for the multilayer. The $\Delta R/R$ values of these films were found to be very sensitive to the thickness of the Au spacer layer. The highest $\Delta R/R$ was obtained at a Au space layer thickness of 24Å, which is close to the second exchange oscillation peak observed in the MBE films (6). As shown in Fig. 2, a $\Delta R/R$ of as high as 4.0% with a H_s of ≈25 Oe and a full width at half maximum (FWHM) of ≈9.5 Oe was observed at 295 K for the film consisting of 10 layers of 24 Å Au/13 Å $Ni_{0.82}Fe_{0.18}$ deposited on a 3 Å Ta-overcoated glass at 50 °C. This gives a sensitivity of 0.4 % per Oe, using the same criteria reported previously (7). A H_s value of as low as ≈20 Oe with a 15 % smaller $\Delta R/R$ has been observed for similar films with a thicker (e.g., 50 Å) Ta underlayer. The best sensitivity we have achieved so far is ≈0.6 % per Oe (2.5 % $\Delta R/R$ with a FWHM of 4.2 Oe) for a film deposited on a 50 Å thick Ta. It is worth to point out that the $\Delta R/R$ value

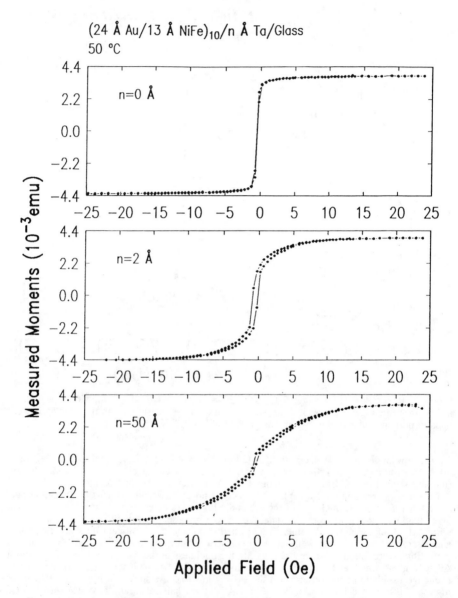

Figure 3. Magnetic hysteresis loops for (24 Å Au/13 Å $Ni_{0.82}Fe_{0.18})_{10}$ films deposited on 0, 2 and 50 Å overcoated glass substrates at 50 °C. The moments were measured with the magnetic fields applied parallel to the film.

is very reversible and reproducible with respect to magnetic fields, judging from the small scatterings of the data for 2 and half cycles of measurements shown in Fig. 2. Magnetic hysteresis loops of these films indicate a saturation magnetization of ≈600-700 emu at an applied field of >3-15 Oe. The magnetization of these films decreases to ≈10 % of the saturation value at zero field, indicating the presence of antiferromagnetic exchange coupling between the $Ni_{0.82}Fe_{0.18}$ layers (3). This exchange coupling is much smaller for the films without the Ta underlayer, as can be seen in Fig. 3 where the measured moments vs applied field data is shown for multilayers consisting of 10 layers each of 24 Å Au/13 Å $Ni_{82}Fe_{18}$ with 0, 2 and 50 Å thick Ta underlayer. (The results for the multilayer with 3 Å Ta underlayer is close to those for the one shown with 2 Å Ta underlayer.) The smaller exchange coupling results in a ≈6x smaller ΔR/R and 10x larger H_s observed for the films without Ta underlayer. High angle X-ray diffraction patterns for a multilayer consisting of 10 layers each of 24 Å Au/13 Å $Ni_{82}Fe_{18}$ is given in Fig. 4 as a function of Ta underlayer thickness. These results show only clusters of satellite at 2θ=37-46 ° from (111) reflections of the multilayer film. No evidence of diffraction from other crystallographic orientations was detected. The film with Ta underlyer is thus better crystallized and has a much stronger (111) texturing than the films without Ta underlayer. This results in an enhanced antiferromagnetic coupling between the $Ni_{0.82}Fe_{0.18}$ layers for the film with Ta underlayer shown in Fig. 3. The larger H_s observed for the film without Ta underlayer is apparently due to its poorer crystalline quality, not to the stronger coupling along the (100) orientation reported previously (8). Similar enhanced exchange coupling was also observed for the multilayers using permalloy doped with 2-10 % Co. These films show a moderately higher ΔR/R (up to 5.7 %) but a much larger H_s (up to ≈200 Oe). Detailed results of these studies and the effect of substrate temperature will be presented in a separate paper.

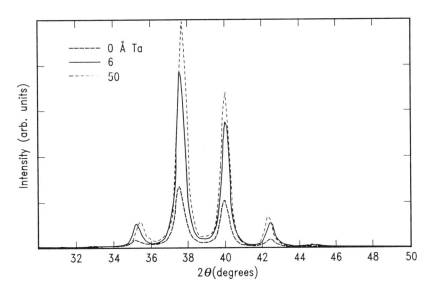

Figure 4. X-ray diffraction patterns of (24 Å Au/13 Å $Ni_{0.82}Fe_{0.18})_{10}$ films deposited on 0, 6 and 50 Å overcoated Si substrates at 50 °C.

SUMMARY

Giant magnetoresistance with low saturation fields in as-deposited $Au/Ni_{0.82}Fe_{0.18}$ multilayer films is reported. A $\Delta R/R$ of as high as 4.0% with ≈ 25 Oe H_s was observed at 295 K for the film consisting of 10 layers of 24 Å Au/13 Å $Ni_{0.82}Fe_{0.18}$ deposited on a 3 Å Ta-overcoated glass at 50 °C. The presence of Ta underlayer is shown to enhance the (111) texturing of the films and the exchange coupling between $Ni_{0.82}Fe_{0.18}$ layers. This results in a much larger $\Delta R/R$ value and smaller saturation field for the multilayer films with a Ta underlayer. The addition of 2-10 % Co to permalloy increases moderately the $\Delta R/R$ value, but also substantially increases the saturation fields.

ACKNOWLEDGMENT

The authors are indebted to R. Farrow and S.S.P. Parkin for many stimulating discussions and for providing a preprint of their paper concerning the oscillatory interlayer exchange coupling in MBE prepared Au/permalloy multilayer films.

REFERENCES

1. M. N. Baibich, J. M. Broto, A. Fert, F. Nguyen Van Dau, F. Petroff, P. Etienne, G. Greuzet, A. Friederich, and J. Chazelas, Phys. Rev. Lett. 61, 2472 (1988).

2. G. Binasch, P. Grünberg, F. Saurenbach, and W. Zinn, Phys. Rev. B39, 4828 (1989). (1992).

3. A. Barthélémy, A. Fert, M. N. Baibich, S. Hadjoudj, F. Petrof, P. Etienne, R. Cabanel, S. Lequien, F. Nguyen Van Dau, and G. Creuzet, J. Appl. Phys. 67, 5908 (1990).

4. S.S.P. Parkin, N. More and K. P. Roche, Phys. Rev. Lett. 64, 2304 (1990).

5. T. C. Huang and W. Parrish, Advances in X-Ray Analysis 35, 137 (1992).

6. S. S. P. Parkin, R. F. C. Farrow, R. F. Marks, A. Cebollada, G. R. Harp, and R. Savoy, Submitted to Phys. Rev. Lett., January 1994.

7. T. L. Hylton, K. R. Coffey, M. A. Parker, and J. K. Howard, Science 261, 1021 (1993).

8. R. Nakatani, T. Dei, and Y. Sugita, J. Appl. Phys. 73, 6375 (1993).

GIANT MAGNETORESISTANCE STUDIES OF NiFe-BASED SPIN-VALVE MULTILAYERS

Chien-Li Lin[*], John M. Sivertsen[*], and Jack H. Judy[**]
The Center for Micromagnetics and Information Technologies (MINT)
[*]Department of Chemical Engineering and Materials Science
[**]Department of Electrical Engineering, University of Minnesota, Minneapolis, MN 55455

ABSTRACT

The giant magnetoresistance in FeMn exchange-biased NiFe-based multilayer spin-valve structures prepared by rf-diode sputtering technique were studied. Experiments were performed on samples with different thicknesses of each layer in these multilayers. The magnetic properties were measured using a vibrating sample magnetometer and the giant magnetoresistance was measured using an in-line four-point magnetoresistance probe. A magnetoresistance of 6.5% in a magnetic field of less than 15 Oe was obtained in a Cu(30Å)/FeMn(150Å)/NiFe(50Å)/Co(15Å)/Cu(20Å)/Co(15Å)/NiFe(60Å) multilayer structure at room temperature. Annealing experiments of these multilayers were performed to study the thermal stability during the recording head fabrication processes. No degradation in the magnetoresistance has been found for annealing these films at 230°C up to four hours.

INTRODUCTION

Since the discovery of the giant magnetoresistance (GMR) in antiferromagnetic coupled Fe/Cr superlattice in 1988 [1], multilayers consisting of magnetic layers separated by nonmagnetic metal spacers have attracted wide attention. Several multilayers exhibiting GMR such as Co/Cu [2], NiFe/Cu [3], and CoFe/Cu [4] have been found. In these systems, the magnetizations of two successive magnetic layers are antiparallel at zero field, due to their antiferromagnetic coupling through the thin spacer layers. Unfortunately, the magnetic fields required to overcome the antiferromagnetic coupling for these multilayers range from several hundred oersted up to several thousand oersted. Such a large saturation field does not permit application to MR read heads and other low field sensors. In order to obtain a significant advantage in terms of sensitivity it is necessary to produce a large magnetoresistance in small magnetic fields.

Recently different types of GMR multilayer structures have appeared that do not depend on an antiferromagnetic exchange coupling to drive the magnetizations into the antiparallel state. GMR structures exhibiting the so-called spin-valve effect were proposed by Shinjo et al. [5] using multilayers composed of uncoupled magnetic layers having different anisotropies and by Dieny et al. [6] using FeMn exchange-biased NiFe/Cu/NiFe sandwiches. GMR ratios of 5~10% have been obtained in a weak field about 100 Oe. From the standpoint of pratical applications, these results showed a very promising research direction for obtaining high-sensitivity GMR materials.

The primary focus of this paper is on the GMR effect of FeMn exchange-biased NiFe-based spin-valve structures since the relatively large signals available with these materials can

411

potentially permit construction of a high-sensitive reading head for future magnetic recording disk drives.

EXPERIMENTAL

All the films used in this study were prepared by a two-step deposition using a three-target Perkin-Elmer RF diode sputtering system. The system pressure prior to deposition was better than 4×10^{-7} Torr and the argon gas pressure during sputtering was 3 mTorr. The target-to-substrate spacing was kept at 3.5 inches enabling us to minimize the interaction of the plasma with the deposition surface. First, the sandwich layers of NiFe, Co, Cu, Co, and NiFe were sequentially deposited on glass and on SiN_x coated Si substrates (which have R_{rms} about 2.5Å verified ex-situ by an atomic force microscope) at ambient temperature. Then, the chamber was vented and a FeMn target was put into the chamber for depositing the exchange bias layer. The top NiFe was slightly etched about 20 Å by rf plasma right before sputtering the FeMn layer to eliminate the possible oxidation of the NiFe surface during the target change. Finally, the multilayers were protected by a 30 Å copper capping-layer. The thickness of each layer was controlled by deposition time, following calibration of the deposition rates of individual materials using a DEKTAK profilometer. A uniaxial in-plane anisotropy of multilayers was induced by applying a magnetic field during the deposition.

Resistivity measurements were performed using a standard in-line four point probe dc method in magnetic field up to 150 Oe at room temperature where the field is applied transverse to the current direction and in the plane of the film. The current and voltage leads were attached to the film surface by pressure contact. The magnetoresistance was defined as the difference between the maximum resistance and the saturation resistance, normalized to the saturation resistance: $MR = (R_{max} - R_{sat}) / R_{sat}$. The magnetization hysteresis curves were measured with a vibrating sample magnetometer (VSM). Annealing experiments of selected samples were carried out in an argon atmosphere in a temperature-controlled tube furnace.

RESULTS AND DISCUSSIONS

Figure 1 shows the schematic diagram of the multilayer structure used in this study. Figure 2 summarizes the magnetoresistance results of a series of samples in which the thickness of the Cu interlayer was varied between 12Å to 32Å at room temperature. For thin Cu layer(12Å), the two magnetic components are strongly ferromagnetically coupled together which might be due to pinhole formation during the sputter-etch process. If no antiparallel state was formed in the magnetization reversal process, the GMR effect is small. For thicker Cu interlayer(>16Å), a well defined two-step reversal process was formed, a significantly increase in GMR was obtained.

Typical magnetization and magnetoresistance hysteresis curves for an as-deposited Cu(30Å)/FeMn(150Å)/NiFe(50Å)/Co(15Å)/Cu(20Å)/Co(15Å)/NiFe(60Å) are shown in Fig. 3. The magnetization curve consists of two separated hysteresis loops. The low field loop results from the magnetization reversal of the "free" bottom NiFe/Co layers, while the high field loop shifted by FeMn exchange bias to about 55 Oe. The plateau region between these two loops indicates the antiparallel alignment of magnetization directions of "free" and "pinned" magnetization layers. Therefore, the resistivity of this film reaches a maximum. A MR ratio of 6.5% was

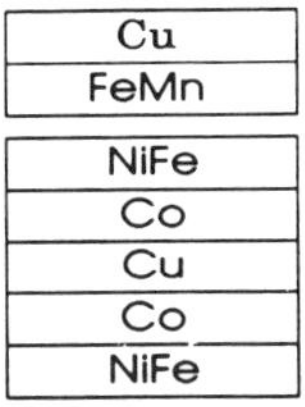

Cu
FeMn

NiFe
Co
Cu
Co
NiFe

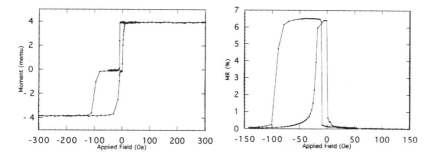

Figure 1: Schematic diagram of the multilayer structure used in this study.

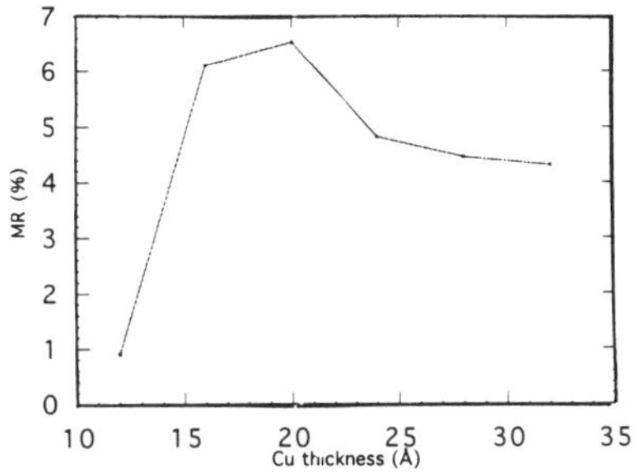

Figure 2: Magnetoresistance versus Cu interlayer thickness for FeMn(150Å)/NiFe(50Å)/Co(15Å)/Cu(xÅ)/Co(15Å)/NiFe(60Å) multilayers.

Figure 3: Typical magnetization and magnetoresistance hysteresis curves for an as-deposited Cu(30)/FeMn(150Å)/NiFe(50Å)/Co(15Å)/Cu(20Å)/Co(15Å)/NiFe(60Å) multilayers.

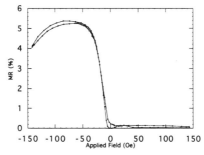

Figure 4: The magnetoresistance hysteresis loop of a transverse biased FeMn/NiFe/Co/Cu/Co/NiFe multilayers.

413

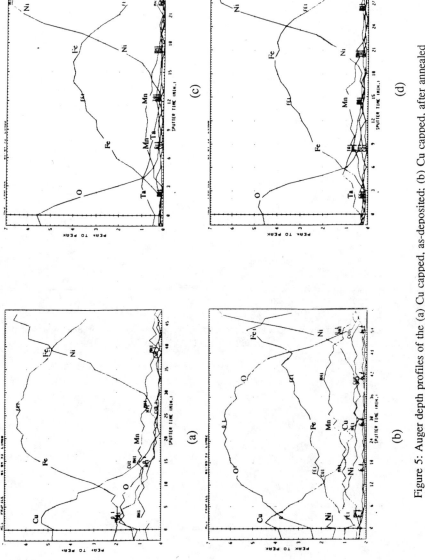

Figure 5: Auger depth profiles of the (a) Cu capped, as-deposited; (b) Cu capped, after annealed at 230°C for 1 hour; (c) Ta capped, as-deposited; (d) Ta capped, after annealed at 230°C for 1 hour NiFe/Co/Cu/Co/NiFe multilayers.

observed in a field of 15 Oe for this structure which is 2~3 times larger than the anisotropic magnetoresistance (AMR) used in today's MR reading heads.

To study the feasibility of using these multilayers as a reading element for disk drives [6][7], selected films were annealed at 200°C in a transverse magnetic field of 3 kOe to reset the "pinned" layer perpendicular to the "free" layer in order to test the signal output of the multilayers with both positive and negative fields. Figure 4 shows the MR hysteresis loop by applying a magnetic field parallel to the "pinned" direction (perpendicular to the easy axis of the "free" layer). A nice linear region was obtained in the field between 10 to 30 Oe with field sensitivity of about 0.25% (i.e. 5% change in a range of 20 Oe) resistance change per oersted.

To study the thermal stability of these films, selected samples were annealed at 230°C for 1 to 4 hours in argon gas to simulate the temperature raise during the head fabrication process. Slightly increasing in MR ratios (~15%) were observed in all the multilayers protected by 30Å Cu layers. To understand the reason, a series of similar multilayers protected by a 50Å Ta capping layer were made. After the same annealing condition, the multilayers capped by 50Å Ta show no increase in MR ratios. But, both multilayers show an increase in the exchange coupling field. Figure 5(a)-(d) show the Auger depth profiles of the O_2, Cu, Ta, Fe, Mn, and Ni distributions in both Cu and Ta capped multilayers before and after the annealing. For multilayers capped with Cu layers, the oxidation of the Cu protective layer and the FeMn layer reduce the shunting effect which caused a slightly increase in MR. On the other hand, for multilayers capped with Ta layers, the oxidation process was inhibited by the Ta capping layer. Table I summarizes the magnetic and transport properties measured on these samples.

Table I - Magnetic and transport properties of Cu-capped and Ta-capped FeMn/NiFe/Co/Cu/Co/NiFe multilayers.

	As-deposited			Annealed at 230°C for 1 hour		
	R (Ω)	ΔR / R (%)	H_{ex} (Oe)	R (Ω)	ΔR / R (%)	H_{ex} (Oe)
Cu-capped	12.79	6.5	70	13.48	7.9	100
Ta-capped	12.61	5.29	60	10.41	5.23	90

SUMMARY

The magnetoresistance and magnetic properties were investigated for Cu/FeMn/NiFe/Co/Cu/ Co/NiFe multilayers prepared by a two-step rf sputtering process. The results obtained in this study show that Cu/FeMn/NiFe/Co/Cu/ Co/NiFe multilayers have a high MR sensitivity and good thermal stability making them good candidates for future magnetic sensors.

ACKNOWLEDGMENTS

The authors would like to thank Drs. J.M. Daughton and Y. Chen (Nonvolatile Electronics, Inc.) for many helpful discussions and suggestions and use of their equipments.

REFERENCES

[1] M.N. Baibich, J.M. Broto, A. Fert, Nguyen van Dau, F. Pertroff, P. Eitenne, G. Creuzet, A. Friederich, and J. Chazelas, Phys. Rev. Lett., Vol. 61, 2472(1988).

[2] S.S.P. Parkin, R. Bhadra, and K.P. Roche, Phys. Rev. Lett. 66(16), 2152(1991); S.S.P. Parkin, Z.G. Li, and D.J. Smith, Appl. Phys. Lett. 58(23), 2710(1991).

[3] R. Nakatani, T. Dei, T. Kobayashi, and Y. Sugita, IEEE Trans. on Magn., Vol. 28, No. 5, 2668(1992).

[4] K. Inomata, Y. Saito, and S. Hashimoto, J. Magn. Magn. Mater. 121(1993) 350.

[5] T. Shinjo and H. Yamamoto, J. Phys. Soc. of Japan, Vol. 59, No. 9, 3061(1990); H. Yamamoto, T. Okuyama, H. Dohnomae, and T. Shinjo, J. Magn. Magn. Mater. 99(1991) 243; H. Yamamoto, T. Okuyama, H. Dohnomae, and T. Shinjo, IEEE Transl. J. on Magn. in Japan, Vol. 7, No. 3, 207(1992).

[6] B. Dieny, V.S. Speriosu, B.A. Gurney, S.S.P. Parkin, D.R. Wilhoit, K.P. Roche, S. Metin, D.T. Peterson, and S. Nadimi, J. Magn. Magn. Mater. 93(1991) 101; B. Dieny, V.S. Speriosu, S. Metin, S.S.P. Parkin, B.A. Gurney, P. Baumgart, and D.R. Wilhoit, J. Appl. Phys. 69(8), 4774(1991); B. Dieny et al., U.S. Patent 5,206,590(1993).

[7] D.E. Heim, R.E. Fontana, Jr., C. Tsang, V.S. Speriosu, B.A. Gurney, and M.L. Williams, TMRC93 Paper, to be published in IEEE Trans. on Magn.

TEXTURE AND GRAIN SIZE OF PERMALLOY THIN FILMS
SPUTTERED ON SILICON WITH Cr, Ta AND SiO_2 BUFFER LAYERS

P. GALTIER, R. JEROME AND T. VALET
Laboratoire Central de Recherches, THOMSON-CSF, Domaine de Corbeville, 91404 Orsay
cedex, France

ABSTRACT

We have investigated the structural properties of $Ni_{80}Fe_{20}$ thin films sputtered on silicon with
Cr, Ta and SiO_2 buffer layers using transmission electron microscopy. We observe a decrease of
the grain size when Ta and SiO_2 underlayers are used instead of Cr. Permalloy films deposited on
Ta layers are strongly (111) textured while those grown on Cr and SiO_2 are mostly randomly
oriented. The results are discussed with respect to the nanostructure of both Ta, Cr and SiO_2
underlayers and in relation to the variation of the magnetic softness observed in this system.

INTRODUCTION

The recent interest in thin magnetic thin films and multilayers has stimulated the need for a
precise knowledge and control of their structural characteristics like texture, interface roughness
and grain size. Permalloy ($Ni_{80}Fe_{20}$) thin films and heterostructures are of great interest for
magnetoresistive devices [1-3]. In particular, the low field magnetoresitive behavior of these
layered systems may be advantageous for high density magnetic recording. In this system, in
plane coercivity as low as 0.1 Oe have been measured under optimized conditions. However this
property is strongly dependent on the growth conditions. Parameters like film thickness, texture
and grain size can be invoked in order to explain the variation of coercivity observed in that kind
of system [4-7]. For films with thicknesses below 100 nm, the nature of the underlayer deeply
affects the structural characteristics responsible for the observed magnetic properties.

In order to clarify the structural aspects connected to their magnetic properties, we present a
Transmission Electron Microscopy study of $Ni_{80}Fe_{20}$ films deposited on various buffer layers.
For this purpose, we have investigated Cr and Ta metallic buffers. In these systems the coercive
field, Hc, can be reduced from 1.0 to 0.1 Oe when Permalloy is sputtered on a Ta instead of Cr
[8]. Although Ta and Cr crystallize both in the bcc phase, they exhibit different melting points.
This should affect the atomic surface mobility during the growth and influence the microstructure.
In addition, we have studied the influence of a well known amorphous underlayer like SiO_2
which has already been used to generate randomly oriented metallic films [7]. An interesting
feature of this underlayer is that it leads to an intermediate value of the coercive field (Hc≈0.5
Oe).

EXPERIMENT

$Ni_{80}Fe_{20}$ films, 250 Å thick, were deposited on Cr, Ta or SiO_2 buffer layers on (100)Si in a
high vacuum sputtering system with a base pressure of 5.10^{-8} Torr [8]. The Silicon substrate was
chemically etched in HF then rinsed in deionized water prior to the growth. RF magnetron was
used for deposition of Cr and Ta at a deposition rates of 2.2 Å/s. The SiO_2 buffer layer, 1200 Å
thick, was grown ex-situ by thermal annealing of silicon at 1050°C under dry oxygen
atmosphere. It was degreased, but not etched, prior to the growth. The $Ni_{80}Fe_{20}$ films, 250 Å
thick, were then deposited using the RF diode configuration at a deposition rate of 1Å/s. A
magnetic field of ≈ 50 Oe was applied in the plane of the substrate during the deposition, in order
to induce a uniaxial anisotropy in the permalloy layer. A cap layer of Cr or Ta was then

Mat. Res. Soc. Symp. Proc. Vol. 343. ©1994 Materials Research Society

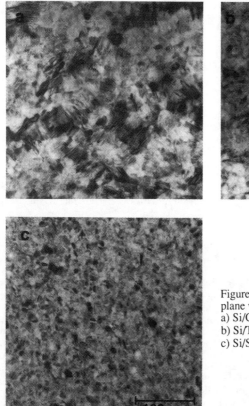

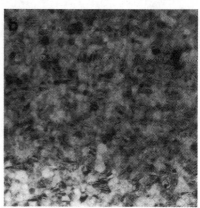

Figure 1: Bright field images obtained on
plane views of:
a) Si/Cr(50Å)Ni$_{80}$Fe$_{20}$(250Å)/Cr(50Å)
b) Si/Ta(50Å)Ni$_{80}$Fe$_{20}$(250Å)/Ta(50Å)
c) Si/SiO$_2$(1200Å)/Ni$_{80}$Fe$_{20}$(250Å)/Cr(50Å)

deposited to prevent the oxidation of the permalloy layers. The substrate temperature was close to 300°C for all the investigated films.

TEM experiments were performed on plane views and cross sections prepared by polishing and dimpling followed by argon milling performed with a beam incidence of 15° and using a liquid nitrogen cooled stage equipped with a sector speed control. The observations were performed using a Topcon 002B microscope operated at 200 kV. It was fitted with a Cs=0.4 mm pole piece giving a resolution of 1.8 Å. Nanodiffraction experiments where performed using a probe size of about 30 Å and a half convergence of 4 mrd.

RESULTS

Grains size and texture of the Permalloy films

In both cases the Ni$_{80}$Fe$_{20}$ films are found polycristalline with the fcc structure. For Permalloy films deposited on Cr, the lateral size of the grains measured on plane views is about 300-500 Å (Figure 1.a). Some columnar growth is visible in the bright field images obtained on cross sections as previously reported [10]. The ring related to the (111) reflection observed in the electron diffraction pattern show that most of the grains are randomly oriented (Figure 2.a).

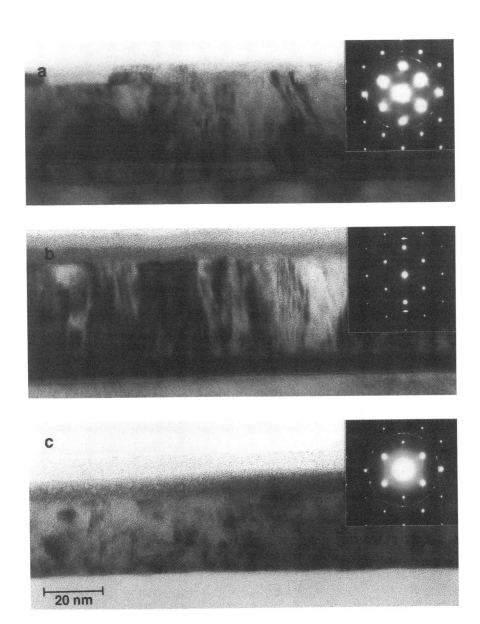

Figure 2: Bright field images and electron diffractions obtained on cross-sections of
a) Si/Cr(50Å)/Ni$_{80}$Fe$_{20}$(250Å)/Cr(50Å), b) Si/Ta(50Å)/Ni$_{80}$Fe$_{20}$(250Å)/Ta(50Å) and
c) SiO$_2$(1200Å)/Ni$_{80}$Fe$_{20}$(250Å)/Ta(50Å)

However, some residual preferential orientations are not excluded. Similar results are observed when Cr is replaced by Fe.

The films deposited on a Ta buffer layer appear to be very different in terms of grain size and texture. Their lateral dimensions are smaller ($\approx$ 100-200 Å), with no particular lateral shape. They exhibit a clear columnar structure as shown in cross-section observations (Figure 2.b). A strong (111)fcc texture is observed with the texture axis parallel to the growth direction (Figure 2.b). The top surface is apparently wavier than the one on Cr and SiO_2 buffers presumably due to the columnar growth.

In the case of a SiO_2 buffer layer, grains with small lateral dimensions are observed ($\approx$ 50-150 Å)(see Figure 1.c). Although their sizes are almost comparable to the one observed with Ta buffers, they do not exhibit the kind of columnar shape noticed previously (Figure 2.c). Furthermore, like for Cr buffer layer, no clear texture is observed.

Structure of the Cr and Ta buffer layers

Electron diffractions on plane views show the bcc structure of the Cr buffer layer. The (200)bcc and (211)bcc reflections are clearly seen whereas the (110)bcc contribution was not resolved from the (111)fcc of $Ni_{80}Fe_{20}$. The grain size, measured on high resolution images on cross sections (Figure 3.a), is larger than 40 Å. A close examination of the high resolution images shows a clear structural coherency between the Cr layer and the upper permalloy layer. This is illustrated in Figure 3.a where we can see the continuity of the lattice fringes from the Cr grains into the upper Permalloy (the lattice spacing measured in that case is about 2.05 Å which corresponds to both the (110) and (200) reflections of Cr an $Ni_{80}Fe_{20}$). Thus in the case of a Cr buffer layer the growth mode of $Ni_{80}Fe_{20}$ is monitored by the different orientations of the grains in the buffer.

The case of the Ta buffer layer appears somewhat complicated. Electron diffractions on plane views show only one broad ring related to Ta with a lattice spacing of 2.35-2.45 Å. This can be attributed to either the (110) reflection of the bcc phase (2.34 Å), either to the (202) reflection of the ß phase (2.35 Å) [11,12]. We have not detected other reflections. High resolution observations performed on cross sections show small grains, $\approx$ 15 Å large, embedded in an "amorphous like" layer (Figure 3.b). This is confirmed by nano-diffraction experiments which show, in addition to diffraction spots related to small Ta grains and to residual contributions from the Si substrate, a ring centered around 2.4 Å (insert of Figure 3.b). This suggests that the Ta buffer layer is partially amorphous. However, it is difficult to definitively conclude about the exact crystallization state of this buffer. A close inspection of the images reveal that the (111) oriented Permalloy grains are nucleated at the interface with the underlayer and apparently without any relationship with the Ta crystallites.

DISCUSSIONS

Our results illustrate the importance of nucleation films on the structure of Permalloy thin films. We show that Cr and Ta buffer layers exhibit very different crystallization states. This is probably in relation with the different melting point values noticed for these metals (1875°C for Cr and 300°C for Ta). A lower atomic mobility is expected far below the melting point which, combined with the relatively low temperature of the growth (300°C), inhibits the crystallization process and probably explain the quasi amorphous structure of the Ta underlayer [13]. On the other hand, the atomic mobility is higher for Cr and this favors the nucleation and growth of well crystallized grains.

In the case of Cr, the properties of $Ni_{80}Fe_{20}$ films are directly monitored by the buffer. However, Cr and $Ni_{80}Fe_{20}$ crystallize in their bulk stable structure. The texture observed with Ta suggests that it is somehow correlated to the "amorphous like" structure of the buffer. However, very different results are observed on SiO_2. The growth mode of thin films deposited on amorphous surfaces has been extensively studied in the past [14]. The orientation of the

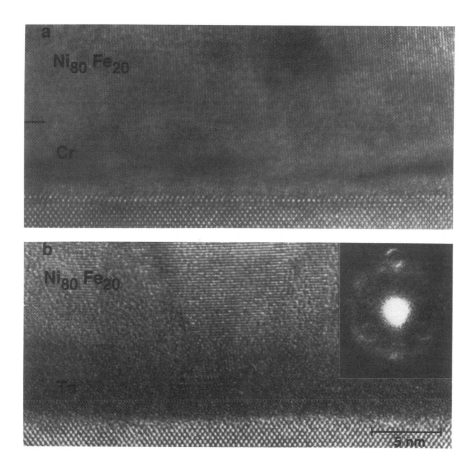

Figure 3: High resolution images of the interface $Si/buffer/Ni_{80}Fe_{20}$ for a) Cr and b) Ta. A nanodiffraction pattern of the Ta layer is shown in insert of b).

crystallites are found to be due (i) to formation of nuclei with special orientations (nucleation orientation) or (ii) to the preferred growth of crystals with special orientations (growth orientation). An tentative explanation of our results can be found in the preparation processes which strongly differ for Ta and SiO_2. The thermal oxide is fabricated *ex-situ* and this probably favors the creation of a lot of nucleation sites, due to contamination, with different orientations of nucleation which lead to randomly oriented grains (Figure 2.c). On the other hand, these nucleation sites are not present in the case of a Ta buffer grown *in-situ*. Thus, the (111) texture observed on the Ta buffer layer should be more likely related to the kinetic of the growth (growth orientation). This is consistent with previous reports of a preferential growth of face-centered cubic metals along the [111] axis [14]. However, the adsorption of atomic species on metals is known to give rise to long range ordering due to the delocalization of the electronic bond and this could also be the origin of the preferential growth observed on Ta. Further investigations are necessary to clarify that point.

Both grain size and preferential orientations of the films have been assumed to influence the magnetic softness. Weak coercivities are generally observed for small crystals where the exchange interaction between the neighboring crystals reduces the effective anisotropy [15]. The growth with preferential orientations is also known to deeply affect the magnetocrystalline anisotropy and magnetostriction [7]. In our study, the highest coercive field (Hc≈1 Oe) is obtained with large and mostly randomly oriented grains (Cr case). The lower coercive field (Hc≈0.1 Oe) is obtained on films with a preferential orientation (Ta case) whereas smaller grains, but randomly oriented (SiO$_2$ case), lead to higher coercivity (Hc≈0.5 Oe). The lowest coercivity obtained on Ta buffers is thus mostly related to the (111) preferential orientation observed with this underlayer. In that case, the magnetization lies in the (111) plane a configuration which reduce the magnetocystalline anisotropy and suppress the sources of local anisotropy fluctuations [8].

CONCLUSIONS

Our results show that the grain size and texture of Ni$_{80}$Fe$_{20}$ thin films are strongly affected by the underlayer. Cr and Ta buffer layers are found to exhibit very different crystallization states. Cr buffer is shown to monitor both the grain size and the absence of texture of the upper Permalloy layer. This contrasts with the results obtained with Ta which is found mostly amorphous and leads to columnar and textured permalloy grains whereas no preferential orientations are observed with SiO$_2$. Both grain sizes and texture are found to affect the coercive field of the Permalloy. However, the preferential orientation observed on Ni$_{80}$Fe$_{20}$ sputtered on Ta appears to be the dominant parameter responsible for the observed weak coercivity.

This work was supported in part by a Brite Euram grant from the European Economic Community. The authors would like to thank C. Chenu for the preparation for the TEM sample, Dr. P. Alnot for stimulating discussions and Dr. F. Plais for the fabrication of the SiO$_2$/Si films

REFERENCES

1. S.S.P. Parkin, Appl. Phys. Lett. **60**, 512 (1992).
2. B. Dieny, V.S. Speriosu, B.A. Gurney, S.S.P. Parkin, D.R. Wilhoit, K.P. Roche, S. Metin, D.T. Peterson and S. Nadimi, J. Magn. Magn. Mater. **93**, 101 (1991).
3. T. Valet, J.C. Jacquet, P. Galtier, J.M. Coutellier, L.G. Pereira, R. Morel, D. Lottis and A. Fert, Appl. Phys. Lett. **61**,3187 (1992)
4. I. Hashim and H.A . Atwater in *Magnetic Ultrathin Films*, Edited by B.T. Jonker, S.A. Chambers, R.F.C. Farrows, C. Chappert (Mater. Res. Soc. Proc. **313**, Pittsburgh, PA, 1993) pp. 749-754.
5. K.Y. Ahn and J.F. Freedman, IEEE Trans. Magn. **MAG-3**, 157 (1967)
6. R. M. Valletta, C. Anderson and H. Lefakis, J. Vac. Sci. Technol. A 9, 2107 (1991)
7. A. Hosono and Y. Shimada, J. Appl. Phys. **67**, 6981 (1990).
8. R. Jerome, T. Valet and P. Galtier, 6th Joint MMM-Intermag Conference, Albuquerque, June 20-23, 1994.
9. T. Valet, P. Galtier, J.C. Jacquet, C. Meny and P. Panissod, J. Magn. Magn. Mater. **12**, (1993).
10. P. Galtier, T. Valet, O. Durand, J.C. Jacquet and J.P. Chevalier in *Magnetic Ultrathin Films*, Edited by B.T. Jonker, S.A. Chambers, R.F.C. Farrows, C. Chappert (Mater. Res. Soc. Proc. **313**, Pittsburgh, PA, 1993) pp. 749-754.
11. L.G. Feinstein and R.D. Hutteman, Thin Solid Films, **16**, 129 (1973).
12. S. Sato, Thin Solid Films, **94**, 321 (1982).
13. K.L. Chopra, in *Thin Film Phenomena* (McGraw-Hill, New York, 1987).
14. E. Bauer, in *Single-Crystal Films*, Edited by M. H. Francombe and H.Sato (Pergamon Press, Oxford, 1964) pp.43-67.
15. H. Hoffman, IEEE Trans. Magn. **MAG-9**, 17 (1973).

EFFECT OF ION IRRADIATION ON THE ARTIFICIAL STRUCTURE AND THE INTERLAYER COUPLING OF COPPER-COBALT MULTILAYERS

J-F BOBO*, E. SNOECK**, M. PIECUCH*, AND M-J CASANOVE***
*LPS-UMR37-URA155-CNRS, BP 239, 54506 Vandœuvre FRANCE
**Dept of Mat. Sci. and Eng., Stanford University, Stanford CA 94305 USA
***CEMES-LOE-CNRS, 29, rue J. Marvig, 31055 Toulouse FRANCE

ABSTRACT

Structural defects like local ferromagnetic short-circuits are responsible for large reductions of the intrinsic interlayer coupling in the case of antiferromagnetically (AF) coupled multilayers (ML). We have investigated the role of such point defects in AF-coupled Cu-Co ML's with $t_{Cu}=8$Å and $t_{Co}=12$Å which have been irradiated with various doses of 200 keV xenon ions. A significant decrease of the giant magnetoresistance and of the saturation field is observed when the density of defects is increased. These results are compared with a micromagnetic model that takes into account local discontinuities of the AF coupling in an ideally AF-coupled ML. Evidence for the presence of such defects is investigated by TEM on cross-sectional specimens.

INTRODUCTION

It is now well established that the preparation of giant magnetoresistance multilayers requires a perfect control of the growth conditions : the presence of too many defects in the artificial structure impairs good AF coupling. The planarity and continuity of the layers are particularly important parameters. As underlined for instance by Johnson [1], local discontinuities of the spacer layer can induce ferromagnetic bridges between two ferromagnetic layers and hide intrinsic interlayer AF coupling. We have already studied [2,3] the influence of these so-called pin-holes. Their effect on antiferromagnetic interlayer coupling is supposed to be strong as there is a ratio 100 between the intensity of the F coupling through a pin-hole and the AF coupling of an equivalent area. We confirm this assumption in what follows by two different approaches : from a theoretical point of view, we have developed a micromagnetic model that takes into account the influence of pin-holes. Then, we have tried to create such columnar defects by irradiating AF-coupled Cu-Co ML's with various doses of 200 keV Xe ions. Structural evidence for the existence of pin-holes has been investigated by TEM while the magnetic properties of these samples have been checked by bulk magnetization and magnetoresistance measurements.

In the first section of our paper, we present our micromagnetic description of the effect of pin-holes on an ideally AF-coupled ML. In the second part we discuss the effect of irradiation on AF coupled ML's and at last, experimental results are compared to our model.

MICROMAGNETIC DESCRIPTION OF THE INFLUENCE OF PIN HOLES

A summary of our model can be given as follows : let us consider two AF-coupled ferromagnetic layers (thickness t_m, magnetization M_S and stiffness constant A) separated by a nonmagnetic layer (thickness t_{nm}). We neglect magnetocrystalline anisotropy. The magnetizations in the two layers will be symmetrical with respect to the direction of the applied field with two opposite angles $\theta_1(r)=\theta$ and $\theta_2(r)=-\theta$. Demagnetizing field effect along the pin-hole is also neglected (it would tend to align the magnetization perpendicularly to the layers). Then we can write the total energy of a pin-hole of diameter d located at the origin of the plane :

$$E_{tot} = \int_{r=0}^{\infty} \left\{ 2t_m A \left(\frac{d\theta}{dr} \right)^2 - 2t_m M_s H \cos\theta - J(r)\cos 2\theta \right\} 2\pi r dr \qquad (1)$$

with $J(r) = J_{12} < 0$ out of the pin-hole ($r>d/2$) and $J(r) = J_{pin-hole}= 2A_{Ferro}/t_{nm} > 0$ in the pin-hole.

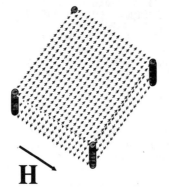

figure 1 : schematic viewof
the AF-coupled trilayer
in the zero field configuration
Local magnetizations are
represented by arrows and
pin-holes as grey columns.

H

The resulting configurations $\theta(r)$ of the moments are obtained by a minimization of E_{tot} with two methods : the first way consists in searching for analytical solutions of $\theta(r)$. Unfortunately, no analytical solution can be found in the realistic 2-dimensional case. However, if we project our problem to 1D, it is possible to integrate an equivalent Euler equation :

$$l_i^2 . \frac{d^2\theta}{dx^2} - 2.\sin\theta\cos\theta + h.\sin\theta = 0 \qquad (2)$$

with two parameters : $l_i = \sqrt{\frac{A.t_m}{J}} = l\sqrt{\frac{t_m}{J}}$ and $h = \frac{H}{2J}$

l_i is the characteristic length of our system (l =15.8Å for Fe and 11.5Å for Co if t_m is expressed in Å and J in erg/cm^2) ; h is the reduced field. We will see further how crucial a role the characteristic length l plays in this problem. We will not give more details about the analytical solution. One must just know that it gives the same results than the numerical approach.

The numerical micromagnetic model can be described as follows : we assume a periodic distribution of pin-holes on a square lattice of cells labeled by i and j indexes (see fig. 1 ; a is the size of the mesh). The total energy of a configuration of the local spins θ_{ij} is the summation over all the cells of the local energy :

$$\frac{E_{ij}}{a^2} = -J_{ij}\cos 2\theta_{ij} - 2M_s tH\cos\theta_{ij} - \frac{2At_m}{a^2}\sum_{i'j'}^{neighbour-cells}\cos(\theta_{ij} - \theta_{i'j'}) \qquad (3)$$

The convergence to the equilibrium configuration is obtained with the relaxation method [4]. Various zero field configurations of the spins and magnetization curves can be obtained with this model : depending on the ratio between the size of a pin-hole (d), the distance between two of them (D) which is equivalent to their concentration per unit of area ($x=d^2/D^2$) and of course the characteristic length l_i we can define some rules (with standard J, M and t_m):

1. $d \ll l_i$ and $D \ll l_i$: ferromagnetic homogeneous solution.
2. $d \ll l_i$ and $D \gg l_i$: antiferromagnetic homogeneous solution.
3. $d \ll l_i$ and $D \approx l_i$: intermediate solution (fig. 2-a), the configuration of the spins is close to 90° biquadratic coupling, similar situations are found for $d \approx l_i$ (see fig. 2-c and 2-d).
4. $d \gg l_i$: inhomogeneous configuration with pin-holes aligned along the applied field direction and orthogonal to the remaining AF-coupled area. This case does not really correspond to pin-holes but rather to large areas where the magnetic layers are in contact separated by AF-coupled domains (see fig. 2-b).

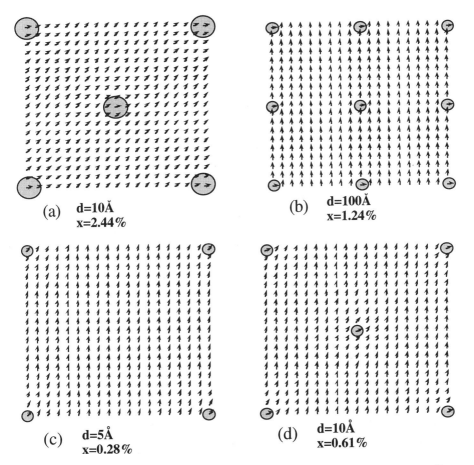

figure 2 : zero-field configurations of our system for a [Co$_{12Å}$-Cu$_{8Å}$] ML (J=-0.25 erg/cm^2) for various distributions of the pin-holes (their area is symbolized by grey circles and their size is d). The field axis is horizontal and only the top layer is represented, the spins in the lower layer are directed downwards and symmetrically to the ones in the top layer.

For each of these spins configurations, the shape of the magnetization curves will evolve from the ideally AF-coupled ML for $d >> l_i$ and $D >> d$ (infinitely diluted and independent pin-holes) with a very low remanence and a susceptibility :

$$\chi = \chi_o = \frac{M_s t_m}{4J} \qquad (4)$$

When D decreases, the remanent magnetization is expected to increase linearly with the pin-hole density x. If the size of the pin-holes is smaller than l_i we predict a strong increase of the initial susceptibility χ_o accompanied by a non negligible remanence (see fig. 3). At the same time, the M(H) curves are no more linear but with a rounded shape which is very similar to the experimental magnetization curves. The saturation field of the magnetization, which is no longer linearly related to J is also reduced when x increases.

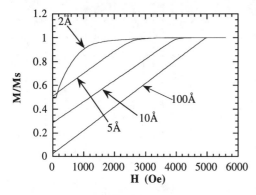

figure 3 : dependence of the magnetizatio: curves of a [Co₁₂Å- Cu₈Å] ML (J=-0.22 erg/cm²) with the size d of the pin-holes (d= 2, 5, 10, and 100Å) and for the same effectiv pin-holes concentratior (x=0.61%). We clearly see the increase of th remanence and of the rounding of the curves when d decreases.

These results shed light on the problem : there is a transition of the behavior of the AF-coupled system when the size of the pin-holes becomes of the same order of magnitude than the characteristic length l of the ferromagnetic material. For AF-coupled ML's, Cu-Co for example, the maximum of the AF coupling intensity will be obtained at the first AF peak. It corresponds to $t_{nm} \approx 8$Å. Therefore, for so thin layers, one cannot avoid the presence of pin-holes which are expected to have a size of the same order of magnitude than t_{nm}. The assumption of $d \approx t_{nm}$ is coherent with TEM observations as attested by figure 4.

figure 4 : Cross sectional TEM micrograph (def=150 nm) of a non irradiated (111) textured [Co₁₂Å- Cu₉Å]₁₅ sample. Large underfocus values are required to enhance compositional contrast, see [7] for instance. The light bands correspond to Co layers, the dark ones to Cu layers. Pin-holes are indicated by small arrows.

ION IRRADIATION OF Cu-Co MULTILAYERS

This technique is well adapted for creating point defects in ML's. It has the advantage of making localized defects compared to thermal annealings (see ref.[5] for example) which affect the totality of the ML. Furthermore, ion irradiation allows us to control the density of defects via the dose of ions that the ML receives. We can even expect to create defects of various sizes by changing the kinetic energy or the atomic number of the beam.

However, a clear description of the effect of irradiation is very difficult to find. As underlined by Dunlop et al. [6], damaging is caused by electronic excitations between the incident ion and the material. Various types of behavior are found for bulk materials and it is a challenge to adapt them so that we can understand what really happens in metallic multilayers. Indeed, in the scope of our theory of pin-holes, we cannot decide whether or not irradiation defects are effective ferromagnetic bridges. Nevertheless disordered tubular volumes are created over several interatomic distances along the latent tracks of the incident ions.

We have chosen doses of 10^{10} and 10^{12} cm^{-2} Xe which correspond to distances between pin-holes of 1000Å and 100Å. These irradiations have been performed on Cu$_{8Å}$-Co$_{12Å}$ AF-coupled ML's prepared during the same sputtering batch for the best accuracy. As shown in fig. 5, the effect of irradiation is a shift of magnetization curves to a lower saturation field and an increase of remanent magnetization. This evolution of the magnetization curves is accompanied by a decrease of the giant magnetoresistance from 70% to 50% at 4.2K and from 35% to 20% at 300K (10^{12}cm^{-2} irradiated ML).

This result is coherent with an increase of the pin-hole concentration as shown in fig. 6 where calculated magnetization curves are plotted. Two runs of simulations with two different sizes of pin-holes have been performed : $d=5$Å and $d=10$Å. The interlayer coupling constant J has been set to a constant value of 0.25 erg/cm^2. The results of our simulations are summarized in table I, see also fig. 5 and 6 for magnetization curves.

dose (cm^{-2})	$d=5$Å	$d=10$Å
untreated	0.67±0.36% (61Å)	1.7±1.5% (76Å)
10^{10}	0.81±0.41% (56Å)	2.0±1.8% (71Å)
10^{12}	0.87±0.36% (54Å)	2.2±1.8% (67Å)

table I : pin-hole densities ($x=d^2/D^2$) deduced from our simulations of magnetization curves for irradiated Cu-Co ML's. Notice the large Δx. The resulting distance D is also reported (Å).

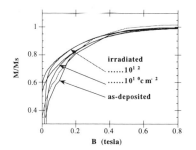

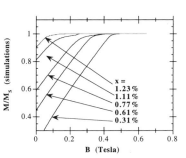

fig. 5 : experimental magnetization curv for untreated and irradiated [C$_{f2Å}$-Cu$_{8Å}$]$_{20}$ ML's. Notice the importan remanent magnetization.

fig. 6 : calculated magnetization curve with a pin-holes size of 5Å and various concentrations with the same nominal characteristics than fig. 5.

These results show that the effect of ion irradiation is small compared to the initial density of defects of the untreated ML as the average distance between pin-holes is found to be already 61Å ($d=5$Å) or 76Å ($d=10$Å) for the untreated sample, which is less than the 100Å caused by a 10^{12} cm^{-2} dose.

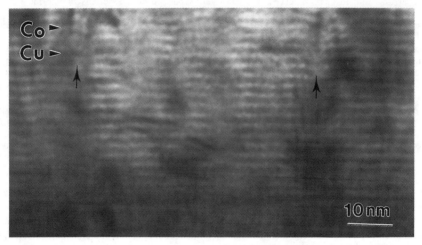

figure 7 : Cross-sectional TEM micrograph of an irradiated sample. The Si substrate and Cr buffer layer are also visible in the figure. Small arrows show bright pin-holes in the otherwise dark copper layers.

TEM observations of the same samples do not provide clear evidence that the pin-holes are more numerous (see fig. 7). This is however coherent with the low required density of pin-holes for our fits (~1%). Small angle X-ray scattering measurements on the same specimens do not reveal any change in the ML structure. All these facts make us think that irradiation has a very small effect on the structure of ML's but nevertheless a non negligible influence on the interlayer exchange coupling.

CONCLUSION

This work is the first study of irradiation effects on AF-coupled multilayers in order to create pin-holes. A decrease of the saturation field is observed for increasing doses on the samples, it is interpreted in terms of the presence of pin-holes with a micromagnetic model. Unfortunately, untreated samples do not present a perfect AF coupling : we have estimated that the effect of irradiation is smaller than 30% of the total number of defects.

According to our model, the fits of the magnetization curves of the irradiated ML's show an increase of the density of defects. TEM observation on the same samples does not clearly evidence for the presence of pin-holes. However, this absence of microscopic confirmation of our hypothesis by TEM is coherent with the low density of pin-holes which is required for an alteration of the AF coupling.

REFERENCES
[1] M.T. Johnson, S.T. Purcell, N.W.E. Mc Gee, R. Coehoorn, J. aan de Stegge and W. Hoving, Phys. Rev. Lett. 68 (1992) 2688
[2] J-F Bobo, M. Piecuch and E. Snoeck, Jour. of Mag. and Mag. Mat. 126 (1993) 440
[3] J-F Bobo, H. Fischer and M. Piecuch, Mat. Res. Soc. Symp. Proc. 313 (1993) 467
[4] A S Arrott and B Heinrich, J. of Mag. and Mag. Mat. 93 (1991) 571
[5] B. Rodmacq, G. Palumbo and Ph. Gérard, J. of Mag. and Mag. Mat. 118 (1993) L11
[6] A. Dunlop, P. Legrand,, D. Lesueur, N. Lorenzelli, J. Morillo, A. Barbu and S. Bouffard, Europhys. Lett. 15 (1991) 765
[7] A. R. Modak, S.S.P. Parkin and D.J. Smith, Ultramicroscopy 47 (1992) 375

PART IV

Polycrystalline Dielectric and Related Thin Films

INTERFACES IN FERROELECTRIC METAL OXIDE HETEROSTRUCTURES

R.Ramesh*, J.Lee*, V.G.Keramidas*, D.K.Fork**, S.Ghonge***, E.Goo*** and
O.Auciello****
*Bellcore, Red Bank, NJ 07701.
**Xerox Palo Alto Research Center, Palo Alto, CA 94305.
***Department of Materials Science and Engineering, University of Southern California
University Park, Los Angeles, CA 90089.
****MCNC Electronics Technology Division, Research Triangle Park, NC 27709-2889.

ABSTRACT

Realization of a viable nonvolatile ferroelectric thin film memory technology hinges on the successful solution of the reliability problems associated with the ferroelectric capacitor concurrent with the integration of these materials with the appropriate Si-CMOS based drive electronics. This integration process introduces a variety of structural, chemical, ionic and electronic interfaces in the memory elements. In this paper, the influence of some of the interfaces on the ferroelectric properties and on the process integration is discussed.

INTRODUCTION

There is currently a strong research and development effort directed at producing a commercially viable solid state, nonvolatile ferroelectric memory (FRAM) technology. Many laboratories are focused on integrating sub-micron thin ferroelectric capacitors of, for example, lead zirconate titanate (PZT), with the mature silicon based transistor technology to yield capacitor-transistor based memory architectures[1-5]. Realization of a commercially viable ferroelectric memory technology has been hampered by one or a combination of problems related to either the reliable performance of the PZT ferroelectric capacitor or to the growth and processing of capacitors that translate to high density memory elements. In Fig.1, the 1transistor - 1 capacitor architecture is schematically illustrated along with the ferroelectric capacitor structure. There are many types of interfaces in this stacked structure which can be classified based on their structural, chemical, ionic and electronic characteristics. If we focus our attention specifically on the electrode-ferroelectric-electrode capacitor stack, several types of interfaces can be identified. The crystal structure, orientation and distribution of crystallographic axes (i.e.,

Mat. Res. Soc. Symp. Proc. Vol. 343. ©1994 Materials Research Society

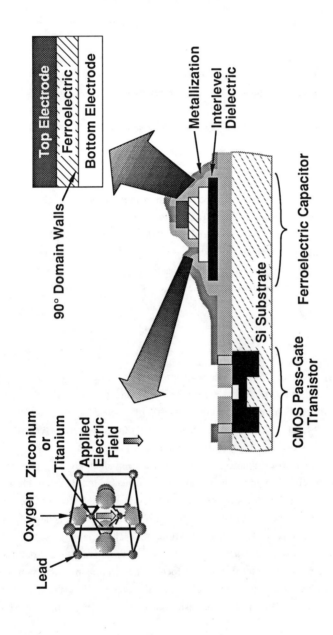

Figure 1: A schematic illustration of the 1 Transistor - 1 Capacitor architecture for the ferroelectric non-volatile random access memory technology. Also shown in this figure are the typical unit cell of the perovskite lead zirconate titanate structure and a schematic of the electrode-ferroelectric-electrode stack.

perfection) and chemistry of the bottom electrode determine to a large extent the corresponding properties of the ferroelectric film and the relative phase stability within it (e.g., perovskite versus pyrochlore). The electrode-ferroelectric interface is very crucial in determining the reliability characteristics of the ferroelectric capacitor, such as fatigue, which is the loss of switchable polarization when subjected to repeated bipolar pulses. This fatigue problem can be overcome for all practical purposes by replacing **metallic** Pt electrodes with **metal oxide** electrodes such as RuO_2, or any of the large number of perovskite metal oxides such as Y-Ba-Cu-O, La-Sr-Co-O, etc.[6-13]. Similarly, internal interfaces in the ferroelectric, such as domain walls and grain boundaries influence the ferroelectric properties.

In order to fabricate a commercially viable ferroelectric memory, these capacitors have to be integrated with Si-based CMOS technology. This raises an important problem, that of growing the complex metal oxides such as PLZT on Si. Direct growth of these lead-based compounds on Si (or SiO2/Si) at high temperatures leads to the formation of very stable binary silicates such as lead silicate. Indeed, this is generally true if the film contains any of the highly reactive Group I or II cations (such as Ba, Ca, K, etc.). Consequently, integration of such metal oxides on Si requires novel approaches to provide chemical isolation of the overlayers from the substrate and also yield the desired crystallographic phase and orientation in the film.

In our research program we are addressing these two important issues related intimately to the engineering of interfaces in the capacitor stack and to the Si wafer through two novel approaches. As described earlier, we are using perovskite metallic oxides as top and bottom electrodes instead of Pt ; this, under optimized processing conditions, yields capacitors that show very little bipolar fatigue at least up to 10^{12} cycles. Two different types of oxide electrodes have been discussed, namely, the layered superconducting perovskites such as Y-Ba-Cu-O (YBCO) [9] and the cubic metallic perovskites such as La-Sr-Co-O (LSCO) [10-13]. These perovskite oxides have room temperature resistivities of the order of a few hundred μohm.cm (for example, epitaxial LSCO has a resistivity of 90μΩ.cm at room temperature). In comparison, Pt has a resistivity in the range of 5-20μΩ.cm at room temperature. In an attempt to lower the effective sheet resistance of the bottom electrode while maintaining the good fatigue characteristics of the LSCO based capacitors, we grew the LSCO/PLZT/LSCO capacitor stack on epitaxial or highly [001] oriented Pt. This can be achieved by growing the Pt layer on a single crystal [100] MgO substrate. The integration of these capacitors with metal oxide electrodes on Si can be accomplished using a novel approach that involves the growth of a thin layer of an anisotropic perovskite **template** layer such as bismuth titanate on the thermally oxidized Si wafer prior to the deposition of the Pt and subsequently the LSCO/PLZT/LSCO capacitor stack[14]. This bismuth titanate template layer grows with a preferred c-axis texture, even on an amorphous surface such as SiO2/Si. It not only provides a good chemical barrier between the overlayers and the Si wafer, but also provides a perovskite-like surface (since it is c-axis oriented and hence the a-b plane is exposed) for the subsequent growth of the cubic perovskite layers. The fact that this template layer itself has a perovskite structure and crystal chemistry is taken advantage of, to nucleate the [001] orientation of the perovskite phase preferentially over other competing orientations of the perovskite phase and other competing phases. Details of the growth and testing of the reliability characteristics of test capacitors with perovskite metal oxide electrodes and the growth process using bismuth titanate template layers are described in earlier papers[10,14]. This paper provides an overview of some of the progress so far in our understanding of the structural, chemical, ionic and electronic properties of the interfaces in the ferroelectric capacitor structure and our attempts to engineer them to obtain the desired integrated capacitor structures.

RESULTS

The films grown using LSCO top and bottom electrodes show a strong c-axis texture with the degree of in-plane texture becoming stronger as the substrate is altered from SiO2/Si to single crystal LaAlO3. We have carried out detailed x-ray diffraction experiments on the heterostructures grown on LaAlO3, YSZ/Si and SiO2/Si (both with the bismuth titanate template layer). Although in all three cases we find that the LSCO/PLZT/LSCO stack is c-axis oriented, the degree of in-plane registry depends on the type of substrate. Fig.2 illustrates x-ray pole figure scans from LSCO/PLZT/LSCO heterostructures grown on these three substrates. The degree of in-plane registry is comparable in the case of heterostructures grown on LaAlO3 and BTO/YSZ/Si , but the heterostructure grown on BTO/SiO2/Si shows very weak in-plane registry, indicating that there is a large density of rotational and/or tilt boundaries in the stack, although the individual grains are all c-axis oriented. High resolution electron microscopy (HREM) shows that the interfaces between the electrode and the ferroelectric in these all-oxide heterostructures are free of second phases as demonstrated in Fig.3 [15].

PLZT ferroelectric capacitors grown using either Pt or LSCO as the top and bottom electrode show excellent hysteresis properties as illustrated in Fig.4. Under appropriate processing conditions, switched polarization values approaching that theoretically expected

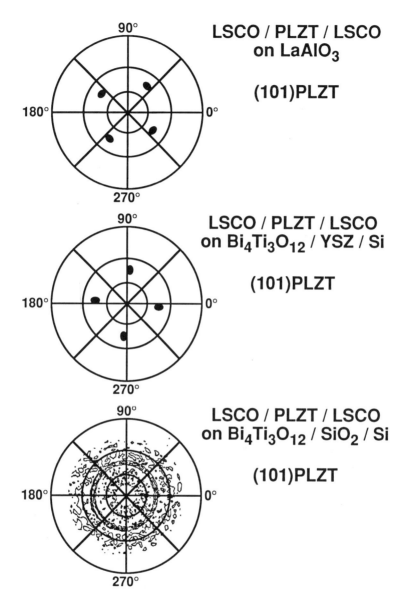

Figure 2 : X-ray pole figure scans for LSCO / PLZT / LSCO heterostructures grown on [001] LaAlO$_3$, Bi$_4$Ti$_3$O$_{12}$ / YSZ / Si and Bi$_4$Ti$_3$O$_{12}$ / SiO$_2$ / Si substrates. The scans were obtained about the [101]PLZT direction.

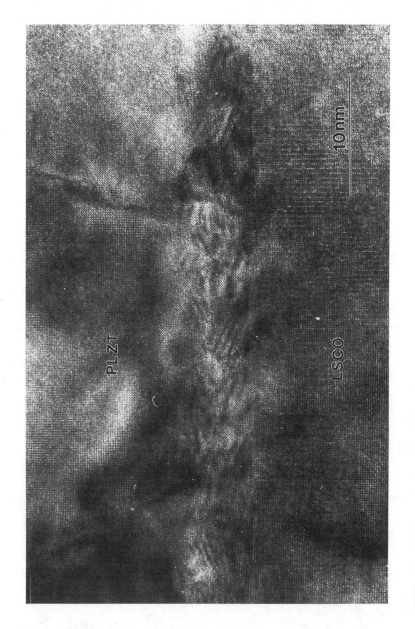

Figure 3 : A high resolution transmission electron micrograph of the LSCO / PLZT interface showing that it is free of second phases. The strain contrast is due to the defects created by the lattice mismatch at the interface.

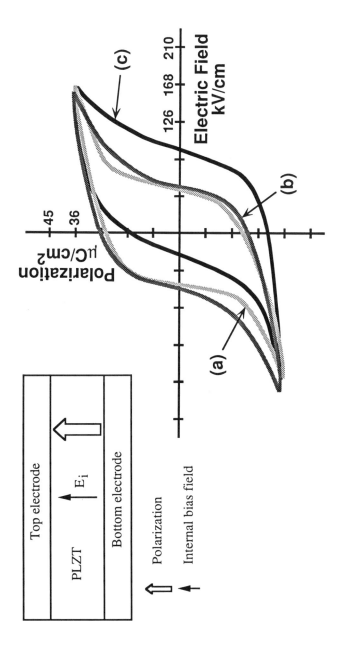

Figure 4 : Pulsed hysteresis loops for films grown with : (a) Pt top and bottom electrodes and PLZT grown by sol-gel processing ; (b) LSCO top and bottom electrodes and the PLZT grown by laser deposition and cooled at 1 atmosphere of oxygen ; (c) LSCO top and bottom electrodes, but cooled in 10^{-6} Torr of oxygen after deposition, showing an internal bias field. The inset shows the capacitor stack with the direction of the internal bias field and the polarization direction.

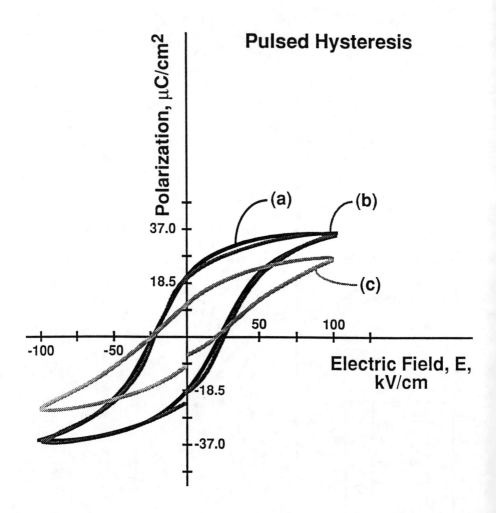

Figure 5 : Comparison of pulsed hysteresis loops for LSCO / PLZT / LSCO heterostructures grown on : (a) [001] LaAlO$_3$; (b) Bi$_4$Ti$_3$O$_{12}$ / YSZ / Si and (c) Bi$_4$Ti$_3$O$_{12}$ / SiO$_2$ / Si substrates.

can be obtained. In general, the hysteresis loops are quite symmetric, but shifts in them (i.e., built-in internal fields) can be induced either through band-gap optical irradiation [16] or by deviations in the processing conditions from the optimum. For example, in the case of capacitors with LSCO electrodes, the post-deposition ambient plays an important role in producing this shift. Cooling under optimum conditions (i.e., one atmosphere of oxygen) yields generally a symmetric loop; cooling in 10^{-6} Torr of oxygen yields a loop that is shifted dramatically, as shown in Fig.4. The origin of the shift can be traced back to internal fields that are generated in the capacitor. Based upon the polarity of the applied field, the direction of the internal field can be deduced as illustrated schematically in the inset to Fig.4. However, the microscopic cause for the internal field is still unclear and needs further study. The differences in the in-plane structural properties induced by the substrate (viz., LaAlO3, YSZ/Si or SiO2/Si) also manifest themselves in the ferroelectric hysteresis properties which is shown in Fig.5. As expected, the capacitors with the highest crystalline quality (i.e., the ones grown on LaAlO3 and YSZ/Si) show better hysteresis and remnant polarization values compared to the capacitors fabricated from the heterostructures grown on SiO2/Si. Nevertheless, all the three capacitor stacks have ferroelectric properties suitable for ultimate integration into nonvolatile memories.

When the metal (i.e., Pt) electrode is replaced by metal oxide electrodes such as RuOx or LSCO, the bipolar fatigue characteristics improve dramatically [10]. This difference is comparatively illustrated in Fig.6 for capacitors with Pt or LSCO as the top and bottom electrodes. The main question is : what is the origin of this difference in fatigue behaviour as a function of electrodes ? It is very likely related to the nature of the electrode-ferroelectric interface since replacing even one of the metal oxide electrodes by Pt induces fatigue in the capacitor [10]. It is also very likely that charged point defects such as oxygen vacancies play an important role in influencing the fatigue behaviour. For example, systematically removing oxygen from the heterostructure leads to a corresponding degradation in the fatigue characteristics eventhough the electrode in contact with the ferroelectric is still a metal oxide.

As illustrated in Fig.1, during the processing of integrated ferroelectric capacitors, many other types of interfaces are created. Specifically, to fabricate capacitors suitable for high density memories (i.e., size in the range of 1-5μm), full wafer etching (preferably dry etching) is required to delineate the capacitor and an interlevel dielectric (ILD) is required to isolate the top and bottom electrode. This ILD, which is typically glass (i.e., SiO_2) or a high temperature polyimid, is in direct contact with the PLZT layer, as can be seen from Fig.1. Consequently, the potential for a chemical reaction between SiO_2 and PLZT exists, especially, if the stack is exposed to elevated temperatures after the ILD deposition. In our

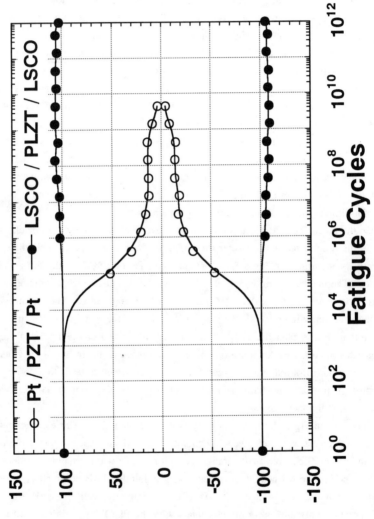

Figure 6 : Comparison of the bipolar fatigue characteristics of PLZT with Pt and LSCO electrodes. Note the rapid drop in remnant polarization in the case of the capacitor with Pt electrodes.

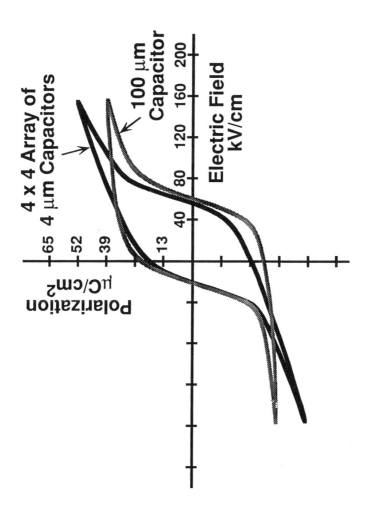

Figure 7 : Pulsed hysteresis loops of 100 μm diameter and 4μm diameter (4x4 array) test capacitors illustrating scaling of the ferroelectric properties with size.

program, we are exploring different routes for the SiO_2 ILD deposition, namely, sputtering and electron beam evaporation as well as other ILD materials. Our experiments so far using SiO_2 as the ILD material show no significant deleterious effect on the ferroelectric properties. These ILD deposition experiments, carried out simultaneously with ion beam etching of the test capacitors, show ferroelectric polarization values in 4μm diameter test capacitors commensurate with those of 100μm diameter test capacitors, as shown in Fig.7. The fully processed test capacitors (i.e., ion milled and with ILD deposition) also show very desirable fatigue, aging and retention characteristics.

SUMMARY AND CONCLUSIONS

Over the past few years, considerable progress has been made in the integration of PLZT-based ferroelectric thin film capacitors with Si-CMOS based drive electronics. Solutions to reliability problems such as fatigue have become possible through suitable modifications of the ferroelectric capacitor stack through the use of metallic oxide electrodes. The use of an anisotropic perovskite template such as bismuth titanate provides a novel route to the growth of highly oriented cubic perovsite thin films (such as PLZT and LSCO) on a thermally oxidized Si surface. Initial results of full wafer processing and the use of SiO2 as an interlevel dielectric show that capacitors as small as 2-4μm can be fabricated without prohibitive loss of ferroelectric properties and reliability characteristics.

REFERENCES

1. See for example, Proc. of MRS Spring 1993 Meeting, San Francisco, CA, April 1993, Eds. E.R.Myers, B.A.Tuttle, S.B.Desu and P.K.Larsen, Materials Research Society, Pittsburgh, PA (1993); Proc. of Fifth Int. Symp. on Integrated Ferroelectrics, Colorado Springs, CO., Ed. C.A. Paz de Araujo, University of Colorado, Colorado Springs, CO.,April 1993.
2. J.F.Scott and C.A.Paz de Araujo, Science, 246, 1400(1989); M.Sayer and K.Sreenivas, Science 247, 1056(1990); G.H.Haertling, J. of Vacuum Science and Technology, 9, 414(1991).
3. S.Sinharoy, H.Buhay, D.R.Lampe and M.H.Francombe, J. of Vac. Science and Technology, A10, 1554(1992).

4. J.T.Evans and R.D.Womack, IEEE J. of Solid State Circuits, 23, 1171(1988).

5. S.K.Dey and R.Zuleeg, Ferroelectrics, 108, 37(1990).

6. R.Ramesh, A.Inam, B.Wilkens, W.K.Chan, D.L.Hart, K.Luther, and J.M.Tarascon, Science, 252, 944(1991).

7. N.E.Abt, P.Misic, D.Zehngut and E.Regan, Proc. of Fourth Int. Symp. on Integrated Ferroelectrics, Monterey, CA, Ed. C.A. Paz de Araujo, University of Colorado, Colorado Springs, CO., March 1992, pp 533-540; S.B.Desu and I.K.Yoo, ibid, pp640-656.

8. R.M.Wolf, in Materials Research Society Fall Meeting Symposium on "Ferroelectric Thin Films II", Boston, MA, Dec., 1991, Eds. A.Kingon, E.R.Myers and B.Tuttle, Materials Research Society, Pittsburgh, PA; J.T.Cheung, and R.R.Neurgaonkar, in Proc. of Fourth Int. Symp. on Integrated Ferroelectrics, Monterey, CA, March 1992, Ed. C.A. Paz de Araujo, University of Colorado, Colorado Springs, CO.

9. R.Ramesh, W.K.Chan, B.Wilkens, H.Gilchrist, T.Sands, J.M.Tarascon, V.G.Keramidas, D.K.Fork, J.Lee and A.Safari, Appl. Phys. Lett., 61, 1537(1992); J.Lee, L.Johnson, A.Safari, R.Ramesh, T.Sands, H.Gilchrist and V.G.Keramidas, Appl. Phys. Lett., 63, 27(1993).

10. R.Ramesh, H.Gilchrist, T.Sands, V.G.Keramidas, R.Haakenaasen and D.K.Fork, Appl. Phys. Lett., 63, 3592(1993).

11. C.B.Eom, R.B.VanDover, J.M.Phillips, R.M.Fleming, R.J.Cava, J.H.Marshall, D.J.Werder, C.H.Chen and D.K.Fork, in Proc. of MRS Spring Meeting Symposium, "Ferroelectric Thin Films III", San Francisco, CA, April 1993, pp145-150, Eds. E.R.Myers, B.A.Tuttle, S.B.Desu and P.K.Larsen, Materials Research Society, Pittsburgh, PA (1993).

12. R.Ramesh, J.Lee, B.Dutta, T.S.Ravi, T.Sands and V.G.Keramidas, in Proc. of Fifth Int. Symposium on Integrated Ferroelectrics, Colorado Springs, CO, March 1993, Ed. C.A. Paz de Araujo, University of Colorado, Colorado Springs, CO, Integrated Ferroelectrics, in press.

13. D.J.Lichtenwalner, R.Dat, O.Auciello and A.I.Kingon, in Proc. of the Eighth Int. Meeting on Ferroelectricity, August 1993, Maryland, USA, in press,1993.

14. R.Ramesh, T.Sands, V.G.Keramidas, J. of Electronic Materials, 23, 19(1994); R.Ramesh, J.Lee, T.Sands, V.G.Keramidas and O.Auciello, Appl. Phys. Lett., in press.

15. S.Ghonge, E.Goo, R.Ramesh, R.Haakenaasen and D.K.Fork, Appl. Phys. Lett., in press.

16. D.Dimos, W.L.Warren, M.B.Sinclair, B.A.Tuttle and R.W.Schwartz, submitted to J. of Appl. Phys.

CHARACTERIZATION OF SOL-GEL DERIVED (Pb,Ba)(Zr,Ti,Nb)O$_3$ THIN FILMS FOR OPTICAL WAVEGUIDE DEVELOPMENT

P. F. Baude, J. S. Wright†, C. Ye, L. F. Francis†, and D. L. Polla
Microelectromechanical Systems Center
Electrical Engineering Department
†Chemical Engineering and Materials Science Department
University of Minnesota
200 S. E. Union Street
Minneapolis, MN 55455

ABSTRACT

(PbBa)(ZrTiNb)O$_3$ thin films and powders have been prepared using the sol-gel technique. Solutions were synthesized in 2-methoxyethanol based upon our previous PZT solution preparation. Three different approaches were used for incorporating barium into PZT alkoxide solutions. Thermal analysis and x-ray diffraction results indicated that barium methoxypropoxide gave the best results. PBZTN (71% Pb and 71% Zr) was deposited onto sapphire substrates as well as oxidized silicon substrates. Optical transmission measurements showed greater than 80% transmission for wavelengths longer than 400 nm. Films with thickness of 3000 Å on sapphire exhibited a refractive index of 2.19 at λ=633 nm.

INTRODUCTION

Ferroelectric thin films have recently attracted attention for use in optical and electrooptical devices. Perovskite, ABO$_3$, ferroelectric films have been demonstrated to be viable candidates for a variety of device applications including passive optical waveguides, optical waveguide modulators, pyroelectric detectors, and optical memories [1,2]. The ability to integrate active optical waveguide devices with silicon CMOS control circuitry in a Micro-Electro-Mechanical System (MEMS) is important for the realization of a family of optical based sensor and actuator devices. The concept of monolithic integration of surface micromachined ferroelectric pressure sensors with high voltage CMOS and optical waveguides, as an example, is depicted below in Fig. 1. Materials that fulfill the requirements of both devices would be attractive for processing. As discussed below, materials in the (PbBa)(ZrTiNb)O$_3$ (PBZTN) system may fulfill such requirements.

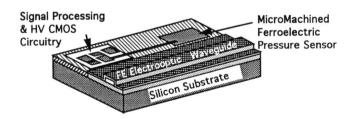

Fig. 1.   Schematic of integrated micromachined ferroelectric pressure sensor with CMOS circuitry and an electrooptic waveguide.

Fig. 2 shows the phase diagram for (Pb,Ba)(Zr,Ti)O$_3$ as reported for bulk ceramics by Ikeda [3]. Ikeda's work identified (Pb,Ba)(Zr,Ti)O$_3$ ceramics as an attractive material system for ferroelectric and piezoelectric device applications. Thacher [4] reported birefringence measurements on bulk ceramics and demonstrated the large electrooptic g_{ij} coefficients in (Pb$_{0.71}$Ba$_{0.29}$)$_{0.99}$[(Zr$_{0.71}$Ti$_{0.29}$)$_{0.98}$Nb$_{0.02}$]O$_3$ [PBZTN (71/29)]. Based on Thacher's results, we chose the PBZTN (71/29) composition for our thin film study. This composition, in bulk form, is rhombohedral and ferroelectric and is characterized by a high electrooptic coefficient, relaxor behavior, and high dielectric constant (ε_r=2200) and electromechanical coupling constant.

In addition to the attractive material properties, PBZTN may offer advantages in processing. Much of the our previous research on thin film optical waveguides has concentrated on the (Pb,La)(Zr,Ti)O$_3$ (PLZT) system. Swartz *et al* first showed that PLZT (9/65/35) films deposited on a SiO$_2$ bottom optical cladding layer [5], require a buffer layer of (Pb,La)TiO$_3$ or PbTiO$_3$ for film adhesion and crack free films. We later confirmed this result [6]. This interlayer usually has a refractive index greater than that of PLZT (9/65/35) which acts to move the guided optical mode into the buffer layer. This reduces the electrically induced Δn in the active layer. Our initial results with the PBZTN material indicate that such a buffer layer is not required.

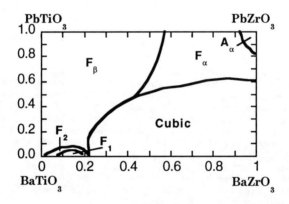

Fig. 2. Phase diagram of bulk PbBaZrTiO$_3$ showing a rhombohedral ferro-electric phase at 71% Pb and 71% Zr. From Ikeda [3].

SOLUTION PREPARATION

Sol-gel solutions of PBZTN (71/29) were prepared using a 2-methoxyethanol based system similar to that used in our Pb(Zr,Ti)O$_3$ solution processing [7]. More recently Haertling [8] discussed the successful preparation of a water soluble PBZT solution. By using such a system Haerlting prepared several compositions and showed that the thin film dielectric properties closely matched that of bulk ceramic. Our 2-methoxyethanol processing of PBZTN was motivated largely by our experience with PZT solution synthesis.

The precursors used were lead acetate trihydrate [Pb(CH$_3$COO)$_2$-3H$_2$O], zirconium propoxide [Zr(OCH$_2$CH$_2$CH$_3$)$_4$], titanium isopropoxide [Ti(OC$_3$H$_7$)$_4$], and niobium isopropoxide [Nb(OC$_3$H$_7$)$_5$]. The processing of this system follows that developed by Budd *et al* [9] for PZT. All procedures were carried out in dry N$_2$. We examined three methods for incorporating the barium into the PZTN solution. Barium hydroxide dissolved in acetic acid was used to prepare barium acetate in addition to the use of barium acetate powders. The former required the mixing of the barium hydroxide monohydrate powder with 2-methoxyethanol and then adding the stoichiometric amount of acetic acid and distilled water with thorough stirring. The barium acetate powder route required the addition of water for dissolving the barium precursor into the 2-methoxyethanol solvent. Just prior to spin coating the barium solution was mixed in proper amounts with the prepared PZTN solution with no additional water added. The mixed solution had a short shelf life and gelled within 36 hours.

Powders and thin films were prepared for thermal analysis and x-ray diffraction and are discussed in the following section. The powders were processed by placing the solution in a glass container and placing inside a oven at 140 °C for 6 hours. As a result of the x-ray diffraction and thermal analysis results an alkoxide approach was chosen for the barium precursor. Barium II methoxypropoxide [C$_8$H$_{18}$O$_4$Ba] obtained from Gelest was dissolved into 2-methoxyethanol and added (under the same dry N$_2$ system) to the prepared PZTN solution. The solution viscosity was modified with the addition of the 2-methoxyethanol solvent. After hydrolysis the solutions were spin coated on a variety of substrates. Metallized (Pt/Ti) and oxidized silicon substrates were used in addition to polished sapphire disks for optical tranmission measurements.

THERMAL ANALYSIS and X-RAY DIFFRACTION

Differential thermal analysis (DTA) and thermal gravimetric analysis were used to determine the crystallization and pyrolysis behavior of the PBZTN solutions. Powders were obtained by placing the hydrolyzed solution into a 130 °C oven for 8 hours. XRD of PBZTN powders was used to determine the crystal structure of the annealed powders.

DTA results for the three synthesis methods are shown in Fig. 3. For the barium acetate precursors (b and c) several exotherms occur. Included are the expected organic removal at temperatures between 200 and 400 °C and peaks at higher temperatures. The three exotherms observed in (c), the barium acetate powder, indicate the possibility of crystallization of multiple phases. This was confirmed by XRD which showed diffraction peaks for a lead barium oxide phase and other unidentified peaks. The same type of multiphase development was observed for the precursor made from barium hydroxide in acetic acid (data labelled b in Fig. 3). No perovskite formation was seen in the x-ray diffraction for both methods (b and c). Although the thin films from these precursors on sapphire, metallized silicon and oxidized silicon substrates were crack free and exhibited excellent adhesion, the dielectric and optical properties were expected to suffer greatly from the non-PBZTN phases.

The PBZTN powder prepared with the Ba methoxypropoxide precursor (c) had a more homogeneous phase development. After the organic decomposition exotherms, a sharp exothermic peak occurs at 550 °C followed by a smaller exotherm at 650 °C. XRD analysis on powders removed from the DTA shows the formation of the perovskite phase with a characteristic peak at $2\theta = 31.5$ ° (Fig. 4.). This peak first appears at 580 °C. Crystallinity is enhanced after heating to 800 °C. Other peaks have been attributed to Al$_2$O$_3$ powder which was used to cover the sample in the DTA analysis.

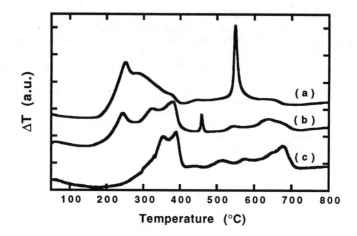

Fig. 3.   Differential thermal analysis of PBZTN powders processed using three different methods for incorporating barium. (a) used barium methoxypropoxide, (b) barium hydroxide in acid and (c), barium acetate powder in water.

Based on XRD and DTA results thin films were prepared using the PBZTN system with the Ba alkoxide precursor. After spin coating, samples were annealed at 250 °C and 475 °C for 15 minutes and then heated again at 700 °C for 30 minutes for crystallization. XRD of films deposited on sapphire and SiO_2 showed presence of the characteristic perovskite peak. More research is underway to further characterize the phase development.

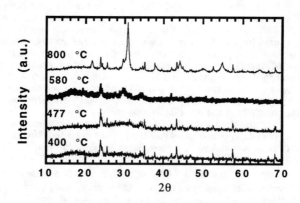

Fig. 4.   X-ray diffraction spectra for PBZTN powder processed using barium methoxypropoxide. Samples were taken from the DTA analysis

448

OPTICAL PROPERTIES

The refractive index and optical transmission properties of thin film PBZTN (71/29) deposited on sapphire were examined. The tranmission spectrum for a 3000 Å thick film annealed at 700 °C is shown in Fig. 5. As with most perovskite oxides the transmission is high for wavelengths in the visible region. The transmission is greater than 84% for $\lambda > 380$ nm. The absorption becomes significant at wavelengths less than about 350 nm. It is clear that radiation sources including Ar-ion and HeNe gas lasers and commercially available laser diodes operating at wavelengths between 500 and 1200 nm will show low loss due to absorption. Of course the main contribution to optical loss in these ferroelectric oxides is expected to arise from grain and defect scattering although these measurements have not been made yet in our films.

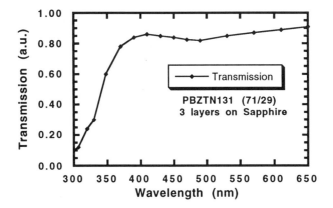

Fig. 5.   Optical transmission through PBZTN thin film on sapphire substrate fired at 700 °C using barium methoxypropoxide. Film thickness was approximately 3000 Å.

The refractive index was measured for several films deposited using barium methoxypropoxide. The refractive index varied slightly with the annealing temperature as expected. For films fired at 700 °C the refractive index of PBZTN (71/29) on sapphire was measured to be 2.19 ±0.07 (at λ=633 nm). This value is significantly lower than our PLZT (9/65/35) and PLT (10% La) thin films which exhibit refractive indices on the order of 2.48 and 2.55 respectively (at λ=633 nm). As noted by Thacher [4] there is a significant dispersion in the PBZTN refractive index. Our measured refractive index agrees well with the value observed by Thacher for the same composition at 633 nm. It should be noted that although it is not expected that the thin film polycrystalline PBZTN (71/29) have an index equal to the bulk, it is possible that interfacial layers will contribute a lower refractive index to the experimental result.

SUMMARY

$(Pb,Ba)(Zr,Ti,Nb)O_3$ thin films have been prepared by sol-gel on sapphire, metallized silicon and oxidized silicon substrates. Barium acetate precursors using both barium acetate powder and barium hydroxide in acetic acid showed did not form the perovskite phase due to inhomogenieties. The use of a barium methoxypropoxide improved the formation of a perovskite phase as shown in DTA and x-ray diffraction. Thin films spin deposited exhibited good adhesion and no cracking. The measurement of the dielectric and electrooptic coefficient of the PBZTN thin films is on-going.

ACKNOWLEDGEMENTS

The authors would like to acknowledge the helpful technical discussions with P. Schiller and T. Cooney and the University of Minnesota's Center for Interfacial Engineering and Microelectronics Laboratory for Research and Education, MLRE. This work was supported in part by the National Science Foundation.

BIBLIOGRAPHY

1.　　C. Ye, T. Tamagawa, P. Schiller, and D. L. Polla, Sensors and Actuators A, **35**, 77-83, 1992
2.　　P. F. Baude, C. Ye, T. Tamagawa, and D. L. Polla, J. Appl. Phys. **73**, 7960 (1993)
3.　　T. Ikeda, J. Phys. Soc. Japan **14**, 168 (1959)
4.　　P. D. Thacher, J. Appl. Phys. **41**, 4790 (1970)
5.　　S. L. Swartz, P. J. Melling, S. J. Bright, and T. R. Shrout, Ferroelectrics **108**, 71 (1990)
6.　　P. F. Baude, C. Ye, and D. L. Polla, Ferroelectric Thin Films II. edited by A. I Kingon, E. R. Myers, and B. Tuttle (Mater. Res. Soc. Proc. **243**, Pittsburgh, PA (1992) pp. 275-280
7.　　D. L. Polla, C. Ye, and T. Tamagawa, Appl. Phys. Lett. **59**, 3539, 1991
8.　　G. Haertling, presented at the 6th International Symposium on Integrated Ferroelectrics, Monterey, CA, 1994
9.　　K. D. budd, S. K. Dey, and D. A. Payne, Brit. Ceram. Proc. **36**, 107 (1989)

TEXTURE VARIATIONS IN SOL-GEL DERIVED PZT FILMS ON SUBSTRATES WITH PLATINUM METALLIZATION

J.G.E. GARDENIERS*, M. ELWENSPOEK* AND C. COBIANU**
* MESA Research Institute, P.O. Box 217, NL-7500 AE Enschede, The Netherlands
** Center of Microtechnology, Bucharest, Romania.

ABSTRACT

Metalorganic precursor solutions of composition Zr : Ti = 0.53 : 0.47 were used to spin-cast PZT layers on sputtered Pt films. After annealing at temperatures of 550 °C - 800 °C, the PZT films of tetragonal perovskite structure reproducibly showed different textures and surface morphologies, depending on whether or not a Ti layer was used as an adhesion layer for the Pt film. The texture differences were found to be independent of annealing treatment. It is argued that the observed texture differences are caused by a change in Pt-PZT interface composition, resulting from the diffusion of Ti into the Pt film during annealing; X-ray diffraction of an annealed Pt/Ti/SiO$_2$/Si film combination provided evidence for a compound Pt$_3$Ti. Annealing at 850 °C caused severe diffusion of Ti from the metal layer into the PZT film, leading to a tetragonal PZT layer with lattice constants corresponding to a Zr : Ti ratio of 30 : 70.

INTRODUCTION

In recent years the interest in thin films of perovskite-type materials like PbZr$_x$Ti$_{1-x}$O$_3$ (PZT) has increased considerably because of the applications in e.g. ferroelectric non-volatile memory devices and piezoelectrically driven micro electromechanical systems. An important issue in the use of PZT films for these and other applications is the choice of the (bottom) electrode material. Because of the relatively high temperatures involved during deposition or annealing (500°C and higher) and the reactivity of PZT, interdiffusion of layers may occur, leading to undesired, irreproducible electrical changes in the films. Prevention of these effects demands the use of chemically inert materials like Pt, which, because of the inertness, in most cases also requires the addition of an adhesion layer (e.g. Ti). In some cases an additional diffusion barier is needed to prevent reaction of the metal package with the (silicon) substrate.

In this paper it will be shown that the use of Pt-Ti layer combinations has important implications for the structural properties (in particular the texture) of sol-gel PZT films. The experimental work will mainly consist of an X-ray diffraction (XRD) study, while additional information is gathered from scanning electron microscopy (SEM) and Energy Dispersive X-ray spectroscopy (EDX).

EXPERIMENTAL

A solution of organometallic precursors in 2-methoxyethanol with mole fraction ratios Pb:Zr:Ti = 1.05:0.53:0.47 and a total Zr + Ti concentration of 0.50 M was prepared, to which a mixture of HNO$_3$-H$_2$O-2-methoxyethanol was added as a catalyst for gel formation (final concentrations of HNO$_3$ and H$_2$O in the solution: 1.4·10^{-3} M and 0.51 M, resp.). Preparations of the solutions were carried out as described by Udayakumar et al. [1]. After filtering, the solution was spin-cast (30 sec. at 2000 rpm.) on metallized substrates. The films were dried for 20 min. on a hotplate at 120 °C in air, and subsequently fired for 20 min at 400 °C in an oven in a flow of pure O_2. To promote crystallization of the films, a high temperature annealing step in pure O_2 was performed at a temperature between 400 °C and 850 °C; different annealing times and temperature ramping procedures were investigated. The substrates used were 3" thermally oxidized silicon wafers coated with either a Pt layer or a Pt layer with a Ti adhesion layer underneath. These metal films were deposited by DC magnetron sputtering from pure metal targets in Ar gas. Full details of the deposition and annealing of the metal films will be given elsewhere [2].

The PZT films were characterized with XRD, SEM, and EDX. Film thicknesses were determined with ellipsometry. XRD was performed with a Cu source; all diffraction angles 2θ mentioned in this paper are related to a wavelength of 0.154186 nm (CuKα). In all measurements the Si (004) substrate peak was found at a 2θ value of within 0.01° of the theoretical value.

RESULTS

Fig.1 shows XRD spectra of a 60 nm thick PZT film, resulting from a single spin-casting step on a substrate containing a Pt film on Ti. The film was fired at 400 °C for 20 min. and subsequently annealed at increasing temperatures up to 850 °C. Annealing at a particular temperature was performed for 30 min, while in between the annealing steps the samples were heated up and cooled down (to 25 °C) as fast as possible. The spectra show that after firing at 400 °C an amorphous film results. Crystallization does not occur below 500 °C; the corresponding spectrum shows faint PZT diffraction peaks at 22.0° and 44.8°, and somewhat stronger diffractions at 14.5°, 29.7° and 35.8°. The latter diffraction peaks correspond well with those expected for a pyrochlore-type oxide, $Pb_2(Zr,Ti)_2O_{(7-x)}$ [3,4]. The pyrochlore diffraction peaks remain visible until 650 °C. Their relative intensity with respect to those of the PZT diffractions decreases considerably in this temperature range. The film which results at 650 °C shows PZT diffractions for 2θ-values of 21.95°, 38.37° (this is not the exact value, since the peak is positioned on the "foot" of the Pt (111) peak at 39.91°) and 44.75°, corresponding to the (100)/(010), (111) and (200)/(020) diffractions, resp., of tetragonal PZT with lattice parameter $a = 0.4049$ nm. The literature value for a for the chosen Zr:Ti ratio of 53:47 is 0.4040 nm [5], which implies that the PZT film either is in a state of compressive stress (giving a strain of 0.23 % perpendicular to the film surface) or has a composition with a slightly higher Zr : Ti ratio (which was also found in the EDX measurements). In addition to these changes, it is observed that the α-Ti (002) peak, which is present in the spectra at 2θ = 38.22° at temperatures up to 600 °C, has disappeared in the spectra for temperatures of 650 °C and higher, while from 550 °C on a new diffraction peak is observed, at 46.50°.

Subsequent heating of the sample does not give structural changes in the PZT film, until a temperature of 850 °C is reached. At this temperature new diffraction peaks appear, at 22.46° and 45.88°, while the PZT (111) peak seems to have disappeared from the spectra; these peaks are probably due to a tetragonal PZT phase with a Zr:Ti ratio of 30:70. Furthermore, a broad diffraction peak arises at 30.47°, which is either due to a pyrochlore phase $Pb(Zr,Ti)_3O_7$ [4] or to yellow-PbO (002) [6]. Additionally, peaks due to TiO_2 (rutile) are observed at 27.50°, 36.19° and 54.48°.

Fig. 2 shows an XRD spectrum of a PZT film deposited on Pt without Ti adhesion layer, which was annealed at 700 °C (after the drying and firing steps mentioned before). It can be seen

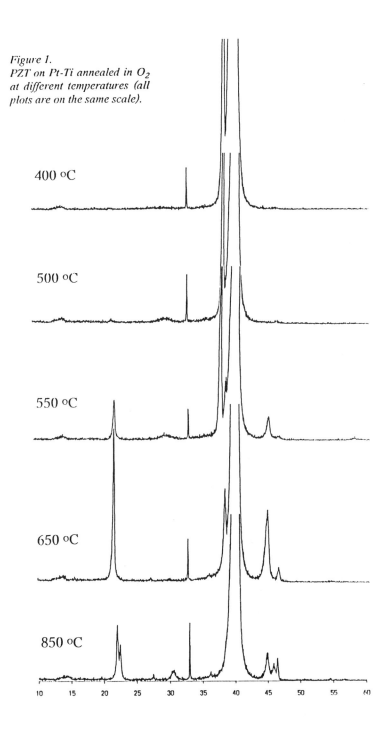

Figure 1.
PZT on Pt-Ti annealed in O_2
at different temperatures (all
plots are on the same scale).

400 oC

500 oC

550 oC

650 oC

850 oC

that the texture of this film is different from that in fig. 1: peaks are found for PZT (101)/(011) at 31.04° and for PZT (121)/(211) at 55.40°. At 29.48° a broad peak is found, which is probably due to pyrochlore (222).

Fig. 2.
PZT on Pt (without Ti)
annealed at 700 °C

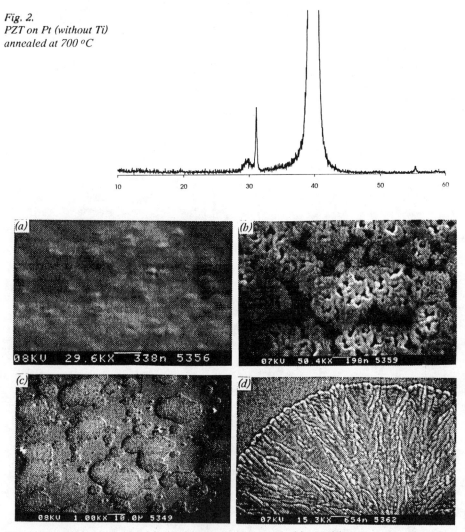

Fig. 3. SEM photographs of: (a) and (b) PZT on Pt with Ti; (c) and (d) PZT on Pt without Ti; films of (a), (c) and (d) were annealed in O_2 at 700 °C, (b) in O_2 at 850 °C; (d) is an enlarged view of part of (c).

The above-mentioned texture differences between PZT on Pt with and without Ti adhesion layer turned out to be reproducible, and independent of the heating treatment (annealing temperature and time, ramping procedures) of the films.

The surface morphology of the two films was also markedly different (fig. 3): a film on Pt without Ti shows the "rosettes" often encountered in PZT films crystallizing from an amorphous matrix [7], while a PZT film on Pt with Ti underneath shows a smooth surface. When the latter film was annealed at 850 °C, it became porous, probably because of extreme PbO evaporation.

Fig. 4 shows XRD spectra of Pt films, with and without Ti underneath, as-deposited and annealed at 700 °C in N_2. All Pt films show a pronounced {111} texture. Table 1 summarizes the results. After annealing, the XRD spectrum of Pt on Ti shows a diffraction at 46.56°, while the Ti (002) peak at 38.20° has disappeared. Although some authors have attributed the former peak to Pt(200) [8], we think that it must be due to the (200) planes of the alloy Pt_3Ti, which forms during annealing at temperatures at or above 550 °C (fig.1). The Pt (200) diffraction is expected at 46.44°, considering the 2θ value of 39.93° for Pt (111) in this film.; a peak position difference of 0.14° is significant, considering the experimental error of 0.02°.

Table 1. XRD results of as-deposited and annealed Pt films on Si-SiO₂.

sample	2θ Pt (111)	FWHM Pt (111)	2θ other	FWHM other
Pt, as-deposited	39.58°	0.36°	-	-
Pt, annealed	39.93°	0.15°	-	-
Pt-Ti, as-dep.	39.55°	0.38°	38.20°	0.40°
Pt-Ti, annealed	39.93°	0.18°	46.56°	0.35°

(a)   (b)

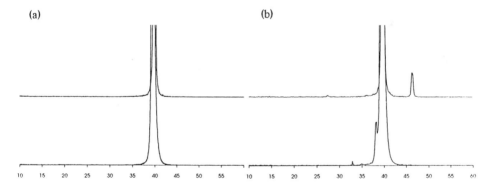

Fig.4. (a) Pt film; bottom: as-deposited, top: annealed at 700 °C; (b) Pt on Ti; bottom: as-deposited, top: annealed at 700 °C; all plots have the same vertical scale.

The XRD results indicate that as-deposited Pt films have a compressive residual stress (strain perpendicular to surface 0.53 %), which changes to tensile stress after annealing (perpendicular strain 0.31 %); the decreased line widths indicate an increased grain size after annealing. No differences between Pt with or without Ti are found in this respect.

In agreement with other authors (e.g. ref. [8]), we have found evidence for the presence of TiO_2 (rutile) after annealing of our Pt-Ti-SiO_2-Si layer package, although the corresponding XRD peak at 26.85° has only a very low intensity with respect to the Pt and Pt_3Ti peaks.

DISCUSSION AND CONCLUSIONS

The results for PZT on Pt agree with those reported by others (e.g. refs. [4,7]): PZT crystallization is impeded, until at a temperature the already present nanocrystalline pyrochlore phase becomes so unstable with respect to PZT, that PZT nuclei are formed, which expand in an explosive fashion to fractal-like dendritic crystals ("rosettes"). It was observed that after a certain fraction of the film had crystallized, prolonged annealing did not give further growth of the rosettes; this is probably caused by PbO loss from the film, which impedes PZT crystallization. The higher Pb content of the rosettes with respect to the matrix was confirmed by EDX.

For PZT on Pt+Ti it was found that crystallization occurs readily, giving almost featureless films. XRD results indicate that in this case Ti diffuses into the Pt layer (giving a compound Pt_3Ti) and probably also through the Pt. At 850 °C this results in a film with two PZT phases: one with the expected ratio Zr:Ti = 53:47, and one with Zr:Ti = 30:70. We think that at lower temperatures (550 °C) a Ti-rich layer is formed at the Pt-PZT interface, which lowers the barrier for PZT nucleation and results in a high density of small crystallites in the film.

A remarkable observation was that after annealing at 850 °C the PZT (111) diffraction has disappeared. We think that this indicates that the nucleation layer on the Ti-rich interface is (111) oriented; on this layer new PZT crystals nucleate, which have (100) texture. When Ti diffusion becomes very severe, new (100)-oriented crystallites form at the expense of the (111)-oriented initial layer.

ACKNOWLEDGEMENTS

The authors thank B. Otter for SEM and EDX measurements, J. Sanderink for sputtering depositions. C. Cobianu acknowledges the support of the EC-program "TEMPUS" and wishes to thank profs. J. Fluitman and D. Dascalu for their efforts in establishing the joint research project.

REFERENCES

1.. K.R. Udayakumar, J. Chen, S.B. Krupanidhi and L.E. Cross, Proc. 7th Int. Symp. Appl. Ferroel., Urbana-Champaign, June 6-8, 1990, p.741.
2. J.G.E. Gardeniers and C. Cobianu, submitted to J. Appl. Phys.
3. F.W. Martin, Phys. Chem. Glasses **6**, 143 (1965).
4. B.A. Tuttle, R.W. Schwartz, D.H. Doughty and J.A. Voigt, Ferroelectric Thin Films, Mater. Res. Soc. Proc. **200**, 159 (1990)
5. B. Jaffe, R.S. Roth and S. Marzullo, J. Res. Nat. Bur. Stds. **55**, 239 (1955).
6. B.G. Hyde and S. Andersson, Inorganic crystal structures, J. Wiley & Sons, New York (1989).
7. L.N. Chapin and S.A. Myers, Ferroelectric Thin Films, Mater. Res. Soc. Proc. **200**, 153 (1990)
8. K. Sreenivas, I. Reaney, T. Maeder, N. Setter, C. Jagadish and R.G. Elliman, J. Appl. Phys. **75**, 232 (1994).

GROWTH AND CHARACTERIZATION OF $Ba_{0.5}Sr_{0.5}TiO_3$ THIN FILMS ON Si (100) BY 90° OFF-AXIS SPUTTERING

S. Y. Hou, J. Kwo, R. K. Watts, D. J. Werder, J. Shmulovich, and H. M. O'Bryan
AT&T Bell Laboratories, Murray Hill, NJ 07974

ABSTRACT

We have studied $Ba_xSr_{1-x}TiO_3$ (BST) thin films (x=0.5) grown on Si (100) with and without a Pt/Ta barrier layer using 90° off-axis RF sputtering. The growth conditions were optimized according to film crystallinity, stoichiometry, and dielectric properties. Polycrystalline BST films with strong (100) texture were obtained via growth on Si (100). The measured dielectric constant from these films was low, presumably because of the parasitic effect of native oxide at BST/Si interface as revealed by TEM. On the other hand, BST films grown on Si with Pt/Ta barrier layers have crystallinity inferior to that on bare Si as determined by X-ray diffraction. Nevertheless, the best BST films on Pt/Ta layers still have good dielectric properties with dielectric constant exceeding 330, leakage current density < 1×10^{-6} A/cm^2 (±1 V), and loss tangent 0.05 at 1 MHz.

INTRODUCTION

$Ba_xSr_{1-x}TiO_3$ (BST) is one of the oxide compounds in the perovskite family that are attractive for a variety of microelectronic applications such as dynamic random access memories (DRAM) [1,2], filters, and capacitors [3]. Since $BaTiO_3$ and $SrTiO_3$ form a complete series of solid solution, the lattice parameters, phase transition temperatures, and dielectric properties of BST may be tuned continuously by varying x between 0 and 1. For bulk BST with x=0.5, it has a Curie temperature about −50 °C, a dielectric constant (ε_r) about 1300 at room temperature, and a dissipation factor or loss tangent 0.05 at microwave frequencies [4], making it a potential candidate for applications requiring materials with high dielectric constant and low temperature coefficient. However, direct growth of BST thin films on Si substrates has been found problematic. The presence of low ε_r SiO_x at the BST/Si interface significantly degrades the overall dielectric properties. Various metal layers have been used as barrier layers between Si and BST to obtain improved dielectric properties [5]. Among them Pt/Ta bilayer has been identified for its relatively good chemical stability and compatibility with BST.

In this paper, we report growth and characterization of BST thin films by 90° off-axis sputtering technique. This technique has been developed for depositing high quality epitaxial high-Tc superconductor thin films [6]. The growth conditions that are essential to BST thin film formation are discussed. Results from Rutherford Backscattering (RBS), X-ray diffraction (XRD), and transmission electron microscopy (TEM) are reported for BST films grown on Si (100) with or without a Pt/Ta barrier layer in different growth conditions. The dielectric properties of the BST films are discussed.

Mat. Res. Soc. Symp. Proc. Vol. 343. ©1994 Materials Research Society

THE EXPERIMENT

BST thin films were grown by 90° off-axis RF sputtering using a 2-inch ceramic $Ba_{0.5}Sr_{0.5}TiO_3$ target. Bare Si (100) as well as Pt/Ta-coated (500 Å-thick each) Si (100) were used as substrates for each deposition. The Pt/Ta layers were deposited in a separate chamber at ~200 °C by on-axis sputtering [7]. Films were optimized by varying the following sputtering parameters: substrate block temperature 500–750 °C; sputtering pressure 2–180 mTorr, O_2/Ar ratio 1/9–1/19, and rf power 30–100 W. A 180°-rotation of the substrate block was performed in the middle of the deposition in order to average out the thickness non-uniformity. After deposition, the samples were vented to 300 Torr O_2 during cool down to ensure full oxygenation.

The stoichiometry and film thickness were determined by Rutherford Backscattering (RBS) using 1.8 MeV He^+. Film crystallinity of the BST films was examined with X-ray diffraction (XRD). For dielectric measurements, samples were coated with 2000 Å Ti/Au top electrodes by evaporation. Sample capacitance was recorded using an impedance analyzer at room temperature and 1 MHz.

RESULTS AND DISCUSSION

Film Stoichiometry

The film stoichiometry has been monitored by RBS to ensure a one-to-one correspondence of composition between films and target. The sputtering pressure has been found to be an important parameter in governing the film stoichiometry. Figure 1 shows the dependence of film stoichiometry on total sputtering pressure during growth. The (Ba+Sr)/Ti ratio changes monotonically from Ti-rich in 70–150 mTorr range to nearly stoichiometric in a pressure at and below 25 mTorr. The individual Sr and Ba content also follow the same trend, with Sr content always slightly higher than Ba. The film stoichiometry is insensitive to other variables such as Ar/O_2 ratio, substrate block temperature, or sputtering power.

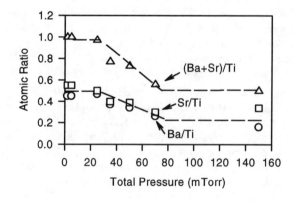

Fig. 1. The dependence of film stoichiometry on total sputtering pressure for off-axis sputtered BST films.

Film Crystallinity

Figure 2 shows XRD results for BST grown directly on Si substrates. Peaks which can be assigned as BST occur in a wide temperature range. Highly (100)-oriented BST was obtained at a deposition temperature of 700°C. However, an in-plane azimuthal scan using a four-circle diffractometer indicated the absence of epitaxial growth between the film and the substrate. At 750°C, a strong impurity peak is observed at 44.65°, presumably due to interaction between constituents in BST and Si as evidenced from RBS.

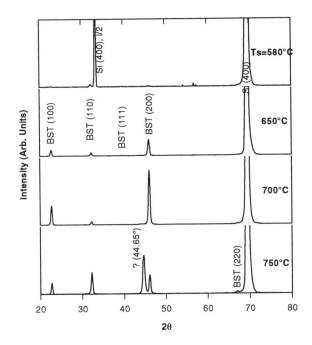

Fig. 2 XRD of BST films grown directly on Si (100) at different substrate temperatures. Note the evolution of (100) texture at 700°C.

Compared to films grown directly on Si, BST films on Pt/Ta/Si substrates have poorer crystallinity under the same growth conditions as judged from the weaker and broader X-ray peaks (Figure 3), although the formation of BST phase is evident. Post annealing at 650°C for 30 min. in air did not result in noticeable improvement in the XRD pattern. This has also been observed in laser ablated BST films grown on Pt/Ti/SiO2/Si substrate [8], manifesting the non-ideal nature of polycrystalline Pt layer supporting BST thin films. However, XRD has revealed an enhanced (110) peak intensity on a sample grown at a pressure as low as 5 mTorr (Figure 3b). Further experiments are in progress to understand this pressure effect.

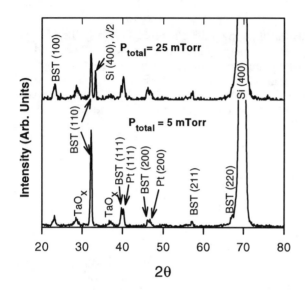

Fig. 3 XRD of BST films grown directly on Pt/Ta/Si (100) at different sputtering pressures.

The interface between BST and its sublayers has been studied by TEM. Figure 4 shows a cross-sectional TEM micrograph taken from a film grown at 650 °C with Pt/Ta layers on Si. The BST/Pt interface shows a greater roughness than that for Pt/Ta, possibly due to the recrystallization or grain growth of Pt during BST deposition. On the other hand, the Pt/Ta and Ta/Si interfaces are relatively smoother. This is probably due to the complete oxidation of Ta during BST film growth, as suggested by our XRD and RBS results.

Dielectric Properties

Table 1. Dielectric properties of BST films on Pt/Ta/Si at a frequency of 1 MHz.

T_s (°C)	Thickness (Å)	J_L (A/cm^2)	ε_r	tan δ
500	1200	1×10^{-5}	167	0.20
600	5400	3×10^{-7}	152	0.04
650	1000	7×10^{-7}	323	0.05
650	1800	3×10^{-8}	333	0.10
690	1900	7×10^{-8}	330	0.16
750	1900	6×10^{-8}	245	0.54

The dielectric properties of BST films on Pt/Ta/Si substrates are summarized in Table 1. A dielectric constant of ~330 was routinely obtained on 1000 to 2000 Å-thick

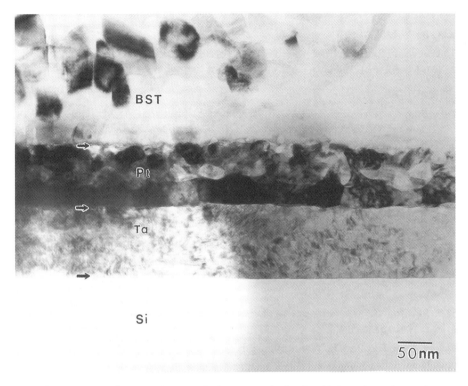

BST

Pt

Ta

Si

50nm

Fig. 4 Cross sectional TEM micrograph showing BST/Pt/Ta/Si interfaces formed at a growth temperature of 650°C.

films grown at 650-700 °C. Films with ε_r exceeding 500 have been found occasionally. This is consistent with the XRD results that BST phase has been attained. In contrast, the best ε_r measured on films on bare Si is ~110 despite the relatively good crystalline quality of these films, presumably due to the shunting effect from the parasitic oxide layer at the BST/Si interface. The loss tangent seems to be optimized at lower temperatures than that for ε_r. The leakage current density under a bias of ±1 V is consistently lower than 1×10^{-6} A/cm^2 for a growth temperature between 600 and 700 °C.

CONCLUSIONS

We have demonstrated growth of BST thin films with good dielectric properties on Pt/Ta/Si substrates by using 90° off-axis sputtering. While Pt/Ta are effective barrier layers in isolating the parasitic oxide layer on Si, they also make difficult the formation of well-crystallized BST phase as evidenced from the much weaker BST peak intensities from XRD. Continual improvement of the dielectric properties is expected by further optimization on sputtering conditions or choosing alternative conducting barrier layers.

ACKNOWLEDGMENT

We would like to thank L. Gomez for technical assistance and Y. H. Wong, Julia M. Phillips (AT&T-BL), and C. S. Chern (Emcore) for helpful discussions.

REFERENCES

[1] E. Fujii, Y. Uemoto, S. Hayashi, T. Nasu, Y. Shimada, A. Matsuda, M. Kibe, M. Azuma, T. Otsuki, G. Kano, M. Scott, L. D. McMillan, and C. A. Paz de Araujo, IEDM Technical Digest, 267 (1992).

[2] T. Eimori, Y. Ohno, H. Kimura, J. Matsufusa, S. Kishimura, A. Yoshida, H. Sumitani, T. Maruyama, Y. Hayashide, K. Morizumi, T. Katayama, M. Asakura, T. Horikawa, Y. Hayashide, K. Moriizumi, T. Katayama, M. Asakura, T. Horikawa, T. Shibano, H. Itoh, K. Sato, K. Namba, T. Nishimura, S. Satoh, and H. Miyoshi, IEDM Technical Digest, 631 (1993).

[3] D. Ueda, in Abstracts of 6th International Symposium on Integrated Ferroelectrics, Mar. 14–16, 1994, CA, 1994, p. 2.

[4] V. K. Varadan, D. K. Ghodgaonkar, V. V. Varadan, J. F. Kelly, and P. Gilkerdas, Microwave J. **35**, 116 (1992).

[5] T. Sakuma, S. Yamamichi, S. Matsubara, H. Yamaguchi, and Y. Miyasaka, Appl. Phys. Lett. **57**, 2431 (1990).

[6] C. B. Eom, J. Z. Sun, K. Yamamoto, A. F. Marshall, K. E. Luther, S. S. Laderman, and T. H. Geballe, Appl. Phys. Lett. **55**, 595 (1989).

[7] Custom deposited by ThinFilms, Inc., NJ.

[8] D. Roy and S. B. Krupanidhi, Appl. Phys. Lett. **62**, 1056 (1993).

THIN FILMS OF SrFeO$_{2.5+x}$ - EFFECT OF
PREFERRED ORIENTATION ON OXYGEN UPTAKE

BRIAN W. SANDERS, JIANHUA YAO, AND MICHAEL L. POST
Institute for Environmental Chemistry, National Research Council Canada, Montreal Road,
Ottawa, Canada, K1A 0R6.

ABSTRACT

Pulsed laser ablation has been used to deposit thin films of SrFeO$_{2.5+x}$ (x = 0 to ≈ 0.5).
Previous work has shown that the orientation of the films, determined by powder x-ray
diffraction depended strongly upon the deposition temperature. Films grown below 770 K
showed little or no orientation. A growth temperature of 900 K resulted in films oriented (200).
Growth temperatures of > 1000 K produced films oriented predominantly (110). At 673 K in
an oxygen atmosphere, oriented films readily converted from the oxygen deficient brownmillerite
form (x=0) to the oxygen rich cubic (or distorted cubic) perovskite form (x $\approx$ 0.3). Films which
exhibited no initial orientation did not react with oxygen under these conditions. Cycling non-
oriented films between 230 and 800 ppm of oxygen in 101.3 kPa of nitrogen at 673 K resulted
in weak (110) orientation. Once oriented, the films reacted readily with oxygen and exhibited
measurable resistance changes. The conversion from oxygen deficient to oxygen rich form was
monitored by x-ray diffraction and the DC resistance of the films.

INTRODUCTION

Current environmental protection concerns require industry to decrease pollution
whenever they can. Accurate sensing of chemical processes should minimize their environmental
impact. Oxygen concentration plays an important role in many processes, and monitoring of
oxygen levels will provide information on potential problems. More specifically, oxygen sensing
has an important role to play in control of engine emissions, metallurgy, food preservation, and
the measurement of ground level ozone.

Perovskite type oxides have some unique chemical and electromagnetic properties which
can be exploited to create an oxygen sensor. Recent studies have shown that thin films and
powders of SrFeO$_{2.5+x}$ ($0 \leq x \leq 0.5$) undergo property changes (due to a change in x) that are
specific to their oxygen exposure level[1,2,3]. To optimize a sensor based on this material the effect
of many parameters on performance need to studied. One such parameter is film orientation.
This paper deals with DC resistance changes that occur when thin films of SrFeO$_{2.5+x}$ of
differing orientations are exposed to various partial pressures of oxygen.

EXPERIMENTAL

Detailed information on the thin film deposition system and procedure can be found
elsewhere[1]. A Lumonics KrF excimer laser was used for the ablation. The power density on the
target (a pressed powder pellet of SrFeO$_{2.5}$ or SrFeO$_3$) was typically 2.5 Jcm^{-2}. Films were
deposited for 20 min on (1$\bar{1}$02) sapphire held at temperatures of 300, 900, or 1130 K. Ablation
atmospheres were vacuum (1.3x10^{-4} Pa) and 26 Pa of oxygen. Two cooling atmospheres were
used: vacuum and 53.3 kPa oxygen.

Mat. Res. Soc. Symp. Proc. Vol. 343. ©1994 Materials Research Society

Once the films were grown, two 0.5 cm² gold electrodes were evaporated 1 cm apart on the surface. Gold wires bonded to these pads allowed two point DC resistance measurements to be made. Using the apparatus shown in figure 1 resistance changes were monitored under flowing conditions of various oxygen concentrations (<30 ppm to 100%, balance nitrogen). Oxygen concentrations of the gas flowing into and out of the chamber were monitored with a micro-gas chromatograph (MTI;M200D). Gas flow was maintained at a rate of ≈200 mlmin⁻¹ using mass flow controllers. The gases were directed over the substrate with a pyrex "umbrella" to ensure rapid exposure of the sample to oxygen concentration changes. The substrates were mounted with silver paint to a resistively heated ceramic plate. A type K thermocouple was then clamped to the film's surface to measure temperature. Resistance measurements and temperature control (constant 670 K) were accomplished with a personal computer interfaced to a Keithley 705 scanner, an HP34401 multimeter (maximum resistance scale - 120 MΩ), and a programmable Sorenson QRD 20-4 power supply.

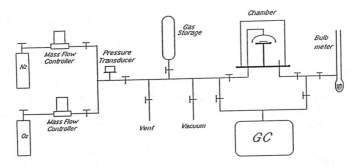

Figure 1. A schematic of the apparatus used to measure film resistance as a function of oxygen concentration.

RESULTS/DISCUSSION

Details of the film's orientation dependence on temperature are given elsewhere[1]. Films grown at 900 K or 1000 K have preferential orientations of (200) or (110), respectively. Films grown below 770 K exhibit no preferential orientation. Since the diffraction peaks of preferentially oriented films are always sharp irregardless of intensity, films grown below 770 K are probably microcrystalline and not amorphous. Figure 2 shows typical diffraction patterns of films grown at 900 and 1000 K. Room temperature resistances of the films are > 120 MΩ. At 670 K, the resistance of $SrFeO_{2.5}$ films were also above 120 MΩ, but the resistance fell to 4 MΩ when the films were exposed to as little as 130 ppm of oxygen. Resistance measurements on pellets[4] of $SrFeO_{2.5+x}$ also show reduced resistance with increasing x. Film resistance continued to fall as the oxygen concentration was increased. Figure 3 shows a typical resistance versus time plot for a thin film of $SrFeO_{2.5+x}$ as the atmosphere is changed from nitrogen (oxygen <30 ppm, moisture <10 ppm) to 820 ppm oxygen. The resistance fell dramatically during the first addition of oxygen. Subsequent additions did not produce as large a change in resistance as the first addition, but resistance plateaus are reached after each addition. Kinetics are not rapid, with the time taken to approach an equilibrium resistance being approximately 1000 sec.

In an effort to reduce $SrFeO_{2.5}$ film resistance, the strontium was partially replaced with lanthanum. Lanthanum substitution has been shown to reduce the resistance of perovskites of this

Figure 2. X-ray diffraction patterns of films grown in 26 Pa O$_2$ and cooled in 53 kPa O$_2$. Growth temperatures are as indicated in the figure.

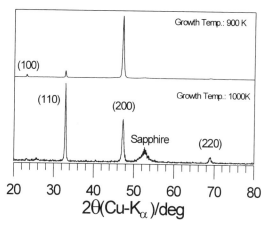

Figure 3. Resistance versus time for a film of SrFeO$_{2.5}$ as the atmosphere is changed from nitrogen to 820 ppm of oxygen in nitrogen.

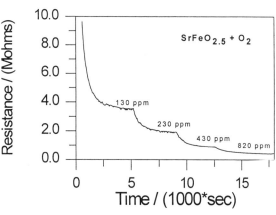

Figure 4. Resistance versus time for lanthanum substituted films with the indicated orientations as the atmosphere is changed from nitrogen to 820 ppm of oxygen in nitrogen

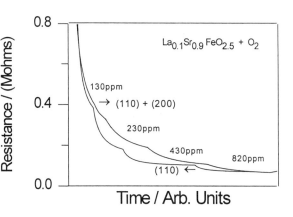

465

type[5]. Figure 4 shows the resistance change for two $La_{0.1}Sr_{0.9}FeO_{2.5}$ films of differing orientation exposed to various oxygen levels. When a comparison is made with figure 3 it is evident that films containing La have lower resistances at a given oxygen exposure level. Whether the film is oriented solely (110) or predominantly (200) almost identical resistance values are reached for a given oxygen exposure level.

Films having no orientation (grown at <770 K) do not exhibit any measurable change in resistance with increasing oxygen level. The resistance remains unmeasurably high (>120MΩ), even after several hours of oxygen exposure at levels that produce measurable resistances in oriented samples. However, at 670 K cycling the films several times between 230 and 820 ppm of oxygen in nitrogen induced a slight orientation. Figure 5 shows x-ray diffraction patterns of a film grown at room temperature, exhibiting no measurable peaks, and the same film after such cycling where a (110) peak is clearly present. Once a noticeable (110) peak is present, resistance changes with increasing oxygen concentration are comparable to initially oriented films (figure 6), although the reaction rate is slower.

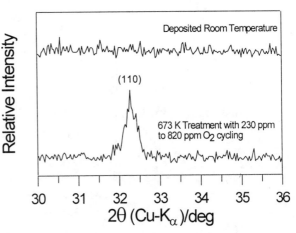

Figure 5. X-ray diffraction patterns of an initially non-oriented film (upper curve), and the same film after cycling between 230 and 820 ppm of oxygen (lower curve).

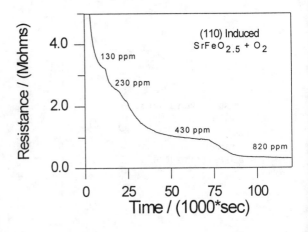

Figure 6. Resistance versus time for a film with an induced (110) orientation as the atmosphere is changed from nitrogen to 820 ppm oxygen in nitrogen.

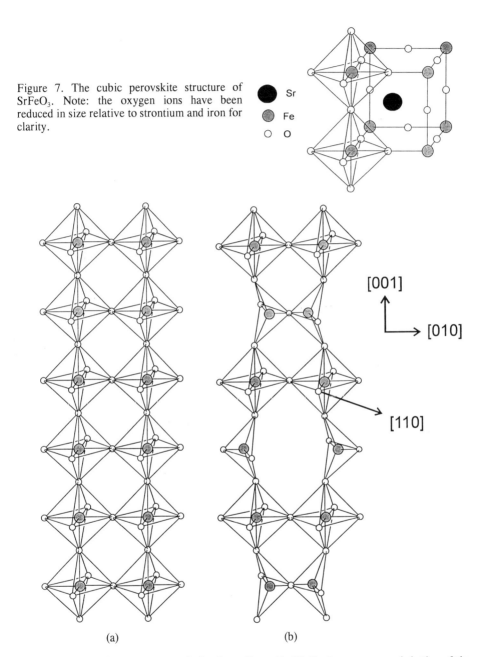

Figure 7. The cubic perovskite structure of SrFeO₃. Note: the oxygen ions have been reduced in size relative to strontium and iron for clarity.

Sr

Fe

O

[001]

[010]

[110]

(a)   (b)

Figure 8 (a) The iron-oxygen octahedra from figure 7. (b) the iron-oxygen sub-lattice of the brownmillerite structure of SrFeO₂.₅.

An explanation of the increased kinetics when a (110) orientation is present can be found by examining the crystal structure of $SrFeO_{2.5}$ and comparing it to $SrFeO_{\approx 3}$. Figure 7 is a schematic of the cubic perovskite crystal structure of $SrFeO_{\approx 3}$[6]. $SrFeO_{2.5}$ assumes an orthorhombic brownmillerite structure. While similar to the cubic perovskite lattice, it has different unit cell and therefore differently defined axes[6]. To avoid confusion, direction and planes mentioned in the following discussion relate to the "equivalent" cubic cell orientations.

Figure 8a is an expanded view of just the Fe-O octahedra from figure 7. The Sr ions have been removed for clarity. As oxygen is removed from the structure a vacancy ordering occurs along the [110] directions. When a stoichiometry of $SrFeO_{2.5}$ is reached, alternating octahedra and tetrahedra of oxygen coordination to iron is found, as shown in figure 8b. There are definite "channels" through the lattice in the [110] direction. These channels are perpendicular to the (110) planes, and could allow a more direct path for penetration of oxygen into the lattice if the surface had a (110) orientation. This might be the reason a (110) orientation is necessary for the films to exhibit higher rates of oxygen uptake. The induction of a (110) orientation during cycling may also be a result of the vacancy ordering upon oxygen removal from the film. Further work will be necessary in order to confirm these hypotheses.

CONCLUSIONS

The resistance of thin films of $SrFeO_{2.5+x}$ is proportional to the amount of oxygen the films are exposed to. To have a measurable resistance at 670 K the films must have some degree of orientation ((110) or (110) + (200)). Films with no initial orientation do not have measurable resistance changes upon oxygen exposure. A (110) orientation can be induced by alternately exposing the film to nitrogen and oxygen at 670 K. After this preconditioning, the films exhibit resistance changes comparable to films initially oriented. The requirement for a (110) orientation might be a result of a vacancy ordering forming channels along the [110] direction which allows for easier penetration of oxygen into the lattice from the (110) oriented surface.

REFERENCES

1. Brian W. Sanders, and Michael L. Post in Laser Ablation in Materials Processing: Fundamentals and Applications, edited by B. Braren, J.J. Dubowski, and D.P. Norton (Mater. Res. Soc. Proc. **285**, Pittsburgh, PA, 1993) pp. 427-432.

2. M.L. Post, B.W. Sanders and P. Kennepohl in Chemical Sensors, (Tech. Digest Fourth Int. Meeting Chemical Sensors, Tokyo, Japan 1992) pp. 322-325.

3. Y. Takeda, K. Kanno, T. Takada, O. Yamamoto, M. Takano, N. Nakayama and Y. Bando, J. Solid State Chem. **63**, 237 (1986).

4. Jukichi Hombo, Yasumichi Matsumoto, and Takeo Kawano, J. of Solid State Chem., **84**, 138 (1990).

5. Y. Yamamura, Y. Ninomiya, and S. Sekido in Chemical Sensors, edited by T. Seiyama, K. Fueki, J. Shiokawa and S. Suzuki (Proc. Int. Meet. Chem. Sensors, Anal. Chem. Symp. Series **17**, Fukuoka, Japan 1983) pp. 187-192.

6. Donhang Liu, Xi Yao, and L.E. Cross, J. Appl. Phys. **71** (10), 5115 (1992).

MICRORAMAN STUDY OF PbTiO₃ THIN FILM PREPARED
BY SOL-GEL TECHNIQUE

E. Ching-Prado, A. Reynés-Figueroa, R.S. Katiyar, S.B. Majumder*, and D.C. Agrawal*
Department of Physics, University of Puerto Rico, San Juan, P.R. 00931.
*Indian Institute of Technology, Kanpur, India.

ABSTRACT

A PbTiO₃ thin film prepared on silicon substrate by sol-gel technique has been studied by micro-Raman spectroscopy. The spectra, in comparison to the single crystal work, show high background in the low frequency region and Raman lines are broader, thus revealing the polycrystalline nature of the film. The frequencies of the Raman bands in the film are clearly shifted to lower frequencies compared to the corresponding ones in the single crystal or powder forms. This phenomenon is similar to the hydrostatic pressure effect on the Raman lines of PbTiO₃ single crystal. The film, therefore, has grains under stress. This stress is caused by non-equilibrium defects and diffusion at the interface. Measurements at different film positions showed variation in the frequency and width of the Raman bands which are associated with the stress and grain size inhomogeneities. The measured shift in the Raman frequencies suggest grain sizes ≤1μm. XRD indicates grain size around 22 nm.

INTRODUCTION

Ferroelectric materials in thin film form have received increasing attention for microelectronic devices, in particular the oxygen octahedral class with perovskite structure. In bulk form these materials exhibit high permitivities, large electromechanical coupling coefficients, and pyroelectric and electro-optic effects[1]. Lead titanate (PbTiO₃) is one of these materials currently under investigation. Thin film of PbTiO₃ (PT) has been prepared by a number of different techniques, such as RF sputtering [2] metalorganic chemical vapor deposition (MOCVD)[3], and sol-gel[4]. In most cases the substrates used were MgO and Al₂O₃, or crystals of perovskite structure, such as SrTiO₃, KTaO₃, etc[5]. In recent years, there has been great interest in combining ferroelectric thin films with semiconductor materials, such as silicon, which results in a wide variety of applications including microactuators, membrane pressure sensors, microvalves, micromotors, etc.

Thin films of PT have been characterized by different techniques, such as x-ray diffraction, scanning electron microscopy (SEM), and electrical measurements. However little work in Raman spectroscopy of thin films has been found in the literature. In particular, Raman scattering of the perovskite phase of PT thin films, on Si substrate, prepared by sol-gel route has not yet been reported. In most of the studies in ferroelectric materials, size and stress effects have received great attention in order to understand physical properties in the ferroelectric phase. Size effects have been investigated by Jaccard et al.[6], where different dielectric properties were found in small ferroelectric particles, in which the long-range coulomb forces play an important role. In BaTiO₃ the Curie-temperature has been found to decrease as the particle size decreases[7]. Also, study of PbTiO₃ ultrafine particles revealed that the ferroelectric phase transition decreases from its bulk value as the size decreases[8]. Size and stress effects have been studied by Nakamura et al. [9] in crystallized amorphous PT. They found that the stress is relaxed for grain size ≥ 1μm. They also observed using Raman spectroscopy, that the effect of increasing annealing temperature on physical properties was approximately analogous to that of the decreasing of hydrostatic pressure in PT single crystal. It has been found that the stress is partially associated with imperfections in the material [4].

VIBRATIONAL MODES

$PbTiO_3$ has five atoms in a primitive cell, giving rise to three acoustic and twelve optic modes. The cubic phase has O_h symmetry and the optical modes are distributed as follows: $3T_{1u}$ + T_{2u} modes. In the tetragonal phase, whose symmetry is C_{4v}, each T_{1u} modes splits into A_1 + E symmetry, while the T_{2u} transforms as B_1 + E. Due to long range electrostatic forces the A_1 and E modes are split into transverse (TO) and longitudinal (LO) optical modes. This effect is also expected in B_1 + E; however, it is not observed. T_{2u} modes are silent. The B_1 and A_1 modes are Raman active. The E modes derived from T_{1u} are both Raman and infrared active[10].

EXPERIMENTAL

The precursor materials used for preparing the sol were lead acetate and titanium tetrabutoxide. The precursor sol was stabilized by acetic acid. Isopropanol was used where dilution was required to reduce sol viscosity and hence the film thickness. Ethylene glycol was used as drying control chemical additives. All the above materials were thoroughly mixed to prepare the sol having nominal composition $PbTiO_3$. The silicon substrate was thoroughly cleaned using standard cleaning procedure. Precursor sol was filtered through 0.2 µm filter just before deposition. Spin coating technique was adopted for coating the sol onto the substrate. The r.p.m varied between 5000-8000. After coating, the sample was dried at 80°C/20 min. to remove solvent followed by a heat treatment of 400°C/30 min. to decompose the residual organics. It was then annealed at 600°C for 6 hrs. in ambient atmosphere for crystallization. The thickness of the samples were measured by a surface profilometer (α step scan 100, Tencor Instruments) and ranged between 0.6 to 1.2 µm. Raman spectra were collected using a Raman microprobe S3000 system, from Instruments S.A, equipped with a diode array from Princeton Instruments as the detection system. The excitation source was the 514.5 nm laser radiation from an argon ion laser.

RESULTS AND DISCUSSIONS

Figure 1 shows the Raman spectra of PT thin film in comparison to that in powder form. A reasonably good agreement between the two spectra is observed, which indicates the tetragonal phase of the film., However, small differences exist. The peaks in the film are broader and most of them are shifted to the lower frequencies, particularly the bands in the high frequency. A similar result is found when we compare the PT film with the PT single crystal. In addition, the background in the film is higher in the low frequency region. These results clearly show the polycrystalline nature of the thin film. In structural terms, the Raman spectrum seems to indicate a very good PT thin film. Raman measurements were made in many different places on the film and the peaks were fitted using a damped harmonic oscillator. These spectra show different frequency shifts and halfwidth in some places. In particular , if we consider only the E(TO) soft mode (Fig. 2), an increase in the halfwidth is observed with decreasing frequency. Figure 3 shows the correlation between the frequency and the halfwidth of this soft mode.

Table I shows the frequencies of the bands in different places on the film. In order to understand the shift to lower frequency, the data was compared with that of the hydrostatic pressure dependence in crystalline PT, obtained by Sanjurjo et al. [10]. Figure 4 shows the frequency of the E(TO) soft mode as a function of pressure, where the solid line corresponds to the best fit using a polynomial equation. The frequencies for the others modes can therefore be estimated from their pressure dependence data. The good agreement found between the data on the films and the pressure dependence data of the single crystal indicates that the film is subjected to a high stress. In

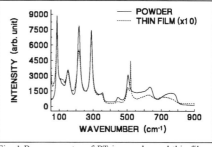

Fig. 1 Raman spectra of PT in powder and thin film.

other words, the most likely cause of the shift of the PT Raman peaks to lower frequencies appears to be due to strain caused by stress, ranging from ~0 in some cases to ~1.2 GPa in most places. Ishikawa et al. [8] and Nakamura et al. [9] found that PT fine particles of 1 μm have the same Raman spectrum as that of a bulk crystal.

But, fine particles smaller than 1 μm present broadening and shifting to lower frequency. This suggests a grain size of ≤1 μm for our PT film. Figure 5 shows the XRD measurements, where the tetragonal structure is confirmed. Using Scherrer's relationship the estimated average grain size is around 22 nm. The grain size was calculated with the integral width of the (100) x-ray peak, taking into account apparatus parameters and stress contributions[4]. This grain size is in agreement with that found by Qu et al.[4] in the PT thin film grown on Si substrate and sintered at 600°C. Scanning electron micrographs of the sample are unclear and not well understood, with some isolated granules in the range of 0.1 to 0.5 μm. The disagreement between XRD and SEM measurements can be explained since the peak broadening of x-ray diffraction patterns is mainly due to small particles, while electron microscope observations tend to overlook the smaller particles. Figure 6 shows the pressure dependence of the lattice parameters measured by Nelmes et al.[11], showing that the c-parameter is more susceptible to pressure change than the a-parameter. The c-parameter of our PT thin film is shown in Fig. 6 by a cross, which corresponds to a compressive

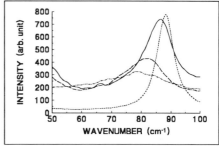

Fig. 2 E(TO) soft mode in various places on the film. Short dash line is the mode in the single crystal.

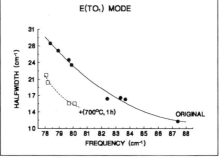

Fig. 3 Correlation between the halfwidth and the frequency of the E(TO) soft mode.

pressure of 1.3 GPa, in good agreement with that obtained by Raman spectroscopy (1.2 GPa). In order to explain the high stresses in our material, the work of Nakamura et al.[9], about crystallization process from amorphous $PbTiO_3$, can be used. They found that annealing of

471

TABLE I: PHONON FREQUENCIES IN PbTiO$_3$ SINGLE CRYSTAL AND PbTiO3 THIN FILM
(frequencies in cm-1)

	SINGLE CRYSTAL			THIN FILM		
PHONON	P~0 GPa	P~0.7 GPa	P~1.2 GPa	POSITION 1	POSITION 2	POSITION 3
E(TO$_1$)	88	82	80	87	82	80
E(LO$_1$)	128	-------	-------	124	126	121
A$_1$(TO$_1$)	149	-------	-------	153	147	145
E(TO$_2$)	220	206	200	219	210	209
B$_1$+E	289	290	290	290	289	289
A$_1$(TO$_2$)	359	347	338	350	340	334
E(LO$_2$)	440	--------	-------	444	444	441
E(TO$_3$)	506	502	504	505	503	515
A$_1$(TO$_3$)	644	617	608	620	616	610
E(LO$_3$)	687	--------	--------	721	703	689
A$_1$(LO$_3$)	794	--------	--------	764	746	743

amorphous PbTiO$_3$ in the range of 600-900°C produces a polycrystalline material, whose Raman frequencies change with the annealing temperature in similar form as the hydrostatic pressure of PT single crystal. Thus, one possible explanation for the shift of the frequencies is as follows: Annealing the thin film at 600°C causes the nucleation rate of the crystalline phase to be very high, while the growth rate is slow. Hence, very fine grains are formed belonging to the paraelectric cubic state, but when the sample is cooled down to room temperature, the cubic to tetragonal deformation should take place at the

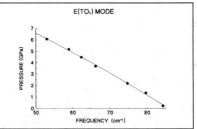

Fig. 4 Pressure dependence of the frequency of the E(TO) soft mode.

Curie temperature (490°C). However, if the grain size is too small to form ferroelectric domains, such deformation is presumably prohibited because each grain boundary is clamped by neighboring grains of different orientation, thus resulting in pronouncedly high stresses. Another possible explanation was presented by Qu et al.[4] in x-ray and Auger studies of PT thin films. They found that the stress in the PT film first decreases and then increases with increasing annealing temperature. The former is due to the elimination of non-equilibrium defects in the grains and the latter is caused by the extension of non-stoichiometric regions at the surface and interface, which increases the defects accumulated at the grain boundary and the interface film-substrate. The total stress in the system must contain contributions due to the difference in the thermal expansion coefficient between the PT film and the Si substrate.

Two mechanisms are possible in order to understand the broadening of the Raman bands in the PT thin film in relation to that of PT single crystal or in powder form. The first is the lifetime broadening. The Raman halfwidth of

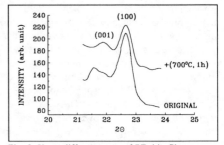

Fig. 5 X-ray diffractograms of PT thin film.

natural PT is related to the lifetime of the phonon created in the Raman process. The primary broadening mechanism in a perfect crystal is decay of the optical phonon into two acoustic phonons with opposite wave vectors of equal magnitude. However, in imperfect crystal such as polycrystalline PT film, the phonon created in the Raman process can also decay at grain boundaries and at defect sites, further reducing the lifetime and broadening the Raman peaks. If phonon lifetime determines the halfwidth, one would then expect the Raman line in nanocrystalline PT film to broaden with increasing defect density

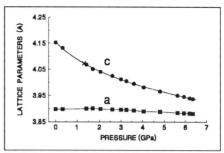

Fig. 6 Pressure dependence of the lattice parameters. The cross corresponds to the c-parameter in the film.

and with decreasing crystallite size. The second mechanism is that the observed broadening arises from averaging over many different nanocrystallites, each of which might have a different amount of internal stress and hence a different Raman frequency[12].

The hydrostatic pressure study of PT single crystal, showed that the halfwidth of the E(TO) soft mode is independent of the frequency between 88 cm^{-1} and approximately 33 cm^{-1}. So that, in this range of frequency no correlation with the width of the E(TO) soft mode is observed. We present two possible explanations for the observed correlation of the halfwidth and frequency shift in the film studied. One possibility is that the halfwidth of the stress distribution in the large number of nanocrystallites sampled by the laser increases with the maximum stress on any nanocrystallite. That is the laser spots with Raman peaks close to that of natural PT might be relatively homogeneous and have a narrow internal stress distribution, while laser spots with large shift of Raman peak to lower frequency might be more heterogeneous and have much wider stress distribution. Another possibility is that a high defect density is present in the grains, which proportionally increases the halfwidth and produces high internal stress. In this case, laser spots with small amount of internal stress must have a low concentration of defects and correspondingly narrow halfwidth, while laser spots with large amounts of internal stress must have a higher defect density and larger bandwidth.

Another possible explanation for the variation in the Raman frequency shift and the halfwidth has to do with the grain size in the laser spots of the PT thin film. In an infinite crystal, momentum conservation limits the first order Raman scattering processes to the zone center (q=0) optical phonons. In a particle of size D, momentum conservation is partially relaxed, and phonons of wave vectors extending from zero to $|q| \sim 1/D$ can contribute to the scattering. However, this explanation is less likely, because the computed dispersion curve of the E(TO) soft mode, along (100) and (001) directions, indicate that the lowest E(TO) frequency is obtained for q=0. Therefore, if the contribution to the Raman measurements of the E(TO) phonon frequency with q≠0 takes place due to small grain size, it is expected broadening and shifting to higher frequency for the E(TO) soft mode, which is opposite to that found experimentally.

In order to increase the grain size, an additional sintering at 700°C for 1 hour was made to the film. SEM micrographs show well defined pictures with grain size around 0.1 μm. Figure 5 also presents the XRD measurements of the sintered film, where the tetragonal structure is confirmed. The grain size, calculated from the XRD data, revealed a value of 0.08 μm, which is in good agreement with that obtained by SEM. The Raman spectra of the PT film with additional annealing show strong signal characteristic of the tetragonal structure. The E(TO) soft mode is found around 79cm^{-1}, which reflects a pressure in the film around 1.2 GPa. In addition, the halfwidth is observed to decrease significantly as shown in Fig. 3. The decrease in halfwidth with increasing grain size is consistent with other reports[8,9]. Therefore, the different halfwidths in the $E(TO_1)$ Raman peaks can be associated with a distribution of grain size in the film. However, Raman works in PT fine particles indicate relaxation of the stress with increasing grain size, which is opposite to that found in the re-annealed film. In addition, our x-ray measurements show somewhat higher pressure in the film re-annealed than the original. This suggests that the original film contains a distribution of grains size, where the concentration of non-equilibrium defects decrease with grain size. In the re-sintered film most of the non-equilibrium defects are removed from within the grain, but more defects are accumulated at the grain boundary. In addition, re-annealing at a higher temperature must produce a higher diffusion between the film and the substrate when compared to the original film. All of these factors give rise to higher stress in the re-annealed sample.

CONCLUSIONS

1- The PT thin film is found to be subject to a high stress of about 1.2 GPa.
2- Stress is mainly created by non-equilibrium defects in the grain and diffusion at the interface.
3- The $E(TO_1)$ peak halfwidth is related to grain size and therefore to the lifetime broadening.
4- The correlation between Raman peak halfwidth and frequency stems from the fact that smaller grains contain a higher amount of non-equilibrium defects, which increases the compressive pressure and decreases phonon lifetime.

ACKNOWLEDGEMENTS

The authors wish to thank J. Cordero and W. Muñoz for technical support. This research is supported by NSF-EPSCoR grant EHR-9108775, DOD-ONR grant N00014-93-1-1266, and ARO grant DAAHO4-93-2-0008

REFERENCES

1- S.K. Dey and R. Zu Leeg, *Ferroelectrics*, **112** , 309 (1990).

2- I. Takuchi, A. Pignolet, L. Wang, M. Proctor, F. Levy, and P.E. Schmid, *J. Appl. Phys.*, **73** 394 (1993).

3- B. S. Kwak, E. P. Boyd, and A. Erbil, *Appl Phys. Lett.*, **53** 1702 (1988).

4- B. Qu, D. Kong, P. Zhang, and Z. Wang, *Ferroelectrics*, **145** 39 (1993).

5- H. Adachi, T. Mitsuya, O. Yamazaki, and K. Wasa, *J. Appl. Phys.*, **60** 736 (1986).

6- A. Jaccard, W. Känzig, and M. Petex, *Helv. Phys. Acta* 26 521 (1953).

7- T. Kanata, T. Yoshikawa, and K. Kubota, *Solid State Comm.*, **62** 765 (1987).

8- K. Ishikawa, K. Yoshikawa, and N. Okada, *Phys. Rev. B* , **37** 5852 (1988).

9- T. Nakamura, M. Takashige, H. Terauchi, Y. Miura, and W. Lawless, *Jpn. J.Appl. Phys.* **23**, 1265 (1984).

10- J. A. Sanjurjo, E. López Cruz, and G. Burns, *Phys. Rev. B*, **28** 7260 (1983).

11- R. J. Nelmes and A. Katrusiak, *J. Phys. C: Solid St. Phys.*, **19** L725 (1986).

12- J.W. Ager III, D. K. Veirs, and G. M. Rosenblatt, *Phys. Rev. B*, **43** 6491 (1991).

PULSED LASER DEPOSITION OF $Bi_4Ti_3O_{12}$ THIN FILMS ON INDIUM TIN OXIDE COATED GLASS

H-J. Cho, William Jo, and T. W. Noh
Department of Physics, Seoul National University, Seoul, 151-742, Korea

ABSTRACT

$Bi_4Ti_3O_{12}$ thin films have been grown on indium tin oxide coated glass by pulsed laser deposition. The films are rapidly thermal annealed at 650 °C in various kinds of ambients. X-ray diffraction and scanning electron microscopy are used to investigate crystallization and microstructures, respectively. Using Auger electron microscopy, chemical compositions and depth profiles are examined. Optical and current-voltage characteristics measurements of the films show that their transmittance and leakage current behaviors are strongly dependent upon the microstructures. O_2 partial pressure in the rapid thermal annealing process is found to be an important parameter which determines crystallization, microstructures, and leakage current behaviors of the $Bi_4Ti_3O_{12}$ thin films.

INTRODUCTION

The ever-growing sophistication of the ferroelectric thin film community is marked by recent developments in deposition techinques, electrode materials, and integration studies for complex device structures.[1] The applications for nonvolatile ferroelectric memories are rapidly evolving.[2] $Bi_4Ti_3O_{12}$ is a good candidate for the applications, since it has a relativley low coercive field along the c-axis.[3,4] For such applications, the film should have a low leakage current and a large dielectric breakdown voltage.[5] In a thin film structure, high electric fields are produced even for low applied voltages. These high electric fields cause nonlinear current-voltage characteristics. Krupanidhi and his colleagues have studied the electrical properties of $Bi_4Ti_3O_{12}$ thin films on Si and Pt coated Si.[6,7] But, understanding of the leakage currents in $Bi_4Ti_3O_{12}$ thin films has not been fully accomplished yet.

For electrical characterization of ferroelectric thin films, various materials including metal, cuprate superconductors, and transitional metal oxides have been adopted as electrodes.[8-10] Recently, non-destructive readout mode using optical birefringence of the ferroelectric thin films is emerging out as a possible access method.[11] So indium tin oxide (ITO) becomes another candiadate for an electrode, since it is optically transparent and conducting. In this paper, leakage current behaviors of $Bi_4Ti_3O_{12}$ thin films on ITO coated glass are presented. Emphasis is placed on understanding how microstructural changes occurred during postannealing process affect leakage current behaviors in the $Bi_4Ti_3O_{12}$ thin films.

EXPERIMENTAL

$Bi_4Ti_3O_{12}$ thin films were grown on ITO coated glasses using pulsed laser deposition. Details of the deposition setup were described earlier.[12] In the present study, an ultra-violet

Mat. Res. Soc. Symp. Proc. Vol. 343. ©1994 Materials Research Society

laser (355 nm, third harmonics of a Q-switched Nd:YAG laser) was used. The laser was pulsed at a rate of 10 Hz, and fluence of the beam was estimated to be 2.5 J/cm^2. O$_2$ pressure and substrate temperature during the deposition were 200 mtorr and room temperature, respectively.

In order to crystallize the as-grown films, rapid thermal annealing (RTA) process was used. RTA process has been widely used in the semiconductor industry. The great advantages of this method are a short annealing time and its relative simplicity as compared with the conventional furnace annealing. The short rising time to the desired annealing temperature minimizes the film-substrate interface reaction. The Bi$_4$Ti$_3$O$_{12}$ films were annealed at 650 °C for 80 seconds in three kinds of ambients, such as O$_2$, air, and N$_2$ at 1 atm.

Phase formation and crystallization were examined by x-ray diffraction analysis using Cu $K\alpha$ radiation. Microstructures were investigated with a scanning electron microscope (SEM). Depth profiles of chemical compositions were looked into by Auger electron spectroscopy. Optical transmission between 320 nm and 620 nm was measured. The leakage currents were measured using a Keithley 236 ammeter with a metal-shielded sample box at 293 K. They were carried out in a metal-insulator-metallic oxide configurations, where the top electrodes of Au with a diameter of 0.5 mm had been prepared by thermal evaporation. Contacts with probe electrodes were made with a wire-bonding method.

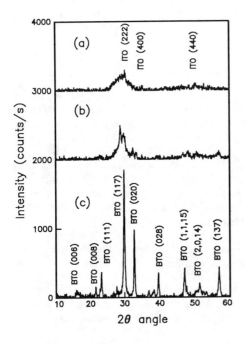

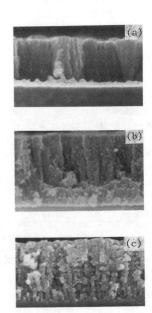

FIG. 1. X-ray diffraction patterns of Bi$_4$Ti$_3$O$_{12}$(BTO) films on ITO coated glass. The samples are rapidly thermal annealed at 650 °C for 80 seconds in atmospheres of (a) O$_2$, (b) air, and (c) N$_2$, respectively.

FIG. 2. Cross-sectional SEM pictures of Bi$_4$Ti$_3$O$_{12}$ thin films annealed in atmospheres of (a) O$_2$, (b) air, and (c) N$_2$, respectively.

RESULTS AND DISCUSSION

The O_2 partial pressure during the RTA process is found to be a very important parameter to determine crystallization of the as-deposited films, which are amorphous $Bi_4Ti_3O_{12}$.[13] Figure 1 shows the x-ray diffraction patterns of the films. The broad peak around $2\theta = 30°$, shown in Fig. 1(a), suggests that the film annealed in O_2 atmosphere is amorphous. In Fig. 1(c), the reflections of the film annealed in N_2 are those of $Bi_4Ti_3O_{12}$, suggesting that the film is polycrystalline. Figure 1(b) indicates that crstallization occurs slightly in the film annealed in air.

The cross-sectional SEM images show that the microstructures of the $Bi_4Ti_3O_{12}$ films have close relation with their crystallization. Figure 2(a) shows that the film annealed in O_2 has a columnar structure in which each layer exhibits a discrete and sharp interface. Figure 2(c) shows that the film annealed in N_2 has a porous structure, which might come from coalescence of the amorphous phase during the annealing process. In the film annealed in air, both of the above mentioned structures are observed. As shown in Fig. 2(b), the top region of the film has the columnar structure and the region close to the ITO layer has the porous structure. From the case of the film annealed in air, it is suggested that crystallization of the $Bi_4Ti_3O_{12}$ phase begin at the interface region.

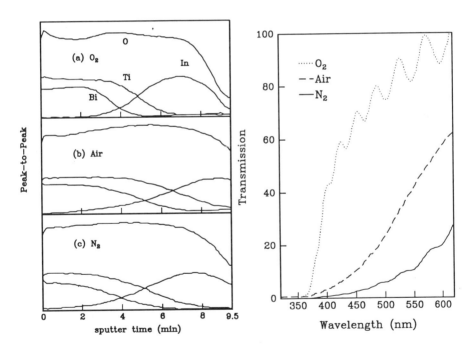

FIG. 3. Auger electron spectroscopy results of the $Bi_4Ti_3O_{12}$ thin films annealed in atmospheres of (a) O_2, (b) air, and (c) N_2, respectively.

FIG. 4. Optical transmission spectra of the $Bi_4Ti_3O_{12}$ thin films annealed at different ambient.

477

According to the Auger electron spectroscopy results, all the films have the same chemical composition within experimental errors. Depth profile results are given in Fig. 3. Although the annealing time was 80 seconds, some indium ions are diffused into the films. In the film annealed in O_2, the indium interdiffusion is relatively lower than those in the films annealed in air and N_2. Compared with the SEM result in Fig. 2(a), it can be proposed that a more dense microstructure in the film annealed in O_2 plays a role to prevent in-diffusion of indium.

Optical transmission spectra between 320 nm and 620 nm of the $Bi_4Ti_3O_{12}$ thin films are given in Fig. 4. For the film annealed in O_2, the interference pattern is shown due to the clean and sharp interface of the film. Moreover, its transmission is larger than those of the other films. The small transmission of the film annealed in N_2 is due to light scattering from its porous microstructure.

The J-E characteristics of the $Bi_4Ti_3O_{12}$ films are shown in Fig. 5. At the low electrical field (E) region, the leakage current density (J) is nearly linear to the applied field. The slopes of the linear fitting curves in the log J vs. log E plot are 1.13, 0.94, and 0.82 for the films annealed at O_2, air, and N_2, respectively. As the O_2 partial pressure in the annealing atmosphere decreases, J in the low E field decreases. This behavior might be related to the formation of the porous structures observed in Figs. 2(b) and (c). It demonstrates that the grain boundary of $Bi_4Ti_3O_{12}$ has considerable effects in suppressing J at the low E field region. A similar behavior has been observed in rf-sputtered $SrTiO_3$ films.[14]

At the high E field region, the electrical behaviors become very nonlinear. As shown in Fig. 5, the log J vs. log E curve for the film annealed in O_2 deviates from the ohmic behavior around 85 kV/cm. The J versus E plot in Fig. 6(a) clearly shows a nonlinear behavior in high E field. A conduction mechanism based on space-chrage-limted emission predicts that J-E characteristics can be fitted with polynomials.[15] The data for the O_2 annealed film can be fitted reasonably well with the modified Langmuir-Child law[5,16], that is, $J(E) = aE + bE^2$. The dashed line in Fig. 6(a) is the best fitting curve where the coefficients a and b are 1.00×10^{-17} A/cm·V and 1.46×10^{-16} A/V^2, respectively. Interestingly enough, the experimental data can also be fitted with $J(E) = cE + dE^3$. The solid line in Fig. 6(a) is the best fitting curve with this new equation, where $c = 1.12 \times 10^{-11}$ A/cm·V and $d = 4.14 \times 10^{-22}$ A·cm/V^3. More studies are required to understand why this cubic equation provides better fitting than the modified Langmuir-Child equation. The fact that J-E characteristics of the O_2 annealed film can be fitted well with polynomials suggests that the role of traps is not significant for this film.[5,15]

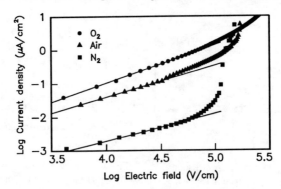

FIG. 5. log J - log E characteristics of the $Bi_4Ti_3O_{12}$ thin films. The linear solid lines represent the nearly ohmic conduction at low electric field.

The film annealed in N_2 shows a drastic change in the log J versus $E^{1/2}$ plot, shown in Fig. 6. At the low E field region, the value of J for N_2 annealed film is smaller than that of the O_2 annealed film. Around 100 kV/cm, the value of J increases abruptly, and it becomes larger than that for O_2 annealed film. This leakage current behavior can be explained in terms of carrier traps. As shown in Fig. 2(c), this film has a porous microstructure which might have lots of carrier traps. In the low E field region, carriers are localized around the traps reducing J. At the high E field region, the localized carriers start to come out of the trap, resulting in a large increase of J. Such a thermoionic emission is called as the Poole-Frenkel hopping, and the current density for this conduction model is given by

$$\log J = \beta_{PF}\, E^{1/2}/k_B T, \tag{1}$$

where $\beta_{PF} = (e^3/\pi\varepsilon_o\varepsilon)^{1/2}$. T, k_B, e, and ε_o are temperature, the Boltzmann constant, the electron charge, and the permittivity in vacuum, respectively. And, ε is high-frequency dielectric constant, which is equal to the square of the refractive index, n. The fitting curve of the experimental data to Eq. (1) is shown in Fig. 6(b). The evaluated value of β_{PF} is 3.92×10^{-5} eV $(m/V)^{1/2}$, which suggests that $n \approx 1.94$. This value is lower than that of single crystal $Bi_4Ti_3O_{12}$, i.e., 2.4-2.6. The small value of n can be understood from the fact that the porous structure in the N_2 annealed sample contains lots of voids.

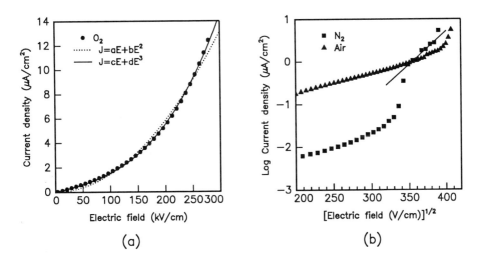

FIG. 6. (a) J vs. E plot of the $Bi_4Ti_3O_{12}$ thin film annealed in O_2. Space-charge limited conduction in models are applied to explain the leakage current behavior at the high electric field. (b) Log J vs. $E^{1/2}$ plot of the $Bi_4Ti_3O_{12}$ thin films annealed in air and N_2. A thermoionic model, Poole-Frenkel hopping, is a possible explanation to the leakage current behavior at the high electric field.

The significant differences of J-E behaviors in the O_2 and N_2 annealed films most likely come from microstructural differences. In the columnar structure observed in the O_2 annealed film, carrier injection into the insulator looks very important, so space charge can distribute along the columnar grain boundaries. In the porous grain structure of the films annealed in N_2, the thermoionic emission from traps in high E fields looks like the most effective origin in the nonlinear behavior.

CONCLUSIONS

$Bi_4Ti_3O_{12}$ thin films were prepared on ITO coated glass substrates by pulsed laser deposition. O_2 partial pressure in rapid thermal annealing has significant effects on both phase formation and microstructures of the films. The nonlinear leakage current behaviors observed in the columnar structure and the porous structure are explained by the space-charge-limited conduction and the Poole-Frenkel hopping, respectively.

ACKNOWLEDGMENT

This work was financially supported by Korea Telecom Research Center, KIST 2000 program, and the Science Research Center of Excellence Program.

REFERENCES

1. Refer to Materials Research Symposium Proceedings, (MRS, Pittsburgh, 1993), Vol. 310.
2. J. F. Scott and C. A. Paz. de Araujuo, Science **246**, 1400 (1989).
3. S. Sinharoy, H. Buhay, W. H. Kasner, M. H. Francombe, D. R. Lampe, and E. Stepke, Appl. Phys. Lett. **58**, 1470 (1991).
4. K. Sugibuchi, Y. Kurogi, and N. Endo, J. Appl. Phys. **46**, 2877 (1975).
5. J. F. Scott, C. A. Araujo, B. M. Melnick, L. D. McMillan, and R. Zuleeg, J. Appl. Phys. **70**, 382 (1991); J. F. Scott, B. M. Melnick, C. A. Araujo, L. D. McMillan, and R. Zuleeg in Proceedings of the 3rd International Symposium on Integrated Ferroelectrics (Colorado Springs, Colorado, 1991), pp. 176-84.
6. Maffei and S. B. Krupanidhi, J. Appl. Phys. **72**, 3617 (1992); J. Appl. Phys. **74**, 7551 (1993).
7. P. C. Joshi and S. B. Krupanidhi, Appl. Phys. Lett. **62**, 1928 (1993).
8. I. K. Yoo, S. B. Desu, and J. Xing in Materials Research Symposium Proceedings, (MRS, Pittsburgh, 1993), Vol. 310, pp. 165-77.
9. R. Ramesh, A. Inam, W. K. Chan, B. Wilkens, K. Myers, K. Remshnig, D. L. Hart, and J. M. Tarascon, Science, **252**, 944 (1991).
10. S. G. Ghonge, E. Goo, R. Ramesh, T. Sands, and V. G. Keramidas, Appl. Phys. Lett. **63**, 1628 (1993).
11. D. H. Reitze, E. Halton, R. Ramesh, S. Etemad, D. E. Leaird, T. Sands, Z. Karim, and A. R. Tanguay, Jr., Appl. Phys. Lett. **63**, 596 (1993).
12. W. Jo, G-C. Yi, T. W. Noh, D-K. Ko, Y. S. Cho, and S-I. Kwun, Appl. Phys. Lett. **61**, 1516 (1992); W. Jo, H-J. Cho, T. W. Noh, B. I. Khim, D-Y. Kim, and S-I. Kwun, Appl. Phys. Lett. **63**, 2198 (1993).
13. The $Bi_4Ti_3O_{12}$ films which were annealed in argon atmosphere showed the properties like as the N_2 annealed thin films. The detailed results are to be published elsewhere.
14. K. Abe and S. Komatsu, Jpn. J. Appl. Phys. **32**, 4198 (1993).
15. K. C. Kao and W. Hwang, Electrical Transport in Solids (Pergamon, Oxford, 1981).
16. R. H. Tredgold, Space Charge Conduction in Solids (Elsevier, Amsterdam, 1966).

THE EFFECT OF LAYER THICKNESS ON POLYCRYSTALLINE ZIRCONIA GROWTH IN ZIRCONIA-ALUMINA MULTILAYER NANOLAMINATES

C.M. SCANLAN,*M.D.WIGGINS,*M.GAJDARDZISKA-JOSIFOVSKA,**AND C.R.AITA*
*Materials Department and the Laboratory for Surface Studies,
University of Wisconsin-Milwaukee, P.O. Box 784, Milwaukee, WI 53201
**Department of Physics and the Laboratory for Surface Studies,
University of Wisconsin-Milwaukee, P.O. Box 413, Milwaukee, WI 53201

ABSTRACT

The mechanical properties of zirconia are known to be a function of phase composition. We show here that a nanolaminate geometry can be used to control the phase composition of zirconia films. The experiment consisted of growth of nanoscale multilayer films (nanolaminates) of polycrystalline zirconia and amorphous alumina by reactive sputter deposition on Si (111) and fused silica substrates. The films were characterized using x-ray diffraction and high resolution electron microscopy. The results show that both monoclinic (m) and tetragonal (t) zirconia polymorphs were formed in the zirconia layers. Most crystallites are oriented with either close-packed {111}-t or {11$\overline{1}$}-m planes parallel to the substrate. The volume fraction of tetragonal zirconia, the desired phase for transformation-toughening behavior, increases with decreasing zirconia layer thickness. Nanolaminates with a volume fraction of tetragonal zirconia exceeding 0.8 were produced without the addition of a stabilizing dopant, and independent of the kinetic factors that limit tetragonal zirconia growth in pure zirconia films.

INTRODUCTION

The bulk zirconia-alumina system is the classic transformation-toughening ceramic composite [1]. Stress triggers a martensitic transition from tetragonal zirconia (t-ZrO_2), a metastable polymorph at standard temperature and pressure (STP), to monoclinic zirconia (m-ZrO_2), the equilibrium STP phase. The t→m transition involves an anisotropic volume expansion. Alumina, with twice the elastic modulus of zirconia (390 vs. 207 GPa) is an effective elastic constraint, providing structural stability to the transformed composite. Bulk zirconia-alumina composites are fabricated at high temperature where t-ZrO_2 is stable. Retaining t-ZrO_2 at STP is usually accomplished by adding a dopant such as yttrium and minimizing the grain size.

The concept of a transformation-toughening ceramic composite can be extended to a polycrystalline thin film coating as well. As in the case of bulk material, the critical coating fabrication step involves retention of t-ZrO_2 at STP. We previously showed [2-6] that in sputter deposited biphasic m+t-ZrO_2 films grown near room temperature, up to 0.5 volume fraction t-ZrO_2 can be retained by crystallite size control alone.

Using a diffusion-limited growth model, we showed [7] that crystallite size in pure zirconia films was determined by the amount a nucleus expanded in the time it took to deposit one monolayer of sputtered flux. The model puts a very narrow window on the process parameters that can be used to obtain biphasic zirconia growth. When the crystallite size exceeds ~13 nm, the films are entirely monoclinic.

The present paper discusses a new approach to producing t-ZrO_2 in thin films. Here, multilayers of polycrystalline zirconia and amorphous alumina were grown by reactive sputter deposition. The layer spacing was scaled to insure nanosize crystallites in the zirconia layers. The result is a "nanolaminate" with a high volume fraction of t-ZrO_2, grown without the addition of a stabilizing dopant, independent of the kinetic factors that limit tetragonal zirconia production in pure zirconia films. In addition to acting as growth termination and restart surfaces, the alumina layers also serve as an elastic constraint, analogous to the function of the alumina matrix in the bulk composite.

481

EXPERIMENTAL PROCEDURE

Film Growth

Sputter deposition was carried out in a multiple target rf diode system. Unheated Si (111) wafers and vitreous fused silica substrates were placed on a rotary table and moved sequentially under Zr and Al targets. The targets were sputtered using 10^{-2} Torr O_2 discharges operated at -1.4 kV (p-p) for Zr and -1.1 kV (p-p) for Al. The deposition started and ended with a zirconia layer. The zirconia layer growth rate was 1.5 nm/min and the alumina layer growth rate was 0.5 nm/min. Zirconia layer thickness was varied in the four different structures grown in this experiment. The Al layer thickness was kept constant at 3.7 nm. The number of zirconia layers, zirconia layer thickness, and total film thickness are recorded in Table I.

Film Characterization

X-ray and electron diffraction techniques were used for crystallographic analysis. Films on fused silica were analyzed by double angle x-ray diffraction (XRD) using unresolved Cu Kα x-radiation (λ =0.1542 nm). Peak position (2Θ), maximum intensity (I), and full width of the peak at half of the maximum intensity (FWHM) were measured. The average crystallite dimension (D) perpendicular to the substrate plane was calculated using the Scherrer relation [8]: D=0.94 λ/BcosΘ, where B is the FWHM after correction for instrument broadening. The volume fraction of each phase was calculated from the integrated intensities using the polymorph method [9].

High resolution electron microscopy (HREM) was performed in a JEM 4000EX transmission electron microscope with a point resolution of 0.17 nm for the 400 keV electrons used in this experiment. Cross-sections of the nanolaminate on Si (111) were prepared using standard sandwiching, cutting, grinding, dimpling and room-temperature Ar^+ ion milling procedures [10]. Care was taken to avoid excessive forces and temperatures during specimen preparation to minimize the possible transformation of metastable t-ZrO_2. HREM images were digitized with an optical CCD camera. Lattice spacing measurements were performed using DigitalMicrograph, an image acuisition and processing software by Gatan.

RESULTS AND DISCUSSION

X-ray diffraction results show that for all film structures studied here, the zirconia layers are polycrystalline and have a t-ZrO_2 component. This latter result is significant because pure zirconia films, as well as nanolaminates with thicker (25-30 nm) zirconia layers, are entirely m-ZrO_2. The alumina layers are amorphous.

The major XRD peaks from films A and B are shown in Fig. 1. These peaks are attibuted to diffraction from the (11$\bar{1}$) planes of m-ZrO_2 (2Θ=28.2^o in an unstressed bulk material at room temperature) and the (111) planes of t-ZrO_2 (2Θ=30.3^o in bulk material at 1020 ^{o}C, the temperature at which reliable data is available [11]). For each phase, these orientations represent growth of the closest-packed planes parallel to the substrate. Bauer has shown this to be the lowest free energy growth form of a crystallite on an amorphous substrate [12]. Very faint, broad peaks in the 2Θ=58-63° range were observed and are attributed to combined higher order diffraction from m- and t- ZrO_2 planes.

TABLE I
Structural Parameters of Zirconia-Alumina Nanolaminates.

Film	Number of ZrO_2 layers	ZrO_2 layer thickness [nm]	Total film thickness [10^2 nm]
A	9	18.0	1.6
B	13	9.0	1.4
C	17	4.5	1.9
D	50	4.5	4.1

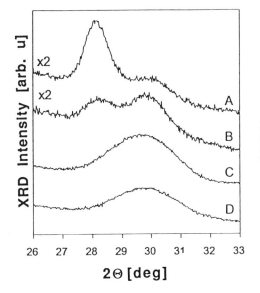

Figure 1: Major x-ray diffraction peaks of zirconia-alumina nanolaminates grown on fused silica under conditions given in Table I.

TABLE II

Zirconia Crystallite Dimension (D) and t-ZrO$_2$ Volume Fraction (VF) in Zirconia-Alumina Nanolaminates on Fused Silica Calculated from X-ray Diffraction Measurements

Film	D, m-ZrO$_2$ [nm][a	D, t-ZrO$_2$ [nm][a	VF, t-ZrO$_2$[b
A	9.8-11.1	5.7-6.6	0.24
B	7.6-8.0	5.5-6.2	0.6
C		4[c	~1[c
D		4[c	~1[c

a) Range represents uncertainty in measurement.
b) Remaining VF is m-ZrO$_2$.
c)Analysis assumes peak is entirely due to t-ZrO$_2$.

The XRD patterns of films C and D consist of a single broad peak tentatively attributed to (111) t-ZrO$_2$ planes. However, the broadness of this peak precludes exclusion of diffraction from (11$\bar{1}$) m-ZrO$_2$ planes. This issue was resolved using HREM, discussed below.

The crystallite dimension perpendicular to the substrate and the volume fraction of each phase in the films are recorded in Table II. Data for films A, B, and C, show that t-ZrO$_2$ volume fraction increases with decreasing zirconia layer thickness. A comparison of data for films C and D shows that the t-ZrO$_2$ volume fraction is dependent only on zirconia layer thickness and not on the total number of zirconia layers in the film.

In films B, C, and D, the largest average crystallite dimension [(11$\bar{1}$) m-ZrO$_2$ in film B and (111) t-ZrO$_2$ in films C and D] is comparable to the zirconia layer thickness. This result indicates that the nanolaminate structure in films B, C, and D truncates zirconia crystallite growth and insures a large t-ZrO$_2$ volume fraction. In film A, which has the thickest zirconia layers, the crystallite dimension is about one-half of the zirconia layer thickness. This result indicates that crystallite growth is terminated by the impinging sputtered flux, as proposed in a growth model for sputter deposited pure zirconia films [7], and not by the nanolaminate structure.

High resolution electron microscopy was used to resolve the ambiguity presented by the single broad XRD peak typical of films with a small zirconia layer thickness, such as films C and D.

Previously reported HREM results [13] on film D', the counterpart of film D grown on Si (111), showed that zirconia was polycrystalline throughout the 99-layer nanolaminate. The typical crystallite dimension perpendicular to the laminate plane was ~4 nm, comparable to both the crystallite dimension calculated from XRD data for film D and to the zirconia layer thickness.

A detailed analyisis of lattice spacings of individual crystallites was carried out using digitized HREM images. Single crystallite regions of these images were selected and Fourier transformed. The spatial frequencies were measured using the Hanning window method, giving typical accuracies between 0.3% and 1% for the selected crystallites with image pixel size between 64 px and 128 px [14]. The {111} Si lattice fringes were used for calibration.

Figure 2a shows a digitized image of an area of the zirconia-alumina bilayer adjacent to the substrate in film D'. The Si substrate (not shown here) is oriented with the [110] zone parallel to the electron beam, resulting in edge-on orientation of the multilayer. Alumina and the native silicon oxide on the substrate (designated SiO_x) display random contrast typical of amorphous structures. Lattice fringes can clearly be seen in the zirconia layer. The boxed area was used to calculate the numerical diffractogram shown in Fig. 2b.

The zirconia phase composition of the boxed area is not obvious from the image. However, the diffractogram clearly shows that this region is biphasic. The lattice spacings obtained from the diffractogram indicate that the top part of the boxed area is tetragonal while the bottom part is monoclinic. The monoclinic area has a twinned structure observed in most m-ZrO_2 crystallites in this nanolaminate.

The diffractogram is indexed in Table III. Comparison with bulk standards shows that the interplanar spacing of the twinned monoclinic region is ~2% smaller. In bulk materials, such behavior is typical of semicoherent interfacial regions between transformed monoclinic grains and the adjacent untransformed t-ZrO_2 parent phase [15]. In the present study, the combination of twinning and difference in $d_{m\{11\bar{1}\}}$ from the bulk standard suggests that a $t \rightarrow m$ transformation has occurred at some point in the film's history.

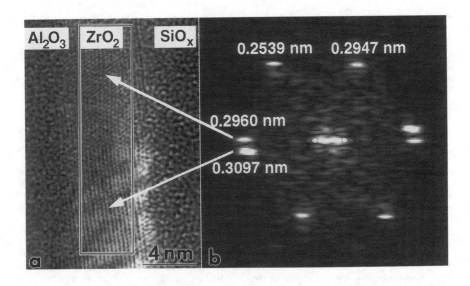

Figure 2: Lattice spacing measurements from digitized HREM images: a) region of first bilayer of film D'; b) numerical diffractogram of boxed region in (a).

TABLE III
Indexing of the Zirconia Diffractogram Shown in Fig. 2

d_{hkl} of film [nm]	Plane	d_{hkl} of bulk standard [nm][11]
0.3097	m (11̄1)	0.3165
0.2960	t {111}	0.296
0.2947	t {111}	0.296
0.2539	m (200), t (200), t (020)	0.254

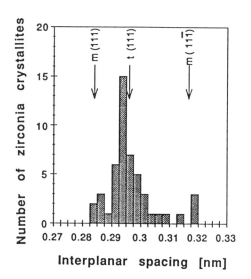

Figure 3: Histogram of lattice spacings in zirconia crystallites obtained from digitized HREM images. Bulk values for the monoclinic (m) and tetragonal (t) spacing are marked with an arrow (Ref. 13).

Approximately fifty zirconia crystallites from throughout the multilayer were analyzed using the method demonstrated in Fig. 2. The obtained interplanar spacings are given in the histogram in Fig. 3. The interval between d=0.27 nm and 0.33 nm is shown for comparison with the complementary XRD data in Fig. 1 for film D. The arrows in the histogram denote the values of d for bulk material [11]. The interplanar spacing for the majority of (111) t-ZrO$_2$ crystallites is slightly smaller than the bulk value. The bulk standard measurement was done at 1020 °C whereas our measurement was done at room temperature.

HREM data were used to estimate the volume fraction of the two zirconia phases in film D'. The data show that film D' contains a volume fraction of 0.72±0.06 t-ZrO$_2$. The remainder was identified as m-ZrO$_2$, although we cannot determine whether this phase was present during growth or is a product of post-deposition transformation. The histogram in Fig. 3 is our definite and reliable confirmation that a large volume fraction of tetragonal zirconia is stabilized in the nanolaminate structure.

Acknowledgements

Support under US Army Research Office Grant No. DAAH04-93-G-0238 and by a gift from the Johnson Controls Foundation to the Wisconsin Distinguished Professorship of CRA is acknowledged. Electron microscopy was performed at the Center for High Resolution Electron Microscopy at Arizona State University, which is supported under NSF Grant No. DMR-9115680. We are grateful to Dr. R. Graham, Dr. F. Shaapur, and Dr. M.R. McCartney from ASU for their assistance.

REFERENCES

1. For a short review, see: M. Ruhle, J. Vac. Sci. Technol. A **3**, 749 (1985).
2. C.-K. Kwok and C.R. Aita, J. Vac. Sci. Technol. A **7**, 1235 (1989).
3. C.-K. Kwok and C.R. Aita, J. Appl. Phys. **66**, 2756 (1989).
4. C.-K. Kwok and C.R. Aita, J. Vac. Sci. Technol. A **8**, 3345 (1990)
5. C.R. Aita and C.-K. Kwok, J. Amer. Ceram. Soc. **73**, 3209 (1990).
6. C.R. Aita, J. Vac. Sci. Technol. A **11**, 1540 (1993).
7. C.R. Aita, Nanostruct. Mater. **4**, (1994) in press.
8. See, for example, L.V. Azaroff, Elements of X-ray Crystallography (McGraw-Hill, New York, NY, 1968) pp. 551-2. The Scherrer equation gives the limiting case of broadening due to size effects with no contribution from random lattice strain, i.e., gives the minimum possible value of the average crystallite dimension.
9. R.C. Garvie and P.S. Nicholson, J. Amer. Ceram. Soc. **55**, 303 (1972).
10. J.C. Bravman and R. Sinclair, J. Electron Micros. Tech. **1**, 53 (1984).
11. ASTM Joint Committee on Powder Diffraction Standards, 1974, File Nos. 13-307 and 17-923.
12. E. Bauer, in Single Crystal Films (edited by M.H. Francombe and H. Sato, Macmillan, NY, NY, 1964) pp. 43-67.
13. C.M.Scanlan, M.Gajdardzisha-Josifovska, and C.R. Aita, Appl. Phys. Lett. **64**, (1994) in press.
14. W.J. deRuijter, M. Gajdardziska-Josifovska, M.R. McCartney, R. Sharma, D.J. Smith and J.K. Weiss, Scanning Microsc **6**, 347 (1992).
15. C.M. Wayman, in Science and Technology of Zirconia (edited by A.H. Heuer and L.W. Hobbs, Advances in Ceramics, Vol. 3, American Ceramic Society, Columbus, OH, 1981) pp. 64-81.

TEXTURE AND PHASES IN OXIDE FILMS ON Zr-Nb ALLOYS

Y.P. LIN*, O.T. WOO** AND D.J. LOCKWOOD***
*Ontario Hydro Technologies, Materials Unit, 800 Kipling Ave., Toronto M8Z 5S4, Canada.
**AECL Research, Chalk River Laboratories, Chalk River K0J 1J0, Canada.
***National Research Council, Institute for Microstructural Sciences, Ottawa K1A 0R6, Canada.

ABSTRACT

Oxide films, 0.2-2.0 μm in thickness on Zr-2.5Nb and 11 μm thick on Zr-20Nb alloys, formed in steam at 673 K, have been examined using TEM, XRD and Raman spectroscopy. Columnar grains of mostly monoclinic ZrO_2 in oxide films on Zr-2.5Nb exhibit a dual texture: a fibre mode with an axis close to the $10\bar{2}_m$ pole and a $[001]_m$ growth mode with an orientation relationship $[100]_m$ // $[4\bar{5}10]_\alpha$ and $(010)_m$ // $(0001)_\alpha$ with the α-Zr metal. In both modes, "tetragonal" (and/or cubic) ZrO_2 was present. Raman spectroscopy differentiated two non-cubic "tetragonal" forms of ZrO_2 within the $[001]_m$ growth texture. In thin oxides (0.5 μm or less), this corresponds to the tetragonal ZrO_2 observed in ceramic zirconia and is characterised by a Raman band near 260 cm^{-1}. The 278 and related 438 cm^{-1} Raman bands observed here in some oxide films (and in other Zr corrosion oxides) are attributed to a separate, non-cubic phase structurally related to the tetragonal ZrO_2. The intensities of the 278 and 438 cm^{-1} bands are dependent not only on the amount of this modified-tetragonal phase but also on the oxide texture (related to the metal texture) and the beam orientation. The lack of Raman response from the "tetragonal" ZrO_2 within the fibre mode of texture indicates either a low volume fraction or a cubic-like structure. For oxide on Zr-20Nb, XRD and Raman spectroscopy show a mixture of monoclinic and "tetragonal" ZrO_2; the Raman results indicate the "tetragonal" ZrO_2 has a high crystal symmetry or nearly cubic structure.

INTRODUCTION

Pressure tubes for CANDU reactors are made from Zr-2.5wt%Nb alloys by a process involving extrusion and cold drawing. Thin oxide films are formed during a stress-relieving treatment at 673 K for 24 hours in a steam atmosphere and serve as barriers against hydrogen ingress. The microstructure of the alloy consists of the hexagonal α-Zr (1%Nb) and the cubic β-Zr (20%Nb) phases. The highly elongated α-Zr grains are generally 0.3-0.5 μm thick and are usually surrounded by thin layer of β-Zr. The barrier characteristics of the oxide film are dependent on its microstructure. Monoclinic (m) ZrO_2 is known to be the major phase in oxide films formed on Zr alloys. Information on the tetragonal (t) ZrO_2, the minor phase, is significant to the understanding of the oxide growth mechanism and has implications on stresses needed to stabilise the t-ZrO_2. Due to the structural relationship between the monoclinic, tetragonal and cubic (c) forms of ZrO_2, ambiguities exist in distinguishing the t- and c-ZrO_2 in the presence of m-ZrO_2 using diffraction methods. The term "tetragonal" (including quotes) ZrO_2 will be used here for simplicity and indexing of t-ZrO_2 will be based on the C-centred cell. By comparison, Raman spectroscopy is sensitive to the nature of chemical bonds and can readily distinguish the three polymorphs. The present work addresses the texture and phases present in oxide films on Zr-2.5Nb and Zr-20Nb alloys using a combination of transmission electron microscopy (TEM), x-ray diffraction (XRD) and Raman spectroscopy (RS).

EXPERIMENTAL

Polished coupons cut from a Zr-2.5Nb pressure tube were oxidised in steam at 673 K for various times to produce oxide thicknesses of 0.2, 0.25, 0.5, 0.7 and 2.0 μm. The oxide films examined correspond to oxide on the radial-normal section of the pressure tube. For the Zr-20Nb alloy (simulating β-Zr), the material was annealed at 1123 K prior to oxidation at 673 K in steam to produce a 11 μm thick oxide. Cross-section and plan-view samples for TEM were produced using ion-milling techniques. XRD were conducted on the oxidised coupons using Cu-K_α irradiation. An Ar laser operating at 457.9nm was used for Raman spectroscopy. In the normal-incidence scattering geometry, a microprobe of 1μm diameter was used; for the 12.3° glancing-incidence arrangement, the probe size was about 1x0.1mm^2.

RESULTS

The general microstructure of the 0.7 μm thick oxide on Zr-2.5Nb consisted of columnar grains of predominantly m-ZrO$_2$, 20-30 nm in width. The axis of the columnar grains is approximately in the growth direction and the length is related to the original metal grain size in that direction, but allowing for the volume increase due to oxidation. Electron diffraction from plan-view samples of the oxide showed a spatial variability on the scale of the metal (α-Zr) grain size and that two texture modes were present. In the commonly observed ring patterns, Fig. 1a, the $11\bar{1}_m$, 002_m and particularly $10\bar{2}_m$ rings were weak or absent, indicating a fibre texture. In contrast to this fibre mode, clusters of columnar grains with similar orientations could be found which required only a minimum amount of sample tilt to bring the $[001]_m$ orientation to the beam direction, Fig. 1b, i.e. an alignment of $[001]_m$ with the growth direction. From suitable cross-section samples, an orientation relationship with the metal for this $[001]_m$ growth mode was determined: $[100]_m$ // $[4\bar{5}10]_\alpha$ and $(010)_m$ // $(0001)_\alpha$; such an orientation relationship yields $[001]_m$ close to $[11\bar{2}0]_\alpha$. In both modes of oxide texture, evidence for the presence of a crystallographically distinct, viz "tetragonal", form of ZrO$_2$ was noted; from the 111_t ring in the case of the fibre mode, Fig. 1a, and from the apparent "splitting" of high order reflections in the $[001]_m$ patterns, an example is arrowed in Fig. 1b.

XRD 2θ scan of the 0.7 μm thick oxide, Fig. 2a, showed predominantly m-ZrO$_2$ peaks with a minor 111_t peak. Due to the presence of texture, the observed relative intensities deviate

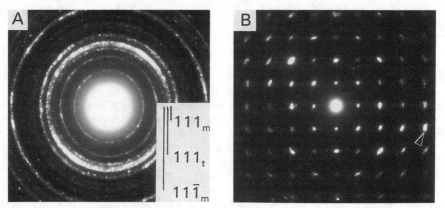

Fig 1   a) Electron diffraction ring pattern and b) m-ZrO$_2$ [001] pattern

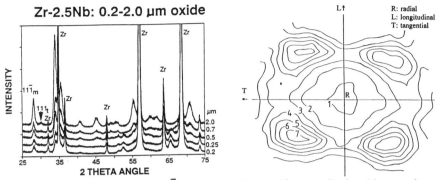

Fig 2 a) XRD 2θ scan and b) $11\bar{1}_m$ partial pole figure (70° max. tilt) in arbitrary units.

considerably from those for random polycrystals or from structure factor expectations. Of the two texture modes, only the fibre mode is expected to affect significantly the observed intensities, as the $[001]_m$ growth mode contribute mainly to the $(002)_m$ peak which already contains contributions from the fibre mode. It is significant that the 111_m peak, which has a strong structure factor, is virtually absent in Fig. 2a. The presence of the $[001]_m$ growth mode is more readily deduced from $11\bar{1}_m$ partial pole figures, Fig. 2b. Two trends are evident from Fig. 2b. The first is the increase in $11\bar{1}_m$ pole density as the angle from the surface normal increased, but it then drops off at higher angles (above ~ 50°); this corresponds to the fibre mode of texture. The second is the high concentrations of $11\bar{1}_m$ pole intensities in all four quadrants of the partial pole figure; this corresponds to two distinct variants of the above orientation relationship. Compared with oxide of other thicknesses, Fig. 2a shows that m-ZrO$_2$ was the dominant phase for all Zr-2.5Nb oxide thicknesses examined here and that trace amounts of the 111_t peak were always present. The variations in relative intensities were also similar indicating that the fibre texture was not altered significantly as the oxide thickened.

Raman spectroscopy of Zr-2.5Nb oxide films readily showed bands due to m-ZrO$_2$. In 0.7 μm thick samples, a band at 278 cm^{-1} was present, Figs. 3a and b. Using a 1 μm probe, a variability of the 278 cm^{-1} signal with location was noted. The 278 cm^{-1} band was related to a band at 438 cm^{-1}, Fig. 3. Both bands and their relative intensities exhibit a dependency on the orientation of the incident laser beam. With the incident beam normal to the surface (i.e. along

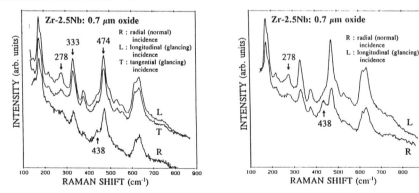

Fig 3 Raman spectra for 0.7 μm thick oxide on Zr-2.5Nb

oxide growth direction or pressure tube radial direction) the 438 cm^{-1} band is generally more pronounced. For glancing beam incidence, the 278 cm^{-1} band was more noticeable and the 438 cm^{-1} band can appear as just a broad shoulder to the 474 cm^{-1} m-ZrO$_2$ band. The prominence of one band over the other also applies when the glancing beam direction is varied between nearly parallel with the longitudinal and with tangential directions, Fig. 3a. In thinner Zr-2.5Nb oxides (0.5 μm or less), m-ZrO$_2$ Raman bands and a 260 cm^{-1} band (rather than 278 cm^{-1}) are seen for the glancing incident beam with little evidence for a band at 438 cm^{-1}, Fig 4. In these thin oxides, a spatial variability of the 260 cm^{-1} band was noted; however, the most noticeable 260 cm^{-1} band was in the thinnest oxide film examined (0.2 μm). For normal beam incidence, the Raman signal was low due to the small probe size and thinness of the oxide; only the 0.5 μm thick film provided sufficient signal above background for a 260 cm^{-1} band to be discerned. In thicker oxide (2.0 μm), Fig. 5, a broad shoulder to the 333 cm^{-1} m-ZrO$_2$ band extending to about 250 cm^{-1} was noted using a glancing incidence beam but a band at 278 cm^{-1} was not resolved; however, a band at 300 cm^{-1} was resolved using a normal incidence beam, Fig. 5.

For the 11 μm oxide on Zr-20Nb, an XRD 2θ scan, Fig. 6, in addition to indicating a texture effect, showed the presence of m-ZrO$_2$, although the strongest peaks could also be assigned to "tetragonal" ZrO$_2$. The Raman spectrum, Fig. 7, showed the presence of some weak m-ZrO$_2$ bands above the background. A broad band at 300 cm^{-1} is evident; however, bands at 260, 278 or 438 cm^{-1} were not resolved. The background consisted of a relatively sharp drop at around 190 cm^{-1} with a broad shoulder between 220 and 310 cm^{-1} followed by a drop to a relatively flat level between 450 and 760 cm^{-1}. Such a background is typical of other oxides on Zr-20Nb formed under different conditions.

DISCUSSION

Electron diffraction, Fig. 1a, and XRD, Fig. 2a, indicate that the axis for the fibre mode of oxide texture is at a large angle to the 111$_m$ pole but at a small angle to the 10$\bar{2}_m$ pole. The 10$\bar{2}_m$ pole, at 73.5° to the 111$_m$ pole, is a likely possibility and is consistent with an axis in the vicinity of the 10$\bar{2}_m$ pole for oxide formed in oxygen on Zr and Zircaloy-2 [1]. The second type of oxide texture is based on the alignment of [001]$_m$ with the growth direction and has a specific orientation relationship with the metal. The propensity of such mode will thus depend on the availability of suitably oriented α-Zr grains. The texture of a pressure tube material has high concentrations of [1$\bar{1}$00]$_\alpha$ in the axial and (0001)$_\alpha$ in the tangential directions. A significant

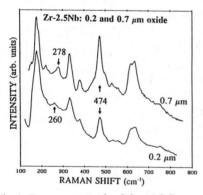

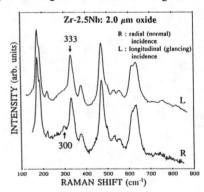

Fig 4. Raman spectra for 0.2 and 0.7 μm oxide.   Fig 5. Raman Spectra for 2.0 μm oxide

490

Zr-20Nb: 11 μm oxide

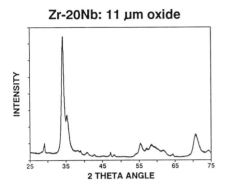

INTENSITY

2 THETA ANGLE

Fig 6. XRD 2θ scan for Zr-20Nb oxide

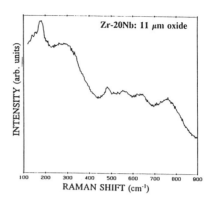

Fig 7. Raman spectra for Zr-20Nb oxide

fraction of the α-Zr grains would be thus expected to develop such an oxide texture, as shown by the spatial variability and by the $11\bar{1}_m$ partial pole figure, Fig. 2b.

A Raman band near 280 cm⁻¹ is commonly observed in corrosion oxide films in Zr alloys [2] and is generally attributed to tetragonal ZrO_2. In stabilised zirconia ceramics, however, a band at 260-264 cm⁻¹ is normally associated with the tetragonal ZrO_2 [3]. In the present work, a Raman band at 278 cm⁻¹ with a related 438 cm⁻¹ band have been observed in 0.7μm thick oxide, while a band at 260 cm⁻¹ was observed in thinner oxide films. These results indicate two structurally distinct forms of non-cubic "tetragonal" ZrO_2. The 260 cm⁻¹ band observed here is taken to be due to the same tetragonal ZrO_2 as in ceramic zirconia. The beam orientation dependency of the 278/438 cm⁻¹ bands implies an origin in a crystallographically aligned phase, which is readily identified with the oriented second phase within the $[001]_m$ texture mode, Fig. 1b. The spatial variability of the observed 278/438 cm⁻¹ bands can also be related to that of the $[001]_m$ oxide texture. As the 260 cm⁻¹ tetragonal ZrO_2 band was observed in thin oxide films, it would seem plausible that the 278 cm⁻¹ band is the result of a transformation to another structurally related phase. A derived structure based on tetragonal ZrO_2 [3] is likely; recent work in ceramic zirconia have shown the existence of more than one "tetragonal" ZrO_2 structure [4]. The observed intensities of the 278 and 438 cm⁻¹ bands will thus be dependent not only on the amount of this modified-tetragonal phase but also the texture of the oxide, texture of the metal and the orientation of the incident laser beam. The association of the 260 and the 278 cm⁻¹ bands implies that the 260 cm⁻¹ tetragonal ZrO_2 band likewise originated from within the $[001]_m$ texture mode in thinner oxide films. Such an origin for the tetragonal ZrO_2 and its derivative means that the 111 reflections from them are not suitably oriented to contribute to the XRD 2θ scans, Fig. 2a, or electron diffraction ring patterns, Fig. 1a; both figures arise from the fibre mode of oxide texture. Yet, both Figs. 1a and 2a showed 111_t reflections from "tetragonal" ZrO_2. RS on the present set of oxide films thus appears to be not sensitive to the "tetragonal" ZrO_2 within the fibre mode of oxide texture. It may be that the volume fraction of the "tetragonal" ZrO_2 is quite low there, or that the "tetragonal" ZrO_2 there exhibits very few Raman bands, i.e. has a higher crystal symmetry closer to that of the cubic ZrO_2. Comparison of results on Zr-2.5Nb oxide films suggests that Raman spectroscopy is sensitive to a two-stage and possibly a three-stage structural modification process in the "tetragonal" ZrO_2 (within the $[001]_m$ mode of oxide texture) such that the 260 cm⁻¹ band from the tetragonal ZrO_2, strongest in the thinner oxide films, first shifts to 278 cm⁻¹ and later to 300 cm⁻¹. Such a scenario could be related to stress relaxation effects as the oxide thickens. However, there has been only limited observations regarding the 300 cm^{-1} band, and further work would be needed to substantiate such a three-stage sequence of structural modifications of the "tetragonal" ZrO_2.

For the oxide on Zr-20Nb, RS showed that m-ZrO_2 was not the major phase; however, bands for the non-cubic "tetragonal" ZrO_2 were not resolved. As an oxide texture is present, distinct from that of oxidised Zr-2.5Nb material, it may be that the incident beam was not suitably oriented to resolve the various "tetragonal" ZrO_2 bands. However, the presence of a broad band and the shape of the background are indicative of a high crystal symmetry for the "tetragonal" ZrO_2, the major phase. The nature of such an oxide is currently being investigated.

CONCLUSIONS

Texture and phases in oxide films, 0.2 - 2.0 μm in thickness on Zr-2.5Nb and 11 μm thick on Zr-20Nb alloys formed in steam at 673 K, have been investigated. Columnar grains of predominantly m-ZrO_2 in oxide on Zr-2.5Nb exhibit a dual texture: a fibre mode with an axis in the vicinity of $10\bar{2}_m$ pole and a $[001]_m$ growth mode with an orientation relationship $[100]_m$ // $[4\bar{5}10]_\alpha$ and $(010)_m$ // $(0001)_\alpha$ with the α-Zr metal. In both modes, presence of "tetragonal" ZrO_2 was found. "Tetragonal" ZrO_2 here refers to the cubic, tetragonal and a modified-tetragonal ZrO_2. Raman spectroscopy differentiated two non-cubic "tetragonal" forms of ZrO_2 in the $[001]_m$ texture mode. In thin oxides, $\leq 0.5\mu$m, t-ZrO_2 is characterised by a Raman band at 260 cm^{-1} and corresponds that observed in ceramic zirconia. In other oxide films (also in other Zr corrosion oxides) related Raman bands at 278 and 438 cm^{-1} are attributed to a separate non-cubic phase, structurally related to the tetragonal ZrO_2. The amount of this phase, the orientation of the incident beam, the textures of the oxide and metal all affect the observed intensity of the 278 and 438 cm^{-1} Raman bands. Within the fibre texture mode, evidence for "tetragonal" ZrO_2 from TEM and XRD but not from RS indicate either a low volume fraction or presence of cubic-like ZrO_2. XRD and RS of oxide on Zr-20Nb show a mixture of monoclinic and "tetragonal" ZrO_2; Raman results indicating a nearly-cubic "tetragonal" ZrO_2.

REFERENCES

1. C. Roy and G. David, J. Nucl. Mat. **37**, 71 (1970).

2. J. Godlewski, J.P. Gros, M. Lambertin, J.F. Wadier and H. Weidiger in Zirconium in the Nuclear Industry, edited by C.M. Eucken and A.M. Garde (ASTP STP 1132, Philadelphia, PA, 1991) pp.416-436.

3. G. Teufer, Acta Cryst. **15**, 1187 (1962).

4. A.H. Heuer and M. Rühle in Advances in Ceramics, Vol. 12, edited by N. Claussen, M. Rühle and A.H. Heuer (American Ceramic Soc, 1983) pp. 1-13.

ACKNOWLEDGEMENTS

This work was funded by CANDU Owners Group, Working Party 35. Thanks are due to N. Ramasubramanian for providing oxide thickness measurements using FTIR, V.F. Urbanic for oxide of Zr-20Nb, and J. De Luca, H.J. Labbé, D. Oad and A. Audet for technical assistance.

PREPARATION OF PZT THIN FILMS BY ECR PECVD

J. W. KIM, S. T. KIM, S. W. CHUNG, J. S. SHIN, K. S. NO*, D. M. WEE
and W.J. LEE
Department of Electronic Materials Engineering
*Department of Ceramic Science and Engineering
Korea Advanced Institute of Science and Technology, Taejon, 305-701 South
Korea

ABSTRACT

Lead zirconate titanite (PZT) thin films have been prepared on $Pt/Ti/SiO_2/Si$ and $Pt/Ta/SiO_2/Si$ substrates by electron cyclotron resonance plasma enhanced chemical vapor deposition (ECR PECVD) at low temperatures using metal-organic (MO) sources. One of the advantages of this method is the easy control of the cation concentrations in the PZT films through the adjustment of the flow rates of each MO sources. The film compositional ratio Ti/Zr was found to be linearly proportional to the flow rate ratio of the input Ti and Zr MO sources. The Pb/(Zr+Ti) ratio was not so much affected by the Ti source flow rate but was inversely proportional to the Zr source flow rate. Pb concentration in the deposited PZT films increased with increasing Ti source flow rate and with decreasing substrate temperature. Perovskite single phase PZT films were successfully fabricated at low temperatures below 500℃. The perovskite structure was obtained when the Pb/(Zr+Ti) ratio in the deposited film was close to 1.

INTRODUCTION

Lead zirconate titanate (PZT) ferroelectric films with perovskite structure have been extensively studied for the applications to pyroelectric infrared sensor, SAW device and memory device, because they have excellent pyroelectric, piezoelectric and dielectric properties[1 4]. Sol-gel[5 6] or sputtering methods[7 8] have been largely used for the fabrication of PZT films because of their simplicity in processing. Recently a great deal of attention has been paid to the metal-organic chemical vapor deposition (MOCVD)[9 11] method which provides good step coverage, large area uniformity and good film quality.

Since the deposition reaction in the thermal MOCVD occurs at the thermal equilibrium condition, the fabrication of the film requires a high process temperature. This makes the stoichiometric control of the multi-component film difficult and the vapor-phase homogenous reactions hardly avoidable. These difficulties can be avoided by lowering the deposition temperature and pressure. Electron cyclotron resonance (ECR) plasma[12] which has a high ionization efficiency decomposes the precursors so powerfully that the films is fabricated at low temperatures. The homogeneous reactions is also significantly inhibited in ECR plasma enhanced CVD (ECR PECVD) because the processing pressure can be lowered to 0.1 mTorr. In the present work, PZT films were fabricated at low temperatures below 500℃ using the ECR PECVD method. The composition and the structure of the films prepared at various source input ratios were examined and the deposition conditions for obtaining perovskite phase were discussed.

EXPERIMENTAL

PZT films were deposited on Pt(111)/Ti/SiO$_2$/Si and Pt(111)/Ta/SiO$_2$/Si substrates at 470~500℃ and 3 mTorr. The schematic diagram of the ECR PECVD system used in the present work is given in Fig.1. The precursors are lead β-diketonate (Pb(DPM)$_2$, Pb(C$_{11}$H$_{19}$O$_2$)$_2$), zirconium t-butoxide (ZrTB, Zr(t-OC$_4$H$_9$)$_4$), tianium iso-propoxide (TiIP, Ti(i-OC$_3$H$_7$)$_4$) and O$_2$. Pb(DPM)$_2$, ZrTB and TiIP are introduced into the chamber by passing Ar gas through the bubblers where constant temperatures are kept : Pb(DPM)$_2$ at 155℃ (P$_{eq}$=0.7Torr), ZrTB at 70℃ (P$_{eq}$=2Torr) and TiIP at 85℃ (P$_{eq}$=1.73Torr[13]). The flow rates of O$_2$ and Ar carrier gas are controlled by mass flow controllers. The flow rates of MO sources are calculated from the vapor pressures of the sources and the conductances of the fine metering valves.

The thicknesses of the deposited films were in the range of 100~150nm and the deposition rate was about 3nm/min. The uniformity of the film which is the deviation in thickness from the mean value was within 5% for the substrate with 3″ diameter. The structural phase of the PZT thin films were characterized using a X-ray diffractometer with CuKα radiation source. Wavelength dispersive spectroscopy (WDS) and Auger electron spectroscopy (AES) were employed to obtain the chemical compositions and the depth profiles of the films. The characteristic X-rays detected in WDS were Pb Mα (2.394keV), Zr Lα (2.042eV) and Ti Kα (4.511keV). The incident electron beam with a low energy of 8 KeV and a incident angle of 30° tilted from the surface normal was used to retain the characteristic X-ray generation volume within the film. The compositions obtained from WDS and AES were calibrated by using Rutherford back-scattering spectroscopy(RBS).

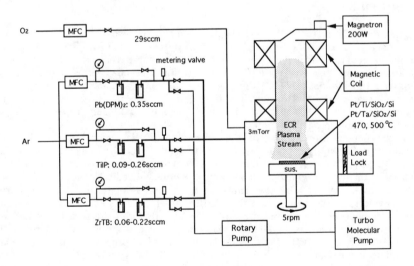

Fig.1. The schematic diagram of the ECR PECVD system.

RESULTS AND DISCUSSION

Fig.2 shows the AES composition depth profiles of the PZT films deposited on (a) Pt/Ti and (b) Pt/Ta substrates at 500 ℃. Both films show uniform composition profiles through the depth, except that Ti-rich interface layer is obsered for the PZT films deposited on Pt/Ti substrate which is formed by the out-diffusion of Ti through the Pt layer.

Fig.3 shows the dependence of the film compositional ratio Ti/Zr on the input flow rate ratio of Ti source to Zr source TiIP/ZrTB. Pb(DPM)$_2$ flow rate was fixed at 0.35 sccm. The film compositions were determined by WDS. It can be seen that the Ti/Zr ratio changes linearly with the flow rate ratio of input MO sources. The linear relationship is satisfied at both temperatures of 470 ℃ and 500 ℃. Therefore, the Ti/Zr ratio can be easily controlled by varying the input sources ratio. The Ti concentrations in the films deposited on Pt/Ti substrates are a little higher than those on Pt/Ta substrates, which is ascribed to the Ti-rich interface layer formed on Pt/Ti substrates.

Fig.4 shows the dependence of the compositional ratio on the flow rate of Ti source TiIP. Pb(DPM)$_2$ flow rate was fixed at 0.35sccm. Pb(Zr+Ti) ratio in the films is not so much affected by the TiIP flow rate, meaning that the increase of the flow rate of Ti source causes the increase of Pb content as well as Ti content in the PZT films. The reason for this is, we believe, that Pb is incorporate into the PZT films through the reaction with Ti to form lead titanate as suggested elsewhere[14,15]. The Pb conentration in films deposited on Pt/Ti substrates are higher than those on Pt/Ta substarates. It is also ascribed to the Ti-rich interface layer which enhances the incorporation of Pb in the film.

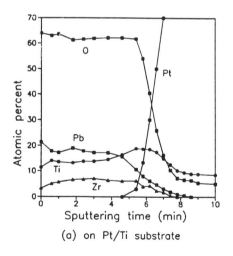

(a) on Pt/Ti substrate

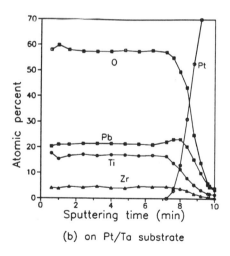

(b) on Pt/Ta substrate

Fig.2. AES depth profiles of the PZT films deposited on (a) Pt/Ti/SiO$_2$/Si and (b) Pt/Ta/SiO$_2$/Si substrates at 500 ℃.

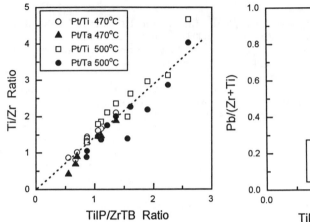

Fig.3. Ti/Zr atomic concentration ratio in the PZT films as a function of TiIP/ZrTB source flow rate ratio. Pb(DPM)₂ flow rate is 0.35 sccm.

Fig.4. Pb/(Zr+Ti) atomic concentration ratio in the PZT films deposited at 500℃ as a function of TiIP flow rate. Pb(DPM)₂ flow rate is 0.35 sccm.

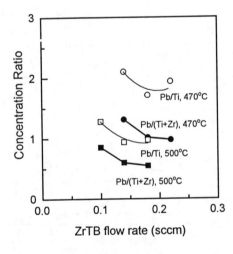

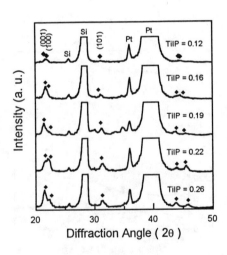

Fig.5. Atomic concentration ratio in the PZT films as a function of ZrTB flow rate. Pb(DPM)₂ and TiIP flow rates are 0.35 and 0.14 sccm.

Fig.6. XRD patterns of the PZT films deposited on Pt/Ti substrates at 500 ℃ with TiIP flow rate as a parameter. ZrTB flow rate is 0.1 sccm.

Fig.5 shows the dependence of the film compositional ratios Pb/(Zr+Ti) and Pb/Ti on the flow rate of Zr source ZrTB. Pb(DPM)$_2$ and TiIP flow rates were fixed at 0.35sccm and 0.14sccm respectively. It can be seen that both Pb/(Zr+Ti) and Pb/Ti ratios decrease with increasing ZrTB flow rate, and Pb is incorporated less at the higher deposition temperature. Since the Ti concentration in the film is not largely affected by the ZrTB input flow rate, it is believed that the Zr MO source interferes with the reaction between Pb(DPM)$_2$ and TiIP and thus the incorporation of Pb in the film is reduced as the Zr source flow rate increases. The decrease of the Pb concentration with increasing temperature is thought to be caused by the vaporization of Pb or PbO from the film or by the decrease in the sticking coefficient of the Pb source. Perovskite single phase PZT films were obtained at the following conditions : ZrTB 0.18 sccm at 470℃ and ZrTB 0.1 sccm 500℃. Note that the Pb/(Zr+Ti) ratio is close to 1 at these deposition conditions.

Fig.6 shows the X-ray diffraction patterns of the PZT films deposited on Pt/Ti substrates at 500℃ with varying TiIP flow rate. Pb(DPM)$_2$ and ZrTB flow rates were fixed at 0.35sccm and 0.1sccm respectively. As the TiIP flow rate increases, 22° and 44.5° peaks are split. As the TiIP flow rate (and thus the Ti content in the films) increases, the film structure changes from rhombohedral to tetragonal, resulting in the splitting of (100) peak into (001) and (100) peaks.

In the present work, we successfully fabricate perovskite single phase PZT films at low temperatures below 500℃. The perovskite phase was obtained when Pb/(Zr+Ti) ratio was close to 1 and the Ti/Zr ratio was higher than 1. As the Pb/(Zr+Ti) ratio deviated from 1 or as the Ti/Zr ratio is less than 1, other phases such as ZrO_2, PbO_2 and pyrochlore appeared to be mixed with perovskite phase. When the stoichiometry deviated very much no perovskite phase was observed.

CONCLUSIONS

PZT films were deposited on Pt/Ti/SiO$_2$/Si and Pt/Ta/SiO$_2$/Si substrates by ECR PECVD at low temperatures below 500℃. For a given substrate and deposition temperature, the cation concentrations in the PZT films was easily controled by adjusting the flow rates of MO sources. Perovskite single phase PZT films were obtained when the film compositional ratio Pb/(Zr+Ti) was close to 1.

The Pb/(Zr+Ti) ratio in the films was almost constant regardless of the Ti source flow rate, but it decreased rapidly as the flow rate of Zr source increased. The Pb concentration in the deposited films increased as the TiIP flow rate increased since Ti enhances the Pb incorporation. Also, the Pb concentration in the films on Pt/Ti substrates was higher than that on Pt/Ta substrates because Pb incorporation was increased by the out-diffused Ti of the Pt/Ti substrate. Pb/(Zr+Ti) and Pb/Ti ratios in the films decreased with increasing temperature. This was ascribed to the increase of the Pb vaporization rate and/or to the decrease of sticking coefficient of Pb precursor with increasing temperature. The Ti/Zr ratio in the films changed linearly with the flow rate ratio of Ti and Zr sources. This linear relationship satisfied at both deposition temperature of 470 ℃ and 500℃.

ACKNOWLEDGMENTS

The authors acknowledge the support of SAMSUNG ELECTRONICS Co. for this work.

REFERENCES

1. R. Takayama and Y. Tomita, J. Appl. Phys., 65, 15(1989).
2. K. Suzuki and M. Nishikawa, Jpn. J. Appl. Phys., 13, 240(1974).
3. K. Tominaga and M. Okada, J. Appl. Phys., 69, 15(1991).
4. Y. Ohya, T. Tanaka and Y. Takahashi, Jpn. J. Appl. Phys., 32, 4163(1993).
5. S. A. Myers and E. R. Myers, Mat. Res. Soc. Symp. Proc., 243, 107(1992).
6. H. Watanabe, T. Mihara, and H. Yoshimori, 4th Intern. Symp. on Integrated Ferroelectrics, 346(1992).
7. T. Hase, T. Sakuma Y. Miyasaka, K. Hirata and N. Hosokawa, Jpn. J. Appl. Phys., 32, 4061(1993).
8. K. D. Gifford, H. N. Al-Shareef, S. H. Rou, P. D. Hren, O Auciello, and A. I. Kingon, Mat. Res. Soc. Symp. Proc., 243, 191(1992).
9. W. C. Hendricks, S. B. Desu, J. Si, and C. H. Peng, Mat. Res. Soc. Symp. Proc., 310, 241(1993).
10. T. Gatayama, M. Shimizu, and T. Shiosaki, 4th Intern. Symp. on Integrated Ferroelectrics, 382(1992).
11. M. Shimizu, M. Fujimoto, T. Katayama, T. Shiosaki, K. Nakaya, M. Fukagawa and E. Tanikawa, Mat. Res. Soc. Symp. Proc., 310, 255(1993).
12. T. Fukuda, M. Ohue, N. Momma, K. Suzuki and T. Sonobe, Jap. J. App. Phys. 28, 1035(1989).
13. Measured experimentally in this laboratory.
14. A. Fujisawa, M. Furihata, I. Minemura, Y. Onuma and T. Fukami, Jpn. J. Appl. Phys. 32 4048(1993).
15. K. Abe, H. Tomita, H. Toyoda, M. Imai and Y. Yokote, Jpn. J. Appl. Phys. 30 2152(1991).

PREPARATION AND PROPERTY CHARACTERIZATION OF ORIENTED PLZT THIN FILMS PROCESSED USING SOL-GEL METHOD

Dae Sung Yoon, Chang Jung Kim, Joon Sung Lee, Chaun Gi Choi, Won Jong Lee*, and Kwangsoo No
Department of Ceramic Science and Engineering
*Department of Electronic Materials Science and Engineering, Korea Advanced Institute of Science and Technology, Taejon, Korea.

ABSTRACT

Highly preferentially oriented lead lathanum zirconate titanate(PLZT) thin films were fabricated on various substrates using the spin coating of metal organic solutions having the composition of (9/50/50) and (10/0/100). The substrates used in this study were $SrTiO_3(100)$, $MgO(100)$, r-plane sapphire, PLT-coated glass, and Pt/Ti/MgO substrates. The films were heat-treated at 600 °C and 700 °C using the direct insertion method. The phases and the orientation of the PLZT thin films were examined using X-ray diffraction(XRD). Pole figure and X-ray rocking curves were measured to study the film orientation. The films were grown with (100), (110), and (001) plane being parallel to the surfaces of $SrTiO_3$, sapphire, and Pt/Ti/MgO, respectively. The dielectric and optical properties of both the oriented films and the polycrystalline films were measured and discussed.

INTRODUCTION

Much attention has been recently paid to lanthanum-modified lead zirconate titanate(PLZT) solid solution systems which are well-known ferroelectric materials. Because PLZT is transparent in the visible and the near infrared region of the electromagnetic waves and has excellent dielectric and electrooptic properties, it is a candidate material for electronic and electrooptic applications. Recently, PLZT thin films have attracted much interest for fabricating novel functional devices, such as thin film capacitors, ferroelectric random access memories(FRAM), optical waveguide modulators, display and other electrooptic devices[1-10].

Recently, we have succeeded in the oriented growth of the PLZT thin films prepared by sol-gel method[10]. This paper describes a procedure for the fabrication of the oriented PLZT thin films. The dielecric and optical properties of both the oriented and the polycrystalline thin films are discussed. We supply a guideline to the property difference between oriented and polycrystalline thin films.

EXPERIMENTAL PROCEDURE

A precise procedure for preparing the precursor solution is shown in Ref. 8. The solution composition was PLZT(9/50/50) and PLZT(10/0/100) consisting of 5 mole % excess Pb. The substrates used in this study were MgO(100), $SrTiO_3(100)$, r-plane sapphire, PLT(10/0/100)-coated Corning 7059 glass, and Pt/Ti/MgO substrates. We also used polycrystalline Pt-coated MgO substrates to fabricate randomly oriented PLZT thin films. A PLT thin layer(50 nm) of the PLT-coated glass substrate acts as a seeding layer to accelerate the nucleation of the oriented PLZT grains. The thin films were deposited by spin-coating the precursor solutions at 3000 rpm for 30 s and then dried at 390 °C for 5 min. The films(with the thickness of about 0.5 μm.) were directly inserted in a furnace to 700 °C and annealed for 30 min[8, 10].

Mat. Res. Soc. Symp. Proc. Vol. 343. ©1994 Materials Research Society

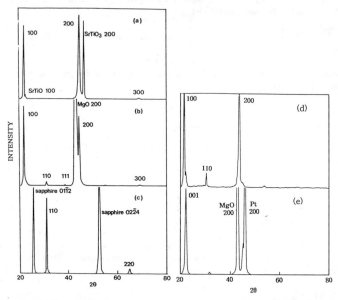

Fig. 1. The X-ray diffraction patterns of the PLZT(9/50/50) films on (a) SrTiO₃,
(b) MgO, (c) sapphire, and (d) PLT-coated glass, and the PLZT(10/0/100)
films on (e) Pt/Ti/MgO

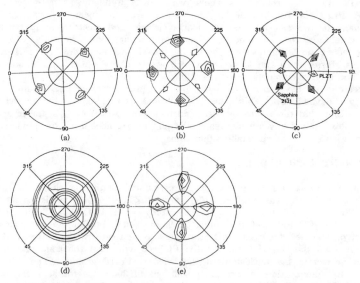

Fig.2. Pole figures of the PLZT(9/50/50) films on (a) SrTiO₃, (b) MgO,
(c) sapphire, and (d) PLT-coated glass, and the PLZT(10/0/100) film on
(e) Pt/Ti/MgO

RESULTS AND DISSCUSSION

X-ray Diffraction patterns of the PLZT thin films fabricated on various substreates are shown in Figure 1. From the XRD patterns, it is known that the PLZT(100) for the SrTiO₃, the PLZT(110) for the sapphire, and the PLT(001) for the Pt/Ti/MgO are parellel to the surface of the substrates. The XRD patterns of the films on an MgO and PLT-coated glass show small extra peaks of (110) and (111) with main (100) and (200) peaks, indicating that the PLZT films on the substrates were grown with preferential orientation.

A relation between the crystal axes of PLZT and the in-plane vectors of the substrates can be obtained using X-ray pole figure. The Schultz reflection method was used in these measurements. The α and β rotations are coupled so that a 360° rotation of β corresponds to 5.0° decrease of α. Figure 2 shows the X-ray pole contour distributions of the PLZT thin films fabricated on various substrates. The pole figures were obtained at 2θ = 38.2°[(111) plane] for the films on SrTiO₃, MgO, and sapphire, and 2θ = 31.8° and 32.8° [(110) plane] for the films on glass and Pt/Ti/MgO. The thin films fabricated on SrTiO₃ and MgO substrates have the strong four (111) reflections coupled with their narrow intensity distributions. This indicate that the normal vectors of the (111) plane is aligned in the x-y plane of the substrates with the polar angle of about 54.7° from the surface normal of the substrates. Figure 2(c) shows that the thin films on sapphire has the sharp contours of PLZT (111) and sapphire (21̄3̄1). The measured angle between two PLZT (111) poles and the surface normal of the substrate is about 35.3°, corresponding to the calculated value. These pole figure data indicate that the films on SrTiO₃, MgO, and sapphire consist of a x-y orientation being parallel to the surface of the substrates. On the other hand, Figure 2(d) shows that the (110) pole distribution of the films on PLT-coated glass are located at β = all degrees with α = 20.0° ~ 63.0°. This feature looks like a ring or a donut which has maximum pole intensity at β = all degrees with α = about 45°. This indicate that the PLZT thin film on PLT-coated glass have complete z-axis alignment and random x-y plane orientation. Therefore, the films of this type have oriented polycrystalline structure consisting of both low and high angle grain boundaries. A X-ray pole figure of the PLZT thin film on Pt/Ti/MgO has the strong four (101) reflections. This indicate that the normal vectors of (101) plane are aligned in the x-y plane of the substrate with the polar angle of about 45° from the surface normal of the substrate. From above results, we know that the films grew epitaxially with a x-y plane orientation.

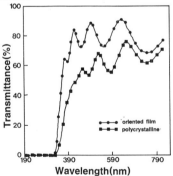

Fig. 3. Transmittance spectra of the (001)-oriented PLZT film on MgO, and the polycrystalline PLZT film on glass.

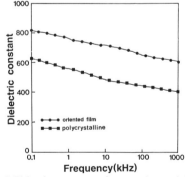

Fig. 4. Dielectric constant vs. frequecy characteristics of the PLZT thin films.

The preferred alignments of the films fabricated on various substrates were further investigated using X-ray rocking curve measurements. The small full-widths-at-half-maximum(FWHM) and the high intensity levels at the extreme values through which θ is rocked suggest that the alignments are limited. The FWHM of the x-ray rocking curves are 1.2°, 6.0°, 1.8°, 6.5°, and 1.4° for the films fabricated on SrTiO₃, MgO, sapphire, PLT-coated glass, and Pt/Ti/MgO, respectively. From above results, it may be speculated that most of the grains in the films are aligned with their oriented direction within 0.6° ～ 3° to the surface normal of the substrates.

The UV transmittance spectra were measured using a Hewlett-Packard 8452A diode array spectrophotometer. Figure 3 shows the transmittance spectra of the PLZT films. The oriented PLZT thin film has an average transmittance of about 80 % in the visible region whereas the transmittance of the polycrystalline film is below 60 %. Generally, polycrystalline films consist of many plane defects such as grain boundaries and interphase boundaries between the perovskite and pyrochlore phases. These defects serve as light scattering sources and cause the low transmittance. The interference oscillation is due to the phase retardation between the reflected light beam at the surface and at the interface between the film and the substrate. The large interference oscillation of the oriented film indicates that it has a smooth surface. The absortion edge in both cases is observed at the wavelength of 340 nm which corresponds to an optical band gap of 3.6 eV.

Figure 4 shows dielectric constant-frequency characteristics of the PLZT thin films. We used Hewlett-Packard 4192A impedance analyzer to measure the dielectric constant vs. frequency. We could not measure the dielectric constants of both films above 10 MHz because of the limitation of the measuring equipment. The PLZT thin films in both cases showed slow decrease throughout all frequencies measured in this study. The (001)-oriented thin film has a dielectric constant of about 738 at 1kHz. The dielectric constant of the polycrystalline films is about 540 at 1kHz. The dielectric constant of the oriented film is about 200 higher than that of the polycrystalline film in low frequency region.

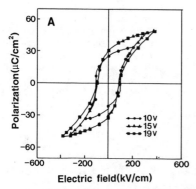 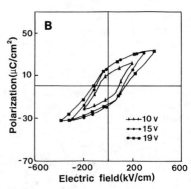

Fig. 5. Hysteresis loops of (a) the (001)-oriented and (b) the polycrystalline PLZT(10/0/100) film on Pt/Ti/MgO.

The ferroelectricity of the (001)-oriented and the polycrystalline PLZT(10/0/100) thin films was investigated by observing the polarization hysteresis using RT66A ferroelectric testing system. Figure 5 shows a family of hysteresis loops measured with applying various maximum voltages(10, 15, and 19 V). The spontaneous polarization(P_s) and the remanent polarization(P_r) of the oriented thin film under maximum voltage of 19 V, are 51.2 μC/cm² and 35.5 μC/cm², respectively. The coercive field(E_c) of the film is 91.8 kV/cm. The hysteresis loop(19 V applied) of the polycrystalline film shows a spontaneous polarization of 33.2 μC/cm², a remanent polarization of 18.2 μC/cm², and a coercive field of 124.2 kV/cm. The oriented PLZT thin films has better squareness and much higher polarization values(P_s and P_r) than the polycrystalline thin film. Note that the coercive field of the oriented thin film seems to be independent of the maximum applied voltage value. The coercive field of the polycrystalline film decreases as the maximum applied voltage decreases.

Figure 6 shows the capacitance-voltage charateristics of the (001)-oriented thin film measured at the various maximum applied voltages(10, 15, and 19 V). Generally, the polarization induced capacitance is caused by many factors such as domain motion, the ionic polarization, the space charge, and some defects[6]. The total capacitance due to all mechanisms corresponds to upper part of the capacitance-voltage curve. The difference between upper and lower value at the same electric field may be affected mainly by the domain motion. The separation between the maxima of the capacitance can be regarded as the double coercive field($2E_c$). The E_c determined in this way is somewhat higher than E_c determined from the hysteresis loop measurements, which may due to the difference of the measuring condition. The oriented thin film has higher maximum capacitance as the maximum appied voltage decreases. This phenomena may indicate that the domain switching of the oriented film occurs dramatically near the coercive field at relatively low maximum applied voltage.

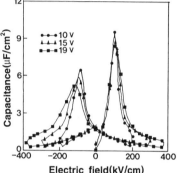

Fig. 6. Capacitance-voltage characteristics of the (001)-oriented PLZT(10/0/100) film.

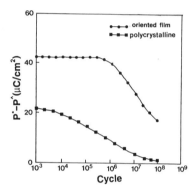

Fig. 7. Fatigue characteristics of the (001)-oriented and the polycrystalline PLZT(10/0/100) films on Pt/Ti/MgO.

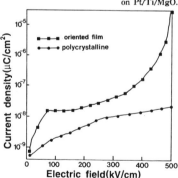

Fig. 8. Current density-voltage curves of the (001)-oriented and the polycrystalline PLZT(10/0/100) films on Pt/Ti/MgO.

We used the pulsed polarization method to measure the fatigue of the PLZT thin films. Because the switching polarization(P^*) and non-switching polarization($P^\wedge$) are important in the memory operation, the fatigue effects were analyzed in terms of these polarizations. Figure 7 shows the fatigue curves of the two films measured under the maximum applied voltage of ±10 V. Bipolar cycling decreases the difference between the switching polarization(P^*) and the non-switching polarization($P^\wedge$) rapidly after 10^6 cycles in the case of the (001)-oriented film. The difference value of the polycrystalline film decreases readily throughout all cycles. The value of ($P^* - P^\wedge$) after 10^8 cycling is about 18.5 $\mu C/cm^2$ for the oriented thin film whereas it is about 2.1 $\mu C/cm^2$ for the polycrystalline film. From above results, it may be speculated that the oriented thin films have slightly better fatigue characteristics than the polycrystalline films.

The current-voltage characteristics of the PLZT thin films were also investigated. Figure 8 shows the leakage current densities of the (001)-oriented and polycrystalline films as a function of applied electric field. The oriented thin films has a leakage current of 2.0 X 10^{-8} A/cm^2 whereas the polycrystalline films of 6 X 10^{-9} A/cm^2. The leakage current of the oriented thin films increases rapidly above 360 kV/cm. But, the polycrystalline films shows small increase of the leakage current as the the applied electric field increases. From above results, it is known that the leakage current property of the oriented thin film is worse than that of the polycrystalline film.

SUMMARY

Highly preferentially oriented lead lathanum zirconate titanate[PLZT(9/50/50) and (10/0/100)] thin films were fabricated on various substrates using the sol-gel method. The thin films fabricated on $SrTiO_3$, r-plane sapphire, and Pt/Ti/MgO grew epitaxially to (100), (110), and (001) orientations, respectively. The transmittance of the (001)-oriented thin films is an average value of 80 % in the visible range. The (001)-oriented PLZT thin film had higher dielectric constant and leakage current, and better ferroelectricity and fatigue characteristics than the polycrystalline films.

REFERENCE

1. G. Yi, Z. Wu and M. Sayer, J. Appl. Phys. **64**, 2717 (1988).
2. R. W. Vest and J. Wu, Ferroelectrics **93**, 21 (1989).
3. C. H. Peng, J. Chang and S. B. Desu in *Ferroelectric Thi. Films* II, edited by A. I. Kingon, E. R. Myers, and B. Tuttle (Mater. Res. Soc. Symp. Proc. **243**, Boston, Massachusetts, 1991), p.21.
4. P. J. Borrelli, P. H. Ballentine and A. M. Kadin in *Ferroelectric Thi. Films* II, edited by A. I. Kingon, E. R. Myers, and B. Tuttle (Mater. Res. Soc. Symp. Proc. **243**, Boston, Massachusetts, 1991), p.417.
5. A. B. Wegner, S. R. J. Brueck and A. Y. Wu, Ferroelectrics **116**, 195 (1991).
6. T. Mihara, H. Watanabe, H. Yoshimori, C. A. Paz de Araujo, B. Melnick, and L. D. Mcmillan, Integrated Ferroelectrics 1, 119 (1992).
7. C. K. Baringay and S. K. Dey, Appl. Phys. Lett. **61**, 1278 (1992).
8. D. S. Yoon, C. J. Kim, J. S. Lee, W. J. Lee and K. No, J. Mater. Res. 9, 420 (1994).
9. S. H. Rou, T. M. Graettinger, A. F. Chow, C. N. Soble, D. J. Lichtenwalner, O. Auciello and A. I. Kingon in *Ferroelectric Thi. Films* II, edited by A. I. Kingon, E. R. Myers, and B. Tuttle (Mater. Res. Soc. Symp. Proc. **243**, Boston, Massachusetts, 1991), p.81.
10. J. S. Lee, C. J. Kim, D. S. Yoon, and Kwangsoo No, submitted to Jpn. J. Appl. Phys. (1994).

IMPROVED SURFACE ROUGHNESS AT POLY-OXIDE/POLY-Si INTERFACE BY A NOVEL OXIDATION METHOD

M.C. JUN, J.W. KIM*, K.B. KIM*, B.C. AHN**, AND M.K. HAN

Dept. of Electrical Engineering, Seoul National University, Seoul 151-742, Korea
*Dept. of Metallurgical Engineering, Seoul National University, Seoul 151-742, Korea
**Anyang Research Lab., Goldstar Co., Ltd., Anyang-Shi, Kyungki-Do 430-080, Korea

ABSTRACT

We present a novel oxidation method to improve the surface roughness at the poly-oxide/poly-Si interface. Instead of directly oxidizing the poly-Si to the desired thickness of the SiO_2, a thin oxide layer is thermally grown on the poly-Si layer and then an a-Si layer is deposited on the top of the oxide layer. The a-Si layer is used as a silicon-source during next step of oxidation. The a-Si layer is fully oxidized until the poly-oxide/poly-Si interface advances below the initial interface. For comparison, the poly-oxide/poly-Si interface is also obtained by the conventional oxidation method. The surface roughness at the interface is investigated using transmission electron microscopy (TEM) and atomic force microscopy (AFM). For the novel oxidation method with the 50 Å thick intermediate oxide, the rms surface roughness at the poly-oxide/poly-Si interface is 30 Å, whereas that is 120 Å for the conventional method.

INTRODUCTION

Polycrystalline silicon (poly-Si) films are widely used in various applications such as silicon based integrated circuits and thin film transistors (TFT's) for active matrix liquid crystal displays (AMLCD's). During the fabrications of poly-Si devices, thermal oxidation of poly-Si is performed to make interlayer dielectrics in silicon based integrated circuits and to form gate insulators in TFT devices. For many applications, thermal oxides grown from poly-Si, referred to as poly-oxides, are required to have low leakage current and high dielectric strength. The TFT devices for AMLCD's also require high field effect mobility. However, it is well known that the thermally grown poly-oxides have higher leakage current and lower dielectric strength than thermal oxides of similar thickness grown on single crystalline silicon [1,2]. The inferior properties of the poly-oxides are correlated with the rough poly-oxide/poly-Si interface, which leads to the higher local electric field than the average field [3,4]. It is also reported that the field effect mobility of TFT devices is affected by the roughness of the gate insulator [5]. In order to reduce these effects, many efforts have been introduced to improve the interface roughness. A stacked thermal/CVD oxide technology is one method being generally used [6,7].

In this paper, we propose a novel oxidation method to improve the surface roughness at the poly-oxide/poly-Si interface. Thin oxide layer was thermally grown from the poly-Si film and amorphous silicon (a-Si) film was deposited on that oxide by low pressure chemical vapor deposition (LPCVD). The thermal oxidation was performed on the a-Si film until the poly-oxide with the desired thickness was formed. We examined the surface roughness at poly-oxide/poly-Si interface with various intermediate oxide thickness. We investigated the surface roughness at the

505

interface using transmission electron microscope (TEM) and atomic force microscope (AFM) with nanometer resolution.

EXPERIMENTAL

The a-Si films of 800 Å thickness were deposited on the thermally oxidized silicon wafer substrates by LPCVD at 550°C. The a-Si films were crystallized by solid phase crystallization at 600°C for 36 hours. Thin oxide layers (intermediate oxides) were formed from poly-Si films by thermal oxidation in a dry oxygen ambient. The thicknesses of the intermediate oxides were 50 Å, 100 Å, and 200 Å. The 400 Å thick a-Si films were deposited on the intermediate oxide by LPCVD at 550°C. The thickness of a-Si film was determined in order to form about 800 Å thick poly-oxide. The 1000 Å thick poly-oxides were subsequently grown at 950°C in dry O_2. It should be noted that the a-Si films were fully oxidized during thermal oxidation and new poly-oxide/poly-Si interface was formed in the lower poly-Si layers since only the 400 Å thick a-Si is deposited. For comparison, the poly-oxide/poly-Si structure was fabricated by the conventional oxidation method. In the conventional oxidation method, the 1000 Å thick poly-oxide layer was directly grown from poly-Si film by dry thermal oxidation.

The surface roughness at the poly-oxide/poly-Si interface was investigated by cross-sectional TEM and AFM with nanometer resolution. For the AFM measurement of the surface roughness, the poly-oxide layer was removed by wet etching using buffered HF acid. The true replica of the interface is preserved by such treatment, because the poly-Si is not attacked by HF-based solutions.

RESULTS AND DISCUSSION

Fig. 1 shows the cross-sectional TEM micrographs of the poly-oxide/poly-Si structures processed (a) by a novel oxidation method with the 50 Å thick intermediate oxide and (b) by the conventional oxidation method. In this figure, the upper SiO_2 is the poly-oxide, and the lower SiO_2 is the oxide grown from single crystalline silicon. It is observed that the surface of poly-Si formed by the novel method is smoother than that by the conventional method. It is clearly seen in the AFM images of Fig. 2. For the novel oxidation method with the 50 Å thick intermediate oxide, the rms surface roughness of the interface is 30 Å, while that is 120 Å for the conventional method. Fig. 3 shows the contribution of the intermediate oxide to the surface roughness at the poly-oxide/poly-Si interface. From this figure, it is easily seen that the surface roughness is dramatically improved by the novel oxidation method. In addition, for the novel oxidation method, the surface roughness increases as does the intermediate oxide thickness. This is due to the intermediate oxide/poly-Si interface which is formed while the thermal oxidation for the intermediate oxide is processed.

In thermal oxidation of poly-Si, grain boundaries are oxidized more rapidly than the center of the grains. The enhanced oxidation of grain boundaries makes V-grooves at the poly-oxide/poly-Si interface [8]. Longer oxidation increases the size of the grooves [8] and the surface roughness at the poly-oxide/poly-Si interface [4]. The rough surface of poly-Si causes the local electric field greater than the average field. This problem can be reduced by the intermediate oxide used in this novel method. Fig. 4 shows the schematic representations of the novel oxidation method. The a-Si film, which is on the intermediate oxide, is crystallized and changed to poly-Si (upper poly-Si) during the oxidation at 950°C. The grain size of the upper poly-Si differs from that of lower poly-

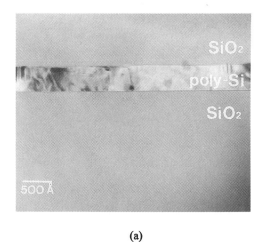

(a)

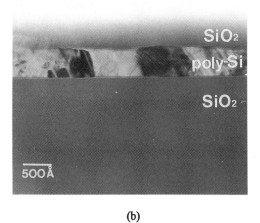

(b)

Fig. 1   The cross-sectional TEM micrographs of the poly-oxide/poly-Si structures
processed (a) by a novel oxidation method with the 50 Å thick intermediate
oxide and (b) by the conventional oxidation method.

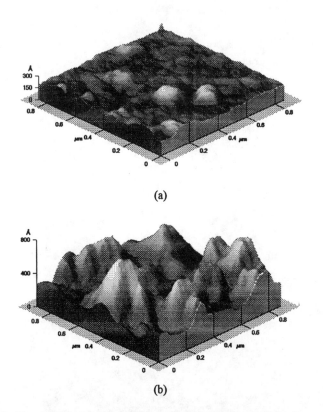

(a)

(b)

Fig. 2   The AFM images of the poly-oxide/ploy-Si processed (a) by a novel
oxidation method with the 50 Å thick intermediate oxide and (b) by the
conventional oxidation method.

Si due to higher annealing temperature, and its grain boundaries do not coincide with that of the lower poly-Si due to the intermediate oxide. Therefore, the enhanced oxidation of the grain boundaries of the upper poly-Si does not directly affect the oxidation of the grain boundaries of the lower poly-Si. The poly-oxide/poly-Si interface along the grain boundaries reaches the interface of the initial a-Si/poly-oxide interface faster than that along the center of grains. After that, oxygen penetrates the intermediate oxide and the lower poly-Si is oxidized from the lower point of the upper poly-Si grain boundaries, whereas the remaining upper poly-Si is oxidized during that time. The peak of the initial poly-oxide/poly-Si interface under the upper poly-Si grain boundaries is lowered during that oxidation, and the surface of the lower poly-Si is smoothened. The oxidation is finished, when the remaining upper poly-Si is fully oxidized and new poly-oxide/poly-Si interface is formed below the initial poly-oxide/poly-Si interface. Thus, a smoother poly-oxide/poly-Si interface is obtained by the novel oxidation method than by the conventional oxidation method.

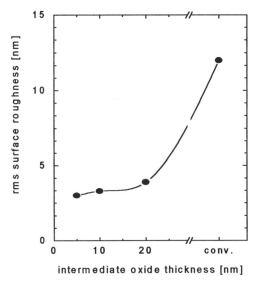

Fig. 3 The rms surface roughness at the poly-oxide/poly-Si interface as a function of the thickness of the intermediate oxide. The 'conv.' means the conventional method.

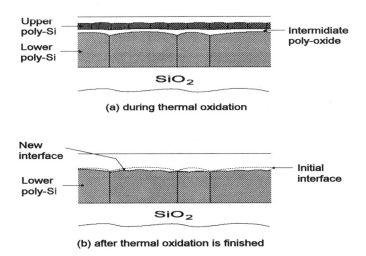

Fig. 4 The schematic representations of the novel oxidation method (a) during thermal oxidation (b) after thermal oxidation is finished. The a-Si film is crystallized and changed into poly-Si (upper poly-Si).

CONCLUSION

We have proposed a novel oxidation method to improve the surface roughness at the poly-oxide/poly-Si interface. The poly-oxide/poly-Si interface by the novel oxidation method is much smoother than that by the conventional oxidation method. For the novel oxidation method, the rms surface roughness of the interface is 30 Å, while that is 120 Å for the conventional oxidation method. The smoother interface by our oxidation method may improve the device characteristics such as dielectric strength, leakage current, field effect mobility.

REFERENCES

1. D.J. DiMaria and D.R. Kerr, Appl. Phys. Lett. 27, 505 (1975).
2. R.M. Anderson and D.R. Kerr, J. Electrochem. Soc. 48,4834 (1977).
3. P.A. Heimann, S.P. Muraka, and T.T. Sheng, J. Appl. Phys. 53, 6240 (1982).
4. L. Faraone and G. Harbeke, J. Electrochem. Soc. 133, 1410 (1986).
5. K. Takechi, H. Uchida, and S. Kaneko in Amorthous Silicon Technology-1992, edited by M.J. Thompson, Y. Hamakawa, P.G. LeComber, A. Madan, and E. Schiff (Mat. Res. Soc. Proc. 258, Pittsburgh, PA, 1992) pp. 955-960.
6. W.K. Park, H.K. Kim, M.S. Yang, S.C. Kim, I.G. Lim, and S.W. Rhee, Proc. Asian Symp. on Inform. Display. 1, 1 (1993).
7. R. Moazzami and C. Hu, IEEE Electron Device Lett. 14, 72 (1993).
8. J.C. Bravman and R. Sinclair, Thin Solid Films 104, 153 (1983).

RuO₂ THIN FILMS AS BOTTOM ELECTRODES FOR HIGH DIELECTRIC CONSTANT MATERIALS

K.Yoshikawa, T. Kimura, H. Noshiro, S. Ohtani, M. Yamada, and Y. Furumura
Process Developmnt Div., Fujitsu Ltd., 1015 Kamikodanaka, Nakahara - ku, Kawasaki 211, Japan

ABSTRACT

Ruthenium dioxide (RuO_2) thin films are evaluated as bottom electrode for dielectric $SrTiO_3$. It was found that a RuO_2 (50nm) / Ru (20nm) barrier layer on a Si substrate is effective as an oxygen barrier layer and as a metal diffusion barrier layer for sputter deposited $SrTiO_3$ films at substrate temperature of 450℃. To test suitability for high temperature processes, RuO_2/ Ru electrodes were annealed in air at 600 ℃. 100nm-thiick RuO_2 was sufficient to prevent oxygen diffusion. After annealing in the same condition, the leakage current of sputter deposited $SrTiO_3$(150nm) on RuO_2(50nm) / Ru(50nm) was 7.6 $\times$ 10^{-9} (A/cm²) at 2V.

INTRODUCTION

Extensive works have been done on oxide ferroelectric and paraelectric materials for capacitor applications used in non-volatile memories [1] and dynamic random access memories [2](DRAMs). Using these oxide materials, a high charge storage density is realized in small cell area. In capacitor fabrication the most troublesome problem is the choice of an appropriate electrode material, because these capacitors must be formed directly on Si ; when oxide dielectric films are deposited directly onto Si, formation of SiO_2 reduces the effective dielectric constant of the capacitor, therefore a barrier layer between the dielectric layer and Si is required. Pt is a well known electrode / barrier material for oxide high dielectric materials. Pt lattice matches well with $SrTiO_3$ and Pt is less reactive with $SrTiO_3$. However, direct deposition of Pt onto Si forms PtSi, and the PtSi is easily oxidized. Takemura et al. used double barrier layers ; they reported the barrier mechanism of Pt / Ta and Pt / Ti bilayer for $SrTiO_3$ or $(Ba,Sr)TiO_3$ thin film capacitors on Si during annealing.[3] - [4] An insulating TaO_x layer is formed after annealing in oxygen atmosphere at temperatures higher than 600℃ in the Pt / Ta bilayer. Insulating TiO_2 is formed at temperatures higher than 500 ℃ in the Pt / Ti bilayer. These results indicate that Pt's ability as a barrier layer to oxygen is poor. These oxidized layers deterionate the electrical contact between the bottom Pt electrode and Si. Other major problem with Pt as the electrode is hillock formation at higher temperatures [5] and difficulty in etching. In this study, we evaluated RuO_2 as a bottom electrode for dielectric materials and in particular we focused on barrier ability to oxygen.

EXPERIMENTAL

Thin films of RuO₂ were reactively sputtered in oxygen / argon (O₂/Ar=4:1) atmosphere at a total pressure of 15 mTorr onto Ru. Ru was also sputtered in Ar atmosphere onto Si substrates at a pressure of 3 mTorr. 10 Ω·cm p-type Si wafers cleaned with a HF solution were used as the substrates. The substrates were kept at a temparature of 350 ℃ during deposition of RuO₂ and Ru.

The SrTiO₃ thin films were reactively sputtered in oxygen / argon (O₂ / Ar = 1:9) atmosphere at a total pressure of 20mTorr onto the RuO₂ / Ru electrodes. The substrate temperature during depositon was 450 ℃. The crystallinity of the films was measured using X - ray diffraction (XRD) method. Reactivity of each layer and oxygen distribution was measured using Auger Electron Spectrometory (AES). Surface morphology was examined by scanning electron microscope (SEM). For electrical measurement, Au films were deposited as a top electrode through a metal mask with a round hole of 0.5mm in diameter. The capacitance and leakage current were measured with an LCR meter and picoammeter / DC voltage source, respectively. Dielectric constants of the films were calculated from capacitance measurement at 100kHz.

RESULTS AND DISCUSSION

To prevent oxidation of the Si surface, an intermediate metallic Ru layer is indispensable before deposition of RuO₂. XRD pattern for SrTiO₃/ RuO₂ / Ru / Si is shown in Fig.1. Poly-crystalline SrTiO₃ films with a perovskite structure were formed on the RuO₂ / Ru electrode. Figure 2 shows AES spectra of as grown (a) RuO₂(50nm) / Ru (20nm) film and (b) after deposition of SrTiO₃ (120nm) film. The interface between RuO₂ and SrTiO₃ is very sharp. Oxygen diffusion through RuO₂ into Ru was also completely inhibited. 50 nm-thick RuO₂ was sufficient to prevent oxygen diffusion during SrTiO₃ deposition at 450 ℃. Electrical measurement were done using two types of contacts ; to the bottom of SrTiO₃ from the RuO₂ directly and from the backside of Si wafer (25mm²) sputtered with Pt (100nm). Resultant dielectric constant (ε_r) were 165 to 174. The I-V characteristics were shown in Fig. 3. There is no difference in the electrical propeties between two types of contact. This shows that no uniform insulating phases or low dielectric phases were formed between SrTiO₃ and Si.

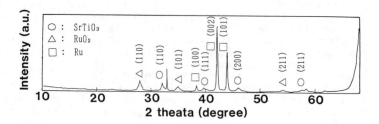

Fig. 1 XRD pattern for SrTiO₃ / RuO₂ / Ru / Si.

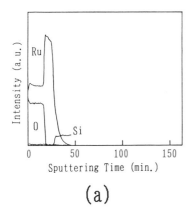 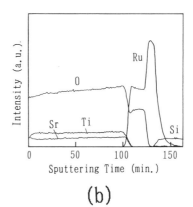

(a)   (b)

Fig. 2 AES depth profiles of (a) $RuO_2(50nm)$ / $Ru(20nm)$ and (b) $SrTiO_3(120nm)$ / RuO_2 (50nm) / Ru (20nm).

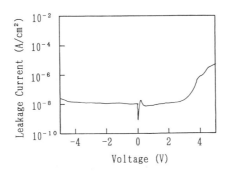

Fig. 3 Leakage current density vs voltage characteristics for $SrTiO_3(120nm)$/ $RuO_2(50nm)$/Ru(20nm).

To get good conformal coverage, $SrTiO_3$ films are required to be deposited by chemical vapor deposition (CVD) method. To test suitability for high temperature processes such as CVD, RuO_2/Ru films with various thickness were annealed in air. Figure4 shows an AES spectra of as-grown RuO_2(50,100,200nm) / Ru(50nm) films and that after annealing in air at 600 ℃ for 1 hour. For as-grown films, the RuO_2 / Ru interface was very sharp and there was no interdiffusion of oxygen. For annealed films, oxygen diffusion through RuO_2 into Ru is almost completely inhibited for RuO_2 thickness above 100nm. This indicates that no SiO_2 is formed at the Ru / Si interface even after high temperature annealing in oxygen atmosphere.

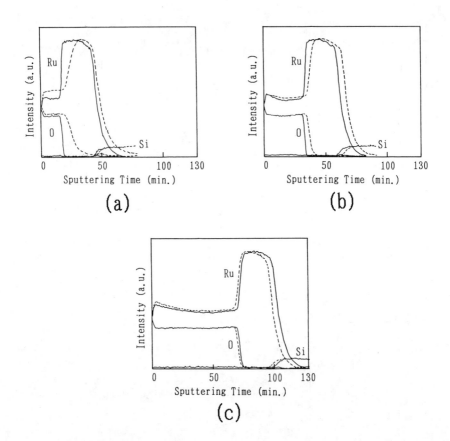

Fig. 4 AES depth profiles of as-grown RuO_2/Ru films(shown by solid lines) and those annealed at 600 °C for 1 hour (shown by dotted lines), (a) RuO_2 (50nm) / Ru (50nm), (b) RuO_2 (100nm) / Ru (50nm), and (c) RuO_2 (200nm) / Ru (50nm)

To investigate reactivity of $SrTiO_3$ with RuO_2 and to check if there is any change in the leakage current after high temperature annealing, sputter deposited $SrTiO_3$(120nm) on RuO_2(50nm) / Ru(20nm) were annealed in same annealing condition as above. AES depth profile of before and after annealing is shown in Fig. 5. This profile did not change comparing before and after annealing. Figure 6 shows I-V characteristics of sputter deposited as-grown films and those of annealed film. Leakage current did not degradate so much after annealing. Figure 7 shows SEM photographs of the surface of RuO_2 film (Fig.7 (a)) and that of $SrTiO_3$ deposited on the RuO_2(Fig.7 (b)). Both surfaces of RuO_2 and $SrTiO_3$ were smooth and did not show hillock formation unlike Pt films.

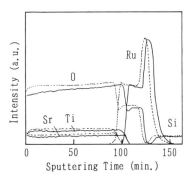

Fig. 5 AES depth profiles of as-grown   SrTiO₃(120nm)/RuO₂(50nm)
/Ru(20nm) film (shown by solid line) and those annealed at 600℃
for 1 hour (shown by dotted lines).

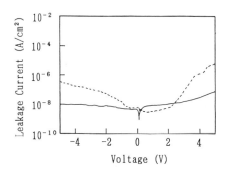

Fig. 6 Leakage current density vs voltage characteristics for
as-grown SrTiO₃(150nm)/RuO₂(50nm)/ Ru(50nm) (shown by solid
lines) and those of annealed at 600 ℃ for 1 hour(shown by
dotted lines).

CONCLUSION

   A RuO₂(50nm)/Ru(20nm) layer on a Si substrate is effective as
an oxygen barrier layer and as a metal diffusion barrier layer for
sputter deposition of SrTiO₃ films at 450℃. AES results show
that 50 nm-thick RuO₂ is sufficient to prevent oxygen diffusion
in sputter prosesses at 450℃. Dielectric constants and leakage
current of SrTiO₃ on RuO₂ were 165 to 174 and 1.0 × 10 ⁻⁸ (A /
cm²) at 2V respectively, independent of measuring from RuO₂ or
backside of Si. This indicates that the bottom electrode
electrically contacts with the Si without any insulating layer
and low dielectric constant material.

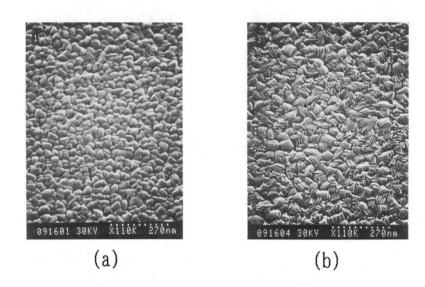

(a)   (b)

Fig. 7 SEM photographs (a) the surface of RuO_2 (200nm) / Ru (50nm) and (b) that of $SrTiO_3$(100nm)/ RuO_2(200nm)/Ru(50nm).

    100 nm-thick RuO_2 is required to prevent oxygen diffusion at high process temperatures (eg. 600 °C) in air. The leakage current of sputter deposited $SrTiO_3$(150nm) film on RuO_2(50nm) / Ru (50nm) was 7.6 $\times$ 10^{-9} (A / cm^2) at 2 V after annealing at 600 °C in air. The RuO_2 / Ru electrode must be a possible candidate for bottom electrodes of CVD grown $SrTiO_3$ films.

ACKNOWLEDGEMENT

    The authors would like to thank Mr.Y. Kudo for AES analysis.

REFERENCES

1. M. Okuyama and Y. Hamakawa, Int. J. Eng. Sci. 29 (3), 391 - 400 (1991)

2. K. Koyama, T. Sakuma, S. Yamauchi, H. Watanabe, H. Aoki, S. Ohya, Y. Miyasaka, and T. Kikkawa, (1991) IEDM Technical Digest, 1991, Piscataway, New Jersey, U.S.A., PP 823

3. K. Takemura, Proc. of 4th Int. Symp. on Integrated Ferroelectrics, March 9 - 11, 1992, Montery, CA, U.S.A. PP 481.

4. K. Abe, H. Tomita, H. Toyoda, M.Imai, and Y.Yokote, Jpn. J. Appl. Phys., 30, 2152 (1991).

5. Husam N. Al-Shareef, K. D. Gifford, P. D. Hren, S. H. Rou, O. Auciello and A.I. Kingon, Proc. of 4th Int. Symp. on Integrated Ferroelectrics, March 9 -11, 1992, Montery, CA, U.S.A. pp 187

GROWTH OF ROCK-SALT AND SPINEL STRUCTURED OXIDES ON α-Al$_2$O$_3$, c-ZrO$_2$ and MgO

SUNDAR RAMAMURTHY, PAUL G. KOTULA AND C. BARRY CARTER
Department of Chemical Engineering and Materials Science, University of Minnesota,
421 Washington Ave. SE., Minneapolis, MN 55455

ABSTRACT

Pulsed-laser deposition has been used to grow thin films of the rock-salt and spinel structured oxides NiO, CoO and Co$_3$O$_4$ on single-crystal substrates of α-Al$_2$O$_3$, MgO and c-ZrO$_2$. The resultant microstructures were characterized in plan-view by transmission electron microscopy and by low-voltage scanning electron microscopy. In all the depositions, the parameters could be controlled to grow predominantly single-crystal films. In the NiO/(0001)α-Al$_2$O$_3$ and CoO/(100)c-ZrO$_2$ systems, {111}-oriented films were observed which were found to be twinned close to 60° and 90° about <111> respectively. Growth of cobalt oxide films on (100) MgO at 800°C and 15 mTorr of oxygen produced {100}-oriented domains of Co$_3$O$_4$ meeting at antiphase boundaries. These observations and other recent studies re-emphasize the role of substrate crystallography in governing the orientation relationships in the overlayer.

INTRODUCTION

One potential application of thin oxide films is as buffer layers in the production of heteroepitactic systems in which the role of interfaces on the properties cannot be under-emphasized. Transmission electron microscopy (TEM) is a common characterization tool for studying the atomic-level structure and chemistry of oxide-oxide interfaces; improvements in electron-optics design with field-emission guns in scanning electron microscopes (SEM) have enabled spatial-resolution of surfaces that approaches the resolution obtainable from a conventional TEM.[1] Although oxide-interfaces have been studied in detail for many years, the kinetics of thin-film growth and the evolution of microstructure are not completely understood, either due to the large number of different growth-techniques used or simply because the specimen geometry is not suitable for a detailed study of the interfaces. TEM investigations of interfaces may be hampered primarily due to two reasons. Firstly, TEM sample preparation procedures may damage the interface and produce artifacts which make image interpretation difficult. Secondly, a suitable specimen geometry is necessary for studying the interface and associated defect structure. With the use of pulsed-laser deposition (PLD), both these difficulties can be overcome. Plan-view, well-characterized TEM-ready foils can be used directly as substrates onto which films are grown. By this approach, post-deposition specimen preparation is eliminated.

PLD has evolved as a common technique for growing oxide films with controlled stoichiometry by simply fabricating a target of suitable composition and altering the deposition parameters.[2] The possibility of growing high-quality films of mixed oxides such as superconductors and ferroelectric materials has spurred more research interest in PLD.[3] In the present study, rock-salt and spinel structured oxides have been grown onto single-crystal substrates of α-Al$_2$O$_3$, MgO and c-ZrO$_2$. The choice of these substrates stems from the fact that they are probably the most extensively studied oxides and their crystal structures are distinctly different. Thus, a study involving the evolution of microstructures in thin films of rock-salt and spinel structured oxides on these substrates serves as a model for the understanding of oxide interfaces in general.

Mat. Res. Soc. Symp. Proc. Vol. 343. ©1994 Materials Research Society

EXPERIMENTAL PROCEDURE

Substrates

Single-crystals of α-Al$_2$O$_3$, c-ZrO$_2$ (yttria stabilized) and MgO were solvent-cleaned, acid-cleaned and then annealed separately at different temperatures. The substrate preparation procedures that were adopted are described elsewhere.[4-6] Depositions were made on TEM-ready foils and on bulk substrates for plan-view TEM and SEM studies. All the substrates were characterized prior to the deposition to offer a comparison with the film microstructure. The orientations of the substrates were; (0001) α-Al$_2$O$_3$, (100) c-ZrO$_2$ and (100) MgO.

Film deposition

NiO and CoO films were deposited by PLD using an excimer laser operating with KrF (λ=248 nm). The target materials used were sintered pellets of NiO (80% of theoretical density) and Co$_3$O$_4$ (96% of theoretical density). A detailed description of the experimental setup for PLD is described elsewhere.[7] NiO films, approximately 100 nm thick for bulk substrates and 8 nm thick for TEM-ready foils, were grown onto (0001) α-Al$_2$O$_3$. CoO films of thicknesses varying between 50-100 nm were grown on (100) c-ZrO$_2$ and (100) MgO. Depositions were carried out with the laser operating at a pulse repetition rate varying from 1 Hz to 10 Hz, with an energy of about 200 mJ per pulse. All the depositions were carried out at substrate temperatures between 700°C and 950°C. Prior to the deposition, the chamber was evacuated to about 1 μTorr, and then backfilled with 5-20 mTorr of oxygen. After deposition, the substrates were cooled to room temperature in the same oxygen pressure as used for deposition.

Microstructural Characterization

The films were characterized in a Philips CM30 TEM operating at 300 kV and a Hitachi S900 low-voltage FESEM (field-emission SEM). TEM samples were prepared by conventional polishing and dimpling techniques. Final thinning to electron transparency was done by ion milling from the dimpled side using 5 kV Ar ions at an angle of 18°.

RESULTS

Prior to any deposition, all acid-cleaned and annealed substrates had surfaces with well-characterized crystallographic steps and ledges.

NiO on (0001)α-Al$_2$O$_3$[8,9]

The growth of NiO on α-Al$_2$O$_3$ produced two twin variants which have a close-packed plane parallel (e.g., (111)) to that of the substrate (i.e., (0001)). Figure 1 is a plan-view bright-field (BF) TEM image recorded on the common zone axis, [0001]α-Al$_2$O$_3$ and [111]NiO, of an 8 nm thick NiO film grown on a TEM foil at a pulse-repetition rate of 10 Hz at 750°C. The moiré fringes in this image are due to the interference of {2$\bar{2}$0} and {30$\bar{3}$0}-type reflections. In the inset of this image is a selected-area diffraction (SAD) pattern showing the orientation relationship between the NiO film and α-Al$_2$O$_3$ substrate. The film was found to be twinned about the [111] direction (60 or 180° rotation) producing Σ=3 twin boundaries which are seen in this image (arrow indicates one such boundary). Figure 2 is a 2kV secondary-electron image of a 100 nm NiO film grown on a bulk substrate at 900°C and 1 Hz. In this case, the twins were found to be larger (about 0.5 μm wide) and the twin boundaries were found to run parallel to the surface steps in the underlying substrate. The location of the bright lines corresponds to the twin boundaries as has been confirmed by TEM.[8] The reason for this contrast in the secondary-electron image will be discussed elsewhere.

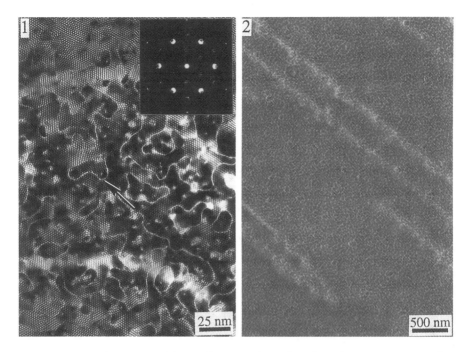

Figure 1. Plan-view BF TEM image from a region oriented on the common zone axis; [0001]α-Al$_2$O$_3$ and [111]NiO as seen in the SAD pattern in the inset. Three sets of moiré fringes seen in this image indicates that the film is continuous. The bright curved lines correspond to the twin boundaries separating the two twin variants (as indicated by an arrow, for example).

Figure 2. 2kV secondary-electron SEM image of a NiO film grown under optimized conditions. The twin variants are found to be ≈ 0.5µm wide and the twin boundaries (seen by enhanced secondary-electron contrast) run parallel to the surface steps in α-Al$_2$O$_3$.

CoO on (100)c-ZrO$_2$[10,11]

CoO films were deposited onto (100) c-ZrO$_2$ at a nominal substrate temperature of 700°C. TEM study of plan-view samples showed the film to grow epitactically with [111] CoO parallel to [100] c-ZrO$_2$. Figure 3 shows the morphology of the film; the microstructure is characterized by inclined grain boundaries which are seen as "stacking-fault-like" fringes, and moiré fringes. SAD studies and multiple dark-field imaging in the TEM revealed the presence of two variants of CoO grains rotated by 90° with respect to each other.[10] Figure 3 is a strong-beam bright-field (BF) image, with $\mathbf{g}=\{2\bar{2}0\}$ of one variant strongly excited, showing a particular set of grains in dark contrast. The SAD pattern in figure 4 was taken from the film in a region where the underlying substrate was milled away. The apparent 12-fold symmetry of the {220} reflections of CoO (due to the two variants) seen in this pattern is essentially a convolution of the 4-fold symmetry of the substrate and the 3-fold symmetry of the film.

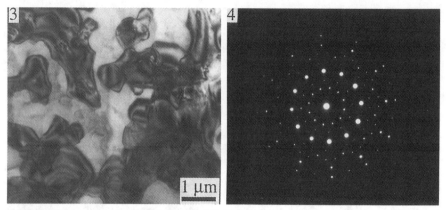

Figure 3. Strong-beam BF TEM image of CoO film grown on c-ZrO$_2$. The widely spaced fringes are due to the inclined boundaries where two CoO grains rotated by about 90° meet.
Figure 4. An SAD pattern from a region where the substrate was milled away, shows the apparent 12-fold symmetry of the {220} reflections of CoO.

Co$_3$O$_4$ on (100)MgO

Cobalt oxide films were deposited onto (100) MgO using the same target material (Co$_3$O$_4$) that was used for growing CoO films on c-ZrO$_2$. Under similar growth conditions (800°C, 5 Hz and 15 mTorr oxygen), a continuous Co$_3$O$_4$ film was obtained which showed a near "cube-on-cube" relationship with the MgO. Figure 5 is a BF image taken from the common [100] pole of the film and substrate. Moiré fringes extend continuously over the entire image indicating a uniform continuous film. The fringes are due to the interference of (002)$_{MgO}$ and (004)$_{spinel}$ reflections. The dark-field (DF) TEM image in figure 6 was recorded using a (022)-type reflection from Co$_3$O$_4$. The dark lines seen in this image are most likely antiphase boundaries (APB's) where domains of Co$_3$O$_4$ spinel meet. The morphology of these boundaries is similar to those observed in films of Mg$_2$TiO$_4$ spinel[12] and the defect spinel γ-Fe$_2$O$_3$[13] grown on MgO.

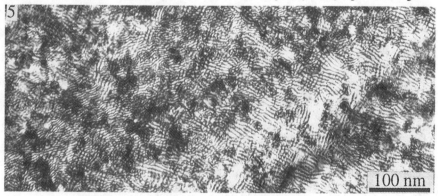

Figure 5. BF TEM image showing two sets of moiré fringes which are characteristic of the "cube-on-cube" orientation relationship between Co$_3$O$_4$ and MgO.

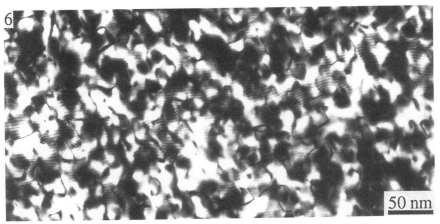

Figure 6. DF TEM image recorded with a {220}-type reflection of Co₃O₄ showing APB's separating the different domains of the spinel.

DISCUSSION

The observations made in the three different systems considered in this paper attest to the importance of the substrate in governing the crystallography and quality of the films grown on them. Depending on the substrate surface and the deposition parameters, the microstructure of the films can be significantly altered as observed in the growth of $YBa_2Cu_3O_{7-\delta}$ films on MgO.[14,15] Annealed single-crystals of oxides often have relaxed surfaces containing steps and ledges which influence the nucleation and growth of thin films, i.e. promote grapho-epitaxy.[14,16]

In the NiO/(0001)α-Al₂O₃ system, the NiO twin variants have a close-packed plane (e.g., (111)) parallel to the (0001) plane of α-Al₂O₃ with [1$\bar{1}$0] and [$\bar{1}$10] of the twins parallel to [10$\bar{1}$0] α-Al₂O₃ respectively. This relationship has also been observed for MgO films grown on (0001) α-Al₂O₃.[17] Films grown at a higher temperature (900°C) and lower deposition rate (1 Hz) produced twins conforming to the surface steps in alumina (see figure 2). The two twins are rotated by 60° (or equivalently 180°) about the [111] direction presumably due to the equal probability for nucleation of both these variants on the (0001) surface. However, the influence of surface steps in alumina on the nucleation of these two variants cannot be ignored. Based on the SEM image in figure 2, it appears that specific ledges and kinks may favor the nucleation of a particular orientation and then growth follows along the terraces. Such an observation where inclined surface steps influence the growth of thin films has been made in TiO₂ films deposited on MgO by PLD.[12] The ability to control the film microstructure by varying the deposition parameters in the NiO/(0001)α-Al₂O₃ system has been successfully exploited to prepare a simple geometry to study the solid-state reaction to form NiAl₂O₄ spinel.[18]

Observation of two variants of <111>-oriented CoO film (rotated by approximately 90°) on c-ZrO₂ is another example where the symmetry of the substrate influences the overall geometry of the grains in the film. This phenomenon has also been observed in Al films grown onto (111) Si.[19] The widely spaced fringes seen in figure 3 are due to the inclined boundaries between the different orientation variants of CoO. The structure of c-ZrO₂ (fluorite structure) in the <100> orientation has alternating planes of cations and anions, i.e. it is a polar surface. Dravid and co-workers have investigated interphase interfaces in the directionally-solidified eutectic (DSE) NiO-ZrO₂(CaO) system and have observed a similar orientation relationship.[20] In that analysis, they found the (111)NiO/(100)c-ZrO₂ to be a low energy interface where the respective planes showed no relaxation at the interface. Following their observations of similar orientation relationships in DSE NiO-Y_2O_3,[21] they proposed that this interface is stable because the cation-anion sequence is maintained across the interface i.e. an oxygen plane is commonly shared by both the lattices at the interface. Growth of c-ZrO_2 on (100) and (111) MgO are topics of current research and will be of use in understanding the stability of other interfaces in this system.

Cobalt oxide films grown on (100) MgO resulted in [100]-oriented Co_3O_4 spinel films which might appear unusual since the lattice mismatch between Co_3O_4 and MgO is much larger ($\approx 4\%$) than that between CoO and MgO ($\approx 1\%$). However, recent studies in this system involving different deposition parameters indicate the possibility of growing epitactic CoO films on MgO. It must be noted that the detection of CoO in a film which is predominantly Co_3O_4, grown on MgO, is difficult using plan-view geometry and standard electron-diffraction analysis because of the small mismatch between CoO and MgO.

The growth of high-quality films of rock-salt and spinel structured oxides on three crystallographically distinct oxide surfaces by PLD has been demonstrated. The influence of substrate structure on the microstructure of the films is as important as the deposition parameters themselves.

ACKNOWLEDGMENTS

The authors wish to thank Prof. Stan Erlandsen for access to the Hitachi S-900 SEM and Chris Frethem for technical assistance. The TEM is part of the Center for Interfacial Engineering, an NSF Engineering Research Center. This research has been supported by the DoE under Grant No. DE-FG02-92ER45465 and by the NSF under Grant No. DMR-8901218. PGK acknowledges a graduate student Fellowship from IBM.

REFERENCES

1. D. C. Joy, Ultramicroscopy **37**, 216-233 (1991).
2. See for example, papers in Mater. Res. Soc. Symp. Proc. **191**, (1990).
3. See for example, papers in Proc. 2nd Int. Conf. on Laser Ablation, AIP Conf. Proc. **288**, (1993).
4. D. W. Susnitzky and C. B. Carter, J. Am. Ceram. Soc. **75**, 2463-2478 (1992).
5. E. L. Fleischer, M. G. Norton, M. A. Zaleski, W. Hertl, C. B. Carter and J. W. Mayer, J. Mater. Res. **6**, 1-8 (1991).
6. M. G. Norton, S. R. Summerfelt and C. B. Carter, Appl. Phys. Lett. **56**, 2246-2248 (1990).
7. P. G. Kotula and C. B. Carter, Mat. Res. Soc. Symp. **285**, 373 (1993).
8. P. G. Kotula and C. B. Carter, edited by G. W. Bailey and C. L. Reider (Proc. **51**st Annual Meeting of MSA Cincinatti, OH, 1993) pp. 1120-21.
9. P. G. Kotula and C. B. Carter, edited by (Proc. 2nd Int. Conf. on Laser Ablation, AIP Conf. Proc. **288**, 1993) pp. 231-236.
10. S. McKernan, S. Ramamurthy and C. B. Carter, edited by G. W. Bailey and C. L. Reider (Proc. **51**st Annual Meeting of MSA Cincinnati, OH, 1993) pp. 926-27.
11. S. Ramamurthy, S. McKernan and C. B. Carter, edited by G. W. Bailey and C. L. Reider (Proc. **51**st Annual Meeting of MSA Cincinnati, OH, 1993) pp. 1154-55.
12. S. L. King and C. B. Carter, edited by J. C. Miller and D. B. Geohegan (Proc. 2nd Int. Conf. on Laser Ablation, AIP Conf. Proc. **288**, 1993) pp. 209-214.
13. I. M. Anderson, L. A. Tietz and C. B. Carter, edited by W. A. T. Clark, U. Dahmen and C. L. Briant (Mat. Res. Soc. Symp. Proc. **238**, 1992) pp. 807-814.
14. M. G. Norton and C. B. Carter, J. Crystal Growth **110**, 641-651 (1991).
15. M. G. Norton, B. H. Moeckley, C. B. Carter and R. A. Buhrman, J. Crystal Growth **114**, 258-263 (1991).
16. L. A. Tietz and C. B. Carter, Phil. Mag. A **67**, 699-727 (1993).
17. D. X. Li, P. Pirouz, A. H. Heuer, S. Yadavalli and C. P. Flynn, Phil. Mag. A **65**, 403-425 (1992).
18. P. G. Kotula and C. B. Carter, edited by E. P. Kvam, A. H. King, M. J. Mills, T. D. Sands and V. Vitek (submitted to Mater. Res. Soc. Symp. Proc. **319**, 1993).
19. U. Dahmen and N. Thangaraj, Mat. Sci. For. **126-128**, 45-54 (1993).
20. V. P. Dravid, C. E. Lyman and M. R. Notis, Ultramicroscopy **29**, 60-70 (1989).
21. V. P. Dravid, V. Ravikumar, G. Dhalene and A. Revcolevschi, edited by W. A. T. Clark, U. Dahmen and C. L. Briant (Mat. Res. Soc. Symp. Proc. **238**, 1992) pp. 815-820.

LOW TEMPERATURE ATMOSPHERIC PRESSURE CHEMICAL VAPOR DEPOSITION OF GROUP 14 OXIDE FILMS

David M. Hoffman,* Lauren M. Atagi,*,¥ Wei-Kan Chu,¶ Jia-Rui Liu,¶ Zongshuang Zheng,¶ Rodrigo R. Rubiano,†,1 Robert W. Springer,¥ and David C. Smith¥
*Department of Chemistry, University of Houston, Houston, TX 77204
¥Los Alamos National Laboratory, Los Alamos, NM 87545
¶Texas Center for Superconductivity, University of Houston, Houston TX 77204
†Department of Nuclear Engineering, MIT, Cambridge, MA 02139

ABSTRACT

Depositions of high quality SiO_2 and SnO_2 films from the reaction of homoleptic amido precursors $M(NMe_2)_4$ (M = Si, Sn) and oxygen were carried out in an atmospheric pressure chemical vapor deposition reactor. The films were deposited on silicon, glass and quartz substrates at temperatures of 250 to 450 °C. The silicon dioxide films are stoichiometric (O/Si = 2.0) with less than 0.2 atom % C and 0.3 atom % N and have hydrogen contents of 9 ± 5 atom %. They are deposited with growth rates from 380 to 900 Å/min. The refractive indexes of the SiO_2 films are 1.46, and infrared spectra show a possible Si-OH peak at 950 cm^{-1}. X-Ray diffraction studies reveal that the SiO_2 film deposited at 350 °C is amorphous. The tin oxide films are stoichiometric (O/Sn = 2.0) and contain less than 0.8 atom % carbon, and 0.3 atom % N. No hydrogen was detected by elastic recoil spectroscopy. The band gap for the SnO_2 films, as estimated from transmission spectra, is 3.9 eV. The resistivities of the tin oxide films are in the range 10^{-2} to 10^{-3} Ω cm and do not vary significantly with deposition temperature. The tin oxide film deposited at 350 °C is crystalline cassitterite with some (101) orientation.

INTRODUCTION

Doped tin oxide, a transparent conductor, and silicon oxide, a dielectric, are important materials in the microelectronics industry [2,3]. In atmospheric pressure chemical vapor deposition (APCVD) processes, silicon oxide is prepared from silane and oxygen or by pyrolysis of tetraethylorthosilicate. Tin oxide is typically deposited by reacting oxygen with tetramethyltin or by hydrolysis of tin tetrachloride. These processes sacrifice quality and/or growth rates at low deposition temperatures.

Homoleptic metal amido complexes, in combination with ammonia, are known to be excellent precursors to nitride films at low temperatures [4]. They are also potential precursors to oxide films because they contain reactive metal-nitrogen bonds as possible sites for oxygen insertion [5,6]. Herein we report the atmospheric chemical vapor deposition of SiO_2 and SnO_2

from the reaction of oxygen with Si(NMe$_2$)$_4$ and Sn(NMe$_2$)$_4$, respectively, at temperatures of 250–450 °C [7].

EXPERIMENTAL

The amido precursors Si(NMe$_2$)$_4$ and Sn(NMe$_2$)$_4$ were prepared using literature methods [8]. Depositions were carried out in an atmospheric-pressure rectangular glass reactor as described previously [7]. Ultra-high purity argon was used as carrier gas, and extra-dry grade oxygen was used directly from the cylinder. The substrates were cleaned by immersing them in H$_2$O$_2$:H$_2$SO$_4$ (1:4) for 10 minutes, rinsing in deionized water for 15 minutes, and drying under a stream of argon. Total flow rates through the system ranged from 1000-1200 sccm. Rutherford backscattering (RBS) analyses and elastic recoil spectroscopy (ERS) were performed using a 2.0 MeV ^{4}He^{2+}, and the data was analyzed and modeled by using the program RUMP. Kapton (C$_{22}$H$_{10}$N$_2$O$_5$) foil was used as the ERS standard. Nuclear reaction analyses (NRA) were performed using a 1.26 MeV ^{2}H$^+$ beam with graphite as a standard. Sheet resistivity was measured with a 4-point probe and converted to local film density using film thicknesses obtained by ellipsometry or RBS. Depth profiling for Auger electron (AES) and X-ray photoelectron spectroscopy (XPS) measurements was carried out using a 3.5-keV Ar$^+$ sputter gun. X-Ray irradiation for XPS used a Mg source ($h\nu$ 1253.6 eV) with a spot size of 5 microns. The XPS binding energies were referenced to a gold standard (Au 4f$_{7/2}$ at 83.95 eV).

RESULTS

Silicon oxide films are deposited from Si(NMe$_2$)$_4$ and oxygen on silicon and quartz at substrate temperatures of 300 to 400 °C. No deposition is observed when the amido precursor is pyrolyzed in the absence of oxygen. The SiO$_2$ films are stoichiometric with O/Si ratios of 2.00 ± 0.10 as determined by RBS. The carbon and nitrogen contents, estimated from XPS survey scans, are less than 0.2 atom % and 0.3 atom % respectively. The hydrogen content determined by ERS is < 9 ± 5 atom % and is uniformly distributed throughout the film. Transmittance IR spectra have characteristic Si–O absorbances at 1070 and 810 cm^{-1} and a peak around 950 cm^{-1}, which may be due to Si-OH (reported at 940 cm^{-1}) [9]. The O-H group is observed as a broad band around 3500 cm^{-1}.

Auger electron spectroscopy depth profiling indicates the silicon oxide films have uniform composition, consistent with the RBS data. The XPS binding energies for the Si 2p and O 1s electrons in the bulk are 106.1 and 535.6 eV respectively. These energies are higher than the reported values of 104 and 531 eV [10], which may be due to sample charging during data collection. The growth rates of SiO$_2$ varied with flow rates and temperature. The conditions were not optimized, but the highest rate was observed at a deposition temperature of 350 °C (700

Å/min) when the bubbler containing the organometallic was heated to 55 °C and the feed lines were maintained at 120 °C.

The films show good adhesion (Scotch tape test) and are chemically resistant except to KOH or HF. The refractive indexes of the films (λ 632.8 nm) are 1.458 ± .003, which agree with the typical value of 1.46 reported for SiO_2 [10]. X-ray diffraction for a SiO_2 film deposited at 350 °C indicates the film is amorphous. Scanning electron microscopy show that the film is smooth with no obvious cracks or other features.

Tin oxide films are deposited on silicon, glass and quartz substrates from $Sn(NMe_2)_4$ and oxygen at substrate temperatures of 250–450 °C. No deposition is observed at 350 °C in the absence of oxygen. The O/Sn ratio is 2.00 ± 0.10 by RBS. Auger depth profiling show the film deposited at 350 °C has uniform composition. The carbon content in the films is less than 0.8 atom % based on NRA, and the nitrogen content is less than 0.3 atom % from XPS survey scans. No hydrogen was detected in three films examined by ERS.

The growth rates of the SnO_2 films are up to 10,000 Å/min. At a substrate temperature of 375 °C the growth rates do not vary significantly with oxygen flow rate and at a low oxygen flow rate (250 sccm/1000 sccm total flow) the rates do not appear to vary with temperature. For depositions with substrate temperatures greater than 375 °C and high oxygen flow rates (400 sccm/1000 sccm total flow), however, the growth rates decrease with increasing temperature. The inverse relationship between the growth rate and temperature at high oxygen flow rates may be due to significant reaction of the precursor and oxygen before reaching the substrate, which decreases the observed deposition rate.

X-Ray diffraction indicates a SnO_2 film deposited at 250 °C is amorphous, while films deposited at 350 °C are crystalline cassitterite with some (101) orientation (see Figure 1). The grain size is estimated from the diffraction study to be approximately 100 Å. Scanning electron microscopy of SnO_2 films deposited at 350 °C reveal nodular grains while the film deposited at 250 °C is smooth and featureless. The films demonstrate good chemical stability to common acids (except HF) and adhere well to the substrates with no delamination observed over a period of months.

The SnO_2 films show good transmittance in the visible range. A 1400 Å SnO_2 film on quartz, for example, is 70-75% transmittant between 400 and 700 nm. The optical band gap was calculated from the absorbance data by plotting $(AE)^2$ vs. E, where A is the absorbance of the film and E is the photon energy. A sample plot is shown in Figure 2 for a 6600 Å SnO_2 film deposited on quartz at 400 °C where the calculated band gap is 3.95 eV. This value is within the range of values reported for SnO_2 films (3.6-4.1 eV) [11]. The films are conductive, with resistivities ranging from 10^{-2}–10^{-3} Ω cm. The resistivities of the films did not appear to vary significantly with deposition temperature.

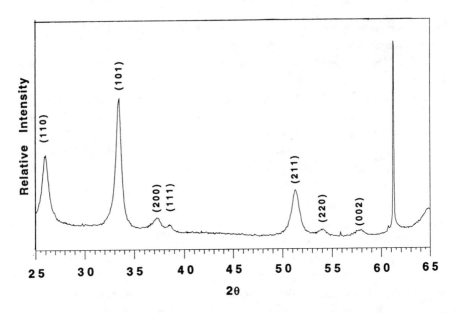

Figure 1.   X-Ray diffraction data for a 4400 Å SnO$_2$ film deposited at 350 °C on silicon.

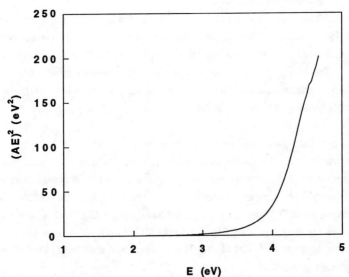

Figure 2.   Plot to determine the band gap for a tin oxide film deposited on quartz at 400 °C. (A) is the absorbance, (E) is the photon energy.

DISCUSSION

The homoleptic amido precursors $M(NMe_2)_4$, (M = Si, Sn) react with oxygen at low temperatures to give oxide films with little or no carbon or nitrogen contamination. The mechanism of the reaction probably involves insertion of dioxygen into the reactive M-N bonds of the precursors followed by decomposition of the unstable peroxo intermediates. The fact that we see little or no carbon or nitrogen indicates that all of the amido ligands are cleanly removed from the precursors.

CONCLUSION

Homoleptic main group amido complexes appear to be viable precursors for the low temperature deposition of oxide thin films. The films adhere well to the substrates and demonstrate promising electrical and optical properties. The reactions to give the oxides are proposed to occur via oxygen insertion into the M-N bonds of the amido precursors.

ACKNOWLEDGMENTS

We thank Norman Elliot for the SEM measurements and Kevin Hubbard for his help with the NRA calculations. DMH is a 1992-1994 Alfred P. Sloan Research Fellow. Support from the Robert A. Welch Foundation and the Texas Advanced Research Program is gratefully acknowledged. WKC acknowledges support from the State of Texas through the Texas Center for Superconductivity.

REFERENCES

1. current address: Raychem Corporation, MS 109/6503, 300 Constitution Dr., Menlo Park, CA 94025.
2. W. Kern, R.S. Rosler, J. Electrochem. Soc. 14, 1082 (1977). W. Kern, G.L. Schnable, A.W. Fisher, RCA Review 37, 3 (1976).
3. J.L. Vossen, Physics of Thin Films 9, 1 (1977). Z.M. Jarzebski, Phys. Stat. Sol. A 71, 13 (1982). K.L. Chopra, S. Major, D.K. Pandya, Thin Solid Films 102, 1 (1983).
4. D.M. Hoffman, Polyhedron, in press. R.M. Fix, R.G. Gordon, D.M. Hoffman, Mat. Res. Soc. Symp. Proc. 168, 357 (1990); J. Am. Chem. Soc. 112, 7833 (1990).
5. Deposition of TaO_x from $Ta(NMe_2)_5$ has been reported: T. Tabuchi, Y. Sawado, K. Uematsu, S. Koshiba, Jpn. J. Appl. Phys. 30, L1974 (1991).
6. The CVD of silicon oxide from $Si(NMe_2)_4$ with oxygen/ozone or oxygen has been reported. Only IR spectroscopy was used to characterize the films. T. Maruyama, T. Shirai, Appl. Phys. Lett. 63, 611 (1993).

7. Portions of this work have been published. L.M. Atagi, D.M. Hoffman, J.-R. Liu, Z. Zheng, W.-K. Chu, R.R. Rubiano, R.W. Springer, D.C. Smith, Chem. Mater., in press.

8. R.G. Gordon, D.M. Hoffman, U. Riaz, Chem. Mater. **4**, 68 (1992). R.G. Gordon, D.M. Hoffman, U. Riaz, Chem. Mater. **2**, 480 (1990). K. Jones, M. F. Lappert, J. Chem. Soc. 1944 (1965).

9. M. Adachi, K. Okuyama, H. Tohge, M. Shimada, J. Satoh, M. Muroyama, Jpn. J. Appl. Phys. **31**, L1439 (1992).

10. S.B. Desu, C.H. Peng, T. Shi, P.A. Agaskar, J. Electrochem. Soc. **139**, 2682 (1992).

11. Z.M. Zarzebski, J.P. Marton, J. Electrochem. Soc. **123**, 333c (1976).

ATOMIC FORCE MICROSCOPY STUDY OF NICKEL OXIDE FILMS MODIFIED BY REACTIVE ELEMENT

F. CZERWINSKI AND J.A. SZPUNAR
Department of Metallurgical Engineering, McGill University, Montreal, Que., H3A 2A7, Canada

ABSTRACT

The evolution of microstructure of thin oxide films during growth has been quantitatively analysed by atomic force microscopy (AFM). The oxide films were formed in pure oxygen atmosphere at temperatures ranging from 873 to 1173 K on polycrystalline nickel substrates. The substrates were both pure and superficially modified by nanometer-sized dispersions of CeO_2. The incorporation of reactive element into the oxide inhibits the growth kinetics and affects the microscopic surface morphology. The extent of this effect depends essentially on the substrate surface microstructure prior to modification and the size of CeO_2 particles. A correlation was found between the surface topography and the reduction of oxide growth rate caused by the reactive element.

INTRODUCTION

The growth of solids from gas phase reactants is a stochastic process responsible for morphology of the film surface [1]. The surface roughness affects the film properties and its control is of great practical importance. Therefore, various analytical techniques are employed to measure the surface roughness and to characterize its relationship with thin film microstructure. The recently developed AFM is a powerful tool, useful for the analysis of any specimen irrespective of its electrical conductivity. Moreover, the digital images having a unique topological perspective can be quantitatively analysed by numerous software packages.

The previous studies [2,3] of oxide formation on the metallic substrates at high temperatures described the evolution of surface topography for films with thicknesses up to a few hundred nanometers. In addition, it was shown that the morphological development of pure oxides was modified by the addition of very small quantities of chemically active elements. Since the growth kinetics control the surface morphology, valuable information about the mechanism of oxide film growth can be obtained from the quantitative analysis of the topography of oxide/gas interfaces.

The objective of this study is to compare the topographical description of oxide surfaces with the kinetics and mechanisms of oxide film growth.

EXPERIMENTAL PROCEDURE

All the topographical measurements were conducted using a Digital Instruments Nanoscope III AFM operated in contact mode and fitted with a 0.12 Nm^{-1} Si_3N_4 cantilever.

529

Data files of 512 x 512 points were collected from the 12.5 x 12.5 μm scan area under condition of the constant force at scan rates between 6 and 10 Hz. Scanning parameters were set to collect the data without filtering or smoothing.

Oxide films were grown at the temperature range of 873-1173 K in pure oxygen. A polycrystalline Ni substrate (A.D. MacKay, Inc.) was used both the pure and superficially modified with 14 nm thick CeO_2 sol-gel coatings [2,3]. The initial size of CeO_2 particles was 5 nm and was increased up to 20-30 nm by vacuum annealing of the coatings before oxidation. Two techniques of Ni surface finish were employed: the mechanical polishing with 1 μm diamond paste to promote the nucleation of the randomly oriented oxide, and the chemical polishing to promote the epitaxial growth. Oxidation was performed in the ultra high vacuum manometric system equipped with a high sensitivity gauge to monitor the oxygen uptake during the reaction [2]. After the first preliminary measurements, the oxidation parameters were adjusted to produce oxide films with thicknesses in the range of 100-1000 nm for all of the substrates. This allowed the comparison of films having the similar thicknesses.

RESULTS AND DISCUSSION

Kinetics of oxide growth

The measurements showed that the growth rate of oxide on mechanically polished Ni was generally higher than that observed for chemically polished samples. The coatings of 5 nm CeO_2 particles diminished the oxide growth rate, however, the effect was dependant on substrate surface finish. For the mechanically polished substrates, the coating decreased the oxidation rate by about 2-3 times. The coatings of 5 nm CeO_2 particles deposited on chemically polished Ni were more effective and the oxide growth rate was about 10 times lower than that measured on pure Ni. Additionally, in the range of film thicknesses studied, this factor had a tendency to increase with oxidation time. The coatings which were composed of CeO_2 particles having the size of 20-30 nm did not inhibit the Ni oxidation and the oxide growth rate was even higher than that observed without modification. Details regarding the growth kinetics are given elsewhere [2,3].

Analyses of the mechanism of oxide growth were conducted using both the oxygen isotopes $^{16}O_2/^{18}O_2$ and the inert marker techniques which revealed the essential differences in diffusion properties of the oxides produced [3,4]. The oxide on pure Ni has grown by outward diffusion of Ni^{2+} ions what is in agreement with other results [5]. The same mechanism was also evaluated for oxide growth on Ni coated with CeO_2 particles of size 20-30 nm. For oxides formed on mechanically polished Ni after modification with 5 nm CeO_2 particles, the inward diffusion of O^{2-} ions was also detected, however, the film still grew predominantly by outward Ni^{2+} transport. The essential change in growth mechanisms from outward Ni^{2+} into inward O^{2-} predominant was observed for oxides formed on chemically polished Ni coated with 5 nm CeO_2 particles.

It is generally known that at temperatures used in this study, the oxide growth on pure Ni is controlled by the short-circuit diffusion of ions via oxide grain boundaries and dislocations [5]. The activation energies evaluated from Arrhenius plot indicate that the same diffusion paths dominate the growth of NiO modified with CeO_2 [2]. Thus, the observed differences in oxide growth mechanisms may be explained by the blocking of these diffusion paths by reactive elements ion segregants. The comparison of kinetic measurements and analysis of the transport properties of oxides indicates, that there is a correlation between the oxide growth reduction caused by the reactive element and the mechanism of oxide formation. The higher the decrease of oxide growth rate caused by the reactive element, the higher the contribution of the inward oxygen diffusion to oxide growth mechanism.

The morphology of oxide surface

The typical morphologies of the growth surfaces of oxides are shown in Fig.1. The pure NiO grown on mechanically polished Ni exhibits faceted surface with the pyramidal grains limited by one dominant type of surface plane. The nucleation of new grains on the outer surface was also observed. The growth of oxide on chemically polished Ni depends on the orientation of substrate grains. In general, however, the oxide surface has crystallographic and faceted character (Fig. 1a). The morphology of NiO modified with CeO_2 depends on substrate surface finish and the size of CeO_2 particles. The coatings deposited onto mechanically polished substrates as well as the coatings of 20-30 nm CeO_2 particles deposited onto the chemically polished substrates forms the oxide which after the initial transient stage has morphology which is very similar to that observed on pure Ni.

A totally different morphology was observed for the oxide growth on the chemically polished substrates coated with 5 nm CeO_2 particles (Fig.1b). Very small, spherical oxide grains uniformly covered the substrate. It is suggested that such surfaces have been formed during very initial stages of growth as a result of nucleation of small oxide grains on the nano-sized CeO_2 dispersions. The subsequent change of the growth place to the oxide/substrate interface inhibited the development of this surface.

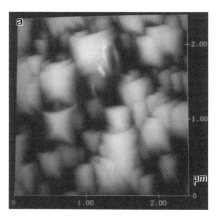 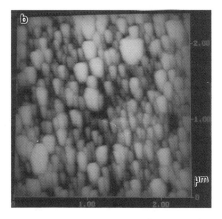

Fig. 1. Top view of oxides formed on chemically polished Ni after 4 h exposure at 973 K: a) pure Ni, b) after coating with 5 nm CeO_2 particles.

In addition to the morphological features discussed above, there are also differences in the surface roughness of oxides formed. An example of oxide produced under the same exposure conditions is shown in Fig. 2. In general the roughness of oxide grown on pure Ni (Fig. 2a) is much higher than after coating with 5 nm CeO_2 particles (Fig. 2b). Once again, the CeO_2 coating with particle size of 20-30 nm produced the oxides having roughness very similar to that on pure Ni (Fig. 2c).

Surface roughness measurement

A full description of the roughness of growth surface requires the characterization of the components parallel to the plane of the interface (lateral roughness) as well as the normal to the plane (vertical roughness). An example of such complex analysis of topographic data is proposed by Williams et al. [1]. The experimental results presented in the literature cover a wide range of film thicknesses and roughnesses. An example of such films are 10 nm thick the gate oxides applied in advanced electronic devices, where the roughness is about 1 nm [6]. In films up to 500-1200 nm thick, chemically vapour deposited glass the roughness is as high as 100 nm [7]. It is important to be able to understand the continuous evolution of surface topography over such a wide range of film thicknesses.

A more detailed description of the surface roughness of NiO grown at high temperatures is presented elsewhere [3]. In this report the mean roughness (R_a) was used as a parameter describing the oxide growth topography [8]. The R_a values reported are an average from AFM analysis of at least 5 regions of area of 12.5x12.5 μm. All the results, plotted as a function of the film thickness, are presented in Fig. 3. In this plot the experimental points exhibit two kinds of behaviour. For all the oxides grown on pure substrates, the R_a values increased with increasing of oxide thickness. The same tendency was observed for mechanically polished Ni substrates and for chemically polished substrates coated with large CeO_2 particles. On the other hand, the mean roughness of oxide formed on chemically polished Ni after coating it with 5 nm CeO_2 particles, remained practically constant for all oxide thicknesses.

For both pure and CeO_2 modified NiO the comparison of the surface morphology changes, the kinetics and the growth mechanism indicates on existing correlation. It is clear that for the films grown by outward diffusion of metal cations, with reaction taking place at oxide/gas interface, the surface roughness increased with time. In the case of growth by inward diffusion of oxygen ions the reaction takes place inside the oxide (beneath the Ce rich region) or at the oxide/substrate interface. During such growth process, the oxide/gas interface remained relatively stable. Thus, the topographic measurements of the growth surface indicate what is the mechanism of oxide formation. There are, however, also other practical aspects of surface roughness measurements. For pure NiO films, they may be useful to design the oxide films with precisely defined topography, which is necessary in solar batteries applications [9]. For NiO films modified by reactive elements, the knowledge of the surface roughness is also crucial for microchemical characterization of the films [3]. Namely, it enables the correct interpretation of the depth-composition profiles measured by many analytical techniques.

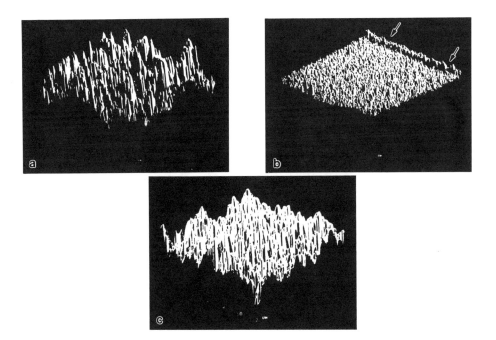

Fig. 2. AFM images (back illuminated) of growth surfaces of oxides formed on chemically polished Ni after 4 h oxidation at 973 K: a) pure Ni, b) coated with CeO_2 (size 5 nm), c) coated with CeO_2 (size 20-30 nm). Arrows indicate the grain boundary in substrate. Vertical scale = 500 nm/div for all films.

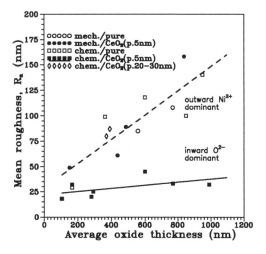

Fig. 3. Mean roughness versus oxide thickness for samples analysed. Mechanisms of oxides growth are indicated.

CONCLUSIONS

The application of AFM proved valuable in measuring the surface roughness of the semiconductor NiO films both pure and modified with CeO_2. The measurements enabled us to establish the factors affecting the oxide morphology evolution during growth. A correlation exists between the topography of oxide/gas interface and the mechanism of NiO growth. The oxide films growing predominantly by the outward diffusion of metal cations exhibit the continuous development of surface morphology as growth proceeds. For these films, the surface roughness shows a nearly linear relationship with thickness. Oxide films growing predominantly by inward diffusion of oxygen anions did not show any marked increase of the surface roughness after the state of steady growth was achieved. The quantitative description of the surface morphology may be used as an indicator of growth mechanism for oxides films modified by reactive elements.

ACKNOWLEDGEMENTS

The financial support of this work by the Natural Sciences and Engineering Research Council of Canada is gratefully acknowledged.

REFERENCES

1. E. Chason, Ch.M. Falco, A. Ourmazd, E.F. Schubert, J.M. Slaughter and R.S. Williams in Evolution of Surface and Thin Film Microstructure, edited by H.A. Atwater, E. Chason, M.H. Grabow and M.G. Legally (Mater. Res. Soc. Proc., 280, Pittsburgh, PA, 1992) p. 203.
2. F. Czerwinski and W.W. Smeltzer, J. Electroch. Soc., 140, 2606 (1993).
3. F. Czerwinski, J.A. Szpunar, R.G. Macaulay-Newcombe and W.W. Smeltzer, Oxid. Met., submitted.
4. F. Czerwinski, G.I. Sproule, M.J. Graham and W.W. Smeltzer, to be published.
5. R.A. Rapp, Metall. Trans., 15A, 765 (1984).
6. M. Offenberg, M. Liehr and G.W. Rubloff, J. Vac. Sci. Technol., A9, 1058 (1991).
7. H. Rojhantalab, M. Moinpour, N. Peter, M.L.A. Dass, W. Hough, R. Natter and F. Moghadam in Evolution of Surface and Thin Film Microstructure, edited by H.A. Atwater, E. Chason, M.H. Grabow and M.G. Legally (Mater. Res. Soc. Proc., 280, Pittsburgh, PA, 1992) p. 147.
8. J.M. Bennett and L. Mattsson, Introduction to Surface Analysis and Scattering, Optical Soc. of America, Washington, D.C. (1989).
9. F. Morin, L.C. Dufour and G. Trudel, Oxid. Met., 37, 39 (1992).

THE MICROSTRUCTURAL CHARACTERIZATION OF NANOCRYSTALLINE CeO₂ CERAMICS PRODUCED BY THE SOL-GEL METHOD

F. CZERWINSKI AND J.A. SZPUNAR
Department of Metallurgical Engineering, McGill University, Montreal, Que., H3A 2A7, Canada

ABSTRACT

CeO₂ ceramics were manufactured in the form of surface coatings deposited onto various substrates by sol-gel technology. The size of the CeO₂ crystallites, dried at room temperature, was about 5 nm and did not change significantly after heating, up to 680 K. Further increase of the temperature resulted in a rapid growth of crystallites. The process of growth depends also on the film thickness and nature of substrate. The results obtained using thermogravimetric analysis (TGA) and infrared spectroscopy (IR) demonstrated that the thermal decomposition of gel was completed at about 750 K. There was no evident texture in both the as-deposited state and after heat-treatment. X-ray diffraction (XRD), the atomic force microscopy (AFM), and transmission electron microscopy (TEM) were used to characterize the structure of coa:.. ;. The examples of application of CeO₂ ceramics as coatings for high temperature corrosion protection are presented. The role of size of CeO₂ particles in modification of grain boundary transport is discussed.

INTRODUCTION

Ceramics produced by the solution-sol-gel technology are new materials with a high chemical purity and high homogenity, achieved by atomic scale mixing of all constituents. Shaping the viscous sol or gel is done to make the nuclear fuel pellets, ceramic fibres and abrasive grains [1]. Sol-gel route is also ideal for deposition of thin coatings on various surfaces, including those having large size and complex geometry. It has been noted by Pfeil [2] that coatings developed from elements with high affinity for oxygen can markedly improve the high temperature oxidation resistance of metals and alloys. The high performance coatings derived from sol-gel exhibit many advantages including the smallest and the best controlled particle size [3]. Therefore these coatings are useful for oxidation inhibition. The latest studies have indicated, also, that protective properties of sol-gel coatings depend strongly on their microstructure which is often transformed at high temperatures [4].

This study is focused on relationship between the microstructure of CeO₂ coatings and the kinetics of growth of NiO and Cr₂O₃ films at high temperatures on substrates modified with CeO₂.

EXPERIMENTAL PROCEDURE

The high purity sol having concentrations of 5-50 g CeO₂/l in 0.1 M HNO₃ was prepared from the 20 wt.% colloidal dispersion supplied by Johnson Matthey Ltd. The 25

535

ml/l of nonionic surfactant (Triton X-100, Aldrich Chemical Co.) was added to prevent the aglomeration of CeO_2 particles. Coatings were produced by dipping a substrate in this sol at room temperature. Following that, the substrate was dried at room temperature for 20 h and calcined at 573 K for 1 h. Polycrystalline Pt and Ni, and single crystal Si were used as substrates. The coatings thicknesses were in the range of (10-100) nm as measured gravimetrically assuming that the CeO_2 has density of 7.1 g/cm^3. To eliminate the effect of substrate, some measurements were made on the 10-20 μm thick platelets of CeO_2 which were obtained by deposition and then stripping the dried layer from the glass substrate. Thermal decomposition of gel was analyzed by TGA and IR spectroscopy. To characterize the microstructure of coatings the XRD, SEM, TEM and AFM were used.

RESULTS AND DISCUSSION

CeO₂ structure after deposition

The CeO_2 deposit in the state of gel, after drying at room temperature, when examined by XRD, already displays all the major reflections of CeO_2 which has fluorite type structure. The direct imaging of the wet gel by TEM is not possible, therefore the film was first calcined at 573 K for 1 h and then stripped from the substrate. According to the XRD analysis, this procedure did not cau.. .ne coarsening of crystallites. We also observed (Fig. 1) that the crystallites are of uniform size in the range of 4-6 nm. The selected area electron diffraction (SAD) pattern of film region in Fig. 1 shows continuous rings for CeO_2, indicating the random orientation of crystallites. Some additional rings are present in SAD pattern, these are not marked by arrows and were formed by thin oxide film supporting the brittle CeO_2 deposit during TEM examination.

Modification of the deposits structure by heat-treatment

Calcination changes the deposit's weight. TGA measurements made on CeO_2 platelets during continuous heating at a rate of 5 deg/min revealed two stages characterized by different rates of weight loss. The higher weight rate loss was during the first stage up to 450 K. The IR analysis proved that this weight loss was accompanied by loss of water content. During the second stage, the lower rate of weight changes was associated with removal of the other species from the coatings. At about 750 K the thermal decomposition was completed with total weight loss of 17.5 % compared to the sample weight after drying.

The characteristic feature associated with the densification process was the formation of microcracks in coatings. The density of microcracks increased with coating thickness. For coatings having the thickness greater than 100-150 nm the network of microcracks covered the whole surface area of the specimen.

The grain growth was observed after the heat-treatment of the CeO_2 films. The grain sizes obtained from XRD measurements of line broadening are shown in Fig. 2. Powder specimens were prepared from the CeO_2 platelets calcined at high temperatures for 5 h. The crystallites were initially very fine and remained fine after calcination up to about 680 K. However, at temperature higher than 680 K they coarsen rapidly. In Fig 2 the measurements by Chen and Chen [5] of CeO_2 particles in powder obtained from sol-gel are also shown. Although, the initial size of CeO_2 crystallites was larger than that in our experiment, they

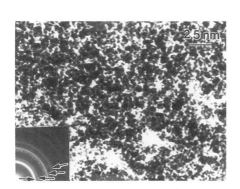

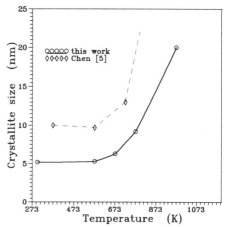

Fig. 1. TEM micrograph and SAL . :ern of CeO$_2$ coating after calcination at 573 K for 1 h. Arrows indicate the diffraction rings for CeO$_2$.

Fig. 2. Effect of temperature on the size of CeO$_2$ crystallites as determined by XRD.

have also found that after a certain temperature the rapid growth of crystallites took place.

The microstructure of coatings obtained from the sol-gel is affected by gel-substrate interactions. The existance of the solid-state epitaxy in nanocomposite xerogel powders and sol-gel coatings has been reported by Roy, Selvaraj et al. [1,6]. To investigate the influence of substrate on the microstructure of CeO$_2$ ceramics, 100 nm thick coatings were deposited on substrates having different structure and different oxidation resistance.

The heat treatment of CeO$_2$ films deposited on polycrystalline Pt substrate resulted in the coarsening of CeO$_2$ crystallites and the sharpening the XRD peaks (Fig. 3a,b). The temperature of rapid growth of CeO$_2$ crystallites is, however, about 100 deg higher than that for separated CeO$_2$ platelets. This suggests the inhibition of coarsening of CeO$_2$ crystallites by interaction with the Pt substrate. The comparison of the peaks intensities in the XRD pattern with the JCPDS standard does not indicate the existance of texture.

To investigate the possibility of epitaxial growth from a sol-gel precursor solution, CeO$_2$ was deposited onto a polished Si(100) single crystal. Despite the fact that the pure oxygen atmosphere was used, the XRD pattern obtained after heat-treatment at 1173 K (Fig. 3c) does not reveal the presence of silicon oxide. The intensities of CeO$_2$ peaks are also similar to those of a random specimen, what indicates that the substrate does not have influence on orientation of CeO$_2$ grains. The heat-treatment of CeO$_2$ deposited on polycrystalline Ni was accompanied by the growth of NiO film, easily detected by XRD (Fig. 3d). The oxide formation, however, did not affect the crystallographic orientation of CeO$_2$. Once again, the XRD peaks intensities are characteristic for the random orientation, similarly as for Pt and Si substrates. The formation of epitaxial relationships in sol-gel

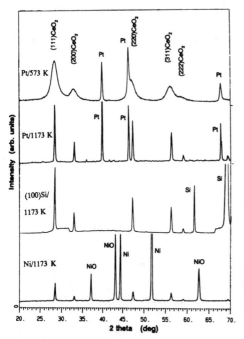

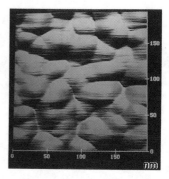

Fig. 4. AFM top view of CeO_2 coating deposited on Pt substrate and calcined at 973 K for 15 min.

Fig. 3. XRD patterns of 100 nm thick CeO_2 coatings calcined on various substrates at temperatures indicated.

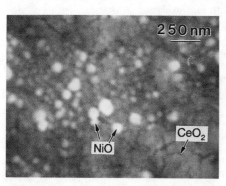

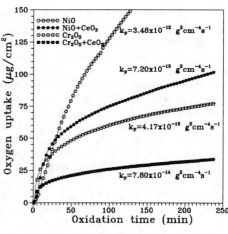

Fig. 5. SEM image of 14 nm thick CeO_2 coatings on Ni substrate after 5 min exposure at 973 K in oxygen.

Fig. 6. Effect of CeO_2 coatings on the growth rate of oxides at 1073 K on Ni and Cr substrates.

coatings reported by Roy, Selvaraj et al. [1,6] was always associated with the growth of new crystals in coating from amorphous state or due to phase transformation. Our results indicate that without the crystallization process in deposits, the deposit-substrate interaction at high temperatures leads only to inhibition of the coarsening of crystallites caused by process of sintering and does not affect their crystallographic orientation.

Essential microstructural differences were observed for coatings deposited onto various substrates. The heat-treatment caused the coarsening of CeO_2 crystallites deposited on Pt substrate. On this substrate the oxides are not formed (Fig. 4.). For Ni substrate which forms a protective oxide film, the small grains of NiO nucleate at high temperature on the surface of the coating (Fig. 5). Nucleation rate of NiO depends on the CeO_2 coating thickness. The thinner the coating, the faster the nucleation of NiO grains. Also, the incorporation of coating into growing oxide film is fast. Although, CeO_2 is highly insoluble in bulk NiO, computer simulation studies [7,8] have indicated that its energy of solution is considerably reduced at grain boundaries by interaction energy between the large Ce^{+4} ions and the boundary sites. The presence of Ce ions segregants in NiO grain boundaries has been proved experimentally [3,4]. The CeO_2 crystallites, when incorporated in oxide, markedly modify the subsequent growth process.

Application of CeO_2 coatings at high temperatures

The effect of CeO_2 coatings on growth rate of NiO and Cr_2O_3 films is shown in Fig. 6. For both oxides, the presence of coatings reduced the growth rate by an order of magnitude. Despite the practical importance of this fact, we do not have a generally accepted theory to explain the mechanism of growth inhibition. The blocking by agglomeration of reactive elements oxide which form the barrier layer at the substrate/oxide interface has been proposed by some authors [9]. However, this barrier has not been observed in our studies [4,10]. The highly effective, very thin CeO_2 coatings are more likely to disperse than aggregate during oxide growth. Recent analyses [3,4,11,12] suggest that a decrease in the oxide growth rate is caused by the reactive element ion segregants which occupy the oxide grain boundaries. Thus, the CeO_2 crystallites are the source of Ce ion segregants. It has also been found that CeO_2 efficiency depends on the size of CeO_2 crystallite. The inhibition of oxide growth (Fig. 6) was achieved for CeO_2 crystallites having a mean diameter of 5 nm. The coating of CeO_2 with larger grain size of 20-30 nm were ineffective and may even increase the oxide growth rate [4].

The sol-gel precursors for CeO_2 coatings are most effective if the size of crystallite is small. This study shows that nano-sized crystallites obtained after deposition are affected by the heat-treatment. We found that above certain critical temperature the rapid coarsening of crystallites takes place. To have the high corrosion protection, the appropriate thermal cycle should be applied to prevent the coarsening of crystallites and to incorporate the extremely small, nano-sized CeO_2 crystallites into native oxide film.

CONCLUSIONS

The CeO_2 coatings manufactured by sol-gel technique on various substrates are characterized by randomly oriented CeO_2 crystallites having the size of 5 nm. The size of CeO_2 crystallites is affected by heat-treatment. For separate CeO_2 platelets, the critical temperature of rapid coarsening of crystallites is 680 K. However, in the case of coatings this temperature is higher and depends on coating thickness and nature of the substrate. Heat treatment of coatings on substrates which do not form oxides is associated with the coarsening of CeO_2 crystallites only, while coatings deposited on oxidizing substrates are incorporated at high temperatures into growing oxide film. The gel-substrate interaction leads to inhibition of the coarsening of the crystallites and does not affect their crystallographic orientation.

The thin films of CeO_2 are very effective in reducing the growth rate of NiO and Cr_2O_3 on various metallic substrates. The size of CeO_2 crystallites is the major factor affecting the protective properties of coatings at high temperatures.

ACKNOWLEDGEMENT

The financial support of this research by the Natural Sciences and Engineering Research Council of Canada is gratefully acknowledged.

REFERENCES

1. R. Roy, Science, **238**, 1664 (1987).
2. L.B. Pfeil, U.K. Patent 459,848 (1937).
3. D.P. Moon and M.J. Bennett, Mater. Sci. Forum, **43**, 269 (1989).
4. F. Czerwinski and W.W. Smeltzer, Oxid. Met., **40**, 503 (1993).
5. P.L. Chen and I.W. Chen, J. Am. Ceram. Soc., **76**, 1557 (1993).
6. U. Selvaraj, A.V. Prasadarao, S. Komarneni and R. Roy, ibid, **75**, 1167 (1992).
7. D.M. Duffy and P.W. Tasker, J. Phys. (Paris), **C-4**, 185 (1985).
8. D.M. Duffy and P.W. Tasker, Phil. Mag. A, **54**, 759 (1986).
9. A. Galerie, M. Gaillet and M. Pons, Mater. Sci. Eng., **69**, 329 (1985).
10. F. Czerwinski and J.A. Szpunar, to be published.
11. K. Przybylski and G.J. Yurek, Mater. Sci. Forum, **43**, 1 (1989).
12. B.A. Pint, Ph.D. thesis, Massachusetts Institute of Technology (1992).

LOW DIELECTRIC CONSTANT, HIGH TEMPERATURE STABLE COPOLYMER THIN FILMS BY ROOM TEMPERATURE CHEMICAL VAPOR DEPOSITION

JUSTIN F. GAYNOR AND SESHU B. DESU*
Department of Materials Science and Engineering, Virginia Polytechnic Institute and State University, Blacksburg, VA 24061-0237, USA

ABSTRACT

Polyxylylene thin films grown by the chemical vapor deposition (CVD) process have long been utilized to achieve uniform, pinhole-free conformal coatings. They have recently been cited as possible low dielectric constant films for intermetal layers in high-speed ICs. Homopolymer films are highly crystalline and have a glass transition temperature around room temperature. We have demonstrated that room temperature copolymerization with previously untested comonomers can be achieved during the CVD process. Copolymerizing chloro-p-xylylene with perfluorooctyl methacrylate results in the dielectric constant at optical frequencies being lowered from 2.68 to 2.19. Copolymerizing p-xylylene with vinylbiphenyl resulted in films which increase the temperature at which oxidative scission occurs from 320 to 450C. Copolymerizing p-xylylene with 9-vinylanthracene resulted in a brittle, yellow film.

INTRODUCTION

Vacuum polymerization of cyclo-di-p-xylylene is a unique reaction which requires no catalyst and results in no reaction byproducts, resulting in extremely pure material with exceptional conformity. Here, we report on a copolymerization method which leads to increased thermal stability, reduction in refractive index and lowered crystallinity. This may increase the suitability of this method for producing intermetallic dielectrics for the electronics industry or low-loss planar waveguides.

The formation of polymer films from the spontaneous polymerization of thermally cleaved cyclo-di-p-xylylene was first reported in 1947[1]. Perfection of the deposition method led to the commercial use of parylene films as coatings for circuit boards by the late 1960's. The first report of room-temperature copolymerization during CVD was with maleic anhydride as the comonomer[2]. In addition, several vinylic monomers have been copolymerized at low temperatures, but when deposition occurs below the glass transition temperature (13C) of the polymer, poor morphology results[2,3].The reader is referred to excellent reviews to supplement this brief introduction[3,4]. Here, only details relevent to the copolymerization process and copolymer properties are presented.

Parylene films grow by a free-radical mechanism. The initiation scheme is shown in Figure 1. First, the paracyclophane is sublimated, then thermally cleaved into two paracyclophane molecules. Upon reaching a surface below 30C, they

---

*To whom correspondence should be addressed.

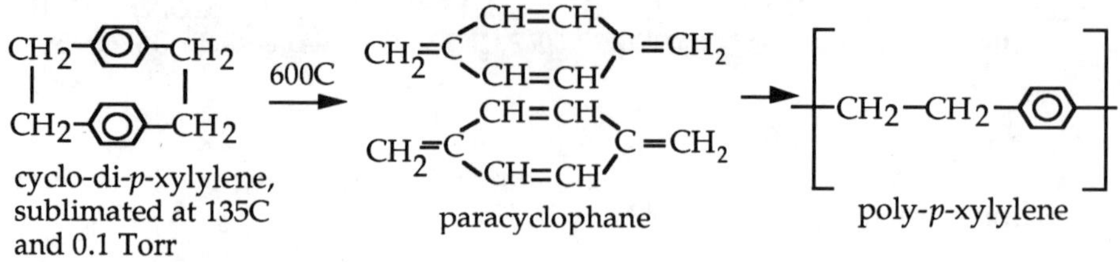

CH$_2$—⬡—CH$_2$ 600C   CH$_2$=C⟨CH=CH / CH=CH⟩C=CH$_2$ CH$_2$=C⟨CH=CH / CH=CH⟩C=CH$_2$ →   [—CH$_2$—CH$_2$—⬡—]
CH$_2$—⬡—CH$_2$

cyclo-di-*p*-xylylene,
sublimated at 135C
and 0.1 Torr

paracyclophane

poly-*p*-xylylene

Figure 1: Formation of poly-p-xylylene from cyclo-di-p-xylylene

condense upon the substrate. By condensing, they are sufficiently concentrated to form the more stable three-unit diradical. Once the diradical is formed, additional monomer adds efficiently[5,6].

Because the rate of initiation is proportional to the [monomer concentration]^3, polymerization in the gas phase is suppressed. However, once initiation has occurred on the substrate, propogation of the polymer chain occurs at a rate proportional to the [monomer concentration]. This polymerization mechanism, unique to the cyclo-di-paraxylylenes, is largely responsible for the utility of these films; the conformity and purity of films grown by this method exceed any other type of thin film polymer processing technique.

The major drawback to this process is the limited number of compounds which can be polymerized, and the consequent limits to property tailoring. Many substituted monomers have been synthesized and studied, and most form linear polymers whose properties are in accordance with prediction. However, only three compounds are commercially available: cyclo-di-p-xylylene, shown in Figure 1; cyclo-di-*p*-chloroxylylene, which has an average of one chlorine atom substituted on each carbon ring; and cyclo-di-*p*-dichloroxylylene, which has two chlorine atoms substituted on each carbon ring. The goal of this research is to significantly alter the properties of the films with commercially available materials while retaining the high purity and conformality of CVD processing.

The commercial uses of these films are based upon their low dielectric constants, high degradation temperatures, conformal deposition, low processing temperatures, low stresses, hydrophobicity, biocompatibility and solvent resistance. Many of the properties listed vary with percent crystallinity of the films. Because the degree of crystallinity in a polymer is greatly dependent on its thermal history, the properties may change significantly during use. The "doping" of the backbone with comonomers, as discussed here, often inhibits the ability to crystallize. In applications where crystallinity is unneccesary or undesirable, the copolymerization method results in properties which are less temperature or thermal history dependent.

Experimental

Fabrication All samples were prepared according to the Gorham method[7] in a homemade reactor modified to allow for the addition of comonomers to the gas

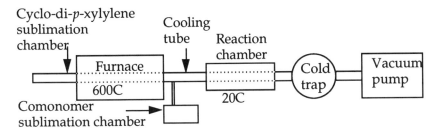

Figure 2: Reactor for chemical vapor copolymerization

mixture. This reactor is shown in Figure 2. Sublimation temperatures for cyclo-p-xylylene and cyclo-p-chloro-xylylene were 135C when homopolymers were grown; however, when a comonomer was used, the sublimation temperature was reduced to 128C to prevent clouding. The furnace was held at 620C. The cooling tube was maintained at 130C for homopolymers and for copolymers with perfluorooctyl methacrylate; this was increased to 180C when copolymerizing with vinylbiphenyl and 9-vinylanthracene. The reaction chamber was maintained between 20-25C for homopolymers and copolymerization with vinylbiphenyl and 9-vinylanthracene, and between 15-20C for copolymerization with perfluorooctylmethacrylate. When copolymers with varying composition were desired, a temperature gradient was set up in the reaction chamber, varying approximately linearly from 30C at the warm end to 18C at the cooler end. Generally, the warmer end led to thicker films with a lower concentration of comonomer.

Cyclo-di-p-xylylene, trade name DPXN, and cyclo-di-p-chloroxylylene, trade name DPXC, were purchased from the Nova Tran corporation. Perfluorooctyl methacrylate, 9-vinylanthracene and vinyl biphenyl were purchased from Lancaster Synthesis.

Characterization FTIR spectra were obtained on a Perkin-Elmer 1710 with a MCT detector. Samples were grown on single crystal CaF_2, purchased from International Crystal Laboratories. Spectra were averaged over 64 measurements, with a bare CaF_2 crystal run as a background. Peak heights were determined from software which performed baseline corrections.

Refractive indices and thicknesses were determined from a variable-angle spectroscopic ellipsometer purchased from J. Woollam Company. Samples were deposited on <111> Si with a 1.3nm layer of SiO_2. Samples were scanned from 400 to 1000nm wavelength light in 2nm increments and at θ = 70, 75 and 80 degrees. Signals were the average of 25 measurements per wavelength. Dielectric constant was assumed to be the square of refractive index.

TGA analysis was performed with a Seiko Thermal Analysis System. Samples were deposited on a detergent-smeared silicon wafer, and removed with running tap water. Samples were then thoroughly rinsed in alcohol and distilled water before being dried in a vacuum oven for 12 hours at 105C. Sample weights were between 10 and 20 mg. The heating rate was 10C/min. Samples were tested in an air environment.

X-ray diffraction scans were performed with a Scintag Automated X-Ray Diffraction System. Samples were deposited on CaF_2 and annealed at 105C under vacuum for sixteen hours. The incident beam was held at 5° above the plane of the film, and the position of the detector scanned.

Absolute compositions of the perfluorooctyl methacrylate/cyclo-di-p-chloro xylylene samples were determined by wavelength dispersive analysis with a Camerac SX-50 electron probe. Samples were grown on B-doped <111> Si.

Results and Discussion

Thermal Stability   It has been reported previously[8,9] that poly-p-xylene and poly-p-chloroxylene begin losing weight at temperatures around 270C in air, and is nearly completely charred at 500C. When heated in nitrogen, however, it is stable to about 480C, where it rapidly decomposes. When the hydrogen bonds in the CH_2 groups are replaced by fluorine, the polymer is stable above 500C in both oxygen and nitrogen. It has been proposed, therefore, that the decomposition of poly-*p*-xylylene occurs through the formation of peroxy radicals which attack the CH_2 bonds, resulting in chain scission[9]. Finally, when poly-p-dichloroxylylene is heated in air, it does not gain weight before decomposing, and the onset of decomposition does not occur until about 320C. Thus, the formation of peroxy radicals appears to be inhibited. It has been speculated that the additional chlorine atom on each ring may sterically hinder oxygen from attacking the CH_2 bonds, or that the electron may be withdrawn from the carbon atom by the strongly electronegative Cl atom, stabilizing the bond[9]. A similar phenomenon is noticed in the poly-p-xylylene-co-vinylbiphenyl samples. In Figure 3, it can be seen that the undoped sample decomposes in the same manner reported earlier. However, as the amount of vinylbiphenyl in the sample is increased, the thermal stability is increased. At the

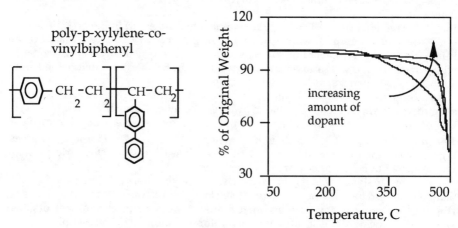

Figure 3: Chemical structure of vinylbiphenyl and thermogravimetric analysis of copolymer films.

maximum concentration, the sample behaves almost the same as the homopolymer in the absence of oxygen. The structure of vinylbiphenyl suggests that it could act efficiently to either sterically shield the CH_2 groups or electronegatively stabilize them. Unfortunately, the chemical similarity of vinylbiphenyl to poly-p-xylylene did not allow for the determination of actual chemical composition in these samples.

Dielectric Constant The dielectric constant of poly-p-chloroxylylene is about 3 in the range 60Hz-1MHz, and 2.7 in the gigaherz region[10]. Two strategies to reduce this would be to lower the density of the material by making it amorphous, and lowering it chemically by the introduction of carbon-fluorine bonds. With the copolymerization method, both of these strategies can be implemented simultaneously. Cyclo-di-p-chloroxylylene was copolymerized with perfluoro-octyl methacrylate. The reaction chamber was held at a thermal gradient to achieve a series of samples with differing compositions. As expected, an increase in the amount of perfluorooctylmethacrylate resulted in a reduction in crystallinity (Figure 4). The major x-ray diffraction peaks cited in the literature are the 020 (16.7°), -111 (17.7°) and 110 (22.7°). A decrease in refractive index was observed; in the optically nonabsorbing region, the dielectric constant is simply the square of the refractive index. The concentration of perfluorooctyl methacrylate in each sample was determined by wavelength dispersive analysis and confirmed by peak height ratios in the FTIR spectrum. In the wavelength dispersive analysis, the ratio of chlorine to fluorine in each sample was determined. The FTIR peaks considered were at 1730/cm, which is assigned to the C=O stretch in the perfluorooctyl methacrylate,

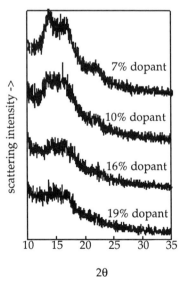

Figure 4: X-ray scattering
vs. composition

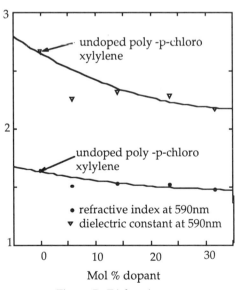

Figure 5: Dielectric constant
and refractive index vs. composition

and at 1052/cm, characteristic of carbon rings in polymer backbones. Figure 5 shows the dielectric constant, as determined by ellipsometry, as a function of perfluorooctyl methacrylate concentration.

Conclusions

The use of comonomers to change the properties of cyclo-p-xylylenes has been demonstrated. The conditions for growing different types of copolymers are described. Copolymers which can be grown at room temperature include p-xylylene with vinylbiphenyl and vinylanthracene and p-chloroxylylene with perfluorooctyl methacrylate. The addition of vinylbiphenyl to p-xylylene significantly increases its oxidative stability; the undoped material starts to decompose around 270C, while the copolymer remains stable to above 450C. The crystallinity and refractive index of p-chloroxylylene can be lowered; the lowest refractive index achieved was 1.47.

Acknowledgements

The authors wish to thank the Innotech Corporation of Roanoke, Virginia for financial support; Todd Solberg of the Virginia Tech Department of Geological Sciences for performing x-ray and wavelength dispersive analysis; and Srivastan Srivinas of the Virginia Tech Department of Chemical Engineering for performing thermal analysis tests.

References

1. M. Swarc, *Disc. Faraday Soc.*, **2**, 48, (1947).
2. V. Sochilin, K. Mailyan, L. Aleksandrova, A. Nikolaev, A. Pebalk and I. Kardash, Plenum Publishing document 0012-5008/91/0007-0165, translated from Doklady Akademii Nauk SSSR, Vol. 319, No. 1, pp. 173-176, July 1991.
3. W. Beach, C. Lee, D. Bassett, T. Austin and R. Olson, *Polymer Encyclopedia*, Vol. 17, pp. 990-1024, Wiley and Sons.
4. W. Beach and T. Austin, *2nd International SAMPE Electronics Conference,* pp. 25-35, June 1988.
5. M. Swarc, *Disc. Faraday Soc.*, **2**, 46, (1947).
6. L. Errede and M. Swarc, *Quarterly Review of the Chemical Society*, Vol. 12, No. 4, 301-320, 1958.
7. W. Gorham, *J. Poly. Sci. A-1*, Vol 4, pp. 3037-3039, 1966.
8. B. Bachman, *1st International SAMPE Electronics Conference,* pp. 431-40, 1987.
9. B. Joesten, *J. Appl. Poly. Sci.*, Vol 18, pp. 439-448 (1974).
10. W. Beach, C. Lee, D. Bassett, T. Austin and R. Olson, p. 1007.

PART V

Mechanics and Mechanical Properties
of Polycrystalline Thin Films

ENHANCED MECHANICAL HARDNESS IN COMPOSITIONALLY MODULATED Fe/Pt AND Fe/Cr EPITAXIAL THIN FILMS

B.J. DANIELS, W.D. NIX, and B.M. CLEMENS
Department of Materials Science and Engineering, Stanford University, Stanford, CA 94305-2205

ABSTRACT

The hardnesses and elastic moduli of sputter-deposited Fe/Pt and Fe/Cr multilayers grown on MgO(001) are evaluated as a function of composition wavelength, Λ. Structural determination by x-ray diffraction showed these films to be oriented in the plane as well as out of the plane. The mechanical behavior of these films was evaluated by nanoindentation. The combination of nanoindentation and x-ray diffraction is an attempt to determine the structural underpinnings of the mechanical behavior of these metal multilayer systems. For both systems there is no observed enhancement in the elastic modulus (the so-called supermodulus effect) across a wide range of bilayer spacings. Nanoindentation results show that for Fe/Pt multilayers, the hardness is enhanced over that expected from a simple rule of mixtures by a factor of approximately 2.5, with a maximum enhancement of 2.8 times this value at a wavelength of 25 Å. This enhancement in hardness occurs for bilayer spacings from 20 Å to 100 Å and is not a strong function of Λ over this range. Results for Fe/Cr multilayers show a hardness enhancement over a similar wavelength range of approximately two times the rule of mixtures value, with a maximum enhancement of 2.2 times this value at a wavelength of 40 Å. The larger hardness enhancement in the Fe/Pt system may be due to the structural barrier (FCC/BCC) to dislocation motion between the two materials. The dominant mechanism responsible for the hardness enhancement in Fe/Pt and Fe/Cr multilayers is not yet known, however three models for dislocation interactions which could account for the hardness enhancement in these multilayers are discussed.

INTRODUCTION

Recent studies of multilayered structures fabricated by thin film deposition techniques such as evaporation, sputter deposition, and electrodeposition have revealed interesting changes in the mechanical, electrical, magnetic, and optical properties of these materials. These property changes are a direct result of the compositionally modulated structure that is created upon deposition. Previous investigations of the mechanical behavior of these artificial superlattice structures have centered on the supermodulus effect[1-5] and the superhardness effect[6-9]. Although the supermodulus effect resulted from errors in the interpretation of bulge test data[4, 5], the superhardness effect appears to be a real material phenomenon. The strength enhancement at small bilayer spacings for compositionally modulated materials has been modeled in various ways. In this work we report on ultrahigh strength Fe/Pt and Fe/Cr multilayers which have been characterized by nanoindentation and x-ray diffraction (XRD). The nanoindentation results for both systems are compared with the predictions of three different models which attempt to explain the hardness enhancement for multilayer thin films with small bilayer spacings.

EXPERIMENTAL

Organically-cleaned, polished MgO(001) substrates were placed in a UHV chamber which was then evacuated to a base pressure of approximately 5×10^{-9} Torr. All films were grown by DC magnetron sputtering in a 3 mTorr Ar environment with a target-to-sample distance of

Mat. Res. Soc. Symp. Proc. Vol. 343. ©1994 Materials Research Society

2.25 inches. Deposition rates at the sample were 1-3 Å/sec as determined by tooling factor-corrected quartz crystal rate monitor rates. A 5 Å Fe + 200 Å Pt seed layer was deposited at approximately 360°C to promote the growth of epitaxial Fe(001)/Pt(001) multilayers on MgO(001). This seed layer was found to be critical in achieving epitaxial growth in this and other systems[10, 11]. A similar seed layer consisting of 200 Å of Cr also grown at 360°C was used to achieve epitaxy in the Fe(001)/Cr(001) system. After deposition of the seed layer the substrate was permitted to cool to 50-60°C. At this point a multilayer film consisting of equal thicknesses of each material was deposited. Although the bilayer spacing, Λ, was varied from film to film, the total thickness of each multilayer film was held at 1 μm. Samples in which the final layer (Pt or Cr) was less than 30 Å received a capping layer, increasing the thickness of the final layer to 30 Å in order to inhibit Fe oxidation.

The structure of these films was determined by XRD. A conventional powder diffractometer using Cu K_α radiation and equipped with an exit beam monochromator was used to obtain low-angle and high-angle symmetric XRD data. Low-angle superlattice lines were analyzed to determine the composition modulation present in the fabricated films. This allowed for the correction of errors in the tooling factors and variations in the sputtering rates due to changes in the target geometry during sputtering. High-angle scans were used to determine if any unwanted (i.e. (110), (111)) types of growth had occurred, eliminating these samples from further testing.

In order to determine the epitaxial orientation of the samples, asymmetric x-ray scans were performed with a three-circle pole figure diffractometer which also uses Cu K_α radiation. ϕ scans for Fe{110} and Pt{220} peaks revealed that both Fe and Pt grew epitaxially in the (001) orientation with a relative rotation around the [001] axis of 45°. This is known as the Bain orientation and is designated as (001) || (001), [010] || [110]. XRD data for Fe{110} and Cr{110} showed that these multilayers were epitaxial with the [001] normal to the direction of growth and no relative rotation between the two films. This orientation is known as cube-on-cube epitaxy and is designated as (001) || (001), [010] || [010].

The mechanical properties of these films were evaluated using a commercially-available depth-sensing indentation machine known as the Nanoindenter. The Doerner and Nix analysis method and the Nanoindenter itself are discussed in detail elsewhere[9, 12, 13]. For each of the samples tested using the Nanoindenter, a six by six array of indentations spaced 5 μm apart was made with six indentations each at nominal depths of 25, 50, 75, 100, 125, and 150 nm. Since analysis of the data showed that hardness was not a function of indentation depth over this range, we present only the data for indentation depths of 150 nm. This clarifies the discussion and allows us to focus on the data from the largest indentations, where the scatter is smallest. Analysis of test data results in the determination of both the "indentation modulus" ($E/(1-\nu^2)$) and the hardness for each film.

RESULTS

The indentation modulus was measured as a function of Λ for both Fe(001)/Pt(001) and Fe(001)/Cr(001) films. Data obtained for a pure Fe (solid line), a pure Pt (solid line), and many Fe/Pt multilayer films (markers) are shown in Figure 1. As expected, we do not find a supermodulus effect. In fact, a trend towards lower indentation modulus at smaller bilayer spacings is apparent, and for bilayer spacings below about 70 Å the indentation modulus of the film is significantly lower than that expected from a simple rule of mixtures. This type of behavior is consistent with bonding at the interface (Fe-Pt) which is expected to be weaker than bonding in the bulk materials (Fe-Fe and Pt-Pt). As Λ is decreased, the number of interfaces per unit volume is increased, increasing the proportion of interfacial bonds. Although the scatter in the data is considerably larger, this drop in modulus is not seen in the Fe/Cr system (Figure 2). Since the structure and lattice parameter of Fe and Cr are similar, we expect the bonding between the two to be more similar than in the Fe/Pt system. Once again, there is a small variation in the measured indentation modulus, however, there is no evidence of a supermodulus effect.

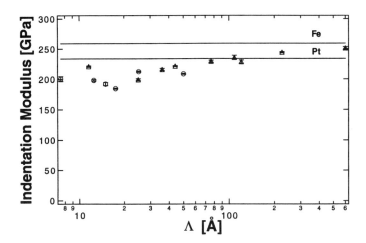

Figure 1: Indentation modulus as a function of composition wavelength for Fe(001)/Pt(001) multilayer films. Note that no supermodulus effect is observed, but rather the films become more compliant at smaller bilayer periods. This effect results from an increased number of interfaces per unit volume at smaller Λ.

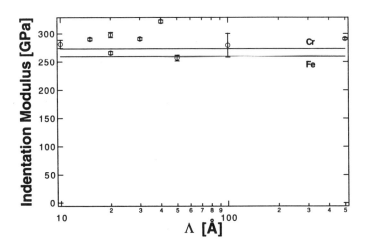

Figure 2: Indentation modulus as a function of composition wavelength for Fe(001)/Cr(001) multilayer films. In contrast to the Fe/Pt system, the modulus stays rather constant with Λ. This occurs because the bonding between Fe and Cr atoms is expected to be similar, while the bonding between Fe and Pt atoms differs.

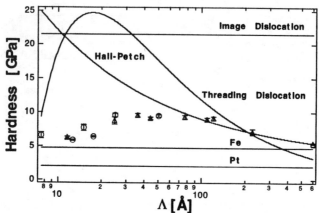

Figure 3: Nanoindentation hardness as a function of composition wavelength for Fe(001)/Pt(001) multilayer films. Indentation depths are nominally 150 nm, and markers represent the mean value for six indentations while error bars are plus and minus the standard deviations for these measurements. Theoretical fits for the threading dislocation, image dislocation, and Hall-Petch models are also shown.

Hardness measured by the Nanoindenter as a function of Λ for pure Pt (solid line), pure Fe (solid line), and Fe/Pt multilayer films (markers) is shown in Figure 3. The hardness is enhanced over that expected from a simple rule of mixtures by a factor of approximately 2.5, with a maximum enhancement of 2.8 times the rule of mixtures value at a wavelength of 25 Å. This enhancement in hardness occurs for bilayer spacings from 20 Å to 100 Å and is not a strong function of Λ over this range. The decrease in hardness enhancement at low values of Λ is due to the absence of sharp interfaces between the Fe and Pt layers. It has been reported by Visokay, et.al. that below a bilayer spacing of 23 Å, a compositionally modulated FCC alloy, rather than a distinctly multilayered BCC/FCC Fe/Pt structure, is formed[14, 15].

Results for Fe/Cr multilayers show a hardness enhancement of approximately two times the rule of mixtures value over a similar wavelength range, with a maximum enhancement of 2.2 times this value at a wavelength of 40 Å (Figure 4). The fact that the hardness enhancement is smaller in the Fe/Cr system indicates that structure modulated strengthening may be a small effect. Clearly some other strengthening mechanism dominates the yield behavior of these multilayer thin films.

DISCUSSION

Three different models have been evaluated in an attempt to explain the hardness enhancement in both multilayer systems. We have developed a model for deformation of FCC/BCC multilayers which treats the creation and movement of threading dislocations trapped within one layer of the multilayer (assuming the structural barrier is the dominant strengthening mechanism). The predictions of that model are compared with the hardness measurements for the Fe/Pt system in Figure 3. The potential importance of the structural barrier is further supported by the observation of a smaller hardness enhancement for Fe/Pt multilayers with wavelengths in the range where the FCC/BCC structural barrier is known to disappear in this system[14, 15]. While this model appears to fit the experimental data at large values of Λ, another deformation mechanism clearly dominates for smaller bilayer spacings. The drawback of this model is that it does not explain the strengthening observed

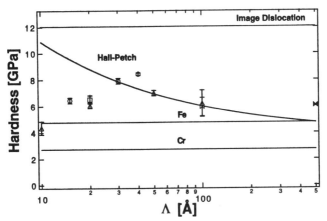

Figure 4: Nanoindentation hardness as a function of composition wavelength for Fe(001)/Cr(001) multilayer films. Indentation depths are nominally 150 nm, and markers represent the mean value for six indentations while error bars are plus and minus the standard deviations for these measurements. Theoretical fits for the image dislocation and Hall-Petch models are also shown.

in the Fe/Cr system and other systems that lack a structural barrier.

A second approach involves the use of the Hall-Petch model for grain size-dependent strengthening. For multilayer thin films, interfaces are treated as grain boundaries, with half of the bilayer spacing analogous to the grain size. Thus we obtain an equation for the hardness, H, of the form:

$$H = H_{in} + k\Lambda^{-\frac{1}{2}} \qquad (1)$$

where H_{in} is the intrinsic hardness and k is a parameter analogous to the locking parameter in polycrystalline materials. This use of this model requires the assumption that the lateral grain size is larger than the bilayer spacing in these multilayer films. Although the validity of this assumption has not yet been verified, we present the results here for comparison. By letting H_{in} be the average hardness for each system and picking k to give the best fit, we obtain the curves shown in Figures 3 and 4. While this model explains the data at large bilayer spacings fairly well, it fails for wavelengths below about 50 Å.

The third model takes into account the image forces on a dislocation that glides from a layer with a low shear modulus to a layer with a higher shear modulus. This approach, which was discussed first by Koehler[16], uses the results of Pacheco and Mura[17] to calculate the stress required to move a dislocation to the interface between the two materials. For the range of bilayer spacings in which we are interested, this model predicts a constant hardness as shown in Figures 3 and 4. Although it consistently overestimates the hardness, this model predicts the proper functional form for measured data in the 10-100 Å regime for both Fe/Pt and Fe/Cr multilayers. It seems plausible that the deformation of multilayers with Λ less than 100 Å is dominated by image dislocation effects, while for films with larger bilayer spacings the plastic behavior may be dominated by dislocation pile-ups at the interfaces.

CONCLUSIONS

Epitaxial Fe(001)/Pt(001) and Fe(001)/Cr(001) multilayers were fabricated via sputter deposition and characterized by XRD and nanoindentation. Elastic anomalies were seen in both systems, but no supermodulus effect was observed in either. In the Fe/Pt system the hardness was enhanced by a factor of approximately 2.5 over the rule of mixtures value for Λ ranging from 20 Å to 100 Å. The hardness in the Fe/Cr system was enhanced by a factor of approximately two over the rule of mixtures value for a similar range in Λ. Three models for the dislocation interactions that could account for a hardness enhancement in these systems were discussed. Although it consistently overpredicts the hardness, it appears that the image dislocation model most accurately explains the hardness behavior for films with wavelengths from 10-100 Å. The yielding behavior for films of larger wavelength may be explained using the Hall-Petch model.

ACKNOWLEDGMENTS

B.J.D. would like to thank G.R. English and G.F. Simenson for many fruitful discussions involving this research. Funding for this investigation was provided by NSF under contract DMR-9100271.

REFERENCES

1. W.M.C. Yang, T. Tsakalakos, and J.E. Hilliard. *J. Appl. Phys.*, **48**:876, 1977.

2. G.E. Henein and J.E. Hilliard. *J. Appl. Phys.*, **54**:728, 1983.

3. T. Tsakalakos and J.E. Hilliard. *J. Appl. Phys.*, **54**:734, 1983.

4. M.K. Small, B.J. Daniels, B.M. Clemens, and W.D. Nix. *J. Mater. Res.*, **9**:25, 1994.

5. S.P. Baker, M.K. Small, J.J. Vlassak, B.J. Daniels, and W.D. Nix. *Proceedings of the NATO Advanced Study Institute: "Mechanical Properties and Deformation Behavior of Materials Having Ultra-Fine Microstructures."* Praia do Porto Novo, Portugal, June 29 - July 10, 1992.

6. A.F. Jankowski. *J. Magn. Magn. Mater.*, **126**:185, 1993.

7. T. Foecke and D.S. Lashmore. *Script. Met. et Mat.*, **27**:651, 1992.

8. S.R. Nutt, K.A. Green, S.P. Baker, W.D. Nix, and A.F. Jankowski in *Thin Films: Stresses and Mechanical Properties Symposium.* Mater. Res. Soc. Proc. 130, Pittsburgh, PA, 1989.

9. M.F. Doerner. *Ph.D. Dissertation.* Stanford University, 1987.

10. P. Etienne, J. Massies, S. Lequien, R. Cabanel, and F. Petroff. *J. Crystal Growth*, **111**:1003, 1991.

11. B.M. Lairson, M.R. Visokay, R. Sinclair, and B.M. Clemens. *Appl. Phys. Lett.*, **61**:1390, 1992.

12. M.F. Doerner and W.D. Nix. *J. Mater. Res.*, **1**:601, 1986.

13. S.P. Baker. *Ph.D. Dissertation.* Stanford University, 1993.

14. M.R. Visokay, B.M. Lairson, B.M. Clemens, and R. Sinclair. *J. Magn. Magn. Mater.*, **126**:136, 1993.

15. B.M. Lairson, M.R. Visokay, R. Sinclair, and B.M. Clemens. *J. Magn. Magn. Mater.*, **126**:577, 1993.

16. J.S. Koehler. *Phys. Rev. B*, **2**:547, 1970.

17. E.S. Pacheco and T. Mura. *J. Mech. Phys. Solids*, **17**:163, 1969.

INTERFACIAL STRUCTURE AND MECHANICAL PROPERTIES OF COMPOSITIONALLY-MODULATED Au-Ni THIN FILMS.

Shefford P. Baker, Max-Planck-Institut für Metallforschung, Institut für Werkstoffwissenschaft, Seestr. 92, 70174 Stuttgart, GERMANY.

James A. Bain, Data Storage Systems Center, Carnegie Mellon University, Pittsburgh, PA 15213, USA.

Bruce M. Clemens and William D. Nix, Department of Materials Science and Engineering, mc 2205, Stanford University, Stanford, CA 94305, USA.

ABSTRACT

Several characteristics of compositionally-modulated Au-Ni thin films have been observed to vary with composition wavelength for wavelengths between 0.9 and 4.0 nm. The average lattice parameter normal to the film plane displays a maximum, the elastic stiffness shows a minimum and the substrate interaction stress in the film goes through a compressive peak in this regime. These variations are all consistent with a model in which deviations from bulk behavior are confined to the interfaces in the material. In this paper, we present the results of microstructural analyses of these films. Symmetric and asymmetric θ - 2θ x-ray diffraction scans were conducted with scattering vectors oriented at a variety of angles to the film normal. Rocking curve analyses were also conducted. Features arising from the composition modulation in symmetric scans are quite sharp. However, asymmetric scans and rocking curves indicate that these films have a relatively poor {111} texture. Data from all scans provide clear evidence that Au intermixes preferentially into Ni. These results are supported by computer simulations of the diffraction spectra and the results of electron-image and -diffraction analyses. These measurements provide a consistent explanation for the mechanical properties of these films.

INTRODUCTION

Interesting mechanical properties have been observed in thin Au-Ni compositionally-modulated films (CMF's) when the composition wavelength, λ, becomes small. These include a minimum in elastic stiffness and a maximum in compressive stress near $\lambda = 2$ nm as shown in Figure 1 [1,2]. A substantial number of investigations of the elastic stiffness of metal CMF's have been conducted, primarily motivated by reports of very large stiffness enhancements. Although such enhancements have been discredited [1,2], λ-dependent minima in elastic stiffness similar to that shown in Fig. 1 have been observed in a number of metal CMF systems (*e.g.* references in [3]). Explanations for this type of stiffness variation include the effects of large strains introduced by interlayer coherency or interface stresses on the elastic constants of the constituents, various electronic effects, and the properties of the interfaces between the constituents themselves. Variations in stress with composition wavelength have been much less frequently reported. However, stresses which become more tensile [4] and stresses which become more compressive [1,2] with decreasing composition wavelength have been observed. This type of stress variation has been attributed to interface stresses [4]. Correlated with the behavior shown in Fig.1 is a maximum in the average lattice parameter measured in the direction normal to the plane of the film, $\bar{d}$ [5]. Variations in stiffness with λ in other CMF systems are strongly correlated with this type of lattice parameter variation in the cases in which $\bar{d}$ has been reported [3].

The Au-Ni samples were prepared by first depositing Cr adhesion layers about 50 nm thick onto oxidized silicon substrates. Following this, Au and Ni were alternately

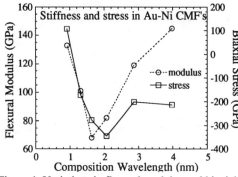

Figure 1. Variations in flexural modulus and biaxial stress with composition wavelength in Au-Ni films.

deposited until the total film thickness was between 610 and 860 nm. The substrates were maintained at room temperature (< 40°C) during planar magnetron sputter deposition of the metals. The composition wavelengths of the Au-Ni films were 0.91, 1.30, 1.63, 2.06, 2.86 and 3.98 nm. A more detailed description of the fabrication of these films is available elsewhere [1,2].

This paper is a preliminary report of our investigations of the microstructure of these samples and its relation with the mechanical behavior shown in Fig. 1.

X-RAY DIFFRACTION

A variety of x-ray diffraction (XRD) techniques were used to investigate the microstructure of these Au-Ni films. Symmetric θ-2θ scans have been reported previously [5]. Additional symmetric scans, as well as rocking curves, were made using Cu $K\alpha$ radiation and a four-circle diffractometer. Asymmetric scans were made using a Huber four-circle goniometer on beam line 7-2 of the Stanford Synchrotron Radiation Laboratory. A double crystal monochromator was used to provide a beam energy of 8003 eV

Symmetric θ - 2θ scans

Diffraction scans in the standard θ-2θ mode were performed with the scattering vector parallel to the film normal in each case. Sample scans are shown in Figure 2. In a homogeneous specimen, such scans are used to sample the population density of atomic planes as a function of the plane spacing (lattice parameter) in the direction of the scattering vector. However, in the range of λ shown, the periodicity of the composition modulation in this direction provides the dominant features. A typical scan includes a sharp central peak whose full width at half of maximum intensity, or FWHM, is typically 0.6 - 0.8 degrees. The position of the central peak is determined by the *average* lattice parameter in the scattering direction, $\bar{d}$. In addition, "satellite peaks" whose spacing is proportional to the inverse of λ arise from the composition modulation. The location of the central peak indicates that these films have <111> texture. No diffraction corresponding to other orientations was

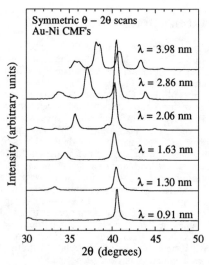

Figure 2. Symmetric x-ray diffraction scans of Au-Ni modulated films.

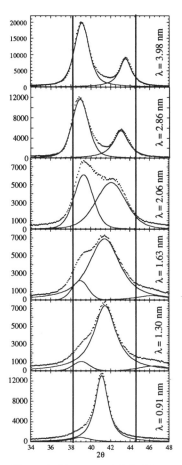

Figure 3. "Asymmetric" θ - 2θ scans of {111} planes having normals oriented 70.53° from the film normal. The vertical lines to the left and right of the peaks indicate the expected positions of 111 peaks from bulk Au and Ni, respectively. (Intensity shown on the vertical axis in arbitrary units.

observed in symmetric scans. The position of the central peak varies systematically with λ and the variation of $\bar{d}$ obtained is in agreement with earlier results [5].

Asymmetric θ - 2θ scans

Diffraction scans in the θ - 2θ mode were also made with the scattering vector oriented at a variety of angles, ψ, with respect to the film normal. These "asymmetric" scans were made at ψ = 35.26, 54.74, 70.53 and 90.00° (grazing incidence), which would correspond to {220}, {200}, {111} and {220} type planes in a film with pure (111) texture. Scans taken at ψ = 70.53° from samples at each composition wavelength are shown in Figure 3. These scans show diffraction from {111} type planes. The vertical lines at 2θ = 38.2 and 44.5° represent the expected peak positions for diffraction from strain-free {111} planes in bulk Au and Ni, respectively. The data were fit with peaks which were a linear combination of Lorentzian and Gaussian distributions plus a linear background. Three important features are evident: First, the peaks observed do not occur at the 2θ values expected for bulk Au and Ni, but are shifted towards each other from these values. Second, the position of the peak closest to Au {111} is relatively constant with composition wavelength, while the peak which is nearest to Ni {111} for the largest values of λ, steadily shifts to lower 2θ values as the composition wavelength decreases. Finally, the peak widths are rather broad when compared with the peaks obtained from the symmetric scans. Similar behaviors have also been reported recently by other authors [6].

Rocking curves

For a "rocking curve," the geometry of the incident beam and the detector is fixed and the sample is rotated about some axis. Thus, the length of the scattering vector is fixed while its direction is varied. This type of scan samples the population density of planes with fixed spacing as a function of direction within the crystal. Two different rocking geometries were used. The central peaks in the symmetric scans were examined by rotating the sample around the θ - 2θ axis. In this configuration, θ is decoupled from 2θ by the rotation of the sample and the rocking angle, ω, is just the difference, $\omega = 2\theta - 2(\theta)$. The 200 reflections of both Au and Ni at ψ = 54.74 were scanned by rotating the sample in ψ. The FWHM values obtained from these scans are shown in

	ω 111	ψ	200
λ (nm)	$\bar{d}$	Au	Ni
3.98		9	8
2.86	9	11	12
2.06	10	13	15
1.63	13	15	18
1.30	14	20	23
0.91	14	20	21

Table 1. Full Width at Half Maximum (FWHM) in degrees for ω and ψ rocking curves.

Table 1. We observe that the <111> texture of these films is relatively poor and decreases with decreasing composition wavelength.

COMPUTER SIMULATIONS OF SYMMETRIC X-RAY SCANS

The data from the symmetric scans was simulated using a modification of the algorithm of Fullerton et al. [7]. In this model, the composition is represented as a square wave of statistically distributed period. Our modification allows the amplitude of the composition modulation to be a fittable parameter as well. We describe here briefly only two results. We observe that (1) the difference between the lattice parameters of the two constituents in the film normal direction and (2) the difference between the composition amplitudes of the two constituents both *decrease* with decreasing composition wavelength. The difference in lattice parameter decreases by about 20% and the difference in composition amplitudes decreases to about 40 atomic % Au. The difference parameters are reported (as opposed to absolute values) because they avoid effects which arise from stresses and minimize systematic errors.

TRANSMISSION ELECTRON MICROSCOPY and ATOMIC FORCE MICROSCOPY

Both images and electron diffraction patterns were generated for these films in plan view using a transmission electron microscope (TEM) operated at 120keV. In addition, images of the surface topology were generated using atomic force microscopy (AFM). TEM images are similar to results for similarly prepared films [8] and indicate that these films contain a columnar structure of grains having diameters between about 30 and 60 nm. Images obtained by AFM indicate that the top of each grain is rounded with the center of each grain being on the order of 10 nm higher than its boundaries [1,2].

Selected area electron diffraction was conducted on samples having $\lambda = 1.63$ and 3.98 nm as well. Diffraction patterns containing smooth pairs of rings corresponding to diffraction from 220 planes with their normals in the plane of the film were obtained. The plane spacings obtained from the diameters of these rings agree with the plane spacings obtained by XRD in grazing incidence and are qualitatively similar to the behavior shown in Fig 3; that is, one ring corresponding to Au {220} which is slightly compressed and essentially independent of λ and a second ring which corresponds to Ni {220} which is slightly expanded at $\lambda = 3.98$ nm but which moves to nearly halfway between Au {220} and Ni {220} by $\lambda = 1.63$ nm.

ANALYSIS AND DISCUSSION

Symmetric x-ray scans and plan view imaging reveal features which are commonly observed in compositionally-modulated films. The sharp features in the symmetric scans indicate <111> textured films having consistent composition modulations and substantial coherence lengths along the growth direction. TEM and AFM images reveal columnar microstructures of fine grains which are randomly oriented in the film plane. Each grain has a convex surface.

Asymmetric scans and rocking curves avoid the convoluting effects of the composition modulation and reveal further information about the structures of these films. The <111>

texture is revealed to be fairly broad—consistent with atomic plane curvature corresponding to the curvature of the grain surfaces—and dependent on λ. Diffraction from Au and Ni can be resolved in asymmetric scans. The lattice parameters attributed to Au and Ni converge with decreasing λ. A similar convergence has been reported for other Au-Ni films and was attributed to coherency strains [8].

To better understand the distribution of stresses and strains in these modulated films, a "$\sin^2\psi$ analysis" was performed on the asymmetric diffraction data. It can be shown [9] that, for a <111> fcc film subject to an equal-biaxial stress, σ, in the plane of the film, the variation in the measured plane spacing of a particular set of crystallographic planes with $\sin^2\psi$ can be used to obtain the *unstrained* lattice parameter, d_o—that is, the plane spacing that this particular set of planes would have if the applied stress were removed—and the stress itself. This analysis was performed using the data from the {220} peaks scanned at $\psi = 35.26°$ and at grazing incidence ($\psi \approx 90°$). The unstrained lattice parameters of the layers which we have attributed to Au and Ni are shown as a function of $1/\lambda$ in Figure 4 along with bulk Au and Ni {220} lattice parameters. We observe that with the exception of the sample with $\lambda = 0.91$ nm, the unstrained lattice parameter of the Au layer remains close to the bulk {220} lattice parameter of Au, independent of composition wavelength. However, although d_o of the second constituent layer starts out near to that of bulk Ni {220} for the largest wavelength sample (to the left in this plot), it moves to larger values as λ decreases until it appears to saturate at a value about midway between the bulk {220} spacings of Au and Ni for $\lambda \leq 1.63$ nm. This indicates that the behavior shown in Fig. 3 is *not* due to an increase in coherency stresses, but that the *equilibrium* lattice parameter of the "Ni" phase is becoming larger with decreasing λ.

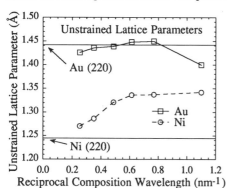

Figure 4. Unstrained {220} lattice parameters from asymmetric scans using the "$\sin^2\psi$ analysis."

Au and Ni are virtually immiscible at room temperature. Nonetheless, one plausible explanation for the lattice parameter behavior shown in Figures 3 and 4 is preferential intermixing of Au into Ni. This correctly accounts for the shifting intensities and positions of the Au and "Ni" peaks as exemplified in Fig. 3. Further evidence that the shift in the "Ni" peak position does not arise from, say, increasing coherency, is provided by the simulations. If the constituents became more coherent with decreasing λ, the difference in their out-of plane lattice parameters would *increase* whereas we calculate a decrease. The fact that the difference in calculated composition amplitudes decreases with decreasing λ provides direct support for intermixing.

We observe that d_o of the intermixed phase varies linearly with $1/\lambda$, at least for $1/\lambda$ less than the saturation value. Quantities which vary linearly with $1/\lambda$ in such a plot scale simply with the number of interfaces in the material. Furthermore, this line points, in the limit of an interface-free material ($1/\lambda = 0$), to the bulk Ni {220} lattice parameter. This suggests a composition profile like that shown in Figure 5. If a thin layer of intermixed material of constant thickness (taken to be about 0.5-0.8 nm from the saturation λ) forms at each Au-on-Ni interface, then as λ decreases, the relative amount of intermixed phase increases. This is supported by the fact that the FWHM of the "Ni" phase peaks goes through a maximum at about $\lambda = 2$ nm for all asymmetric scans (at which point this layer would be about half intermixed).

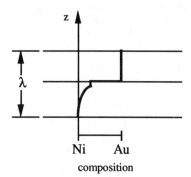

Figure 5. Possible composition profile in one wavelength.

We note that for composition wavelengths larger than $\lambda = 1.63$ nm, the stiffness of these films as well as $\bar{d}$ also vary linearly with $1/\lambda$ [1,2].Thus, an intermixed layer with both a larger {111} spacing and a lower stiffness than a simple rule of mixtures average of these values based on bulk Au and Ni, can account for these λ-dependent phenomena as well.

Regarding stresses, the general trend is that the stress in both layers varies in the same way (analogous to Fig 1) with the stress in Au being more compressive than that in "Ni". Again, there is no evidence of in-plane interlayer coherency. However, these results suggest that interface stresses could account for the stress behavior shown in Fig 1. (*e.g.* see [4]).

The fact that Au is more compressive than Ni agrees with observations that, for sputtering under identical conditions, a larger mass species will be deposited in a more compressive stress state than lighter elements [10]. · Mechanical and/or structural energies might then provide the driving force required to overcome any chemical energy barrier and generate the somewhat counter-intuitive asymmetric mixing of a larger species into a smaller one.We note that similar behavior has been observed in Co-Pt modulated films [11].

CONCLUSIONS

Asymmetric XRD, electron diffraction and computer simulations of symmetric diffraction scans provide evidence of preferential intermixing of Au into Ni in these films. The data are consistent with the existence of a thin intermixed layer of constant composition at Au-on-Ni interfaces. The existence and properties of this intermixed layer could account for variations in stiffness and average lattice parameter normal to the film plane with composition wavelength.

ACKNOWLEDGEMENTS

The authors would like to thank Dr. Alan F. Jankowski for preparation of the Au-Ni films and Dr. Wolf-Michael Kuschke for additional x-ray work and useful discussions.

REFERENCES

1.   S.P. Baker, PhD Dissertation, Stanford University, (1993).
2.   S.P. Baker and W.D. Nix, submitted to*J. Mater. Res.* (1994).
3.   I.K. Schuller, K. Fartash and M. Grimsditch, *MRS Bulletin* **XV**, 33 (1990).
4.   J.A. Ruud, A. Witvrouw and F. Spaepen, *J. Appl. Phys.* **74**, 2517 (1993).
5.   A.F. Jankowski, *Superlat. Microstruct.* **6**, 427 (1989).
6.   N. Nakayama, L. Wu, H. Dohnomae, T. Shinjo, J. Kim and C.M. Falco, *J. Magn. Magn. Mater.* **126**, 71 (1993).
7.   E.E. Fullerton, I.K. Schuller, H. Vanderstraeten and Y. Bruynseraede, *Phys. Rev. B* **45**, 9292 (1992).
8.   M.A. Wall and A.F. Jankowski, *Thin Solid Films* **181**, 313 (1989).
9.   B.M. Clemens and J.A. Bain, *Mat. Res. Soc. Bull.* **XVII**, 46 (1992).
10. J.A. Thornton and D.W. Hoffman, *Thin Solid Films* **171**, 5 (1989).
11. J.A. Bain, B.M. Clemens, H. Notarys, E.E. Marinero and S. Brennan, *J. Appl. Phys.* **74**, 996 (1993).

CALCULATION OF THE [111]-TEXTURE DEPENDENCE
OF THE ELASTIC BIAXIAL MODULUS

MARTHA K. SMALL* AND WILLIAM. D. NIX**
*Laboratoire de Métallurgie Physique, Université de Poitiers (URA CNRS 131), 86022 Poitiers
Cedex, France
**Department of Materials Science and Engineering, Stanford University, Stanford, California
94305

ABSTRACT

The elastic biaxial modulus of a polycrystalline material as a function of [111] fiber texture was
determined by imposing an equibiaxial strain and calculating the resulting stress assuming a plane-
stress condition. The situation of a such a texture in a film under plane-stress is commonly seen in
deposited thin films of face centered cubic metals. The calculation was performed by determining
the average biaxial modulus of a single crystal as a function of tilt angle off the [111] plane normal
and then using these values in a Gaussian distribution of crystal tilts off the [111] to simulate
varying degrees of texture. Results are presented for several fcc materials with different elastic
anisotropy factors.

INTRODUCTION

It is quite a common phenomenon for materials to develop a preferred crystallographic
orientation, or texture, in the course of growth or processing. As very few single-crystal materials
are elastically isotropic, a textured material will generally have directions of enhanced stiffness. In
the case of polycrystalline thin films, the film tends to grow by the addition of close-packed planes,
which creates a fiber texture perpendicular to the plane of the film: [111] for fcc and [110] for bcc
materials. Film texture can be verified by x-ray methods. Comparing the relative intensities of
peaks in a θ-2θ scan to their polycrystalline values gives an indication of the perfection of the
texture and a more precise measure can be obtained from a pole figure scan. However, the
stiffness of such a textured film depends on the direction dependence of individual elastic constants
and is difficult to estimate precisely.

In this paper we present a calculation of the [111]-texture dependence of the elastic biaxial
modulus. This is the relevant elastic modulus for thin films on substrates and results are presented
for several fcc metals.

TEXTURE-DEPENDENCE CALCULATION

The modulus of a single crystal is determined from Hooke's Law,

$$\sigma = C\varepsilon . \tag{1}$$

where σ and ε are the stress and strain tensors, respectively, and C is the stiffness matrix of the
material. There are two major steps in calculating the texture-dependence of an elastic modulus for
a polycrystalline material from single crystal elastic constants. The first is to determine the relevant
elastic modulus of the single crystal as a function of rotation relative to the preferred orientation.
This can be calculated exactly by transformation of the stiffness matrix, C, from the cube
coordinate system (CCS) to the rotated coordinate system (RCS). The second step is to determine
how the stiffnesses of an agglomeration of single crystals combine to give the modulus of a

textured polycrystal. For the latter one must make certain simplifying assumptions about the distribution of grains of a given orientation and about the stresses and strains in individual grains.

For this study we assume that each grain in the film is under the same unit equi-biaxial strain. This is the Voigt average. Though the actual strain state is undoubtedly more complicated, this meets the compatibility requirements and allows one to calculate the stresses from Eq. (1). Another simplifying assumption is that a thin film on a substrate is in a plane-stress condition, i.e., $\sigma_3=0$. The strain and stress tensors can thus be written as follows:

$$\varepsilon = \begin{pmatrix} 1 & 0 & 0 \\ 0 & 1 & 0 \\ 0 & 0 & \varepsilon_o \end{pmatrix} \quad \text{and} \quad \sigma = \begin{pmatrix} \sigma_1 & 0 & 0 \\ 0 & \sigma_2 & 0 \\ 0 & 0 & 0 \end{pmatrix}. \tag{2}$$

The biaxial modulus, Y, is defined as the biaxial stress divided by the biaxial strain. When the material is not isotropic in the plane then Y is given by its average in the principal stress directions, or, in this case,

$$Y = \frac{\sigma_1 + \sigma_2}{2}. \tag{3}$$

The solution to Eq. (3) is found by combining Eq. (1) in the RCS with Eq. (2). The calculation is in principle quite simple, but becomes cumbersome in practice due to the complexity of the resultant tensor manipulations. Mathematics software [1] developed for solving analytical expressions has made this problem tractable, particularly for rotations of specific coordinate systems in specific directions.

Stiffness Matrix in the Rotated Coordinate System

Using the four-index notation for the stiffness tensor, the stiffness in an arbitrary coordinate system can be calculated from the stiffness in the CCS using the following relationship:

$$C_{ijkl} = \ell_{im}\ell_{jn}\ell_{kp}\ell_{lq}C_{mnpq} \tag{4}$$

where ℓ_{ab} are the direction cosines between the axes in the two coordinate systems. Material elastic constants are reported in terms of the (6x6) stiffness matrix in the CCS, and these must be transformed into the fourth rank stiffness tensor using the following rules:

$$
\begin{aligned}
C_{iikk} &= C_{ik} \\
C_{ijkk} &= C_{9-i-j,k} \quad &\text{for} \quad & i \neq j \\
C_{iikl} &= C_{i,9-k-l} \quad &\text{for} \quad & k \neq l \\
C_{ijkl} &= C_{9-j-i,9-k-l} \quad &\text{for} \quad & i \neq j \text{ and } k \neq l
\end{aligned}
\tag{5}
$$

where each subscripted letter can take on values from 1 to 3. Using the above relations the biaxial modulus has been calculated for the (111), (110), and (100) planes [2]. Results for several fcc materials are listed along with polycrystalline biaxial moduli in Table 1.

Ideally, one would like to solve the problem of a texture film completely generally. However, in retaining all of the required parameters the solutions quickly increase in complexity until they are no longer readily manipulable. We therefore derive solutions for three distinct cases of rotation of the coordinate system, and these are illustrated in Fig. 1. For each case the initial coordinate system is the $[\bar{1}10]$, $[\bar{1}\bar{1}2]$, $[111]$, to be referred to as the [111]CS. In the figures,

562

Table 1. Biaxial moduli, in GPa, calculated for different crystal orientations with the polycrystalline value listed for comparison. Elastic constants found in ref's [3-5].

Table 1. Biaxial moduli, in GPa, calculated for different crystal orientations with the polycrystalline value listed for comparison. Elastic constants found in ref's [3-5].

Material	Y(100)	Y(110)	Y(111)	$Y = \dfrac{E}{1-v}$
Al	100.0	87.4	114.5	107.3
Ag	76.7	154.2	173.6	130.6
Au	78.0	120.3	188.6	139.3
Cu	114.8	182.7	261.0	197.6
Ni	217.8	290.0	389.3	290.0
Pd	130.2	255.1	287.8	198.4

the [111]CS and the CCS are indicated by the indices of their directions, and the RCS is denoted
by the directions α, β, and γ. Case 1 (Fig. 1a) is a rotation of the [111]CS about the [$\bar{1}$10] axis
with γ rotating towards the [110]. Case 2 is a rotation about the same axis, but in the opposite
sense (i.e., γ towards the [001]). The third case, shown in Fig. 1b, is rotation about the [$\bar{1}\bar{1}2$],
which is equivalent in either direction due to mirror symmetry. The solutions for the direction
cosines for Case 1 can be obtained from simple geometry and written as follows:

$$\ell_{\alpha x} = -\sqrt{2} \qquad \ell_{\alpha y} = \sqrt{2} \qquad \ell_{\alpha z} = 0$$

$$\ell_{\beta x} = \ell_{\beta y} = \frac{\sqrt{2}}{2}\sin[\theta - \delta] \qquad \ell_{\beta z} = \cos[\theta - \delta] \qquad (6)$$

$$\ell_{\gamma a} = \ell_{\gamma b} = \frac{\sqrt{2}}{2}\cos[\theta - \delta] \qquad \ell_{\gamma z} = -\sin[\theta - \delta]$$

where $\qquad \delta = \tan^{-1}\left(\sqrt{2}/2\right).$

The direction cosines for Case 2 are obtained from those of Case 1 by simply replacing θ with $-\theta$
in Eq's (6). The direction cosines for Case 3 are slightly more complicated and can be found in
ref. [6]. Using Eq's (4) to (6) one obtains the stiffness tensor in the RCS as a function of angle of
rotation. The 6x6 stiffness matrix in the RCS is then calculated by reversing the rules listed in Eq.
(5). These values can then be substituted into Eq. (1) to find the biaxial modulus as a function of
rotation.

Figure 2 shows plots of modulus variation with angle of rotation off the [111] calculated
using the elastic properties of Ag for each of the three cases. Case 2 is simply the reverse of Case
1, and Case 3 is almost exactly the average of the other two. To understand the shape of the
curves, one need only follow the RCS as γ rotates away from the [111]. Take Case 1 as an
example. At 0° the RCS coincides with the [111]CS and Y is at a maximum. At $\theta=\delta$, 35.3°, γ has
rotated to the [110] direction and the biaxial modulus is equal to the (110) value. The same rotation
again brings us back to [111], and an additional 54.7° brings the RCS into coincidence with the
[00$\bar{1}$] and to a sharp minimum in biaxial modulus. Case 3 rotation takes place on a plane which is
at equal angles from those for Cases 1 and 2. It is therefore not surprising that this solution is
close to the average of the other two.

An exact solution to the texture calculation would require knowing the integrated average of
biaxial moduli at all angles about the [111] of crystals rotated an angle θ off the [111]. As noted
above, such an exact solution is very computationally intense. However, the solutions we obtain

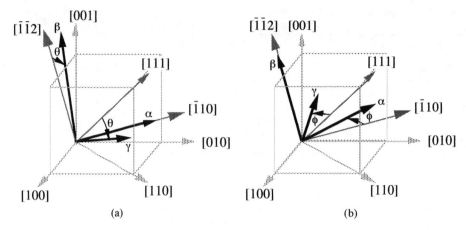

Figure 1. (a) Case 1 rotation (black lines) about the α-axis, shown in relation to the CCS (light gray) and RCS (dark gray). Case 2 rotation is the same, but in the opposite direction of Case 1. (b) Case 3 rotation, about the β-axis.

for a limited number of specific rotations indicate that the average of Cases 1 and 2 is representative of the average over all angles. We therefore use this average, shown in Fig. 3 for Ag, in subsequent steps of the calculation.

Texture Dependence of Biaxial Modulus

Having determined the orientation dependence of biaxial modulus for a single crystal, what remains to be found is the probability of finding a grain with a particular orientation. The probability function used to describe the distribution of planes in the film, the texture, is a

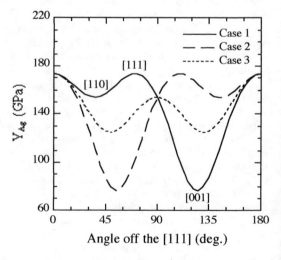

Figure 2. Biaxial Modulus of silver as a function of rotation of the plane normal, γ, off the [111] for Cases 1 through 3. The direction of γ at the peaks and valleys of the curve for Case 1 is also indicated.

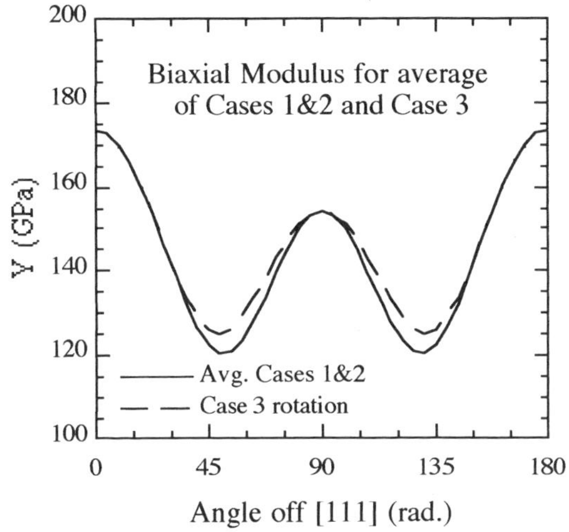

Figure 3. Biaxial Modulus of Ag as a function of rotation off the [111] for the average of Cases 1 & 2 with Case 3 plotted for comparison.

Gaussian normalized to a total integrated probability of 1 between zero and 90° of rotation of the surface normal off of the [111]:

$$f(\theta,\kappa) = \frac{\sqrt{2}}{\kappa\sqrt{\pi}\; Erf\left(\dfrac{\pi}{2^{3/2}\kappa}\right)} \exp\left(-\frac{\theta^2}{2\kappa^2}\right) \qquad (7)$$

where θ is the angle of rotation away from the [111] and κ is the standard deviation of the probability curve, which can be obtained from a pole figure x-ray scan. The biaxial modulus as a function of this deviation is simply the product of $f(\theta,\kappa)$ and the biaxial modulus integrated over the range of angles from zero to 90°:

$$Y(\kappa) = \int_0^{\pi/2} Y_{avg}(\theta) f(\theta,\kappa)\,d\theta \qquad (9)$$

where $Y_{avg}(\theta)$ is the average of the biaxial moduli calculated for Cases 1 and 2. A plot of $Y(\kappa)$ normalized by the polycrystalline value (see Table 1.) is shown in Fig. 4 for several fcc metals. The curves start at the (111) value for a κ of 0° and change very little over the first 5°. Most of the decrease occurs between 5 and 30°, and above 45° the modulus remains essentially constant.

The biaxial moduli should converge to the polycrystalline value at high values of κ (untextured film), but, as is usual for a Voigt average calculation [3], they are overestimated by between 5 and 20%. At higher κ's the grains are nearly randomly oriented and the variation in stiffness from grain to grain is accommodated by more complicated stress and strain states than are accounted for in the simple model. The solution increases in accuracy at smaller κ's as the (111)-plane normals align themselves in a narrow distribution of grain orientations about the preferred axis. In that case the assumption that all of the grains experience the same strain is a good one.

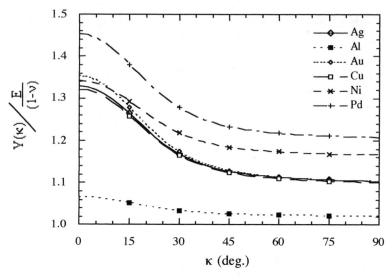

Figure 4. Texture-dependence of biaxial modulus normalized by the polycrystalline value for several fcc metals.

DISCUSSION AND CONCLUSIONS

In this paper we have described a relatively simple method for calculating the texture dependence of the biaxial modulus and calculated the results for the (111) texture dependence of several fcc metals. The results indicate that for narrow distribution of plane orientations ($\kappa < 5°$) the biaxial modulus is essentially equal to the (111) value. It then decreases to the polycrystalline value at a standard deviation of roughly 45°. The curves provided in Fig. 4 can be used to estimate the biaxial modulus as a function of texture when κ is known.

ACKNOWLEDGEMENTS

Financial support for this research was provided by DOE grant DE-FG03-89ER45387.

REFERENCES

1. Mathematica v. 2.0.3 enhanced, Wolfram Research, Inc. (1991).
2. W. D. Nix, *Met. Trans. A* **20A**, 2217-2245 (1989).
3. G. Simmons and H. Wang, *Single Crystal Elastic Constants and Calculated Aggregate Properties: A HANDBOOK* M.I.T. Press, Cambridge, MA, (1971).
4. M. A. Meyers and K. K. Chawla, *Mechanical Metallurgy: Principles and Applications* Prentice-Hall, Eaglewood Cliffs, (1984).
5. C. J. Smithell, *Smithells Metals Reference Book* (edited by E. A. Brandes). Butterworth & Co., London, (1983).
6. M. K. Small, Ph.D. Dissertation, Stanford University (1992).

ELASTIC PROPERTIES OF *FeNi/Cu* and *FeNi/Nb* METALLIC SUPERLATTICES INVESTIGATED BY BRILLOUIN LIGHT SCATTERING

G. CARLOTTI*, G. SOCINO*, HUA XIA**, AN HU** AND S. S. JIANG**
* Dipartimento di Fisica, Unità INFM, Università di Perugia, Via Pascoli, 06100 Perugia, Italy
** National Laboratory of Solid State Microstructures, Nanjing University, 210008 Nanjing P.R. China

ABSTRACT

The Brillouin light scattering technique has been exploited in order to investigate the elastic properties of periodic superlattices consisting of alternating layers of $Fe_{20}Ni_{80}$ (permalloy) and either Cu or Nb. These multilayers, with total thicknesses ranging between 0.2 and 0.7 mm and with periods of typically 3-5 nm, present a polycrystalline structure with (110) texture for the Nb layers and (111) texture for both the FeNi and Cu layers. Measurement of the frequency position of the Brillouin peaks corresponding to the Rayleigh and Sezawa acoustic modes allowed the effective elastic constants of these structures to be determined. The values obtained are compared with those calculated from the elastic constants of the bulk materials, taking into account the polycrystalline nature of the superlattices and the crystallographic orientation of the layers.

INTRODUCTION

The elastic properties of metallic multilayers and superlattices have been extensively investigated in the last decade, by use of both conventional elastic techniques and Brillouin light scattering. These studies have indicated the presence of deviations and anomalies with respect to predictions based on the effective medium description within the elastic continuum approximation [1]. In particular, a marked softening of some of the effective elastic constants has been observed in several superlattice systems whose constituents do not form solid solutions, such as Nb/Cu [2,3], V/Ni [4], Mo/Ni [5], Mo/Ta [6], Co/Au [7] and Ag/Ni [8]. More recently, an opposite behaviour, i.e. an appreciable increase of some of the effective elastic constants, has been observed in structures with partially miscible constituents, such as Ag/Pd [9] and Ta/Al [10]. These findings have stimulated the development of theoretical models which, in turn, attribute these anomalies to different physical mechanisms, such as electronic effects [11], coherency [12] or interface [13] strains, difference of the Fermi energy of the constituents [14] and atomic disorder at the interfaces [15]. However, the problem of a satisfactory understanding of the elastic response of metallic multilayers and superlattices is still open and further experiments are required in order to clarify this subject [16].

In this paper we present the results of a BS investigation of the elastic properties of FeNi/Cu and FeNi/Nb superlattices, which are of current interest in the field of magnetic media. Measurements were made on thin film superlattices supported by glass substrates and characterized by different periodicities. We could thus evaluate the effective elastic constants of these structures and compare them with the values expected from calculations based on the effective medium approximation, taking into account the microstructural properties of the films.

567

EXPERIMENTAL

The samples studied here consist of alternate layers of $Fe_{20}Ni_{80}$ (permalloy) and either Cu or Nb, whose elemental thicknesses will be denoted as d_{FeNi}, d_{Cu}, and d_{Nb}. They were deposited by dc sputtering on glass substrates, with periods p ranging between about 3 and 5 nm. X-ray diffraction experiments have shown that these superlattices are polycrystalline with <111> texture for $Fe_{20}Ni_{80}$ and Cu layers and <110> texture for Nb layers. A detailed list of the structural parameters of the specimens investigated, including the period p, the total thickness h and the mass density ρ, are reported in Tables I and II.

TABLE I Structural characteristics of the FeNi/Cu specimens investigated

sample	d_{FeNi} (nm)	d_{Cu} (nm)	p (nm)	h (nm)	ρ (gm/cm^3)
FNC34	1.5	2.0	3.5	640	8.849
FNC35	2.5	2.0	4.5	660	8.845
FNC30	3.0	2.0	5.0	440	8.804

TABLE II Structural characteristics of the FeNi/Nb specimens investigated

sample	d_{FeNi} (nm)	d_{Nb} (nm)	p (nm)	h (nm)	ρ (gm/cm^3)
FNN1	1.5	2.5	4.0	235	8.618
FNN2	1.5	1.3	2.8	300	8.640
FNN3	2.0	1.8	3.8	190	8.638

Brillouin scattering experiments were performed in air, at room temperature, by means of a tandem triple pass, Sandercock-type Fabry-Perot interferometer [17, 18], characterized by a contrast ratio higher than 10^{10} and a finesse of about 120. Both the incident and the scattered light were polarized in the plane of incidence (p-p scattering). Measurements have been taken in the back-scattering interaction geometry; in this condition the wave vector Q of surface phonons coming into the scattering process is fixed by the conservation of momentum $Q=2k\sin(\theta)$, where k is the optical wavenumber and θ is the angle of incidence of light. From this relationship the phase velocity v of surface phonons can be expressed in terms of the frequency shift f of the corresponding Brillouin peaks as follows: $v = \dfrac{\pi f}{k\,\sin(\theta)}$.

ACOUSTICS OF SUPERLATTICE FILMS

Polycrystalline metallic films, exhibiting a preferential orientation, are modelled as hexagonal acoustic media, with the c-axis parallel to the growth direction. In fact, due to the random orientation of the crystallites in the surface plane, the structure is isotropic in this plane and five independent effective elastic constants are required to describe the system. Referring to a coordinate system with the x_3 axis perpendicular to the superlattice layers, these independent constants are c_{11}, c_{12}, c_{13}, c_{33} and c_{44}. Their expected values can be calculated from the single crystal elastic constants of the two superlattice constituents by means of a two step procedure. First, the Reuss or Voigt [19] average of the elastic constants around the crystallographic orientation axis (the (111) axis for FeNi and Cu, the (110) axis for Nb) is performed; then the effective elastic constants of the superlattice are obtained from those of the separate constituents [20], by imposing the continuity of stress and strain across the interfaces [21]. One thus obtains direct expressions of the effective elastic constants of the multilayer in terms of those of the two constituents A and B. Specifically, for the shear constant c_{44} the relationship is particularly simple:

$$c_{44} = p \left[\frac{d_A}{c_{44}^A} + \frac{d_B}{c_{44}^B} \right]^{-1} \qquad (1)$$

In Table III we report the calculated values of the shear effective elastic constant c_{44} for the FeNi/Cu and FeNi/Nb superlattices analysed here.

An experimental determination of these constants can be achieved by means of Brillouin spectroscopy. In BS experiments one measures the surface phonon spectrum, in the long wavelength limit, at a fixed value of the wave vector [18]. This spectrum strongly depends on the thickness of the superlattice film under investigation. When the total thickness h is comparable with the value of the phonon wavelength Λ, discrete guided acoustic modes are present in the spectrum, whose number increases with the ratio h/Λ [22]. One can thus measure their phase velocity dispersion curves and obtain information about the values of the effective elastic constants. To this respect, previous investigations on thin metallic films have shown that the contribution of different constants to the guided modes can be very similar, so that they cannot be determined unambiguously or are affected by a rather large error. We notice, however, that for values of the ratio h/Λ higher than about 1.5, the shear elastic constant c_{44} can be selectively determined because it is well decoupled from the other constants, its influence on the Rayleigh wave velocity being much stronger than the influence of the other three constants [8].

TABLE III Values of the shear effective elastic constant of the FeNi/Cu and FeNi/Nb superlattices analysed, as calculated by means of the Voigt (Reuss) average, within the effective-medium approximation.

	FNC34	FNC35	FNC30	FNN1	FNN2	FNN3
c_{44} (GPa)	49.7 (38.0)	53.2 (40.9)	54.5 (42.1)	49.9 (43.1)	53.9 (45.9)	53.7 (45.7)

RESULTS AND DISCUSSION

Figure 1 presents typical Brillouin spectra relative to both a *FeNi/Nb* and a *FeNi/Cu* specimen; in the former case (*FNN2*) the value of the ratio h/Λ is about 1.08 and only the Rayleigh and the first Sezawa mode are revealed in the spectrum; in the latter case (*FNC35*) the film thickness is larger so that $h/\Lambda=2.37$ and several discrete guided modes are revealed, up to the fifth Sezawa mode (note that in this case the first Sezawa mode merges into the Rayleigh peak). In order to achieve a determination of the effective elastic constants of the FeNi/Cu superlattices, we have measured the phase velocity of the guided acoustic modes for different

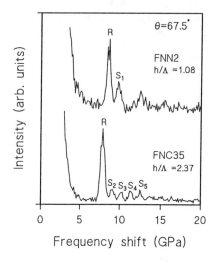

Fig. 1 Brillouin spectra relative to a FeNi/Nb (FNN2) and a FeNi/Cu (FNC35) specimens, for an angle of incidence $\theta=67.5°$

ratios h/Λ, by changing the angle of incidence of light. Two spectra relative to the specimen *FNC30* are shown in Figure 2a. The experimental values of these velocities are reported in Figure 2b, together with the dispersion curves obtained by a best fit procedure, using the effective elastic constants of the superlattice as free parameters. To this respect, we notice that in the range of h/Λ investigated the shear effective constant c_{44} can be easily determined, because it is decoupled from the other constants, due to its almost exclusive influence on the Rayleigh mode; on the other hand, c_{11}, c_{13}, and c_{33} are affected by a rather large error (about 10%), becouse they influence the acoustic modes in a similar way. In the case of the FeNi/Nb systems, due to the reduced total thickness of the superlattice films, only the Rayleigh and the first Sezawa mode were revealed, so that there was not enough information to determine the whole set of elastic constants. Only the shear constant c_{44} could be determined in this case, due to its dominant influence on the Rayleigh mode. The experimental values of the elastic constants for both FeNi/Cu and FeNi/Nb are shown in Table IV. On comparing the experimental values of c_{44} with the theoretical values of Table III, obtained from the effective medium approximation, it turns out that there is a substantial agreement between theory and experiment. In particular, there is an increase of this constant, as expected, when the fraction of FeNi in the structure increases with respect to the Cu (or Nb) fraction, while a change of the value of the superlattice period seems to be uneffective on modifying the elastic constants. A similar ideal behavior, characterized by the independence of the elastic muduli from the multilayer periodicity, has been previously observed in the systems Fe/Pd [23], Cu/Ni [24], Cu/Pd [25], and Cu/Co [26].

CONCLUSIONS

In summary, we have investigated the elastic properties of sputter deposited *FeNi/Cu* and *FeNi/Nb* superlattices, using the Brillouin light scattering technique. The effective elastic constants of these structures have been determined from measurement of the frequency position of the Brillouin peaks corresponding to the Rayleigh and Sezawa acoustic modes. The values obtained are in fairly good agreement with those calculated from the elastic constants of the bulk materials, taking into account the polycrystalline nature of the superlattices and the crystallographic orientation of the layers. We thus conclude that, different from a number of metallic superlattices previously investigated, the elastic behavior of the *FeNi/Cu* and *FeNi/Nb* systems shows no evidence of appreciable elastic anomaly.

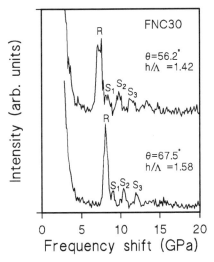

Fig. 2a Brillouin spectra relative to the FNC30 specimen, for two different angles of incidence of light.

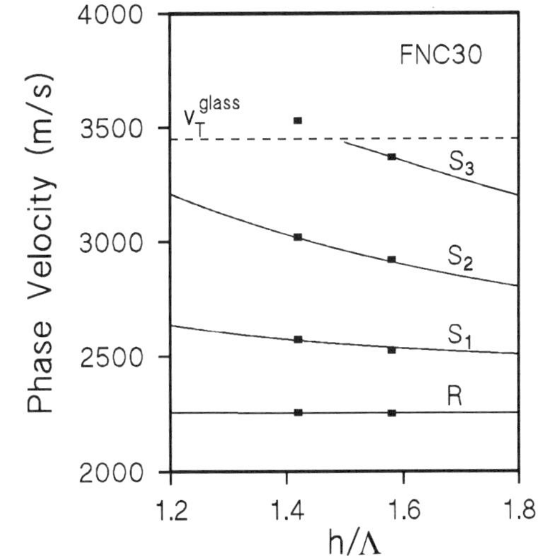

Fig. 2b Calculated dispersion curves (solid lines) and experimental values of the phase velocity of the Rayleigh and Sezawa modes for the specimen FNC30.

TABLE IV Experimental values of the effective elastic constants of the superlattices analyzed. The uncertainty in the determination of c_{11}, c_{13} e c_{33} is of the order of 10%, because of the critical interdependence among them.

sample	c_{11} (GPa)	c_{13} (GPa)	c_{33} (GPa)	c_{44} (GPa)
FNC34	188	98	207	47±1
FNC35	190	98	210	48±1
FNC30	193	98	213	51±1
FNN1	....	....	....	52±1
FNN2	....	....	....	53±1
FNN3	....	....	....	54±1

ACKNOWLEDGMENTS

This research was supported by the Consorzio Interuniversitario di Fisica della Materia (Italy) as well as by the National and Jiangsu provincial Natural Science Foundations of China.

REFERENCES

[1] M. Grimsditch, in *Light Scattering in Solids V,* edited by M. Cardona and G. Guntherodt (Springer, Berlin, 1989), p. 285

[2] J.A. Bell, W.R. Bennet, R. Zanoni, G.I. Stegeman, C.M. Falco and C.T. Seaton, Solid State Commun. **64**, 1339 (1987)

[3] G. Carlotti, D. Fioretto, L. Palmieri, G. Socino, L. Verdini, H. Xia, A. Hu, X.K. Zhang, Phys. Rev. B, **46**, 12777 (1992)

[4] R. Danner, R.P. Huebener, C.S.L. Chun, M. Grimsditch and I.K. Schuller, Phys. Rev. B, **33**, 3696 (1986)

[5] R. Khan, C.S.L. Chun, G.B. Felcher, M. Grimsditch, A. Kueny, C.M. Falco and I.K. Schuller, Phys. Rev. B, **27**, 7186 (1983)

[6] J.A. Bell, W.R. Bennet, R. Zanoni, G.I. Stegeman, C.M. Falco and F. Nizzoli, Phys. Rev. B, **35**, 4127 (1987)

[7] B. Hillebrandt, P. Krams, K. Sporl and D. Weller, J. Appl. Phys., **69**, 938 (1991)

[8] G. Carlotti, D. Fioretto, G. Socino, B. Rodmacq and V. Pelosin, J. Appl. Phys., **71**, 4897 (1992); G. Carlotti, D. Fioretto, L. Giovannini, G. Socino, V. Pelosin and B. Rodmacq, Solid State Commun. **6**, 487 (1992)

[9] J.R. Dutcher, S. Lee, J. Kim, G.I. Stegeman and C.M. Falco, Phys. Rev. Lett., **65**, 1231 (1990)

[10] Hua Xia) X.K. Zhang, An Hu, S.S. Jiang, R.W. Peng, Wei Zhuang, Duan Feng, G. Carlotti, D. Fioretto, G. Socino and L. Verdini, Phys. Rev. B, **47**, (1993); G. Carlotti, G. Socino, An Hu, Hua Xia and S.S. Jiang, J. Appl. Phys., **75** (1994)

[11] W.E. Pickett, J. Phys. F: Met. Phys., **12**, 2195 (1982)

[12] A.F. Jankowski, J. Phys. F **18**, 413 (1988)

[13] R.C. Cammarata and K. Sieradzki, Phys. Rev. Lett., **62**, 2005 (1989)

[14] M.L. Huberman and M. Grimsditch, Phys. Rev. B, **62**, 1403 (1989)

[15] J.A. Jaszczak, S.R. Phillipot and D. Wolf, J. Appl. Phys., **68**, 4573 (1990)

[16] A comprehensive phenomenological model has been recently proposed by M. Grimsditch, E.E. Fullerton and I.K. Schuller, Mat. Res. Soc. Proc., Vol. 308, (1993), p. 685

[17] G. Carlotti, D. Fioretto, L. Palmieri, G. Socino, L. Verdini and E. Verona, IEEE Transaction on Son. Ultrason. and Freq. Control, **38**, 56 (1991)

[18] F. Nizzoli and J.R. Sandercock, in *Dynamical Properties of Solids,* edited by G.K. Horton and A.A. Maradudin (North-Holland, Amsterdam, 1990), Vol. 6, p 307

[19] M.J.P. Musgrave, <u>Crystal Acoustics</u>, (Holden Day, San Francisco, 1970), p.177

[20] Landolt-Bornstein, *Numerical Data, New Series, Group III* (Springer, Berlin, 1979), Vol.11, p.79

[21] M Grimsditch and F. Nizzoli, Phys. Rev. B, **33**, 5891 (1986)

[22] G.W. Farnell and E.L. Adler, in *Physical Acoustics,* edited by W.P. Mason and R.N. Thurston (Academic, New York, 1972), Vol. 9, p.35

[23] P.Baumgart, B. Hillebrands, R. Mock, G. Guntherodt, A. Boufelfel and C.M. Falco, Phys. Rev. B., **34**, 9004 (1986)

[24] J. Mattson, R. Bhadra, J.B. Ketterson. M. Brodsky and M. Grimsditch, J. Appl. Phys., **67**, 2873 (1990)

[25] A. Moreau, J.B. Ketterson and B. Davis, J. Appl. Phys., **68**, 1622 (1990)

[26] J.R. Dutcher, S. Lee, C.D. England, G.I. Stegeman and C. M. Falco, Mat. Sci. Eng. A**126**, 13 (1990)

INTERNAL STRESS CHANGE OF PHOSPHORUS-DOPED AMORPHOUS SILICON THIN FILMS DURING CRYSTALLIZATION

HIDEO MIURA AND ASAO NISHIMURA
Mechanical Engineering Research Laboratory, Hitachi, Ltd., 502 Kandatsu,Tsuchiura, Ibaraki 300, Japan

ABSTRACT

Internal stress change of phosphorus-doped silicon thin films during crystallization is measured by detecting substrate curvature change using a scanning laser microscope. The films are deposited in an amorphous phase by chemical vapor deposition using Si_2H_6 gas. The deposited films have compressive stress of about 200 MPa. The internal stress changes significantly to a tensile stress of about 800 MPa at about 600°C due to shrinkage of the films during crystallization. The high tensile stress can be relaxed by annealing above 800°C. The phosphorus doping changes the crystallization process of the films and their final residual stress.

INTRODUCTION

Polycrystalline silicon thin films are widely used in semiconductor devices. In metal oxide semiconductor (MOS) transistors, these films are used as a gate electrode material. Recently, polycrystalline films have been formed by crystallizing phosphorus-doped amorphous films to obtain thin films of about 100 nm, in which phosphorus is doped uniformly. However, dislocations were often observed at the gate edges in the single crystal substrates [1]. Once dislocations are introduced in a transistor structure, the device reliability decreases significantly because of increased leakage current. Therefore, dislocations must be eliminated in transistor structures to improve device reliability.

During film crystallization, it was found that the internal stress of silicon thin films without phosphorus doping changed from compressive to tensile [2,3]. The magnitude of the stress change sometimes exceeds 1000 MPa [4]. Such a stress change may cause dislocations in the substrate. This is because the strength of the single crystal silicon substrate decreases drastically at temperatures above 600°C and crystallization occurs at that temperature. Therefore, it is very important to make clear the internal stress change of the phosphorus-doped silicon thin films. In this paper, the internal stress change of the phosphorus-doped amorphous silicon thin films was measured during annealing, especially during crystallization of the film.

EXPERIMENTAL METHOD

Phosphorus-doped amorphous silicon films about 550 nm thick were deposited on thermally oxidized single crystal silicon substrates and on substrates without an oxide layer by chemical vapor deposition using Si_2H_6 gas at about 525°C. The dopant concentration was varied from 0 to $4 \times 10^{20}/cm^3$. Test strips were cut from the substrates, 30 mm long and 4 mm wide, along the

573

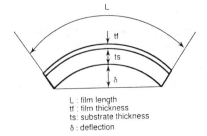

L : film length
tf : film thickness
ts: substrate thickness
δ : deflection

Fig. 1 Stress measurement method

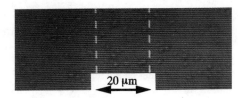

20 μm

Fig. 2 Scanning laser micrograph of
a partially crystallized film

<110> crystallographic direction of the substrate.

The internal stress of the film was determined by measuring the deflection of the test piece using a scanning laser microscope [5], as shown in Fig. 1. The microscope measures the deflection of a sample by detecting the change in the focal point of the scanned incident laser beam. Measurement resolution was about 0.1 μm when an Ar-ion laser of 488 nm was used. Assuming uniform film stress, the internal stress of the film (σ_f) was calculated using the following equation.

$$\sigma_f = 4E_s t_s^2 \delta / \{3L^2 (1-v_s) t_f\} \tag{1}$$

Here, E_s and v_s are Young's modulus and Poisson's ratio of the substrate, t_s and t_f are the thicknesses of the substrate and the film, L is the length of the test strip, and δ is the deflection of the test piece. In this study, L was 30 mm, t_s was 0.55 mm, and t_f was 550 nm. By substituting Young's modulus of 170 GPa, Poisson's ration of 0.06 and the δ measurement resolution of 0.1 μm, into Eq. (1), the stress measurement resolution was determined to be about 15 MPa.

The test piece was annealed in an infrared furnace in flowing argon gas for 30 minutes at temperatures between 400°C and 900°C. The deflection of the test piece was measured continuously during annealing. Crystallization was also observed during annealing by detecting the reflex index change of the silicon against the incident laser beam from the amorphous phase to the polycrystalline phase. An example scanning laser micrograph is shown in Fig. 2. The white particles are the newly grown single crystal particles. During observation of the crystallization process, the internal stress change of the film was measured.

RESULTS

Internal stress change in the phosphorus-doped silicon films during annealing is summarized in Fig. 3. The internal stress of the amorphous films was compressive. The stress value of 200 MPa was almost constant, regardless of the dopant concentration. The internal stress of the films did not change even when the film was heated up to 550°C. With continued heating, crystallization started at various temperatures, depending on the dopant concentration. The crystallization temperature decreased from about 630°C to 580°C as the dopant concentration was increased. However, the magnitude of the internal stress change was almost constant at 1000 MPa, i.e., the internal stress reached a tensile stress of about 800 MPa. The tensile stress decreased with

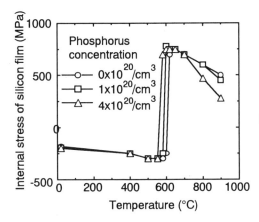

Fig. 3 Internal stress changes of
phosphorus-doped silicon films

annealing at temperatures above 800°C. The stress relaxation rate also depended on the dopant concentration. The final residual stress of the film without the dopant after 900°C annealing was about 500 MPa, while that with the dopant of $4\times10^{20}/cm^3$ was about 250 MPa.

Stress development process during crystallization was caused by volume shrinkage of the films. The shrinkage rate reached about 0.6%. Assuming that the volume shrinkage causes the stress, the predicted stress value, which is calculated by multiplying the volume shrinkage rate by Young's modulus of the film, reaches about 1000 MPa. This value agrees very well with the measured data.

Stress relaxation with high temperature annealing was caused by the viscoelastic or plastic characteristics of the silicon films. Since the critical resistance of silicon against plastic deformation at high temperatures (above 700°C) decreases with phosphorus doping, the internal stress of the phosphorus-doped films is relaxed due to these characteristics.

The reason that the crystallization temperature decreases with increases in the dopant concentration is another issue. The doped phosphorus probably plays an important role. To analyze the crystallization process in detail, the internal structure of partially crystallized films was observed using a transmission electron microscope. A cross-sectional micrograph of the film with a dopant concentration of $4\times10^{20}/cm^3$ annealed at 580°C for 10 minutes is shown in Fig. 4. The nucleation and growth of single crystal particles started at the interface of the amorphous film and the base SiO_2. Even if the doped phosphorus affects the crystallization process, only the phosphorus existing at the interface has the most important role. To confirm this assumption, a very thin undoped layer about 5 nm thick was deposited on the thermal oxide and then a phosphorus doped film of about $4\times10^{20}/cm^3$ was deposited on this undoped layer. The internal stress change of this film with annealing is compared with that of the film without the undoped layer in Fig. 5. The crystallization temperature of this film with the undoped layer was about 620°C, while that of the film without the undoped layer was about 580°C. The crystallization temperature of the film with the undoped layer agreed well with that of the film without phosphorus doping. Besides, the crystallization-induced stress of the film with the undoped layer increased to about 1200 MPa. This stress increase was due to the increase in the average grain size of the crystallized film. The average grain size of the film with the undoped layer was about 5 μm,

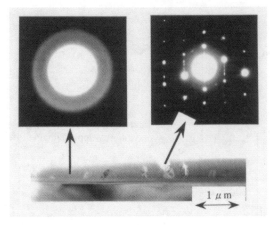

Fig. 4 Cross-sectional transmission micrographs of a partially crystallized films

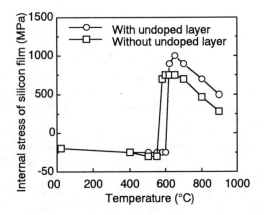

Fig. 5 Effect of phosphorus existing at the substrate interface on the internal stress of the films

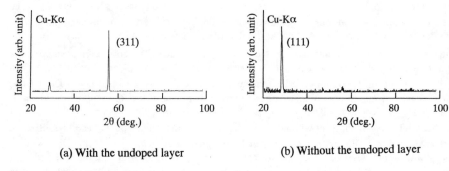

(a) With the undoped layer

(b) Without the undoped layer

Fig. 6 X-ray diffraction profile of fully annealed phosphorus-doped amorphous silicon thin films

Fig. 7 Transmission electron
micrographs of grown
particles in the
phosphorus-doped films

1 μm

0.5 μm

Fig. 7(a) without the undoped layer   Fig. 7(b) with the undoped layer

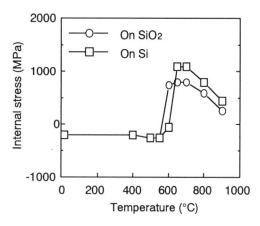

Fig. 8 Substrate effect on internal
stress change of phosphorus-doped
silicon thin films. Phosphorus
concentration is 4×10^{20}/cm^3.

while it was about 2 μm in the film without the undoped layer. The former film showed about 0.7% volume shrinkage during crystallization. According to the stress development process mentioned before, this shrink rate causes about 1200 MPa. This predicted value also agrees well with the measured data. This result clearly shows that the crystallization temperature decreases with phosphorus doping.

In addition, x-ray diffraction analysis showed that a (311) crystallographic plane mainly grew in the film with the undoped layer, while a (111) plane mainly grew in the film without the layer as shown in Figs. 6(a) and 6(b), respectively. The grown particles were oval in the film with the undoped layer, as shown in Fig.7(a). Each oval particle was single crystal. On the other hand, they were flower-like in the film without the layer, as shown in Fig. 7(b). The flower-like particle was polycrystalline. However, each "petal" of the particle was single crystal. Energy dispersive x-ray (EDX) analysis showed that phosphorus was highly concentrated at the center of the flower-like particle. The maximum concentration was about 50% higher than the average concentration. This result indicates that phosphorus played an important role on the initial nucleation of the particle.

The crystallization process and the internal stress of the film were also affected by the base material. Figure 8 shows the internal stress changes of the films deposited on SiO_2 and the chemically rinsed substrate surface. The dopant concentration was $4 \times 10^{20}/cm^3$ in both films. The initial internal stress of the amorphous films was almost constant at -200 MPa. This compressive stress did not change before crystallization. Crystallization started at about 610°C in the film deposited on the rinsed substrate, while it started at 580°C in the film deposited on SiO_2. The internal stress change of the former film reached about 1400 MPa and was about 1000 MPa in the latter film. This was because the average grain size of the film directly deposited on the substrate was about three times that of the film deposited on SiO_2. The grown polycrystalline film on the rinsed substrate showed strong x-ray diffraction in the (111) crystallographic direction of silicon. Thus, it is concluded that the base material on which the amorphous silicon film is deposited changes the crystallization process and the internal stress of the film.

Further study is necessary to clarify the effects of the base material and phosphorus existing at the interface of the film and the base on the crystallization process of phosphorus-doped amorphous silicon films. However, our results have shown that it is important to use films with a higher dopant concentration and to anneal the films fully in order to minimize their internal stress. To control the crystallographic direction of the film, the dopant concentration must be controlled at the interface of the film and the base material.

SUMMARY

Internal stress change of phosphorus-doped amorphous silicon film during annealing was measured. The initial internal stress of the amorphous film was compressive at about 200 MPa. It changed significantly to a tensile stress of about 800 MPa during crystallization at about 600°C. This stress change occurred because of shrinkage of the film during crystallization. The crystallization process was determined by the doped phosphorus, especially that existing at the interface of the film and the base material. The final residual stress of the polycrystalline film after full annealing decreased with increases in the dopant concentration.

ACKNOWLEDGEMENTS

The authors would like to thank Dr. Eiji Takeda, Dr. Toru Kaga and Mr. Takashi Kobayashi of the central research laboratory, Hitachi, Ltd. for their helpful discussions and sample preparation.

REFERENCES

1. H. Miura, N. Saito and N. Okamoto, in Simulation of Semiconductor Devices and Processes Vol. 5, edited by S. Selberherr and H. Stippel (Springer-Verlag, Wien, 1993), p. 177.
2. S. P. Murarka and T. F. Retajczyk, J. Appl. Phys., 54, 2069 (1983).
3. J. Admaczewska and T. Budzynski, Thin Solid Films, 113, 271 (1984).
4. H. Miura, H. Ohta, N. Okamoto and T. Kaga, Appl. Phys. Lett., 60(22), 2746 (1992).
5. D. Awamura, T. Ode and M. Yonezawa, SPIE, 765, 53 (1987).

INTRINSIC STRESS AND MICROSTRUCTURAL EVOLUTION IN SPUTTERED NANOMETER SINGLE AND MULTILAYERED FILMS

TAI D. NGUYEN

Center for X-Ray Optics, Lawrence Berkeley Laboratory, University of California, Berkeley, CA 94720.

ABSTRACT

The relationship of intrinsic stress and microstructural evolution in nanometer thick Mo and Si films, and Mo/Si multilayers deposited by magnetron sputtering at low working pressure (2.5 mTorr) is studied. The stress depends strongly on the microstructure which evolves with the film thickness. Transition from tensile to compressive films is observed in the metal films, in which nucleation and columnar grain growth occur. Deposition of layered Mo films by time-delayed sequential sputtering of thin layers results in smaller grains that do not extend through the film thickness, and in more tensile stress state than thick films of the same thickness. The Si films are highly compressive at all thicknesses studied. The multilayers in this study show compressive stresses, with higher compressive stress at longer periods, and decreasing stress at shorter periods. The interface stress in amorphous Mo/Si multilayers is determined to be 1.1 J/m^2. Comparison with values in other systems is made.

INTRODUCTION

Sputtered films and multilayers usually exhibit the presence of intrinsic stress in the structures. Stress in multilayer mirrors are undesirable in many x-ray optical applications where flat mirrors, or precise control of curved mirrors, are designed. Controlling of the stress in these multilayers requires understanding of stress and microstructure, and their relationship, in multilayers and thin films. Stress and microstructure changes in thin films with working pressure have been studied extensively (for example, references 1-5). In this paper, intrinsic stress and microstructural evolution with thickness in single and multilayered thin films at low working pressure (2.5 mTorr), which supposedly provides smooth interfaces for optimum performance in nanometer period x-ray mirrors, are studied.

Stress usually arises from the strain or misfit that must be accommodated elastically when a film is deposited on a substrate. Various types of strains include growth strains, thermal strains, and epitaxial strains. If the density of the film changes after it has been deposited on the substrate, an intrinsic or growth stress develops in the film. Thermal strains result from the different thermal expansions between the film and the substrate when there is a change in temperature. Subsequent cooling from deposition of the film at high temperatures, for example, results in a stress in the film if the film and substrate have different thermal expansion coefficients. Other processes that lead to volume changes due to densification, such as phase transformations, composition changes, and annihilations of excess vacancies, dislocations, and grain boundaries, also produce stress in the film. Epitaxial strains arise from the natural misfit between the film and the substrate lattices in their stress-free state.

In the case of sputtered films and multilayers in this study, the films and the nanometer-thick layers in the multilayers are predominantly amorphous or polycrystalline, and thus epitaxial strains are not a source of stress in these structures. Different microstructures in different thicknesses of the films, and phase transformation and interdiffusion between the layers in the multilayers, produce changes in the stress state of the films and multilayers.

DETERMINATION OF STRESS IN THIN FILMS

Stress in thin films can be determined by the amount of bending or curvature of the substrate caused by the films. For a film deposited on a flat substrate, stress in the film is measured by using the laser scanning technique.[6] In this technique, reflection of a laser light determines the curvature of the film and substrate as the laser beam scans through the sample.

For the usual case of a substrate that is elastically isotropic in the plane of the film, the biaxial stress in the film is given by the Stoney equation:[7]

$$\sigma = \frac{E_s t_s^2}{6(1-\gamma)t} K \tag{1}$$

for the case in which the total thickness of the films is much less than that of the substrate, where E and γ are the substrate biaxial modulus and Poison ratio, t_s and t_f are the substrate and film thicknesses, and K is the substrate curvature. For a substrate that has an initial curvature before the film is deposited, then the substrate curvature in the equation is modified to ΔK, the change in the curvature before and after the film is deposited.

STRESS IN PERIODIC MULTILAYER STRUCTURES

Many analyses have been given for the case of a thin film,[7] thick film,[8-9] bilayer strip,[10] thin film on a plate,[11] and multilayer thin films.[12] A detail analysis of the elastic relation for a composite of layers with equal and different Young's moduli has also been given by Townsend et al.[13] In a multilayer stack, the thin film approximation assumes that the interaction between the layers is negligible. The average stress in a multilayer stack is:

$$\sigma = \frac{E_s t_s^2}{6(1-\gamma)t} \cdot \sum_i K_i = \frac{E_s t_s^2}{6(1-\gamma)t} \cdot \sum_i \frac{6(1-\gamma)t_i}{E_s t_s^2} \sigma_i = \frac{1}{t} \cdot \sum_i t_i \sigma_i , \tag{2}$$

in which t is the total thickness of the multilayer structure. The equation above assumes that the interactions between the layers are negligible. Each layer is assumed to cause a fixed bending amount K_i to occur, independent from the bending caused by adjacent layers. For a periodic multilayer stack of N bilayers of thickness $(t_A + t_B)$, then $t = N \cdot (t_A + t_B)$, so:

$$\sigma = \frac{1}{N \cdot (t_A + t_B)} \cdot N \cdot (t_A \sigma_A + t_B \sigma_B) = \frac{t_A \sigma_A + t_B \sigma_B}{t_A + t_B} . \tag{3}$$

The average stress thus is independent of the bilayer period, d, and the number of bilayers N, and depends on the layer stress and the relative thickness of the layers.

The derivation above assumes that the layer characteristics or microstructures and interfaces are identical for all thicknesses of the layers. It ignores the different stress of the first layer on the oxidized substrate, which is different from the stress in subsequent layers, and the strain relief that subsequent layers have on the already deposited layers. The contribution of the interfaces to the stress of the multilayers has also been neglected, which can be included by an additional term to equation 3. The interface stress contribution is proportional to the number of interfaces in a multilayer structure, or to the density or frequency of the interfaces in the structure, or equivalently, is inversely proportional to the period of the multilayers. Assuming that the bulk stress does not depend on the thickness of the layers, or in a thickness (or multilayer period) range that the stress is almost constant, the contribution of stress from the interfaces can be determined from the slope of the average stress vs. inverse period plot.[14]

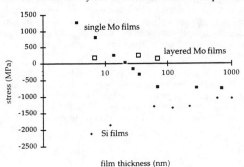

Figure 1. Stress evolution in single Mo (filled squares), layered Mo (open squares), and Si films (diamonds).

580

EXPERIMENTAL TECHNIQUES

Thin films and multilayers were prepared by magnetron sputtering at floating temperature onto oxidized (111) 3-inch Si wafer substrates. The sputtering pressure was 2.5 mTorr. The thickness of the single films studied ranged between a few nanometers and one micron. The multilayers were prepared such that the ratio of the metal layer thickness to the bilayer period is approximately 0.35. They were made by rotating the substrates sequentially over two materials targets, while different thickness single films were prepared by leaving the substrates stationary over the targets for multiple periods of time. For layered Mo films, thin layers of Mo were deposited with the delay time of 30 seconds, which is similar to the time required for the substrates to make one revolution in the deposition of the multilayers. The film thicknesses and the multilayer periods were determined by low angle x-ray diffraction.

Characterization of the microstructures in the multilayers was performed by High Resolution Transmission Electron Microscopy (HRTEM). The TEM specimens were prepared by mechanical thinning, followed by ion beam milling.[15] They were then studied in a Topcon 002B microscope operating at 200kV. Film stresses were measured in a commercial Tencor FLX stress measurement system. In this experiment, the thickness of the single films and multilayers is significantly less than the thickness of the substrate (381 μm) that the flexural modulus of the film and substrate is dominated by the properties of the substrate, and thus formulation for thin films can be applied.

RESULTS

The evolution of stress in Mo and Si films is shown in Figure 1 for thicknesses ranging from 5 nm to approximately one micron. Thin Mo films are tensile and become compressive as the thickness increases and remains constant at -750 MPa when the thickness reaches 70 nm. Transition between tensile and compressive occurs near 30 nm thickness. The Si films are compressive at all thicknesses, with the stress highly compressive at small thickness, and approach an asymptote at $\sigma = -1200$ MPa. The evolution of stress in Mo is typical of that found in elemental metal films studied (W, Ru, and Cr), and that in Si is typical of that found in amorphous non-metallic films (C, SiC, and B_4C). Each film however has different rates of stress change with thickness, and different asymptotic stress values at thick films.

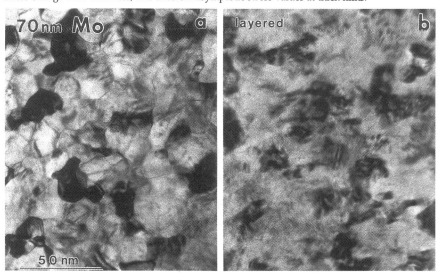

Figure 2. Plan-view TEM images of a) single, and b) layered 70 nm thick Mo films. The layered film contains 40 layers of approximately 1.8 nm thick.

Also shown in Figure 1 are the stress values of 3 Mo layered films prepared by 4, 20, and 40 sequential layers of approximately 1.8 nm thickness. All the layered films exhibit tensile stress, in contrast to the compressive stress found in single Mo films having comparable thicknesses to the 20- and 40-layer Mo films. Transmission electron microscopy of the 40-layer Mo film, and of the single film having similar thickness, was performed to study the microstructure in the films. Plan-view TEM images of the samples show that the grains in the single Mo films are much larger than those in the layered Mo films (Figure 2). Cross-sectional HRTEM image of the single film show the columnar structure of the Mo grains that grow typically from the substrate to the surface of the film, while most grains in the layered films do not extend from the substrate to the film surface.

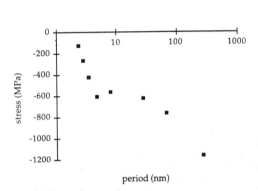

Stress variation with period in Mo/Si multilayers is shown in Figure 3. The total thickness of the multilayers is kept constant at 280 nm, while the period and number of bilayer periods vary from 2.3 to 280 nm, and 120 to 1, respectively. As can be seen from the figure, the compressive stress in the multilayers remains relatively constant near -600 MPa for periods between 7 and 28 nm. The compressive stress, however, increases as the period increases above 28 nm, or N decreases below 10, and decreases as the period decreases below 7 nm, or N increases above 40.

Figure 3. Stress variation with period in Mo/Si multilayers. Total thickness = 280 nm.

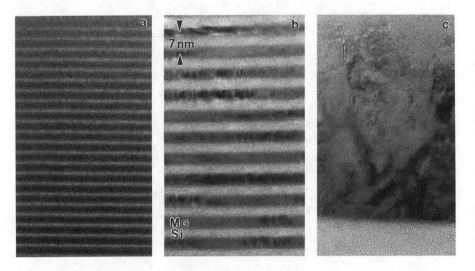

Figure 4. Cross-sectional HRTEM images of a) 3.5, b) 7, and c) 280 nm period Mo/Si multilayers (corresponding to those arrowed in Figure 3).

The evolution of stress in these multilayers with period is attributed to the different microstructures in the multilayers. Cross-sectional HRTEM images of three different period (3.5, 7, and 280 nm) multilayers from Figure 3 (arrowed) are shown in Figure 4a, b, and c, respectively. The images show that the layered structure in all samples is apparent, as also evidenced from the enhanced Bragg's peaks in their low angle x-ray diffraction. The Si layers in all the samples show a typical amorphous structure of sputtered films. The Mo layers are amorphous at short period, and become polycrystalline as the period, or effectively the Mo layer thickness, increases. A columnar growth structure can be seen in the 280 nm period multilayer, which has approximately 80 nm thick Mo layers (Figure 4c).

DISCUSSION

The results of the single and layered Mo films suggest that stress in polycrystalline films depends strongly on the microstructure or growth of the films. At low working pressure, tensile stress is observed as the amorphous film is formed. As the film thickness increases, the tensile stress decreases and the films become compressive as the voids at the grain boundaries in the columnar structure close up. The microstructure in the layered films suggests that grain growth has been interrupted during film growth. Possible mechanisms for this growth interruption include surface relaxation of the deposited layers, low mobility of the Mo atoms on the film surface, and contamination of residual gas on the surface of the deposited layered films that inhibit continued growth of the Mo grains.

Equations 2 and 3 suggest that the tensile stress values found in the layered films are representative of the stress in a 1.8 nm Mo film that makes up the layered films, assuming the microstructures in each layer in the layered films and in the 1.8 nm films are the same, and that the effects from the substrate interface and strain relief from the underneath layers are negligible. A value of near 200 MPa for a 1.8 nm Mo film is reasonable since the high tensile stress eventually will decrease to zero as the film thickness decreases (Figure 1). The thinnest film in this study indicates a thickness limit for the available commercial wafer curvature measurement system for Mo films. Stress in thinner films can be determined by employing the layered film structures, as long as the microstructures in the layered films are identical to each other.

Changes in the stress in different period Mo/Si multilayers shown in Figure 3 can be explained by the different microstructures in the layers. Equation 3 assumes that the layer characteristics or microstructures and interfaces are identical for all thicknesses of the layers. Changes in the thickness of the layers, however, produce different microstructures in the layers and at the interfaces, and thus result in a different stress state. In thin films and multilayers of nanometer thickness, nucleation and grain growth are possible, and result in densification of the materials and volume changes of the layers. In practice then, stress in periodic multilayers can vary with period where microstructural evolution occurs. Higher compressive stress at longer period multilayers seen in Figure 3 probably results from more compressive stress in the polycrystalline Mo layers, and stress relaxation observed in shorter period multilayers is attributed to increased number of interfaces at short periods, and to changes in the microstructure and stress in the Mo and Si layers. Previous studies have shown that the amorphous Mo_xSi_y interfaces in Mo/Si multilayers range between 0.6 and 1.0 nm.[16-17] As the period approaches this interfacial layer thickness, effects of the interfaces become dominant. Contribution of the stress from the silicide layers at the interfaces, which can be different from the stresses in Mo and Si layers, becomes significant.

Since stress in multilayers supposedly is independent of the number of bilayer periods, stress of single bilayer pairs of Mo and Si films can be compared with the values measured in the multilayer structures. The curve of the stress sums follows the curve of the measured multilayer stress quite well in the range of period between 7 and 280 nm, where the stress data for single Mo and Si films are available.[18] The stress difference in this range of period is approximately 700 MPa, which can be crudely attributed to contributions from the interfaces.

Stress contribution from the interfaces was determined using the method similar to that described in reference 14. While volume stress represents the force acting on a unit area (Pa or N/m^2), interface stress is the force necessary to extend a unit length of interface or equivalently the work required to increase a unit area of interface by a unit strain (N/m or J/m^2). Different values of interface stress, however, are found in different ranges of Mo/Si periods, in which the microstructure of the Mo layers evolves from amorphous to crystalline structure.[19] For the Mo/Si multilayer period between 2.3 and 4.6 nm in this study, where the microstructure is amorphous in both the Mo and Si layers, this work per unit area was found to be 1.1 J/m^2. This value is close to that found in (111) Au/amorphous Al_2O_3 multilayers in the period range between 3.4 and 16.2 nm (1.13 J/m^2),[14] in Ti/Si multilayers of period between 1 and 6 nm (1.05 J/m^2),[20] and in Cr/C multilayers of period between 3.5 and 15 nm.[19] Although the interface stresses in these multilayer systems are similar to each other, the interfacial structures in the multilayers however are very distinct. The interfaces in the (111) Au/amorphous Al_2O_3 multilayers were found to be very sharp by TEM (no interfacial phases),[14] while that in the Ti/Si multilayers is attributed to the formation of an interfacial phase.[20] The interfaces in Mo/Si[16-17] and Cr/C[21] multilayers also show the presence of an interfacial layer which results from intermixing of the layers during deposition.

CONCLUSIONS

The relationship between intrinsic stress and microstructure in sputtered nanometer Mo and Si thin films and Mo/Si multilayers has been studied. Stress in the polycrystalline metal films depends strongly on the microstructure or film growth. Deposition of layered Mo films results in different microstructure and stress from those in single films of same thickness. Stress in periodic multilayers depends on the microstructure of the component layers at thickness where phase transformation occurs. Interface stress in Mo/Si multilayers is measured to be 1.1 J/m^2, similar to values reported in other multilayers.

ACKNOWLEDGMENT

This work was supported by the Director, Office of Energy Research, Office of Basic Sciences, Materials Sciences Division, of the U.S. Department of Energy under Contract No. DE-AC03-76SF00098.

REFERENCES

1. J.A. Thornton, J. Tabock, and D.W. Hoffman, Thin Solid Films 64 (1979) 111.
2. D.W. Hoffman and J.A. Thornton, J. Vac. Sci. Tech. 20, 3 (1982) 355.
3. J.A. Thornton, J. Vac. Sci. Tech. A4, 6 (1986) 3059.
4. M. Itoh, M. Hori, and S. Nadahara, J. Vac. Sci. Tech. B9, 1 (1991) 149.
5. T.J. Vink, M.A. J. Somers, J.L.C. Daams, and A.G. Dirks, J. Appl. Phys. 70, 8 (1991) 4301.
6. J.T. Pan and I. Blech, J. Appl. Phys. 55 (1984) 2874.
7. G.G. Stoney, Proc. Roy. Soc. Lon. A82 (1909) 72.
8. A. Brenner and S. Sendroff, J. Res. Natl. Bur. Stand. 42 (1949) 105.
9. N.N. Davidenkov, Sov. Phys.-Solid State 2 (1961) 2595.
10. S. Timoshenko, J. Opt. Soc. Am. 11 (1925) 23.
11. R.W. Hoffman, Phys. Thin Films 3 (1966) 211.
12. J. Vilms and D. Kerps, J. Appl. Phys. 53 (1982) 1536.
13. P.H. Townsend, D.M. Barnett, and T.A. Brunner, J. Appl. Phys. 62 (1987) 4438.
14. J.A. Ruud, A. Witvrouw, and F. Spaepen, in Defects in Materials (MRS Proc. 209, 1991) 737.
15. T.D. Nguyen, R. Gronsky, and J.B. Kortright, J. Elec. Microsc. Tech. 19 (1991) 473.
16. K. Holloway, K.B. Do, and R. Sinclair, J. Appl. Phys. 65 (1989) 474.
17. D.G. Stearns, M.B. Stearns, Y. Cheng, J.H. Stith, and N.M. Ceglio, J. Appl. Phys. 67 (1990) 2415.
18. T.D. Nguyen, X. Lu, and J.H. Underwood, in Physics of X-Ray Multilayer Structures (OSA Tech. Digest Series 6, 1994) 102.
19. T.D. Nguyen, to be published.
20. S. Radelaar, in Advances in Phase Transitions, ed. by J.D. Embury and G.R. Purdy (Pergamon Press, Oxford, 1988) 178.
21. T.D. Nguyen, Ph.D. thesis, University of California at Berkeley (1993).

Microstructure - Stress - Property Relationships in Nanometer Ge/C Multilayers

Xiang Lu[1,2], Tai D. Nguyen[1], and James H. Underwood[1]

[1]Center for X-ray Optics, Lawrence Berkeley Laboratory,
One Cyclotron Road, Berkeley, CA 94720

[2]Applied Science and Technology Graduate Group,
University of California, Berkeley, CA 94720

Abstract
A series of sputtered as-prepared and annealed Ge/C multilayer structures with periods ranging from 2 to 8 nm has been studied with high resolution transmission electron microscopy (HRTEM), x-ray scattering, and stress measurement techniques. Ge/C multilayers have potential applications as normal incidence reflective mirrors near 4.4 nm wavelength. The reflectivity and stress in these structures depend on the microstructural evolution of the component layers. The as-prepared structure of both Ge and C layers appear amorphous from TEM imaging and diffraction. Annealing at 500 °C for 60 minutes leads to crystallization of the Ge layers. As the phase diagram indicates, no carbide compound has been found. X-ray scattering reveals that the multilayer period expands by as much as 10% after annealing. Both TEM images and x-ray profiles suggest that the layer structures remain well-defined upon annealing. *In-situ* stress-temperature measurements directly show the Ge/C multilayer microstructure evolution path. X-ray measurements show that the structures with periods near 2 nm undergo a significant improvement on optical performance with annealing. The physical mechanisms that may have caused the optical enhancement are discussed. Correlation of the stress evolution in the multilayers and in individual layers during annealing, and their relationships to the microstructures and optical properties are examined.

I. Introduction
Nanometer multilayer structures, consisting of alternating high and low atomic number materials, have been widely used in optical applications as dispersive and reflecting elements for radiation of ultraviolet to x-ray wavelengths. Multilayers using C as the low atomic number material, such as W/C, WC/C and Ru/C systems, have shown good optical performance[1]. Ideally, the optimum multilayer performance is obtained in perfect structures with atomically sharp interfaces. However, actual fabricated multilayers are far from ideal, and their performance strongly depends on the microstructure, especially the interfacial structure of the multilayers. In general, the interfacial structure depends both on the microstructure inside the layers, and on the interaction between the layer materials. Presumably, in systems in which the components do not react with each other, such as Ru/C and Ge/C, minimum intermixing and diffusion should allow relatively sharp and well-defined interfaces. For this reason, Ge/C was studied as a potential candidate for the x-ray multilayer mirrors. Early materials studies showed that no compounds of Ge and C were detected up to 3170 °C[2], and also indicated virtual insolubility of C in Ge, or Ge in graphite [3]. These properties suggest that Ge/C should form sharp multilayer interfaces and hence may provide significant specular x-ray reflectance, especially at wavelengths just above the carbon K edge (4.4 nm), where interfacial microstructure and morphology play a critical role in the determination of reflectivity. In this paper, we study the effects of the microstructure and interfacial roughness, and their thermal evolution on the performance of Ge/C using HRTEM, x-ray scattering and reflectance measurement. We also investigate the multilayer intrinsic stress and its evolution, and the correlation with the microstructure and optical performance.

II. Experimental Techniques
Ge/C multilayers with periods ranging from 2 to 8 nm were prepared by a magnetron sputtering system [1] on semiconductor grade Si (111) wafers. The argon sputter gas pressure

was 2.5 mTorr. The period of the multilayers was determined using grazing angle x-ray reflectance measurements of the first several multilayer Bragg peaks with Cu K_α x-rays. The same x-rays were used in rocking scan [4] experiments to study the multilayer interfacial roughness. In the rocking scan, the detector angle 2θ was fixed at the multilayer first Bragg perk, while the sample angle θ was scanned at 0.005° gradient around the Bragg peak. Normal incidence reflectance of the multilayers was measured at $\theta = 85°$ using a reflectometer with a laser produced plasma source [5]. Microstructural characterization of the multilayers was performed by cross-sectional and plan-view HRTEM in a JEOL JEM 200CX microscope operating at 200 kV [6]. Thermal treatment of the multilayers was performed in a tube furnace under a vacuum of 10^{-6} torr. Unless specified, the samples were heated at 500 °C for 60 minutes.

Stress - temperature measurements were made in a nitrogen atmosphere on a Tencor FLX thin film stress measurement system. The mechanism of this stress measurement system is as follows: the substrate is bent under the stress of the deposited film. The Stoney formula [7] relates the curvature of the substrate and the in-plane stress of the thin film. By laser scanning, the system measures the curvature of the substrate before and after the film deposition, and uses the Stoney formula to calculate the intrinsic stress in the film.

III. Results and discussion

Microstructure observation

We studied the microstructure of Ge/C multilayers with periods of 2.5 nm, 5.0 nm and 8.0 nm respectively. Γ value, the ratio of the Ge thickness over the multilayer period ($\Gamma = \dfrac{d_{Ge}}{d_{Ge} + d_C}$), is 0.47 for all as-prepared samples. The microstructural phase information is obtained from both plan-view and cross-sectional HRTEM bright field imaging and electron diffraction. As in other metal/carbon systems reported [1], the C layers in Ge/C are amorphous before and after annealing. The Ge layers in as-prepared samples appear amorphous, while annealing results in crystallization in the 5.0 nm and 8.0 nm period multilayers. The Ge layers in 2.5 nm period are amorphous after annealing. However, observed fuzzy rings in diffraction patterns indicate tendency of Ge structure evolution towards crystallization at atomic level. Figure 1(a) shows a cross sectional bright field image and corresponding diffraction pattern of an as-prepared 8 nm period sample. Both image and diffraction pattern indicate amorphous structures in C and Ge layers. Figure 1(b) shows image and diffraction pattern of the same multilayer after annealing. Lattice fringes can be clearly seen in the Ge layers, as pointed by the arrows. The crystallite size is a few nanometers, at the same range as each Ge layer thickness. As the phase diagram [8] indicates, no carbide compound has been found in either as-prepared or annealed samples. HRTEM show that Ge/C layer structures are more stable upon annealing at 500 °C than other immiscible systems, such as Ru/C, Cu/C [1].

Stress measurements

The multilayer intrinsic stress is directly related to the multilayer microstructure. We find that the multilayer biaxial stress strongly depends on Γ. As shown in Figure 2, the as-prepared Ge/C multilayers may be compressive or tensile, with biaxial stress varying from -780 MPa to 70 MPa at different Γ values. The smaller Γ, the more compressive the multilayer is. Figure 2 shows both pure Ge and pure amorphous C films to be very compressive, with stress values at -650 MPa and -780 MPa respectively. In-situ stress-temperature measurements show that the C single film becomes more compressive after annealing, mainly due to thermal expansion[1, 9]. The stress - temperature measurements on the Ge single film indicate that the stress in Ge films becomes less compressive and eventually turns to tensile upon annealing, caused by Ge crystallization and grain growth. The stress - temperature behavior of the Ge/C multilayer is shown in Figure 3. The as-prepared multilayer ($\Gamma = 0.45$) has a compressive stress value of -120 MPa. During early heating from 25 °C to 150 °C, the greater thermal expansion of the multilayer compared to the Si substrate causes the multilayer to be more compressive. As the

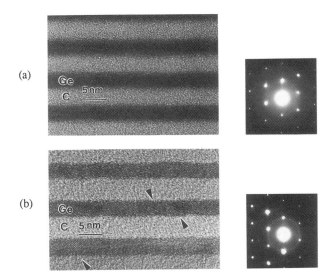

Figure 1. (a) Cross-sectional HRTEM images of 8 nm period as-prepared Ge/C multilayer show amorphous structure; (b) cross-sectional HRTEM images of the same 8 nm multilayer after annealing at 500 °C for 60 minutes. The area in arrows shows crystalline Ge lattice fringes. Spots in the diffraction patterns come from the Si substrate.

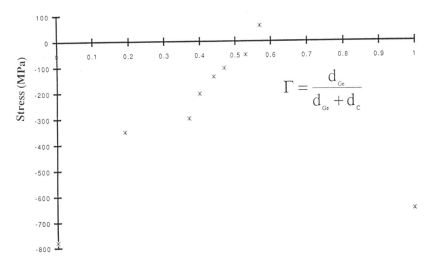

$$\Gamma = \frac{d_{Ge}}{d_{Ge} + d_{c}}$$

Figure 2. Biaxial stress strongly depends the multilayer Γ value. All the multilayers have 2.3 nm period and 345 nm total thickness. The pure C film ($\Gamma = 0$) thickness is 310 nm, and the pure Ge film ($\Gamma = 1$) thickness is 92 nm.

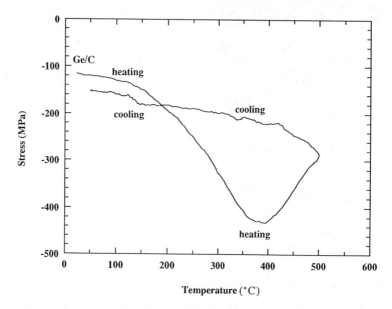

Figure 3. *In-situ* stress-temperature measurements imply the microstructure evolution path. The multilayer has a period of 2.3 nm, a total thickness of 345 nm, and $\Gamma = 0.44$.

sample is heated from 150 °C to 400 °C, the C thermal expansion becomes another major factor, and compressive stress buildup becomes even faster. Stress relaxation starts at about 400 °C. Above this temperature, densification of Ge becomes the dominant factor. The HRTEM microstructure results show that the Ge layers in 2.3 nm multilayers do not crystallize after annealing. Nevertheless, we do see the change of interatomic structure of Ge after annealing, which directly causes the stress relaxation. The stress relaxation continues throughout the subsequent heating and cooling process. After temperature is below 400 °C during cooling, the microstructure evolution becomes less active, and the stress relaxation is dominantly driven by the difference of the thermal expansion coefficients between the multilayer and the substrate. From then on, the cooling curve has the same slope as the early heating curve (from 25 °C to 150 °C). The stress - temperature behavior gives us a detailed picture of the microstructure evolution process.

Optical properties

We studied the optical performance of Ge/C multilayers with period near 2.3 nm, which may have potential applications as normal incidence x-ray mirrors near the carbon edge. Figure 4 shows the normal incidence reflectance results from an as-prepared multilayer, and the same multilayer after annealed at 400 °C, 450 °C and 500 °C respectively for 60 minutes. Interestingly, there is a significant increase of the normal incidence reflectance after annealing. Compared with the as-prepared, the 500 °C annealed sample shows a 53% increase of the integrated (over the wavelength) peak reflectance. The bandwidth of the normal incidence peak remains at 0.05 nm, approximately the same before and after annealing. The normal incidence measurements also show a shift of the peak from 4.58 nm to 5.02 nm after annealing at 500 °C,

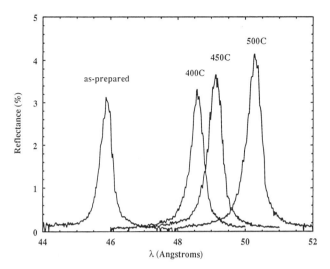

Figure 4. Normal incidence reflectance of an as-prepared multilayer, and after annealing at 400 °C, 450 °C and 500 °C respectively for 60 minutes. The 500 °C annealed sample shows 53% increase of integrated peak reflectance. The as-prepared sample has a period of 2.27 nm, a total thickness of 454 nm, and $\Gamma = 0.47$.

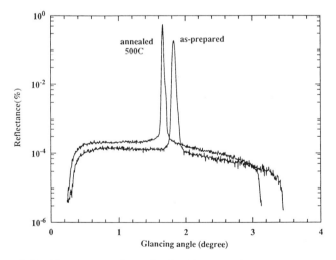

Figure 5. Rocking scan results at the first Bragg peaks using Cu K_{α} x-rays. The 500 °C annealed sample shows 60% increase of the integrated peak reflectance and background nonspecular scattering. The as-prepared sample has a period of 2.27 nm, a total thickness of 454 nm, and $\Gamma = 0.47$.

which indicates a 9.6% expansion of the multilayer period. The theoretical Ge/C peak reflectance at 50 nm wavelength is 17% assuming perfect interfacial structures, so the experimental reflectance is 24% of the ideal value, a reasonably good result. Figure 5 also reveals that the significant increase of the reflectance occurs when annealing temperature is above 400 °C. Notice that the stress relaxation also starts at 400 °C as shown in Figure 3.

X-ray rocking scan is performed to compare the interfacial roughness before and after annealing. Figure 5 shows the rocking scan results at the first Bragg peak using Cu K_α x-rays. Similar to the soft x-ray normal incidence reflectance results, a 60% increase of the integrated (over the scanning angle) peak specular reflectance is observed after annealing. The background nonspecular scattering is also enhanced by 60%, which indicates the ratio of specular to nonspecular reflectance is the same before and after annealing. We conclude that the increase in the x-ray reflectance does not result from a decrease of the interfacial roughness after annealing.

Theoretical calculations show that carbon expansion and subsequent Γ value decreasing can not dramatically improve the optical performance. Other possible mechanisms for the increasing of the reflectance include sharping of the chemical composition at the interface, which results from phase separation between Ge and C, and change in density inside the layers. Stress relaxation may be another possible reason. Further work is underway to solve the puzzle.

IV. Summary

The interrelationships among the microstructure, stress and optical properties have been investigated using HRTEM, stress measurement technique, x-ray scattering and reflectance. All as-prepared Ge/C multilayers studied are amorphous, and annealing results in crystallization of the Ge layers in longer period multilayers. Intrinsic stress strongly depends on the multilayer Γ value, and the stress-temperature behavior directly indicates the microstructure evolution process. Significant reflectance increase upon annealing was observed from both Cu K_α and soft x-ray measurements. The best normal incidence peak reflectance we obtained so far is about 4 - 5% near the carbon edge, and the ratio of experimental to ideal reflectance is about 24%. Ge/C multilayer is quite stable and its stability can be a very valuable property in many applications.

Acknowledgments

The authors like to thank R. Soufli and E. M. Gullikson for helping the normal incidence measurements. Valuable discussions with J. B. Kortright are acknowledged. This work was supported by the Director, Office of Energy Research, Office of Basic Science, Materials Sciences Division, of the U.S. Department of Energy under Contract No. DE-AC03-76SF00098. HRTEM were performed at National Center for Electron Microscopy, Materials Sciences Division, Lawrence Berkeley Laboratory, University of California.

References

[1] T. D. Nguyen, R. Gronsky, and J. B. Kortright, Mat. Res. Soc. Symp. Proc. 139 (1989) 357; 187 (1990) 95; 280 (1993) 161.
[2] R. I. Scace, and G. A. Slack, J. Chem. Phys., 30 (1959); and R. I. Scace, and G. A. Slack, in *Silicon Carbide*, Pergamon Press, New York (1960) 24.
[3] A. Taylor, and N. J. Doyle, Scripta Met. 1 (1967) 161.
[4] J. B. Kortright, J. Appl. Phys. 70 (1991) 3620.
[5] E. M. Gullikson, J. H. Underwood, P. C. Batson, and V. Nikitin, J. X-ray Sci. Tech., 3 (1992) 283.
[6] T. D. Nguyen, R. Gronsky, and J. B. Kortright, Elec. Microsc. Tech. 19 (1991) 473.
[7] G. G. Stoney, Proc. Roy. Soc. London, A82, 172 (1909).
[8] M. Hanse, and K. Anderko, In *Constitution of Binary Alloys*. 2nd ed. (1989).
[9] C. A. Lucas, T. D. Nguyen, and J. B. Kortright, Appl. Phys. Lett. 59 (1991) 2100; X. Jiang, D. Xian, and Z. Wu, Appl. Phys. Lett. 57 (1990) 2549.

THE EFFECT OF INCIDENT KINETIC ENERGY ON STRESS IN SPUTTER-DEPOSITED REFRACTORY-METAL THIN FILMS

T.J. Vink, J.B.A.D. van Zon, J.C.S. Kools and W. Walrave

Philips Research Laboratories Eindhoven, Prof. Holstlaan 4, 5656 AA Eindhoven, The Netherlands

ABSTRACT

In magnetron sputter deposition the intrinsic stress in refractory-metal films changes from compressive to tensile on increasing the working-gas pressure. This pressure dependence is linked to the particle transport process from target to substrate during deposition. In this work we apply a Monte Carlo (MC) technique to simulate the transport of sputtered atoms and reflected neutrals in a background gas. Specific examples of Cr, Mo and W thin film growth in Ar and Ne gas ambients are presented. Trends in thermalization of the depositing atoms coincided with the observed trends in the compressive-to-tensile stress curves, for the different target and working-gas combinations studied. Furthermore, a quantitative correlation between the stress transition pressures and the incident kinetic energy of both sputtered atoms and reflected neutrals during film growth was found. In this case the contribution of the latter species was weighted with a relatively low factor.

INTRODUCTION

It is well known that in sputter-deposited films tensile or compressive stresses can be generated, depending on the energetics of the deposition process. Modeling of film growth and extensive experimental evidence show an important effect of energetic bombardment onto the film and its evolving microstructure, which in turn determines the (growth-induced) stress. A comprehensive review on this matter has recently been published by Windischmann.[1] High-energy transfer to the growing film generally results in dense films which experience compressive stress. The origin of this stress state is interpreted in terms of the atomic peening mechanism.[2] Modeling of the atomic peening process has provided simple stress scaling laws,[3,4] which depend on the level of the normalized energy flux. Tensile stress is often associated with open, porous films, grown under conditions of limited energetic bombardment and low adatomic mobility. Short-range attractive forces across atomic scale gaps can act effectively in such microstructures and provide a mechanism for the generation of tensile stress. Simulations of both the microstructure and the stress evolution as a function of the kinetic energy of the depositing atoms and fast neutralized inert gas atoms were carried out by Müller.[5] The calculated tensile stress profile was shown to mimic the extensive experimental data. More recently, in a similar molecular-dynamics approach by Fang et al.[6] these calculations were extended to the compressive regime as well.

591

With respect to the above mentioned energy-stress relationships, it is clear that the particle energy and flux striking the developing film are fundamental quantities. To this end, we have carried out Monte Carlo (MC) simulations of the transport of sputtered atoms and reflected neutrals in a background gas. Examples of Cr, Mo and W thin film growth in Ar and Ne gas ambients are presented. From these calculations the effects of pressure and target/working-gas mass ratio on the energetics of the deposition flux can be estimated, and are related to the observed compressive-to-tensile stress transitions of the deposited films.

MONTE CARLO SIMULATION

Various authors[7-11] have used MC simulations to study the particle transport process in a neutral background gas. The method has been well developed and therefore the set-up of the present calculations will be described only briefly.

Sputter particle energies are generated from a Thompson distribution[12] in which the high energy tail is modified in order to incorporate the effect of low discharge voltages.[13] This distribution does not account for the difference in projectile mass, although initial energies with Ne will be lower than in the case of Ar.[14] Therefore several computations are performed with mono-energetic particles. Their average energy is scaled according to the ion energy[15] and according to the projectile mass.[14] Measured energy distributions have been used as generator and have been compared with mono-energetic sources with respect to the average kinetic energy on the substrate. Only minor differences occurred. In the case of reflected neutrals mono-energetic sources have also been used. The energy is based on the discharge voltage and reflection co-efficients for energy.[16] Average starting energies for sputtered particles ($<E_s^i>$) and reflected neutrals ($<E_n^i>$), and their respective yields (Y_s and Y_n), are given in Table I. Both particles are assumed to follow a cosine distribution of emission intensity. The energy is taken independent on the emission angle (defined with respect to the surface normal), although in general the energy increases with increasing angle.[17] Background particles are generated from a Maxwellian velocity distribution.

Energy dependent cross-sections have been calculated from Ne–Ne, Ar–Ar and Kr–Kr potentials, using the elastic collision integral.[18] These cross-sections are scaled with the atomic diameter, which resulted in an average universal cross-section. This universal cross-section is scaled by $(d_{sput} + d_{gas})^2$. The mean free path is generated from a Poisson distribution based on the energy-dependent cross-section and a factor which corrects the distribution when the velocity of the sputtered particle comes close to the velocity of the background particle.

The elastic collision is carried out 3–D by transformation to the centre-of-mass system and using Euler rotation. The model assumes an infinite target and substrate surface. Only the trajectory of the target atom is tracked. However, for reflected neutrals the recoil atom also can be taken into consideration. In that case the average energy of the reflected neutrals on the substrate decreases less steeply with increasing pressure, as observed in a simplified 2–D calculation.

In the thermalization calculations a particle is considered *thermalized* when its energy is less than twice the average energy of the background particles. The temperature of the background is set at a value of 500 K. In principle the temperature is dependent on the deposition rate, the energy of the sputtered particles and the gas pressure itself.[11,19]

EXPERIMENTAL

Cr, Mo and W thin films were deposited on glass substrates in a load locked modular magnetron-sputter system, described in more detail elsewhere.[20] Ar and Ne were used as working-gases. The distance between the planar magnetron sputtering cathode with a target and the substrates was 70 mm. A direct-current (dc) power supply was used, and was driven in the constant power mode. A rotatable substrate carrier allowed rotation up to 10 rpm. All films were deposited in the static deposition mode (the substrate placed opposite the target) at a constant input power of 0.8 kW, except for Cr in Ar where the dynamic deposition mode was used (10 rpm) and an input power of 2.4 kW was applied. Film thicknesses (Cr and Mo: 200 nm; W: 150 nm) were constant within the pressure range used.

Thin film stresses were determined using the bending-beam method,[21] applying the following techniques: i) determination of the force required to straighten a bent, coated sample (thin glass substrates: Schott type D263-1; dimensions were $60 \times 10 \times 0.15$ mm^3), and ii) optical measurement of the radius of a coated glass strip (Flexus F5200). In the case where films were deposited in the dynamic deposition mode, the glass strips were aligned such that their long axis was parallel to the transport direction. The temperature induced stress component is considered to be negligible (approximately 0.05 GPa), since the temperature increase of the substrate during sputter-deposition was estimated to be less than 50 K. Therefore, in the present work the reported stress values, averaged over the film thickness, are conceived as incorporated during film growth.

RESULTS AND DISCUSSION

Thermalization of sputtered atoms

The calculated fraction of thermalized Cr, Mo and W atoms arriving at the substrate surface are depicted in Fig. 1, as a function of Ar and Ne pressure. From this graph, it is seen that the thermalization curves shift to higher pressure for increasing mass ratio, $m_{t(target)}/m_{g(as)}$, and are very much alike for target/working-gas combinations with a similar mass ratio: $m_{Cr}/m_{Ne} \sim m_{Mo}/m_{Ar} \sim 2.5$ and $m_{Mo}/m_{Ne} \sim m_{W}/m_{Ar} \sim 4.7$. The latter observation suggests that thermalization is primarily dominated by the average energy reduction per collision, which goes with $[(1-\mu)/(1+\mu)]^2$ [where $\mu = m_g/m_t$]. For combinations with equal m_t/m_g mass ratio, the effects of initial energy of the sputtered atoms and cross-section on thermalization apparently compensate each other.

Incident kinetic energy and film stress

Growth-induced stresses in Cr, Mo and W thin films, sputter-deposited in Ar and Ne ambients, are depicted in Fig. 2 as a function of pressure. This is the same series of target/working-gas combinations used in the above mentioned MC calculations on the thermalization of the sputtered atom flux. A well known feature of the stress state in these refractory-metal films is a transition from compression to tension with increasing pressure. The transition pressure increases with increasing m_t/m_g mass ratio. This trend compares well with previously published data on the dependence of transition pressure on target-to-gas mass ratio for planar magnetron deposited films.[22,23]

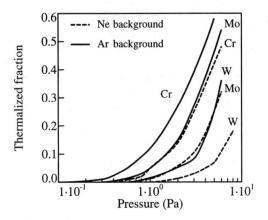

Fig. 1. The fraction of thermalized Cr, Mo and W atoms arriving at the substrate surface as a function of Ar and Ne pressure.

From a comparison of Figs. 1 and 2 it is seen that the thermalization curves reflect the observed trend in the stress transitions. Both the sequence in stress curves along with m_t/m_g and also the similarity in stress reversals at the same m_t/m_g mass ratio are predicted accurately by the thermalization curves. Film growth can be expected to be dominated by the thermalized atoms at relatively high fractions, where the films experience high tensile stress or relaxation of stress occurs. The role of the thermalized atoms at the stress reversals is probably of minor importance since their fractions are observed to be fairly low in all cases.

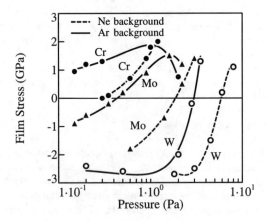

Fig. 2. Compressive-to-tensile stress transition curves for Cr, Mo, and W thin films sputter-deposited in Ar and Ne.

The energy-stress relationship for the present films has been studied on more quantitative aspect also. An estimate of the particle bombardment of the growing film at the stress transition pressures (p_t), as observed in Fig. 2, is depicted in Table I. Here, the average energy deposited per atom is given for both the sputtered atoms $(<E_s^f>)$ and the reflected neutrals $(<E_n^f>)$. The energy carried by these species differ considerably for the different target/working-gas combinations studied. For instance in the case of Cr/Ar the incident energy flux is primarily carried by the sputtered atoms, whereas for W/Ne the energy is largely transferred by the reflected neutrals. The energies of the sputtered atoms and reflected neutrals do not simply add up. Molecular dynamics modeling of microstructure and stresses in sputter-deposited thin films[5,6] show that the effect of incident kinetic energy of sputtered atoms (Ni) and fast neutrals (Ar) are not the same. In these studies it was found that stress varies in a similar fashion as a function of incident kinetic energy for both species, however, the required energy deposited per atom was roughly a factor of 10 higher in the case of bombarding reflected neutrals. Applying this weight factor for $<E_n^f>$, the following "effective" values for the average energy deposited per atom can be generated, as given in Table I. The values thus obtained are more or less constant for a given metal, e.g. independent of whether Ar or Ne is used (note that this is not the case when no weight factor is applied: c.f. 39 eV/at for W/Ne and 15 eV/at for W/Ar). However, even with these corrected values for the average energy deposited per atom the definite sequence in stress transition points is not as easily predicted as with the thermalization curves. In particular the estimated energy values for W are relatively low compared to the other materials [note the difference with the increase in binding energy values (U_b) along with the metal series Cr, Mo and W]. The latter might be related to the fact that a constant background gas temperature is used in the calculations, whereas gas rarefaction can be expected to be more severe at these high transition pressures.[19]

From the above energy-stress relationships it is seen that it is not straightforward to correlate the incident average energy of both sputtered atoms and fast neutrals with observed film properties. The average energy deposited per atom can be related to film stress, but whether it is the universal parameter is not clear (see also Ref. 24). Furthermore, thermalized deposited atoms might play an important role in film growth as well, most probably at relatively high fractions.

Table I
Calculated results at the observed stress transition pressures

	Y_s	Y_n	$<E_s^i>$ eV	$<E_n^i>$ eV	p_t Pa	$<E_s^f>$ eV/at	$<E_n^f>$ eV/at	$<E_s^f>+0.1\cdot<E_n^f>$ eV/at	U_b eV
Cr/Ne	0.74	0.19	5.0	63	0.27	3.7	13.3	5.0	4.1
Mo/Ne	0.37	0.26	7.3	105	2.0	3.2	40.9	7.3	6.8
W/Ne	0.21	0.35	12.0	140	6.0	1.8	37.2	5.5	8.9
Cr/Ar	0.93	0.06	8.7	25	0.07*	7.7	1.0	7.8	4.1
Mo/Ar	0.62	0.15	12.7	60	0.45	6.9	7.9	7.7	6.8
W/Ar	0.44	0.26	21.0	98	3.0	5.0	10.1	6.0	8.9

*This value is extrapolated from the tensile stress data (see Fig. 2).

CONCLUSIONS

MC simulations have been used to calculate the energetics of the deposition flux during sputter-deposition of Cr, Mo and W thin films in Ar and Ne working-gases. From these calculations the effect of pressure and target/working-gas combination on the energetic bombardment during film growth was estimated, and related to the observed stress properties of the deposited films. The following energy-stress relationships were found:

i) Trends in the compressive-to-tensile stress curves coincided with the trends in thermalization of the sputtered atom flux.

ii) Also a quantitative correlation between the stress transition pressures and the average energy per deposited atom was found. In this case the contribution of the reflected neutrals, with respect to the sputtered atoms, was weighted with a low factor of $\sim$0.1.

REFERENCES

[1] H. Windischmann, Critical Reviews in Solid State and Materials Sciences **17**, 547 (1992).
[2] F.M. d'Heurle, Metall. Trans. **1**, 725 (1970).
[3] H. Windischmann, J. Appl. Phys. **62**, 1800 (1987).
[4] C.A. Davis, Thin Solid Films **226**, 30 (1993).
[5] K.-H. Müller, J. Appl. Phys. **62**, 1796 (1987).
[6] C.C. Fang, F. Jones, and V. Prasad, J. Appl. Phys. **74**, 4472 (1993).
[7] T. Motohiro and Y. Taga, Surface Sci. **134**, L494 (1983).
[8] G.M. Turner, I.S. Falconer, B.W. James, and D.R. McKenzie, J. Appl. Phys. **65**, 3671 (1989).
[9] M.A. Vidal and R. Asomoza, J. Appl. Phys. **67**, 477 (1990).
[10] A.M. Myers, J.R. Doyle, and D.N. Ruzic, J. Appl. Phys. **72**, 3064 (1992).
[11] H.M. Urbassek and D. Siebold, J. Vac. Sci. Technol. A **11**, 676 (1993).
[12] M.W. Thompson, Philos. Mag. **18**, 377 (1968).
[13] H.M. Urbassek, Nucl. Instrum. Methods Phys. B **4**, 356 (1984).
[14] R.V. Stuart and G.K. Wehner, J. Appl. Phys. **35**, 1819 (1964).
[15] R.V. Stuart, G.K. Wehner, and G.S. Anderson, J. Appl. Phys. **40**, 803 (1969).
[16] W. Eckstein and J.P. Biersack, Z. Phys. B **63**, 471 (1986).
[17] W. Eckstein, Nucl. Instrum. Methods Phys. B **18**, 344 (1987).
[18] J.A. Barker, R.A. Fisher, and R.O. Watts, Molecular Physics, **21**, 657 (1971).
[19] M. Rossnagel, J. Vac. Sci. Technol. A **6**, 19 (1988).
[20] T.J. Vink, M.A.J. Somers, J.L.C. Daams, and A.G. Dirks, J. Appl. Phys. **70**, 4301 (1991).
[21] R.W. Hoffman, in *Physics of Thin Films*, edited by G. Hass and R.E. Thun (Academic, New York, 1966), p. 211.
[22] J.A. Thornton, in *Semiconductor Materials and Process Technology Handbook*, edited by G.E. McGuire (Noyes, Park Ridge, New York, 1988), p. 329.
[23] D.W. Hoffman and R.C. McCune, in *Handbook of Plasma Processing Technology*, edited by S.M. Rossnagel, J.J. Cuomo, and W.D. Westwood (Noyes, Park Ridge, New York, 1990), p. 483.
[24] I. Petrov, F. Adibi, J.E. Greene, L. Hultman, and J.-E. Sundgren, Appl. Phys. Lett. **63**, 36 (1993).

MICRO-MECHANICAL CHARACTERIZATION OF TANTALUM NITRIDE THIN FILMS ON SAPPHIRE SUBSTRATES

SHANKAR K. VENKATARAMAN*, JOHN C. NELSON*, NEVILLE R. MOODY+, DAVID L. KOHLSTEDT** AND WILLIAM W. GERBERICH*
* Department of Chemical Engineering and Materials Science, University of Minnesota, Minneapolis, MN 55455.
** Department of Geology and Geophysics, University of Minnesota, Minneapolis, MN 55455.
+ Sandia National Laboratories, Livermore, CA 94551.

ABSTRACT

The adhesion of Ta_2N thin films -- often used as thin film resistors -- to sapphire substrates has been studied by continuous microindentation and microscratch techniques. Ta_2N films, 0.1-0.63μm in thickness, were sputter deposited onto single crystal substrates. Continuous microscratch experiments were performed by driving a conical diamond indenter simultaneously into and across the film surface until stresses high enough to delaminate the film were developed. Continuous microindentation experiments were performed to induce film spallation by normal indentation. From both of these experiments, interfacial fracture toughness was determined as a function of film thickness. The interfacial fracture toughness obtained from continuous microscratch experiments is 0.53±0.17 MPa√m, independent of film thickness. This observation indicates that there is almost no plastic deformation in the film prior to fracture so that a 'true' interfacial fracture toughness is measured. For the 0.63 μm thick film, continuous microindentation data yielded a fracture toughness of 0.61±0.08 MPa√m, which matches closely the value obtained from the microscratch test. Hence, the continuous microscratch and microindentation techniques are viable methods for determining the interfacial fracture toughness in such bi-material systems.

INTRODUCTION

Thin films are used in a variety of electronic, optical and biomedical applications. For example, they provide resistance to abrasion, wear, corrosion and oxidation and provide special magnetic or dielectric properties [1]. Their properties, structure, functional characteristics and performance all depend on the adhesion between the film and the substrate. Hence, a number of experimental techniques have been developed to measure adhesion between films and substrates. However, most of these techniques are only qualitative and serve as relative comparisons between samples. The peel test, bulge test and scratch tests, however, have been made quantitative and yield reasonable estimates of thin film adhesion strengths [1-4].

Thin tantalum nitride films are of interest to the microelectronics community. They are used as resistors and capacitors because of their long term stability and low thermal coefficients of resistance [5-7]. However, the effects of high heat generation on the contact adhesion has been a major concern, as has been the need for reliable techniques to measure the adhesion strengths of these brittle films. In this paper, continuous microindentation and microscratch techniques [8-10] are employed to directly measure the interfacial toughness of Ta_2N films deposited onto sapphire substrates. The effect of film thickness is investigated, and the interfacial fracture toughnesses determined from continuous microindentation and continuous microscratch techniques are compared.

EXPERIMENTAL METHODS

Tantalum nitride films, 0.10 and 0.63 μm in thickness, were reactively sputtered onto (11$\bar{2}$0) oriented single crystal sapphire substrates using a d.c. magnetron sputtering system. Prior to sputtering, the substrates were heated to 170° C in vacuum for two hours to drive off moisture, and then RF backsputtered for 120 s to remove surface contaminants. During sputtering, the vacuum was maintained to less than 10^{-7} Torr. The films were deposited using a tantalum target with argon plus controlled amounts of nitrogen as a carrier gas.

Continuous microscratch tests were performed by driving a conical diamond indenter, with a 1 μm tip radius and a 90° included angle, simultaneously into and across the sample surface at rates of 150 nm/s and 0.5 μm/s, respectively. Once high-enough stresses develop, a section of the film ahead of the indenter tip spontaneously delaminates. Continuous microindentation tests were performed by driving the same conical indenter into the sample at a rate of 150 nm/s, until film spallation was induced by normal indentation. In both of these experiments, the applied normal and tangential load, depth of penetration and scratched distance (for microscratch experiments) were continuously recorded.

RESULTS

a) Continuous microscratch experiments

A typical scratch loading curve for the 0.10 μm Ta_2N/Al_2O_3 sample is shown in Fig. 1. For each scratch loading curve, the load increases as the indenter goes into and across the film. Stresses are built up at the interface; at a critical load, L_{cr}, the applied stress equals the interfacial shear stress. At this instant, the film delaminates from the substrate, and a drop in load is observed. The critical load for film delamination increases as the thickness of the film is increased and is 51±19 mN and 100±12 mN for the films of thicknesses 0.10 and 0.63 μm, respectively. Film delamination is clearly seen in scanning electron micrographs of scratch experiments performed on the 0.10 and 0.63 μm thick films, as shown in Fig. 2 for the 0.1 μm film. The results indicate an extremely brittle kind of interfacial failure. For the 0.63 μm -thick film, the delaminated region was large (Table 1) but for the 0.1 μm -thick film, delamination was localized to within the scratch track itself (Fig. 2). Also, in each scratch experiment the film delaminated well before the indenter reached the interface. For the 0.63 μm -thick film for example, the film delaminated when the indenter tip had gone only ~ 0.1 μm into the film. The critical load and area of delamination (obtained from magnified views of the delaminated regions) for the scratch experiments performed on each sample are averaged over five independent experiments and summarized in Table 1. Note that although the critical load varies only by a factor of two, the area of delamination varies by almost two orders of magnitude, indicating a much higher stress state in the 0.10 μm -thick film.

b) Continuous microindentation experiments

A typical load-displacement curve for indentation experiments performed on the 0.63 μm thick film is shown in Fig. 3. The load increases to about 0.3 N at which point a sharp change in

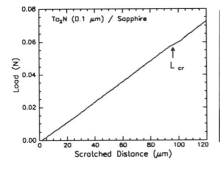

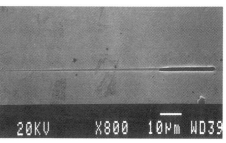

Fig. 1 Typical scratch loading curve for the 0.1μm -thick Ta₂N film on sapphire. At the critical load, the film delaminates.

Fig. 2 SEM micrograph of a typical scratch experiment on a 0.1 μm -thick Ta₂N film. The film delaminates at the critical load.

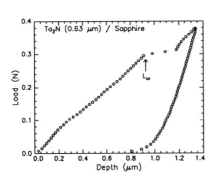

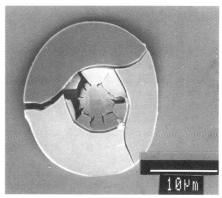

Fig. 3 Typical indentation experiment showing a sharp load (depth) transition at about 0.3 N. The film delaminated at this load.

Fig. 4 SEM micrograph of the indentation shown in Fig. 3. Note the film spallation region as well as the actual indentation size.

Table 1 Determination of the interfacial fracture toughness for Ta₂N thin films on Al₂O₃ by the continuous microscratch technique (data summarized over five different tests)

Sample	t (μm)	L_{cr} (mN)	A_d (μm²)	K (MPa√m)
1	0.10	51	2.6	0.56±0.08
2	0.63	100	178.5	0.50±0.17

the depth of penetration occurs. The film delaminates at this point, as is seen clearly from the SEM micrograph of the indentation, Fig. 4. The nature of the delamination process was clearly mapped out by doing a series of indentations starting at loads just below the critical load for delamination, to loads just above the delamination load [11]. The film delaminates and mushrooms up like a donut at the critical load. At loads beyond the critical value, the delaminated film also fractures as seen in Fig. 4. For loads below the critical values, the loading and unloading portions of the indentation are almost elastic, indicating that very little plastic deformation occurred before delamination of the film [11]. The indentations were also imaged using atomic force microscopy (AFM). From the AFM images and data, the amount of 'up-thrust' of film material around the indentations was estimated. With the up-thrust volume and the volume of the indentation, the volume fraction of the film material that was compressed into the film was calculated. The need for such a calculation will be brought out in the next section and in more detail in a forthcoming publication [11].

DISCUSSION

a) Continuous microscratch experiments

The critical load for film delamination has been used as a qualitative measure of adhesion for a number of years [12-13]. However, in the last few years, various researchers have proposed models for the calculation of properties such as interfacial shear strength, practical work of adhesion and interfacial fracture toughness [12-16]. In this paper, the analysis of Venkataraman et al. [10] is used to calculate the interfacial fracture toughness for the Ta_2N/Al_2O_3 system. In this analysis, the average shear and normal stresses in the delaminated region are determined as a function of the critical load, area of delamination and various shape parameters; and the interfacial fracture toughness is determined from the shear stress and film thickness. To obtain the average shear stress, the friction coefficient must be known. Since both normal and tangential loads were recorded simultaneously during the microscratch experiments, the friction coefficient was determined to be 0.3 from the ratio of the tangential load to normal load at film delamination. The interfacial toughnesses for the 0.10 and 0.63 μm thick Ta_2N films deposited on sapphire are 0.56±0.08 MPa√m and 0.50±0.17 MPa√m, respectively, as summarized in Table 1.

The adhesion strength of the Ta_2N/sapphire system is almost independent of film thickness. Typically, for metal/ceramic systems, the practical work of adhesion and interfacial toughness increase as the thickness of the film increases [18]. This behavior is mainly due to the fact that the energy spent in plastic deformation of the film increases with increasing film thickness. As the film thickness approaches zero, the contribution of the energy involved in plastic deformation of the film approaches zero. An extrapolation of a plot of the practical work of adhesion versus thickness to zero film thickness yields the *true* work of adhesion. Since this trend of adhesion strength with film thickness is not seen for the as-deposited Ta_2N films on sapphire, the true adhesion strength is the same for the film thicknesses investigated in this study. This result indicates that the Ta_2N/sapphire is nearly a perfect brittle-brittle system with almost no plastic deformation of the film prior to interfacial failure.

b) Continuous microindentation experiments

From the load-displacement indentation data shown in Fig. 3 and the corresponding SEM micrograph of the indentation in Fig. 4, it is clear that there is also a critical load for film delamination by normal indentation. An estimate for the strain energy release rate and interfacial fracture toughness for interfacial crack growth can be obtained by applying an indentation-induced delamination model of Marshall and Evans [19]. In this model, volume is assumed to be conserved and the volume of the displaced material from the indentation is assumed to go radially into the film. In such a case, the plastic zone exerts an outward pressure on the film material around the indentation region, causing the film to expand. Marshall and Evans determined an expression for a stress σ_o that has to be applied to prevent the film from expanding out [19]. They determined expressions by including and excluding an additional film residual stress. For stress-free films, σ_o can be written as

$$\sigma_o = \frac{E V_o}{2\pi(1-v)a^2 t} \tag{1}$$

where E is the elastic modulus of the film, V_o is the indentation volume, a is the radius of the circular crack (delamination region) seen in Fig. 3, for example, and t is the thickness of the film. The strain energy release rate, G, is given by

$$G = \frac{(1-v^2)\sigma_o^2 t}{E} \tag{2}$$

The interfacial fracture toughness, K, is obtained from the fracture mechanics result for plane strain

$$K = \sqrt{\frac{E G}{(1-v^2)}} \tag{3}$$

The interfacial toughness of the Ta_2N/Al_2O_3 system can be calculated from equations 1-3. The crack length is obtained from the SEM micrograph of the indentation. The indentation volume is determined from the depth of penetration and the indenter tip shape. The modulus of the Ta_2N film was calculated to be 350 GPa from the unloading portions of the indentation load-displacement data using the model of Doerner and Nix [2]. All of these data, as well as the calculated interfacial fracture toughness, are averaged over four different indentations and are summarized in Table 2. An important point to note is that the above model assumes that the entire volume of the indentation, as calculated from the load-displacement data, is conserved and goes radially into the film. However, as is shown in a forthcoming publication [11], this is not the case. Atomic force microscopy observations and analysis indicate that only 58% of the volume of the indentation goes into the film and the remaining 42% goes into a pile up or up-thrust above and around the indentation. The fracture toughness determined from the Marshall-Evans model is hence corrected to account for this volume difference; the corrected values for the interfacial fracture toughness, K_{corr}, are also shown in Table 2.

The interfacial fracture toughness determined from the continuous microindentation technique agrees, within experimental error, with values obtained from the continuous microscratch experiments. Not only does this comparison indicate that the true interfacial fracture toughness of the Ta_2N/sapphire system has been obtained, but the comparison also shows that the continuous microscratch technique is a viable technique for determining adhesion and toughness of bi-material systems. Even if the residual stress in the Ta_2N film is assumed to be 1 GPa, the fracture toughness only increases to about 0.71 MPa$\sqrt{m}$, and the comparison between the two techniques is still quite good [11].

Table 2 Determination of the interfacial fracture toughness for 0.63 μm thick Ta$_2$N films on Al$_2$O$_3$ by the continuous microindentation technique

a (μm)	V$_o$ (μm^3)	σ$_o$ (GPa)	G (J/m^2)	K (MPa√m)	K$_{corr}$ (MPa√m)
8.66	1.06	1.89	2.82	1.06±0.12	0.61±0.08

ACKNOWLEDGEMENTS

This research was supported by the Center for Interfacial Engineering at the University of Minnesota under Grant No. NSF/CDR-8721551. N.R. Moody would like to acknowledge the support of the U.S. DOE through Contract DE-AC04-94AL85000. The authors thank Erica Lilleodden for performing the AFM. The authors also thank Drs. P.S. Alexopoulos, T.W. Wu and T.C. O'Sullivan, of IBM Almaden Research Center, and Dr. Ridha Berriche of the National Research Council, Canada, for building the microindentation apparatus.

REFERENCES

[1] K.L. Mittal, Electrocomponent Science and Technology, **3**, 21 (1976)

[2] A.J. Perry, Thin Solid Films, **107**, 167 (1983).

[3] B.N. Chapman, in *Aspects of Adhesion* (D.J. Alner, Editor, CRC Press, Cleveland, Ohio) p.43 (1971).

[4] H. Dannenberg, J. Appl. Polymer Sci. **5**, 125 (1961).

[5] J.R. Adams and D.K. Kramer, Surface Science, **56**, 482 (1976).

[6] C.L. Au, W.A. Anderson, D.A. Schmitz, J.C. Flassayer, and F.M. Collins, J. Mater. Res., **5**, 1224 (1990).

[7] N.R. Moody, S. Venkataraman, J. Nelson, W. Worobey, and W.W. Gerberich, presented at the 1993 MRS Fall meeting, proceedings to be published.

[8] T.W. Wu, J. Mater. Res., **6**, 407 (1991).

[9] M.F. Doerner and W.D. Nix, J. Mater. Res. **1**, 601, (1986).

[10] S. Venkataraman, D.L. Kohlstedt and W.W. Gerberich, J. Mater. Res. **7**, 1126 (1992).

[11] S. Venkataraman, J.C. Nelson, N.R. Moody, D.L. Kohlstedt and W.W. Gerberich, in preparation, 1994.

[12] P. Benjamin and C. Weaver, Proc. Roy. Soc., London, **A254**, 163 (1960).

[13] C. Weaver, J. Vac. Sci. Technol., **12**, 18 (1975).

[14] M.T. Laugier, Thin Solid Films, **117**, 243 (1984).

[15] P.J. Burnett and D.S. Rickerby, Thin Solid Films, **148**, 41 (1987).

[16] S.J. Bull, D.S. Rickerby, A. Matthews, A. Leyland, A.R. Pace and J. Valli, Surface and Coatings Tech., **36**, 503 (1988).

[17] K.L. Johnson, *Contact Mechanics* (Cambridge University Press 1985), pp. 50-51, 68-70, 171-179, 427-428.

[18] S.K. Venkataraman, D.L. Kohlstedt, and W.W. Gerberich in, *Thin Films: Stresses and Mechanical Properties, IV* (MRS Symposium Proceedings, Volume 308).

[19] D.B. Marshall and A.G. Evans, J. Appl. Phys., **56**, 2632 (1984).

THE EFFECTS OF HIGH TEMPERATURE EXPOSURE ON THE FRACTURE OF THIN TANTALUM NITRIDE FILMS.

N. R. MOODY, S. K. VENKATARAMAN*, J. C. NELSON*, W. WOROBEY**, AND W. W. GERBERICH*
Sandia National Laboratories, Livermore, CA 94551-0969
*University of Minnesota, Minneapolis, MN 55455
**Sandia National Laboratories, Albuquerque, NM 87185-0957

ABSTRACT

Continuous microscratch testing was used in this study to determine the effects of elevated temperature exposure on the adhesion and toughness of thin tantalum nitride films. These films were sputter-deposited at room temperature on sapphire substrates to a nominal thickness of 600 nm with some films heated to 600°C in vacuum while others were heated to 600°C in air. The films heated in vacuum exhibited no changes in composition or structure while the films heated in air completely transformed to tantalum pentoxide. Comparison of the results shows that the interfacial fracture toughness increases from 0.5 MPa-m$^{1/2}$ for as-sputtered films to 0.8 MPa-m$^{1/2}$ for films heated in air. However, the toughness increases to more than 3.0 MPa-m$^{1/2}$ when the films are heated in vacuum. The increase in toughness values follows the reduction in deposition defect content where formation of an oxygen deficient tantalum oxide layer in air from the as-sputtered film increases interfacial toughness slightly while full densification of the tantalum nitride films in vacuum increases toughness to very high levels.

INTRODUCTION

Exposure of thin films to elevated temperatures can strongly affect film durability through changes in composition and structure. [1-5] In these films, the interface structure can be defined on the atomic level through composition and the nature of bonding. [1,3] However, the relationship between interface structure and durability is not well-defined due to limitations in test techniques.

Thin tantalum nitride films are of particular interest. These film systems are used extensively in microelectronic applications such as thin film resistors because of their long term stability and low thermal coefficients of resistance. [6,7] Elevated temperature exposure is often employed during processing to form a protective oxide overlayer and to stabilize the resistance of the thin film elements [8,9] However, the effects of heat during elevated temperature exposure and during subsequent service on contact adhesion are unknown. Therefore, submicron indentation [10-14] and continuous microscratch testing [15-17] are employed in this study to directly measure the effects of elevated temperature exposure on the properties and adhesion of tantalum nitride films on single crystal sapphire substrates in a first step to understanding elevated temperature exposure effects.

MATERIALS AND PROCEDURE

In this study, thin tantalum nitride films were reactively sputtered onto smooth single crystal sapphire substrates to a nominal thickness of 600 nm using a d. c. magnetron sputtering unit. The substrates had been prepared by heating to 170°C in vacuum for two hours to drive off moisture followed by an RF backsputter for 120 s to remove contaminants and expose fresh material. With a vacuum maintained to less than 1.3x10^{-5} Pa (10^{-7} torr), the films were deposited on the substrates using a tantalum target, argon as a carrier gas, and controlled additions of nitrogen to form Ta$_2$N. Deposition was at a rate of 0.3 nm/s onto (11$\overline{2}$0) surfaces of the single crystal sapphire substrates.

The samples were divided into three groups. One group consisted of as-sputtered samples. A second group was heated to 600°C for 100 hours in air, while the third group was heated to 600°C for 100 hours in vacuum. The thickness of each film was established by direct cross-section measurement in a JEOL 840 SEM.

The elastic moduli of the as-sputtered, air-annealed, and vacuum-annealed films were determined from a series of microindentations on each film employing a microindentation test system developed by IBM. [14,15] The tests were conducted at a constant displacement rate of 15 nm/s to depths ranging from 100 to 600 nm using a Berkovitch indenter. Indentation loads and corresponding depths were recorded continuously throughout each test. The area for each indentation in the as-sputtered film was measured directly in a JEOL 840 SEM. The relationship between area of indentation and plastic indentation depth from this series of tests was then used to determine the areas of indentation in the air and the vacuum annealed films. Stiffness was determined from the slopes of the unloading curves following the method of Oliver and Pharr [12] from which the elastic modulus of each film as a function of depth was obtained. [10-12]

Fracture toughness was then determined on the same microindentation test system configured to control normal and lateral indenter displacements. These tests employed a conical diamond indenter with a 1μm tip radius and a 90° included angle that was simultaneously driven into the films at a rate of 15 nm/s and across the films at a rate of 0.5 μm/s until a portion of the film spalled from the substrate. During each test, normal and tangential loads, depth, and lateral position were continuously monitored.

The work of adhesion and interfacial fracture toughness values of the thin Ta_2N films were determined from the data using the elastic approach of Venkataraman et al. [16,17] where the practical work of adhesion is defined as the total elastic strain energy released into the film and the substrate when the film delaminates. As a zero-order approximation for films and substrates of similar elastic properties as found in this study, the strain energy released from the substrate can be equated to the strain energy released when the film delaminates. [16,17] The work of adhesion is then given as,

$$W_a = G_i = G^f_i + G^s_i = 2G^f_i \tag{1}$$

where W_a is the practical work of adhesion and G_i is the strain energy for interfacial fracture composed of the strain energy released upon fracture into the substrate, G^s_i, and into the film, G^f_i. The strain energy released into the film, G^f_i, is a function of the average elastic shear and normal stresses in the delaminated region. These stresses are determined from the critical normal and tangential loads at fracture, the area of delamination, the length of delamination, the track width at the start of delamination, the shear modulus, the elastic modulus, and the thickness of the film. [16,17] The corresponding fracture toughness is a function of the strain energy for interfacial fracture and is given by equation (2), where E^f and v are the elastic modulus and Poisson's ratio of the film respectively.

$$K_i = \sqrt{\frac{E^f G_i}{(1-v^2)}} \tag{2}$$

RESULTS

Heating the tantalum nitride films in vacuum at 600°C had no effect on structure or composition of the films or their interfaces as measured with scanning auger and x-ray diffraction techniques consistent with previous studies of tantalum and tantalum nitride films on sapphire and alumina substrates. [18] However, film thickness decreased from 600 nm to 530 nm indicating that the as-sputtered films are less dense than the vacuum-annealed films.

In contrast, heating the tantalum nitride films in air at 600°C completely transformed them to tantalum pentoxide, Ta_2O_5. This transformation was accompanied by a growth in film thickness from 600 nm to 1380 nm and follows previous observations on elevated temperature exposure of tantalum and tantalum nitride films. [19] Moreover, scanning auger microscopy reveals that the oxide consists of two regions defined by a small variation in the ratio between oxygen and tantalum as observed in previous work [20] on tantalum and niobium pentoxides. The outer oxygen deficient layer extended approximately 1000 nm in from the surface while the inner layer with a

Table I. Thickness (t), elastic moduli (E), work of adhesion (W_a), and interfacial fracture toughness (K_i) values along with average areas of spallation (A) and critical loads at fracture (P_{cr}) for thin tantalum nitride and tantalum oxide films on single crystal sapphire.

Film/Substrate	t (nm)	E (GPa)	A (um^2)	P_{cr} (N)	W_a (J/m^2)	K_i (MPa-m$^{1/2}$)
Ta$_2$N/As-sputtered	635	350	180.4	0.091	0.7±0.5	0.5±0.2
Ta$_2$N/600°C/100h air (Ta$_2$O$_5$)	1380	160	146.6	0.067	3.6±1.5	0.8±0.2
Ta$_2$N/600°C/100h vacuum	530	400	10.3	0.033	24±11	3.2±0.8
(Buckled Fracture)			75	0.041	0.9	0.6

composition approaching stoichiometric proportions extended almost another 400 nm to the substrate. As for the vacuum annealed films, there was no indication of a reaction between the film and the substrate.

Nanoindentation testing revealed that the film properties differed with exposure conditions as shown in Table I. The as-sputtered film had an elastic modulus of 350 GPa while the vacuum annealed film had an elastic modulus of 400 GPa. The as-sputtered values were independent of depth from 100 to 400 nm while the vacuum annealed values exhibited a slight decrease with increasing depth of indenter penetration. Greater depths of indenter penetration caused film failure. Applying the simple rule of mixtures shows this difference in elastic moduli scales directly with film density.

In contrast, the air-annealed Ta$_2$O$_5$ film had a modulus of 160 GPa, quite close to the value of 186 GPa for bulk tantalum but much lower than the moduli of the tantalum nitride films. These values were also independent of depth to 600 nm. As with the nitride films, greater depths of indenter penetration caused film failure.

A series of five microscratch tests were run on each of the three film systems to determine the work of adhesion and fracture toughness. These tests revealed a significant difference in fracture behavior between each type of film. Failures of the as-sputtered Ta$_2$N film occurred consistently at loads between 0.05 and 0.15 N with an average critical load to failure of 0.091 N and were characterized by a sharp change in indenter depth at the point of delamination as shown in Figure 1a. There were typically two large, well-defined (Figure 1b) spalls occurring sequentially during each scratch test with each spall corresponding to a single fracture event. The first spall provided the data needed to determine adhesion and toughness.

Failures of the tantalum oxide film occurred at loads between 0.05 and 0.10 N (Figure 2a) with a somewhat lower average critical load of 0.067 N than the as-sputtered film. However, the areas of spallation were also somewhat smaller suggesting that the work of adhesion and fracture toughness values for the oxide film are higher than the values for the as-sputtered tantalum nitride film. Within the relatively small range of critical loads for oxide film fracture are differences in fracture behavior not observed in the as-sputtered samples. Fracture can occur along the internal oxide interface as clearly shown by the first spall in Figure 2a and indicated by the discontinuity in the load trace. These failures occur on one side of the scratch track and initiate at a preexisting spall. The final failures were always centered on the scratch track whether or not a preexisting spall was included in the fracture process. This indicates that film failure was not dependent on the presence of these preexisting spalls.

In all tests on vacuum annealed samples, fracture occurred at critical loads between 0.03 and 0.04 N. The areas of spallation for most scratch tests were also much smaller as shown in Figure 2b. Although these loads were considerably lower and the areas of spallation much smaller than in the as-sputtered and air annealed samples, the fracture events were characterized by the same abrupt changes in the load-versus-distance traces. In one test the film buckled, leading to a very large spall that wrapped its way back along the scratch trace. It could not be determined if buckling preceded fracture, if it occurred as part of the fracture process, or if it was a response to the initial fracture event. In all other tests fracture occurred along the interface without buckling.

The average work of adhesion and fracture toughness for each film are given in Table I. In all calculations, Poisson's ratio was assumed to equal 0.35, the value for tantalum as well as for many

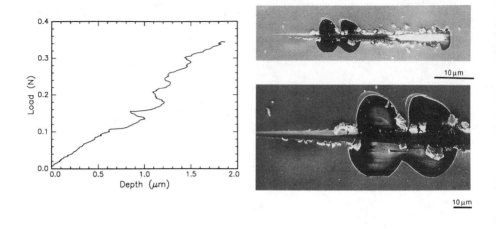

Figure 1. Failures in Ta$_2$N films on single crystal sapphire (a) occurred near 0.1 N and were accompanied by (b) large, well-defined spalls as shown at low and high magnification.

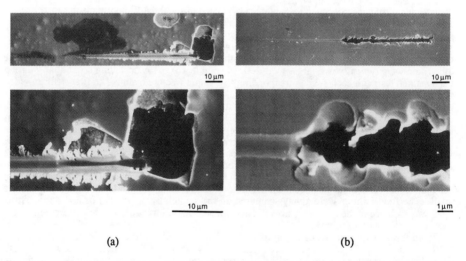

Figure 2. Failures in (a) air-annealed tantalum oxide and (b) vacuum annealed tantalum nitride films occurred at lower loads with smaller spalls than observed on as-sputtered films indicating higher works of adhesion.

other metals. This table shows that the works of adhesion and corresponding fracture toughness values for the as-sputtered and air annealed films that fractured along the substrate interface were similar.

DISCUSSION

The values of adhesion measured in practice are usually from fracture experiments and include contributions from internal stresses, surface roughness, and plastic energy dissipation processes as well as the reversible work done to create two free surfaces. [1,3,4,21,22] In contrast to the more traditional measures of fracture energy, continuous microscratch techniques sample very small volumes of material which minimizes the contributions from plastic energy dissipation. [22] These contributions decrease with film thickness to levels that are near the true work of adhesion [5,23]

Recent work on as-sputtered Ta_2N films shows that values measured for the as-sputtered tantalum nitride films in this study exhibit a lower limit for room temperature fracture. [24] These values significantly exceed van der Waals forces and result from a number of contributary sources. Sputter deposition creates a high density of structural defects that includes pores, vacancies, and dislocations. [25,26] It also creates a high compressive stress through an atomic peening mechanism when deposited under conditions of low working gas pressure and temperatures as used in this study. [27] However, even when residual stresses exceed one GPa, which is near the average internal compressive stress observed for sputter deposited tantalum films [28], there is only a 10 to 15 percent variation in the critical load to fracture. [17] The predominant contribution to adherence is therefore attributable to bond formation between the oxygen active tantalum nitride film and the sapphire substrate. [23] This contribution more than offsets any reduction in adhesion that entrapped argon, pore, and vacancy concentrations induce.

On heating in air at 600°C for 100 hours, the tantalum nitride completely transforms to tantalum pentoxide, Ta_2O_5 with a significant expansion in volume. The large expansion in volume leads to high compressive stresses. [5] Even though these high stresses induce crater-like fractures along the internal oxide interface, they increase the measured critical load to fracture. When coupled with increases in adhesion from release of entrapped argon and shrinkage of any deposition pores, the increased compressive stresses increase the work of adhesion and fracture toughness of the tantalum oxide films from those of as-sputtered tantalum nitride films. Of particular interest is the observation that fracture along the internal oxide interface only occurs when a preexisting spall is struck. This indicates that within the limits of test resolution the strength of oxygen bonds exceeds that of bonds across the film-substrate interface.

Vacuum annealing the tantalum nitride films at 600°C produced the most profound changes in interfacial adhesion and fracture toughness. The high temperatures led to a release of all entrapped argon with an associated increase in adhesion. [26] This release of entrapped argon coupled with annihilation of excess vacancies produced a 10 percent decrease in film volume. Although the effect of these changes on compressive stress is unknown in these films, the magnitude of the measured work of adhesion and corresponding fracture toughness casts doubt that the increase in vacuum annealed over as-sputtered film adhesion values can be attributable to a reduction in defect concentration or to an increase in compressive stress. This suggests that the character of bonding between film and substrate has changed. However, the change is not uniform as there are regions where adhesion exhibits only a small and statistically insignificant increase over as-sputtered films. The size scale of these changes exceeds the resolution of many widely used techniques to study the composition and structure of interfaces. Advances in understanding interfacial adhesion necessitate employing techniques such as high resolution transmission electron microscopy where structure and composition can be studied on the atomic scale.

SUMMARY

Exposure of thin tantalum nitride films in air and vacuum environments at 600°C had a strong affect on measured values of adhesion. Exposure to air at this temperature completely transformed the film to a much thicker, oxygen deficient film of tantalum pentoxide. The film was comprised of two regions differing in composition by a small change in the tantalum to oxygen ratios with a sharp interface. Exposure to vacuum had no effect on film composition but led to a reduction in film thickness through a densification process. Comparison of the results shows that the interfacial fracture toughness increases from 0.5 MPa-m$^{1/2}$ for as-sputtered films to 0.8 MPa-m$^{1/2}$ for films heated in air. However, the toughness increases to more than 3.0 MPa-m$^{1/2}$ when the films are heated in vacuum. The increase in toughness values follows the increase in film density where formation of a more dense tantalum oxide layer increases interfacial toughness slightly while full densification of the tantalum nitride films increases toughness to very high levels.

ACKNOWLEDGMENTS

The authors thank D. Norwood of Sandia National Laboratories, Albuquerque, NM for his assistance in producing the thin films for this study. We would also like to thank B. Bernal, M. Clift, and D. Boehme for their technical support. N. R. Moody and W. Worobey gratefully acknowledge the support of the U.S. DOE through Contract DE-AC04-94AL85000. S. Venkataraman and J. Nelson gratefully acknowledge the support of the Center for Interfacial Engineering at the University of Minnesota under grant NSF/CDR-8721551 and W. W. Gerberich acknowledges support of the U. S. DOE through Contract DE-FG02-84ER45141.

REFERENCES

1. A. G. Evans and M. Ruhle, MRS Bulletin, **XV** (10), 46 (1990).
2. W. E. Bibeau, W. A. Porter, and D. L. Parker, Proc. Electronic Comp. Conf., **28**, 427 (1978).
3. K. P. Tremble and M. Ruhle, in *Metal-Ceramic Interfaces*, edited by M. Ruhle, A. G. Evans, M. F. Ashby, and J. P. Hirth (Pergammon Press, Oxford, 1990) p. 144.
4. K. L. Mittal, Electrocomponent Science and Technology, **3**, 21 (1976).
5. K. L. Chopra, *Thin Film Phenomena*, (McGraw-Hill Book Company, New York, 1969) p. 267-327.
6. J. R. Adams and D. K. Kramer, Surface Science, **56**, 482 (1976).
7. C. L. Au, W. A. Anderson, D. A. Schmitz, J. C. Flassayer, and F. M. Collins, J. Mater. Res., **5**, 1224 (1990).
8. S. Susumu, K. Murasugi, and K. Kaminishi, Proc. Electron. Components Conf., **26**, 177 (1976).
9. D. P. Brady, F. N. Fuss, and D. Gerstenberg, Thin Solid Films, **66** (1980) 287-302.
10. M. F. Doerner and W. D. Nix, J. Mater. Res., **1**, 601 (1986).
11. W. D. Nix, Metall. Trans. A, **20A**, 2217 (1989).
12. W. C. Oliver and G. M. Pharr, J. Mater. Res., **7**, 1564 (1992).
13. S. K. Venkataraman, D. L. Kohlstedt, and W. W. Gerberich, J. Mater. Res., **8**, 685 (1993).
14. T. W. Wu, C. Hwang, J. Lo, and P. S. Alexopoulos, Thin Solid Films, **166**, 299 (1988).
15. T. W. Wu, J. Mater. Res, **6**, 407, (1991).
16. S. K. Venkataraman, D. L. Kohlstedt, and W. W. Gerberich, J. Mater. Res., **7**, 1126 (1992).
17. S. K. Venkataraman, D. L. Kohlstedt, and W. W. Gerberich, Thin Solid Films, **223**, 269 (1993).
18. A. Zhao, E. Kolawa, and M.-A. Nicolet, J. Amer. Vac. Sci. Technol. A, **4**, 3139 (1986).
19. H. Yamagishi and M. Miyauchi, Japanese J. of Applied Physics, **26**, 852 (1987).
20. P. Kofstad, Corrosion, **24**, 379 (1968).
21. D. S. Rickersby, Surface and Coatings Technology, **36**, 541 (1988).
22. S. K. Venkataraman, D. L. Kohlstedt, and W. W. Gerberich, in Thin Film Stresses and Mechanical Properties IV, edited by P. H. Townsend, T. P. Weihs, J. E. Sanchez, and P. Borgesen (Mater. Res. Soc. Proc. **308**, Pittsburgh, PA, 1993) p. 621.
23. P. Benjamin and C. Weaver, Proc. Royal. Soc., London, **A254**, 163 (1960).
24. S. K. Venkataraman, J. C. Nelson, N. R. Moody, D. L. Kohlstedt, and W. W. Gerberich, "Micro-Mechanical Characterization of Tantalum Nitride Thin Films on Sapphire Substrates," Presented at the Symposium on Polycrystalline Thin Films-Structure, Texture, Properties and Applications, 1994 MRS Spring Meeting, San Francisco CA.
25. D. W. Oblas and H. Hoda, J. Appl. Physics, **39**, 6106 (1968).
26. W. W. Lee and D. Oblas, J. Vac. Sci. Technol., **7**, 129 (1970).
27. J. A. Thornton and D. W. Hoffman, J. Vac. Sci. Technol., **14**, 164 (1977).
28. J. A. Thornton, J. Tabock, and D. W. Hoffman, Thin Solid Films, **64**, 111 (1979).

RAMAN CHARACTERIZATION OF POLYCRYSTALLINE SILICON: STRESS PROFILE MEASUREMENTS

M.S. BENRAKKAD[1], A. PEREZ-RODRIGUEZ[1], T. JAHWARI[2], J. SAMITIER[1], J.M. LOPEZ-VILLEGAS[1], J.R. MORANTE[1]
[1]LCMM. Dept. Física Aplicada i Electrònica. Universitat de Barcelona. Avda. Diagonal 645-647. 08028 Barcelona, Spain.
[2]Serveis Científico-Tècnics. Universitat de Barcelona. C. Lluis Solé i Sabarís, 1-3. 08028 Barcelona, Spain.

ABSTRACT

Raman spectroscopy is applied for the analysis of polysilicon films deposited on SiO_2 sacrificial layers. Different deposition technologies and processing parameters have been studied. The features of the first order Si Raman signal (shape, width and position of maximum) are analyzed in order to evaluate the stress average value in the scattering volume. MicroRaman measurements performed at different points on the edge of the samples allows the estimation of the stress gradient through the polysilicon layers. These measurements are correlated with the structure obtained by using micromachined test structures.

INTRODUCTION

Nowadays, polycrystalline silicon is used for active layers in many applications in the field of micro-electro-mechanical systems. For this, the material is deposited on a silicon dioxide film, which constitutes a sacrificial layer. However, the polycrystalline silicon film, after release, can present serious stress effects. These often cause mechanical device failures, curling or fracture. The origin of these large tensile and compressive intrinsic stresses is straightforwardly related to the polysilicon microstructure. Besides this, to know the average stress value from a mechanical point of view it is very important to know the stress profile across the layer thickness.

The uniformity of the stress through the film thickness becomes an essential feature. Changes in the magnitude and direction of the stress can cause cantilevered structures to curl toward or away from the substrate. For instance, a higher tensile stress near the top surface of a cantilever gives rise to an away bend or, viceversa, a lower tensile stress near the top surface gives rise to a substrate bend. Homogeneous stresses do not cause any bending. In the case of bridge structures, only compressive stresses cause bending, whereas tensile bridges are maintained without bending or become broken if the tensile stress value is too high. The existence of these stress gradients limits the maximum possible length before the buckling appears. Moreover, the stress must be kept within a fixed range in order to get reproducible device sensitivity. Often, low tensile stress values are chosen to avoid buckling problems.

Stress gradients can be eliminated or reduced by annealing. Nevertheless, the standard integration rules limit the high temperature processing; and hence, the realization of long length bridges or cantilevered structures by using surface micromachining technologies constitutes a difficult issue.

609

Stress values are usually determined before polycrystalline release from wafer bow measurements or by using X-ray diffraction techniques. After sacrificial layer etching, specially designed test structures allow one to determine the average stress values. Stress gradients have been reported from X-ray diffraction measurements performed on samples etched at different depths. However, this procedure is destructive and time consuming. Information on the average stress can also be obtained by non contact surface profilometry applied to test structures.

In this work, a detailed analysis of the possibilities of Raman spectroscopy to determine the stress profile of polycrystalline silicon layers is presented. Raman spectra obtained under different conditions and using different polysilicon films are presented and discussed. The results of the Raman peak shifts, shape and full width half maxima (FWHM) are interpreted taking into account the stress profile in the layers, as well as the presence of structural defects and stress gradients in the scattering volume.

EXPERIMENTAL DETAILS

Raman spectroscopy measurements were performed with a T64000 Jobin-Yvon triple spectrometer, working in the multichannel mode. Measurements were made at room temperature. For the microRaman measurements, the spectrometer is coupled with a metallographic Olympus microscope, and excitation and light collection are made through the microscope objective. Excitation was provided by the different lines of an Ar^+ laser. From these, the 457.9 nm line was preferred, as it allows one to obtain better lateral resolution in the microRaman configuration.

Special attention was given to the laser power, in order to avoid thermal effects on the Raman spectra [1]. This problem becomes serious in the case of small structures with a very low polycrystalline volume. To avoid these effects, an excitation power density below $0.4 Mw/cm^2$ was used. This has determined the need to use relatively long integration times in the measurement of the spectra.

Samples were vertically placed on a holder controlled by X-Y micropositioners. In this way, the edge of the sample is observed and the microRaman spectra of the polycrystalline cross section can be performed in different regions in the edge. Microscope objective with a numerical aperture of 0.95 was normally used. For this objective, the spot size on the sample is of about 0.6 μm for the 457.9 nm excitation line [1]. Moreover, the spectra are compared with those obtained under the same experimental conditions on a single crystal Si reference sample.

To study the applicability of MicroRaman scattering for stress profile measurements, the analyzed samples have been obtained with different technological parameters, including high and low growing temperature, as well as atmospheric and low pressure conditions. In all cases, the polysilicon layer was grown on a 1 μm thick SiO_2 film. Polysilicon layer thickness ranges from 2 μm to 12 μm. Likewise, average grain size changes from 0,2 μm to sizes larger than 1μm. Usually, the layers present a (220) preferential crystalline orientation, although samples with (422) or (400) preferential orientation, as observed by X ray diffraction measurements.

Finally, in order to facilitate the interpretation of the data, bridge or cantilever based test structures were made on some of the samples. This has allowed us to evaluate the average stress value from a macroscopic point of view.

RESULTS AND DISCUSSION

The Raman spectra measured on the cross section edge of the different polysilicon films show two different behaviours, according to their shape. From to this, we can distinguish between two clear situations: a) single crystal like spectra and b) asymmetric spectra.

Single crystal like spectra

Figure 1 shows the first order Si Raman spectrum measured on the cross-section of a thick poly-Si layer.

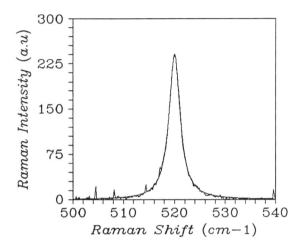

Fig. 1 First order Raman line from polycrystalline Si layer. The spectrum has been fitted with a Lorentzian curve.

This layer was deposited at high temperature ($> 1000°C$). It corresponds to a polysilicon material with large average grain size (about 1 μm) and a defect density of about 10^{11} cm^{-2}. This spectrum has been fitted with a Lorentzian curve. The spectrum is characterized by having FWHM value about 10 % higher than that from the spectrum measured on the Si reference sample. However, the shape of both spectra is very similar, and the main difference between them is given by a shift of the Raman line ($\Delta\omega$). All This rules out any contribution of structural defects in the Raman line for these samples, and the shift in relation to the reference spectrum is determined by the presence of an homogeneous stress (σ) in the scattering volume. Under these conditions, the amount of the shift of the Raman line in relation to the reference one gives in a straightforward way, the stress value. According to Anastassakis [2,3], $\sigma(MPa) = -K\Delta\omega$ being $K = 140$ MPa cm.

Figure 2 shows the stress profile distribution measured in one of these poly-Si layers. As is shown, the distribution is characterized by relatively low stress values. For other samples not shown in fig. 2, the surface region of the layer shows thermal stress and a partial oxidation through the grain boundaries, which causes an additional compressive stress in comparison with the values obseved on the edge. This is bassically due to the high temperature procedure.

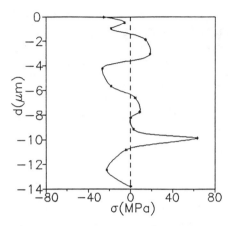

Fig.2 Stress depth profile in the polysilicon layer. d is the distance from the top surface.

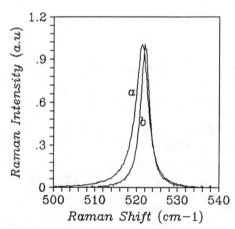

Fig. 3 (a) Raman spectra measured from the thinner layer,(b) is a Lorentzian curve.

Asymmetric spectra

The Raman spectra measured from thinner samples deposited by LPCVD and annealed at various temperatures show a non-Lorentzian shape, as can be seen in figure 3. In this case, the spectra are asymmetrically broadened towards the low frequency side in relation to the reference spectrum, in addition to the frequency shift. FWHM is about 20% higher than that from reference Si. The behaviour suggests the presence of disorder effects in the spectra, determined by structural defects. Then, a deconvolution of these effects has to be made by assuming a correlation length model, which gives an estimation of the length for which translational symmetry and conservation of momentum continue to hold.

Assuming a constant correlation length (L) (and so, a homogeneous distribution of defects in the scattering volume), the shape and broadening of the spectra have been simulated [4]. Figure 4 shows the theoretical Raman line shift versus FWHM according to this model. The dependence of the FWHM on the asymmetry of the spectrum for the simulated spectra is plotted in figure 5, together with the experimental points. The asymmety is calculated as the ratio Γ^-/Γ^+, where Γ^- is the half width at half maximun (HWHM) on the low-energy side to HWHM on the high-energy side. The difference in the position of the maximum of the simulated spectra (plotted in figure 4) from the experimental one gives the stress contribution to the Raman line shift, if an homogeneous stress is assumed in the scattering volume. In this way, an estimation of the average stress value in the scattering volume can be made.

However, we have to remark that for some samples the experimental points tend to have a lower asymmetry than the theoretical simulation, as is shown in figure 5. This could be determined by two facts: i) the presence of a non homogeneous stress in the scattering volume, which would modify the shape of the spectrum, and ii) the existence of regions in the grains with different defect concentrations.

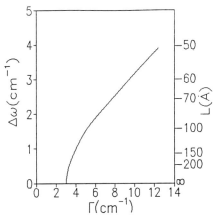

Fig. 4 The relationship between the FWHM (Γ) and peak shift Δω, using the three-dimensional phonon confinement

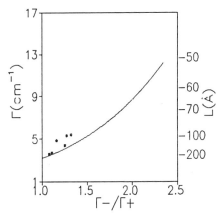

Fig. 5 Relationship between FWHM (Γ) and asymmetry (Γ⁻/Γ⁺) of the Raman line. The points correspond to thinner sample.

In fact, the second situation is to be expected for a polysilicon layer, as the regions close to the grain boundaries are more defective than than inner region of the grains. To take this into account, a length distribution function has to be assumed in the model. Anyway, the deduction of the correlation length from the broadening of the Raman line allows a first estimation of average stress. In this way, values of the same order of magnitude as those measured from micromachined structures are obtained.

So, in general compressive stress value are obtained for samples annealed at high temperature, while a tensile behaviour is found for the samples annealed at lower temperatures ($\leq 1000^\circ$C). These data have been compared with those from the test structures. Although the stress values obtained from both measurements are not the same, they are of the same order of magnitude and character.

Difference observed in the stress values have to take into account the different nature of the measurements. MicroRaman measurements are made at a microscopic level in poly-Si/SiO$_2$/Si structures and reveal non-homogeneous stress distributions across the layer. In fact, the values obtained correspond to a stress estimation, as they assume the existence of homogeneous stress and defect distribution in the scattering volume. Besides this, stress measurements on bridge or cantilever structures are made on a macroscopic scale, as the dimensions of the structures are several orders of magnitude bigger than the Raman scattering volume. Moreover, these are made on processed samples, where the sacrificial SiO$_2$ layer has been removed.

CONCLUSIONS

The first results on the Raman analysis of stress gradients through different polysilicon films are presented, taking into account the possible effects of the structural defects on the spectra. The results obtained already show the potential of this technique for the evaluation of stress distributions, although further development is needed in order to take into account the possible presence of variation of the stress or concentration of defects in the scaterring volume. Anyway, this represents an alternative technique in relation to those using processed test structures.

Likewise, the ability to measure the stress profile through the film allows the analysis of stress evolution as a function of the technological process to be performed. In this sense, the correlation of the stress gradients with the physical processes taking place during the layer deposition or annealing is a critical point.

REFERENCES

[1] E. Martín, J. Jiménez, A. Pérez-Rodríguez, J.R. Morante, Materials Letters 15, 122 (1992).
[2] E. Annastassakis, in Proc. ISPPME'85, Varna, Bulgaria: Physical Problems in Microelectronics, ed. J. Kassabov (World Scientific, Singapore, 1985), 128.
[3] E. Annastassakis, E. Liarokapis, J. Appl. Phy. 62 3346 (1987).
[4] M. Yang, D. Huang, P. Hao, F. Zhang, X. Wang, J. Appl. Phy. 75 651 (1994).

MEASUREMENT OF THERMALLY-INDUCED STRAINS IN POLYCRYSTALLINE Al THIN FILMS ON Si USING CONVERGENT BEAM ELECTRON DIFFRACTION

S.K. Streiffer, S. Bader, C. Deininger, J. Mayer, and M. Rühle
Max-Planck-Institut für Metallforschung, Institut für Werkstoffwissenschaft, Seestraße 92,
D-70174 Stuttgart, Germany

ABSTRACT

Strains in polycrystalline Al films grown on oxidized Si wafers were measured using convergent beam electron diffraction (CBED). CBED patterns were acquired on a Zeiss EM 912 TEM equipped with an imaging energy filter and CCD camera. HOLZ line positions in the (000) CBED disk were matched using an automated refinement procedure. A sensitivity to variations in lattice parameter of approximately 0.00007 nm was obtained. Strong deviations from a simple equibiaxial strain, perfect [111] texture model were observed.

INTRODUCTION

Because of differences in the thermal expansion coefficients of the various materials used in semiconductor heterostructures, thermal processing of such structures can lead to large stresses and strains and even device failure. In particular, the reliability of Al interconnects in integrated circuits is affected by such stresses. Thermally induced stresses in polycrystalline Al thin films on Si have been extensively studied using techniques such as wafer curvature and x-ray diffraction[1]. However, these are large area methods and yield results that are averaged over the sample, rather than the local values which may be important for hillocking, electromigration-induced voiding, or stress relaxation by grain boundary diffusion.

Convergent beam transmission electron diffraction (CBED) offers the possibility of achieving much greater spatial resolution compared to conventional techniques used to measure stress and strain, because of the ability to focus an electron probe to very small size. Despite this promise of improved spatial resolution, the usefulness of transmission electron diffraction for strain determination has been limited because electrons are very strongly scattered. This leads to an effective failure of Bragg's Law due to dynamical diffraction effects, and to a large inelastic background in the diffraction pattern that sharply reduces the signal to noise ratio for samples of reasonable thickness.

Recently, it has proved possible to address these limitations and measure lattice parameters with very high precision and very high spatial resolution in the TEM by measuring the position of higher order Laue zone (HOLZ) lines in CBED patterns, and matching these positions to appropriate computer simulations which include corrections for dynamical diffraction effects[2,3]. A new generation of microscopes equipped with imaging electron energy filters and cooled, slow-scan CCD cameras allows one to remove most of the inelastic background from the patterns and to more easily quantify the intensity and position of diffraction features. Here we report on the application of these techniques to the measurement of local strain variations in [111]-textured Al films deposited onto oxidized Si wafers.

Mat. Res. Soc. Symp. Proc. Vol. 343. ©1994 Materials Research Society

EXPERIMENTAL PROCEDURES

[111] fiber-textured Al films were grown on oxidized Si wafers to a thickness of approximately 0.22 µm by UHV evaporation at room temperature. The oxide thickness was 50 nm. Tensile strains were induced in the Al by cycling the samples to T=450°C for 20 minutes in forming gas. This is expected to generate an equibiaxial stress of approximately +250 MPa in the Al film at room temperature immediately after annealing.

Specimens for transmission electron microscopy were prepared by cutting 3 mm disks from the wafers with an ultrasonic core drill, and thinning these disks from the backside to a thickness of 100 µm. Next, the samples were dimpled to a thickness of approximately 5 µm. They were then attached Al side down to a glass microscope slide using a low temperature wax, and the edges of the disks were coated with acid resistant lacquer. A solution of 20% HF / 80% HNO_3 was then used to etch a window in the Si substrate in the dimpled region, creating a large, uniform-thickness area of Al thin film for examination in the microscope.

Convergent beam electron diffraction patterns were collected on a Zeiss EM 912 transmission electron microscope equipped with an imaging Omega energy filter and Gatan CCD imaging system. The electron probe size used for diffraction was approximately 32 nm. In order to reduce contamination, the specimens were cooled during observation to temperatures ranging from –50°C to –149°C. Electrons suffering thermal diffuse scattering have energy losses too small to be removed by the filter; therefore cooling also improved the visibility of the HOLZ lines even in filtered patterns. After collection, the point spread of the CCD array was removed from the CBED disks by deconvolution.

The Al <233> zone axis, located 10.02° from [111], was chosen as having the best combination of number and contrast of HOLZ lines, proximity to the surface normal, and minimum dynamicity. Measurements were performed on grains of various sizes at this zone axis.

The data were analyzed based on methods developed by Zuo[3]. Line equations were measured for a number of experimental HOLZ lines in a given pattern, and were used to calculate a set of HOLZ line intersections. The distances between intersections were then compared to distances calculated from a simulated HOLZ line pattern for a trial set of lattice parameters. Simulated patterns were generated using a kinematical description[4] for electron diffraction, plus dynamical corrections to the HOLZ line positions evaluated using a modified version of the program TCBED written by Zuo and Spence[5,6]. Least-squares minimization of the difference between simulated and experimental distances as a function of lattice parameters was accomplished using a quasi-Newton algorithm[7]. The lattice parameters obtained from this procedure were then given to TCBED and used to calculate a new set of dynamical corrections, and the best-fit lattice parameters redetermined. This was repeated until convergence, usually requiring of the order of three iterations. Care was taken to include only HOLZ lines near the center of the (000) CBED disk so as to minimize the sensitivity of line positions to thickness[8] and to avoid regions of HOLZ line hybridization, which leads to both abrupt shifts in HOLZ line positions and to strong deviation of the affected lines from the straight line approximation.

Other algorithms for extracting lattice parameters from HOLZ line arrangements have also been demonstrated[9]. The technique described above has the advantages that it is invariant with respect to pattern orientation, may be easily corrected for pattern magnification, and presents a relatively simple situation for error analysis. As discussed by Zuo, the intersection distances may all be normalized to a particular distance, scaled by a factor chosen to minimize χ^2, or scaled absolutely. We have chosen to calibrate the microscope camera length at the CCD based on CBED patterns recorded from a bulk, strain-free polycrystalline aluminum sample, and to allow the minimization routine to choose the best pattern magnification within narrow bounds about this value. These same CBED patterns were also used to evaluate electron-optical distortions, e.g. anisotropic magnification.

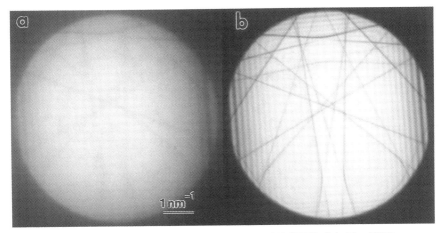

Figure 1: Unfiltered (a) and filtered (b) [233] zone-axis (000) CBED disk. T= -50°C.

The necessity of energy filtering the CBED patterns is demonstrated as follows. The inelastic mean free path λ_{in} for Al at this accelerating voltage is of the order of 100 nm[10]. For the sample thickness t = 0.22 μm used here, the intensity of electrons not suffering energy losses I_{el}, divided by the total intensity I_T, is given by:

$$\frac{I_{el}}{I_T} = \exp\left[-\frac{t}{\lambda_{in}}\right] \sim 0.1. \qquad (1)$$

Therefore the majority of electrons for this sample thickness are inelastically scattered. Inclusion of electrons with energy losses in the diffraction pattern leads to a strong diffuse background and a broadening and shift of specific diffraction features, because many of the inelastically scattered electrons with an effectively incorrect wavelength are subsequently diffracted. Chromatic aberration further degrades the observed pattern. For the present case, the inelastic background is so intense as to render the experiments impossible without filtering, as demonstrated by the comparison of unfiltered (Figure 1(a)) and filtered (Figure 1(b)) [233] zone-axis (000) CBED disks, taken from the bulk aluminum sample from a region of approximately 0.18 μm thickness, at -50°C. Based on consideration of the magnification at the CCD, the HOLZ lines used and their width, and errors in the dynamical corrections and measurement of the accelerating voltage, it is concluded that a change in lattice parameter of approximately 0.00007 nm is detectable, corresponding to a strain resolution of the order of 2×10^{-4}.

RESULTS AND DISCUSSION

A representative unfiltered bright field image is shown in Figure 2. The average grain size of the Al was on the order of 0.5 μm. Some abnormally large grains are clearly visible.

The HOLZ line arrangement near the center of the (000) disk for a 1 μm grain at T=-50°C and a 2 μm grain at T=-50°C and T=-150°C are shown in Figures 3 (a), (b), and (c) respectively. The lattice parameters giving the best fit to Figure 3(c) were used to calculate the

Figure 2: Unfiltered bright-field micrograph of the Al thin film.

full dynamical simulation shown in Figure 3(d). Movements of the HOLZ lines are visible both between the two grains at the same temperature and with temperature for the same grain. The two patterns taken at T=-50°C should also be compared to the pattern from the bulk Al sample taken at the same temperature, shown in Figure 1.

If perfect [111] texture and equibiaxial strain are assumed in the film, the expected lattice distortions in the Al are satisfied by a trigonal unit cell given by $a_{cubic} \Rightarrow a_{trigonal}$ and $\alpha_{cubic}=90° \Rightarrow \alpha_{trigonal} \neq 90°$. The [233] zone axis of such a unit cell still displays mirror symmetry, which is clearly not the case for the patterns presented here, nor for any other [233] pattern observed for these films. Reasons for this deviation from mirror symmetry include tilt of the [111] direction from the grain surface normal, dislocation effects on the strain field, and inhomogeneities in the sample, as well as any other nonequibiaxial strain and nonplane stress sources.

Although deviations from perfect [111] texture are certainly present, the deviations are generally small (<10°). Misorientations of such a magnitude are insufficient to generate the observed deviations from mirror symmetry. Therefore, the lattice parameters were refined according to a model of biaxial (but not equibiaxial) strain, using the bulk Al lattice parameters at the appropriate temperature as reference. The data for four different grains at T=-50 °C, and for one grain (Grain 3) at various temperatures, are shown in Figure 4 (a) and (b). The principal strains deduced from the biaxial model (shown as filled symbols) are compared to the best fits for an equibiaxial model (squares) in these two graphs. The nonequal, biaxial model typically improved the goodness of fit, as measured by χ^2, by approximately two to four over the equibiaxial case, and by several times the measurement error. Grains 1 and 2 had diameters of approximately 1 μm, while Grains 3 and 4 had diameters of 2 μm and 0.5 μm, respectively.

Clearly, significant deviations from the simple equibiaxial strain, perfect [111] texture model are required to produce the symmetry breaking observed in the experimental patterns. No correlation of strain state with grain size is apparent, although this is perhaps not unexpected given the small number of measurements presented and the variations in the environment (exact orientation, orientation of neighboring grains, etc.) of each grain. Finally, no absolute calibration of the unstrained lattice parameter in the Al film was available, so the absolute values of the strains presented in Figure 4 are somewhat uncertain. The relative changes for the different measurements, however, are expected to be very reliable.

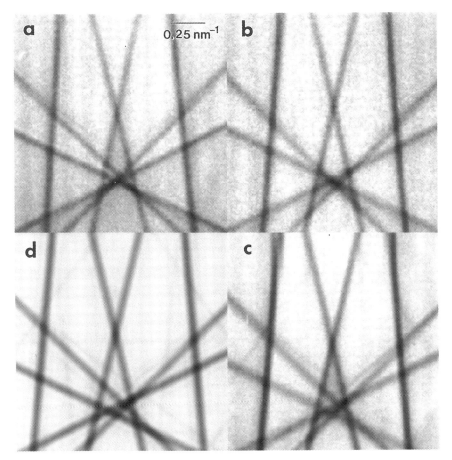

Figure 3: [233] zone-axis HOLZ line arrangements near the center of the (000) disk for (a) grain size 1 μm, T=-50°C; (b) grain size 2 μm, T=-50°C; and (c) grain size 2 μm, T=-150°C. (d) dynamical simulation using best-fit lattice parameters for (c).

SUMMARY AND CONCLUSIONS

Thermally-induced strains in polycrystalline Al thin films deposited onto oxidized silicon wafers were measured for several different grains and at temperatures ranging from –50°C to –149°C. The positions of HOLZ lines in the (000) disk of the [233] zone axis of the Al were matched using an automated refinement procedure that calculated lattice parameters according to two models: one assuming perfect [111] texture and equibiaxial strain, and one assuming perfect [111] texture and nonequal, biaxial strain. Strong deviations from equibiaxial strain are seen in all of the experimental CBED patterns. More work is needed to understand the origin of these deviations, and to correlate the strain measured in individual grains with the surrounding microstructure.

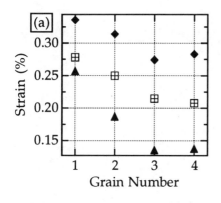 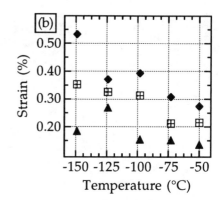

Figure 4: (a) Principal biaxial (filled symbols) strains determined for 4 different grains, compared to best-fit equibiaxial strain (squares). Grain sizes are 1 μm, 1 μm, 2 μm, and 0.5 μm, for Grains 1-4, respectively. (b) Best-fit principal biaxial (filled symbols) or equibiaxial (squares) strains for Grain 3 as a function of temperature.

ACKNOWLEDGMENTS

This research was performed under the auspices of the Max-Planck-Gesellschaft, and with the support of the BMFT contract # NTS 0215/8. S.K.S. wishes to acknowledge the Alexander von Humboldt-Stiftung for financial support. We wish to thank U. Bladeck for assistance with sample preparation and microscope operation. Many helpful discussions with Drs. S. Baker, R. Venkatraman, and P. Besser are gratefully acknowledged.

REFERENCES

1. see, e.g., C.J. Schute and J.B. Cohen, J. Mater. Res. **6**, 950 (1991); M.F. Doerner and S. Brennan, J. Appl. Phys. **63**, 126 (1988); M.F. Doerner and W.D. Nix, J. Mater. Res. **1**, 601 (1986).

2. Y.P. Lin, A.R. Preston, and R. Vincent, *EMAG* 1987, IOP Conference Series No. 90 (Institute of Physics, Bristol) p. 115.

3. J.M. Zuo, Ultramicroscopy **41**, 211 (1992).

4. see J.C.H. Spence and J.M. Zuo, *Electron Microdiffraction* (Plenum Press, New York, 1992).

5. J.M. Zuo, K. Gjønnes, and J.C.H. Spence, J. Electron Microsc. Tech. **12**, 29 (1989).

6. C. Deininger, G. Neckar, and J. Mayer, accepted for publication in Ultramicroscopy.

7. implemented in *IMSL/IDL* , IMSL, Inc. Sugar Land, TX.

8. John Mansfield, David Bird, and Martin Saunders, Ultramicroscopy **48**, 1 (1993).

9. H.J. Maier, H. Renner, and H. Mughrabi, Ultramicroscopy **51**, 136 (1993); S.J. Rozeveld, J.M. Howe, and S. Schauder, Acta Metall. **40**, S173 (1992).

10. R.F. Egerton, *Electron Energy Loss Spectroscopy in the Electron Microscope* (Plenum Press, New York, 1986) , p. 294.

TRIBOLOGICAL PROPERTIES OF LASER DEPOSITED SIC COATINGS

T. ZEHNDER*, A. BLATTER*, J. BURGER**, C. JULIA-SCHMUTZ**, R. CHRISTOPH**
*Institute of Applied Physics, University of Bern, Sidlerstr. 5, 3012 Bern, Switzerland
**Swiss Center for Electronics and Microtechnology, Inc., P.O. 41, CH-2007 Neuchâtel, Switzerland

ABSTRACT

We studied on a nanometer scale the tribological properties of thin silicon carbide films on Si(100) wafers and stainless steel. The coatings were fabricated from a sintered SiC target by pulsed ArF laser deposition at substrate temperatures between 20 °C and 1000 °C. Amorphous films resulted at low deposition temperatures while nanocrystalline structures developed at high deposition temperatures. An atomic force/lateral force microscope was employed to characterize the film topography and the friction behavior. The microhardness was determined from measurements utilizing a depth-sensing nanoindentation instrument. The SiC films on Si(100) exhibit a smooth surface with an average roughness R_a of a few nanometer, the amorphous films being even an order of magnitude smoother. No appreciable differences were found in microhardness and friction coefficient between amorphous and nanocrystalline films. On stainless steel, amorphous SiC coatings were obtained for deposition temperatures up to 500 °C. Their surface relief portrayed the grain boundaries of the underlying steel substrate, reflecting the ballistic nature of the deposition process. No stoichiometric films were obtained above 500 °C as the silicon from the growing film quickly dissolved in the steel substrate.

INTRODUCTION

Tribological coatings must be not only wear resistant, but also strongly adherent to the bulk material and, usually, smooth to minimise mechanical friction. Silicon carbide is a chemically inert high temperature semiconductor with interesting potential in the fabrication of powerful electronic devices. SiC is also among the most abrasion resistant materials, which makes it an excellent coating material for tribological applications [1]. There exist some difficulties, however, in the fabrication of SiC thin films. One of the major problems is the high decomposition temperature, which hinders the fabrication of stoichiometric films by conventional techniques like thermal evaporation. Chemical vapor deposition methods have been developed that allow deposition temperatures below 1000 °C [2]. The technique of

Mat. Res. Soc. Symp. Proc. Vol. 343. ©1994 Materials Research Society

pulsed laser deposition (PLD) allows temperatures as low as 25 °C [3]. The short thermal cycle in PLD results in congruent ablation and the evolving supersaturated vapor promotes the growth of smooth films with the stoichiometry of the target. The high kinetic energy of the ablated species results in enhanced adhesion and density even for low-temperature deposition.

In this paper we study the mechanical properties of SiC films deposited by PLD on Si(100) wafers and stainless steel. We compare amorphous with nanocrystalline films which were obtained by variation of the deposition temperature. The film microhardness was derived from nanoindentation measurements with a depth-sensing instrument. The surface roughness and the submicron film friction behavior were characterized using an atomic force microscope (AFM).

EXPERIMENTAL PROCEDURES

The experimental procedure of SiC deposition by PLD is described in detail elsewhere [4]. Briefly, 20 ns pulses from an ArF excimer laser ($\lambda = 193$ nm) were focused onto a sintered SiC target placed in a deposition chamber. The laser irradiance on the target was about 10^8 W/cm^2 and the chamber pressure 10^{-2} Pa. The distance between target and substrate was 21 mm. The polished Si(100) wafers and soft stainless steel substrates (316L [5]) were radiatively heated during deposition up to 1000 °C. All films were usually deposited under the same laser conditions at a pulse repetition rate of 20 Hz during one hour and during a quarter of one hour for the Si substrates and the steel substrates, respectively. The resulting film thicknesses were 2 μm and 0.5 μm, respectively. A 7 μm thick film was deposited on Si as well.

Films were amorphous at deposition temperatures up to 800 °C whereas nanocrystalline ß-SiC with a prominent (100) texture grew at 900 °C and above as revealed by X-ray diffractometry [4]. Photoelectron Spectroscopy (XPS) verified dominant Si-C bonding for both the amorphous and crystalline structures, along with some oxygen contamination (~8%, most probably SiO_x) due to residual gas pressure in the deposition chamber. Some C-C bonds existed in the films (as was most obvious from Raman spectroscopy). These likely indicate the presence of amorphous carbon. Films deposited on steel at temperatures above 500 °C were not transparent. At 700 °C, for instance, X-ray diffraction showed extra Bragg peaks, which could not be exactly identified, and XPS revealed only a thin SiC surface layer of less than 5 nm. This showed that Si quickly dissolved in the steel substrate when heated to temperatures above 500 °C [1].

Nanoindentation measurements were performed with a depth-sensing instrument with a Vickers diamond indenter under a load of 20 mN (2 g). Topography and roughness were measured over several scan ranges (1 x 1 μm^2, 5 x 5 μm^2 and 10 x 10 μm^2) with a stand-alone AFM using high-quality Ultralever tips with curvatures below 10 nm. The same AFM was operated in the lateral-force mode to evaluate the friction coefficient at a submicron scale. Upon varying the applied normal force from 1 nN to 20 nN, the resulting torsion of the cantilever was simultaneously measured and the lateral forces therefrom calculated [6]. The ratio of lateral force to normal force is the friction coefficient.

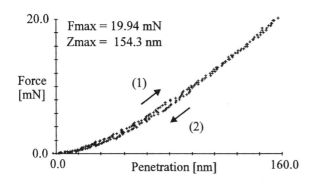

Fig. 1. Typical loading (1) - unloading (2) curves during indentation of an amorphous, 7 μm thick SiC film on Si(100). The film exhibits almost complete elastic deformation up to loads of at least 20 mN.

RESULTS AND DISCUSSION

A typical loading/unloading curve during indentation of a SiC film deposited on a Si(100) substrate is shown in Fig. 1. The displayed data are taken from an amorphous (700 °C deposition temperature) film of thickness 7 μm. The recorded depth is composed of both elastic and plastic displacement. The film shows almost complete elastic deformation behavior as the unloading curve closely follows the loading curve. The maximum penetration depth at 20 mN (2 g) load is 154.3 nm. Assuming that during initial unloading the area in contact with the indenter remains constant, the plastic and elastic contributions can be separated by fitting a tangent to the unloading curve at F_{max}. The intercept of the tangent with the abscissa gives the plastic displacement at F_{max}. From these values and an experimentally determined indenter shape calibration the microhardness can be calculated [7]. A microhardness of 42.6 GPa is derived from the curve in Fig. 1. After unloading, there was no residual indentation visible in an optical microscope because of complete elastic recovery. All other films grown between 300 °C and 950 °C had comparable microhardnesses, i.e. no marked correlation between film hardness and microstructure exists. The scattering of the data for an individual film lies in the range of 10% to 20% which is on the order of the difference between the highest and the lowest hardness value of all the films. However, there seems to be a weak trend of slightly increasing hardness with increasing deposition temperature, i.e. when going from amorphous to crystalline microstructures. Compared with the hardness of 12 GPa tabulated for Si [8] we definitely observe a substantial, 3 to 4 times increase of the microhardness after coating with a thin SiC film.

The steel-SiC system showed a strongly different behavior in the same range of loads. The smaller film thickness and the softer substrate (1.7 GPa [5]) clearly had their effect on the indentation. Unloading curves did not show any elastic recovery. Cracks were formed around the rim of the impressions, which is a manifestation of plastic deformation of the underlying soft substrate. The microhardness of the steel-SiC system is affected by the softer substrate and is load dependent [9].

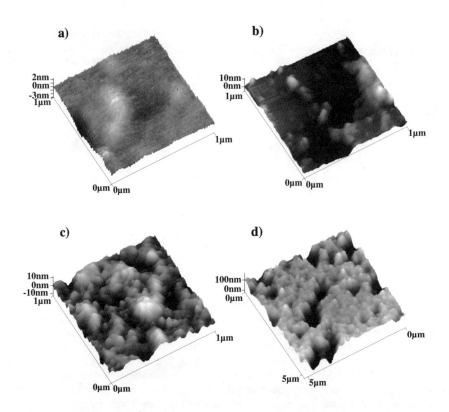

Fig. 2. AFM images of the topography of SiC films grown a) at 300 °C on Si(100), b) at 700 °C on Si(100), c) at 950 °C on Si(100) and d) at 200 °C on steel.

AFM images of the topography of SiC films on Si(100) substrates, grown at three different deposition temperatures, are pictured in Fig. 2. a) to c). The amorphous film grown at 300 °C is smooth (a) with an average roughness of 0.35 nm. The roughness parameters for the different scan ranges are summarized in Table I. It is seen that the values are larger than the

Table I. Average roughness of SiC films determined for square areas with 1 μm, 5 μm and 10 μm side and friction coefficient from Atomic Force Microscope measurements.

Substrate	Si(100)	Si(100)	Si(100)	Si(100)	Steel
Deposition temperature	bare	300 °C	700 °C	950 °C	200 °C
Microstructure	Monocryst.	Amorphous	Amorphous	Nanocryst.	Amorphous
R_a(1 μm) [nm]	0.09	0.35	1.76	2.51	-
R_a(5 μm) [nm]	0.08	0.30	2.10	2.38	24.70
R_a(10 μm) [nm]	0.14	0.68	6.37	2.82	28.00
Friction coefficient	0,22	0.031	-	0.28	0.68

values of the bare silicon substrate, i.e. the deposition leads to surface roughening. The film grown at 700 °C has a rougher surface, due to the presence of some hilltype features of lateral dimensions of several 10 nm (b). The film is still X-ray amorphous. However, these features might well be nanocrystallites indicating the beginning of crystallization, the crystalline fraction being too small to be detected by X-ray diffractometry. The step in the right corner is an embedded large particle. Such inclusions are the result of splashing which is an undesired but characteristic effect of PLD [10]. Embedded particles may drastically increase the measured film roughness as was observed when increasing the scan area from 1 x 1 μm^2 (R_a = 1.76 nm) to 10 x 10 μm^2 (R_a = 6.37 nm). The nanocrystalline film grown at 950 °C (c) exhibits a pronounced topography. The microroughness of nanocrystalline films was always less than 3 nm. The nanocrystallites closely resemble the features in b), supporting the idea that the film in (b) is partly crystalline.

An amorphous SiC film grown at 200 °C on the polished steel substrate exhibits deep pores (Fig. 2 d)). These are the consequence of the steel grain boundaries. This demonstrates that the roughness of the amorphous films is determined by the surface finish of the substrates. The reduced surface diffusion of the monomers at the low deposition temperatures and the uniform spatial distribution of impinging particles tend to conserve features of the substrate surface.

The measured lateral force on the cantilever versus the employed normal force, i.e. the friction coefficient, was determined for amorphous SiC films grown at 300 °C on Si(100) and at 200 °C on steel, for the nanocrystalline film grown at 950 °C and both substrates. The friction coefficients at a submicron scale are listed in Table I. The corresponding value for the bare steel substrate is 0.31. The friction coefficient of the smooth amorphous SiC film on Si(100) is nearly one order of magnitude lower than that of the rougher nanocrystalline film. This difference may be due to the different material properties of the amorphous and the crystalline phase of SiC, or it may be solely a consequence of the accentuation of the surface topography. A hint that points to a topography effect is the much higher friction coefficient of the amorphous SiC film on rough steel substrate. This film had a microstructure and composition equal to those deposited on Si(100). On the other hand the friction as measured with our technique should not depend on roughness. In comparison with the friction coefficients of the substrate materials we note that the rough coatings have higher friction coefficients, whereas the smooth amorphous film on silicon exhibits substantially lower friction.

CONCLUSIONS

No appreciable differences in the microhardnesses were found between the amorphous and the nanocrystalline SiC films deposited on Si. The microhardnesses of both types of films were in the range of 40 GPa. The topography was more accentuated in crystalline films than in amorphous films. This is due to nanometer-sized grains which give rise to a larger roughness value and result in a higher friction coefficient compared to the amorphous films. Only amorphous SiC films were obtained on steel because chemical reactions took place between the coating and the substrate at deposition temperatures above 500 °C. Due to the low surface diffusion at growth temperatures below 500 °C, the topography of the amorphous films is a portrait of the substrate. Compared to films deposited on Si, we also observe a dramatic increase of the friction coefficient for the coated steel substrate.

In conclusion, amorphous SiC coatings prepared by low-temperature PLD show at least as good nanotribological behavior as crystalline ones. Adhesion and hardness are similar but the amorphous films are smoother and have a lower friction coefficient. The influence of the substrate material seems to be more prominent than that of the microstructure. The possibility of low-temperature deposition broadens the class of materials which can be protected by a SiC overcoat as demonstrated with steel.

ACKNOWLEDGEMENTS

We are particularly grateful to Professor H. P. Weber for helpful discussions and O. Küttel of the Institute of Physics, University of Fribourg for the XPS measurements. This work was supported by the Swiss Commission for the Encouragement of Scientific Research.

REFERENCES

1.   Bharat Bhushan and B. K. Gupta, Handbook of Tribology, 1st ed. (McGraw-Hill Inc., 1991), p. 14/73.
2.   M. Ignat, M. Ducarroir, M. Lelogeais, J. Garden, Thin Solid Films **220**, 271 (1992).
3.   J. Krishnaswamy, A. Rengan, J. Narayan, K. Vedam, C. J. McHargue, Appl. Phys. Lett. **54**, 2455 (1989).
4.   T. Zehnder, A. Blatter, A. Bächli, Thin Solid Films **241**, 138 (1994).
5.   Notz Stahl AG, 2555 Brügg, Switzerland.
6.   O. Marti, Physica Scripta **T49**, 599 (1993).
7.   M. F. Doener and W. D. Nix, J. Mater. Res. **1** (4), 601 (1986).
8.   Landolt Börnstein NS III/17a, (Springer-Verlag, Heidelberg, 1982), p. 77.
9.   B. Jönsson and S. Hogmark, Thin Solid Films **114**, 257 (1984).
10. H. Dupendant, J. P. Gavignan, D. Givord, A. Lienard, J. P. Rebouillat, Y. Souche, Appl. Surf. Sci. **43**, 369 (1989).

PART VI

Polycrystalline Metallization

POLYCRYSTALLINE COMMUNICATION LINKS:
MICROELECTRONIC CHIP INTERCONNECTS

TIMOTHY D. SULLIVAN
IBM Microelectronics, 1000 River St., Essex Junction, VT 05452

ABSTRACT

All microelectronic chip interconnects are composed of polycrystalline films with properties that vary discretely with grain orientation. Techniques for dealing with this variation are developing as diminishing dimensions accentuate the effects of grain orientation. Variations in metallization microstructure affect mass transport and median failure time for metallization wearout mechanisms. Stability of adhesion layers, barrier layers and redundant conduction layers is influenced by the microstructure of the films in contact with each other. Intermetallic reactions, which are of no concern with single-level metallization, can become significant for multi level metallizations. Ti/Al systems provide typical examples of these trends.

INTRODUCTION

In integrated circuit microelectronics, all on-chip device interconnects are fabricated from polycrystalline films. Multi-level metallization layers, such as shown in Fig. 1, employ a variety of Al alloys vertically connected to each other with studs made from CVD tungsten films. Thin films of Ti/W and Ti are used as adhesion layers and redundant conductors for Al-alloy films, and as barrier layers at metal/Si contacts. TiN films provide antireflective coatings and diffusion barriers, while polysilicon, refractory silicide films, and CVD W films are employed for local interconnects. The polycrystalline nature of these films is commonly known, but they are usually treated as though they were homogeneous. Resistivity, composition, stress levels, etc., are measured over large areas, covering hundreds or thousands of grains. Although this approach has served the industry well in the past, decreasing feature dimensions have transformed the boundary effects of past technologies into volume effects in present technologies, and are requiring techniques to deal with the granular, or locally inhomogeneous, properties of these films. For purposes of this paper, polycrystalline films used in microelectronic chip wiring are classified into three catagories: simple films, fabricated with a single element (e.g., Al, Cu, W); complex films, fabricated from metal alloys (e.g., AlCu, AlSi); and composite films composed of layered films (e.g., W/AlSi, Ti/AlCuSi).

Polycrystalline simply means the film is composed of a mosaic of crystallites (grains) packed together in the shape of the film. The internal structure of the film is highly varied on a microscopic scale and is termed 'microstructure,' which includes features ranging from a few Angstroms to several microns. This idea is familiar to most technical people working with these materials, but some of the implications may not be so obvious. What distinguishes any given grain from its neighbor is orientation of the grain's crystal lattice, which is important because many physical properties of crystals, such as conductivity and stiffness, are directionally dependent. Grain orientation (i.e., film texture) can be measured by X-ray diffraction which reveals the volume fraction of grains oriented along specific directions. Interfaces (grain boundaries) exist between the grains and abrupt changes in properties occur across them. The presence of grain boundaries is

629

Mat. Res. Soc. Symp. Proc. Vol. 343. ©1994 Materials Research Society

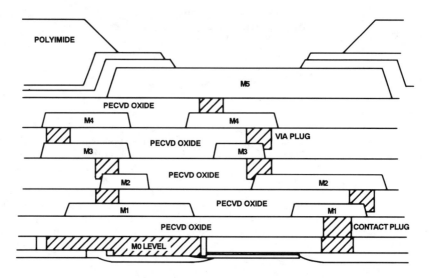

Figure 1.   Cross-sectional representation of fully planarized five-level metallization.

largely responsible for the range of properties exhibited by polycrystalline films, in contrast to their single crystalline counterparts. For example, atomic diffusivity is orders of magnitude greater along the grain boundaries than through the grains themselves, and allows the mass transport through the film to be controlled by variation of the grain size.

The sheer variety of grain boundaries found in a polycrystalline film makes exact characterization of any specific film very difficult. Considered from the perspective of a metal interconnect (line) on a microelectronic chip, a grain boundary can be simplistically defined by orientation parameters given with respect to a Cartesian coordinate system aligned with the principal axes of the line (Fig. 2). These parameters define three quantities: the orientations of grain one, grain two, and the normal to the plane of their common boundary. In Fig. 2, the x-axis is in the direction of the the line width, the y-axis in the direction of the line length, and the z-axis in the direction of the line height (film thickness). It should be readily apparent, after consideration, that a wide variety of possibilities exist for grain boundary orientation, even after the relative orientations of the two grains have been fixed. Those familiar with such structures will recognize that grain boundaries in fine Al lines are generally oriented perpendicular to the major axis of the line after annealing, as indicated by the dashed box in the figure. This does little to diminish the variety of grain boundaries encountered, because grains are distributed over all orientations.

The distribution of possible boundary orientations is, however, dependent on film texture. A film with strong texture (high degree of alignment of a high percentage of grains) means that most grains have nearly the same base plane, limiting the variations in grain boundaries to those produced by rotations of the two grains relative to each other. Two of the three parameters for each grain are severely restricted in range, which substantially reduces orientational possibilities for boundaries.   In wide lines, grain boundaries intersect at 'nodal points' and form a network around the grains.

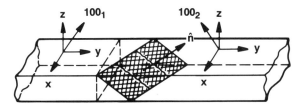

Figure 2. Describing a grain boundary in a narrow metal line. Reference coordinates are fixed to the line, with the y-axis parallel to the longitudinal axis of the line.

A variety of defects can exist within the grains themselves, including dislocations, and vacancy and impurity (unintended) concentrations. Dislocations provide alternate diffusion paths through the grain and nucleation sites for void formation. Vacancies and impurities impact diffusivity. Although these defects exist in single crystal materials as well, they are usually uniformly distributed throughout the material, whereas in a polycrystalline film, they can vary substantially from grain to grain. It is evident that self-diffusion of metal atoms within a metal line can vary widely depending upon the details of film microstructure. In past technologies, the widths of these conductors was great enough that average properties varied little along the line length, so the crystallinity of the film could often be overlooked. However, in the past several years, deposition cleanliness and reduced linewidths have produced the "bamboo" structure in fine lines. The lines have literally become chains of rectangular solid grains laid end to end to carry the electrical signals used by the chip to communicate internally and with the external world: polycrystalline communication links.

Introduction of alloying elements to a single component film to produce a complex film adds variety to each of the properties mentioned above. In concentrations below the solubility limit, alloying elements may alter diffusivities, strengthen or weaken grain boundaries, alter surface energies, slow or pin dislocations, getter impurity elements, etc. In concentrations above the solubility limit, alloying elements form precipitates internally, on grain boundaries, and at grain boundary nodal points. Depending upon the film deposition technique, alloying elements can also influence film texture. These effects are further complicated by fabrication process anneals because the metal is left at a different stage of microstructural evolution (typically non-equilibrium) after each anneal. Not only does this alter the resistivity of the metal, but it also alters the mass-transport characteristics and affects metal-line lifetime.

Composite films are formed by depositing a layer of a simple or complex film on or between other simple or complex films. Since all of these films are annealed, the neighboring metals begin to form alloys to some degree, transforming the simple films to complex films. Elements from adjacent films often react to form compounds as well as alloys, and the reaction process is affected by the polycrystalline nature of both films. Grain size is an important property in controlling reaction uniformity.

As a general rule, the polycrystalline nature of a film is important whenever details of the microstructure affect a property of importance to the product being made. In fine metal lines, however, the microstructure becomes particularly important when line width approaches median grainsize.

EFFECTS ON MICROELECTRONIC RELIABILITY

Electromigration

Chip interconnect reliability is strongly affected by film grainsize. The median lifetime of an Al interconnect due to wiring wearout mechanisms such as electromigration depend upon the interconnect width, even though the current density is kept constant. The lifetime has a minimum at a critical conductor width, which depends upon the specific grain size distribution in the lines[1,2,3]. When the conductor width is greater than the median grain size, atomic transport occurs by grain boundary diffusion, which is relatively rapid. The probability of developing a hole large enough to sever the Al decreases as the line gets larger, hence the lifetime increases with increasing width. For lines much smaller than the median grain size, mass transport occurs by lattice diffusion, which is orders of magnitude slower than boundary diffusion. The lower diffusivity offsets the greater probability of developing a hole large enough to sever the smaller line. Clearly, variability in the grainsize distribution can introduce variability in electromigration lifetime.

The presence of strong film texture has also been reported to increase electromigration lifetime, even when film median grain sizes are smaller than for films of comparable composition with weak structure[4]. This suggests that the restriction on grain boundary types, discussed earlier, is a stronger factor than grain size.

The variability of electromigration lifetime with linewidth highlights the importance of the distribution of grain sizes in a metal film. Because clusters of grains smaller than the median grain size are always present in a film, fine metal lines rarely exhibit a pure bamboo structrue. Rather, the structure is mostly bamboo with segments which contain at least one grain boundary within the line width. Failure is most likely to begin with a void nucleated at such a site. Because of this, the physics of grain growth and the statistics of grain distribution in as-deposited films have drawn increased attention[5].

Examination of grain growth in fine lines, as opposed to larger areas of metallization, is important. For a one-micron thick film, and over an area of 10 microns square (about 10,000 grains), the distribution of grain boundaries might be adequately sampled. But one would need a submicron line more than 40,000 microns long to come close to sampling the same variety of configurations; each grain in the contiguous area has several sides in contact with other grains, while those in the narrow line interface with other grains on only two sides.

Stress-Induced Voiding

Failure by stress-induced voiding is also extremely dependent on the polycrystalline nature of metal lines. Contrary to electromigration, for which trans-granular failure has been observed, virtually all stress-induced voids nucleate on grain boundaries. Some boundaries appear to be more suitable for voiding than others, and thus the statistics of boundary distributions and grain size are of primary interest.

In addition to line width reduction to dimensions on the order of film granularity, an additional issue of edge effects has emerged. For example, in the case of stress-induced voiding, notches have been present for years along line edges, where passivation induced stress in the metal is greatest. These notches were largely ignored in earlier technologies because the size of the notch was small compared to the line width. Metal voiding becomes a problem in narrow interconnects because the stress in the center of narrow lines is nearly as great as at the edge. The edge effect has become a volume effect, and notches have not scaled down with line width.

In bamboo or near-bamboo structures, where grain boundaries are nearly perpendicular to the line axis, the probability of a void occurring which will extend completely across the line depends very strongly on the orientation of the grains adjoining the boundary. Voids in AlSi lines have been shown to prefer having (111) planes as void sides[6]. If (111) planes in each grain adjoining the boundary are at large angles to the line axis, a void will have to grow very large before it severs the line. Alternately, if the (111) planes of both grains are closely aligned and normal to the line axis, the line can be cut by a "slit-like" void requiring very little volume, and will have a much shorter lifetime.

Variables such as deposition chamber vacuum, sputter target structure or length of use, alloying inhomogenieties or substrate temperature (which can influence grain size and alter diffusivities), can produce sufficiently different films in a nominally controlled deposition process to produce measurably different performances in chip lifetime.

Cu alloying with Al is a popular measure taken to improve stress-induced voiding and electromigration lifetime by developing Al_2Cu precipitates in the grain boundaries to impede grain boundary diffusion. However other results can occur. Shown in Fig. 3 is a scanning electron micrograph of Al.5%Cu lines which have been held at 225 °C for 7000 hr. and then stripped of passivation. A number of Al_2Cu precipitates are visible (light contrast) across the lines, presumably at Al grain boundaries. In many cases, small voids are visible immediately adjacent to the precipitates. Whether precipitate formation is accompanied by vacancy accumulation, or the precipitate was there first and attracted the void (or vice versa), or both nucleate at the same kind of boundary is unknown. The incidence of paired voids and Cu precipitates is too high to ignore, suggesting an additional factor at work in void nucleation.

Void Nucleation

Void nucleation is critically important to conductor reliability. If nucleation can be prevented, failure will not occur. If nucleation can be controlled, failure can be delayed, perhaps beyond the end of product life. Both the location and density of

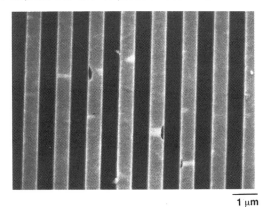

1 μm

Figure 3.   Cu precipitates and stress-induced voids in Al0.5%Cu lines after 7000 hr at 225°C.

nucleation sites might be controlled. But location is difficult to predict or control unless the vulnerable grain boundaries can be placed in insensitive locations, and this is unlikely at present. Control of the density of nucleation sites is more likely. The best situation occurs when the void density is high enough that no single void can grow large enough to sever a line. However, this is risky for a bamboo line because slit-like voids can have very small volumes, and nucleation densities may be too low to effectively stall growth because of the large separation between grain boundaries.

Void nucleation depends both upon microstructure and stress in the metallization[7] (Fig. 4). Dependence upon microstructure is illustrated by the line width dependence and by different nucleation behavior of the metal deposited at different temperatures. Dependence on stress in the Al is indicated by the dependence on passivation stress. However, it needs to be noted here that passivation stress is generally not a good indicator of stress in the conductor[8]. The experiment in Ref. 7 is probably an exception because the passivation geometry was held constant while the proportion of nitride and oxynitride films were varied systematically. Thus, the properties of the enclosing passivation varied with the film stress, and varied the stress experienced by the conductor.

Void nucleation threshold stress is a property of the metallization and is expected to vary with both composition and microstructure. Difficulties arise in predicting local stresses which depend very strongly upon structural variations (Fig. 5). One such variation is produced by the presence of a W stud, which has a much higher thermal expansion coefficient than the surrounding glass, and a greater stiffness than the Al. The combination increases stress in the Al over the level experienced away from the stud. Another variation is produced by the presence of sharp protrusions into the Al as a result of an etch step.

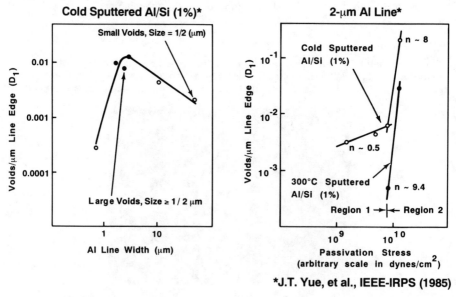

*J.T. Yue, et al., IEEE-IRPS (1985)

Figure 4. Dependence of stress-induced void nucleation density upon passivation stress and linewidth.

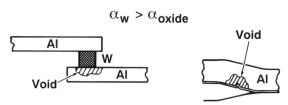

$\alpha_w > \alpha_{oxide}$

Al

W

Void

Al

Void

Al

Figure 5.   Structural variations, which can generate high local stresses and nucleate voids, include W studs and sharp protrusions left from selective etch steps.

A good model for nucleation must include both structural variations and the probability of occurrence of susceptible boundary configurations. Analytical solutions are impractical or impossible for most complex geometries, and finite element modeling (FEM) must be used in these cases. Not much has been done in this arena to date, largely due to the complexity of dealing with microstructural details. Yet it is exactly these grain properties which need to be included in modeling detail to refine predictive capabilities.

COMPOSITE FILMS

Ti/Al System

Rather than grapple with development of complex modeling routines, the approach taken by several manufacturers has been to combine suitable alloying elements in the Al to retard void nucleation and growth. Another method to protect wiring from electromigration or stress-induced voiding is the use of refractory redundant conductors. A void in the Al film is then bridged by the refractory layer, preventing open circuit failure. A refractory underlayer such as Ti has a much higher melting temperature and elastic modulus than Al, enabling it to withstand higher temperature and stress than the Al, and making it relatively immune to electromigration and stress voiding. From one perspective, such a refractory layer has a much finer grain structure than the Al, and behaves in a more homogeneous manner than the larger-grained Al. This is an attempt to buffer the discrete nature of the Al with the more continuum-like refractory layer.

Multilevel wiring using vias or studs to connect the different levels may require an additional redundant layer on top of the Al to protect against the formation of holes under the studs. Multiple high-temperature processing steps cause the Ti layers to react with the Al, and the resulting reaction causes a net decrease in the interconnect volume and Al thickness, affecting both void formation and conductivity. Full reaction of the Ti layers before depositing interlevel dielectric provides a stable metallization with the required conductivity.

Ti/Al reaction kinetics as a function of time at temperature for different temperatures are often reported by RBS analysis[9]. This analysis is an averaging technique, since the spot diameter for the incident beam is hundreds of microns in diameter, while the Ti grains are tenths of the film thickness in diameter. The RBS data suggest a uniformly growing reaction front because of this averaging, while in actual fact, the reaction can proceed at different rates over each Ti grain. Transmission electron microscopy (TEM) of the reacted couple (Fig. 6) show considerable roughness   at the reaction interface, particularly at grain boundaries. The granularity of the interface is comparable to that of the Ti, indicating that Ti granularity substantially influence the reaction. As metallizations become thinner, this irregularity increasingly affects line conductivity.

200 nm

Figure 6. Ti/AlCu/Ti after 1-hr anneal at 400°C (left) and Ti/AlCu/TiN after 20 hr at 400°C. Note incursions into Al grain boundaries. (Photos courtesy of K. Rodbell, T. Domenicucci.)

Fabrication processes used to build these structures often depend upon performance characteristics required of final structures, which in turn depend upon properties of the materials used. For example, circuits designed to achieve a specified speed require interconnect resistances below a specified level. Annealing each metallization level is required to achieve stable metallization, but over-annealing can lead to excessive incursion into the Al grain boundaries by barrier metallization, with a resultant increase in sheet resistance. Barrier thickness, stack height and sheet resistivity thus become competitors.

Ti-AlCuSi Reaction

A refractory layer has also been used with Al in some applications to contact Si junctions[10]. Ti reacts with both Si and Al; with Si it forms $TiSi_2$, while with the Al it forms $TiAl_3$. Since the $TiSi_2$ reaction is generally induced at temperatures around 600 °C or higher, it was generally thought to be negligible at usual processing temperatures for interconnects and interlevel dielectrics. However, the $TiAl_3$ reaction can proceed at a measurable rate in the vicinity of 400 °C. The disadvantages of having the $TiAl_3$ reaction occur is that low resistivity Al alloy is consumed to form much higher resistivity $TiAl_3$, and second, that $TiAl_3$ allows interdiffusion of both Si and Al to produce spiking. Before the use of the refractory metal underlayer, Si was used as an alloying element in the Al to prevent Al penetration of the junctions at the Al-Si interface (spiking). When used in conjunction with Ti, the presence of Si in the Al alloy was found to substantially retard $TiAl_3$ formation, and thus still act as a protection against spiking.

However, increases in circuit density not only drive line dimensions down, but also require additional levels of wiring. This means that the first levels of wiring will experience longer times at elevated temperatures, such that structures which were sufficiently stable for single level metal processing become vulnerable after repeated anneals. A study of Ti/Al0.5%Cu2%Si metallization at the contact level was conducted to define the width of the process window for $TiAl_3$ formation. Surprisingly, although $TiAl_3$ formation occurred, Ti barrier failure appeared somewhat unexpectedly to occur by simultaneous formation of $TiSi_2$ at the Ti/Si interface (Fig. 7).

Examination of spiked samples showed that spike locations appeared to align with specific Al grains. Fig. 8 shows a bright-field TEM cross section of a film after spiking occurred. A portion of the spike can be seen on the left of the picture. The Ti layer in the region of the Al spike is thicker than that over an unspiked region, probably indicating a more complete reaction of Ti with Al and Si. The edge of an Al grain is visible immediately above the right end of the spike. The occurrance of this alignment was observed in several cases. A correspondence of position between selected Al grains and spikes would be reasonable if the Ti layer had been fully converted to $TiAl_3$, but not so obvious given the conversion of the bottom of the layer to $TiSi_2$.

Since Ti was deposited first, it could have influenced the orientation of the Al grains. But this would have to occur over a much larger area than could be accounted for by any one grain; therefore, it must be due to a collection of grains. Ti films are usually highly textured and may be arranged in texture "domains" which share basal plane orientations and extend over several tens of grains. The spiked region would then correspond to the position of the Al grain because both the Al grain and the spiked region would have been produced by the same Ti layer texture domain. These grains may react somewhat faster with the underlying Si and the overlying Al to produce a layer incorporating Al, Si and Ti. After all the pure Ti is consumed by either the $TiSi_x$ phase on the bottom or the Ti/AlCuSi on the top, the Si-Al is free to produce spiking.

This study was performed on metallization with an 0.1 μm-thick Ti layer and 1.2 μm-thick Al. The process window at 400 °C appeared to be greater than 23 hours. Athough quite adequate for single level metallization, one must bear in mind that decreasing metallization thickness drives Ti thickness down in proportion. For an application requiring six levels of metal and 0.3 μm line widths, the Ti layer would be reduced to about 300 Angstroms at the same time as the total high-temperature processing time is increased. The time needed to breach a Ti layer this thin would be comparable to the total processing time at 400 °C.

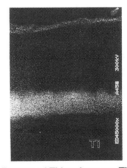

Figure 7.   X-ray dot maps of Si, Al and Ti in region of spike. Bottom of Ti has become $TiSi_2$; top of Ti is $TiAl_3$.

Figure 8.   Cross-sectional TEM image of Ti/AlCuSi film after formation of Al spike (bright contrast, lower left). Grain boundary is visible in the Al above edge of spike.

SUMMARY

An overview of properties of polycrystalline films relevant to microelectronic interconnect metallization has been presented. Effects of these properties were described especially with respect to chip interconnect wearout mechanisms, electromigration and stress-induced voiding. In these cases, the polycrystalline character of the metallization becomes important when interconnect line widths are comparable to the median grain size of the films. Less obvious is the introduction of a substantial barrier to both analytical and computer-aided modeling of interconnect lifetime due to the complexity of describing the properties of these films at the microscopic level.

Some effects of film microstructure on the stability and uniformity of composite metallizations such as Ti/AlCu and Ti/AlCuSi have also been described. These effects, though subtle, nevertheless control the end performance of the films and would not occur without the presence of grain boundaries and multiple crystal orientations. The microstructure of these films can thus be seen to influence a wide range of product-related phenomena, from the rather obvious to the rather obscure.

ACKNOWLEDGEMENTS

The author thanks Dave Thomas for views of multilevel metallization; Dennis Bouldin for voiding data; and Ken Rodbell and Tony Domenicucci for TEM micrographs of layered metallization.

REFERENCES

1. T. Kwok, in <u>Proc. of 1st Int'l ULSI Science and Technology Symposium,</u> ECS. ed., 593 (1987).

2. E. Kisbron, Appl. Phys. Lett.,**36**, 968 (1980).

3. J. Cho and C. V. Thompson, Appl. Phys. Lett., **54**, 2577 (1989).

4. T. J. Licata, et al., in <u>Mat. Res. Soc. Symp. Proc. Vol. 309,</u> MRS. ed., 87 (1993).

5. See, for example, D, T. Walton, H. J. Frost, and C. V. Thompson, Appl Phys. Lett., **61**, 40 (1992).

6. K. Hinode, et al., J. Vac. Sci. Technol. **B 5**, 518 (1987).

7. J. T. Yue, W. P. Funsten, H. V. Taylor, in <u>Reliability Physics 1985 Annual Proceedings,</u> IEEE, NY, (1985), p.126.

8. K. Hinode, et al., J. Vac. Sci. Technol. **B 8**, 495 (1990).

9. I. Krafcsik, et al., Appl. Phys. Lett., **43**, 1015 (1983).

10. R. M. Geffken, J. G. Ryan, and G. J. Slusser, IBM J. Res. Develop., **31**, 611 (1987).

EFFECTS OF CRYSTALLOGRAPHIC ORIENTATION
ON FILM MORPHOLOGICAL EVOLUTION

John E. Sanchez, Jr.
Advanced Film Development, Advanced Micro Devices
One AMD Place, Sunnyvale, CA 94088

ABSTRACT

The factors that determine the thermal strains, stresses and plastic yielding during the annealing of deposited Al thin films on Si substrates are reviewed. The effect of film texture orientation on dislocation glide is described and is shown to lead to variations in film strain energy with texture orientation. The strain energy difference between adjacent grains of different texture is proposed to account for several morphological changes during film annealing. If grain boundaries are mobile (i.e., with small initial diameter) film stresses and strain energies may induce the (secondary) growth of weakly oriented (110), (112), etc., grains at the expense of the normally favored (111) strongly oriented grains. If grain boundaries are stagnant, the strain energy differences between grains may induce the hillock growth of weakly oriented (110) grains in order to reduce film compressive stresses, and may also induce the depletion "sinking" of strongly oriented (111) grains in order to relieve film tensile stresses. This model generally accounts for these experimental findings of others.

INTRODUCTION

The properties of deposited thin films are determined by aspects of the film microstructure such as grain size, grain size distribution, and crystallographic texture. These are in turn dependent upon the details of the deposition conditions (including the substrate material) and the film thickness. For example, thin film grain growth is limited by the film thickness due to grain boundary surface grooving[1,2], and the texture of metal films is controlled to a large degree by the nature of the substrate surface. The understanding and control of the factors that determine film microstructure is often critical in applications such as metallization interconnects in integrated circuit (IC) devices. For example, the long term service reliability of metallization interconnects is determined in part by grain size and texture[3]. Further, changes in film morphology, such as hillock[4] and "sunken grain" formation[5], limit the manufacturing yield during IC fabrication. In addition, it has been shown that the thermal expansion mismatch stresses applied during IC processing significantly affect the morphological and microstructural evolution of the metallization films and interconnects. Therefore a better understanding of the interactions between film microstructure, mechanical stress and morphological evolution is required in order to design more robust and reliable metallization systems.

Several categories of phenomena which occur which during annealing of Al films on Si substrates are of interest. In the situation where grain boundary motion (i.e., grain growth) may occur, it has been shown that secondary (abnormal, extensive) grain growth of Al[6,7] and Cu[8] films under certain conditions of film thickness, grain size and initial texture produces a change in the film crystallographic texture, from the typically initial (111) to (110), (311), (200), etc. This is of interest for several reasons. IC interconnects fabricated from strongly (111) textured Al films provide increased resistance to electromigration failures[9], so that the secondary growth of the off-(111) grains may decrease device reliability. Secondly, the evolution of texture away from (111) during secondary grain growth is contrary to the effect of surface energy minimization[10] on grain growth. This model predicts the enhancement and growth of the preferred low surface energy (111) texture component through grain growth which may become "secondary" or "abnormal" in extent, leading to grains exceptionally larger (i.e., 50-100 times film thickness) than accounted for in a typical lognormal size distribution.

In the situation where grain boundaries are stagnant, both hillock growth and sinking/depletion of Al grains occur during the annealing. Hillocking is generally described as the diffusive out-of-plane growth of grains due to applied compressive thermal stresses during heating, whereas the "sunken grain" phenomena may be described as the diffusive depletion of material from a grain induced by the tensile thermal stresses during cooling from high temperature. Recent experimental evidence has shown that both hillock and sunken grains exhibit a preferred orientation, (110) or off-(111) for hillock grains[11] and (111) for the sunken-depleted grains[12].

The model for plasticity in thin films on rigid substrates will be reviewed, and the effects of film thickness, grain size and crystallographic texture on film flow stress will be described. It is proposed that the variation of flow stress between adjacent grains of different crystallographic texture may affect the film microstructural and morphological evolution during annealing. It is shown how this effect[13] accounts for the loss of (111) texture orientation during secondary grain growth, the hillocking of "weakly oriented" (110) grains, and the sinking-depletion of "strongly oriented" (111) grains during annealing, as described below.

THIN FILM PLASTICITY

The processes used during IC fabrication often involve heating of the (usually Al alloy) metallization layers to approximately 425°C. These cycles induce significant thermal strains in the thin Al films due to the large thermal expansion mismatch between the Al and thick Si substrate. The thermal strain (ε) in the continuous film for a given change in temperature (ΔT) is given by $\Delta \alpha \, \Delta T$, where $\Delta \alpha = \alpha_{Si} - \alpha_{Al}$ = the difference between the Si and Al thermal expansion coefficients. The film stress (σ_{el}) at temperature (T) induced by thermal expansion mismatch, *prior* to any yielding behavior in the Al, is given by

$$\sigma_{el} = M \, \Delta \alpha \, \Delta T = M \, \Delta \alpha \, (T - T_o), \tag{1}$$

where M is the biaxial elastic modulus for aluminum given by (E/1-v), E is Young's modulus, v is Poisson's ratio, and T_o is the temperature at which $\sigma_{el} = 0$ on heating. The subscript (el) is chosen to emphasize that this is the stress prior to yielding. We assume that the moduli (M or E) are isotropic in the plane of the film, however they may vary for different fiber texture orientations of the film, M_{ijk} and E_{ijk}. The moduli variations with texture orientation are determined by the elastic anisotropy of the material under consideration; Al is nearly isotropic, while Cu varies significantly from isotropic elastic behavior. The thermal expansion coefficients (α_{Al}, α_{Si}) are isotropic, given the temperature stability of each lattice.

During typical thermal cycles the films undergo sufficient straining to induce yielding. Typically the initial stress state (at room temperature) is somewhat tensile, and the compressive strain induced as the temperature is increased produces a compressive stress which eventually reaches the yield stress, Figure 1. It has been shown that the compressive yield stress (σ_y) is comprised of both film thickness (h) and grain size (d) dependent contributions[14]

$$\sigma_y (T) = \sigma_h + \sigma_d, \tag{2}.$$

The tensile yielding is somewhat analogous to the compressive yielding, but it is interesting that there exists a persistent asymmetry between the compressive and tensile yield behaviors. We are initially concerned with the film yielding during first heating into the compressive regime, so that we will focus on the compressive behavior. These σ_h and σ_d are of the form

$$\sigma_h (T) = m (T) / h, \tag{3a},$$

$$\sigma_d (T) = k (T) / d^n, \tag{3b}.$$

where n is approximately 1. There is only a weak temperature dependence of σ_y, and we assume for the following that the factors k and m (and therefore σ_y) are independent of temperature, assuming of course that the thermal straining has been sufficient to induce yielding in the film. Both k and m are similar to "Hall-Petch" constants. The grain size dependence has also been noted by several other reports[15]. The result is that films of varying grain size and/or varying thickness will yield in compression at different stress levels, as illustrated in Figure 1.

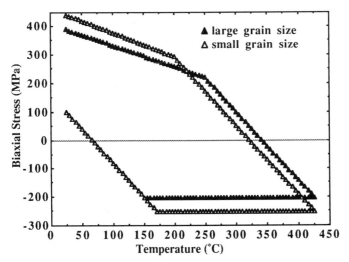

Figure 1. Schematic of the stress-temperature behavior of polycrystalline films on Si substrates, showing the elastic and plastic regimes for both compressive and tensile stresses. Given the grain size and film thickness dependence of yield stress, a film with a smaller (stable) grain size (or equivalently a thinner film) will have a higher yield stress.

The dependence of film flow stress on thickness and grain size may be generally described as the constraining effect of both the film (top and bottom) interfaces and grain boundaries on dislocation motion along glide planes. Thicker films provide longer glide planes inclined in the film and therefore allow easier dislocation motion between the "pinning" effects of the interfaces as the gliding dislocations leave training segments near the top and bottom interfaces, Figure 2. Grain boundaries impede continued dislocation glide, and dislocation pile-ups at boundaries may have a strengthening effect as the grain size decreases, Figure 3. Nix[16] and others have considered in detail the energetics of gliding dislocations in oxidized/capped thin films rigidly attached to thick substrates, Figure 2. This analysis includes the geometry of the glide plane in the biaxially stressed film and the elastic energy stored in the system (the oxide cap, the film and the substrate) as "threading" dislocations glide and leave the trailing segments near to the interfaces. Nix provides an analytical expression for σ_h of the form

$$\sigma_h = C_{ijk} \, G^* \, b \, / \, h \qquad (4),$$

where $C_{ijk} = \sin \phi / \cos \phi \cos \lambda$ and is described in more detail below, G^* is the effective shear modulus of the capped film-substrate system, b is the Burgers vector, and ϕ and λ are the included angles between the glide plane normal direction and Burgers vector and the film normal direction, respectively, as shown in Figure 2. The effective shear modulus (G^*) is given by the term[16] $\{(1/(2\pi[1-v]))\}\{(\mu_f \mu_s/(\mu_f + \mu_s)) \ln (\beta_s h/b) + \mu_f \mu_o/(\mu_f + \mu_o) \ln \beta_o t /b)\}$, where β_s and β_o are constants, t is the oxide capping layer thickness, and μ_f, μ_s, and μ_o are the shear moduli of the film, substrate and oxide capping layer, respectively. G^* will be a constant for a given film/substrate system. The effect of geometry of the glide plane in the film is expressed in the term $C_{ijk} = \sin \phi / \cos \phi \cos \lambda$ and is made up of two contributions. The "length" of the glide plane, or the distance between the interface pinning points, is given by $(\sin \phi)/h$, assuming that the grain structure is typically columnar in the film, that is, the grain height is equal to the film thickness. The relation between the applied biaxial stress and the resolved stress on the glide plane involves the "Schmid factor" $(\cos \phi \cos \lambda)^{-1}$. These geometric factors are grouped into the term $C_{ijk} = \sin \phi/(\cos \phi \cos \lambda)$ which is obviously determined by the texture orientation (ijk) of the film grains, where the (ijk) lattice planes are oriented parallel to the film plane.

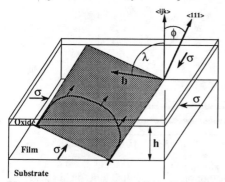

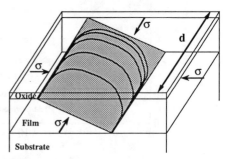

Figure 2. Model of dislocation glide along (111) planes in a grain of texture orientation (ijk) in a film of thickness h. The thermal applied stress σ is equal-biaxial in the film plane. (After Nix[16].)

Figure 3. Illustration of dislocation pile-up at a grain boundary as a possible mechanism for the "Hall-Petch" grain size strengthening behavior observed in the flow stress of thin Al films.

TEXTURE ORIENTATION DEPENDENT PLASTICITY

Typically, as-deposited Al films have principally (111) texture orientations. In this case $C_{111} = 3.46$. The C_{ijk} vary and may be calculated for other grain orientations. The significance is that the film thickness-dependent flow stress (σ_h) varies with film texture orientation through the factor C_{ijk}. Further, the previous analysis suggests that the grain size dependent flow stress (σ_d) should also vary with film texture in the same way, and an analytical expression for σ_d should also include the factor C_{ijk}. However in the following we consider only the effect of C_{ijk} on σ_h; the effects of C_{ijk} on σ_d will only emphasize the effects described below.

The values for C_{ijk} are calculated for each available {111} <110> slip system in the various low index texture orientations (100), (110), (111), (012), (112), etc., since the C_{ijk} for each slip system will vary for a given texture orientation, Table I. We assume that only 4 systems are required for general slip in each grain, and use therefore the average of the 4 lowest C_{ijk} as listed in Table II. We note that films with (110) orientation have the lowest value, $C_{110} = 1.42$, while (111) films have the highest value, $C_{111} = 3.46$, etc., Table II.

Table I. Calculation of C_{ijk} for the various slip systems and (ijk) orientations of Al films.

Grain Orientation	Slip system {111}	<110>	C_{ijk}	Grain Orientation	Slip System {111}	<110>	C_{ijk}
(100)	(111)	$[\bar{1}10]$	2.0	(211)	(111)	$[\bar{1}10]$	1.22
		$[\bar{1}01]$	2.0			$[\bar{1}01]$	1.22
	$(\bar{1}11)$	$[110]$	2.0		$(1\bar{1}1)$	$[110]$	2.16
		$[101]$	2.0			$[011]$	3.25
	$(1\bar{1}1)$	$[110]$	2.0			$[\bar{1}01]$	6.48
		$[\bar{1}01]$	2.0		$(11\bar{1})$	$[101]$	2.16
	$(11\bar{1})$	$[\bar{1}01]$	2.0			$[011]$	3.25
		$[\bar{1}10]$	2.0			$[\bar{1}10]$	6.48
(110)	(111)	$[\bar{1}01]$	1.42				
		$[0\bar{1}1]$	1.42	(221)	(111)	$[\bar{1}01]$	1.2
	$(11\bar{1})$	$[101]$	1.42			$[0\bar{1}1]$	1.2
		$[011]$	1.42		$(\bar{1}11)$	$[110]$	5.41
(111)	$(\bar{1}11)$	$[110]$	3.46			$[101]$	7.2
		$[101]$	3.46			$[0\bar{1}1]$	21.6
	$(1\bar{1}1)$	$[110]$	3.46		$(1\bar{1}1)$	$[110]$	5.41
		$[011]$	3.46			$[011]$	7.2
	$(11\bar{1})$	$[\bar{1}01]$	3.46			$[\bar{1}01]$	21.6
		$[011]$	3.46		$(11\bar{1})$	$[101]$	2.0
(210)	(111)	$[\bar{1}10]$	2.58			$[011]$	2.0
		$[\bar{1}01]$	1.29				
		$[0\bar{1}1]$	2.58	(311)	(111)	$[\bar{1}01]$	1.33
	$(\bar{1}11)$	$[110]$	3.93			$[110]$	1.33
		$[101]$	5.91		$(\bar{1}11)$	$[110]$	6.65
		$[0\bar{1}1]$	11.8			$[101]$	6.65
	$(1\bar{1}1)$	$[110]$	3.93		$(1\bar{1}1)$	$[110]$	1.91
		$[011]$	11.8			$[011]$	3.83
		$[\bar{1}01]$	5.91			$[\bar{1}01]$	3.83
	$(11\bar{1})$	$[\bar{1}01]$	1.29		$(11\bar{1})$	$[101]$	1.91
		$[011]$	2.58			$[011]$	3.83
		$[\bar{1}10]$	2.58			$[\bar{1}10]$	3.83

Table II. Average of the lowest 4 C_{ijk} for the available slip systems in each of the low index (ijk) texture orientations considered.

orientation	average of 4 lowest C_{ijk}
(100)	2.0
(110)	1.42
(111)	3.46
(210)	1.94
(211)	1.73
(221)	1.6
(311)	1.62

An example of this behavior is shown in Figure 4. The stress-temperature behavior of both quasi-epitaxial Al (111) single crystal films on Si (111) substrates and Al (110) bicrystal films on Si (100) substrates was determined by x-ray methods[17]. The significantly higher compressive σ_y of the "stronger" Al (111) oriented film in comparison to the Al (110) "weak" film is evidently due to the difference between the C_{111} (= 3.46) and the C_{110} (= 1.42) for the two films. The key conclusion is that the yield stress is determined by the film texture orientation in addition to the grain size and thickness.

Deposited polycrystalline films are not in general perfectly textured. Al films are usually principally (111) textured but also contain components of (110), (112), etc., and randomly oriented grains. We make the key assumption that individual grains differing in texture orientation yield independently as determined by their respective size (d) and orientation (ijk). This assumption is justified since the local strain in each grain in a continuous film is imposed by

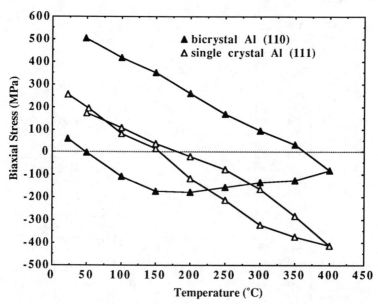

Figure 4. Stresses in Al (111) single crystal and Al (110) bicrystal 0.3 μm films during the second thermal cycle as determined by x-ray diffraction methods (courtesy Paul Besser[17]).

the common substrate. While film continuity provides common constraint for each grain, any displacements at grain boundaries due to differences between stress levels in adjacent grains are assumed to be small. Further, the glide processes described above consider only the orientation and size of each grain, so that yielding behavior will occur independently unaffected by adjacent grain behavior. As an example, consider the annealing of an Al film of uniform grain size made up of both (111) and (110) oriented grains. Prior to yielding, the stress level is continuous and equal throughout the film but changing with temperature. Eventually the weakly oriented (110) grains will yield first at a stress $\sigma_{y,110}$ approximately 40% ($C_{110}/C_{111} = 1.42/3.46 = 0.40$) of that required for yielding of the (111) grains, $\sigma_{y,111}$. The stress differences between the (111) and (110) grains continues to increase, since the (110) stresses remain at the yield level, until the (111) grains eventually yield, at which point the stress difference between the different components is at the maximum. The absolute magnitude of these stress differences will depend upon the film thickness and grain size, so that it is conceivable that, for a given thermal cycle, either or all texture components in the film remain elastic through the anneal. In this case the stress differences are expected to be small, or zero. Practically, small grain sizes and thin films lead to higher stresses (equation 1), so that the maximum stress differences will occur in such films as yielding occurs. Note that, in the case of extremely small grains/thin films, all components of the film may remain elastic, until grain growth "weakens" the film and yielding occurs. In general, however, the grain/film flow stresses are dependent on the texture orientation ($\sigma_{y,ijk}$) through the factors C_{ijk}.

The differences in stress levels lead to differences in strain energy density W between grains[13]. For a biaxially stressed film W is given by

$$W = 1/2 \, (\sigma_1 \, \varepsilon_1 + \sigma_2 \, \varepsilon_2) = 1/2 \, (\sigma_1^2 / M_1 + \sigma_2^2 / M_2) \tag{5},$$

where the biaxial moduli, stresses and strains (M_i, σ_i, and ε_i, respectively) are measured or resolved in orthogonal directions in the film plane. For the biaxial situation considered here, M is isotropic and the stresses $\sigma_1 = \sigma_2 = \sigma_{el}$ are uniform in each grain prior to any yielding, and differ $\sigma_1 = \sigma_2 = \sigma_{y,ijk}$ after yielding in each grain of orientation (ijk). The W is given by (σ_{el}^2 / M) prior to yielding and by ($\sigma_{y,ijk}^2 / M$) in yielding (ijk) oriented grains thereafter. Therefore the difference in biaxial strain energy density ($\Delta W_{ijk-abc}$) between grains of different texture orientations is a maximum when the strongest grains have yielded, and is given by

$$\begin{aligned}
\Delta W_{ijk-abc} &= 1/M \, (\sigma_{y,ijk}^2 - \sigma_{y,abc}^2) \\
&= 1/M \, \{(\sigma_{h,ijk} + \sigma_d)^2 - (\sigma_{h,abc} + \sigma_d)^2\} \\
&= 1/M \, \{\sigma_{h,ijk}^2 - \sigma_{h,abc}^2 + 2 \, \sigma_d(\sigma_{h,ijk} - \sigma_{h,abc})\}
\end{aligned} \tag{6}.$$

Using equations 3 and 4 we can write

$$\Delta W_{ijk-abc} = 1/M \, \{(\, G^* \, b/h)^2 (C_{ijk}^2 - C_{abc}^2) + 2 \, (G^* \, b \, k/d \, h) \, (C_{ijk} - C_{abc})\} \tag{7}.$$

Estimates for the possible magnitudes of $\Delta W_{ijk-abc}$ may be made using the experimental results of Venkatraman[14]. Those results for (111) textured Al and Al-Cu films in compression in the range $\approx 300°C$ found $m_{111} \approx 25$ MPa-μm, and $k \approx 100$ MPa-μm. From equations 3a and 4, $m_{ijk} =$

C_{ijk} G*b. Since C_{111} = 3.46, we can find G*b ≈ 7.2 MPa-μm. Thus, $\Delta W_{ijk-abc}$ may be estimated for Al and Al-Cu films of known grain size d and thickness h using the tabulated C_{ijk} and the factors (G*b) and k found above.

The $\Delta W_{ijk-abc}$ discussed here is due to orientation dependent plasticity for the case of an ideally isotropic material. We consider below the possible effects of significant ΔW variations between adjacent grains on film morphology. However, similar orientation dependent mechanical behavior may also be due to elastic anisotropy, such that the M_{ijk} may produce significant stress and strain energy differences in differently textured grains and films *prior* to yielding. Cu films are very anisotropic, and we note that several effects to be discussed below have also been observed in Cu films during annealing.

STRAIN ENERGY INDUCED SECONDARY GRAIN GROWTH

Several pervious investigations[6,7] found that Al and Al-Cu films layered with other intermediate films (Cu, Cr, etc.) underwent a change in texture from the initial (111) orientation to (133)[6] and (110) and (112)[7] as a result of secondary grain growth during annealing. The effect of layering is to produce effectively thinner films with initially small grain size, on the order of < 0.2 μm in diameter, Figure 5. We can estimate the magnitude of the $\Delta W_{110-111}$ for this case. Using d = 0.2 μm, h = 0.4 μm from the previous experiments and the factors C_{110}, C_{111}, (G*b), and k from above, we find $\Delta W_{110-111}$ ≈ 3.5 10^5 J/m³. This can be compared to the driving force for grain growth (F_{gg}) for these films. F_{gg} is given by γ_{gb}/r^*, where γ_{gb} is the grain boundary energy and r* is the in plane boundary curvature for grains of diameter d, r* ≈ 5-6 d. For d = 0.2 μm, r* ≈ 1 μm (10^{-6} m). Typically, γ_{gb} is ≈ 0.3 J/m³, so that F_{gg} ≈ 3 10^5 J/m³. Thus, the strain energy difference between "weak" and "strong" grains in sufficiently thin films at yielding is of the same magnitude as the grain growth driving force.

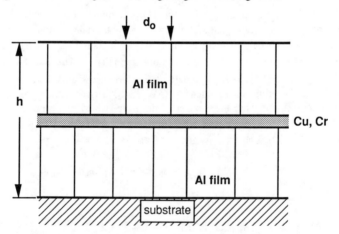

Figure 5. Cross-section schematic drawing of as-deposited thin Al films layered with an intermediate Cu or Cr layer. The effect of the intermediate layer is to maintain small grain size and high stresses during first annealing, leading to strain energy induced secondary grain growth (εEISGG) and a change in film texture from (111) to (133)[6] and (110) and (112)[7].

It is proposed that under certain initial conditions of film h, d and thermal stress the strain energy difference between strongly and weakly oriented grains provides sufficient driving force for the preferred growth of the weakly oriented grains. This *strain energy induced secondary grain growth* (εEISGG) produces a textural evolution opposite that described as surface energy driven secondary grain growth[10], which predicts the growth of low surface energy (111) grains at the expense of off-(111) texture components in the film.

εEISGG is expected to be most likely for small initial grain sizes and for thin films, so that the stress levels will be greater during annealing. As discussed earlier, it is possible that for small enough d and h the film initially remains entirely elastic (no yielding) at high temperature, with $\Delta W \approx 0$ for elastically isotropic materials. However, if grain growth is allowed to occur then the film may "weaken" enough as the grain size increases (equation 3b) to allow yielding, which in turn will generate a significant ΔW and the subsequent effects. This sequence is the likely situation for the Al-Cu films[6,7], where small precipitates pinned boundaries from moving until they dissolved at high temperature and high stress. Note that the εEISGG process produces an increasing driving force advantage for the weakly oriented grains as they grow and become weaker with size (equation 3b). Therefore it is likely that the large weakly oriented grains continue to consume the strongly oriented (111) grains without limit, leading to the secondary growth observed.

The final grain size after εEISGG has occurred is on the order of 50-100 μm[6,7] for the films of initial grain size ≈ 0.2 μm. Therefore only one grain in $(0.2/50)^2 = 60,000$ grains is required to be available as a suitable weakly oriented "seed' grain to drive the εEISGG process. Given the typical log-normal initial grain size distribution of the film, it is probable that there always exists a grain weak enough (due to orientation and size effects) to drive a εEISGG process. However, thicker films, thermal cycles and competing normal grain growth will limit the opportunities for εEISGG to occur in general.

HILLOCK AND SUNKEN GRAIN FORMATION

The well-known hillock growth is the out-of-plane growth of grains in order to relieve the compressive thermal stresses during annealing, while grain sinking is the depletion of grains to relieve tensile thermal stresses. Both phenomena are believed to involve diffusional mass transport, but the details of their formation are as yet unclear. However recent work has shown that both hillock and sunken grains exhibit preferred orientations. Hillocks have been shown[11] to be principally (110) or significantly off-(111) in "weak" orientations as described above, while sunken grains (Figure 6) are generally strongly oriented (111) grains[12,18]. We propose that orientation dependent plasticity and the resultant strain energy effects account for these observations, as described below.

Typically, the first annealing of Al films after deposition induces compressive stresses, grain and hillock growth at elevated temperatures. Careful measurements have shown that the rapid hillock formation (as a result of the compressive stresses) does not relieve a significant portion of the applied compressive strain. The atypical (110) orientation of hillocks in a matrix of normally (111) oriented grains suggests that material diffuses from the high strain energy-strongly oriented (111) grains to the adjacent weakly oriented (110) grains, inducing their out-of-plane growth and reducing the biaxial film stress in a region surrounding the hillock. (This mechanism will be

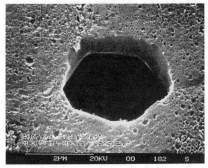

Figure 6. Example of a sunken grain in an Al 1.5% Cu annealed film. The orientation of the remaining material in the sunken region is of (111) texture orientation. (courtesy of H. Shin[12].)

described in more detail below.) Similarly, sunken grains are formed when a (111) high stress-high strain energy grain is surrounded by weak (110) or random grains, inducing the diffusion of material from (and depletion of) the (111) grain to the surrounding weak matrix, as shown schematically in Figure 7. An example of a sunken (111) grain surrounded by weakly oriented off-(111) grain is presented in Figure 8. The texture orientations of the grains in Figure 8 were determined by electron back-scattered diffraction analysis[18] and are listed in Table III. An asymmetry exists between the frequency of hillocking and grain sinking. Since Al films are principally (111) textured, it is likely that any weakly oriented grains will be surrounded by a "strong" (111) matrix, and that (110) hillocking will be commonly observed. However, for the same films it is unlikely that a (111) grain will be surrounded by a "weak" (a random or (110) etc.) matrix, so that the occurrence of sunken grains will be less frequent. This asymmetry between the frequency of hillocking and grain sinking is generally observed.

Lattice and interface diffusion are normally much slower than boundary diffusion at the temperatures of interest. Therefore, diffusional grain hillocking and sinking occurs via boundary diffusion along the network of boundaries surrounding the growing/depleting grain in a radius (R) which grows with time as $R \approx (D_{gb} t)^{1/2}$, where D_{gb} is the effective boundary diffusivity and t = time, Figure 9. Stress is relieved only near the boundary of grains in this region (shaded boundaries) since removal or addition of boundary material will not relieve stresses within the grain interiors.

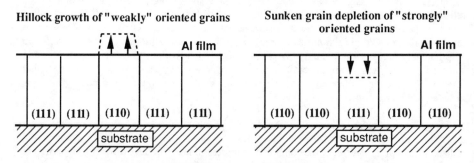

Figure 7. Schematic of hillock growth of "weakly" oriented (random or (110)) grains to relieve compressive thermal stresses in the film (left), and the sunken grain depletion of strongly (111) oriented grains in a matrix of "weak" grains to relieve tensile thermal stresses in the film (right).

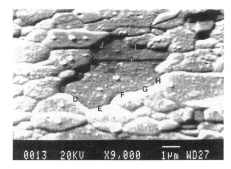

Figure 8. SEM micrograph[18] of a sunken grain in an annealed Al film. The texture of orientation of labeled grains is shown in Table III. Sunken grain is (111) while surrounding grains are significantly off-(111) texture

Table III. Orientation of the grains in Fig. 8 determined by EBSD analysis[18].

Grain	Texture Index
A	111
B	748
C	546
D	102
E	537
F	919
G	-----
H	407
I	435
J	----
K	435

A more detailed analysis of hillock and sunken grain formation as a boundary diffusional stress relaxation mechanism is being developed. This model includes the effects grain size, film thickness and thermal cycle on the relaxation behavior.

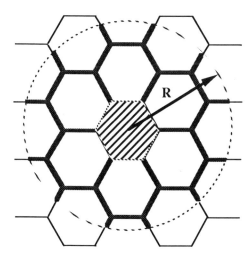

Figure 9. Plan view of the mechanism for diffusional grain hillocking or depletion (striped grain) where material is redistributed in a region defined by the radius $R \approx (D_{gb} t)^{1/2}$, where D_{gb} is the effective boundary diffusivity and t = time. The stress is relieved only in material near to the boundaries in the grains (shaded boundaries) since removal or addition of boundary material will not relieve stresses within the grain interiors.

SUMMARY

The role of grain and film texture orientation on dislocation plasticity in biaxially stressed thin films on rigid substrates was described. An expression for the orientation dependent strain energy density (W) in grains and films was derived. The magnitude of the differences in W for variously textured film components was shown to be approximately equal to the driving force for grain growth for thin films with small (pinned) grains during annealing. The strain energy induced secondary grain growth (εEISGG) model for the preferential growth of favored weakly oriented (110), etc., grains at the expense of strongly oriented (111) grains was presented. This model accounts for the changes in texture (from the normally preferred (111)) as a result of secondary grain growth observed by others.

The effects of stress and strain energy differences between differently textured adjacent grains was also shown to account for the formation of preferentially oriented hillock and sunken grains, which form via boundary diffusion to partially relieve compressive and tensile stresses, respectively. This model predicts weakly oriented (i.e., (110), etc.) grains will hillock, and strongly oriented (111) grains will deplete, as observed experimentally.

ACKNOWLEDGMENTS

The author wishes to acknowledge fruitful discussions with A. Wanner (Max-Planck Institut, Stuttgart, Germany) and the support of the Max-Planck Society during his tenure in Stuttgart where this work was initiated. The collaboration with P. Besser (Stanford University and AMD) on the single crystal/bicrystal Al film work is also acknowledged. Appreciation is extended to K. Rajan (RPI) and H. Shin (Motorola) for the use of sunken grain micrographs and analysis.

REFERENCES

1. W. W. Mullins, *Acta Metallurgica* **6**, 414-427 (1958).
2. J. E. Sanchez, Jr., E. Arzt, *Scripta Metallurgica et Materialia* **26**, 1325-1330 (1992).
3. M. J. Attardo, R. Rosenberg, *J. Applied Physics* **41**, 2381-2386 (1970).
4. P. Chaudhari, *J. Applied Physics* **45**, 4339-4346 (1974).
5. N. Kristensen, et al, *J. Applied Physics* **69** (**7**), 2097- 2104 (1991).
6. A. Gangulee, F. M. d'Heurle, *Thin Solid Films* **12**, 399-402 (1972).
7. H. P. Longworth, C. V. Thompson, *J. Applied Physics* **69** (**7**), 3929-3940 (1991).
8. D. P. Tracy, D. B. Knorr, *J. Electronic Materials* **22**, 611-616 (1993).
9. S. Vaidya, A. K. Sinha, *Thin Solid Films* **75**, 253-259 (1981).
10. C. V. Thompson, *J. Applied Physics* **58** (**2**), 763-772 (1985).
11. D. Gerth, D. Katzer, R. Schwarzer, *Materials Science Forum* **94-96**, 557-562 (1992)
12. H. Shin, *ProceedingsVLSI Multilevel Interconnect Conference* (IEEE, Santa Clara, CA, 1991), vol. **1991** VMIC, pp. 292-294.
13. J. E. Sanchez, Jr., E. Arzt, *Scripta Metallurgica et Materialia* **27** 285-290 (1992).
14. R. Venkatraman, J. C. Bravman, *J. Materials Research* **7**, 2040-2048 (1992).
15. A.J. Griffin, F.R. Brotzen, C. Dunn, *Scripta Metallurgica Materialia* **20**, 1271-72 (1986).
16. W. D. Nix, *Metallurgical Transactions A* **20A**, 2217-2245 (1989).
17. P. Besser, et al, *this proceedings*
18. K. Rajan, Rennselaer Polytechnic institute, personal communication

LOCAL TEXTURE AND ELECTROMIGRATION IN FINE LINE MICROELECTRONIC ALUMINUM METALLIZATION

J. L. Hurd, K. P. Rodbell[A], D. B. Knorr[B] and N. L. Koligman

IBM Analytical Services Group, 1580 Route 52, Hopewell Junction, NY 12542.
[A] IBM Research Division, Yorktown Heights, NY 10598.
[B] Materials Engineering Dept., Rensselaer Polytechnic Institute, Troy, NY 12180.

ABSTRACT

Aluminum films 1 μm thick were deposited on oxidized silicon by sputtering and partially ionized beam evaporation to vary the crystallographic texture. These films were patterned into lines and subsequently annealed at 400 °C for 1 h. A strong correlation between the electromigration behavior and the blanket film texture (X-ray diffraction (XRD) / pole figures) has been reported previously for these films. In this work, an Electron Backscatter Diffraction (EBSD) a.k.a. Backscatter Kikuchi Diffraction (BKD) technique was employed using a scanning electron microscope (SEM) to interrogate individual grain orientations. BKD pole figures were acquired for lines $\geq$0.3 μm wide and for blanket (pad) regions. Identical inverse pole figures were found for blanket films measured using both XRD and BKD (pads). Furthermore, the BKD (111) fiber texture shows a line width dependency, with narrow lines having a slightly improved texture over blanket (pad) regions. Local grain orientations were investigated near and within electromigration testing sites with characteristic void and hillock morphologies. The relationship of neighboring grain orientations to electromigration damage is shown.

INTRODUCTION AND EXPERIMENTAL PROCEDURES

Aluminum line texture in microelectronics has been reported previously for 1 μm wide lines [1,2]. In this paper we report a comparison of XRD and BKD texture measurement techniques for blanket films and BKD for sub-micrometer wide aluminum lines. BKD proves a useful technique to explore the influence of film lateral dimensions on grain texture. Further, we examine the relationship between local grain orientations and electromigration damage.

Pure Al films 1 μm thick were deposited onto oxidized silicon by both sputtering and partially ionized beam (PIB) evaporation [3]. The substrate bias for the sputtered deposition was 2 keV (SP-2). For the PIB depositions the substrate bias was fixed at 2 keV and the ion content varied; either 1% or 2% (PIB-2/1 and PIB-2/2). These films were patterned into 4 point probe structures and then annealed at 400 °C for 1 hour in forming gas. The electromigration results (time to failure, t_{50}, and sigma, σ) and film characteristics have been reported previously [3,4] and are summarized in TABLE I. The median grain size and grain size distribution were determined from transmission electron microscopy (TEM). In this paper we compare BKD derived grain

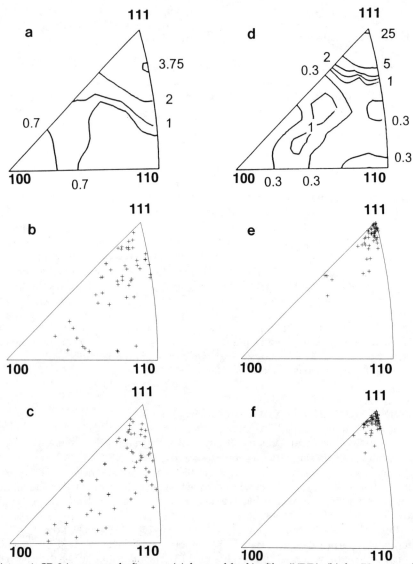

Figure 1. SP-2 inverse pole figures: (a) from a blanket film (XRD), (b) for 50 grains from a 100 x 100 μm² pad region on a chip (BKD) and (c) for 60 grains from a 0.6 μm wide line (BKD). PIB2/1 inverse pole figures: (d) from a blanket film (XRD), (e) for 56 grains from a 100 x 100 μm² pad region on a chip (BKD) and (f) for 60 grains from a a 0.5 μm wide line (BKD).

texture for both the strongest and weakest textured films (PIB-2/1 and SP-2). The crystallographic texture was quantified by XRD on blanket films as the (111) and random volume fractions and the half width of the (111) component ($\omega_{90\%}$). Additionally, (200) and (220) pole figures were determined by XRD on blanket SP-2, PIB-2/1 and PIB-2/2 films, from which inverse pole figures were calculated.

TABLE I. Summary of the blanket annealed Al grain size, grain size distribution and electromigration data (at 225 $^{\circ}$C and 1×10^6 A/cm^2) for three different film textures.

Film	Median Grain Size (μm)	Grain Size Distribution	Volume Fraction (111)	Random	$\omega_{90\%}$ (deg.)	t_{50} (hours)	σ
PIB 2/1	0.75	0.48	0.91	0.09	13.1	772	0.28
PIB 2/2	0.64	0.49	0.84	0.16	14.8	235	1.23
SP-2	0.75	0.52	0.64	0.36	21.2	29	1.60

The BKD experimental technique for determining individual grain orientations is well known [5]. Electron backscatter diffraction of patterned lines were measured using a commercially available BKD system [6] and software [7] on a JEOL JXA 733

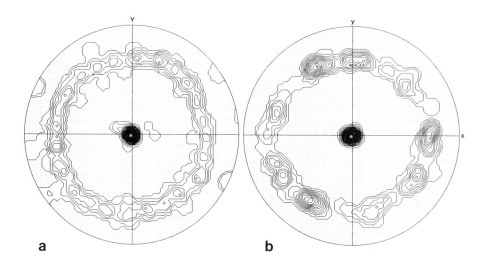

a   b

Figure 2. Two <111> BKD contoured pole figures for PIB2/1 Al (56 grains each) showing a slight texture strengthening with decreasing linewidth; (a) pad and (b) a 1.8 μm wide line. The line transverse and longitudinal directions are represented by x and y respectively.

electron microprobe with a LaB$_6$ filament. Wet or dry etched lines were used to measure successive crystallographic grain orientations in 0.3 to 1.8 μm wide by 1 μm thick patterned Al lines at the two extremes of the <111> surface normal texture (SP-2 and PIB2/1). Successive grains along both electromigration failed and non-damaged lines were measured. BKD texture contour maps were obtained using a 100/n algorithm [8], where n is the number of grains counted and (100/n)% is the corresponding area percent used for determining the contour interval values.

RESULTS

Fig. 1 is a comparison of the two <111> texture extremes. The XRD texture correlates well with the BKD texture obtained from the pad regions. No difference is observed between the pad and line inverse pole figures for weakly textured Al (SP-2); a slight increase in the texture is seen for the already strongly textured Al (PIB2/1) as the line width decreases. Two <111> contoured BKD pole figures for PIB2/1 are shown in Fig. 2. Grain orientation clustering in rotation about the surface normal is noted for the 1.8 μm wide line (Fig. 2b). This grain texturing is local and is most visible for subsets of this texture pattern representing a small number (~10) of adjacent grains. The local texture effect on electromigration damage is shown in Figures 3 and 4 where 6 individual grains are indexed near a damage site from a PIB2/1 sample which failed at 719 hours at 225°C and 1 x 10^6 A/cm^2. Fig. 5 shows the orientation of 78 grains for an undamaged portion of the same line. A region of hillock growth is noted at grain #5, which is misoriented anomolously from the <111> fiber texture of the surrounding grains. Grains 1 & 2 have characteristic <111> orientations, however the orientation of the voided region between these is unknown due to its melting and subsequent recrystallization when the line failed during electromigration.

DISCUSSION & CONCLUSIONS

This paper presents a complete analysis of BKD data on pure Al sub-micrometer wide lines in comparison with XRD and BKD of blanket Al films. Identical inverse pole figures were found for blanket film regions (pads) measured using both XRD and BKD. This is justification that both approaches to texture measurements converge to the same results. Nevertheless, XRD requires the determination of (111), (200) and (220) pole figures from blanket films in order to accurately determine the inverse pole figure, whereas BKD builds the inverse pole figure directly one grain at a time. The determination of individual grain orientations is the most straightforward method for measuring texture [9], however it potentially suffers from two drawbacks (1) in order to obtain sufficient statistical relevance a large number of grains must be measured which is both tedious and time consuming and (2) the data may be compromised by selection effects. For example, if the crystals of a certain orientation are difficult to measure, they may be underrepresented in the set of measured orientations. In this paper the excellent correspondence between the XRD and BKD data is direct evidence that the BKD technique is capable of reproducing the XRD data one grain at a time.

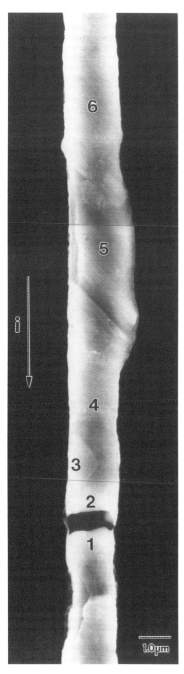

Figure 3. Field emission composite SEM micrograph of a PIB2/1 electromigration damaged region on a 1.8 µm wide line with 6 individual grains indicated.

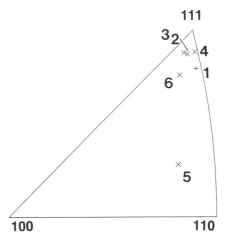

Figure 4. BKD inverse pole figure for the 6 grains labeled in Fig. 3.

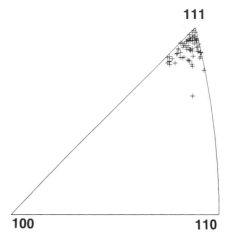

Figure 5. BKD inverse pole figure (78 grains) for undamaged portion of line shown in Fig. 3.

This is most likely due to the small spot size of our BKD system (where the spot size is smaller than the average grain size of the Al). Furthermore, the BKD technique is not limited to blanket film samples and is capable of measuring fine line texture. The slight strenghtening of the PIB2/1 texture with decreasing linewidth requires further verification. However, it suggests that narrow lines may have a tendency to favor certain orientations due to the presence of three free surfaces that limits the number, and degree, of allowed variations from the <111> surface normal orientation. The grain texturing observed in Fig. 2b indicates a local effect during grain growth where minimization of grain boundary energy may cause a preferential grain matching at boundaries on an atomic scale. This results in a slight minimization of the number of random grains in a patterned line as evidenced by the stronger texture observed. As the sample size becomes larger, the grain rotation distribution becomes more diverse due to statistical averaging over the line length. This occurs when many clusters from different regions of the line are overlapped in a pattern. Finally, local grain orientations were investigated near and within electromigration testing sites with characteristic void and hillock morphologies. There is insufficient evidence to establish a mechanism for preferential electromigration based on local grain texture, but electromigration damage is clearly related to local variations in the <111> texture. A useful technique has been established and further investigations are underway to determine the relationship of grain boundary index and orientation to local texture near and away from electromigration failure sites for both strong and weakly textured sub-micrometer wide lines.

REFERENCES

1. K.Z. Troost and M.H.J. Slangen, "EBSP Studies in Microelectronics", European Microscopy and Analysis, September 1993, p. 17.
2. S. Kordic, R.A.M. Wolters and K.Z. Troost, J. Appl. Phys. 74, 5391 (1993).
3. D.B. Knorr, K.P. Rodbell and D.P. Tracy, Proceedings of the First MRS Symposium on Materials Reliability Issues in Microelectronics, (MRS Publications, Pittsburgh, 1991), 225, p.21.
4. D.B. Knorr, D.P. Tracy, and K.P. Rodbell, Appl. Phys. Lett. 59, 3241 (1991).
5. J.A. Venables, C.J. Harland, Phil. Mag. 27, 1193 (1973).
6. SINTEF Metallurgy, Trondheim, Norway.
7. CHANNEL Software, N. H. Schmidt, Randers, Denmark.
8. J. Starkey, Can. J. Earth Sci. 14, 268 (1977).
9. H.J. Bunge, Int. Mat. Rev. 32, 265 (1987).

MECHANICAL BEHAVIOR OF SINGLE CRYSTAL Al (111) AND BICRYSTAL Al (110) FILMS ON SILICON SUBSTRATES

Paul R. Besser*[†], John E. Sanchez, Jr.*, S. Brennan[¶], John C. Bravman[†], G. Takaoka[+] and I. Yamada[+]

*Advanced Process Development, Advanced Micro Devices, Sunnyvale, CA 94088
[†]Department of Materials Science and Engineering, Stanford University, Stanford, CA 94305
[¶]Stanford Synchrotron Radiation Laboratory, Menlo Park, CA 94025
[+]Ion Beam Laboratory, Koyoto University, Kyoto, Japan

ABSTRACT

Single crystal Al (111) films and bicrystal Al (110) films have been deposited on bare Si (111) and (100) wafers using ion-cluster beam deposition. The stress in the films was determined using X-rays diffraction as the films were thermally cycled from room temperature to 400C. The (111) film exhibited nearly ideal elastic behavior as the stress was essentially linear with temperature. The (110) Al film yielded in compression at a lower temperature and stress than the (111) film and exhibited broad hillocks not found in the Al (111) film. During the second thermal cycle, both films behaved in a nearly ideal elastic fashion. Measurement of the strain relaxation in the films during the second thermal cycle showed that the (110) film relaxed significantly while the (111) did not relax.

INTRODUCTION

Al films deposited by the ion-cluster beam (ICB) method[1] have been shown to have a quasi-epitaxial relationship to the single crystal Si substrates. For ICB depositions onto Si (100) substrates, two variants of the Al (110)//Si(100) are formed; Al (110)//Si (100) and either Al [001]//Si [011] or Al [-110]//Si [011]. These two variants are essentially the same orientation relationship varying only by a 90° rotation. Earlier[2] studies of the Al (111)//Si (111) system showed that during the earliest stage of Al film formation on Si(111) the relationship was of the type Al (100)//Si (111) with the three variants of Al <011>//Si <211>. However all results have shown that for complete fully formed ICB deposited Al films the orientation relationship on Si (111) substrates is Al (111)//Si (111) and Al [-110]//Si [-110].

The TEM studies[3] of the microstructure of these films have shown that the Al (111) films on Si (111) are essentially single crystals, with few low angle grain boundaries and dislocation loops in the as-deposited condition. Annealing at ≈ 400°C eliminated these defects, leaving "perfect" single crystal films. The Al (110) films on Si (100) showed a bicrystal grain morphology, with low angle meandering grain boundaries separating the two Al orientation variant grains in a mosaic morphology without grain triple points. Annealing allowed grain coarsening and growth but preserved the essential bicrystal grain morphology.

A model for the plastic deformation of thin films on rigid substrates[4] has recently been described in terms of dislocation motion threading through the film and being "pinned" at the film surface and lower interface. In this description the film flow stress is determined in part by the film thickness and grain size. Further, it has been pointed out that the film crystallographic orientation in relation to the applied biaxial strain may also help determine the resolved shear stress on the film glide planes[5]. This effect has been proposed[5] as playing a key role in

659

processes which alter film morphology during annealing, such as hillock or sunken grain formation and secondary grain growth which also alters the film crystallographic texture. Recent experimental work has illustrated the film thickness/flow stress effect[6] for Al (111) films on oxidized Si substrates. However the present Al (111) single crystal and Al (110) bicrystal films provide an opportunity to experimentally determine the effects of film crystallographic texture and grain size on film flow stress. X-ray determination of biaxial film strain is an effective method for film stress measurement. This paper presents the results of x-ray strain and stress measurements during annealing of Al (111) single crystal and Al (110) bicrystal films on Si (111) and Si (100) substrates, respectively.

EXPERIMENTAL

The 3000Å thick (111) and (110) epitaxial Al films were prepared under UHV conditions by ion-cluster beam (ICB) deposition at room temperature onto cleaned Si (111) and Si (100) single crystals, respectively.[3] The ICB source was operated with an accelerating voltage of 3 kV. The electron voltage and current were 200 V and 100mA, respectively. Reflected high-energy electron diffraction (RHEED) patterns from the as-deposited films confirmed the orientation of the films: Al (111) on Si (111) and Al(110) on Si (100). After thermal cycling, the films were characterized using cross-sectional transmission electron microscopy (TEM). The (111) Al was indeed still a single crystal, with no grain boundaries over a 30 mm area. The fact that there were no rings in the diffraction pattern (only diffraction dipoles) indicates that the (110) Al film had an epitaxial relationship with the Si although the film was clearly polycrystalline with a mean grain diameter of ~2 mm. High-resolution TEM revealed an atomistically rough interface between the Al and the Si, which is consistent with previous results of similarly deposited films in the same system.[7]

The strain in the films was measured with X-ray diffraction using the grazing incidence X-ray scattering (GIXS) geometry. The GIXS technique has been used extensively to measure the strain gradients in thin films (see for instance, (8)) For the present work, the GIXS method was chosen since the in-plane strain can be measured directly and rapidly. The diffraction experiments were performed at the Stanford Synchrotron Radiation Laboratory (SSRL) under dedicated conditions (3GeV, 20-90 mA) on a focused wiggler beamline. The X-ray energy of 8700 eV was selected with a double-crystal Si (111) monochromator. The incident beam conditions are described elsewhere.[9] Asymmetric θ–2θ scans were performed on the (422) in-plane reflection in the GIXS geometry, and symmetric scans of the (222) reflection normal to the sample surface were performed. The intensity vs 2θ data was fit to a Pearson VII function with a background to determine the peak position.[10] The unstressed interplanar spacing was calculated from the interplanar spacing determined from the two orientations,[8,11] and used to determine the strain. A heating stage designed for use on a four-circle diffractometer was used for the experiments.[9] The samples were thermally cycled from room temperature to 400°C, and the stress was measured every 50°C. For the second thermal cycle, the samples were heated to, and held at 200°C for ~3 hours, and the strain relaxation was measured. Afterwards, the sample was heated to 400°C, then cooled to 200°C and the strain relaxation was again measured. The stress was calculated directly from the measured strain by rotating the stiffness coefficient matrix to the film coordinate system.[11]

RESULTS

The as-deposited room temperature stress of both films was tensile, with Al (111) films having a stress of ≈ 260 MPa, while the Al (110) films having s stress of ≈ 60 MPa. The stress-temperature behavior for both the first and second cycles is included in Figure 1 and will be described in detail in the next section. During the second thermal cycle, the strain relaxation in the films was measured for several hours at 200°C during heating to and cooling from 400°C. The results are included in Figure 2.

DISCUSSION

The difference between the as-deposited stresses for the Al (111) and Al (110) films is puzzling. Because the lattice constants for Al (a_0 = 4.0496Å) and Si (a_0 = 5.4306Å) vary so much, a true epitaxial lattice-matching relationship in general does not exist. For the case of Al (111)//Si (111), it has been pointed out[3] that $4a_0$ (Al) ≈ $3a_0$ (Si), with a difference of +0.6% for the Al. Thus, in this view a quasi-epitaxial orientation relationship requires a strain of 0.6% in the Al. Considering Al (111) with a biaxial modulus of 115 GPa, this strain will induce a stress of ≈700 MPa, which is greater that observed (Figure 1) by a factor of 2. It could be speculated that relaxation via plastic deformation has decreased the as-deposited Al (111) film stress; however, further work is required to understand the source of the as-deposited Al (111) strain.

In the case of the Al (110) bicrystal, the lattice mismatch must be calculated in the two directions, Al [001]//Si [001] and Al [-110]//Si [001]. For Al [001]//Si [001] the mismatch is calculated as d_{100} (Al) - $d_{110}/2$ (Si) = 4.0496Å - 3.8400Å = 0.2096Å. Lattice matching requires ≈ -0.2096Å/4.0496Å ≈ -5% strain in the Al, which is clearly not feasible. This suggests that there is no lattice strain in the Al as a result of this orientation relationship in this direction. For the Al [-110]//Si [011] case lattice mismatch is the same as for Al (111)//Si (111), ≈ .6% for the Al. The sum of these effects for the Al (110) bicrystal suggests that the as-deposited strain should be lower than for the Al (111) film, as observed, however a better understanding of the observed low (60 MPa) Al (110) stress is required.

The stress as a function of temperature for the (111) single crystal Al film is qualitatively similar to other authors' observations for polycrystalline Al films on Si substrates.[4,12] The film stress is tensile at room temperature and behaves linearly with temperature during heating. The slope of this linear portion ($\Delta\sigma/\Delta\alpha$) was calculated (2.1 MPa/°C). Using Equation 1 and

$$\Delta\sigma = M_{111} (\alpha_{Al} - \alpha_{Si}) \Delta T \qquad (1)$$

knowing the thermal expansion (α) of Al and Si, the measured biaxial modulus (M) was calculated to be 96 MPa. This compares well with the theoretical biaxial modulus in the [111] direction of Al (114 GPa). Above 300°C (320 MPa compressive), the film yields in compression with continued heating to 400°C. During cooling, the stress behaves linearly until 250°C, when it yields in tension. Below 150°C, strain hardening is observed and the film returns to a stress similar in magnitude to that measured prior to thermal cycling. The stress-temperature behavior of this single crystal film differs from polycrystalline films in two ways. First, the magnitude of the stress is larger in compression than measured by other authors. Typically, a polycrystalline film would begin yielding at compressive stresses on the order of 100-150 MPa and temperatures below of 200°C. Secondly, there is little hysteresis in the plot (yielding in the film). This will be discussed later and is not the case for the (110) Al film.

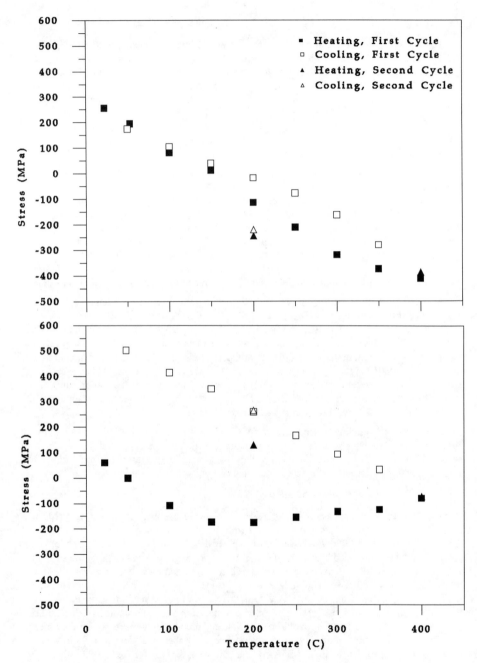

Figure 1. Stress as a function of temperature for a (111) single crystal Al (above) film and a (110) bicrystal Al film. The first and second thermal cycles are shown.

The stress-temperature behavior of the (110) Al film was qualitatively different from the (111) film. As shown in Figure 1, the stress at room temperature is much smaller (60 MPa). Upon heating, the stress is linear with temperature below 100°C. The film yields in compression above 125°C until 400°C. As the temperature is decreased from 400°C, the stress is essentially linear with temperature. From Equation 1, the slope of the stress-temperature line is proportional to the biaxial modulus in the [110] direction in Al. The calculated modulus (71 GPa) is less than the theoretical modulus (87 GPa). This can be justified by considering that strain relaxation will occur during the strain measurement at each temperature, thus reducing the slope of this line and the measured Biaxial Modulus. The stress reaches 500 MPa upon cooling to room temperature.

It is interesting to note that the (110) film exhibits extensive yielding during the first heating. However, even though the (111) is at the same stress level as the (110) and at a higher temperature, it does not. Yielding of the (110) Al film in compression (175 MPa, 150°C) is consistent with the literature for polycrystalline Al films on Si substrates. One physical mechanism suggested to explain this yielding is hillocking in the films. Hillocking typically occurs when thin films are in compression at elevated temperatures. When the films were analyzed in a scanning electron microscope after thermal cycling, hillocks were observed on the surface of the (110) film, but not on the surface of the (111) film. This suggests that the (110) Al film undergoes strain relaxation, while the (111) does not. This was verified during the second thermal cycle. The samples were heated directly from room temperature after the first cycle to 200°C, and the strain relaxation was measured for three hours. The samples were then heated to 400°C and held for 10 minutes. After cooling to 200°C, the relaxation was again measured for three hours. The results are included in Figure 2. The (110) film, which is in tension during both measurements, relaxed 6 MPa at 200°C during heating and 45 MPa during cooling. The (111) film showed no significant relaxation during either heating or cooling.

CONCLUSIONS

The stress in single crystal Al (111) films and bicrystal Al(110) films was determined using X-rays diffraction as the films were thermally cycled from room temperature to 400°C. The (111) film exhibited nearly ideal elastic behavior as the stress was essentially linear with temperature. The (110) Al film yielded in compression at a lower temperature and stress than the (111) film and exhibited broad hillocks that the (111) did not. During the second thermal cycle, both films behaved in a nearly ideal elastic fashion. Measurement of the strain relaxation in the films during the second thermal cycle showed that the (110) film relaxed significantly while the (111) did not relax.

ACKNOWLEDGMENTS

The experiments were performed at SSRL, which is supported by the Dept. of Energy, Div. of Chemical Sciences. One author (PB) acknowledges the support of a Semiconductor Research Corporation Fellowship during collection of these data. We gratefully acknowledge Roger Alvis at AMD for the TEM analysis.

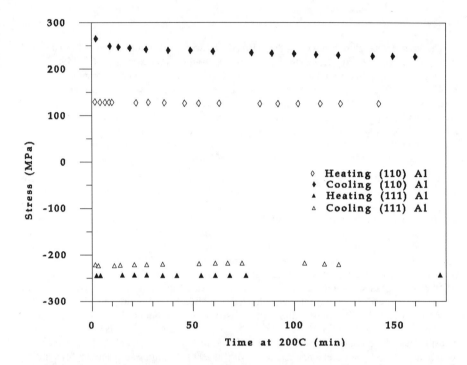

Figure 2. Stress relaxation measured in the (111) and (110) Al films during the second thermal cycle. The samples were heated to and held at 200°C to measure the relaxation. After a subsequent anneal 400°C, the films were cooled to and held at 200°C again.

REFERENCES

1) I. Yamada, H. Inokawa and T. Takagi, J. Appl. Phys. 56(10), 2746 (1984).
2) H. Inokawa, I. Yamada and T. Takagi, Jap. J. Appl. Phys. 24 (3), L173 (1985).
3) I. Yamada, H. Usui, S. Tanaka, U. Dahmen and K.H. Westmacott, J. Vac. Sci. Technol. A 8(3) 1443 (1990).
4) W.D. Nix, Met. Trans. A, 20A, 2217-2245 (1989).
5) J. E. Sanchez, Jr and E. Arzt, Scripta Mettallurgica 27, 284 (1992).
6) R. Venkatraman and J.C. Bravman, J. Mat. Res. 7, 2040 (1992).
7) M. Thangaraj, S. Hingerberger, K.H. Westmacott and U. Dahmen, in D.P. Favreau, Y. Shacham-Diamand, and Y Horiike, Editors, Advanced Metallization for ULSI Applications in 1993, 247 (1993).
8) R. Venkatraman, P.R. Besser, J.C. Bravman, S. Brennan, J. Mater. Res. 9(2), 328(1994).
9) P.R. Besser, S. Brennan and J.C. Bravman, J. Mater. Res. 9(1) 13 (1994).
10) I.C. Noyan, Adv. X-ray Anal. 28, 281-288 (1985).
11) J.A. Bain and B.M. Clemens, MRS Bulletin 17(7), 46 (July, 1992).
12) P.A. Flinn, D.S. Gardner, and W.D. Nix, IEEE Trans. Electr. Dev. 25, 2160 (1987).

CHARACTERIZATION OF TEXTURED ALUMINUM LINES AND MODELLING OF STRESS VOIDING

S.C. Wardle, Department of Mechanical Engineering, Yale University, Newhaven, CT, 06520,
B.L. Adams, Department of Mechanical Engineering, Brigham Young University, Provo, UT84602,
C.S. Nichols, Department of Materials Science and Engineering, Cornell University, Ithaca, NY 14853 and
D.A. Smith, Department of Materials Engineering, Stevens Institute of Technology, Hoboken, NJ 07030.

ABSTRACT

It is well known from studies of individual interfaces that grain boundaries exhibit a spectrum of properties because their structure is misorientation dependent. Usually this variability is neglected and properties are modeled using a mean field approach. The limitations inherent in this approach can be overcome, in principle, using a combination of experimental techniques, theory and modeling. The bamboo structure of an interconnect is a particularly simple polycrystalline structure that can now be readily characterized experimentally and modeled in the computer. The grain misorientations in a [111] textured aluminum line have been measured using the new automated technique of orientational imaging microscopy. By relating boundary angle to diffusivity the expected stress voiding failure processes can be predicted through the link between misorientation angle, grain boundary excess free energy and diffusivity. Consequently it can be shown that the high energy boundaries are the favored failure sites thermodynamically and kinetically.

INTRODUCTION

The limits to the performance of materials are frequently imposed by the behavior of weak links. In the context of grain boundary dominated processes in polycrystalline materials failure is governed by the properties of a critical number of connected boundaries which are susceptible to breakdown in response to some external field [1-5]. This characteristic naturally introduces a statistical element into materials behavior. The elucidation of the failure behavior of a polycrystalline aggregate then requires knowledge of the properties and connectivity of all the grain boundaries present together with an appropriate constitutive law. This paper is a first report of some progress towards this ambitious goal by combining the rapid crystallographic data acquisition possible using Back Scatter Kikuchi Diffraction (BKD) with modeling and applying it to

Mat. Res. Soc. Symp. Proc. Vol. 343. ©1994 Materials Research Society

the relatively simple case of stress voiding in a bamboo microstructure such as is found in narrow interconnects.

EXPERIMENTAL DATA

Unpassivated sub-micron aluminum lines were prepared by conventional processing on oxidized silicon substrates and annealed for 1hr. at 673K. This produces a structure which contains extensive bamboo regions. The misorientations of the grains in a segment of line about 100 microns in length were determined by BKD. The data are displayed in three ways: pole figures, figs 1a and b, a listing of misorientation angles, Table 1 and stereographic representations of the individual grain misorientation axes, figs. 2a-f . The grain boundaries present fall into two crystallographic classes: low angle, misorientation less than ten degrees, and with a broad distribution of rotation axes; high angle, misorientation greater than ten degrees with a strong <111> texture. The sample consisted of more than 140 grains of which almost two thirds were of low angle character. No information was collected concerning the nature of the grain boundary planes but transmission electron microscopy of similarly processed materials indicates that the high angle grain boundaries are predominantly of the asymmetrical tilt type.

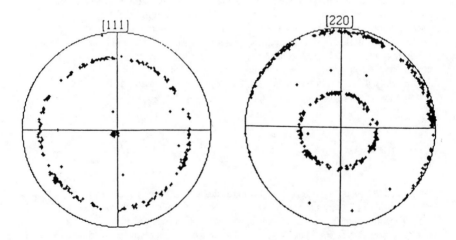

Fig. 1 Pole figures showing the strong 111 fiber texture of the aluminum metallization (a) a 111 pole figure (b) a 220 pole figure.

Table I : Misorientation angles (in degrees) between grains

Range of misorientations	Number of misorientations	Misorientation angles
10-20	13	10,11,11,14,14,15,17,17,18,18,18,18,19
20-30	9	20,23,23,23,25,25,28,28,28
30-40	10	31,31,33,34,36,37,38,38,39,39
40-50	11	40,40,40,43,45,46,47,48,48,49,49
50-62.8	12	51,51,51,52,53,53,56,56,57,58,59,60

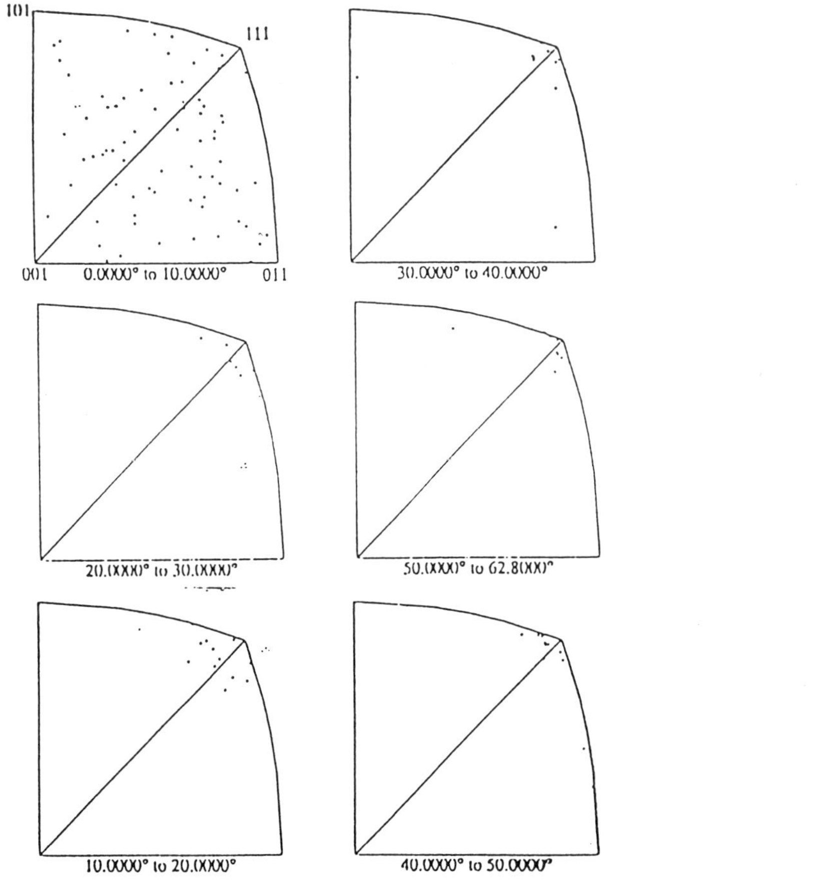

Fig 2 Stereographic projections of the grain rotation axes grouped by rotation angle; note the strong bias towards a [111] rotation axis for misorientation angles greater than ten degrees i.e. the high angle boundaries.

CONNECTION BETWEEN CRYSTALLOGRAPHIC MEASURES OF GRAIN BOUNDARY CHARACTER AND GRAIN BOUNDARY BEHAVIOR

There is a need to establish a fundamental link between experimentally accessible characteristics of grain boundaries, such as their crystallographic parameters, and their properties. At present this task is incomplete. However the grain boundary excess energy γ is a useful global measure of the difference between the coordination at a grain boundary and that in the grain interior. According to the arguments of Borisov [6] there is a quantitative connection between γ, the lattice diffusivity and the grain boundary diffusivity which can be written as follows:

$$\gamma = kT \ln D_b/D_l \tag{1}$$

where D_b and D_l are the grain boundary and lattice diffusivities respectively, k is Boltzmann's constant and T is the absolute temperature. Thus for two grain boundaries with energies γ_1 and γ_2 grain boundary diffusivities are related by a factor $\exp(\gamma_1/\gamma_2)$. The functional form of the γ/misorientation function is known explicitly only for low angle grain boundaries for which Read and Shockley showed that :

$$\gamma = \gamma_0 \theta (A - \ln \theta) \tag{2}$$

where A is a term containing the energy of the dislocation core and γ_0 is an elastic term. For the regime where the Read Shockley equation is valid it would be predicted that:

$$\ln (D_{b1}/D_{b2}) = \{\theta_1(A - \ln\theta_1)\} / \{\theta_2 (A - \ln\theta_2)\} \tag{3}$$

However the diffusion mechanism in tilt boundaries is along the pipes defined by dislocation cores and their density is proportional to the misorientation angle so that the above relationship may be reduced to:

$$\ln(D_{b1}/D_{b2}) = \theta_1 / \theta_2 \tag{4}$$

The equation above describes the misorientation dependence of the grain boundary diffusion coefficient which is governs the kinetics of the growth of grain boundary defects at low homologous temperatures.

The heterogeneous nucleation kinetics of a defect are also related to γ and θ; the free energy of nucleation is decreased by a factor $1/2(2 - 3\cos\alpha + \cos^3\alpha)$ [8] relative to the corresponding value for homogeneous nucleation. This factor appears in the argument of the exponential term describing the nucleation rate. Consequently there is an overwhelming preference for the selection of high energy boundaries as sites for nucleation and growth of defects.

The analysis above has so far only been applied to model systems so far but it clearly has the potential to be used as a diagnostic for grain boundary character or when the relationship between crystallographic measures and properties is established to be a predictor.

GROWTH OF STRESS VOIDS IN A MODEL SYSTEM

A model has been set up in which it is possible to observe the dynamics of void nucleation, growth and coalescence in a segment of a bamboo structured interconnect. The void growth can be modeled as a function of stress, current density and temperature. In accord with data obtained from bicrystal experiments the boundary diffusion coefficient and the sink efficiency for point defects are made misorientation dependent. A simple binary classification is made into (a) low angle and special boundaries and (b) random high boundaries which respectively have low and high diffusivities and sink efficiencies. Each grain consists of an array of particles on a two dimensional square lattice; the line adheres to a substrate but is otherwise free to relax. Each particle represents a group of many atoms. An initial sample with a specified grain size but a random distribution of boundary types is generated by the computer. Vacant sites are also place randomly thoughout the structure in accordance with the desired temperature. Diffusion of particles is treated according to a Monte-Carlo Metropolis [9] algorithm and in the present case a uniaxial stress was applied during the relaxation. Voids were found to accumulate in a higher concentration on random boundaries than on low angle or special boundaries in accord with the theoretical considerations outlined in the previous section.

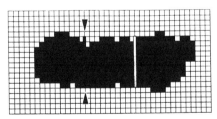

Fig. 3 shows a model bamboo tricrystal with a high energy boundary and a low energy boundary after being stressed in tension at an elevated temperature; note the greater growth of the edge voids at the high energy boundary (indicated by arrows)

DISCUSSION

The work described in this paper highlights the complexity involved in going beyond the mean field approach to the behavior of polycrystalline materials. However in practice failure processes are frequently structure sensitive. The mean field approach is too optimistic for first fails and underestimates the performance which is potentially attainable in the most favorable cases. Consequently the understanding of the upper and lower limits of performance by taking account of, and ultimately controlling, the intrinsic boundary to boundary variability of properties in a polycrystalline aggregate offers considerable technological leverage. To realize this objective it is necessary to have an explicit, experimentally validated and comprehensive understanding of the spectrum of grain boundary properties coupled with an experimentally accessible measure of character such as crystallographic parameters. The grain boundary excess energy is a good fundamental link to properties and BKD related techniques coupled with property measurements offer a way of establishing the necessary database so that relationships like that of Borisov can be validated and fully quantified. The behavior of thin films provides a context which offers practical importance with the simplifications which come from a quasi-two-dimensional grain structure which is strongly textured so that the grain boundary population is limited to low angle boundaries and tilt boundaries with a common fiber axis. The parallel modeling program can potentially develop into a predictive tool which will have been validated by the experimental program and include experimentally determined descriptors of the grain structure (and associated properties) as a function of material, composition and processing.

CONCLUSIONS

A rosetta stone linking properties to crystallographic measures of grain boundary character can be developed by combining BKD with property measurements; this provides the basis for predicting the structure sensitive properties of a polycrystalline aggregate at a level which included boundary to boundary differentiation in properties.

REFERENCES

[1] C.S. Nichols and D.R. Clarke, Acta metall. mater. **39**, 995, (1991).
[2] C.S. Nichols, R.F. Cook, D.R. Clarke and D.A. Smith, Acta metall. mater. **39**, 1657 (1991).
[3] C.S. Nichols, R.F. Cook, D.R. Clarke and D.A. Smith, Acta metall. mater. **39**, 1667, (1991).
[4] T. Watanabe, Materials Forum, **11**, 284 (1988).
[5] G. Palumbo, K.T. Aust, U.Erb, P.J. King, A. M. Brennenstuhl and P.C. Lichtenberger, Phys. Stat. Sol. (a) **131**, 425 (1992).
[6] V. T. Borisov, V.M. Golikov and G.V. Scherbedinsky, Phys. Met. Metall. **17**, 80 (1964).
[7] W.T. Read and W. Shockley, Phys. Rev. **78**, 275 (1950).
[8] J.W. Christian, Transformations in metals and alloys, Pergamon, Oxford, 1965, p412.
[9] N. Metropolis, A.W. Rosenbluth, M.N. Rosenbluth, A.H. Teller and E. Teller, J.Chem. Phys. **21**, 1087 (1953).

PART VII

Polycrystalline Semiconductor Films

NUCLEATION AND GROWTH OF POLYCRYSTALLINE SILICON FILMS IN AN ULTRA HIGH VACUUM RAPID THERMAL CHEMICAL VAPOR DEPOSITION REACTOR USING DISILANE AND HYDROGEN

Katherine E. Violette[1], Mehmet C. Öztürk[1], Gari Harris[2], Mahesh K. Sanganeria[1], Archie Lee[2], and Dennis M. Maher[2]   [1])North Carolina State University, Department of Electrical and Computer Engineering, Box 7911, Raleigh, NC 27695-7911, [2])North Carolina State University, Department of Materials Science and Engineering, Box 7916, Raleigh, NC 27695-7916

A study of Si nucleation and deposition on SiO_2 was performed using disilane and hydrogen in an ultra high vacuum rapid thermal chemical vapor deposition reactor in pressure and temperature ranges of 0.1 - 1.5 Torr and 625 - 750°C. The film analysis was carried out using scanning electron microscopy, transmission electron microscopy and atomic force microscopy. At lower pressures, an incubation time exists which leads to a retardation in film nucleation. At 750°C, the incubation time is 10s at 0.1 Torr and decreases to less than 1s at 1.5 Torr. The nuclei grow and form three dimensional islands on SiO_2, and as they coalesce, result in a rough surface morphology. At higher pressures, the inherent selectivity is lost resulting in a higher nucleation density and smoother surface morphology. For ~ 2000 Å thick films, the root-mean-square surface roughness at 750°C ranges from 110Å at 0.1 Torr to 40Å at 1.5 Torr. Temperature also strongly influences the film structure through surface mobility and grain growth. At 1 Torr, the roughness ranges from 3Å at 625°C to 60Å at 750°C. The grain structure at 625°C/1Torr appears to be amorphous, whereas at 750°C the structure is columnar. The growth rate at 625°C/1.5 Torr is 1200 Å/min provides a surface roughness on the order of atomic dimensions which is comparable to or better than amorphous silicon deposited in LPCVD furnaces.

INTRODUCTION

Polycrystalline silicon (polysilicon) has many applications in Si based microelectronics including self-aligned metal-oxide-semiconductor (MOS) gate technology, thin film transistor (TFT) technology, dopant diffusion source for shallow junction formation[1] and electrically erasable programmable read only memories (EEPROMS)[2]. Each application has different requirements for the polysilicon film structure: MOS gate technology demands smooth films for gate reliability[3], TFT's need large grains for device active areas[1], and diffusion sources require columnar grains for fast dopant diffusion[1]. EEPROMS with a double poly-insulator structure have the most stringent requirements on polysilicon smoothness as the second gate oxide is formed on a polysilicon layer[2]. The wide variety of polysilicon film requirements mandates a thorough understanding of the film growth and factors influencing the basic properties of polysilicon films such as surface morphology and grain structure.

In recent years, single wafer manufacturing using multi-chamber cluster tools has gained in popularity[4]. As a potential technique in such systems, RTCVD has been considered for many processes including polysilicon deposition with successful results[5]. In a typical RTCVD reactor, the wafer is heated optically through a quartz window, and thermal switching is used to initiate and terminate chemical reactions on the wafer surface in a matter of seconds. The technique aims at minimizing the thermal budget, the temperature and time product, instead of simply reducing the process temperature. Satisfactory growth rates for single wafer processing can be achieved at elevated temperatures by accurately controlling the temperature profile to within a few seconds by means of lamp switching and computer control. Thus, thin films can be deposited at high growth rates which is an important concern in single wafer processing. A Si source gas which provides high growth rates is of great importance for the single wafer environment.

A potential source gas for silicon deposition is disilane which offers a significant deposition rate advantage over the more typically used silane and dichlorosilane. Sadamoto, Comfort and Reif show that disilane provides an order of magnitude improvement in the silicon deposition rate over silane in spite of lower process pressures[6]. Several studies reported on LPCVD of silicon from disilane exploit the increase in growth rate from disilane over silane for applications in amorphous Si deposition[7, 8] and in-situ doping[9, 10]. The surface roughness of the amorphous films deposited in these studies, although not quantified, are expected to be similar to amorphous films obtained from silane[1]. As yet, disilane has not been studied in an RTCVD reactor for polysilicon deposition.

There are many mechanisms contributing to the structure of the polysilicon films: total deposition pressure, deposition temperature, hydrogen partial pressure, and the presence of dopants and impurities[1]. These variables strongly influence the film growth in terms of the initial Si nucleation[11] and the subsequent film surface morphology and grain structure[1]. Previous work has suggested that surface morphology in RTCVD films is related to the oxygen levels in the films[5, 12]. Detailed work[13] has recently been performed on LPCVD polysilicon from silane which links the final grain size and surface morphology to Si nucleation and subsequent film coalescence on the oxide surface.

In this work, using disilane for the first time, the influence of pressure and temperature on polysilicon surface morphology and grain structure has been studied in an RTCVD reactor. We have explored the initial nucleation and subsequent growth of polycrystalline silicon films on SiO_2 substrates over the temperature range of 625° to 750°C and the pressure range of 50 mTorr to 1.5 Torr. By using an ultra-clean, ultra-high vacuum (UHV) RTCVD reactor the contamination of the polysilicon films has been minimized, thus allowing separation of temperature and pressure effects from impurity effects.

EXPERIMENTAL

In this study, the depositions were carried out in an UHV-RTCVD reactor described in detail elsewhere.[14] 4" Si (100) wafers with 100 nm thick thermal SiO_2 were cleaned in a batch Huang clean: 5 min. in a 1:1:5 $NH_4OH:H_2O_2$:deionized (DI) water bath at 80°C, 5 min. DI rinse, 5 min. in a 1:1:5 $HCl:H_2O_2$:DI water bath at 80°C followed by a second 5 min. DI water rinse. The wafers were spin-dried in nitrogen and stored in a nitrogen purged box until use. Immediately before insertion into the UHV RTCVD system, a 20 s, 5% HF dip, 30 s water rinse, and nitrogen blow dry were performed to obtain a clean oxide surface. After sample loading and reaching the base pressure (~ 3×10^{-9} Torr) in the cryopumped reaction chamber, the pumping was switched to the turbomolecular / drag combination pump and gas flows were initiated. The gas used in this work was ULSI grade 10% Si_2H_6 in H_2 further purified at point-of-use to obtain oxygen and water vapor levels below 50 ppb. After establishing steady gas flows, an arc lamp was turned on to heat the substrate to the deposition temperature at a ramp rate of approximately 150°C/s. The deposition process was initiated and terminated by temperature switching. When the deposition was complete, the lamp was extinguished, the gas flow stopped and the wafer allowed to cool. The polysilicon surface morphology and film structure were evaluated using atomic force microscopy (AFM), transmission electron microscopy (TEM) and scanning electron microscopy (SEM).

RESULTS

We have previously reported the temperature dependence of the Si growth rate from 10% Si_2H_6 in H_2 at a total pressure of 100 mTorr[14]. At 1 Torr the growth rate ranges from 120 nm/min. at 625°C to 650 nm/min. at 750°C . Over the pressure range of 100 mTorr to 1.5 Torr at 750°C, the growth rate ranges from 330 nm/min. to 890 nm/min with a constant gas flow rate at 250 sccm. The deposition pressure was controlled by a throttle valve which modifies the pumping speed at the chamber exhaust. Closed loop feedback control was established real time using a capacitance manometer for pressure measurement. The polysilicon film thicknesses were measured by interferometry. Comparable growth rates using 10 % silane in argon requires higher pressures and temperatures[5, 12]. Sadamoto, Comfort and Reif at MIT have clearly demonstrated the differences in deposition rates obtained from disilane and silane[6]. Disilane clearly is able to provide high growth rates while reducing pressure and temperature making it a more suitable gas source for single wafer manufacturing.

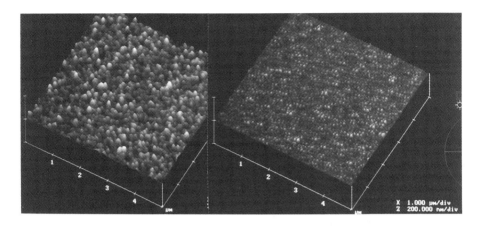

Figure 1: AFM images of films deposited at a) 100 mTorr and b) 1.5 Torr.

Figure 1 shows AFM images of polysilicon samples at a) 100 mTorr and b) 1.5 Torr demonstrating that the surface features are similar but are different sizes: at 100mTorr they are on the order of 250 nm whereas at 1.5 Torr the features are on the order of 50 nm. Figure 2 quantifies the trend observed in surface roughness using both the rms surface roughness and the peak-to-peak surface roughness as a function of pressure at 750°C . As shown, as the pressure is increased from 0.1 Torr to 1.5 Torr the root-mean-square (rms) roughness decreases from 10.5 nm to 3.8 nm. The peak-to-peak roughness decreases with increasing pressure from 85 nm at 100 mTorr to 32 nm at 1.5 Torr. These values are lower than those reported previously for films deposited by RTCVD using SiH_4[5, 12].

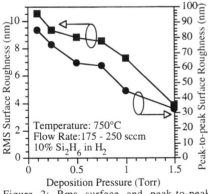

Figure 2: Rms surface and peak-to-peak roughness vs. pressure at 750°C.

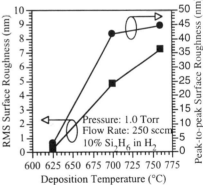

Figure 3: Rms surface and peak-to-peak roughness vs. temperature at 1Torr.

Figure 3 shows the rms surface roughness and peak to peak surface roughness respectively as measured by AFM as a function of temperature at 1 Torr. The surface roughness at 750°C, 1 Torr is 8 nm, and, as the temperature is decreased to 625°C, the rms surface roughness decreases to 0.3 nm which is on the order of atomic dimensions. The peak-to-peak surface roughness also decreases, from 44 nm at 750°C to 33 nm at 625°C. It is our belief that, for these extremely smooth films, the curvature of the wafer surface and artifacts from AFM begin to contribute to, and thus limit, the minimum value of the measurable peak-to-peak surface roughness.

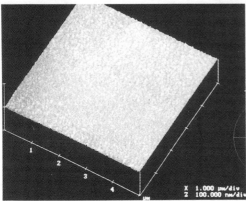

Figure 4: AFM image of polysilicon deposited at 625°C and 1 Torr.

Figure 4 shows and AFM image of a sample deposited at 625°C and 1 Torr. The image shows the extremely smooth surface. The film deposited at 625° appears to have features on the order of 25 nm, extremely small relative to the higher temperature samples. Films which are this smooth tend to be amorphous in the literature and TEM confirms this sample does have an amorphous layer at the surface.

Typical LPCVD of polysilicon from silane results in surface roughness values ranging from 1 nm[13] to 5 nm[15] with some reported data as high as 30 nm[16]. A major difficulty with these data is the range of film thicknesses involved: the smoothest films are 100 nm thick[13] and the roughest are 500 nm thick[16]. Polysilicon surface roughness can be correlated strongly to the film's thickness[17] and can contribute to the variation seen in these different works. In spite of this wide variation in data, the best as-deposited rms roughness is 1 nm[13], a factor of three higher than the as-deposited value obtained at 625°C in this work from disilane. The most notable result here is that the rms surface roughness obtained in this study is smoother than the reported values for amorphous silicon films deposited from silane at temperatures below 580°C[15]. Using disilane also has the significant added advantage of high deposition rate at low temperatures used in this work, allowing for the high throughput required for single wafer manufacturing.

DISCUSSION

Previous data we have obtained indicate that pressure has a strong effect on nucleation in the regime we consider here[18]. The mechanism leading to the pressure dependence of nucleation may be explained as follows. The flux of silicon to the growth surface is dependent on pressure and will increase as pressure increases[19]. The flux to the surface in turn determines the density of adatoms. At lower pressures, the surface density of atoms will be reduced leading to a reduction in the probability of nucleation. Conversely at higher pressures, nucleation occurs more readily leading to higher initial nucleation densities. Once nucleation has occurred, the diffusing adatoms prefer to attach to existing nuclei before establishing new nucleation sites thus leading to a saturation in the nuclei density[11]. The adatom surface mobility is a strong function of temperature[1] where, at high temperatures, the adatoms have high mobilities and therefore diffusion lengths and are able reach and attach to an existing nucleus. With decreasing temperature, the adatoms lose thermal energy which decreases their diffusion length thus leading to an increase in the number of nucleation sites until they are separated by an adatom diffusion length.

At the high growth rates associated with RTP, nuclei tend to grow three dimensionally and form a rough surface upon coalescence. At higher pressures, the flux of silicon species to the oxide surface will be high, thus greatly increasing the nucleation density on the oxide surface. When the nucleation density is high, coalescence occurs early, while the nuclei are still small, and the initial interface will have a smooth morphology. At lower pressures, the nucleation density is reduced and coalescence occurs when the nuclei are large leading to a rough initial surface. This initial film sets the stage for the film grain structure later in the deposition. The nucleation sites provide seed crystallites for the diffusing adatoms and will form the initial grains in the film after coalescence. Since it is energetically favorable for an atom to attach to a crystalline location[1], longer surface diffusion lengths obtained at the reduced pressures lead to larger nuclei and thus larger grains. Because of the high level of cleanliness in the UHV environment, the diffusion length is likely not to be limited on the system impurities. Extensive secondary ion mass spectroscopy data show that the background oxygen levels in the deposited silicon films are below the SIMS background level of 2E18 cm^{-3}.

The adsorption of hydrogen may begin to play a role in the diffusion of adatoms as temperature is reduced below 650°C as the hydrogen surface coverage will be high at high pressures[19]. If hydrogen occupies a large number of surface sites it can impede adatom diffusion and can lead to an extremely high initial nucleation density and in turn an extremely smooth initial surface upon coalescence. After film coalescence the location of adatom bonding will depend on its ability to diffuse to a crystalline location. Since the temperature at 625°C is so low, the surface mobility will be reduced both thermally and by hydrogen coverage preventing adatoms diffusion to the favorable existing crystalline sites. The result will then be short range incorporation and thus an amorphous structure and ultimately an atomically smooth surface morphology.

CONCLUSIONS

In this work we have studied RTCVD of polysilicon using Si_2H_6. Our results indicate that extremely smooth films can be deposited at rates compatible with single wafer manufacturing. Best surface morphology is obtained at higher pressures and lower temperatures. At 625°C / 1Torr, the deposition rate is ~ 100nm/min. and the rms roughness is only 0.3 nm. The film has an amorphous structure at the top and polycrystalline structure at the oxide interface with a smoother surface than that obtained in conventional LPCVD furnaces using SiH_4.

REFERENCES

1. Kamins, T.I., Polycrystalline Silicon for Integrated Circuit Applications. The Kluwer International Series in Engineering and Computer Science, ed. J. Allen. 1988, Norwell, MA: Kluwer Academic Publishers. 280.
2. Cobiani, C., O. Popa, and D. Dascalu, IEEE Electron Device Letters, 1993. **14**(5): p. 213-215.
3. Ibok, E. and S. Garg, Journal of the Electrochemical Society, 1993. **140**(10): p. 2927-2937.
4. Burggraaf, P., . 1992, p. 71-74.
5. Ren, X., et al., Journal of Vacuum Science Technology B, 1992. **10**(3): p. 1081-1086.
6. Sadamoto, M., J.H. Comfort, and R. Reif, Journal of Electronic Materials, 1990. **19**(12): p. 1395.
7. Nakazawa, K., Journal of Applied Physics, 1991. **69**(3): p. 1703-1706.
8. Voutsas, A.T. and M.K. Hatalis, Journal of the Electrochemical Society, 1993. **140**(3): p. 871-877.
9. Nakayama, S., I. Kawashima, and J. Murota, Journal of the Electrochemical Society, 1986. **133**(8): p. 1721-1724.
10. Madsen, L.D. and L. Weaver, Journal of the Electrochemical Society, 1990. **137**(7): p. 2246-2251.
11. Bloem, J., Journal of Crystal Growth, 1980. **50**: p. 581-604.
12. Xu, X.-L., et al., Journal of Electronic Materials, 1993. **22**(11): p. 1345-1351.
13. Voutsas, A.T. and M.K. Hatalis, Journal of the Electrochemical Society, 1993. **140**(1): p. 282-288.
14. Sanganeria, M.K., K.E. Violette, and M.C. Öztürk, Applied Physics Letters, 1993. **63**(9): p. 1225-127.
15. Harbeke, G., et al., Journal of the Electrochemical Society, 1984. **131**(3): p. 675-682.
16. Foster, D., A. Learn, and T. Kamins, Solid State Technology, 1986. (May): p. 227-232.
17. Kamins, T.I., Solid State Technology, 1990. (April): p. 80-82.
18. Violette, K.E., M.K. Sanganeria, and M.C. Öztürk, Journal of the Electrochemical Society,
19. Gates, S.M. and S.K. Kulkarni, Applied Physics Letters, 1992. **60**(1): p. 53-55.

FABRICATION OF POLYCRYSTALLINE $Si_{1-x}Ge_x$ FILMS ON OXIDE FOR THIN-FILM TRANSISTORS

JULIE A. TSAI[†], ANDREW J. TANG[‡], and RAFAEL REIF[‡]
[†] Department of Materials Science and Engineering
[‡] Department of Electrical Engineering and Computer Science
Microsystems Technology Laboratories, Massachusetts Institute of Technology, Cambridge, MA 02139

ABSTRACT

Polycrystalline-$Si_{1-x}Ge_x$ films have been formed by various methods on oxide-coated Si substrates at temperatures $\leq 600°C$. Compared to thermal growth, plasma deposition of poly-$Si_{1-x}Ge_x$ promotes smoother films with smaller grains having a {200}-dominated texture. Poly-$Si_{1-x}Ge_x$ films formed by plasma deposition of amorphous-$Si_{1-x}Ge_x$ followed by a crystallization anneal have an even smoother surface with grain sizes enhanced by an order of magnitude and a weak {111} grain texture. Hydrogen incorporated in amorphous-$Si_{1-x}Ge_x$ evolves completely during crystallization without disrupting the smooth surface morphology. The largest grain sizes ($\sim 1.3\mu m$) are achieved in poly-$Si_{1-x}Ge_x$ films formed by Si^+ ion implantation for amorphization with a subsequent recrystallization anneal.

INTRODUCTION

Polycrystalline-$Si_{1-x}Ge_x$ (poly-$Si_{1-x}Ge_x$) alloy materials are attractive for thin-film transistor (TFT) applications such as active-matrix liquid crystal displays (AMLCDs) and high-density static random-access memory (SRAM) cells. As an alternative to poly-Si, poly-$Si_{1-x}Ge_x$ offers lower temperatures for deposition and processing [1] and can lead to higher carrier mobilities [2] while remaining essentially compatible with exisiting Si processing and integrated circuit technologies.

Deposition of poly-$Si_{1-x}Ge_x$ alloys has been achieved by chemical vapor deposition (CVD) techniques such as conventional low-pressure CVD (LPCVD) [3], rapid thermal CVD (RTCVD) [4], and plasma-enhanced very-low-pressure CVD (PE-VLPCVD) [5]. However, as-deposited films in polycrystalline form tend to have rough surfaces and small grain sizes that degrade carrier conduction in the surface channel of a TFT. Although plasma-enhanced deposition of poly-$Si_{1-x}Ge_x$ has been reported to improve the surface morphology compared to thermal CVD [6], issues of grain size, grain structure, and texture must also be considered.

Research in poly-Si TFTs has explored several technologies to improve device performance by enhancing the poly-Si grain size. One method is the amorphization of LPCVD poly-Si films by Si^+ self-implantation with subsequent solid-phase recrystallization [6],[7],[8]. Other methods involve deposition of LPCVD amorphous Si, with [9] or without [10],[11] subsequent Si^+ implantation, followed by solid-phase crystallization. This work investigates similar methods to improve the quality of poly-$Si_{1-x}Ge_x$ films and compares the resulting films to both as-deposited poly-$Si_{1-x}Ge_x$ and to poly-Si control films. Specifically, we investigate low-temperature deposition of amorphous $Si_{1-x}Ge_x$ with subsequent crystallization, and amorphization of poly-$Si_{1-x}Ge_x$ by Si^+ implantation with subsequent recrystallization.

EXPERIMENTAL PROCEDURES

Amorphous and polycrystalline $Si_{1-x}Ge_x$ thin films with various Ge contents were deposited using a flexible plasma-enhanced very-low-pressure chemical vapor deposition (PE-VLPCVD) re-

a   b

Figure 1: XTEM micrographs of as-deposited poly-$Si_{0.68}Ge_{0.32}$: a) thermal CVD, and b) plasma-enhanced CVD [5].

actor that has previously been described in detail [5]. Substrates were bare (100)-oriented Si wafers coated with 1000Å of thermally grown SiO_2. After a brief ex-situ HF dip clean, each substrate was loaded into the PE-VLPCVD reactor and baked at 600°C in Ar gas until the base pressure reached 10^{-8} Torr. The temperature was then lowered to the desired film growth temperature: 500°C for as-deposited poly-$Si_{1-x}Ge_x$, or 400°C for as-deposited amorphous-$Si_{1-x}Ge_x$. Just prior to film deposition, the oxide surface was briefly Ar-plasma-sputtered to improve nucleation of $Si_{1-x}Ge_x$ on oxide. Thin films of Si and $Si_{1-x}Ge_x$ were plasma-deposited at pressures ≤ 5 mTorr using SiH_4 and GeH_4 process gases of varying ratios, at a total gas flow rate of 30 sccm, and with 10 Watts of RF plasma power, with the exception of a thermal CVD poly-$Si_{0.68}Ge_{0.32}$ film that was grown for comparison to a plasma-deposited poly-$Si_{0.68}Ge_{0.32}$ film.

The Si and $Si_{1-x}Ge_x$ films deposited in amorphous form were 850Å thick and had Ge contents of $x = 0\%$, 18%, and 31%, as measured by Rutherford backscattering spectroscopy (RBS). To evaluate the extent of hydrogen incoporation, a 1000Å-thick amorphous $Si_{0.69}Ge_{0.31}$ film was deposited directly on a double-polished high-resistivity Si wafer for Fourier transform infrared spectroscopy (FTIR) analysis. Annealing of the amorphous films at 600°C in a N_2 ambient for times of 2, 6, and 15 hours was performed to investigate crystallization behavior. The Si and $Si_{0.75}Ge_{0.25}$ films plasma-deposited in polycrystalline form were 1000Å thick and were amorphized by Si^+ ion implantation at a dose of 5e15 cm^{-2} and energy of 55 keV. These films were also annealed at 600°C in a N_2 ambient for times of 6 and 15 hours to investigate re-crystallization behavior. Resulting film properties were characterized by glancing-angle X-ray diffraction analysis (XRD) for crystallinity and grain texture, by plan-view transmission electron microscopy (p-TEM) for grain size estimates, and by cross-sectional TEM (XTEM) for surface morphology.

RESULTS AND DISCUSSION

As-Deposited Poly-$Si_{1-x}Ge_x$ Films

An important advantage of plasma-enhanced CVD poly-$Si_{1-x}Ge_x$ films over thermal CVD is the resulting surface morphology, as shown in Figure 1, where the plasma-deposited film has a smoother surface and a more columnar {220}-oriented grain structure [5]. For TFT applications, smooth surfaces are conducive to uniformity and reproducibility during subsequent device processing steps such as patterning and oxidation. Strong grain texture is desirable for low-angle grain boundaries between aligned grains for reduced carrier scattering in TFTs. However, grain size must also be considered. Figure 2 shows the corresponding grain sizes of the poly-$Si_{0.68}Ge_{0.32}$ films from plan-view TEM analysis. Whereas thermal growth results in an average grain size of about 1200Å,

680

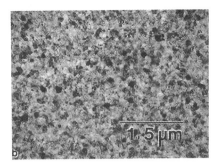

Figure 2: Plan-view TEM micrographs of as-deposited poly-$Si_{0.68}Ge_{0.32}$: a) thermal CVD, and b) plasma-enhanced CVD.

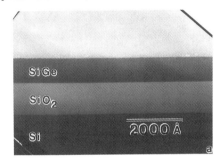

Figure 3: XTEM micrographs of 400°C $Si_{0.69}Ge_{0.31}$: a) amorphous as-deposited, and b) after crystallization (600°C, N_2 ambient, 15 hours).

plasma-enhanced deposition yields an even smaller average grain size of about 400Å. Therefore, despite the smoother surface of the plasma-deposited film, the presence of large grain boundary areas may greatly reduce effective carrier mobilities in TFTs due to excessive grain boundary trap states. For high-performance TFTs, much larger grain sizes are desirable.

Amorphous Deposition with Subsequent Crystallization

One technique for grain-size enhancement involves deposition in the amorphous state followed by solid-phase crystallization annealing. We concentrated on plasma-enhanced deposition rather than thermal growth not only because it results in smoother as-deposited films but also to avoid the very slow growth rates associated with thermal deposition at low temperatures. Figure 3 shows XTEM micrographs of the 400°C amorphously-deposited $Si_{0.69}Ge_{0.31}$ film before and after a 15-hour crystallization anneal. In the amorphous state, a very smooth $Si_{1-x}Ge_x$ surface is evident, but also evident are faint fine columnar features. Since glancing-angle XRD analysis showed the film to be fully amorphous, these columnar features may be ascribed to fluctuations in the density of amorphous material resulting from the growth temperature being lower than that required for adequate surface diffusion to occur. Plasma-deposition of amorphous $Si_{1-x}Ge_x$ at 450° resulted in a very smooth surface with no columnar features. After crystallization of the 400°C amorphous film, the very smooth surface is maintained, as shown in Figure 3b, and the resulting grain size is larger than the figure's field of view.

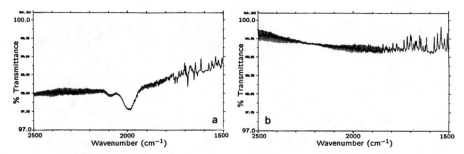

Figure 4: FTIR scans of 400°C $Si_{0.69}Ge_{0.31}$/Si: a) amorphous as-deposited, and b) after crystallization (600°C, N_2 ambient, 15 hours).

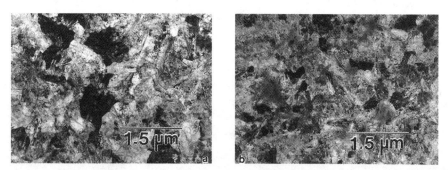

Figure 5: Plan-view TEM micrographs of crystallized amorphous films: a) $Si_{0.69}Ge_{0.31}$, and b) Si.

At low deposition temperatures such as 400°C, H incorporation is a concern in amorphous films deposited using hydride gas chemistries (i.e. SiH_4, GeH_4) because any H present during crystallization or subsequent device processing steps may create voids or bubbles in the film. FTIR analysis in the region of stretching mode frequencies for Si-H [12] and Ge-H [13] was performed on a $Si_{0.69}Ge_{0.31}$ film before and after a 15-hour cryrstallization anneal, and the results are shown in Figure 4. Two peaks are observed for the amorphous film, both of which are intermediate between the peaks for pure Si-H and pure Ge-H: one near 2000 cm^{-1} corresponding to monohydrides (SiH and/or GeH), and a smaller peak near 2100 cm^{-1} for dihydrides (SiH_2 and/or GeH_2). After crystallization, both peaks are no longer detectable by FTIR; thus, H incorporated in the film during deposition is eevolved during crystallization without disruption of the smooth surface quality of the poly-$Si_{1-x}Ge_x$ film.

Analysis by glancing-angle XRD showed that all amorphously deposited Si and $Si_{1-x}Ge_x$ films were crystallized within the shortest anneal time investigated of 2 hours. In contrast, reports in the literature of solid phase crystallization of amorphous Si [11],[12] or amorphized Si [6],[8] films often require tens of hours. Since grain growth during solid-phase crystallization can be retarded by electrically inactive impurities such as O, N, and C [14], our very fast crystallization behavior even of pure Si may be attributable to the low base pressure and very-low operating pressure of our PE-VLPCVD deposition process.

The plan-view TEM micrographs in Figure 5 compare the crystallized $Si_{0.69}Ge_{0.31}$ film to the crystallized pure Si film after 15 hours of annealing. The average grain size of the poly-$Si_{0.69}Ge_{0.31}$ film is $0.87\mu m$ and that of poly-Si is $0.40\mu m$, both of which are much larger than the grain sizes

Figure 6: Plan-view TEM micrographs of Si$^+$-implant-amorphized films after recrystallization: a) Si$_{0.75}$Ge$_{0.25}$, and b) Si.

of either of the as-deposited poly-Si$_{1-x}$Ge$_x$-films (Figure 2). Since the poly-Si$_{0.69}$Ge$_{0.31}$ grain size is more than twice that of poly-Si, the technique of amorphous deposition with subsequent crystallization appears to be even more effective at enhancing grain sizes of Si$_{1-x}$Ge$_x$ films than Si films. Because Ge has a lower melting point than Si, a particular anneal temperature will be more effective for the nucleation and growth processes of a Si$_{1-x}$Ge$_x$ alloy compared to pure Si, resulting in larger Si$_{1-x}$Ge$_x$ grain sizes. From XRD analysis, both films have a weak {111} grain texture, similar to that observed in thermal as-deposited poly-Si$_{1-x}$Ge$_x$ films.

Amorphization by Si$^+$ Implantation with Subsequent Re-crystallization

Another technique used for grain size enhancement is the amorphization of polycrystalline films by implantation followed by recrystallization. Just as all the amorphously deposited films had crystallized within 2 hours of annealing, the poly-Si and poly-Si$_{0.75}$Ge$_{0.25}$ films which were amorphized by Si$^+$ implantation also recrystallized within the shortest anneal time investigated. In Figure 6, the plan-view TEM micrographs show the average grain sizes after a 15-hour anneal: 0.47μm for poly-Si, and 1.33μm for poly-Si$_{0.75}$Ge$_{0.25}$. Final grain size data for poly-Si$_{1-x}$Ge$_x$ films from the different fabrication methods is summarized in Figure 7.

For poly-Si, a slightly larger grain size resulted from fabrication by Si-implant-amorphization with recrystallization compared to the method of amorphous deposition with crystallization; this increase can be attributed to the slightly larger initial film thickness. However, an even greater increase in grain size was observed for poly-Si$_{1-x}$Ge$_x$. The implantation process thus appears to further enhance Si$_{1-x}$Ge$_x$ grain size by reducing the density of nucleation sites more effectively in Si$_{1-x}$Ge$_x$ than in Si. We speculate that ion-bombarded Ge atoms impede the nucleation process due to its heavier mass compared to Si during the reordering of the amorphous alloy film. For both Si and Si$_{1-x}$Ge$_x$, XRD analysis showed that initially {200}-oriented as-deposited polycrystalline films (Figure 1b) which are subjected to implant-amorphization with recrystallization become weakly {111}-textured, similar to the thermal as-deposited film and the amorphous-crystallized films. Therefore {111} grains resulted from random nucleation rather than from growth of existing nuclei.

SUMMARY AND CONCLUSION

Two methods were investigated for fabricating improved polycrystalline-Si$_{1-x}$Ge$_x$ films on oxide at low temperatures ($\leq$ 600°C) suitable for TFTs: plasma deposition of amorphous material with subsequent crystallization, and Si$^+$ implant-amorphization of polycrystalline material with

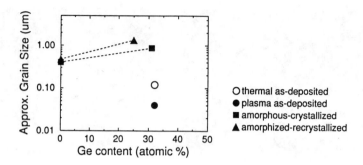

Figure 7: Si and $Si_{1-x}Ge_x$ grain sizes as a function of Ge content for the various fabrication methods.

subsequent recrystallization. Both methods result in high quality poly-$Si_{1-x}Ge_x$ films with very smooth surfaces, enhanced grain sizes, weak {111} grain texture, and no detectable hydrogen content. The $Si_{1-x}Ge_x$ grain sizes are enhanced by an order of magnitude compared to as-deposited poly-$Si_{1-x}Ge_x$ films and by more than a factor of two compared to identically-processed poly-Si films. The presence of Ge in poly-Si promotes grain size enhancement not only because the nucleation and growth process is more effective but also because the density of nucleation sites after implant-amorphization is reduced, resutling in even larger grain sizes. The largest grain size observed was $1.3\mu m$ in the implant-amorphized-recrystallized poly-$Si_{0.75}Ge_{0.25}$ film.

The authors would like to thank Dr. T. Noguchi on leave from Sony Corp. and Prof. C. Thompson for helpful discussions. One author (J. Tsai) gratefully acknowledges the Semiconductor Research Corporation (SRC) Education Alliance for fellowship support.

REFERENCES

1. T-J. King and K.C. Saraswat, IEDM Tech. Dig., 567 (1991).
2. T-J. King, J.P. McVittie, K.C. Saraswat, and J.R. Pfiester, IEEE Trans. Electron Dev., **41** (2), 228 (1994).
3. T-J. King, J.R. Pfiester, J.D. Shott, J.P. McVittie, and K.C. Saraswat, IEDM Tech. Dig., 253 (1990).
4. D.T. Grider, M.C. Ozturk, Y. Zhong, J.J. Wortman, and M.A. Littlejohn, TechCon'90 Extended Abstracts, 293 (1990).
5. J.A. Tsai and R. Reif, Mater. Res. Soc. Proc. **317**, (1993), in press.
6. R. Reif and J.E. Knott, Electron. Lett., **17** (17), 586 (1981).
7. T. Noguchi, H. Hayashi, and T. Ohshima, Jpn. J. Appl. Phys., **25** (2), L121 (1986).
8. K. T-Y. Kung and R. Reif, J. Appl. Phys., **64** (4), 1638 (1987).
9. I.-W. Wu, A. Chiang, M. Fuse, L. Ovecoglu, and T.Y. Huang, J. Appl. Phys., **65** (10), 4036 (1989).
10. W.G. Hawkins, IEEE Trans. Electron Dev., **33** (4), 477 (1986).
11. M.K. Hatalis and D.W. Greve, IEEE Electron Dev. Lett., **8** (8), 361 (1987).
12. T. Noguchi, T. Ohshima, and H. Hayashi, Jpn. J. Appl. Phys., **28** (1), 146 (1989).
13. W.A. Turner, D. Pang, A.E. Wetsel, S.J. Jones, W. Paul, and J.H. Chen, Mater. Res. Soc. Proc. **192**, 493 (1990).
14. E.F. Kennedy, L. Csepregi, J.W. Mayer, and T.W. Sigmon, J. Appl. Phys., **48** (10), 4241 (1977).

DOPANT IMPLANTATION AND ACTIVATION IN POLYCRYSTALLINE-SiGe

W.J. EDWARDS*, YUEN-SHUNG CHIEH**, SAMUEL LIN**, D.G. AST*, J.P. KRUSIUS** AND T. I. KAMINS***
* Department of Materials Science and Engineering, Cornell University, Ithaca, NY 14853-1501
** School of Electrical Engineering, Cornell University, Ithaca, NY 14853-1501
*** Hewlett-Packard, Palo Alto, CA 94303-0867

ABSTRACT

The activation of dopants and the evolution of the microstructure of poly-SiGe films during the activation anneal are of interest for applications such as removable diffusion sources for shallow junctions in CMOS circuit processing, low resistance contact layers and elements for thin film transistors in active matrix flat panel displays. To study these effects, 2300Å poly-SiGe films with Ge content between 20 and 48% were deposited on 600Å poly-Si seed layers by APCVD on thermally oxidized Si substrates. Growth temperatures were between 600 and 700°C. These films were then implanted with B and As at energies of 40 and 80 keV, respectively, at two different doses of $5 \times 10^{13}/cm^2$ and $5 \times 10^{16}/cm^2$. To activate the dopants, the samples were furnace annealed at temperatures from 600 to 1000°C in an N_2 ambient. Additional samples were rapidly annealed at temperatures between 800 and 1050°C for times up to 120 seconds. The sheet resistance, measured using a four-point probe, of $5 \times 10^{16}/cm^2$ arsenic implanted films increased by a factor of two as the Ge content in the film increased from 20 to 48%. The boron doped film, on the other hand, showed no increase in sheet resistance with Ge content. TEM showed that in all cases the grain size increased after anneal. However, the grain sizes of the annealed, arsenic doped samples were on average 5 times larger than those of the boron-doped samples with the same anneal. The sub-grain structure of these films also changed after implantation and annealing. In particular, the twin planes increased in size in boron doped samples and sub-grain twinning virtually disappeared in the arsenic doped samples. In $5 \times 10^{16}/cm^2$ As implanted samples, precipitates were formed which may indicate that the As solubility limit in the films has been exceeded. These precipitates are believed to be As or As-rich and could be responsible for the sheet resistance increase with Ge fraction, since the maximum solubility for As in Ge is lower than in Si by about a factor of ten.

INTRODUCTION

Silicon-germanium alloys have been used in CMOS processing as an ultra-shallow source/drain dopant diffusion source [1]. Implantation into the film with the desired dopant, controlled diffusion of the dopant from the film into the substrate and selective removal allows this film to be used as a sacrificial layer absorbing implantation damage. The use of silicon-germanium alloys for thin film transistors (TFTs) has also been investigated recently [2]. Greater carrier mobilities and lower processing temperatures than those achievable in poly-Si are expected by the addition of Ge to the film. Unpublished results from this laboratory indicate

Mat. Res. Soc. Symp. Proc. Vol. 343. ©1994 Materials Research Society

good device characteristics for p-type devices in agreement with reference 2, but the characteristics of n-type devices have not been promising.

The use of silicon-germanium alloys for the above mentioned applications requires a better understanding on how the properties of these films depend on doping conditions and subsequent thermal treatment. Boron implanted poly-SiGe films have been investigated previously, but n-type implanted films have received little attention to date [3]. Since the diffusivity of dopants from polycrystalline films is strongly dependent on the grain structure of the films [4], an understanding of structural changes during processing is required.

EXPERIMENTAL

Polycrystalline silicon-germanium films were deposited by APCVD (atmospheric pressure chemical vapor deposition) on a 600Å polycrystalline silicon seed layer on thermally oxidized Si substrates at temperatures between 600 and 700°C. Film compositions ranged from 20 to 48% Ge with thicknesses of 2300Å. These films were then implanted with B and As at doses of $5 \times 10^{13}/cm^2$ and $5 \times 10^{16}/cm^2$. Implantation energies were 40 keV for boron and 80 keV for arsenic. Following implantation, the films were capped with a plasma-enhanced CVD oxide at a temperature of 240°C and furnace annealed at temperatures from 600 to 1000°C in flowing nitrogen to activate the dopants. Some samples were also subjected to a subsequent rapid thermal anneal (RTA) to assist in dopant activation. The sheet resistance was then measured using a four-point probe. X-ray diffraction was performed before and after implantation and thermal treatment to determine changes, if any, in the initial texture. Both plan-view and cross-sectional transmission electron microscopy (TEM) were performed on selected samples. TEM specimens were prepared by mechanical thinning and dimpling followed by argon ion milling to electron transparency.

RESULTS AND DISCUSSION

Figure 1a shows the dependence of the sheet resistance, measured with a four point probe, for films implanted with arsenic and boron at a dose of $5 \times 10^{16}/cm^2$ as a function of the Ge fraction in the film. Following implantation, these samples were annealed at 975°C for 60 minutes in flowing nitrogen. Inspection of figure 1a shows that in B implanted films, the sheet resistance is relatively independent of Ge content between 20 and 48% Ge. On the other hand, the As implanted films show an increase of over 120% in sheet resistance as Ge fraction increases from 20% to 48%. It can also be seen that the sheet resistance of the As implanted films exceeds that of the B implanted films for all Ge fractions. King et al., using a dose lower by an order of magnitude to that in this work, found similar results for the resistivity of B implanted films but noted about a 25% decrease in the resistivity with increasing Ge fraction in this composition range [3]. The similar resistivity values obtained here and in the reference, suggest that both films are near the solid solubility limit for boron. As a reference, the resistivity of similarly doped poly-Si films for both As and B doping is on the order of 2×10^{-3} Ω-cm [4] which corresponds to a sheet resistance of ~90Ω/square for the film thickness studied in this work. This value is comparable to that of B doped poly-SiGe films for all Ge concentrations. The sheet resistance of the As doped $Si_{.52}Ge_{.48}$ film, however, is three times as high as that of the corresponding poly-Si films.

686

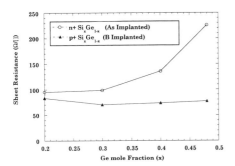

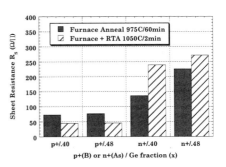

Figure 1a. Plot of sheet resistance, measured with a four-point probe, vs. Ge content of $5 \times 10^{16}/cm^2$ B and As implanted films furnace annealed for 60 min. at 975°C.

Figure 1b. Sheet resistance of 40 and 48% Ge films, implanted with $5 \times 10^{16}/cm^2$ As and B, after furnace anneal (solid bars) and additional 120 sec RTA at 1050°C (striped bars).

All films were then rapidly thermally annealed (RTA) at 1050°C for 120 sec. in dry nitrogen to determine if the dopants were fully activated. Figure 1b shows the results for the 40 and 48% Ge containing films. The additional anneal decreases the sheet resistance by 40% for the B implanted films but increases the sheet resistance for the arsenic implanted films by 77% for the $Si_{.60}Ge_{.40}$ film and 18% for the $Si_{.52}Ge_{.48}$ film.

To measure the crystallographic texture in these films, X-ray diffraction was used. Figure 2a shows the diffracted intensity versus 2Θ for the boron and arsenic implanted $Si_{.52}Ge_{.48}$ films after both the furnace and rapid thermal anneals discussed above. In general, the {111} and {220} peaks dominate the spectra and these peaks are shown in the figure, along with the {311} peak. The higher index peaks were difficult to differentiate from the background and are not shown. The integrated peak intensities, corrected for absorption and normalized to that of a randomly oriented poly-Si powder, are shown in figure 2b. The most striking feature is the nearly five-fold increase in {111} texture in the arsenic implanted and annealed sample relative

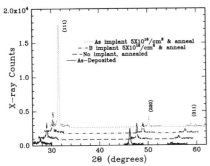

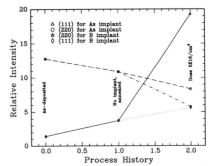

Figure 2a. X-ray diffraction spectra of {111}, {220} and {311} x-ray peaks of $Si_{.52}Ge_{.48}$ film as a function of process history.

Figure 2b. Integrated and normalized intensity of the {111} and {220} x-ray peaks in figure 2a.

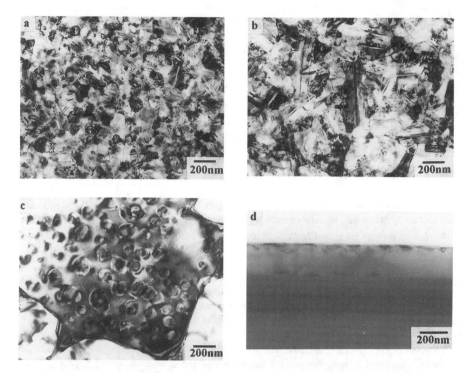

Figure 3. Plan-view and cross-sectional TEM micrographs of 60 min., 975°C furnace and 1050°C, 120 sec. RTA annealed films after: a. no implant species, b. Boron implanted at 40 keV at a dose of $5 \times 10^{16}/cm^2$, c. As implanted at 80 keV and a dose of $5 \times 10^{16}/cm^2$. d. A cross-sectional TEM image of the same sample in c.

to that of a non-implanted sample with the same thermal history. The B implanted film also shows an increase in the {111} texture, making it nearly equal to that of the {220} texture.

To determine what changes in grain size and structure accompany these texture changes, TEM was performed on these samples. Figure 3 shows plan-view and cross-sectional transmission electron micrographs of the 48% Ge film for different implantation conditions. Figure 3a shows a control area on the wafer masked during implantation but annealed under the same conditions as areas implanted with B and As at $5 \times 10^{16}/cm^2$. The activation anneal sequence for these samples was 975°C for 60 minutes in flowing nitrogen followed by the RTA at 1050°C for 120 sec, also in nitrogen. This non-implanted film has a grain size of roughly 1400Å. Figures 3b and 3c show the B and As implanted and annealed films, respectively. A comparison of the three figures shows that both B and As enhance grain growth. In B implanted films the grain size increased to 2300Å, and in As implanted films it increased to ~1.1µm. This is in contrast with poly-Si films which show no increase in grain size after boron implantation [4]. Twin platelets in the grain structure of the B implanted and annealed film appear to accompany the increase in grain size. These platelets accompany most of the individual grains and their length is on the order of the grain size. Films implanted with As show very little subgrain twin structure but appear to contain precipitates about 1000Å in diameter in most grains. This is believed to result from the implant exceeding the solubility limit of As in the alloy. The average As concentration, calculated by dividing the implant dose by the film thickness, is ~1.7×10^{21}/cm^3. Since the maximum solubility limit of As is ~2×10^{20}/cm^3 in Ge and ~3×10^{21}/cm^3 in Si [5], it is reasonable to assume that the solubility of As in a 48% Ge film is less than 1.7×10^{21}/cm^3. Therefore it is likely that the precipitates are As or As-rich regions. The same As implanted and annealed film is shown in cross-section in figure 3d which shows that the precipitates nucleated heterogeneously at the surfaces of the film. This observation indicates that the precipitates are not an artifact of TEM sample preparation and that As can redistribute itself over the distance of half of the total film thickness, about 1500Å including the poly-Si seed layer. Further investigation of the nature and formation of the precipitates is underway.

The observation that the sheet resistance increases with Ge content for the furnace annealed specimens and that an additional RTA at a higher temperature further increases the sheet resistance is consistent with the hypothesis that the sheet resistance increase is caused by the formation of As or As-rich precipitates. Since most thin film processing occurs under non-equilibrium conditions, it is conceivable that for each Ge content, an optimal anneal temperature at which the film remains supersaturated with As exists and precipitation can be avoided. It appears, however, that a higher temperature anneal, in this particular case, the RTA, allows the film to equilibrate and accelerates precipitation. The precipitation, as seen from cross-sectional TEM, occurs heterogeneously at the film surfaces but judging from the TEM, not at grain boundaries. The effect of the loss of As by precipitation on the film resistivity can be determined by the volume fraction of precipitates and the increase in sheet resistance. If one assumes that the precipitates are As, one finds from the precipitate density and average precipitate volume that the electrically active As concentration in the film should fall by about 22%. The resistivity increase after the RTA is 18% as seen in figure 1b. In non-alloyed poly-Si the conductivity of heavily doped (~ mid 10^{20}/cm^3) films is limited not by grain boundary barriers but by the dopant solid solubility. In view of the high average doping level in the films studied here, the conductivity is expected to also be limited by the solid solubility of As. Therefore the 18% increase in resistivity of the Si$_{.52}$Ge$_{.48}$ film is consistent with the 22% loss in As from solution as determined from TEM.

The development of a pronounced {111} texture in doped and annealed poly-SiGe films is of potential importance for the use of these films for diffusion sources for ultra-shallow junction formation. Since lower temperature anneals will be used to drive-in the implanted dopants and since at lower temperatures grain boundary diffusion dominates, any preferred orientation in the film would have an impact on the diffusivities. The most rapid diffusion along grain boundaries occurs in polycrystalline films with a columnar structure characterized by a strong {220} texture [6]. The very large grain size observed in TEM of the As doped films in conjunction with the strong {111} texture indicates that diffusion from the film at lower temperatures would be slower than those from B doped films, which exhibit stronger {220} texture and smaller grains. The B doped films are better suited to out-diffuse dopant by grain boundary diffusion than the As doped films. The effect of implantation and annealing on the texture of films deposited directly on a Si substrate will need to be investigated to determine the rate of diffusion of dopants from the poly-SiGe film for ultra-shallow source/drain regions in ULSI.

689

CONCLUSIONS

This paper has reported on the activation of implanted dopants in poly-SiGe films with applications in CMOS processing and TFTs. The sheet resistance for $5 \times 10^{16}/cm^2$ B implanted films is relatively constant with increasing Ge fraction but in $5 \times 10^{16}/cm^2$ As implanted films sheet resistance increases super-linearly with Ge fraction and is 2.5 times as high for 48% Ge films compared to 20% Ge containing films. An additional RTA was found to degrade activation of the As doped films; the sheet resistance increased after the additional anneal. The texture of the As doped 48% Ge film was found to be more strongly {111} than that of the B doped and non-doped films. The grain size was found to increase after implantation and annealing by almost a factor of two in B and a factor of eight in As implanted $Si_{.52}Ge_{.48}$ films compared to control films that were not implanted but underwent identical thermal treatments. The presence of precipitates observed in TEM in the 48% Ge film is consistent with the hypothesis that the cause of the sheet resistance increase with Ge fraction is the precipitation of As since the solubility of As in Ge is an order of magnitude lower than in Si.

ACKNOWLEDGMENTS

This work was supported by the Semiconductor Research Corporation contract number 93-SC-069. The research made use of the facilities of the National Nanofabrication Facility at Cornell, supported by NSF grant number ECS 861 9049. This work also made use of MRL Central Facilities supported by the National Science Foundation under Award No. DMR-9121654. The authors would also like to thank P.K. Yu and H. Birenbaum for assistance with the SiGe deposition.

REFERENCES

[1] D.T. Grider, M.C. Ozturk and J.J. Wortman, 3rd Int. Symp. on ULSI Science and Technology, 296, (1991).
[2] Tsu-Jae King, K.C. Saraswat, J.R. Pfiester, IEEE Elect. Dev. Lett. 12, 584 (1991).
[3] Tsu-Jae King, J.R. Pfiester, J.D. Schott, J.T. McVittie and K.C. Saraswat, IEDM Tech. Digest, 253 (1990).
[4] T.I. Kamins, Polycrystalline Silicon for Integrated Circuit Applications, (Kluwer Academic Publishers, Boston, 1988), p 91.
[5] F.A. Trumbore, Bell System Tech. J., 39, 205 (1960).
[6] T.I. Kamins, J. Manoliu and R.N. Tucker, J. Appl. Phys., 43, 83 (1972).

POLYCRYSTALLINE SILICON LAYERS FOR SHALLOW JUNCTION FORMATION: PHOSPHORUS DIFFUSION FROM IN SITU SPIKE-DOPED CHEMICAL VAPOR DEPOSITED AMORPHOUS SILICON

D. KRüGER, J. SCHLOTE, W. RöPKE, R. KURPS AND CH. QUICK
Institute of Semiconductor Physics, W.-Korsing-Str.2, PO Box 409
D-15204 Frankfurt (Oder), Germany

ABSTRACT

Shallow and lateral homogeneous delineated n^+p-junctions were formed utilizing solid source diffusion from a deposited amorphous silicon layer with an in situ imbedded ultrathin phosphorus-rich zone. SIMS, AES, and TEM investigations were carried out to analyze the dopant behavior in correlation to morphological and structural changes during subsequent heat treatments. After heat treatments up to 950°C the layer remained flat, surface roughness was found to be less than 3 nm. Dopant pile up at the Si-layer/Si-substrate interface was observed and interpreted on the basis of segregation phenomena. From the time dependence the P segregation-pile-up was found to be diffusion limited except for a small starting period. The dopant concentration in the Si-substrate drops down over more than 2 orders of magnitude in a thickness range less than 20 nm.

1. INTRODUCTION

The development of next generations of high speed silicon bipolar devices requires the fabrication of ultrashallow emitter-basis profiles. In the past dopants were typically added by ion implantation into single crystalline Si, followed by a heat treatment for electrical activation. To obtain shallow high-quality junctions, implantation damage in the silicon substrate must be avoided or removed by post implantation heat treatments. Additionally, ion channeling during implantation and dopant transient diffusion during annealing make difficult the formation of extremely shallow n^+p-junctions. Shallow damage free junctions can be obtained using an out-diffusion process from a solid source layer on the Si substrate. The dopant is introduced into a layer of polycrystalline silicon [1] or a silicide layer [2] grown directly on the substrate either in situ, during layer deposition, or through ion implantation after deposition. The implantation damage in this case remains in the source layer, preserving the high crystal quality in the underlying silicon. Impurity diffusion in polycrystalline silicon, which consists of single crystal grains separated by grain boundaries, is dominated by grain boundary diffusion since diffusivities are 3 to 4 orders of magnitude higher along grain boundaries than in single crystal silicon, while the diffusivities in the intragrain region are comparable to those in single crystal silicon. The dopant redistributes rapidly through grain boundaries inside the source layer and subsequent dopant movement into the substrate can be limited, forming a very shallow junction.

Recently, as-deposited amorphous silicon has been shown to have advantages over as deposited polycrystalline silicon films in terms of surface smoothness, grain morphology, and sheet resistivity [3]. Polysilicon layers with a low level of surface roughness can be obtained using a two-step process consisting of the deposition of in-situ doped amorphous silicon followed by a separate recrystallization annealing. In this work we report the deposition of in situ phosphorus-spike-doped amorphous silicon on silicon as a diffusion source using low pressure chemical vapor deposition (LPCVD). We studied the dopant behavior (redistribution and diffusion) by secondary ion mass spectroscopy (SIMS) and Auger electron spectroscopy (AES) and used transmission electron microscopy (TEM) for structural characterization of the layers.

2. EXPERIMENTAL

The deposition of in-situ doped amorphous silicon was carried out in a conventional horizontal LPCVD reactor for 100 mm wafers using monosilane (SiH_4) and tertiarybutylphosphin $(CH_3)_3 CPH_2$ (TBP) as reactant gases [4]. To achieve more flexibility in adjusting the overall dopant concentration, the TBP flow was switched on only for a defined short period during the deposition. Thus a sandwich structure of undoped/doped/undoped amorphous silicon with an ultrathin heavily phosphorus-doped interlayer was produced. The silicon wafers were cleaned using a procedure similar to the "RCA Standard Clean" [5]. In addition, the wafers were "prebaked" for 5 min in a HCl-atmosphere just before the deposition at 560°C. Different annealing treatments in nitrogen atmosphere were chosen for recrystallization and dopant activation in the temperature range 700°C - 950°C.

For the characterization of the extremely high P content in the as-deposited layers AES was used. Auger depth profiling was carried out with a 10 keV, 10 nA electron beam scanned over a 100 x 100 μm^2 area. For sputtering we used an 4 keV argon ion beam rastered over 1.3 x 1.3 mm^2 at 55° ion beam angle to sample normal. The quantification of the AES data was performed using the intensities of the Si-LVV and P-KLL transitions with calibrated sensitivity factors from heavily ion implanted layers with a known composition.

The SIMS measurements of the phosphorus redistribution and in-diffusion were performed using an ion microprobe ATOMIKA 6500. The experiments were carried out with mass-filtered O ions at energies of 2 and 9 keV. The focused ion beams were scanned over areas of typically 300 μm^2 at normal incidence, an electronic gate of 15 x 15 μm^2 was chosen for secondary ion gating. To analyze possible origins of interface segregation also oxygen content and its distribution was measured by means of 15 keV Cs^+ primary ions using the mass ^{16}O. In some cases the mass $^{42}SiO^-$ was also analyzed. For quantification of the oxygen content reference samples with shallow oxygen implantations were used.

The transmission electron microscope (TEM) technique was employed to characterize the morphology and the defect structure with sufficient high lateral resolution. Plan view TEM samples were prepared by chemical thinning from the backside using a solution containing HNO_3:HF. The samples for cross-section TEM (XTEM) were glued face to face using epoxy, mechanical polished to a thickness of approximately 30 μm, and ion milled to perforation.

3. RESULTS

The crystallization, grain growth, dopant redistribution, interface grain boundary segregation and in-diffusion into the Si substrate are complicated interacting processes. The polycrystalline grain boundaries are fast diffusion paths for dopant redistribution [7]. The dopants can rapidly redistribute along the boundaries before significant diffusion into the substrate takes place. It has been shown [9] that the lateral distribution of the dopant front is uniform, even if high temperature annealing caused a non uniform breakup of the interface and the partial alignment of the grains in the polycrystalline layer.

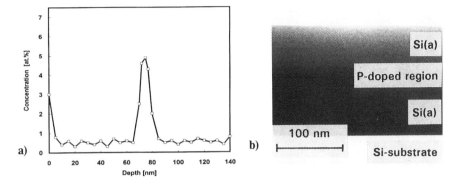

Fig. 1. Amorphous silicon layer with spike-doped P region, as deposited.
a) AES depth profile, b) XTEM micrograph.

A typical phosphorus profile (AES) of an as deposited sample is shown in Fig. 1a. As visible, a concentration of 5 at.% can be reached by means of the described doping. The TEM cross-section in Fig.1b reveals the flat and homogeneous structure of the amorphous silicon. The thickness of the doped region is less than 10 nm. The transition region between the doped and undoped region is very sharp (Fig.1a, b). Quantitatively it was estimated to be in the range of 1 to 2 nm, as concluded from the AES and XTEM measurements. The location of the dopant spike inside the amorphous layer can be adjusted very precisely. This is an additional advantage in comparison to dopant introduction by means of implantation.

Fig. 2a shows a series of SIMS profiles from different heat treatments of a sample with 5 at.% P in a 10 nm layer. The whole thickness of the as-deposited amorphous sandwich is 150 nm. As visible, at a temperature of 700°C (60 min) the phosphorus has not reached the polycrystalline Si/Si-substrate interface while at 800°C (30 min) the interface was reached. A steep in-diffusion profile into the substrate is obtained in the following heat treatments. This was proofed also by high resolution spreading resistance measurements.

A pile up of phosphorus was observed as soon as the dopant reached the interface. In comparison to the case of a diffusion process starting from a polycrystalline silicon layer the P peak shows less broadening effects. This can be explained by a lower surface roughness, leading to reduced sputter profile broadening. Reduced surface roughness was confirmed by XTEM analysis, and is shown in Fig. 2b. Even after heat treatments of 950°C the surface of the layer remains flat, roughness values less than 3 nm were measured. The dopant transition width at the poly-Si/Si interface was estimated to be less than 6 nm (FWHM), this is by a factor of 3 to 5 less than published from the diffusion from polycrystalline deposited silicon layers [10].

The dopant redistribution mechanism can be explained by grain boundary diffusion. However, crystallization and grain growth inside the amorphous layer and normal grain growth have to be taken into account. It has been shown that in the early stages of annealing,

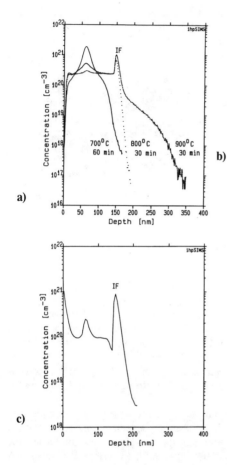

a)

c)

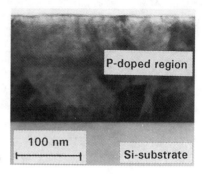

b)

Fig. 2. Characterization of the P redistribution in annealed layers,

a) SIMS profiles for samples annealed at different temperatures, the P pile up at the interface is clearly visible,

b) XTEM micrograph of the grain structure,

c) quantified SIMS oxygen profile.

the grain boundary movement is the major mechanism by which the dopants transfer from the grain to the grain boundaries [9].

The interface between polycrystalline silicon and the substrate also acts as a high diffusivity grain boundary. The pile up of P at the interface may be interpreted as an indication of grain boundary segregation and/or as a trapping of P within the interfacial region caused mainly by oxygen as the dominating contamination at the interface. The interfacial layer influences the segregation and transport of dopants across the interface [8]. The quantified oxygen SIMS profiles are shown in Fig. 2c. From the area under the peak a coverage of about 1.5 monolayer (ML) has been estimated. We suspect that this is due to unremoved oxide rests after substrate preparation before the deposition of the amorphous silicon layer.

To be sure, that really a segregation of phosphorus takes place and that the increase of P is not related to interfacial SIMS artefacts (e.g. increasing ionization probability due to the presence of oxygen) we performed AES depth profiling and verified the presence of P at the interface. A phosphorus level of about 2 at.% has been found in the AES depth profile near the interface. To clarify the segregation process in more detail, the time dependence of the increasing P segregation during first annealing periods and the decreasing as deposited P concentration was measured. This should give some hints concerning the limiting mechanism of the P redistribution. In general, taking into account the sharp interfaces and possible interface contaminations the limiting stages of the redistribution may be as well interface

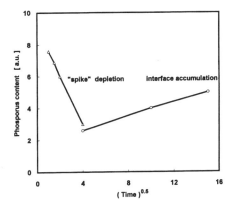

Fig. 3. Time dependence of the P pile up at the polycrystalline Si/Si-substrate interface.

reactions as diffusion. The kinetics of the interfacial P as a function of diffusion time is shown in Fig.3.

As visible, the increase of the interfacial P content can be described with a $(t)^{0.5}$ dependence. Consequently, the limiting mechanism for the discussed behavior seems to be diffusion. However, extrapolating the curve to time zero, it is obvious that at the beginning stages other limiting mechanisms are responsible for the increasing interfacial P content.

XPS measurements reveal a binding energy of 129,5 eV for the $2p_{3/2}$ peak, indicating that phosphorus at the interface is mainly in its elemental state.

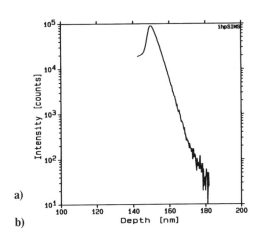

To examine the dopant diffusion into the substrate high resolution SIMS measurements with an 2 keV oxygen primary beam were performed. The SIMS profile in Fig. 4a reveals a decrease of the P concentration of two orders of magnitude over a depth of 15 nm. The steepness of the profile was measured to be 8 nm/dec. Further, the P diffusion front was delineated by chemical etch of the prepared XTEM specimens in a solution consisting of 3 % HF in HNO_3. The etching solution etches faster in the heavily doped substrate regions than in the lightly doped areas.

Fig. 4. Dopant diffusion into the Si(001) substrate.
a) High resolution SIMS profile of the interface near region after diffusion at 850°C (30 min), measured with an oxygen beam of 2 keV after presputtering with 9 keV.
b) XTEM micrograph of a sample after diffusion at 900°C (30 min), chemically etched in HF:HNO_3.

After junction delineation a bright region beneath the polycrystalline Si/Si-substrate interface was observed, running parallel to the interface. This is shown in Fig. 4b for a sample heat treated at 900°C (30 min). The delineated region in the substrate shows the diffusion front and corresponds to a P concentration of about 5×10^{17} cm^{-3}. The front was found to be uniform. For very shallow profiles, however, no sufficient etching contrasts could be measured to determine the junction depth precisely.

The described method of deposition allows also amorphous layer deposition in deep trenches. This is an additional advantage in comparison with the common technology of ion implantation.

4. SUMMARY

In the present work ultrathin (less than 10 nm) phosphorus-rich layers were in situ embedded into thin amorphous Si LPCVD films. We have shown the capability of such layers for the formation of shallow n$^+$p-junctions. SIMS, AES, XPS and TEM investigations were carried out to analyze the dopant behavior in correlation to morphological and structural changes during subsequent heat treatments. Phosphorus shows a strong pile up at the polycrystalline Si/Si-substrate interface due to segregation. The pile up is diffusion limited except for a short starting period. Extremely shallow and laterally uniform delineated junctions were achieved. The dopant slope was about 8 nm/dec. The main interfacial contamination was a thin oxide of about 1.5 ML. No interaction of P with this oxide layer was found. Small surface roughness (3 nm) and homogeneous deposition possibilities in deep trenches offer high potentials for technological application.

References

[1] I. R. C. Rost, P. Ashburn, IEEE Trans. Electron Dev., **ED-39**, 1717 (1992).
[2] V. Probst, H. Schaber, A. Mitwalsky, H. Kabza, B. Hoffmann, K. Maex,
    L. van den hove, J. Appl. Phys., **69** ,3 (1991).
[3] K. Park, S. Batra, S. Banerjee, G. Lux, T. C. Smith, J. Appl. Phys., 70(3), 1397
    (1991).
[4] E. T. Tang, IEDM´ 89, Washington, N.Y., IEEE, p. 39 (1989).
[5] W. Kern, D. Puotinen, RCA Rev., **31**, 187 (1970).
[6] T. Kamins, "Polycrystalline silicon for Integrated circuit applications", Kluwer,
    Boston (1988).
[7] B. Swaminathan, K. C. Saraswat, R. W. Dutton, T. I. Kamins, Appl. Phys. Lett.,
    40, 795 (1982).
[8] B. Swaminathan, E. Demoulin, T. W. Sigmon, R. W. Dutton, R. Reif,
    J. Electrochem. Soc., **127**, 2227 (1980).
[9] A. G. O'Neill, C. Hill, J. King, C. Please, J. Appl. Phys., **64**, 167 (1988).
[10] J. L. Hoyt, E. F. Crabbe, R. F. W. Pease, J. B. Gibbons, A. F. Marshall,
    J. Electrochem. Soc., **135**, 1773 (1988).

CRYSTALLIZATION OF HYDROGENATED AMORPHOUS SILICON THICK FILMS ON MOLYBDENUM SUBSTRATES

Nagarajan Sridhar[*], D. D. L. Chung[*], and W. A. Anderson[+], Center for Electronic and Electro-Optic Materials, State University of New York at Buffalo, NY 14260-4400, and J. Coleman, Plasma Physics Corp., P. O. Box 548, Locust Valley, NY 11650.

[*] Also with Department of Mechanical and Aerospace Engineering
[+] Also with Department of Electrical and Computer Engineering

ABSTRACT

Crystallization of hydrogenated amorphous silicon thick films deposited by dc glow discharge on molybdenum substrates was studied by Raman scattering and x-ray diffraction. Investigation was made as a function of amorphous silicon film deposition temperature. On heating the films at a rate of 5 $^\circ$C/min to 650 $^\circ$C for various times, it was observed that the film deposited at 300 $^\circ$C started crystallization faster than the film deposited at 150 $^\circ$C. The degree of crystallinity increased with increasing annealing time for all the films. However, at all annealing times, the degree of crystallinity for the annealed film deposited at 150 $^\circ$C was higher than that of the annealed film deposited at 300 $^\circ$C, indicating that the crystallization growth rate was higher for the film deposited at a lower temperature. These results were consistent with the dark conductivity measurements. The film deposited at 150 $^\circ$C showed a photoresponse which increased with increasing annealing time whereas no photoresponse was observed for the film deposited at 300 $^\circ$C. This was probably due to the degree of crystallinity and grain size being much larger for the film deposited at 150 $^\circ$C than the film deposited at 300 $^\circ$C.

INTRODUCTION

Polycrystalline silicon (polysilicon) has received a lot of attention as a material for applications such as solar cells [1], thin film transistors [2] and integrated circuits [3]. It can be obtained directly by a variety of chemical vapor deposition techniques such as low pressure chemical vapor deposition (LPCVD) [4], atmospheric pressure chemical vapor deposition (APCVD) [5] and plasma enhanced chemical vapor deposition (PECVD) [6]. However, it has been reported that polysilicon obtained after crystallization of amorphous silicon has better structural (such as a larger grain size [7, 8]) and electrical properties (higher conductivity [8]) than as-deposited polysilicon. Due to this, there is a growing interest in the use of polysilicon obtained by the crystallization of amorphous silicon. However, previous work has focused mainly on thin film transistor applications where film thicknesses less than 0.5 μm have been employed on glass and oxidized wafer substrates [9-12]. For solar cell applications the minimum polysilicon film thickness necessary is at least 5 μm. Moreover, for large area low cost solar cells, it is economical to employ a low cost substrate such as sheet metal. There is a lack of information on the correlation between structural properties (such as grain size and crystallinity) and the photoresponse of the resulting polysilicon film obtained after crystallization of amorphous silicon. This correlation is essential for solar cell applications.

In this work, we have investigated the correlation between the structural properties and the photoconducting properties of polysilicon thick films obtained by the crystallization of PECVD

697

amorphous silicon films through Raman scattering, x-ray diffraction (XRD) and conductivity measurements. This study was conducted as a function of amorphous silicon film deposition temperature.

EXPERIMENTAL

Intrinsic a-Si:H films were deposited at Plasma Physics Corp. on molybdenum substrates by PECVD (or dc glow discharge) in a gradient field from silane with deposition rates up to 1 μm min^{-1}. The deposition temperatures were 150, 225 and 300 $^{\circ}$C; the pressure was 250-300 mtorr; the SiH$_4$ flow was 10 cm^{3}min^{-1}. The film thickness was 7 μm. All the samples had a thin n$^+$ amorphous silicon layer (formed by the addition of 1 % PH$_3$ to H$_2$) deposited on molybdenum prior to the deposition of the intrinsic film to provide ohmic contacts between silicon and molybdenum. The thickness of the n$^+$ layer was about 500 Å.

Isothermal vacuum annealing was conducted to crystallize the amorphous silicon films. For this purpose, the sample was inserted in a Vycor tubing. This tube was evacuated to a pressure of 10^{-6} torr by a turbomolecular pump and then sealed. This ampoule was then heated at a rate of 5 $^{\circ}$C/min to 650 $^{\circ}$C in a three zone furnace and isothermally heated to different times ranging from 0 to 50 hours. After this treatment, the ampoule was removed from the furnace and then air cooled.

After the annealing treatment, the films were characterized for crystallization and conductivity measurements. Crystallization measurements were done by employing Raman sattering and x-ray diffraction. Raman scattering was performed on the sample at room temperature to determine the time for nucleation for crystallization. It was performed in the backscattered mode using a Spex 1877 triplemate Raman spectrometer equipped with a CCD detector. The wavelength of the Ar laser was 4880 Å and the laser power was 100 mW. X-ray diffraction was performed on the sample using a Phillips-Norelco x-ray diffractometer and CuKα radiation. The 2θ scan rate was 0.25 $^{\circ}$/min. The grain size was determined from the full width at half maximum value of the XRD spectrum by means of Scherrer's formula, after compensating for the line broadening [13]. The degree of crystallinity along the hkl direction was determined by normalising the integrated intensity of the XRD peak measured along that direction to the corresponding integrated intensity for a randomly oriented silicon powder along the same direction which was corrected for film thickness using the x-ray absorption coefficient of silicon [13].

Room temperature dark and photoconductivity measurements were measured to determine the photoresponse, which is defined as the ratio of the photoconductivity to the dark conductivity. The ohmic contacts were fabricated by patterning and etching the molybdenum after attaching the front side of the film to the ceramic substrate. This technique does not involve any heating, which is usually required during the evaporation of contacts.

RESULTS AND DISCUSSION

7 μm thick intrinsic a-Si:H films deposited at three different temperatures, namely 150, 225 and 300 $^{\circ}$C were heated at a rate of 5 $^{\circ}$C/min to 650 $^{\circ}$C and isothermally annealed for different times. From gas evolution and thermogravimetric analysis, we observed that the hydrogen evolved from the films when heated to 650 $^{\circ}$C. Thus, there is no effect of hydrogen in the annealed films. Fig. 1 shows the Raman spectra of the films deposited at 150 and 300 $^{\circ}$C, and

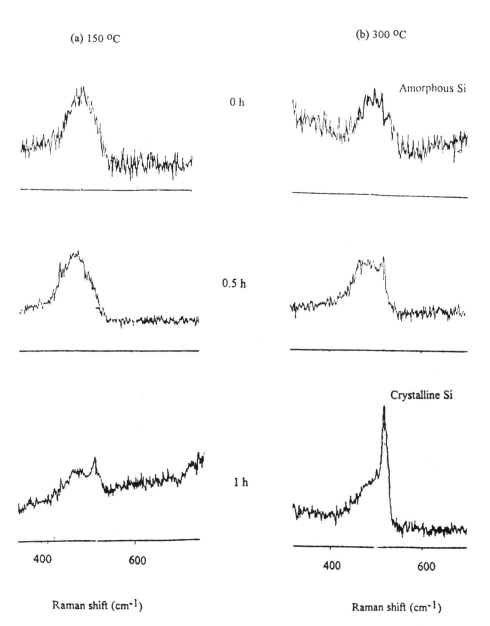

(a) 150 °C   (b) 300 °C

0 h

Amorphous Si

0.5 h

Crystalline Si

1 h

400   600   400   600

Raman shift (cm^{-1})   Raman shift (cm^{-1})

Fig. 1 Raman spectra of amorphous silicon films deposited at (a) 150 and (b) 300 °C and annealed at 650 °C for 0, 0.5 and 1 h.

annealed for 0, 0.5 and 1 h at 650 °C. At 0 h, a peak was observed at 480 cm^{-1} in the Raman spectra for both films (deposited at 150 and 300 °C) indicating that the films were amorphous. At 0.5 h, an additional peak occurred at 520 cm^{-1} for the film deposited at 300 °C, indicating that the presence of crystalline Si, whereas for the film deposited at 150 °C, only a single peak (at 480 cm^{-1}) was observed, indicating the film was still amorphous. This suggests that the nucleation of crystalline silicon occurred faster for the film deposited at a higher temperature. However, it is known that the penetration depth of the Ar laser during Raman scattering measurements is less than 1 µm, so Raman scattering was performed on the back side of the film. The back side of the film was exposed by attaching the front side of the film to a ceramic substrate and then etching away the Mo. A similar trend was observed in the Raman spectra when the measurement was done in the backside of the film, indicating that the crystallization in these films was bulk induced. This trend is consistent with that of the dark conductivity data (Table 1) which increased at an annealing time of 0.5 h more for the film deposited at 300 °C than those deposited at 225 or 150 °C.

Ref. 14 observed a similar increase in dark conductivity during isothermal annealing of doped and undoped APCVD films and they employed this measurement to extract the activation energy of crystallization on these films. Our results are in good agreement with Ref. 10 and 11. Ref. 11 proposed that at higher deposition temperatures, the structural disorder decreases and the local arrangement of tetrahedra may resemble more closely that of crystalline silicon. Hence these regions may act as favorable sites for nucleation during annealing, thus resulting in a higher nucleation rate.

At low annealing times (1 h or less), the crystalline fraction was so small that the presence of any x-ray peak could not be detected. For annealing times greater than 1 h, the crystallinity increased (as evidenced by the increase in the height of the crystalline silicon peak in the Raman spectra), so that we could employ XRD to study the degree of crystallinity, texture (if any) and grain growth in these films. From the XRD spectra, it was observed that for all the three deposition temperatures (150, 225 and 300 °C), the films on annealing showed the presence of only the 111 peak at all times. Our results were different from Ref. 12, where the crystallization of a-Si deposited by LPCVD was studied at 570 °C and they observed the growth of 111, 220 and 311 orientations. However, our results cannot be directly compared with Ref. 12 because, in LPCVD films, the transition from amorphous to polysilicon deposition temperature is much different and this can affect the growth of the various orientations [15]. From Table 1, it can be seen that the degree of crystallinity increased with increasing annealing time in all the films. However, the degree of crystallinity was higher for the film deposited at 150 °C than those deposited at 225 and 300 °C at all annealing times, whereas, from Raman scattering (Fig. 1), the time for the onset of crystallization was delayed for the film deposited at 150 °C. Hence, a higher crystallization growth rate and a lower crystallization nucleation rate were obtained for the film deposited at a lower substrate temperature, in accordance with Ref. 10. This is consistent with the dark conductivity (Table 1), which sharply increased with annealing time for the films deposited at 150 °C. Table 1 shows that the grain size of the annealed films calculated from XRD results was greater than 2000 A (the maximum limit of grain size as measured from XRD) for the annealed film deposited at 150 °C at all annealing times. However, for the annealed films deposited at 225 and 300 °C, the grain size increased from a few hundred angstroms at short annealing times to greater than 2000 Å at long annealing times. Only the annealed film deposited at 150 °C showed a photoresponse (ratio of the photoconductivity to the dark conductivity) which increased with increasing annealing time, whereas the film deposited at 225 °C showed a very low photoresponse and the film deposited at 300 °C showed

Table 1 Effect of amorphous silicon deposition temperature on the structural and electrical properties of the resulting polysilicon films obtained by isothermal annealing at 650 $^{\circ}$C at various times

	Annealing time (h)					
	0	0.5	1	4	30	50
Fractional increase in dark conductivity						
150*	-	3.6	43	1.4×10^5	1.5×10^5	1.5×10^5
225*	-	4	30	480	1.7×10^4	2×10^4
300*	-	10	20	310	1×10^4	1×10^5
Photoresponse						
150*	-	-	0.002	0.02	0.04	0.08
225*	-	-	-	-	0.03	0.04
300*	-	-	-	-	-	-
Degree of crystallinity along 111 direction (%)						
150*	0^a	0^a	$> 0^a$	42^b	48^b	63^b
225*	0^a	0^a	$> 0^a$	32^b	40^b	47^b
300*	0^a	$>0^a$	$> 0^a$	25^b	32^b	30^b
Grain size (A)						
150*	-	-	-	> 2000	> 2000	> 2000
225*	-	-	-	500	1500	> 2000
300*	-	-	-	200	500	> 2000

* Amorphous silicon film deposition temperature ($^{\circ}$C)
a Based on Raman results
b Based on XRD results

no photoresponse at all annealing times. This suggested that the grain size was probably so small for the annealed films deposited at higher temperatures that there were too many defects in these films to show any photoresponse. From Table 1, it can also be seen that the degree of crystallinity was higher for the annealed film deposited at 150 $^{\circ}$C in comparison to the annealed film deposited at higher temperatures, indicating that a high degree of crystallinity could also be a factor in improving the photoresponse of the film. Ref. 10 fabricated thin film transistors from annealed films deposited in the amorphous state at different temperatures. They observed that the field effect mobility increased with decreasing film deposition temperature and they attributed this increase in mobility to the increase in the grain size.

CONCLUSIONS

This work was focussed on the crystallization of hydrogenated amorphous silicon thick films deposited from silane by dc glow discharge. The structural properties measured by Raman scattering and XRD were correlated to the dark conductivity and the photoresponse of the film annealed at 650 °C for various lengths of time. The investigation was made as a function of the amorphous silicon film deposition temperature. The following findings were obtained. The nucleation of crystallization occurred faster for the film deposited at a higher temperature (300 °C) in comparison to the film deposited at a lower temperature (150 °C). This is consistent with the increase of the dark conductivity at the onset of crystallization for the film deposited at 300 °C. At longer annealing times, the degree of crystallinity was higher for the film deposited at 150 °C than those deposited at 225 or 300 °C, indicating a higher crystallization growth rate for the film deposited at a lower temperature. Only the film deposited at 150 °C showed a photoresponse which increased with increasing annealing time, probably due to a higher degree of crystallinity and a larger grain size.

ACKNOWLEDGEMENTS

The authors acknowledge financial support from National Renewable Energy Laboratory. The technical assistance of W. Y. Yu of State University of New York at Buffalo in Raman scattering is greatly appreciated.

REFERENCES

1. T. L. Chu, J. Cryst. Growth 39, 45 (1977).
2. S. Morozumi, K. Oguchi, T. Misawa, R. Araki and H. Ohshima, in SID 84 Dig., 316 (1984).
3. S. D. Malhi, H. Shichijo, S. K. Banerjee, R. Sundaresan, M. Elahy, G. Pollack, W. Richardson, A. H. Shah, L. R. Hite, R. H. Womack, P. K. Chatterjee and H. W. Lam, IEEE Trans. Electron Devices ED-32, 258 (1985).
4. T. I. Kamins, M. M. Mandurah and K. C. Saraswat, J. Electrochem. Soc., 125, 927 (1978).
5. N. Nagasima and N. Kubota, Jpn. J. Appl. Phys. 14, 1105 (1975).
6. T. I. Kamins and K. L. Chiang, J. Electrochem. Soc. 129, 2326 (1982).
7. G. Harbeke, L. Krausbauer, E. F. Steigmeier, A. E. Widmer, H. F. Kappert and G. Neugebauer, J. Electrochem. Soc. 131, 675 (1984).
8. F. S. Becker, H. Oppolzer, I. Weitzel, H. Eichermuller and H. Schaber, J. Appl. Phys. 56, 1233 (1984).
9. R. B. Iverson and R. Reif, J. Appl. Phys. 62, 1675 (1987).
10. K. Nakazawa and K. Tanaka, J. Appl. Phys. 68, 1029 (1990).
11. M. K. Hatalis and D. W. Greve, J. Appl. Phys. 63, 2260 (1988).
12. S. Hasegawa, T. Nakamura and Y. Kurata, Jpn. J. Appl. Phys. 31, 161 (1992).
13. B. D. Cullity, Elements of X-ray Diffraction (Addison- Wesley, Reading, MA, 1967).
14. R. Bisaro, J. Magarino, K. Zellama, S. Squelard, P. Germain and J. F. Morhange, Phys. Rev. B, 31, 3568 (1985).
15. S. Hasegawa, M. Morita and Y. Kurata, J. Appl. Phys. 64, 4154 (1988).

PULSE-TO-PULSE LASER STABILITY EFFECTS ON MULTIPLE SHOT EXCIMER LASER CRYSTALLIZED a-Si THIN FILMS

R. I. JOHNSON, G. B. ANDERSON, J. B. BOYCE, D. K. FORK, P. MEI, and S. E. READY

Xerox Palo Alto Research Center, 3333 Coyote Hill Rd., Palo Alto, CA 94304

ABSTRACT

Laser crystallized amorphous silicon thin films on quartz exhibit a peak in the grain size, electron mobility and the Si (111) x-ray intensity as a function of the laser fluence, substrate temperature, film thickness, and the number of laser shots per unit area. The peak in grain size has also been shown to be dependent on the stability of the pulse-to-pulse laser energy density, particularly at high shot densities. The shape of the distribution profile of the pulse-to-pulse laser fluence can significantly alter the grain growth at higher shot densities. The modified growth can be expressed by a simple model based on the mean and standard deviation of the laser energy density relative to the characteristic fluence at which the grain size, mobility, and Si (111) x-ray intensities are maximized.

INTRODUCTION

Earlier studies have investigated the influence of the substrate bias temperature, shot density, film thickness, and laser fluence on the grain size, mobility and x-ray intensities of laser crystallized a-Si thin films [1-13]. From these studies, it has been shown that the resulting poly-Si films exhibit exceptionally large grain sizes and electron mobilities for a narrow range of laser fluence [1-3]. For shot densities considerably larger than 100, variations in the normal Si (111) x-ray peak energy position and growth pattern begin to appear. These variations in grain growth are caused by normal fluctuations in the pulse-to-pulse laser fluence of the excimer laser.

In this study, we will discuss the different modes of grain growth as seen over a wide range of shot densities and present a simple model which describes a mechanism by which fluctuations in the pulse-to-pulse excimer laser fluence are able to cause the growth variations observed.

EXPERIMENTAL CONSIDERATIONS

The grain size calculations, used in this paper, depend primarily on the strong correlation between the Si (111) x-ray intensity and the grain size for determining the average crystallite dimensions. All x-ray analysis was confined to the Si (111) x-ray peak, because it is the most intense peak. The characteristic laser fluence, F_m, is defined as the laser energy density required to produce the maximum Si (111) x-ray intensities for 100 nm a-Si films, that have been exposed to a shot density of 100 laser pulses per unit area. Shot density is defined as the number of laser pulses at any point on the sample.

The samples were laser crystallized in a static (unscanned) mode with a XMR 5100 excimer laser operating at 308 nm with a 50 ns pulse length. The laser beam was

Mat. Res. Soc. Symp. Proc. Vol. 343. ©1994 Materials Research Society

homogenized and focussed down to a beam spot of 9 mm by 9 mm. All laser crystallization was done in a vacuum at 10^{-6} Torr, or better, on 100 nm of LPCVD deposited a-Si films, implanted with phosphorous (2×10^{15}/cm^2) on 4 inch, fused silica substrates. Characterization of the samples included x–ray diffraction, transmission electron microscopy and Hall mobility in the van der Pauw geometry. The samples were processed at laser pulse rates of both 5 and 10 Hz to rule out the effects of substrate heating.

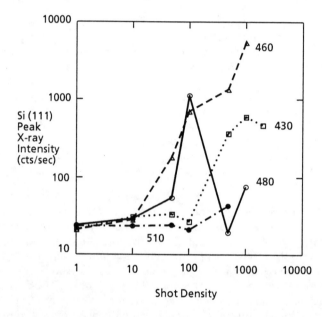

Figure 1
The Si (111) peak x-ray intensities with respect to the shot density for energies of 430, 460, 480, and 510 mJ/cm^2.

RESULTS AND DISCUSSION

The x-ray measurements, shown in Fig. 1, exhibit four distinct patterns of grain growth as a function of shot density with respect to the laser fluence. The four modes of growth are represented by a laser fluence below F_m, above F_m, close to F_d, and above F_d. F_m is the energy density where large grain growth can occur and F_d is the energy density where irreversible substrate interface damage begins. F_m and F_d are estimated to be 445 mJ/cm^2 and 495 mJ/cm^2, respectively, at a shot density of 100, for the materials used in this experiment. Fig. 2 shows three typical pulse-to-pulse excimer laser fluence distribution profiles taken at the operating voltages of 28, 29 and 30 kV.

At 430 mJ/cm^2, where the beam spot energy is below F_m, one sees little change in the Si (111) peak x-ray intensity in the first 100 shots. Above 100 shots, an increase begins that continues up to about 1000 shots, as shown in Fig. 3a. At laser fluences below F_m, most of the shots have insufficient energy to grow large crystals. However, because of fluctuations in the pulse-to-pulse laser fluence about the mean, some pulses may exceed F_m. As a result, while most of the shots below F_m are forming small grains, a small percentage of pulses exceeding F_m will form sparsely

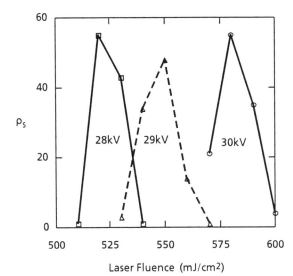

Figure 2
Excimer laser energy density distribution profiles for different laser voltage settings. ρ_s is defined as the number of laser pulses occurring with respect to the laser fluence for a given voltage.

distributed large grain clusters. The accumulation of these larger grains will be made evident by an increase in x-ray intensity observed at the higher shot densities.

At 460 mJ/cm², above F_m, a steady rise in the x-ray signal intensity is observed, starting at about 10 shots, as seen in Fig. 3b. Energy densities above F_m produce a high percentage of shots that efficiently grow large, highly textured crystals. Few, if any, of the shots have reached the interface damage level, F_d, where irreversible interface roughening occurs.

At 480 mJ/cm², close to F_d, an increase in x-ray signal intensity starts again at about 10 shots, but peaks near 100 shots. As the shot density increases above 100, the intensity falls to a value close to the single shot value, as shown in Fig. 3c. At fluences between F_m and F_d, most shots have sufficient energy to grow large crystals, giving rise to the initial increase seen. However, a substantial percentage of shots may exceed the interface damage threshold at F_d, causing a steady irreversible roughening of the interface and a steady increase in the number of nucleation sites. These sites, in turn, produce small crystals that eventually achieve dominance at the higher shot densities. The x-ray intensity increase observed after 500 shots is associated with the ablation of the film. Curiously, the ablated regions of these films produce a large (111) x-ray signal. Optical microscope investigations of these ablated areas show that considerable silicon remains in the form of the weblike structure shown in Fig. 4a. These structures represent highly textured (111) crystalline silicon drawn into a filigree pattern by surface tension. A TEM micrograph in Fig. 4b shows a cross section of one filigree segment. The material has a thickness of almost 1 micron.

At the laser fluence of 510 mJ/cm², above the interface damage threshold, F_d, irreversible interfacial roughening occurs with almost every shot, creating many

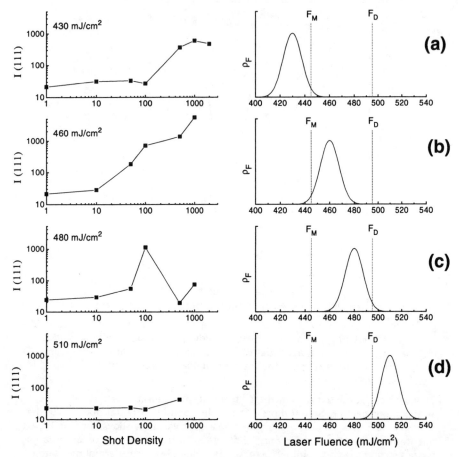

Figure 3. Four crystallization modes are shown, together with their pulse-to-pulse laser fluence distribution profile energy positions relative to F_m and F_d. ρ_F is the laser fluence distribution function. I (111) represents the Si (111) x-ray intensity in counts/sec.

nucleation sites that facilitate the growth of large numbers of small crystals. The x-ray intensity shown in Fig. 3d shows little change with respect to increasing shot density. When a sufficient number of laser pulses exceeds F_d, the growth of nucleation sites overwhelms any large crystal growth that might occur from pulses at lower energy densities, giving rise to the flat intensity observed. The existence of any large crystals from lower energy density pulses will ultimately be reduced to smaller status by irreversible interface roughening, if the shot density is sufficiently high. The increase observed in the x-ray intensity above 100 shots is again associated with the strong x-ray signal from the highly textured residual silicon left on ablated regions of the film.

One can conclude from the consequences of the positional relationships that exists between the laser fluence distribution profile and the energy density thresholds at F_m, and F_d, that the pulse-to-pulse energy stability of an excimer laser is an important parameter for scanned multiple shot laser crystallization. Large variations in fluence make large uniform grain growth difficult to achieve.

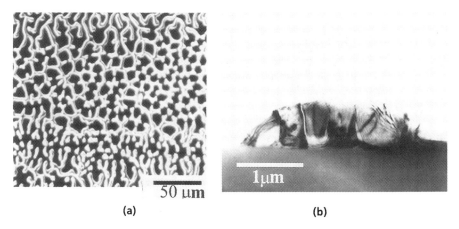

(a)   (b)

Figure 4. (a) A top view of the residual silicon left after laser ablation. (b) A TEM cross section, of one residual silicon segment. This material is almost a micron thick and exhibits a strong Si (111) x-ray signal.

Early studies [8,9] have shown that static or unscanned laser crystallized a-Si films, resulted in samples that showed little change in the x-ray intensity or grain size with increasing shot density. Those studies were done at laser fluences significantly above or below the narrow fluence region where large grain growth is likely.

CONCLUSIONS

We have presented results from laser crystallized amorphous silicon thin films on fused silica substrates. Specifically, we have looked at the effect of shot densities from 1 to 3000 at various laser fluences. The results show that a-Si thin films on quartz exhibit distinctive growth patterns which are determined primarily by the shot density, laser fluence, and the pulse-to-pulse stability of the laser energy density.

We have also introduced a model which provides an explanation for the observed variations in the grain growth seen at different shot densities. It is based on the distribution of the pulse-to-pulse laser energy density position relative to the thresholds at F_m and F_d. The implications of this work strongly suggest that the pulse-to-pulse energy stability of the laser holds considerable importance for multiple shot laser crystallization. Laser systems with small fluctuations in the pulse-to-pulse energy density will enable the production of superior material in contrast to systems with large fluence variations.

ACKNOWLEDGEMENT

We wish to thank S. Chen and S. McNeal at XMR, Inc, Santa Clara, CA, for the use of their laser system. A part of this work was funded under DARPA Contract No F33615-92-C-5811

REFERENCES

1. R. I. Johnson, G. B. Anderson, J. B. Boyce, D. K. Fork, P. Mei, S. E. Ready, and S. Chen, Mat. Res. Soc. Proc. **297,** 533 (1993).
2. J. B. Boyce, G.B. Anderson, D. K. Fork, R. I. Johnson, P. Mei, and S. E. Ready, Mat. Res. Proc. **321,** 671 (1994).
3. J. S. Im, H. K.Kim, and M. O. Thompson, Appl. Phys. Lett. **63,** 1969 (1993).
4. R. I. Johnson, G. B. Anderson, S. E. Ready, J. B. Boyce, Mat. Res Soc. Proc. **219,** 407, (1991).
5. H. Kuriyama, et al, IEEE International Electron Meeting, Wash. D.C. (1991).
6. T. Sameshima and S. Usui, Mat. Res. Soc. Symp. Proc. **71,** 435 (1986).
7. T. Sameshima, M. Hara, and S. Usui, Polycrystalline Semiconductors II, eds. J. H. Werner and H. P. Strunk, Springer Proceedings in Physics, Vol. **54,** Springer–Verlag, Berlin (1991).
8. R. Z. Bachrach, K. Winer, J. B. Boyce, S. E. Ready, R. I. Johnson, and G. B. Anderson, J. Electron. Materials **19,** 241 (1990).
9. R.Z Bachrach, K. Winer, J. B. Boyce, F. A. Ponce, S. E. Ready, R. I. Johnson, G. Anderson, Mat. Res. Proc. **157,** 467 (1990)
10. K. Winer, G. B. Anderson, S. E. Ready, R. Z. Bachrach, R. I. Johnson, F.A. Ponce, and J. B. Boyce, Appl. Phys. Lett. **57,** 2222 (1990).
11. R. I. Johnson, G. B. Anderson, S. E. Ready, D. K. Fork, and J. B. Boyce, Mat. Res.Soc. Proc. **258,** 123 (1992).
12. S. E. Ready, J. B. Boyce, R. Z. Bachrach, R. I. Johnson, K. Winer, G. B. Anderson, and C. C. Tsai, Mat. Res. Soc. Proc. **149,** 345 (1989).
13. J. S. Im, and H. J. Kim, Appl. Phys. Lett, **64,** 2303, (1994)

CHARACTERIZATION OF THE SUBSTRATE INTERFACE OF EXCIMER LASER CRYSTALLIZED POLYCRYSTALLINE SILICON THIN FILMS

G. B. ANDERSON, J. B. BOYCE, D. K. FORK, R. I. JOHNSON ,P. MEI, and S. E. READY
Xerox Palo Alto Research Center, 3333 Coyote Hill Road, Palo Alto, CA 94304

ABSTRACT

Excimer laser crystallized Si thin films on fused silica substrates exhibit a peak in the average grain size as a function of laser energy density. The average grain size increases with increasing laser fluence until a maximum value , approximately 10 microns for a 100 nm thick Si film, is achieved. The peak in grain size is accompanied by a peak in the electron Hall mobility. Further increases in the laser fluence result in a decrease in the Si grain size and an increase in the intragranular defects. A small energy range of 40 mJ/cm^2 exists in which this peak in grain size can be achieved. Cross section TEM has shown that when the peak laser fluence is exceeded , the fused silica substrate can be as rough as 17 nm . Atomic force microscopy ,performed on the substrate surface after the Si has been etched off, also shows that the magnitude and spatial frequency of the roughness increases when the critical laser fluence is exceeded. This degradation of the interface may also produce sites for stacking faults to form during the solidification of the Si. This result and results of simulations of the temperature of the interface during crystallization suggests that the peak energy range exists after the complete melting of the Si thin film and before the silica substrate starts to soften.

INTRODUCTION

Large area Si electronic devices have application in two dimensional electronic imaging and flat panel displays [1,2]. Excimer laser crystallization of amorphous Si on noncrystalline substrates has the opportunity of becoming a useful technique for processing large area Si electronic devices [3-12]. At this time solid phase crystallization of Si is the predominant processing technique. Excimer laser crystallization has several advantages over solid phase crystallization: The fast heating and cooling of laser crystallization produce Si grain sizes greater than those obtained by solid phase crystallization with fewer intragranular defects. The average substrate temperature during laser crystallization is usually below 600 ºC and therefore can be used with temperature sensitive substrates. And finally the laser can crystallize selected regions of a wafer, thereby producing hybrid a-Si and crystalline Si devices.

Most commercially available excimer lasers and optics produce inhomogeneous beams which in turn produce inhomogeneous crystalline Si films. These inhomogeneities are manifested in the films by large variations in the Si grain size [7,13]. This large variation in grain size is undesirable for TFT applications. One method of reducing the effects of beam inhomogeneities is to scan the excimer laser beam over the sample so that each area of the sample is exposed to different parts of the laser beam . The scanning is achieved by stepping the sample in increments smaller than the laser beam size; this has the affect of partially averaging out beam inhomogeneities.

When samples are crystallized by this multi - shot scanning technique, the average grain size increases with increases in the laser fluence until a maximum grain size is reached. Once the laser fluence exceeds this maximum the average grain size is reduced. This behavior of the Si film is also observed in static laser

709

crystallization. Several models have been propose to explain this behavior [13-15]. This study uses cross section transmission electron microscopy (TEM) and atomic force microscopy (AFM) to examine Si films and substrates for laser crystallization energy densities past the peak value.

EXPERIMENTAL CONSIDERATIONS

100 mm quartz wafers were coated with 100 nm of LPCVD a–Si. The wafers were then laser crystallized with an excimer laser operating at 308 nm with a 50 ns pulse length. The beam was homogenized and focussed down to a beam spot size of 4 x 15 mm. The scanned laser beam had a pulse rate of 20 Hz and each area on the substrate received 100 laser exposures with the step size between the exposures being 40 microns. The crystallizations were performed in a vacuum of 10^{-6} torr. The crystallized samples were then characterized by x-ray diffraction using the Si (111) reflection. Plane view TEM samples were prepared using the lift off technique while cross section TEM sample were prepared by mechanically thinning the samples to 20 microns and then ion milling the samples from both sides on a liquid nitrogen cooled stage. These samples were also placed in a saturated KOH solution and the Si films were etched off the substrate. Atomic force microscopy was then performed on the surface of the substrate .

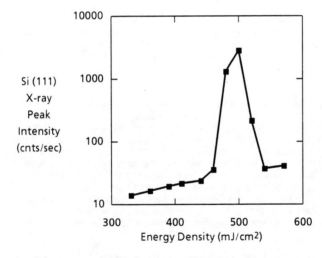

Fig. 1.

Si (111) x-ray peak intensity as a function of laser fluence for 100 nm of LPCVD Si on fused silica.

RESULTS AND DISCUSSION

Figure 1 shows the Si (111) x-ray peak intensity as a function of laser energy density . The Si (111) peak occurs at a laser energy density of about 500 mJ/cm2 for these conditions. Plane view TEM shows that this peak corresponds to a maximum in the average grain size. When this peak laser energy density is exceeded the average grain size is reduced. Figure 2 shows cross section TEM micrographs of Si samples crystallized at the peak laser energy density and well beyond the peak at 560 mJ/cm2. The differences between the two images are the grain size of the Si and the smoothness of the interface. The sample crystallized at 560 mJ/cm2 has smaller Si grains and an undulating interface.

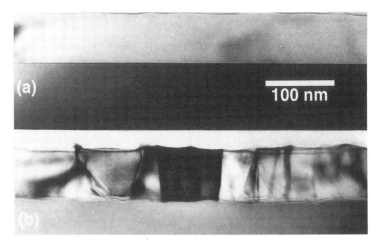

Fig.2. Cross section TEM images of crystallized Si on fused silica (a) at the peak laser energy density (500 mJ/cm2) and (b) past the peak (560 mJ/cm2)

Figure 3 shows a high resolution TEM image of this undulating interface. This figure shows the (111) and (220) planes of the Si lattice. The Si lattice terminates at the amorphous silica forming a flat interface except in the center of the image , here a Si facet has formed , extending 17 nm into a cavity in the silica substrate . This cavity may have formed during the crystallization. Such a feature in the substrate may act as a nucleation site for Si grain growth. At laser energy densities past the peak, the substrate may start to soften and this may roughen the substrate causing many of these nucleation sites and therefore inducing the growth of many small grains.

Fig.3. High resolution TEM image of a 17 nm Si facet extending into the fused silica substrate. Arrows near the Si point to the interface, the arrow in the substrate is the location of bottom of the facet.

Plan view TEM shows that large (up to 4 microns) Si grains do exist in the Si film crystallized past the peak laser energy density. However, these grains are not as large as the grains formed at the peak and typically have many more intragranular defects. The most common extended intragranular defects found in this material are stacking faults. Stacking faults exist in the Si crystallized at the peak laser fluence but the occurrence of these defects drastically increases as the peak laser fluence is exceeded. Figure 4 is a cross section TEM micrograph of Si crystallized past the peak laser fluence showing stacking faults propagating through a large Si grain. This micrograph shows that the stacking faults start from the roughened substrate and terminate on the surface of the Si grain. Electron diffraction shows that these stacking faults are in the (111) slip planes of the Si . This micrograph suggests that as a fast growing grain has nucleated and is growing laterally, the roughness on the silica substrate causes perturbations in the stacking order of the crystal as it is growing and this nucleates a stacking fault.

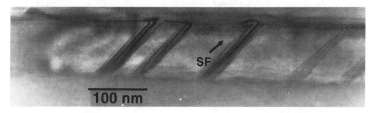

Fig. 4. Stacking faults propagating from the substrate in a large Si grain

Figures 5 shows a pair of three dimensional renderings of atomic force microscopy (AFM) images. The samples for the images were produced by etching the Si off the substrate in a KOH solution and then examining the texture of the resulting sample surface. This surface is the fused silica portion of the interface. Figure 5a is an image produced by the sample crystallized at the peak laser fluence of 500 mJ/cm2. The AFM image shows that the substrate surface is not flat but undulating with height differences as great as 10 nm. Figure 5b is an AFM image of the sample crystallized well past the peak laser fluence. This image shows an increase in the spatial frequency and magnitude of the roughness. Height variations as great as 20 nm exist in the surface of this sample.

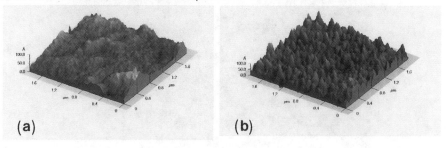

(a) (b)

Fig. 5 Atomic Force microscopy images of Si crystallized at the peak laser fluence (a) past the peak laser fluence (b)

Figure 6 contains a heat flow simulation of the maximum interface temperature in degrees kelvin, vs. laser energy density. The melting point of Si is 1685 K and the softening temperature of fused silica is 1883 K. This simulation suggests that the maximum in the average grain size occurs between the point at which the Si is completely melted (1685 K at about 400 mJ/cm2) and the point at which the silica substrate softens (1813 K at about 500 mJ/cm2). Figure 1 shows that this is the fluence region where the peak develops. Beyond 500 mJ/cm2 the silica softens, the interface roughens and the average grain size declines. Again this suggests that the degradation of the interface is causing many nucleation sites to form.

Maximum Interface Temperature

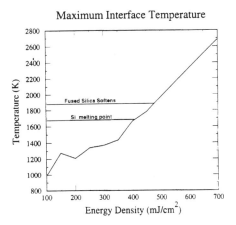

Fig. 6. Simulation of interface temperature vs. laser energy density

CONCLUSION

For laser crystallized Si thin films on silica substrates, the average grain size exhibits a peak as a function of the laser fluence. When this peak laser fluence is exceeded then the average Si grain size is reduced. Cross section TEM has shown that this reduction in grain size coincides with a an increase in the silica substrate roughness. Also this roughness is manifested as pits as deep as 17 nm into the silica substrate. These pits in the substrate may act as nucleation sites for grain growth as well as for intragranular defects such as stacking faults. Atomic force microscopy of the substrate surface performed on samples crystallized at the peak laser fluence and beyond the peak laser fluence show that the magnitude and spatial frequency of the substrate roughness increase as the peak laser fluence is exceeded. Finally , simulations suggest that the interface temperature during crystallization beyond the laser fluence at the peak is well past the melting point of both Si and the silica substrate. This suggest that the Si is completely melted and that the first 10 nm of the substrate my begin to flow under these crystallization conditions. The peak in average grain size may exist at the point where the Si is completely melted and before the silica substrate begins to soften.

ACKNOWLEDGMENT

A part of this work was funded under DARPA Contract No. F33615-92-C-5811.

REFERENCES

1. P. Mei, J.B. Boyce., M. Hack, R. Lujan, S.E. Ready, International Semiconductor D evice Symposium Proceedings, Vol. 1, 47, (1993).
2. P. Mei, J.B. Boyce., M. Hack, R. Lujan, R.I. Johnson, G.B. Anderson, D.K. Fork, and S.E. Ready. To be published in Appl. Phys. Lett., February 28, 1194.
3. T. Samashima and S. Usui, Mat. Res. Soc. Symp. Proc. 71, 435 (1986).
4. S. E. Ready, J. B. Boyce, R. Z. Bachrach, R. I. Johnson, K. Winer, G. B. Anderson, and C. C. Tsai, Mat. Res. Soc. Proc.. 149, 345 (1989)
5. K. Winer, R. Z. Bachrach, R. I. Johnson, S. E. Ready, G. B. Anderson, and J. B. Boyce, Mat. Res. Soc. Proc. 164, 183 (1990).
6. K. Winer, G. B. Anderson, S. E. Ready, R. Z. Bachrach, R. I. Johnson, F.A. Ponce, and J. B. Boyce, Appl. Phys. Lett. 57, 2222 (1990).
7. R. Z. Bachrach, K. Winer, J. B. Boyce, S. E. Ready, R. I. Johnson, and G. B. Anderson, J. Electron. Mat. 19, 241 (1990)
8. R. I. Johnson, G. B. Anderson, S. E. Ready, J. B. Boyce, Mat. Res. Soc. Proc. 219, 407 (1991)
9. H. Kuriyama, et al., IEEE International Electron Meeting, Wash. D.C. (1991)
10. R. I. Johnson, G. B. Anderson, S. E. Ready, D. K. Fork, and J. B. Boyce, Mat.Res. Soc. Proc. 258, 123 (1992)
11. T. Samashima, Mat. Res. Soc. Proc., 283, 679 (1992).
12. H. Iwata, et al., Mat. Res. Soc. Proc., 283 709 (1992).
13. H. J.Kim, J. S. Im, and M. O. Thompson, Mat. Res. Soc. Proc., 321, 558 (1993).
14. J.B. Boyce, G.B. Anderson, D.K. Fork, R.I. Johnson, P. Mei, and S.E. Ready, Mat. Res. Soc. Proc. 321, 671 (1993)
15. H. J.Kim, J. S. Im, Mat. Res. Soc. Proc., 321, 665 (1993)

CHARACTERIZATION OF SPUTTERED Ge FILM

JAESHIN CHO* AND N. DAVID THEODORE**
*Compound Semiconductor-1, Motorola Inc., Tempe, Arizona 85284
**Materials Research and Strategic Technologies, Motorola Inc., Mesa, Arizona 85202

ABSTRACT

The electrical resistivity and microstructure of sputtered germanium film were characterized as a function of anneal temperature from 400 to 700°C. The as-deposited sputtered Ge film had an amorphous structure with resistivity of 165 Ω-cm which was maintained after annealing up to 540°C. After annealing above 550°C, the resistivity dropped by almost four orders of magnitude to ~0.027 Ω-cm. The sharp transition of resistivity at 550°C is believed to be due to the recrystallization of Ge film from the as-deposited amorphous structure. The ohmic contact property on Ge films was also evaluated using sputtered tungsten. Low resistance ohmic contacts were obtained both on as-deposited and annealed Ge films, with typical ohmic contact resistance of 0.05 ± 0.008 Ω-mm on annealed Ge films. The W contacts were thermally stable after annealing up to 650°C.

INTRODUCTION

Polycrystalline semiconductor materials such as polysilicon are being widely used in modern integrated circuits. Polysilicon films can be produced with a wide range of conductivity from 10^{-6} to 10^3 mho cm^{-1} corresponding to doping levels from 10^{15} to 10^{20} cm^{-3}. This feature of polysilicon has been utilized for realizing ultra-small monolithic resistors over a wide range of values from several ohms to several megaohms. The conduction process in polysilicon is also known to be greatly influenced by variations in microstructural properties such as grain structure, average grain size and the nature of grain boundaries which in turn depend on the deposition technique [1-6].

In this study, the electrical resistivity of sputtered Ge film was characterized as a function of anneal temperature by fabricating van der Pauw resistors. The microstructure and crystallinity of Ge film were characterized by transmission electron microscopy and x-ray diffractometry. The ohmic contact property on Ge films using tungsten was evaluated using the transmission line method.

EXPERIMENTAL PROCEDURES

The resistivity of sputtered Ge film was measured by fabricating van der Pauw resistors and transmission line method (TLM) structures. These test structures were fabricated using two photolithography steps. First, 500Å SiN$_x$ and 3500Å SiO$_2$ layers were sequentially deposited on Si wafers by plasma enhanced chemical vapor deposition (PECVD). Then the areas where Ge was going to be deposited were defined by the first photo step, and the SiO$_2$ layer was opened by etching down to SiN$_x$ layer using buffered HF. This isotropic etch generates sidewall profiles in the SiO$_2$ layer that ensures breakage of sputtered Ge film at the steps that will be patterned by lift-off. The Ge films were sputter deposited by DC magnetron with thicknesses from 200 to 3000Å, and patterned by lift-off. The background pressure of the sputtering chamber was below 8x10^{-7} Torr for all cases. The wafers were then capped with 2000Å PECVD SiO$_2$ film deposited at 350°C, and rapid thermal annealed from 400 to 750°C for 1 minute in Ar ambient. The second photo step was performed to define the ohmic contact vias, where the 2000Å SiO$_2$ film was etched open using buffered HF. The wet etching of SiO$_2$ was to prevent the erosion of Ge, if the reactive ion etch of SiO$_2$ was used, and also to enhance the lift-off yield of ohmic metal. 1000Å thick tungsten was then sputter deposited for ohmic contact and patterned by lift-off. The schematic cross section of the fabricated TLM structure is drawn in Figure 1.

Mat. Res. Soc. Symp. Proc. Vol. 343. ©1994 Materials Research Society

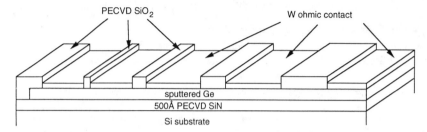

Fig. 1. Schematic cross section of fabricated transmission line method (TLM) structures with gap spacings of 2, 4, 8, 16, and 32 μm.

The sheet resistance of Ge film was measured from the van der Pauw resistor using an automatic DC parametric measurement system. The ohmic contact property on Ge film was evaluated using TLM structure, where the R_{sh} and R_c values were extrapolated from the slope and intercept of resistance as a function of gap spacings [7]. Using this technique, the sheet resistance of Ge as high as several hundred MΩ/sq can be measured accurately. Other techniques such as 4-point probe can be used to measure sheet resistance, but the as-deposited Ge film had R_{sh} beyond the upper measurement limit of our 4-point probe (5 MΩ/sq, Prometrics RS50). For sheet resistance values below 5 MΩ/sq, the 4-point probe values were well correlated with van der Pauw and TLM values within 5%.

The microstructure and crystallinity of the Ge films were characterized using x-ray diffractometry and transmission electron microscopy (TEM). For TEM, a JEOL 200CX transmission electron microscope was used, operating at 200 kV. Plan-view and cross-section TEM samples were made using dimpling and polishing followed by 5 keV argon ion milling. Bright-field and dark-field micrographs were obtained to characterize microstructures of the Ge layers. In-focus and under-focus images were obtained to observe the presence of voids. Selected-area electron diffraction patterns were obtained from the layers to further analyze the extent of crystallinity and the phases present.

RESULTS AND DISCUSSION

Effect of anneal on the resistivity of Ge

Figure 2 shows the sheet resistance of 2000Å thick Ge films as a function of anneal temperature. The as-deposited Ge film had a sheet resistance of 8.57 ± 0.25 MΩ/sq and showed a slight increase when annealed up to 540°C. Above 550°C, the sheet resistance dropped to 1.09 ± 0.04 kΩ/sq, which is almost four orders of magnitude lower than that of as-deposited film. It was found that further annealing at higher temperatures does not lower the Ge resistivity. This abrupt transition of resistance at 550°C was observed for all Ge thicknesses used in this study (200 to 3000Å).

Figure 3 shows the sheet resistance of Ge film as a function of thickness for both as-deposited and 600°C annealed films. For both films the sheet resistance scales with thickness as expected. The resistivity of as-deposited Ge is 165 Ω-cm, three times higher than that of intrinsic single crystalline Ge of 47 Ω-cm [6]. For 600°C annealed samples, the R_{sh} vs. thickness curve deviates for thinner films indicating that the transition becomes more sluggish as film thickness decreases.

The sharp transition of electrical resistivity indicates that there must be major changes in the film properties during annealing. It could be related to the purity of the Ge target or some contamination from the sputtering chamber. The Ge target used in this study was MARZ grade with impurity levels of Al, Ca, Cu, Mg, Fe, and Na less than 1 ppm. SIMS analysis showed no significant changes on the chemical composition of Ge film before and after annealing. Since the

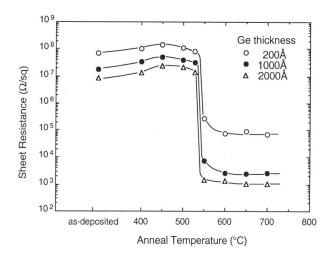

Fig. 2. Sheet resistance of 200Å, 1000Å and 2000Å thick sputtered Ge films as a function of anneal temperatures from 400 to 700°C.

Ge film was patterned by lift-off, the presence of photoresist in the sputtering chamber may cause outgassing and might be the source of contamination. However, a separate study of Ge deposition without photoresist inside the sputtering chamber also shows the same resistivity transition at 550°C. Therefore, it is believed that the contamination from the target or from the sputtering chamber is not responsible for the observed resistivity transition.

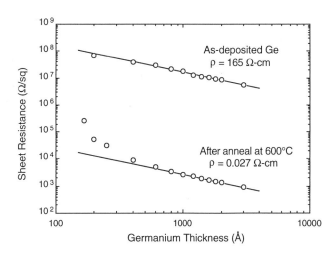

Fig. 3. Sheet resistance of as-deposited (top curve) and annealed Ge film at 600°C (bottom curve) as a function of film thickness.

717

Microstructure of Ge

The microstructure of Ge films was characterized as a function of anneal temperature. Figures 4-5 present representative TEM micrographs and electron diffraction patterns obtained from the Ge films. Figure 4 presents a bright-field TEM micrograph obtained from the as-deposited Ge film. The film appears to be uniformly amorphous. There is no clear evidence of microcrystallinity or nanocrystallinity. Under-focus images show evidence of a slight porosity in the film. Cross-section TEM micrographs of the as-deposited Ge film (not shown here) show some evidence of columnar porosity in the amorphous film. Such porosity can arise from shadowing of growing amorphous Ge islands during deposition. A selected-area electron diffraction (SAED) pattern obtained from the as-deposited Ge film is shown in Fig. 4b. The SAED pattern confirms that the Ge film is amorphous. Diffuse rings are present in the SAED pattern that are typical of amorphous materials. Calibration of such diffraction patterns using Si spots (not shown in Fig. 4b) confirms that the innermost diffuse ring in Fig. 4b matches the 111 reciprocal lattice spacing for Ge. This indicates that the nearest neighbor distance in the amorphous film corresponds to Ge {111} lattice spacings. Short-range order is present in the film, but long-range order is absent.

Figure 5a presents a bright-field plan-view TEM micrograph obtained from the Ge film rapid thermal annealed at 600°C for 1 minute. The film is now clearly crystalline. Extensive local strain is present in the material, resulting in the mottled appearance of the bright-field image. Dark-field plan-view TEM micrographs (not shown here) show that grain-sizes of the Ge film vary from 30Å to 450Å with average grain-size of ~200Å. The film is therefore nanocrystalline in nature. Figure 5b presents a selected-area electron-diffraction pattern obtained from the 600°C-annealed Ge film. Continuous sharp rings are present in the SAED pattern. The rings are much sharper than in the case of Fig. 4b, which was obtained from an amorphous film. The rings also extend out to higher order lattice-plane Miller indices; this indicates a presence of much higher long-range order than in the case of the amorphous film. The rings seen in Fig. 5b do not show clearly discrete diffraction spots as is commonly seen with microcrystalline films. The diffraction rings therefore indicate a presence of a very high density of very small randomly oriented grains in the Ge annealed film. This information correlates well with the nanocrystallinity observed in the plan-view TEM micrograph (Fig. 5a). Under-focus bright-field plan-view TEM micrographs (not shown here) indicate a presence of slight porosity in the Ge films. This porosity could be void-like in nature, or could consist of low-density regions. In either case, for the purpose of this paper, we call these regions 'pores'. Cross-section under-focus bright-field TEM micrographs (not shown here) show that the pores have coalesced to be roughly spherical. The density of such pores has decreased with respect to as-deposited film. Electrical data shown in Fig. 2 show a significant drop in resistivity when as-deposited Ge films are annealed above 550°C. The microstructural analysis reveals two phenomena that can contribute to the observed drop in resistivity: (i) transformation from amorphous to crystalline material, with an associated drop in electron scattering, and (ii) coalescence and spheroidization of pores in the Ge film, with an associated decrease in pore surface-area, with an associated decrease in electron scattering by pore walls.

Contact Property

The I-V characteristics of W ohmic contacts on as-deposited and 600°C annealed Ge films showed that W makes low resistance ohmic contact even without contact anneal. Since Ge has a low bandgap energy of 0.66 eV, it is expected that the ohmic contacts can be made easily on Ge film.

The ohmic contact resistance was evaluated using the TLM method. For the as-deposited Ge film, the extrapolated contact resistance was 1.0 ± 30 Ω-mm. The wide variation of R_c comes mainly from the measurement errors since the determination of R_c from TLM is sensitive to the small variations in the gap spacings between contacts in the TLM test structure [7]. Since the measured resistance between gaps was dominated by the high sheet resistance of as-deposited Ge films, the small variations on the gap spacings which were defined by wet etching can cause large errors in the extrapolated contact resistance. For 600°C annealed Ge films that had several

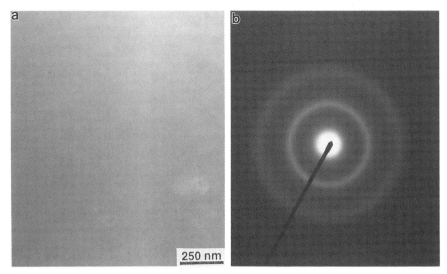

Fig. 4. (a) Bright-field plan-view TEM micrograph, (b) selected-area electron-diffraction pattern obtained from Ge film as-deposited.

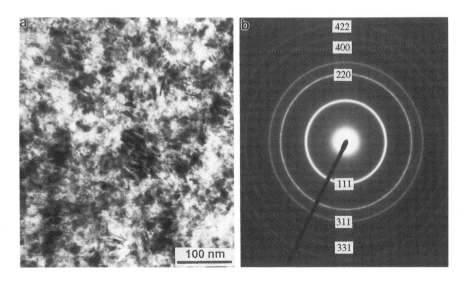

Fig. 5. (a) Bright-field plan-view TEM micrograph, (b) selected-area electron-diffraction pattern obtained from Ge film after 1 minute rapid thermal anneal at 600°C.

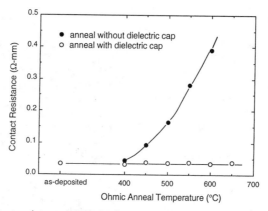

Fig. 6. Ohmic contact resistance of W ohmic on the 600°C annealed Ge films as a function of anneal temperature. Samples were rapid thermal annealed for 1 minute in Ar ambient. When the ohmic contact anneal was performed with a dielectric cap, the contact resistance was thermally stable up to 650°C.

kΩ ranges of sheet resistance, the extrapolated contact resistance was 0.05 ± 0.008 Ω-mm with small variations.

The thermal stability of W ohmic contact on Ge was tested by annealing up to 650°C. It was found that the contact resistance was below 0.05 Ω-mm after annealing at 650°C as shown in Figure 6. It should be noted that when the ohmic anneal was performed without dielectric cap, the contact property was degraded due to the oxidation of tungsten thereby degrading the W/Ge interface. When the dielectric cap layer such as PECVD SiO_2 film was used to prevent oxidation of W, there was no degradation of ohmic contact after annealing up to 650°C.

CONCLUSIONS

The electrical resistivity of sputtered Ge film had an abrupt transition of almost four orders of magnitude after annealing above 550°C. The microstructural analysis reveals two phenomena that can contribute to the observed drop in resistivity: (i) transformation from amorphous to crystalline material, with an associated drop in electron scattering, and (ii) coalescence and spheroidization of pores in the Ge film, with an associated decrease in pore surface-area, with an associated decrease in electron scattering by pore walls. The ohmic contact on Ge film was fabricated using sputtered tungsten. The contact resistance measured from transmission line method was 0.05 ± 0.008 Ω-mm and thermally stable up to 650°C.

REFERENCES

1. N.C.-C. Lu, L. Gerzberg, C.-Y. Lu, and J.D. Meindl, IEEE Trans. Electron Devices, **ED-28**, 818 (1981).
2. N.C.-C. Lu, L. Gerzberg, C.-Y. Lu, and J.D. Meindl, IEEE Trans. Electron Devices, **ED-30**, 137 (1983).
3. T. Ohmi, K. Matsudo, T. Shibata, T. Ichikawa, and H. Iwabuchi, Appl. Phys. Lett. **53**, 364 (1988).
4. T. Ohmi, T. Ichikawa, T. Shibata, and H. Iwabuchi, Appl. Phys. Lett. **54**, 523 (1989).
5. T. Ohmi, H. Iwabuchi, T. Shibata, and T. Ichikawa, Appl. Phys. Lett. **54**, 253 (1989).
6. S.M. Sze, Physics of Semiconductor Devices, 2nd ed. (John Wiley & Sons, New York, 1981), p. 850.
7. A. Scorzoni and M. Finetti, Materials Science Reports, **3**, 79 (1988).

XRD TEXTURE AND MORPHOLOGY ANALYSIS OF POLYCRYSTALLINE LPCVD GERMANIUM-SILICON

CORA SALM, JOS G.E. KLAPPE, JISK HOLLEMAN, JAN BART REM AND PIERRE H. WOERLEE
MESA Research Institute, University of Twente, P.O. Box 217, 7500AE Enschede, The Netherlands.

ABSTRACT

The morphology and texture of polycrystalline LPCVD Ge_xSi_{1-x} alloys have been studied for various deposition temperatures and for x = 0, x = 1 and x ≈ 0.3. The transition temperature of amorphous to polycrystalline growth is much lower for Ge and Ge_xSi_{1-x} than for Si. Just above this transition temperature, we observed a (220) orientation for all polycrystaline layers grown. At elevated temperatures a change to the (004) orientation for poly-Ge_xSi_{1-x} and (331) for poly-Si and poly-Ge is noted.

INTRODUCTION

Polycrystalline silicon (poly-Si) is used in IC processing, e.g. as a gate material, as interconnect and as a diffusion source for shallow junctions [1]. In recent years much attention has been focused on polycrystalline Ge_xSi_{1-x} alloys. This material is compatible with the existing silicon technology and has several advantages compared to poly-Si. Its lower deposition temperature is of interest for the fabrication of thin film transistors on glass substrates. Other advantages are selective deposition and etching and a higher charge carrier mobility, resulting in lower sheet resistance [2,3,4].

The texture and morphology of polycrystalline layers are important properties because the grain size and orientation affect the dopant distribution, charge trapping at grain boundaries and therefore the electrical properties of the material.

EXPERIMENTAL

The substrates we used were 3" oxidized Si(100) wafers. In order to prevent growth incubation, a 50 nm thick amorphous-Si (a-Si) nucleation layer was deposited first. The polycrystalline layers were deposited in the entrance zone of a conventional hot-wall LPCVD reactor at substrate temperatures between 420 and 675 °C. The fast depletion of the GeH_4 makes it difficult to grow in the flat zone of the reactor [5]. All layers were grown at a total pressure of P_{tot} = 1 mbar and partial pressures of SiH_4 and GeH_4 of respectively P_{SiH4} = 0.1 mbar (poly-Si and poly-Ge_xSi_{1-x}) and P_{GeH4} = 0.015 mbar. The carrier gas we used was N_2. This resulted for the Ge_xSi_{1-x} in 0.27<x<0.38 for the different growth temperatures, as determined with EDX [5]. The gradual decrease in the germanium content with increasing temperature is due to an increasing GeH_4 depletion at higher temperature even in the entrance zone [5]. Previous studies have shown that this variation in the germanium content is not critical for the texture of the studied layers [6]. However, the thickness of the layers does influence the texture and therefore layers of equal thickness were grown at different temperatures. The poly-Ge_xSi_{1-x} layers were all approximately 750 nm thick. The poly-Si and poly-Ge layers were 400 and 350 nm respectively.

The texture of the films was determined from X-ray diffraction (XRD) θ-2θ scans. In order to quantify the texture of the samples the orientation fraction f_{hkl} is determined as follows:

$$f_{hkl} = \frac{I_{hkl}}{F_{hkl}} / \sum_{hkl} \frac{I_{hkl}}{F_{hkl}} \tag{1}$$

Mat. Res. Soc. Symp. Proc. Vol. 343. ©1994 Materials Research Society

where F_{hkl} is a scattering correction factor and I_{hkl} is the intensity after background subtraction and corrected for a thickness dependent factor G [1]:

$$G \sim d / \sin(\theta), \qquad (2)$$

where d is the layer thickness and θ the angle of incidence. The correction factors for scattering for poly-Si and poly-Ge were obtained from reference [7], the factor for Ge_xSi_{1-x} was taken as the average between the two, see table 1.

f_{hkl}	poly-Si	poly-Ge_xSi_{1-x}	poly-Ge
111	100	100	100
220	55	56	57
311	30	35	39
004	6	6	7
331	11	11	10

Table 1: Scattering correction factors F_{hkl} for quantitative XRD analysis.

For the quantitative analysis only the peaks that were clearly above the noise level were taken into account.

The surface morphology of the poly-Ge_xSi_{1-x} was studied with a scanning electron microscope (SEM).

RESULTS AND DISCUSSION

The growth behavior of a layer and the amorphization temperature of this layer depends on the substrate material, the substrate temperature and the total and partial pressures of the gasses [8]. We varied the substrate temperature, keeping the gas flows constant, which resulted in an increasing growth rate with increasing temperature. Figure 1 shows the growth rate of poly-Ge, poly-Ge_xSi_{1-x} (0.28<x<0.37) and poly-Si as a function of the substrate temperature.

Figure 2 shows XRD measurements corrected for background for 400 nm thick poly-Si and 350 nm thick poly-Ge layers at different temperatures. Poly-Si was found to have a transition temperature of polycrystalline to amorphous growth at 575 °C. Just above this temperature XRD measurements show small peaks indicating the presence of (111), (220) and (311) oriented grains. Going towards higher temperatures (middle and top figure) these three peaks become larger and at elevated temperatures also a small (331) peak is observed. Poly-Ge shows a less obvious change in the XRD scans. In figure 3 the XRD spectra and SEM micrographs of 750 nm thick poly-Ge_xSi_{1-x} layers are shown at three substrate temperatures. The XRD scans show a very large (220) peak at the lowest temperature. At intermediate temperature this peak decreases rapidly and the (111) and (311) peaks increase. For even higher temperatures the (111) peak become the largest and a (004) peak appears.

From the SEM micrographs shown in figure 3, it can be seen that with an increasing temperature the average grain size increases (120, 200 and 250 nm for 475, 525 and 575 °C respectively) and

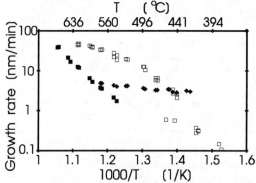

Figure 1: Arrhenius plot of the growth rate [nm/min] for poly-Si (■), poly-Ge_xSi_{1-x} (0.27<X<0.38) (□) and poly-Ge (◆)

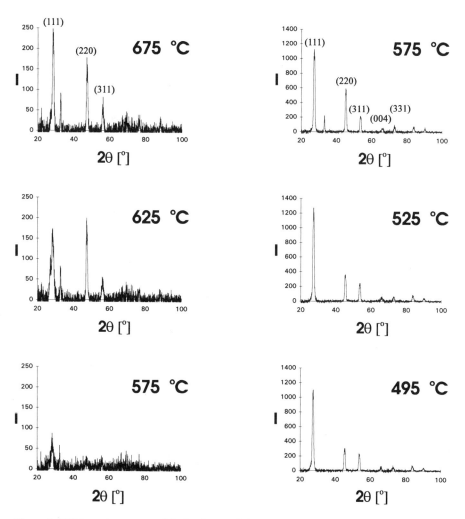

Figure 2: XRD spectra, corrected for background signal, for poly-Si (left) and poly-Ge (right).

the surface roughness increases. We also made micrographs of a layer grown at 475°C of 300 nm thickness. Here the average grain size was 60 nm. In all cases a columnar structure was observed in cross sectional SEM.

For the determination of the fraction of a certain grain orientation present, correction factors for scattering (see equation (1)) and the angle of incidence (see equation (2)) have to be taken into account. As a result, the large (111) peaks that appear in the XRD measurements do not represent the dominating orientation. In figure 4 the orientation fraction f_{hkl} is shown as a function of the deposition temperature of the poly-Si, poly-Ge and poly-Ge_xSi_{1-x} layers.

Increasing the temperature does not have an effect on the relative presence of (111) grains in poly-Si. At about 625 °C the (331) orientation appears at the cost of (311) and eventually (220).

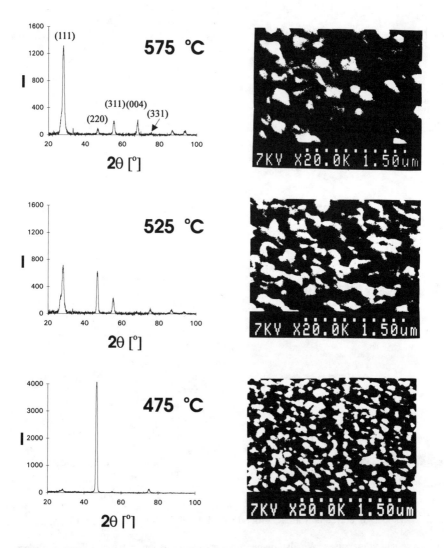

Figure 3: XRD measurements and SEM micrographs of poly-Ge$_x$Si$_{1-x}$ at 3 growth temperatures.

In contrast to poly-Si, poly-Ge shows the presence of the (004) orientation. As for poly-Si, poly-Ge shows at low temperatures that (220) and (331) oriented grains are the most important.

For poly-Ge$_x$Si$_{1-x}$ three areas can be roughly discriminated. Almost all the grains in poly-Ge$_x$Si$_{1-x}$ at low temperatures are (220) oriented. For layers half as thick as those shown in figure 4, this is only about 60% and all the other orientations are slightly more prevalent. This means that the (220) grains grow at the expense of the other orientations. At intermediate temperatures all the orientations are present and at high temperatures the (004) orientation dominates. In the intermediate region the (311) orientation is larger than in the two outer regions. The (331)

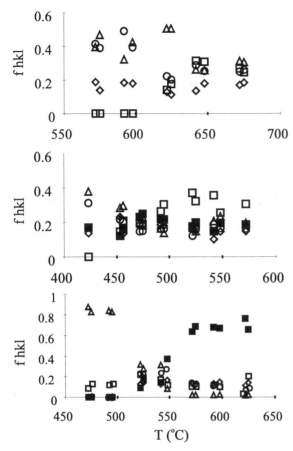

Figure 4: Fraction of each grain orientation (f_{hkl}) as a function of the deposition temperature for poly-Si (top) and poly-Ge (middle) and poly Ge_xSi_{1-x}. ◊=(111), Δ =(220), o=(311), ■=(004) and □=(331).

orientation is equally present over the entire temperature range. Note that at high temperatures the other orientations are small but definitely present, whereas for low temperatures the (004) orientation is not present (or below the detection limit of XRD measurements).

It should be kept in mind that the terminology of low and high temperature regimes are related to completely different regimes, depending on the material studied, e.g. the low temperature regime of poly-Si overlaps the high temperature regime of poly-Ge.

The region defined as the low temperature region starts at the amorphization temperature. A transition range is noted with both polycrystalline and amorphous phases towards a completely polycrystalline layer. The end of the low temperature region is defined at the temperature at which the (220) orientation drops rapidly. This is 650 °C for poly-Si and 510 °C for poly-Ge_xSi_{1-x}. For poly-Ge this temperature is difficult to extract from the present data as it is at the edge of the measured temperature range.

The amorphization temperature was determined at 570 °C for poly-Si and 420 °C for poly-Ge_xSi_{1-x}. For poly-Ge this temperature will be even lower, but cannot be measured with the present LPCVD equipment. For the determination of the amorphization temperature the surface diffusion length as introduced by Voutsas and Hatalis [8] is an important parameter. It gives an indication whether deposited atoms are able to nucleate in a crystalline way, or are forced into amorphous growth if their average surface diffusion length is below a critical value. The surface diffusion length depends on the deposition rate, the surface diffusivity, the hydrogen coverage, the substrate temperature and the surface diffusion activation energy. The activation energy is the most critical parameter, as its magnitude influences the surface diffusion length exponentially. The lower amorphization temperature of poly-Ge_xSi_{1-x} compared to poly-Si might be explained in this manner as the activation energy of poly-Ge_xSi_{1-x} is expected to lie in between the values for poly-Si and poly-Ge.

SUMMARY

We have presented data showing the change from (220) orientation at low temperature to (004) orientation at higher temperatures for poly-Ge_xSi_{1-x} alloys. In a transition range all orientations are present and in this range the (311) orientation is more strongly present than in the outer regions. For poly-Si a drop in (002) orientation can be seen as the temperature is increased, the (311) orientation becomes more important but does not drop again and the (004) orientation is not present at all. For poly-Ge the (004) orientation seems constant. The temperature range should be expanded to both higher and lower temperatures to see if the behaviour over a large range of temperatures of poly-Ge_xSi_{1-x} can be regarded as a weighted average of poly-Si and poly-Ge. The critical temperatures below which the growth is amorphous is 570 °C for poly-Si and 420 °C for poly-Ge_xSi_{1-x} ($x \approx 0.35$).

The grain size of the poly-Ge_xSi_{1-x} layers increases from 120 nm to 250 nm if the growth temperature is increased from 475 to 575 °C. With increasing temperature also the roughness of the layers increases. For all temperatures a columnar structure was observed in cross sectional SEM.

ACKNOWLEDGEMENTS

This work was financed by the Dutch Foundation for Fundamental Research on Matter (FOM) and the Netherlands Technology Foundation (STW), project number 22.2876. We thank Mark Smithers of the Centre for Materials Research (CMO) for making the SEM micrographs.

REFERENCES

[1] T.I. Kamins, Polycrystalline Silicon for Integrated Circuit Applications, (Kluwer Academic Press, Boston, 1988).
[2] T.-J. King and K.C.Saraswat, IEEE El.Dev.Lett. **13**, 309 (1992).
[3] M.C. Öztürk, Y. Zhong, D.T.Grider, M. Sanganeria, J.J. Wortman and M.A.Littlejohn, proceedings SPIE, 260 (1993).
[4] D.J. Godbey, A.H.Krist, K.D.Hobart and M.E.Twigg, J. Electrochem. Soc. **139**, 2943 (1992).
[5] J. Holleman, A.E.T. Kuiper and J.F. Verweij, J. Electrochem. Soc., **140,** 1717 (1993)
[6] J. Holleman (private communication).
[7] Powder Diffraction Data, JCPDS Associatship at the Nat. Bureau of Standards, ed. D.K. Smith, (1976).
[8] A.T. Voutsas and M.K. Hatalis, J. Electrochem. Soc., **139**, 2659 (1992).

ENERGIES AND ATOMIC STRUCTURES OF GRAIN BOUNDARIES IN SILICON:
COMPARISON BETWEEN TILT AND TWIST BOUNDARIES

M. KOHYAMA*, R. YAMAMOTO** AND Y. WATANABE***
*Department of Material Physics, Osaka National Research Institute, AIST,
1-8-31, Midorigaoka, Ikeda, Osaka 563, Japan.
**Institute of Industrial Science, University of Tokyo, 7-22-1, Roppongi,
Minato-ku, Tokyo 106, Japan.
***Cray Research Japan Ltd., 6-4, Ichiban-cho, Chiyoda-ku, Tokyo 102, Japan.

ABSTRACT

    The energies and atomic structures of tilt and twist boundaries in Si have
been examined by using the tight-binding electronic theory, and the reason
why twist boundaries are seldom found in polycrystalline Si has been
investigated. About the frequently observed {122} Σ=9 and {255} Σ=27 tilt
boundaries, the configurations without any coordination defects consistent
with the electron microscopy observations have relatively small interfacial
energies with small bond distortions. About the <111> Σ=7, <011> Σ=3 and
<001> Σ=5 twist boundaries, the configurations contain larger bond
distortions or more coordination defects, and much larger interfacial
energies than those of the tilt boundaries. The <001> twist boundaries have
very complex structures as compared with the other twist boundaries, which
can be explained by the morphology of the ideal surfaces. The stability of
the tilt boundaries in Si can be explained by the viewpoint of the stable
structural units consisting of atomic rings.

INTRODUCTION

    Polycrystalline Si films have been actively investigated for various
applications such as solar cells or thin-film transistors (TFT's). Grain
boundaries very much affect the properties of polycrystalline Si. Thus, it
is of much importance to understand the fundamental properties of grain
boundaries in Si, especially, the structure-property relationships.
    Significant advances have been made in the understanding of the tilt
boundaries in Si. The CSL (coincidence site lattice) tilt boundaries exist
frequently in polycrystalline Si or Ge. Many high resolution transmission
electron microscopy (HRTEM) observations [1-3] have indicated that such tilt
boundaries are constructed by atomic rings without any coordination defects.
And theoretical calculations [4-7] have shown the stability of such
configurations without any electronic states inside the minimum band gap.
    On the other hand, there have been relatively few studies about twist
boundaries in Si or Ge. There remain basic questions such as why twist
boundaries are seldom found in polycrystalline Si or Ge as compared with the
tilt boundaries, or what atomic configurations twist boundaries contain.
Recently, several theoretical calculations have indicated that twist
boundaries in Si or Ge have fairly different features from those of the tilt
boundaries. Tarnow and co-workers [8] have performed energy-minimization
calculations of the <001> Σ=5 twist boundaries in Ge by using the ab-initio
pseudopotential method. The interfaces contain many distorted or weak bonds
and coordination defects, which cause very large interfacial energies.
    In this paper, we examine the energies and atomic structures of various
twist boundaries with different rotation axes, the <111> Σ=7, <011> Σ=3 and
<001> Σ=5 twist boundaries in Si, by using the tight-binding electronic

727

theory. We compare those with the typical tilt boundaries in Si, the {122} Σ=9 and {255} Σ=27 boundaries. And we investigate the reason why twist boundaries are seldom found in polycrystalline Si.

METHOD OF CALCULATIONS

The details of the method are given in [9]. Energy-minimization calculations have been performed by using the transferable semi-empirical tight-binding (SETB) method [10], where the total energy is expressed as a sum of the band-structure energy and the remaining repulsive energy as well as the usual SETB method. However, this method can reproduce the energies and equilibrium volumes of variously coordinated structures of Si, which cannot be attained by the usual SETB method. Recently, we have examined the applicability of this method to graphitic Si as a three-coordinated structure. The calculated equilibrium bond length is 0.2197nm, and the energy increase against diamond structure is 0.63eV/atom. These are in good agreement with the ab-initio results, 0.2249nm and 0.71eV/atom [11].

All the calculations of boundaries are carried out with use of the supercell technique. The intra-atomic electrostatic interactions are included self-consistently through the form of a Hubbard-like Hamiltonian with the value of U used in [12].

ATOMIC MODELS

The {122} Σ=9 and {255} Σ=27 boundaries are constructed by rotating two crystals around the <011> axis by 38.9˚ and 31.6˚, respectively. The boundary planes, {122} and {255}, are parallel to the rotation axis. The atomic models can be constructed by the HRTEM observations [1-3]. These contain a glide-plane symmetry and consist of arrangement of five-membered, seven-membered or six-membered rings without any coordination defects.

The supercells of the Σ=9 and Σ=27 boundaries contain 144 atoms and 192 atoms, respectively, where the distances between the interfaces are about 3.3nm and 1.8nm, respectively. The rigid-body translations only normal to the interface are permitted by the symmetric property. These have been optimized by iterating the relaxations of the supercells of 80 atoms and of 112 atoms for the respective boundaries by using the present method.

About the twist boundaries, we deal with models constructed by two ideal surfaces. The <111> Σ=7 boundary is formed by rotating (111) surfaces around the <111> axis by 38.2˚ or 98.2˚. The <001> Σ=5 boundary is formed by rotating (001) surfaces around the <001> axis by 36.9˚ or 53.1˚. In these boundaries, respective two rotation angles generate different boundaries. The <011> Σ=3 boundary is formed by rotating (011) surfaces around the <011> axis by 70.5˚ or 109.5˚. For this boundary, the two rotation angles generate identical boundaries. Thus, we deal with only the boundary of 109.5˚.

Firstly, we examine the overall features of energies and atomic structures of the twist boundaries against various rigid-body translations parallel to the interface. As discussed in [8], the meaningful translations parallel to the interface are limited within the irreducible part of the DSC (displacement shift complete) unit cell. We have performed the relaxations about the mesh points in the irreducible parts of the DSC unit cells [9]. In these calculations, the small supercells have been used and the translations normal to the interface have been optimized by using an empirical inter-atomic potential [13] for saving the computing time. The small supercells of the Σ=7, Σ=3 and Σ=5 boundaries contain 84, 60 and 60 atoms, respectively.

Secondly, we have performed the relaxations of the large supercells with several selected rigid-body translations parallel to the interface in order to obtain the most stable configurations. In these calculations, the translations normal to the interface have been optimized by iterating the relaxations of the small supercells by using the transferable SETB method. The large supercells of the Σ=7, Σ=3 and Σ=5 boundaries contain 168, 156 and 140 atoms, respectively, where the distances between the interfaces are about 1.9nm, 2.5nm and 2.0nm, respectively.

RESULTS AND DISCUSSION

The tilt boundaries

Fig. 1 shows the relaxed configurations of the tilt boundaries. About the Σ=9 boundary, the optimized translation normal to the interface is +0.006nm. The interfacial energy E_{gb} is 0.32J/m^2. The energy per boundary atom is 0.11eV. The bond-length and bond-angle distortions range from -1.6% to +1.5% and from -16.2° to +20.8°, respectively. There exist no electronic states inside the minimum band gap. These results are in good agreement with the previous theoretical results of the same boundary in Si [4-6].

About the Σ=27 boundary, the optimized translation is +0.005nm. E_{gb} is 0.56J/m^2. The energy per boundary atom is 0.19eV. The bond distortions range from -2.9% to +3.5% and from -20.5° to +24.8°. The energy and the bond distortions are larger than those of the Σ=9 boundary. This can be explained by the relatively complex arrangement of the structural units as compared with the simple arrangement of the units in the Σ=9 boundary [14].

In any case, the present results are consistent with the previous results [4-7]. It can be said that the frequently observed CSL tilt boundaries in Si have relatively small interfacial energies with small bond distortions.

Fig. 1. Stable configurations of the tilt boundaries in Si: (a) the {122} Σ=9 boundary, and (b) the {255} Σ=27 boundary. Atomic positions are projected along the <011> axis, and open and closed circles indicate the atoms with the two kinds of heights along the <011> axis. Atomic rings constituting one period of the structural units [14] are indicated by numerals.

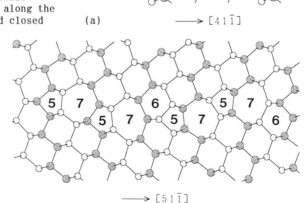

(a)

(b)

The twist boundaries

Fig. 2 shows the most stable configurations of the $\Sigma=7$ boundaries of 38.2°
and 98.2° obtained by the 168-atom cells. The optimized translations normal
to the interface are +0.01nm and zero, respectively. These contain no
coordination defects, which are defined by the radius of 0.28nm. E_{gb} of the
boundary of 38.2° is 1.28J/m², and the energy per boundary atom is 0.51eV.
E_{gb} of the boundary of 98.2° is also 1.28J/m², although this is a little
larger.

These configurations are relatively simple, where the interfacial bonds
are constructed by the dangling bonds of the surface atoms facing each other
at the interfaces. In both the configurations, one kind of interfacial bonds
in the unit cell contain very small distortions at the coincidence sites.
However, the other six interfacial bonds and the neighboring back bonds
inevitably contain larger distortions. The bond distortions in the boundary
of 38.2° range from -1.4% to +4.6% and from -23.5° to +21.3°, and those in
the boundary of 98.2° range from -1.4% to +4.9% and from -25.0° to +22.1°.

About the dependence on the translations parallel to the interface, the
translations introduced to the configurations shown in Fig. 2 induce the
increases of the bond distortions or induce the bond switching and the three-
coordinated defects. E_{gb} by the 84-atom cells ranges from 1.3J/m² to 1.7J/m²
against various rigid-body translations parallel to the interface.

Fig. 3 shows the most stable configuration of the $\Sigma=3$ boundary obtained by
the 156-atom cell. The optimized translation normal to the interface is
+0.01nm. E_{gb} is 1.02J/m², and the energy per boundary atom is 0.33eV. This
configuration contains no coordination defects, although this contains large
bond distortions ranging from -2.4% to +2.0% and from -35.5° to +28.5°.

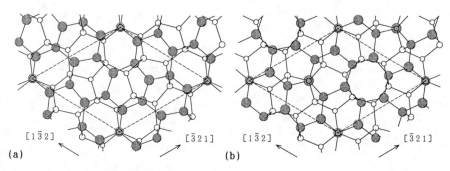

[1$\bar{3}$2]   [$\bar{3}$21]   [1$\bar{3}$2]   [$\bar{3}$21]

(a)   (b)

Fig. 2. Stable configurations of the
<111> $\Sigma=7$ twist boundaries in Si of
(a) 38.2° and of (b) 98.2°. Four
atomic layers are projected along the
<111> axis, and open and closed
circles indicate the atoms of the
lower and upper crystals. Dashed
lines indicate the CSL unit cell.

Fig. 3. Stable configuration of the
<011> $\Sigma=3$ twist boundary in Si. Four
atomic layers are projected along the
<011> axis. Arrows indicate the
four-membered rings in the CSL unit
cell.

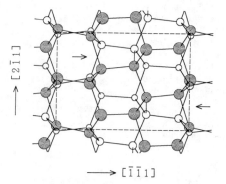

[2$\bar{1}\bar{1}$]

$\longrightarrow$ [$\bar{1}\bar{1}$1]

This configuration contains two four-membered rings in the CSL unit cell, and the large bond-angle distortions are associated with these rings. This point is related to the morphology of the ideal (011) surface. Respective surface atoms have only one dangling bond as well as the ideal (111) surface. However, surface atoms are connected with each other by the bond along the surface, and dangling bonds of neighboring surface atoms have different directions, respectively. Thus, the four-membered rings are easily constructed for the <011> twist boundaries.

About the dependence on the rigid-body translations parallel to the interface, relatively stable configurations without any coordination defects such as that in Fig. 3 can be constructed for the translations that make the atomic rows along the [2,-1,1] direction of the two (011) surfaces coincide with each other at the interface. Large translations along the [-1,-1,1] direction induce very complex structures with five-coordinated defects or large bond distortions. E_{gb} of the 60-atom cells ranges from $1.1 J/m^2$ to $3.1 J/m^2$ against various translations parallel to interface.

About the <111> $\Sigma=7$ and <011> $\Sigma=3$ boundaries, it can be said that the configurations shown in Figs. 2 and 3 exist at the shallow energy minima against the rigid-body translations. However, the absolute values and densities of the bond distortions and the interfacial energies are much larger than those of the tilt boundaries.

Fig. 4 shows the most stable configurations of the $\Sigma=5$ boundaries of $36.9°$ and $53.1°$ obtained by the 140-atom cells. In the same notation with that in [8], the translations parallel and normal to the interface are $\vec{R}_1/20+\vec{R}_2/20$ and $+0.008nm$, respectively, for the boundary of $36.9°$. For the boundary of $53.1°$, those are $\vec{R}_1/10+\vec{R}_2/10$ and $+0.03nm$. $\vec{R}_1$ and $\vec{R}_2$ are the CSL vectors.

Both the configurations contain no coordination defects. However, these are very complex, and the bond distortions and energies are very large as compared with the other twist boundaries. E_{gb} of the boundaries of $36.9°$ and $53.1°$ are $2.05J/m^2$ and $2.34J/m^2$, respectively. The energies per boundary atom are $0.94eV$ and $1.08eV$, respectively. The bond distortions in the boundary of $36.9°$ range from -2.5% to $+16.0\%$ and from $-24.4°$ to $+37.3°$, and those in the boundary of $53.1°$ range from -1.5% to $+14.8\%$ and from $-24.7°$ to $+49.8°$. These bond distortions generate deep states, although the stable configurations of the other twist boundaries contain only shallow states.

About the variations of the structures and energies against the rigid-body

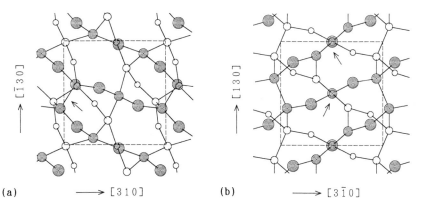

(a) $\longrightarrow$ [310]   (b) $\longrightarrow$ [3$\bar{1}$0]

Fig. 4. Stable configurations of the <001> $\Sigma=5$ twist boundaries in Si of (a) $36.9°$ and of (b) $53.1°$. Four atomic layers are projected along the <001> axis. Arrows indicate the four-membered rings in the CSL unit cells.

translations parallel to the interface, the <001> Σ=5 boundaries reveal very complex features as well as those in Ge [8]. Three-coordinated or five-coordinated defects, or very large bond distortions are frequently observed. The interfacial energies are very large. E_{gb} by the 60-atom cells ranges from 2.2J/m² to 2.9J/m² for the boundary of 36.9° and from 2.5J/m² to 3.0J/m² for the boundary of 53.1°. There exist no sharp or deep energy minima against the rigid-body translations. There seem to exist many metastable configurations with different bonding networks and similar energies. There may exist other configurations with similar energies to those in Fig. 4.

These complex features should be caused by the morphology of the ideal (001) surface, where respective surface atoms contain two dangling bonds pointing to different directions in contrast to one dangling bond per surface atom in the ideal (111) and (011) surfaces. This problem increases the difficulty in constructing stable configurations with as many dangling bonds reconstructed as possible, which should result in the large interfacial energies with coordination defects or large bond distortions. And this problem causes a very high degree of freedom for constructing various configurations through the formation of dimer bonds or interfacial bonds, which should result in the variety of metastable configurations.

Discussion

The present results clearly indicate the possibility that twist boundaries generally contain greater structural disorder and larger interfacial energies than the tilt boundaries in Si. This can be understood from the viewpoint of the structural units. Generally, the stable configurations of the tilt boundaries in Si or Ge are constructed by arranging the structural units consisting of atomic rings without any large bond distortions, as well as the extended defects such as stacking faults or planar defects [15]. For twist boundaries, such stable structural units cannot be easily constructed. This should be the primary reason why twist boundaries are seldom found.

REFERENCES

1. M.D. Vaudin, B. Cunningham and D.G. Ast, Scripta Met. 17, 191 (1983).
2. A. Bourret and J.J. Bacmann, Surface Sci. 162, 495 (1985).
3. M. Elkajbaji and J. Thibault-Desseaux, Phil. Mag. A58, 325 (1988).
4. R.E. Thomson and D.J. Chadi, Phys. Rev. B29, 889 (1984).
5. D.P. DiVincenzo, O.L. Alerhand, M. Schluter and J.W. Wilkins, Phys. Rev. Lett. 56, 1925 (1986).
6. M. Kohyama, S. Kose, M. Kinoshita and R. Yamamoto, J. Phys. Condens. Matter 2, 7809 (1990).
7. M. Kohyama, R. Yamamoto, Y. Watanabe, Y. Ebata and M. Kinoshita, J. Phys. C21, L695 (1988); A.T. Paxton and A.P. Sutton, Acta Metall. 37, 1693 (1989).
8. E. Tarnow, P. Dallot, P.D. Bristowe, J.D. Joannopoulos, G.P. Francis and M.C. Payne, Phys. Rev. B42, 3644 (1990).
9. M. Kohyama and R. Yamamoto, submitted to Phys. Rev. B.
10. S. Sawada, Vacuum 41, 612 (1990); M. Kohyama, J. Phys. Condens. Matter 3, 2193 (1991).
11. M.T. Yin and M.L. Cohen, Phys. Rev. B29, 6996 (1984).
12. D. Tomanek and M.A. Schluter, Phys. Rev. B36, 1208 (1987).
13. F.H. Stillinger and T.A. Weber, Phys. Rev. B31, 5262 (1985).
14. M. Kohyama, R. Yamamoto and M. Doyama, Phys. Stat. Sol. B137, 11 (1986).
15. M. Kohyama and S. Takeda, Phys. Rev. B46, 12305 (1992).

PIEZORESISTIVE PROPERTIES OF BORON-DOPED PECVD MICRO- AND POLYCRYSTALLINE SILICON FILMS

M. Le Berre*, M. Lemiti*, D. Barbier*, P. Pinard*, J. Cali**, E. Bustarret**, J.-C. Bruyère**, J. Sicart°, J. L. Robert°
*LPM-INSA (URA CNRS 358), 20Av. Einstein, 69 021 Villeurbanne, France
**LEPES-CNRS, B.P. 166X, 38 000 Grenoble, France
°GES, Université des Sciences et Techniques du Languedoc, 34 095 Montpellier, France

ABSTRACT

The electrical and piezoresistive properties of in-situ doped PECVD silicon films deposited on oxided silicon wafers have been investigated. One series of films was deposited in the so-called microcrystalline state at 450°C. The other set of samples was deposited in the amorphous state at 320°C and subjected to rapid thermal annealing. Structural properties (grain size, texture, residual stress) were evaluated experimentally through TEM and grazing angle X ray diffraction and related to the measured gauge factor. A maximum longitudinal gauge factor of 28 is measured in the case of advantageously textured microcrystalline material, the magnitude of the gauge factor decreasing sharply for randomly oriented material. For the amorphous deposited and subsequently annealed material, the longitudinal gauge factor is in the range 22-27 depending on dopant concentration. These experimental features are compared to the results of a theoretical approach of piezoresistance in polysilicon. We derive various expressions of the gauge factor according to the assumptions of either constant stress or constant strain within the aggregate. In the case of untextured films, analytical Voigt-Reuss-Hill averages for the elements of piezoresistive and elastoresistive tensors lead to greatly simplified expressions. Theoretical estimates are shown to be in reasonable agreement with the experimental measurements. This confirms the great potential of PECVD microcrystalline and polycrystalline silicon for strain gauges.

1. INTRODUCTION

In many applications the deposition temperature has to be kept as low as possible, whether to minimize damage of the processed wafer or to allow a wider selection of low cost substrate material. PECVD (Plasma Enhanced Chemical Vapor Deposition) process at 450°C yields layers with a fine-crystallized structure (microcrystalline silicon) which enables any crystallization process to be dispensed with. On the other hand thin films may be deposited at lower temperature ($T°_D$=320°C) with the same technique, but then require to be subsequently annealed for crystallization. However the properties of polycrystalline and all the more of microcrystalline silicon are largely influenced by the high number of grain boundaries. Unfortunately only little information is available on the piezoresistive effect of fine-grained layers[1,2,3] (L<30nm). The gauge factor (relative resistance change divided by the strain) is known to have strong dependences on doping concentration, grain size, texture, resistivity and barrier height at grain boundaries. In this paper, the effects of these various factors have been taken into account. A theoretical description of the piezoresistive properties of polycrystalline silicon derived according to the assumptions of constant strain or stress was compared to the measurements.

733

2. PREPARATION OF THE THIN FILMS

Amorphous (a-Si: H) and microcrystalline (μc-Si: H) thin films were deposited in a parallel plate diode reactor by a.c. (50kHz) PECVD of $SiH_4/H_2/B_2H_6$ (1/9/r) gas mixtures[4]. (100)- oriented 2" silicon wafers covered with a 100nm thermally grown oxide were used as substrates.

Microcrystalline thin film deposition occurred at 450°C. The r.f. power density and flow were fixed at 0.2W/cm^2 and 20sccm respectively. In order to obtain a higher degree of crystallization, the total pressure was varied from 1mbar to 1.5mbar and the interelectrode spacing was decreased from 3cm to 2cm for the samples deposited according to condition 1 and 2 respectively[5]. The amorphous films were deposited at a lower temperature (320°C). The power density was lower than in the former case (0.1W/cm^2), and the total pressure was increased to 2.5mbar.

Amorphous and microcrystalline layers were in-situ doped with boron. The diborane to silane volumic flow ratio r is given with the deposition rate in table I. Amorphous deposited layers were crystallized through rapid thermal annealing (RTA: 1100°C/20s). The choice of RTA is due to previous studies that yielded improved properties regarding the electrical activity for rapid thermally annealed samples as compared to conventionally annealed samples[6].

Table I: Deposition conditions (diborane to silane volumic flow ratio r and deposition rate v_d) for microcrystalline deposited samples a) and for amorphous deposited samples b)

a)

Sample	U3	U2	U1	U33	U4	U35
r (B_2H_6/SiH_4)	7.10^{-5}	1.10^{-3}	3.10^{-3}	2.10^{-4}	1.10^{-2}	5.10^{-3}
v_d [nm/min]	2.5	2.6	2.7	0.5	0.9	0.6
	condition1			condition2		

b)

Sample	A23	A11	A10	A24	A27	A5
r (B_2H_6/SiH_4)	5.10^{-3}	5.10^{-3}	5.10^{-3}	1.10^{-2}	2.10^{-2}	2.10^{-2}
v_d [nm/min]	3	2.7	2.7	3.1	2.9	5.6

3. THEORETICAL ANALYSIS

The piezoresistance of polycrystalline silicon has been investigated by several authors[7,8,9,10,11]. The common goal is to explain the influence of the deposition and process conditions on the magnitude of the piezoresistive effect and to optimize this effect. However these approaches lead to various expressions of the gauge factor according to the assumptions on the distribution of strain and stress inside the material. How the polycrystalline texture is accounted for is another key point for the calculation of the gauge factor, as different expressions have been used so far for the calculation of an average value over all possible crystallite orientation[7,8,9].

As mentioned by V. Mosser[11], either local strain or stress has to be considered constant in the material to simplify the calculations (Voigt-Reuss-Hill averaging). Using these assumptions in Hooke's generalized equations and in the equations relating the elastoresistive and the piezoresistive tensor, the longitudinal gauge factor is given in the case of a uniaxial stress by[12]:

$$K_l^v = \frac{\rho_c\left(1-2\frac{W}{L}\right)}{\rho}\left[m_{11}^v + \frac{2\,C_{12}^v\left(1-m_{12}^v\right)}{C_{11}^v+C_{12}^v}+1\right] \qquad (1)$$

$$K_I^R = \frac{\rho_c \left(1 - 2\frac{W}{L}\right)}{\rho} \left[\frac{\pi_{11}^R}{s_{11}^R} - \left\langle \frac{\dot{s}_{21}}{\dot{s}_{11}} \right\rangle - \left\langle \frac{\dot{s}_{31}}{\dot{s}_{11}} \right\rangle + 1 \right] \qquad (2)$$

$$K_I^{R2} = \frac{\rho_c \left(1 - 2\frac{W}{L}\right)}{\rho} \left[\left\langle \frac{\dot{\pi}_{11}}{\dot{s}_{11}} \right\rangle - \left\langle \frac{\dot{s}_{21}}{\dot{s}_{11}} \right\rangle - \left\langle \frac{\dot{s}_{31}}{\dot{s}_{11}} \right\rangle + 1 \right] \qquad (3)$$

where the barrier change produced by the stress at grain boundaries is neglected.

ρ_c stands for the bulk resistivity inside the grain, ρ for the total resistivity, W for the depletion width and L for the mean grain size. The primes mean that the coefficients are expressed in a coordinate system linked to the polysilicon layer. The terms in expressions (1), (2) and (3) with an exponent V or R or in brackets are average terms , as detailed in the following.

In the hypothesis of constant stress, 2 distinct expressions K_I^R and K_I^{R2} of the gauge factor have been derived. K_I^R is obtained by dividing the mean resistivity change by the mean strain. K_I^{R2} on the contrary is the mean value of the resistivity change divided by the strain in each crystallite.

Expressions (1) and (3) are close to the gauge factor expressions proposed by E. Obermeier[7] and P. J. French[8] respectively. A major difference is that average coefficients with an exponent R and V are calculated here in a simple analytical way instead of using a 3 dimensional numerical integral.

The averaging over all possible crystallite orientations of 4th rank tensors has been examined in the scope of elastic and optical properties by Anastassakis[13]. Using tensor invariants, elements of the average tensor can be deduced in a purely mathematical way. We propose to apply a similar approach to determine the average tensor elements of the elastoresistive, piezoresistive tensors as for compliance and elastic constant tensors. Since factors 2 and 4 are introduced when passing from the tensorial to the reduced notation for the piezoresistive and compliance tensors respectively, the final expressions of the elements of the average tensor will differ from one tensor to another[12]. In the hypothesis of constant strain, the mathematical method of averaging is applied to the elastoresistive tensor and to the tensor of elastic constants. Elements of average piezoresistive and compliance tensors are then deduced for consistency. In the hypothesis of constant stress, the averaging method is applied to the piezoresistive and compliance tensors. Elements of average elastoresistive tensor and average tensor of elastic constants are then deduced for consistency (Table II).

Some terms in brackets in expressions (2) and (3) are averages of ratio. For such cases where the mathematical averaging can't be used, the average terms are calculated numerically using Voigt's equivalent integral[7,14]:

$$\left\langle x_{ij}' \right\rangle = \frac{1}{8\pi^2} \int_0^{2\pi} \int_0^{\pi} \int_0^{2\pi} x_{ij}' \sin\theta \, d\phi \, d\theta \, d\psi \qquad (4)$$

When the layers are strongly textured, all average terms in expressions (1), (2) and (3) have to be calculated numerically, according to[15]:

$$\left\langle x_{ij}' \right\rangle = \frac{1}{2\pi} \int_0^{2\pi} x_{ij}' \, d\psi \qquad (5)$$

Table II: Elements of average tensors.

Constant strain	Constant stress
Anisotropy coeffcient: $m = m_{11} - m_{12} - 2\,m_{44}$	Anisotropy coeffcient: $\pi = \pi_{11} - \pi_{12} - \pi_{44}$
Mean terms: $m_{11}^v = m_{11} - \dfrac{2}{5}\,m$	Mean terms: $\pi_{11}^R = \pi_{11} - \dfrac{2}{5}\,\pi$
$m_{12}^v = m_{12} + \dfrac{1}{5}\,m$	$\pi_{12}^R = \pi_{12} + \dfrac{1}{5}\,\pi$
$m_{44}^v = m_{44} + \dfrac{1}{5}\,m$	$\pi_{44}^R = \pi_{44} + \dfrac{2}{5}\,\pi$
Anisotropy coeffcient: $C = C_{11} - C_{12} - 2\,C_{44}$	Anisotropy coeffcient: $S = S_{11} - S_{12} - \dfrac{S_{44}}{2}$
Mean terms: $C_{11}^v = C_{11} - \dfrac{2}{5}\,C$	Mean terms: $S_{11}^R = S_{11} - \dfrac{2}{5}\,S$
$C_{12}^v = C_{12} + \dfrac{1}{5}\,C$	$S_{12}^R = S_{12} + \dfrac{1}{5}\,S$
$C_{44}^v = C_{44} + \dfrac{1}{5}\,C$	$S_{44}^R = S_{44} + \dfrac{4}{5}\,S$
$\pi_{11}^v = m_{11}^v S_{11}^v + 2\,m_{12}^v S_{12}^v$ $\pi_{12}^v = m_{11}^v S_{12}^v + m_{12}^v\left(S_{11}^v + S_{12}^v\right)$ $\pi_{44}^v = m_{44}^v S_{44}^v$	$m_{11}^R = \pi_{11}^R C_{11}^R + 2\,\pi_{12}^R C_{12}^R$ $m_{12}^R = \pi_{11}^R C_{12}^R + \pi_{12}^R\left(C_{11}^R + C_{12}^R\right)$ $m_{44}^R = \pi_{44}^R C_{44}^R$
$S_{11}^v = \dfrac{C_{11}^v + C_{12}^v}{\left(C_{11}^v - C_{12}^v\right)\left(C_{11}^v + 2C_{12}^v\right)}$ $S_{12}^v = \dfrac{-C_{12}^v}{\left(C_{11}^v - C_{12}^v\right)\left(C_{11}^v + 2C_{12}^v\right)}$ $S_{44}^v = \dfrac{1}{C_{44}^v}$	$C_{11}^R = \dfrac{S_{11}^R + S_{12}^R}{\left(S_{11}^R - S_{12}^R\right)\left(S_{11}^R + 2S_{12}^R\right)}$ $C_{12}^R = \dfrac{-S_{12}^R}{\left(S_{11}^R - S_{12}^R\right)\left(S_{11}^R + 2S_{12}^R\right)}$ $C_{44}^R = \dfrac{1}{S_{44}^R}$

4. RESULTS AND DISCUSSION

Microcrystalline thin films deposited according to condition 1 are found by TEM to be (110)-textured whereas thin films deposited according to condition 2 are randomly oriented[5]. The grains appear rather columnar but actually consist of domains of much smaller coherence length ($L \approx 10nm$) (Fig. 1). Internal stress evaluated from Raman spectra[5,16] and grazing angle X-ray diffraction spectra according to the Warren-Averbach method[17] is highly compressive ($\approx 2\text{-}3GPa$), as commonly associated with low frequency PECVD deposition. The gauge factors experimentally measured are greater for (110) textured films than for randomly oriented thin films in agreement with values reported in the literature[1,2,3].

The amorphous and subsequently annealed thin films are also under compressive stress after deposition but the stress becomes tensile after annealing in the range 0.2-0.8GPa[18]. This high compressive stress led in some cases to spalling upon annealing[19]. The thin films are randomly oriented and exhibit a mean grain size about 20nm as determined from grazing angle diffraction spectra with the Warren-Averbach method, though some larger grains may be found (Fig. 2). Such a grain size is much smaller than that determined for amorphous deposited and annealed LPCVD thin films[20] ($L > 100nm$).

Fig. 1: TEM cross section of a
microcrystalline thin film (dark field)

Fig.2: TEM cross section of an amorphous
deposited and annealed thin film
(bright field)

Electrically active carrier concentration has been determined by Hall measurements[18,21] and the barrier height at grain boundaries has been deduced from this data[18]. Using the experimentally determined parameters of mean grain size, active carrier concentration, resistivity and barrier height, a fitted resistivity ρ has been calculated considering thermionic emission as the main conduction mechanism accross grain boundaries[22,7]. A calculated bulk resistivity ρ_c was then extracted from this calculation. Theoretical gauge factors are reported in Fig. 3 and 4 with measured values.

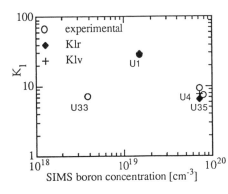

Fig. 3: Experimental and theoretical gauge
factor of microcrystalline PECVD silicon
thin films

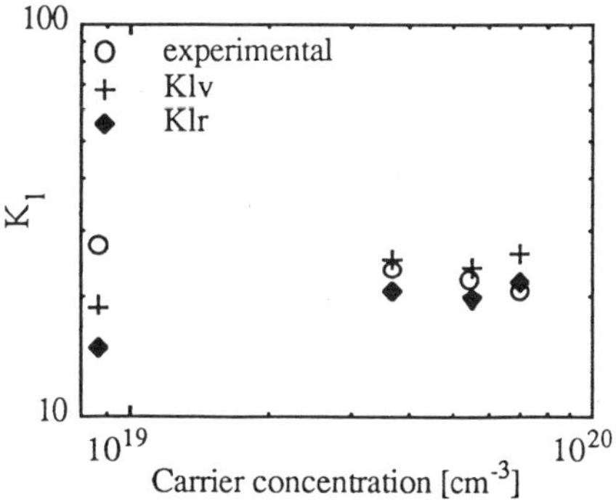

Fig. 4: Experimental and theoretical gauge
factor of amorphous deposited and rapid
thermal annealed PECVD silicon thin films

Obviously experimental values are satisfactorily approximated by the theory. Gauge factors calculated in the approximation of constant strain are always found to be higher than those calculated in the approximation of constant stress. However neither of these hypotheses is totally realistic. In the hypothesis of constant strain forces between the grains wouldn't be in equilibrium whereas in the hypothesis of constant stress the grains wouldn't fit one another. Both approaches should be thus regarded as useful approximations[23].

5. CONCLUSION

We successfully applied an improved model to fit experimental gauge factors in fine-grained polysilicon. This model enables the comparison between gauge factors calculated in the extreme cases of constant stress or constant strain.

For amorphous deposited and annealed thin films, the gauge factor is between 20 and 27, which is much lower than those measured for LPCVD amorphous deposited and annealed thin films. We think this discrepancy is mainly due to a much smaller grain size for the PECVD amorphous deposited and annealed layers.

In spite of a high compressive internal stress and small grain size, PECVD microcrystalline deposited films have gauge factors reaching 28 in the case of advantageous crystalline texture, in agreement with the results in the literature[1,2,3]. When the layers are randomly oriented, the gauge factors are much lower (6-8).

6. REFERENCES

1. W. Germer, Proc. Mater. Res. Soc. Europe Conference, 1984; Strasbourg, France, edited by P. Pinard, S. Kalbitzer (Les Ulis: Les Editions de Physique, 1984) pp. 581-586.
2. W. Germer, Sensors & Actuators 7, 135 (1985)
3. S. W. Guo, S. S. Tan, W. Y. Wang, Mater. Res. Soc. Symp.106, 231 (1988),
4. M. A. Hachicha, J. C. Bruyère, E. Bustarret, A. Deneuville, M. Brunel, IPAT'87 Int Conf Proc (CEP Ltd, Edinburgh UK, 1987), 360
5. M. Le Berre, M. Lemiti, P. Pinard, E. Bustarret, W. Grieshaber, J. C. Bruyère, M. A. Brunel, Mater. Res. Soc. Symp. 283, 573 (1993)
6. P. Jeanjean, J. Sicart, J. L. Robert, M. Le Berre, P. Pinard, V. Conedera, J. Phys. III 3, 47 (1993)
7. E. Obermeier, PhD thesis, University of Munich, 1983
8. P. J. Evans, A. G. R. Evans, Electron. Letters 20, 999 (1984)
9. D. Schubert, W. Jenschke, T. Uhlig, F. M. Schmidt, Sensors and Actuators 11, 145 (1987)
10. J. Suski, V. Mosser, J. Goss, Sensors and Actuators A 17, 405 (1989)
11. V. Mosser, J. Suski, J. Goss, E. Obermeier, Sensors and Actuators A 28, 113 (1991)
12. M. Le Berre, PhD thesis, INSA Lyon, 1993
13. E. Anastassakis, E. Liarokapis, J. Appl. Phys. 62, 3346 (1987)
14. W. Voigt, Lehrbuch der Kristallphysik, Verlag B. G. Teubner, Leipzig, 1910. p. 962
15. Y. Kanda, I.E.E.E. Trans. Elec. Dev. ED-29, 64 (1982)
16. E. Anastassakis, A. Pinczuk, E. Burstein, F. H. Pollak, M. Cardona, Sol. State Comm. 8, 133 (1970)
17. S. Vepreck, F. A. Sarott, M. Rückschloß, J. Non-Crys. Sol. 137&138, 733 (1991)
18. P. Jeanjean, J. Sicart, P. Sellitto, J. L. Robert, E. Bustarret, W. Grieshaber, J. Cali, M. Le Berre, M. Lemiti, P. Pinard, V. Conedera, J. Appl. Phys., accepted for publication
19. Y. Audet, W. Grieshaber, J. C. Bruyère, E. Bustarret, M. Le Berre, D. Barbier, M. Lemiti, C. Dubois, IPAT'91 Int. Conf. Proc.(CEP Ltd., Edinburgh UK, 1991), 190
20. P. J. French, PhD thesis, Southampton, 1986
21. Y. Audet (private communication)
22. N. C. C. Lu, L. Gerzberg, C. Y. Lu, J. D. Meindl, I. E. E. E. Trans. Elec. Devices ED-28, 818 (1981)
23. R. Hill, Proc. Phys. Soc. London Sect. A 65, 349 (1952)

PART VIII

Polycrystalline Nitride Films

MICROSTRUCTURES OF CHEMICAL-VAPOUR-DEPOSITED TiN FILMS

Noboru Yoshikawa and Atsushi Kikuchi
Department of Metallurgy, Faculty of Engineering, Tohoku University,
Aza-Aoba, Aramaki, Aoba-ku, Sendai, Miyagi, Japan 980

ABSTRACT

A gas mixture consisting of $TiCl_4$, H_2 and N_2 was fed into an externally-heated steel tube, and TiN was deposited on the inner wall by CVD. Microstuctures of the films were observed and their relationships with the preferred crystal orientations were studied. Distributions of the film growth rate and gas concentrations along the axial direction were calculated.

By comparing the film microstructures with the calculated local deposition conditions, it is shown that formation of the films with (110) preferred orientation correlated with the conditions at high temperature and low partial pressure of $TiCl_4$ on the substrate.

INTRODUCTION

In order to control the properties of chemical-vapour-deposited films, it is important to investigate the relationships between deposition conditions and film microstructures[1,2].

In this study, TiN was deposited on the inner wall of an externally-heated steel tube. Microstructures of the films deposited at different positions in the reactor under different deposition conditions were observed.

In order to obtain the local deposition conditions for the observed microstructures, numerical analyses of the gas flow and mass transfer were made and the concentration distributions along the axial direction of the reactor were calculated. This paper discusses the microstructural development of chemical-vapour-deposited TiN films, taking into account the relationships between the crystal orientation of films, the wall temperature and the calculated gas concentrations on the substrate.

EXPERIMENTAL

A schematic illustration of the experimental apparatus is shown in Figure 1(Fig. 1). A more detailed description has been given previously[3,4]. A gas mixture consisting of $TiCl_4$, H_2 and N_2 was fed into a steel tube 10mm in diameter and 640mm in length, which was placed horizontally and heated with an infrared image furnace. H_2 was used as a carrier for the vaporized $TiCl_4$. Deposition conditions are listed in Table 1. A temperature distribution existed along the axial direction of the reactor. To account for this, the setup maximum temperature of the distribution is listed as the deposition temperature(T), while the wall temperature at each position is referred to as the substrate temperature.

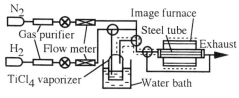

Fig. 1: Schematic illustration of the apparatus

Table 1. Conditions for TiN deposition

Gas flow rate(Q_T)/x10^{-6} m^{3}s^{-1} : 3.0- 30.0
Deposition temperature(T) /K :1123 -1323
Deposition time / ks : 3.6
Process pressure : Atmospheric pressure
Partial pressure of gas mixture / kPa :
P_{TiCl4} ; 1.52 -6.2 , $P_{H2} = P_{N2}$; 49.9-47.56

Mat. Res. Soc. Symp. Proc. Vol. 343. ©1994 Materials Research Society

OBSERVATION OF MICROSTRUCTURES

Microstructures of the films were observed by SEM and TEM. Film composition was analyzed by EPMA. Results indicate that the TiN film had nearly stoichiometric compositions under the conditions described above. Crystal orientation of the films was determined by XRD.

In order to quantify the degree of crystal orientation of films, a parameter suggested by Lotgering[5] was used, which is defined as follows:

$$f(hkl)=\frac{P(hkl)-P_0(hkl)}{1-P_0(hkl)}$$

where P(hkl) is a ratio of the peak intensity of (hkl) reflection to the sum of all peak intensities for the film spectrum, and P_0(hkl) is the same ratio of the peak intensities for the spectrum of a randomly-oriented microstructure.

A MODEL FOR THE NUMERICAL ANALYSIS

In the model for the analysis of flow and mass transfer, following assumptions were made:
1) Flow in the reactor is laminar, steady and axisymmetrical. Gas velocity at the inlet has a parabolic distribution. 2) Gas temperature is not dependent on the radial direction(r). Measured distributions of the temperature on the wall is given. Temperature dependence of the gas properties is taken into account. 3) The gas phase is composed of three components, which are $TiCl_4$, HCl and a H_2 and N_2 mixture. 4) Homogeneous reaction in the gas phase does not occur. Deposition of TiN and generation of HCl occur on the wall according to the following reaction:

$$TiCl_4(g) + 2H_2(g) + 1/2N_2(g) = TiN(s) + 4HCl(g)$$

The flow and mass transfer in the reactor was analyzed by solving the fundamental equations listed below :

$$\nabla \cdot \mathbf{n}_i = 0 \qquad \text{: Equation of continuity}$$

$$\nabla \cdot \rho \mathbf{vv} = -\nabla p - [\nabla \cdot \tau], \tau = -\mu \{\nabla \cdot \mathbf{v} + (\nabla \cdot \mathbf{v})^t\} + \frac{2}{3}\mu(\nabla \cdot \mathbf{v}) \qquad \text{: Equation of motion}$$

$\mathbf{n}_i$ is the mass flux of i th gas species with respect to the stationary axis (r: radial and z: axial), and given by

$$\mathbf{n}_i = - \rho D_{i3}\nabla \omega_i + \rho \omega_i \mathbf{v} \qquad \text{: Equation of mass transfer}$$

where ω_i is the mass fraction of i th component (i=1: $TiCl_4$, 2: HCl, 3: H_2-N_2). Accordingly

$$\omega_1 + \omega_2 + \omega_3 = 1$$

μ, ρ are the viscosity and density of the gas and D_{i3}(i=1,3)is the diffusivity of i th component in in a nearly pure H_2-N_2 mixture. $\mathbf{v}$ is the velocity vector and p is the pressure.
The rate equation obtained in the previous report[6] was incorporated as a boundary condition for the mass transfer equation:

$$\cdot \rho D_{13}\left(\frac{\partial \omega_1}{\partial r}\right)_{substrate} = 1.6 \times 10^4 \cdot \exp(-E/RT) \cdot X_{TiCl_4}^{-0.2} \cdot X_{HCl}^{-0.3} \quad (TiCl_4\, kg/m^2 s^{-1})$$

E=230kJ/mol

where $X_i = (\omega_i/M_i)/(\omega_1/M_1 + \omega_2/M_2 + \omega_3/M_3)$ M_i : Molecular weight of i th component

RESULTS AND DISCUSSION

1. Features of film microstructures

SEM photographs of TiN films deposited at the same positions in the reactor, but at different input partial pressures of $TiCl_4$ (P_{TiCl4}^{in}) are shown in Fig. 2. Film crystals are small and sharp at higher P_{TiCl4}^{in}, while polyhedral grains were formed at lower P_{TiCl4}^{in}.

Changes in microstructure were also observed along the axial direction(z: distance from the gas inlet), as shown in Fig. 3. The film deposited in the upstream region(Fig.3a) also had small, sharp crystals, and the film in the downstream region(Fig.3c) were similar to that deposited at low P_{TiCl4}^{in} (Fig. 2b). Therefore P_{TiCl4} is established as an important factor for formation of the film microstructures.

Temperature variation along the axial direction may also account for the microstructural differences in Fig 3. However, it had been observed[2] that films deposited at different temperatures(T) showed similar crystal shape with only variation in crystallite size with the source gas compositions as in this study. Partial pressures of H_2 and N_2 had much less influence on the formation of microstructures than $TiCl_4$.

a) $P_{TiCl4}^{in} = 6.18kPa$ b) $P_{TiCl4}^{in} = 0.92kPa$

z = 0.28m (1263K)   10μm

Fig. 2:Micrographs of TiN film surfaces deposited at different P_{TiCl4}^{in}.
$Q_T=6.0\times10^{-6}m^3s^{-1}$,T=1273K

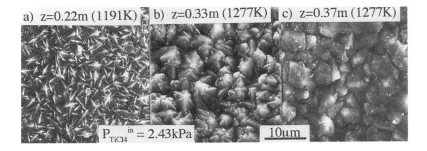

a) z=0.22m (1191K)   b) z=0.33m (1277K)   c) z=0.37m (1277K)

$P_{TiCl4}^{in} = 2.43kPa$ 10um

Fig. 3 : Micrographs of TiN film surfaces deposited at different axial positions in the reactor.
$Q_T=6.0\times10^{-6}m^3s^{-1}$,T=1273K, $P_{TiCl4}^{in}=2.43kPa$

2. Microstructure and crystal orientation of films

In order to investigate the relationship between crystal orientation and film microstructures, the TiN films were observed using TEM. A micrograph of the sharp crystals formed at the upstream region is shown in Fig.4a. Longitudinal direction of the needle-like crystals was determined to be [111]. The grains have grown in different spatial directions, so that the resulting films had either a weak (111) texture or a random orientation.

The polyhedral grain shown in Fig.4b consisted of several crystals having common <110> growth directions. From analysis of the electron diffraction patterns, it was found that the crystals had a (1$\bar{1}$1) twin plane. A cross-section SEM micrograph of the film is shown in Fig. 4c. Films consisted of columnar grains had strong (110) preferred orientations, consistent with Fig.4b. From these micrographs, the deposition conditions favoring such columnar growth were determined.

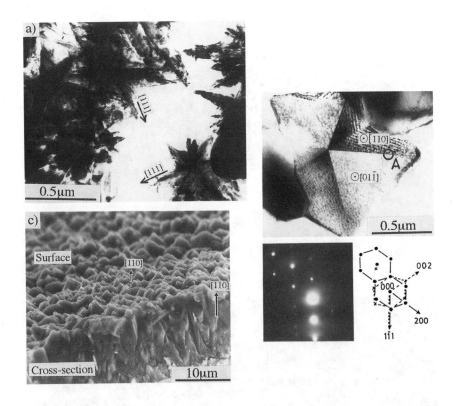

Fig. 4: TEM photographs of films deposited in the upstream region(a), in the down stream region(b) and a SEM photograph of a film consisting of columnar grains(c). Q_T=6.0x10^{-6}m^{3}s^{-1}, T=1273K, P_{TiCl4}^{in}=2.43kPa

3. Correlation between (110) texture of films and local deposition conditions

Calculated distributions of the film growth rate, partial pressure of $TiCl_4$ and HCl as a function of axial position are shown in Fig. 5. The calculated growth rate agreed with the experimental results. Gas velocity at the inlet was uniform throughout the reactor. As the deposition occurred, P_{TiCl_4} decreased and P_{HCl} increased along the axial direction.

(111) and (110)-oriented regions are indicated in the figure. The (111)-oriented region was closer to the inlet and had a higher P_{TiCl_4} than in the downstream region. The position with the highest (110)-orientaion($f(110)=0.11$) is indicated with a circle. The degree of (111) preferred orientation was generally lower than that of (110), as mentioned before.

The degree of (110) preferred orientation of the film was compared with the temperature and the calculated P_{TiCl_4} on the substrate at each position. The results are plotted in Fig. 6. These data were obtained from films deposited under various conditions of $P_{TiCl_4}{}^{in}$ and T, within the range of the gas flow rate shown. Formation of films with (110) preferred orietation correlated with the conditions at high temperature and low P_{TiCl_4} on the substrate.

Similar plots for the films grown at higher gas flow rates are shown in Fig.7a,b. Although the gas mixture had the same composition at the inlet, the degree of (110) preferred orietation is lower than that at the lower Q_T (Fig.6). Concentration distributions at different gas flow rates were calculated, and shown in Fig. 8. It was found that the decrease of $TiCl_4$ partial pressure in the reactor was decreased with gas flow rate. The positions of the highest (110) orientation shifted toward the downstream direction. Together with the results in Fig.6, it is clear that a smaller decrease of P_{TiCl_4} at the higher gas flow rates resulted in less formation of highly (110)-textured films.

T=1273K, Q_T=6.0x10^{-6}m^3/s, $P_{TiCl_4}{}^{in}$=4.05kPa

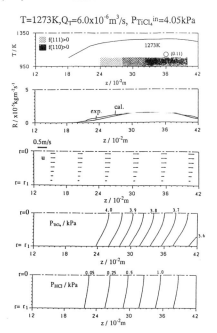

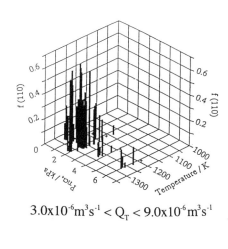

$3.0x10^{-6}m^3s^{-1} < Q_T < 9.0x10^{-6}m^3s^{-1}$

Fig. 6 : Relationships between the degree of (110)preferred orientation, temperature and P_{TiCl_4} on the substrate.

Fig. 5: Distributions of temperature, growth rate of film(TiN/kgm^{-2}s^{-1}), gas velocity, P_{TiCl_4} and P_{HCl}.

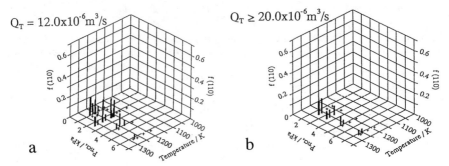

Fig. 7: Relationships among the degree of (110) preferred orientation, temperature and P_{TiCl4} on the substrate at higher gas flow rates.

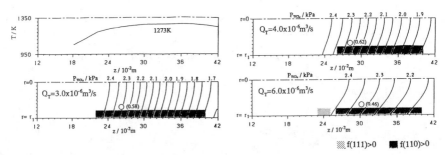

Fig. 8: Calculated distributions of P_{TiCl4} at different gas flow rates. T=1273K, P_{TiCl4}^{in}=2.43kPa

SUMMARY

(110)-oriented films were formed in the region of low P_{TiCl4}^{in} , which occurred in the downstream regions in the reactor. These films were made up of the twinned columnar grains grown in <110> directions. Taking account of the calculated concentration distributions on the substrate, it was shown that formation of (110)-oriented films correlates with the conditions at high temperature and low P_{TiCl4} on the substrate. Highly (110)-oriented films were not formed at the higher gas flow rates, because of small decreases of P_{TiCl4} in the reactor.

REFERENCE

1. W.A. Bryant, J. Mater. Sci. **12**, 1285 (1977).
2. G.Wahl and F.Schmaderer, ibid. **24**, 1141(1989).
3. N.Yoshikawa, H.Aikawa and A.Kikuchi, J. Japan Inst. Met. **55** 571 (1991).
4. N.Yoshikawa, K.Higashino and A.Kikuchi, ibid. **58** 442 (1994).
5. F.K.Lotgering: J. Inorg. Nul.Chem. **9** 113 (1959).
6. N.Yoshikawa and A.Kikuchi, submitted to Mat. Trans. JIM

PULSED LASER DEPOSITION OF NbN$_x$ (0 ≤ x ≤ 1.4) THIN FILMS

RANDOLPH E. TREECE,*† JAMES S. HORWITZ, AND DOUGLAS B. CHRISEY
Naval Research Laboratory, Code 6674, Washington, D. C. 20375-5345. † National Research Council Postdoctoral Research Associate.

ABSTRACT

The structure, morphology, and electrical properties of pulsed laser deposited NbN$_x$ (0 ≤ x ≤ 1.4) thin films have been investigated. Films were deposited from Nb metal targets on oriented MgO (100) and amorphous fused silica substrates as a function of substrate temperature and ambient pressure. A reducing atmosphere (N$_2$ with 10% H$_2$) was used to prevent oxide formation. At elevated temperatures, the N/Nb ratios of films deposited on MgO, as determined from Rutherford Backscattering Spectroscopy, increased from 0 - 1.4 as the ambient pressure was varied from vacuum to 200 mTorr, respectively. NbN$_x$ films (x=1) were deposited at 600 °C and 60 mTorr. Both the structure and electrical properties of the deposited films varied strongly with the substrate type. On fused silica, NbN films were poorly crystalline with low critical temperatures (T$_c$ ~ 8 - 11 K) and low critical currents (J$_c$ (4.2 K) ~ 2 MA/cm^2), whereas on MgO the NbN films were oriented and had better transport properties (T$_c$ ~ 16.4 K and J$_c$ (4.2 K) > 7.1 MA/cm^2).

INTRODUCTION

Pulsed laser deposition (PLD) is a relatively new thin film deposition technique which is currently being used to deposit high quality multicomponent oxide thin films. These materials include complex electronic ceramics such as high temperature superconductors and ferroelectric thin films.[1] Included among the advantages of PLD is that thin films can be deposited in high pressures of reactive gasses. While the vast majority of thin films grown by PLD are oxides there have been several reports recently on the growth of nitride thin films. These reports include the deposition and characterization of metastable CN$_x$[2] and c-BN.[3] With the exception of TiN,[4] the growth of transition metal nitrides by PLD has been virtually unexplored. Transition metal nitrides are an important class of materials which are very hard, have high melting points and are metallic.

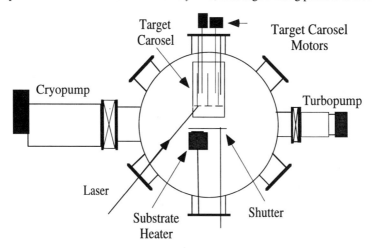

Figure 1. Pulsed laser deposition chamber.

Mat. Res. Soc. Symp. Proc. Vol. 343. ©1994 Materials Research Society

NbN, frequently grown by sputtering techniques, is a refractory nitride with a superconducting transition temperature (T_c) of 16 K.[5] NbN materials are currently being used as superconductors in digital electronics. In this study we have characterized the structure, morphology, and electrical properties of pulsed laser deposited NbN_x ($0 \le x \le 1.4$) thin films.

EXPERIMENTAL PROCEDURE

NbN_x films were grown in a high vacuum chamber equipped with a turbomolecular and a cryopump, shown in Figure 1. A pulsed KrF excimer laser beam (248 nm, 30 ns FWHM) operating at 10 Hz was focused with a 50 cm focal length lens at 45° onto a dithering and rotating (30 RPM) Nb foil target. The ambient gas (N_2 with 10% H_2) input pressure was regulated to a dynamic equilibrium (~10 sccm) by a solenoid-activated leak valve controlled by a capacitance manometer with the chamber under gated or throttled pumping. The substrates were washed with ethanol, attached to the substrate heater with silver paste and maintained ~6 cm from the target. The laser fluence was 6 J/cm^2. The film compositions, as expressed by NbN_x ($x = N/Nb$), and thicknesses were determined by Rutherford Backscattering Spectroscopy (RBS) with 6 MeV He^{2+}. Typical films were ~ 1000Å thick after 5000 shots. X-Ray diffraction patterns were collected using a rotating anode source and a conventional $\Theta-2\Theta$ geometry and indexed using a least squares fit of the data. Γ (FWHM) for a particular reflection was measured by fixing 2Θ at the peak maximum and scanning through Θ. The T_c's, taken where $R = 0$, were determined resistively by the four point probe method and J_c at 4.2 K was measured inductively by pressing a multi-turn pancake-shaped coil against the film surface and inducing shielding currents.[6]

RESULTS AND DISCUSSION

Effect of N_2 Pressure on Film Composition & T_c

The nitrogen content in films deposited from Nb foil targets was found to increase with reactive gas pressure. The H_2 gas was introduced to reduce oxide contamination of the nitride films. Elastic recoil detection (ERD) measurements are currently underway to determine if any hydrogen was incorporated into the films. Shown in Table I are the structural, T_c and phase assignments for films grown from a Nb foil target and deposited on MgO (100) held at 600 °C as a function of gas pressure. At 10^{-7} Torr, films with a cubic lattice constant corresponding to that of niobium were deposited. At gas pressures of 1, 10 and 20 mTorr the deposited films displayed

Table I. Pressure Dependence of NbN Properties for Films Grown on MgO (100) and at 600 °C

Gas Pressure (mTorr)	T_c (R=0) (K)	Lattice Constant(s) (Å)	Material Deposited
2×10^{-4}	9.15	$a_0 = 3.335$	Nb
1	< 4.2	$a_0 = 3.12$ $c_0 = 4.82$	Nb_2N
10	< 4.2	$a_0 = 3.16$ $c_0 = 4.90$	Nb_2N
20	< 4.2	$a_0 = 3.10$ $c_0 = 5.01$	Nb_2N
60	16.4	$a_0 = 4.442$	NbN
100	6.7	$a_0 = 4.351$	NbN + Nb_3N_4
200	< 4.2	$a_0 = 4.343$	Nb_3N_4

Table II. Substrate and Temperature Dependence of NbN Properties for Films Grown at 60 mTorr

Substrate	Sub. Temp (°C)	T_c (R=0) (K)	Lattice Constant (Å)
MgO (100)	25	7.1	$a_0 = 4.378$
Fused Silica	25	7.5	$a_0 = 4.372$
MgO (100)	400	16.1	$a_0 = 4.441$
Fused Silica	400	8.4	$a_0 = 4.362$
MgO (100)	600	16.4	$a_0 = 4.442$
Fused Silica	600	11.3	$a_0 = 4.415$
MgO (100)	800	14.3	$a_0 = 4.408$
Fused Silica	800	11.8	$a_0 = 4.44$

the hexagonal unit cells expected for Nb_2N with $a_0 = 3.10\text{-}3.16$ Å and $c_0 = 4.77\text{-}5.01$ Å. An RBS spectrum of the film grown at 20 mTorr revealed a composition of $NbN_{0.58}$, consistent with the Nb_2N assignment. Films grown at 60 mTorr consistently displayed a cubic lattice with $a_0 \approx 4.44$ Å and a composition, determined by RBS, of $NbN_{1.0}$. Above 60 mTorr a_0 equals 4.351 Å and 4.343 Å at 100 mTorr and 200 mTorr, respectively. This phase is assigned as Nb_3N_4. RBS measurements confirm that $N/Nb > 1.3$ and a decrease in a_0 is expected as the number of Nb vacancies in the cubic lattice increase.

Shown in Figure 2 is T_c for the films deposited on MgO (100) at a substrate temperature of 600 °C as a function of N_2 / H_2 pressure. T_c drops from 9.2 K (that of pure Nb) to < 4.2 K as reactive gas pressure is increased from vacuum to 20 mTorr. The inability to superconduct above 4.2 K has previously been reported for Nb_2N.[5] Between 20 and 100 mTorr, T_c rises to ~ 16 K, consistent with earlier observations of NbN. At 100 mTorr T_c drops to 6.7 K and decreases to < 4.2 K above 200 mTorr.

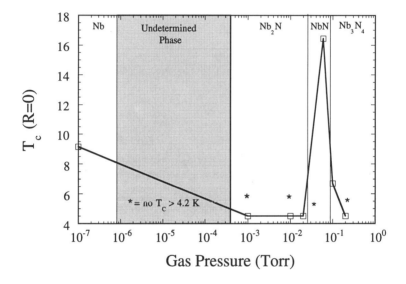

Figure 2. T_c for the films deposited on MgO (100) at a substrate temperature of 600 °C as a function of N_2 / H_2 pressure.

Effect of Substrate Temperature on T_c

The growth of NbN on single crystal and amorphous substrates was examined as a function of substrate temperature. Oriented MgO (100) and fused silica substrates were placed adjacent to each other on the substrate heater so that the films could be deposited under identical conditions. As shown in Table II, the films grown at room temperature showed virtually identical T_c's (~7 K) and lattice constants ($a_0 \approx 4.37$ Å). Above room temperature, the films grown on MgO displayed considerably higher T_c's than those deposited on silica. The T_c's of films grown on MgO rose to a maximum of 16.4 K at a growth temperature of 600 °C, where $a_0 = 4.441$ Å, and decreased to 14.3 K when deposited at 800 °C, where $a_0 = 4.408$ Å. In contrast, the T_c's of films grown on silica rose continuously to a maximum 11.8 K at a substrate temperature of 800

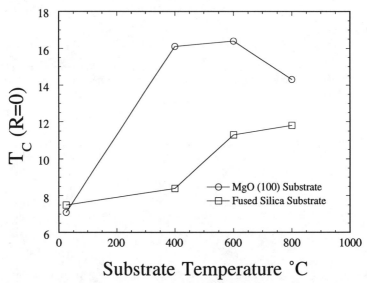

Figure 3. T_c of films grown on MgO (100) and fused silica at 60 mTorr N_2 / H_2
pressure as a function of substrate temperature.

°C, with a_0 also reaching a maximum at 4.44 Å at 800 °C. The XRD pattern of the film grown on
silica at 800 °C showed evidence for a second phase, possibly Nb metal.

A correlation was observed between a_0 and T_c for films deposited on MgO. The film with
the highest T_c also had the largest a_0. This dependence likely is due to N/Nb ratio in the films.
NbN films with nitrogen deficiencies and materials with niobium deficiencies both lead to a
decrease in a_0 and T_c.[7] Detailed compositional analysis is in progress.

Effect of Epitaxy on T_c

In addition to the differences in electrical properties of NbN films deposited on MgO and
fused silica, structural differences were observed. XRD patterns of the films deposited
simultaneously on MgO and silica at a N_2 / H_2 pressure of 60 mTorr and a substrate temperature of
600 °C are presented in Figure 4. The pattern of the NbN film grown on fused silica (top) reveals
only a very broad (111) peak on a descending background caused by the amorphous silica
substrate. X-ray diffraction from the film grown on MgO (100) is highly textured with Γ (200) =
1.0 °.

The structural differences between the films grown on oriented and amorphous substrates
had significant effect on the electrical properties. A plot of resistivity versus temperature, shown in
Figure 5, reveals that the NbN on MgO has higher T_c (15.9 versus 11.3 K) and lower residual
resistivity (60 versus 120 $\mu\Omega$-cm) than the sample grown on fused silica. In addition, J_c
measurements at 4.2 K show that the oriented material can accommodate four times the critical
current of the polycrystalline film.

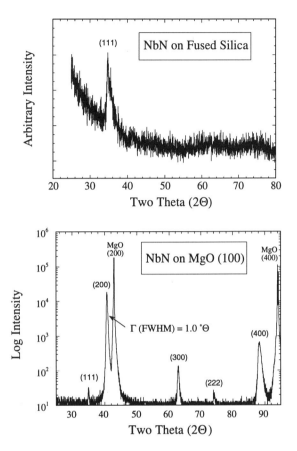

Figure 4. XRD patterns of NbN films deposited simultaneously on MgO and fused silica at a N_2 / H_2 pressure of 60 mTorr and a substrate temperature of 600 °C.

CONCLUSIONS

The N/Nb ratios for NbN_x films grown by PLD at elevated substrate temperatures depends most critically on nitrogen pressure. The ratio varied from 0 to 1.4 for films deposited in vacuum to 200 mTorr. A narrow range of pressures yields films with $T_c > 8$ K. At pressures < 20 mTorr Nb_2N forms and at pressures above 100 mTorr Nb_3N_4 is deposited. The effect of substrate temperature on the transport properties of NbN is pronounced. NbN grown at room temperature on either oriented or amorphous substrates had a $T_c < 8$ K. Films grown on amorphous substrates had T_c's increase with deposition temperature through a substrate temperature of 800 °C, whereas

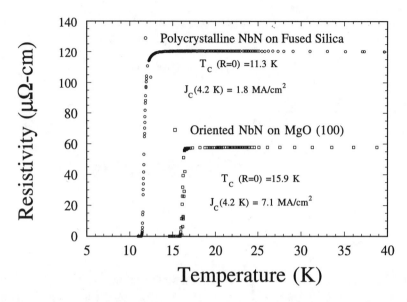

Figure 5. Resistivity versus temperature measurements of NbN films deposited simultaneously on MgO and fused silica at a N_2 / H_2 pressure of 60 mTorr and a substrate temperature of 600 °C.

films deposited on oriented substrates had T_c's increase to a substrate temperature of 600 °C and decrease at higher deposition temperatures. There is a strong correlation between film structure and optimal transport properties. A highly oriented NbN film grown on MgO (100) displayed higher T_c, higher J_c and lower residual resistivity than a poorly crystalline sample grown on fused silica under identical conditions.

REFERENCES

1. Growth of electronic ceramics by PLD is reviewed in J. S. Horwitz, D. B. Chrisey, K. S. Grabowski and R. E. Leuchtner, Surf. Coatings Technol. **51**, 290 (1992).
2. R. E. Treece, J. S. Horwitz and D. B. Chrisey, Mater. Res. Soc. Proc.(1994) in press.;.F. Xiong and R.P.H. Chang, Mater. Res. Soc. Proc. **285**, 587 (1993).; C. Niu, Y.Z. Lu, and C.M. Lieber, Science **261**, 334 (1993).
3. A. K. Ballal, L. Salamanca-Riba, G. L. Doll, C. A. Yaylor II and R. Clarke, Mater. Res. Soc. Proc. **285**, 513 (1993).
4. P. Tiwari, T. Zheleva and J. Narayan, Mater. Res. Soc. Proc. **285**, 349 (1993).; V. Cracium, D. Cracium and I. W. Boyd, Mater. Sci. and Engin. **B18**, 178, (1993).
5. S. A. Wolf, D. U. Gubser, T. L. Francavilla and E. F. Skelton, J. Vac. Sci. Technol. **18**, 253 (1981).; E. J. Cukauskas, W. L. Carter and S. B. Qadri, J. Appl. Phys. **57**, 2538 (1985).
6. J. H. Claassen, M. E. Reeves and R. J. Soulen, Rev. Sci. Instrum. **62**, 996 (1991).
7. W. Lengauer, Surface and Interface Anal. **15**, 377 (1990).

POLYCRYSTALLINE GRAIN STRUCTURE OF
SPUTTERED ALUMINUM NITRIDE FILMS

ASHER T. MATSUDA[1], H. MING LIAW[1], WAYNE A. CRONIN[1], HARLAND G. TOMPKINS[2],
PETER L. FEJES[2] AND KEENAN L. EVANS[3]
[1] Motorola Inc./Semiconductor Products Sector, 2100 E. Elliot Rd., Tempe, AZ 85284
[2] Motorola Inc./Semiconductor Products Sector, 2200 W. Broadway Rd., Mesa, AZ 85201
[3] Motorola Inc./Semiconductor Products Sector, 5005 E. McDowell Rd., Phoenix, AZ 85008

ABSTRACT

Reactively-sputtered, polycrystalline thin film aluminum nitride (AlN) is an attractive material for use in acoustic wave devices, for which it requires a strong preferred orientation, similar to that found in epitaxial films. This investigation evaluated the grain structure including preferred orientation, grain size, and surface morphology of sputtered AlN films. The characterization techniques utilized included x-ray diffraction (XRD), secondary ion mass spectroscopy (SIMS), transmission electron microscopy (TEM), and atomic force microscopy (AFM). The results revealed two types of grain structure: 1) a single-grain columnar structure that is perfectly oriented in the [001] direction throughout the entire film thickness and 2) a multiple-grain columnar structure that possesses a strong [001] orientation at the bottom of the film and a tilted [001] combined with other orientations at the top of the film. Strong correlations between orientation and surface morphology, oxygen content, and grain size were observed, namely higher degrees of c-axis orientation correlated with lower mean surface roughness values, reduced oxygen concentration, and narrower grains.

INTRODUCTION

Aluminum nitride (AlN), a III-V "semiconductor" with a hexagonal wurtzite crystal structure, is an attractive material in thin film form due to its wide-ranging applications in microelectronic devices. In addition to having a high thermal conductivity, high electrical resistivity, and excellent chemical and thermal stability (making it suitable as an insulating or passivating layer), AlN possesses good acoustic properties including a high acoustic wave velocity and a moderately high electromechanical coupling coefficient. For these reasons, AlN has been the subject of extensive investigation for use in high frequency acoustic devices including surface (SAW) and bulk acoustic wave (BAW) resonators and filters[1-5].

AlN thin film has been produced by various methods including chemical vapor deposition (CVD)[6-7], ion plating[8], and reactive sputtering[9-17] on various substrate materials. In order to obtain high electromechanical coupling coefficients, films must be deposited with a high degree of atomic alignment, consisting of either a "single crystal" or well oriented polycrystallites. For use in SAW devices, the film must also possess a smooth surface to minimize propagation losses. Epitaxial AlN films grown via CVD on R-plane[6] and basal-plane[7] sapphire at temperatures above 1000°C have shown fairly good piezoelectric properties, but the high temperatures involved are not compatible with semiconductor IC processing. Reactive sputtering of AlN has received much attention recently

Table I. XRD, SIMS, and AFM data for sputter-deposited AlN on (100) Si samples.

sample ID	film thickness (Å)	(002) intensity (arb. units)	(002) FWHM (°)	$2\theta_B$ (°)	oxygen concentr. (at.%)	R_a (nm)
A	12700	35900	0.245	36.033	4.2	8.0
B	14900	99200	0.209	36.026	2.4	14.1
C	10800	15400	0.304	35.990	4.0	7.5
D	11500	79500	0.218	36.037	1.9	9.1
E	10800	207300	0.189	35.891	1.1	4.3
F	10700	265600	0.183	36.026	0.4	4.0
G	9800	55400	0.218	35.979	3.5	8.0
H	9900	117200	0.196	35.990	2.2	6.9
I	10300	176500	0.194	35.996	1.8	4.2
J	17800	28600	Samples J and K are approximately			
K	21000	216200	$2\mu m$ films deposited on Si_3N_4/(100) Si			

as a viable alternative method to produce well aligned polycrystalline films at low temperatures (100-500°C) Liaw et al have shown recently that the SAW transducer characteristics of sputtered AlN films can approach or even match those of epitaxial AlN films, depending on the degree of orientation[18]. The purpose of this study was to investigate the grain structure of sputtered AlN films, including the oxygen content and surface morphology, for various degrees of grain alignment.

EXPERIMENTAL PROCEDURE

AlN films were reactively-sputtered on (100) Si and CVD Si_3N_4-coated (100) Si wafers using a commercial sputtering system. The AlN films and Si_3N_4 seed films (when used) were approximately 1μm and 2000Å in thickness, respectively. The process conditions were chosen so that the deposited polycrystalline films would possess acoustic properties either approaching or equal to those of epitaxial films. A total of nine samples on (100) Si and two on Si_3N_4 on (100) Si processed under varying conditions of power, gas composition, and gas flow rate were used in the evaluation.

The c-axis orientation of the AlN grains was evaluated by measuring the (002) XRD peak intensities and full-width at half maximum (FWHM) values for $2\theta_B \approx 36.1°$. A Rigaku Rotoflex X-ray Diffractometer with a rotating anode copper x-ray tube was operated at 30kV and 20mA in the $\theta-2\theta$ mode. Under these settings, the Si (004) peak registered counts in the range of 175,000 (437,500 converted) to 200,000 (500,000 converted) counts. The values presented in table I, however, were converted from the actual measured values to stay consistent with previously published data.

Cross-sectional TEM was used to resolve the columnar microstructure in terms of texturing, grain size, and microstructural evolution. A JEOL 200CX side-entry transmission microscope with a double tilt stage operating at 200kV was used in the analysis. The samples were prepared in cross-section by mechanically thinning the films to electron transparency following the technique described by Klepeis et al[19]. Selected area diffraction (SAD) was also employed to further verify the degree of c-axis preferred orientation.

A Digital Instruments Nanoscope III AFM was used in height mode to image the AlN surface morphology and to quantify the mean surface roughness (R_a). The wide-legged, 200μm Si_3N_4 cantilever tip (spring constant $k = 0.12$) was used along with a scan rate of 15.3 Hz, sample number of 512, and scan size between 1 and 15μm.

SIMS depth profiling was utilized to determine the oxygen concentration through the thickness of the films. All data for this evaluation were acquired on the quadrupole based Physical Electronics PHI6300 ion microprobe SIMS system using a

(a)   0.2 μm

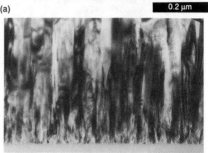
(b)   0.2 μm

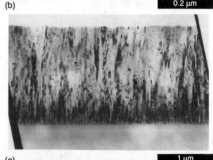
(c)   1 μm

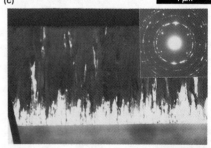
(d)   1 μm

Figure 1. TEM micrographs of sample J. Bright-field images taken from (a) top of film, (b) bottom of film, (c) entire thickness of film. (d) Dark-field image of entire film thickness taken from the Si [002] and AlN [002] beams, along with accompanying SAD pattern.

(a) `0.2 μm`

(b) `0.2 μm`

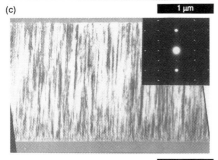

(c) `1 μm`

(d) `1 μm`

Figure 2. TEM micrographs of sample K. Bright-field images taken from (a) top of film, (b) bottom of film, (c) entire thickness of film. (d) Dark-field image of entire film thickness taken from the Si [002] and AlN [002] beams, along with accompanying SAD pattern.

Cs^+ primary ion source and microbeam ion gun for surface sputtering and depth profiling. The auxiliary electron gun was employed for surface charge neutralization during SIMS data acquisition. An AlN reference standard was used in conjunction with the bulk relative sensitivity factor calculation routine in the PHI6300 data processing software to calculate a relative sensitivity factor for oxygen in AlN. The m/e 32 (O_2^-) and the m/e 41 (AlN^-) secondary ion intensities were used as the dopant and the matrix species, respectively.

EXPERIMENTAL RESULTS

X-ray diffraction (XRD)

AlN (002) peak intensity and width data are presented in table I and reveal a wide range of c-axis orientation for the nine samples on (100) Si. FWHM values of the AlN (002) peaks are inversely related with their intensities and have an average value of 0.217°. The lattice constant c_o, calculated from the θ_B values, was found to be between 4.984 and 5.004Å, compared to c_o = 4.979Å for bulk AlN, indicating that the films are in a planar compressive stress state. The diffractograms from $2\theta_B$ = 10 to 80° for samples C and F revealed no significant reflection from crystallographic planes other than (002), thus indicating that both are fairly well oriented in the c-axis direction. Sample pairs C/J and F/K are subjected to further analysis as the worst and best cases of c-axis texturing, respectively, throughout the study.

Transmission electron microscopy (TEM)

Cross-sectional TEM micrographs and SAD patterns for samples J and K are presented in figures 1 and 2, respectively. Both samples are on Si_3N_4 and are approximately 2μm in thickness and allow the observation of grain structure and texture throughout the thickness of the film. Figures 1a and 1b image the structure of the AlN columnar grains at the top (top surface eroded during ion milling process) and bottom of the sample J film, respectively. The top surface micrograph reveals large grains or grain clusters approximately 100nm in diameter with a lack of uniform preferred orientation. The bottom portion of the film, however, appears to consist of small columnar grains approximately 20-30nm in diameter with a high degree of texturing. Low magnification bright- and dark-field images of the entire thickness of the film are shown in figures 1c and 1d, respectively. The dark-field image, taken with the Si [002] and AlN [002] diffracted beams, clearly reveals that AlN c-axis texturing is present in the initial 0.5μm portion of the film, but is eventually lost as the film is deposited. An SAD pattern taken from the entire film thickness (fig. 1d) shows sparse rings from various orientations of AlN as well as both [001] and [103] texturing.

Figures 2a and 2b image the structure of the AlN columnar grains at the top and bottom of the sample K film, respectively. In contrast to sample J, sample K possesses strong [001] texturing throughout most of the thickness of the film as seen in the low magnification bright- and dark-field images (figure 2c and 2d, respectively) with an average grain size of roughly 30nm at the top and 20nm at the bottom of the film. The SAD pattern taken from the entire film thickness (fig. 2d) is spotty (slight arcing is visible), indicative of a strong [001] preferred orientation; tilting of the sample confirmed that a-axis basal orientation is random.

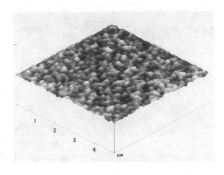

Figure 3. AFM surface image of sample C.

Atomic force microscopy (AFM)

AFM surface roughness (R_a) measurements for the AlN/(100) Si samples are tabulated in table I and range from approximately 4 to 14nm with an average value of 7.34nm. Surface images are shown in figures 3 and 4 for samples C and F, respectively. Sample C (R_a = 7.5nm) is observed to possess a rough surface characterized by a dense aggregation of approximately 0.2μm wide "peaks" which are thought to be clusters of individual columnar grains. Sample F (R_a = 4.0nm), on the other hand, possesses a relatively smooth topography that is only periodically interrupted with the large "peaks" seen in sample C. Samples E and I also possessed surface morphologies similar to that of sample F, while the rest of the samples possessed surface morphologies similar to that of sample C. Figure 5 is a plot of mean surface roughness (R_a) measured by AFM vs. AlN (002)

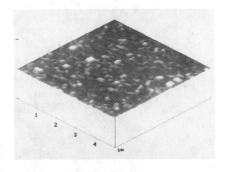

Figure 4. AFM surface image of sample F.

peak intensity. The unfilled points represent the measured R_a values while the filled points represent the measured R_a values normalized to a 1μm film thickness. It is evident that there is a transition to lower values of R_a at approximately 120,000 counts.

Secondary ion mass spectroscopy (SIMS)

Elemental oxygen concentration in the AlN films measured at the midplane ranged from 0.4 to 4.2 atomic percent with an average value of 2.39 atomic percent (table I). SIMS depth profile data reveal a relatively uniform distribution of oxygen throughout the outer half of the films, with some decrease in oxygen content as the interface between the AlN and Si substrate is reached. Figure 6, which is a plot of oxygen concentration vs. AlN (002) peak intensity, shows that there is a strong trend of decreasing oxygen concentration with increasing (002) intensity.

DISCUSSION

The piezoelectric properties of sputtered AlN are highly dependent on its structure, particularly its preferred orientation. Examination of samples J and K in cross-section reveals distinct differences in structure and texture. Sample J nucleates with a dominant [001] preferred orientation, but eventually loses this texturing after about 0.5μm of deposition. Seeds of c-axis oriented AlN nucleate at the substrate surface and continue to grow uniformly with further deposition of the film, but are eventually pinched off and overtaken by other growth orientations. Continued deposition results in the formation of wide, multi-grain aggregates, each composed of mutually aligned, non c-axis oriented grains, which grow non-uniformly through the rest of the thickness of the film. Although the micrographs reveal a columnar structure, the SAD pattern indicates the presence of not only

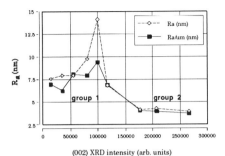

Figure 5. Surface roughness vs. (002) XRD intensity for AlN on (100) Si samples.

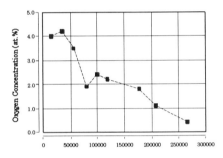

Figure 6. Oxygen concentration vs. (002) XRD intensity for AlN on (100) Si samples.

[001] texturing, but also [103] texturing combined with random orientations as seen in the sparse ring formations. Sample K also nucleates in the same manner as sample J, but continued growth of the film results in preservation of the narrow and well aligned grain structure. Reactively-sputtered AlN, therefore, possesses good initial [001] alignment (first few hundred angstroms) under a wide range of process conditions as well as on various crystalline and amorphous seed materials. (002) basal arrangement appears to be a natural growth habit for sputtered AlN as it nucleates rather readily. Continuation of the [001] preferred orientation with film growth, however, is heavily dependent on process conditions.

The surface morphology of sputtered AlN appears to be significantly affected by the preferred orientation of the film. Examination of figure 5 reveals the existence of two distinct groups of surface morphology for the samples deposited on Si: 1) the first six points below 120,000 counts have an average R_a of approximately 7.5nm and 2) the last three points above 175,000 counts have an average R_a of approximately 4nm. The AFM surface images of the two groups of samples reveal that the surface morphology of group 1 samples are all very similar to that of sample C, characterized by a dense agglomeration of large (~0.2μm diameter) peaks, and the surface morphology of group 2 samples are all similar to that of sample F, characterized by a relatively smooth surface periodically interrupted by the large protrusions seen in sample C. The large surface peaks or protrusions are most likely due to the widening and tapering of the grain aggregates (as seen in figure 1) that lose the preferred orientation and emerge at the surface. These misoriented grains have different growth rates, thus resulting in grains or grain aggregates with varying heights and a rough surface. Sample F consists mainly of perfectly aligned grains throughout most of the film (as in sample K) and, therefore, only a few misaligned grains protrude from the surface.

The mechanism of structural evolution in AlN is not fully understood at this time, but oxygen incorporation into the films appears to be a key factor in the degradation of a preferred orientation. In general, the oxygen content in the films increased with increasing gas flow rates (base pressure was constant for all samples). It is likely, therefore, that oxygen is incorporated into the films from the sputtering gases. Increasing oxygen concentration in the AlN films, as seen in figure 6, in turn results in degradation of the c-axis preferred orientation. Although the oxygen concentration vs. (002) intensity plot does not reveal two distinct groups or a maximum oxygen concentration allowed for high quality AlN films, comparison with figure 5, where 175,000 counts was the threshold value for well-oriented films, produces an approximate critical value of 1.8 at.% oxygen to ensure a strongly c-axis oriented film. From our past experiences, films with an oxygen concentration not exceeding 1.0 at.% typically resulted in strong piezoelectric properties. It is believed that an oxygen concentration surpassing approximately 1.0 at.% sufficiently distorts the AlN lattice to disrupt the continued growth of the c-axis aligned crystallites which, in turn, degrades the acoustic properties of the AlN film.

CONCLUSION

In summary, the polycrystalline grain structure of reactively-sputtered AlN films was investigated. Two types of polycrystalline microstructures in so-called [001] oriented AlN films (defined by the sole presence of {002} diffraction peaks) were observed: 1) a single-grain columnar structure that possesses a strong c-axis preferred orientation throughout the entire thickness of the film and 2) a multi-grain columnar structure that nucleates with a strong c-axis orientation, but eventually loses it with continued deposition of the film. Degradation of a strong c-axis preferred orientation also correlated with a rougher surface, wider grains, and higher values of oxygen concentration. The exact mechanism for the degradation of a c-axis preferred orientation in sputtered AlN is not known at this time, but oxygen incorporation into the film appears to play a significant role.

ACKNOWLEDGMENTS

The authors would like to thank K. Koetz, E. Cull, and J. Lewis for preparing the AlN films; L. Johnston and R. Doyle for TEM sample preparation; and R. Legge, D. Convey, and T. Hopson for assistance with and use of the atomic force microscope.

REFERENCES

1. J.S. Wang and K.M. Lakin, IEEE 1981 Ultrasonics Symp. Proc., 502-505 (1981).

2. K.M. Lakin, J.S. Wang, G.R. Kline, A.R. Landin, Y.Y. Chen, and J.D. Hunt, IEEE 1982 Ultrasonics Symp. Proc., 466-475 (1982).

3. K.M. Lakin, G.R. Kline, R.S. Ketcham, A.R. Landin, W.A. Burkland, K.T. McCarron, S.D. Brayman, and S.G. Burns, 41st Annual IEEE Freq. Ctrl. Symp. Proc., 371-381 (1987).

4. G.R. Kline, K.M. Lakin, and R.S. Ketcham, IEEE 1988 Ultrasonics Symp. Proc., 339-342 (1988).

5. S.V. Krishnaswamy, J. Rosenbaum, S. Horwitz, C. Vale, and R.A. Moore, IEEE 1990 Ultrasonics Symp. Proc., 529-536 (1990).

6. J.K. Liu, K.M. Lakin, and K.L. Wang, J. Appl. Phys. 26, 1555 (1987).

7. K. Tsubouchi and N. Mikoshiba, IEEE Trans. Sonics and Ultrasonics SU-32, 634 (1985).

8. M. Kishi, M. Suzuki, and K. Ogawa, Jpn. J. Appl. Phys. 31 (pt. 1, no. 4), 1153-1159 (1992).

9. C.R. Aita, J. Appl. Phys. 53 (3), 1807-1808 (1982).

10. J.A. Kovacich, J. Kasperkiewicz, and D. Lichtman, J. Appl. Phys. 55 (8), 2935-2939 (1982).

11. C.R. Aita and C.J. Gawlak, J. Vac. Sci. Technol. A 1 (2) 403-406 (1983).

12. A. Kawabata, Jpn. J. Appl. Phys. 23 (supplement 23-1), 17-22 (1984).

13. S.V. Krishnaswamy, W.A. Hester, J.R. Szedon, and M.H. Francombe, Thin Solid Films 125, 291-298 (1985).

14. F.S. Ohuchi and P.E. Russell, J. Vac. Sci. Technol. A 5 (4) 1630-1634 (1987).

15. G.L. Huffman, D.E. Fahnline, R. Messier, and L.J. Pilione, J. Vac. Sci. Technol. A 7 (3), 2252-2255 (1989).

16. M.H. Francombe and S.V. Krishnaswamy, J. Vac. Sci. Technol. A 8 (3), 1382-1390 (1990).

17. H. Okano, Y. Takahashi, T. Tanaka, K. Shibata, and S. Nakano, Jpn. J. Appl, Phys. 31 (pt. 1, no. 10), 3446-3451 (1992).

18. H.M. Liaw, W. Cronin, and F.S. Hickernell, IEEE 1993 Ultrasonics Symp. Proc., 267-271 (1993).

19. S.J. Klepeis, J.P. Benedict, and R.M. Anderson, Mat. Res. Symp. Proc. 115, 179-184 (1988).

EVOLUTION OF THE PROPERTIES OF REACTIVELY SPUTTERED ZIRCONIUM NITRIDES ON SILICON SUBSTRATES.

PIERRE VANDEN BRANDE[*], STEPHANE LUCAS AND RENE WINAND
LUCIEN RENARD[**] AND ALAIN WEYMEERSCH
[*]Université Libre de Bruxelles, Department of Metallurgy-Electrochemistry, CP165, 50 avenue Roosevelt, B1050 Brussels (BELGIUM).
[**]Centre de Recherches et Développements du Groupe Cockerill-Sambre, Campus Universitaire du Sart Tilman, Bd. de Colonster B57, 4000 Liège (BELGIUM).

ABSTRACT

Zirconium nitride films on silicon substrates have been prepared by DC magnetron enhanced reactive sputtering in a N_2/Ar gas mixture under various experimental conditions. The films properties (chemical composition, structure, morphology and optical response) were investigated and related to the experimental conditions. It is shown that these properties are strongly related to the target current density which governs the deposition rate, to the deposit thickness, to the nitrogen partial pressure and to the residual gas contamination. The objective of this paper is to detect the minimum film thickness threshold under which a zirconium nitride coating cannot be used in decorative applications.

INTRODUCTION

Nitrides and carbides of transition metals find increasing use as decorative coatings combining intensive colors, high wear resistance and good corrosion resistance. The gold-like color of the group IVb nitrides and particularly of zirconium nitride has attracted a great deal of interest in decorative applications for large area substrates [1].

It is well known that the optical properties of these nitrides depend very much on the deposition conditions and that various factors can influence the film composition, purity and structure. ZrN films can be produced by magnetron reactive sputtering of a zirconium target in a gas mixture of argon and nitrogen. The most important parameters that govern the composition and the structure of these films are the nitrogen partial pressure, the deposition temperature, the substrate bias voltage and the deposition rate. It was shown [2] that the thickness also plays a critical role in the evolution of the film properties, leading to an evolution of the crystalline structure which induces modifications of the optical properties. Moreover, it is expected that the reflection optical response of the film substrate-system will depend on the possibility of light transmission through the film. The film thickness seems then to be a key parameter in decorative applications. The main objective of this paper is consequently to find out the minimum thickness (threshold) under which the film cannot be used in decorative applications as well as the factors influencing this threshold.

EXPERIMENTAL PROCEDURE

Film preparation.

Zirconium nitride deposits were produced by reactive DC magnetron sputtering with a cylindrical target (90mm in diameter) in a mixture of nitrogen and argon at a total gas flow of 100 standard cm^3 min^{-1}. Under these conditions the total gas pressure was 5 10^{-3} mbar. The gases were mixed before introduction into the vessel. They were injected through a circular gas shower placed halfway between the substrate and the target. Decontamination of the zirconium target was achieved before each experiment by presputtering for 10 min at a 0.5 A target current in pure argon. Prior to each experiment the vacuum chamber was pumped down to a residual vacuum below 10^{-5} Torr. During each experiment, the deposition rate was constant and fixed by the target current, the deposition time was fixed by a clock regulated shutter and the target voltage was controlled relative to ground. The substrates were grounded Si(100) slices kept at room temperature and placed 55mm from the target. They were etched, for 10 min, at 10^{-2} mbar in pure argon before each experiment.

Film analysis

The geometrical thickness of the deposit was measured with a Dektak surface profilometer in order to calculate the deposition rate.

The nitrogen content of deposits produced under various experimental conditions, was determined with the help of a nuclear nitrogen depth profiling technique based on the nuclear resonant reaction $^{15}N(p,\alpha\gamma)^{12}C$. The deposits were made with natural nitrogen and since the isotopic percentage of ^{15}N in natural nitrogen is 0.37%, it is possible to determine the total nitrogen concentration by the determination of the ^{15}N distribution. The nuclear resonant reaction $^{15}N(p,\alpha\gamma)^{12}C$ is very intense, occurs at E_p = 429 keV, is isolated and has a natural width of 120 ev [3]. An automatic energy scan system mounted on a Van De Graaff accelerator was used to obtain the excitation curve at high resolution [4]. Further details about this technique can be found in [4,5]. Once the excitation curve, *e.g.* the gamma yield *vs* the proton energy, is recorded, a further treatment of the information is needed because the recorded spectrum is the convolution of the actual depth with a function including the energy distribution of the beam (depth dependant) and the width of the nuclear reaction. This deconvolution has been done with the help of the program VAVLOV [6]. Fig.1 shows the effect of the deconvolution (fig.1b) on the nitrogen depth profile measured on a 50 nm thick ZrN sample produced at 2.5 standard cm^3 min^{-1} of N$_2$ (fig.1a). The deconvolution procedure has the effect of making the actual nitrogen depth profile sharper. Conversion of the deconvoluted spectrum to nitrogen concentration *vs* depth has been achieved with a TiN standard (fig.1c). From the energy loss of the proton beam through the film, the total number of atoms per unit area and the deposit specific density can be estimated.

The chemical composition profile of the film was derived by Auger electron spectroscopy (AES) under 4keV ion argon erosion. The experimental conditions for AES analysis were the followings: E_p=5 keV and θ_p=30°.

The structure of the deposit was carried out by X ray diffraction (XRD) from 1.4 μm thick films.

Color and gloss, which are the most important properties of decorative coatings, were studied by spectral reflectance spectroscopy with a Micro Color tristimulus colorimeter equipped with an Ulbricht globe coupled to a Xenon flash lamp for diffuse illumination of the sample. The light diffuse reflection from sample was measured at an angle of 8° in accordance with the German industrial standard DIN 5033. The results are given in L*a*b* coordinates were L* specifies the position on the white-black axis, a* the position on the red-green axis and finally b* the position on the blue-yellow axis [7-9]. In order to get more fundamental information about these optical properties, transmission spectra were also recorded between wavelengths of 300 nm and 800 nm through various thicknesses of ZrN films deposited on glass substrates.

EXPERIMENTAL RESULTS AND DISCUSSION

Fig.2 indicates that the deposition rate decreases when the nitrogen partial pressure increases at constant target current (0.5A). The deposition rate decreases from 70 nm min^{-1} under pure argon sputtering of Zr to 15 nm min^{-1} for a nitrogen flow of 15 standard cm^3 min^{-1}. Above this threshold flow, the deposition rate is minimum and constant. This behaviour can be explained by the progressive surface contamination of the target with nitrogen (target poisoning). Above the threshold of 15 standard cm^3 min^{-1}, the target is probably completely poisoned by nitrogen and the deposition rate is low because the sputtering yield of ZrN is low [10-11]. The deposition rate is also an increasing function of the target current (fig.3). Correspondingly and also displayed fig.3, the absolute value of the target potential presents a growing slope *vs* the target current because the zirconium target resistivity increases by Joule heating.

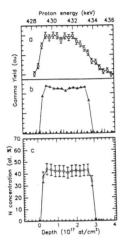

Fig.1 Nitrogen depth profile determined in Zr by the nuclear resonant reaction ^{15}N(p,αγ)^{12}C (a, recorded spectrum; b, deconvoluted spectrum; c, N depth profile).

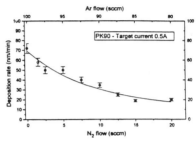

Fig.2 Deposition rate *vs* the nitrogen flow at constant target current (0.5A).

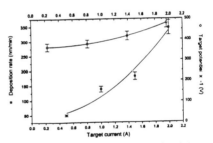

Fig.3 Deposition rate (squares) and target potential (circles) *vs* the target current at constant nitrogen flow (5 sccm)..

Table 1. L*a*b* coordinates *vs* the nitrogen flow (0.5A).

N₂ (sccm)	L*	a*	b*
0	77.6	5.1	2.9
5	70.2	9.9	30.1
10	57.7	16.1	12.0

Table 2. L*a*b* coordinates *vs* the film thickness. (0.5A- 5 % N₂).

Thickness (nm)	L*	a*	b*
0	68.2	4.5	-9.0
25	59.9	5.7	4.0
50	71.2	6.9	20.0
250	79.0	9.9	32.1
500	73.2	10.1	31.5
1500	70.2	9.9	30.1

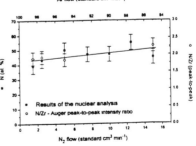

Fig.4 Nitrogen concentration in Zr (squares) and peak-to-peak nitrogen-to-zirconium Auger intensity ratio *vs* the nitrogen flow at constant target current (0.5A).

From similar curves to those presented figure 1, it was possible to derive the evolution of the deposit nitrogen concentration *vs* the N_2 flow at 0.5 A. Fig.4 indicates that the nitrogen concentration increases slowly from 40 at.% to 48 at.% when the nitrogen flow increases from 1.5 to 15 standard $cm^3 min^{-1}$. A similar behaviour is observed for the corresponding peak-to-peak nitrogen-to-zirconium Auger intensity ratio plotted *vs* the N_2 flow (fig.4). When the target current increases from 0.5 A to 1 A, the N/Zr Auger signal decreases from 1.853 to 1.563 (nitrogen flow: 5 standard $cm^3 min^{-1}$). The AES depth profile of a ZrN film obtained at a 0.5 A target current and 1.5 standard $cm^3 min^{-1}$ of N_2 shows that the surface film is always highly contaminated by carbon and oxygen. This contamination is substantially reduced inside the deposit (fig. 5). The peak-to-peak nitrogen-to-zirconium Auger ratio stays constant through the deposit and this confirms the result derived with the nitrogen nuclear depth profiling technique (fig.1c). Fig. 5 indicates the presence of a surface carbon contamination layer and a zirconium carbide contamination phase into the deposit. The observed peaks in the AES depth profile are the following: Zr at 147eV, C at 272eV, N at 379eV and O at 503eV. Fig. 6 shows the effect of an incomplete decontamination of the Zr target by pure argon sputtering on the film composition. The oxygen content of the ZrN deposit is dramatically high.

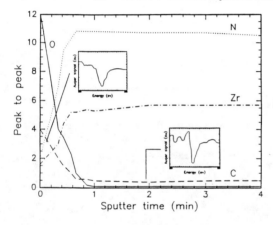

Fig.5 AES analysis of ZrN produced with a 0.5A target current and 5 sccm nitrogen flow.

Fig.6 AES analysis of a ZrN film contaminated by oxygen during sputtering (0.5A, 5 sccm of N_2).

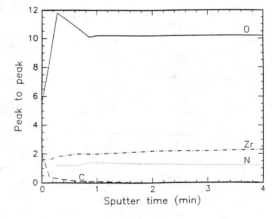

The L*a*b* measurements summarized in Table 1 indicate a strong dependance of the ZrN color on the nitrogen partial pressure. Above a threshold thickness (250nm) the diffuse reflection optical response becomes independent of the film thickness (Table 2). Transmission spectra recorded through different ZrN films display, as expected, an increase of light absorption with the film thickness (between the wavelengths of 300nm and 800nm). This absorption is almost complete above a thickness of 500nm (Table 3). Nevertheless, the oxygen contamination can strongly alter the film optical properties by decreasing the light absorption in the visible wavelength range as shown Table 3.

Table 3. Optical transmission results recorded between wavelengths of 300nm and 800nm (0.5 A). I_p: peak intensity, I_0: incoming intensity through the glass, λ_{min} - λ_{max}: transmission band width

Sample type	λ_p (peak) (nm)	I_p/I_0 (%)	width at $I_p/2$ (nm)	λ_{min} (nm)	λ_{max} (nm)
glass substrate	512.8	100.0	131.9	363.8	691.5
ZrN 5%N₂ 500nm + O₂	519.1	91.7	121.3	363.8	691.5
ZrN 5%N₂ 100nm	521.3	83.3	134.0	363.8	695.7
ZrN 5%N₂ 250nm	512.8	41.7	159.6	374.5	695.7
ZrN 5%N₂ 500nm	544.7	0.6	-	-	-

Table 4. XRD results vs nitrogen flow. 2θ is the diffraction angle, d is the distance between two diffraction planes and I% is the diffraction peak intensity percentage relatively to the maximum peak corresponding to ZrN(111).

2θ	d	N₂ (5sccm) I%	N₂ (10sccm) I%	detected phase
29.88	2.988	11.4	12.0	ZrO₂
32.95	2.716	5.5	5.3	Zr(002)
33.86	2.645	100.0	100.0	ZrN(111)
39.23	2.294	8.0	53.2	ZrN(200)
50.11	1.819	7.4	0.7	ZrO₂
56.85	1.618	5.1	3.3	ZrN(220)
59.72	1.547	3.0	2.4	ZrO₂
67.72	1.382	0.0	13.2	ZrN(311)
71.26	1.322	11.9	0.0	ZrN(222)

The deposits are almost entirely composed of a stoichiometric ZrN phase with small amounts of zirconium, zirconium oxide and zirconium carbide phases as shown by the XRD results registered on 1.4 μm deposits sputtered at 0.5A for 5% N_2 and 10% N_2 (Table 4). The contamination is due to the high chemical reactivity of Zr. Zr contamination by oxygen can strongly alter the film optical properties. In order to increase the optical absorption, it is very important to reduce the oxygen contamination. When the oxygen content is very low the film can be totally absorbing in transmission, in the visible spectrum, for thickness as-thin-as 100nm [12]. Carbon contamination alters also the optical properties of the film by formation of a ZrN_xC_y phase [13]. The color change of the deposit vs the nitrogen flow can be principally explained by structural changes more than composition variations as shown by tables 1 and 2. Table 4 indicates that a high nitrogen flow leads to a strong increase of the ZrN (200) and (311) XRD peaks. The ZrN(111) peak remains the major peak in the two situations. Similar results were observed previously [2,8], but while the ZrN(111) peak was always observed as the highest intensity peak, other secondary peaks than ZrN(200) and ZrN(311) were affected by the variation of the nitrogen partial pressure. This can probably be explained by different deposition conditions (higher substrate temperature and negative bias). It should be noted that recrystallization processes play a very important role in the growth of a ZrN deposit. The crystalline structure can be modified during the film growth [2]. If the films are well crystallized above a thickness of 1μm, it is not the case below 100nm. Observation of the films under transmission electron microscope produced grain growth due to the electron beam heating. This is typical for highly out-of-equilibrium films with high defects densities. This is often usual for films sputtered at high deposition rate on room temperature substrates. The high defect density in the deposit can explain the low energy loss of the proton beam through the film and consequently the low film density of 5.7 g/cm^3 (the calculation is performed using the stopping power of stoichiometric ZrN). This value has to be compared to the specific density of stoichiometric ZrN which is in the range of 7 g/cm^3.

CONCLUSION

Optical properties of ZrN films produced by DC reactive magnetron sputtering on room temperature grounded substrates depend principally of their crystalline structure and their contamination level with carbon and oxygen. The crystalline structure was shown to be affected by nitrogen partial pressure and film thickness. Nevertheless, the nitrogen partial pressure do not affect strongly the film composition above a nitrogen flow of 1.5 standard cm^3 min^{-1}. The films contain a major phase of ZrN and contamination phases constituted by zirconium, zirconium oxide and zirconium carbide. Oxygen contamination was proved to dramatically decrease the optical absorption in the visible wavelength range. For this reason, the gas contamination has to be reduced as-much-as possible in decorative applications. The thickness threshold for decorative applications seems to be 250nm. But a lower threshold value can probably be achieved with a lower oxygen contamination.

REFERENCES

1. H.Randhawa, Surf. and Coat. Tecchnol., 36(1988)829
2   U.Beck, G. Reiners, I.Orban, H.A.Jehn, U.Kopackz & H.Schack, Surf. and Coat. Technol., 62(1993)215
3. B.Maurel & G.Amsel, Nucl. Inst. Meth., 218(1983)159
4. G.Terwagne, M.Piette & F.Bodart, Nucl. Inst. Meth., B19/20(1987)145
5. G.Terwagne, S.Lucas & F.Bodart, Nucl. Inst. Meth., B66(1992)262
6. S.Lucas & M.Piette, VAVLOV: a deconvolution program, LARN, 22 rue Muzet, B-500 Namur(1993), Belgium.
7. ISO 7724/1,2,3 - (1984) E
8. G.Reiners, H. Hantsche, H.A.Jehn, U.Kopackz & A.Rack, Surf. and Coat. Tecchnol., 54/55(1992)273
9. U.Beck, G.Reiners, I.Urban, H.A.Jehn, U.Kopackz & H.Schack, Surf. and Coat. Tecchnol., 61(1993)215
10. J.H.Kim & H.A.Jehn, J. of the Korean Inst. of Met. & Mater., 30(1992)75
11. S.Hofmann, Thin Solid Films, 191(1990)335
12. P.Panjan, A.Zanker, B.Navinsek, A.Demsar, A.Svajger, L.Tanovic & N.Tanovic, Vacuum, 40(1990)161
13. H.Randhawa, Surf. and Coat. Tecchnol., 36(1988)829

The authors thank La Region Wallonne, Belgium, for partial financial support, Le Laboratoire d'Analyses par réactions Nucléaires des Facultés Universitaires Notre Dame de la Paix à Namur, Belgium, for its support in the NRA experiments.

Author Index

Subject Index

accumulation, 217
acoustic
    modes, 561
    wave devices, 753
activation, 685
adhesion, 597
Ag, S-T-1, 217
Al, single crystal, 659
alloy, 375
    Ti-based, 265
Al-Ni, 169
aluminum nitride, 253, 753
amorphization, 679
    reaction, 229
    solid state, 223, 229, 381
amorphous, 101, 715
    alloy, 211
    carbon, 381
    $Si_{1-x}Ge_x$, 679
    silicon, 691, 697, 703, 709
asymptotic behavior, 55
atmospheric pressure chemical vapor deposition,
    523
atom probe/field ion microscope, 223, 229, 309
atomic force microscopy, 119, 529
Au, 77, 405
Au-Ni, 555

backscatter Kikuchi
    diffraction, 653
    patterns, 23
biaxial modulus, 561
$Bi_4Ti_3O_{12}$, 475
boundary
    motion, 71
    sliding, 83
    tilt, 727
    twist, 727
Brillouin light scattering, 567
brownmillerite, 463
buffer layers, 417

C, 585
capillarity, 247
c-axis orientation, 753
CeO_2, 529, 535
chemical
    stability, 253
    vapor polymerization, 541
Co, 285, 315, 345
coalescence, 149
Co-Cr, 309
CoCrPt, 339
CoCrPt/Cr, 321CoCrNi, 381
CoCrPtTaB, 303
CoCrTa/Cr, 297
co-deposit, 271
coercivity, 315, 345, 417
composite, 265

compositional order, 375
computer models, 83
$Co_{1-x}Ni_x$, 351
$Co_{77}Ni_7Cr_4Pt_{12}$, 315
contamination, 759
convergent beam electron diffraction, 615
copolymer, 541
copper, 113
Co-Pt, 387
Co_xPt_{1-x}, 375
$CoPt(L1_0)$, 387
$CoPt_3$-$L1_2$ phase, 375, 387
Cr, 381
cross hatch angle, 303
crystal structure, 345, 399
crystallinity, 697
crystallization, 89, 475, 573, 679, 697
crystallized, 697, 703, 709
crystallographic texture, 327
crystallography, 517
$Cu(In,Ga)Se_2$, 143
$CuInSe_2$, 143
CVD, 113, 673, 741

dark conductivity, 697
decomposition, 309
decoration, 759
defects, 423
detectors, 101
devices, 505
dielectric, 457
    constant, 511
    properties, 499
diffusion, 241, 451
discontinuous, 399
disilane, 673
dopant, 685
DRAMS, 511
DSC, 101

early stages, 223
ECR PECVD, 493
elastic
    constants, 567
    modulus, 549
    properties, 567
electrodeposition, 137
electromigration, 653
electromotive force, 229
electron
    backscatter diffraction, 653
    microscopy, 101
epitaxial relationships, 131
epitaxy, 327, 339
erasable optical data storage, 89
etching, 247
evaporated, 113
excimer laser, 703, 709